METHODS FOR PHASE DIAGRAM DETERMINATION

Edited by

J.-C. ZHAO
GE Global Research, USA

ELSEVIER

Amsterdam • Boston • Heidelberg • London • New York • Oxford
Paris • San Diego • San Francisco • Singapore • Sydney • Tokyo

Elsevier
The Boulevard, Langford Lane, Kidlington, Oxford OX5 1GB, UK
Radarweg 29, PO Box 211, 1000 AE Amsterdam, The Netherlands

First edition 2007

Copyright © 2007 Elsevier BV. All rights reserved

No part of this publication may be reproduced, stored in a retrieval system
or transmitted in any form or by any means electronic, mechanical, photocopying,
recording or otherwise without the prior written permission of the publisher

Permissions may be sought directly from Elsevier's Science & Technology Rights
Department in Oxford, UK: phone (+44) (0) 1865 843830; fax (+44) (0) 1865 853333;
e-mail: permissions@elsevier.com. Alternatively you can submit your request online by
visiting the Elsevier web site at http://elsevier.com/locate/permissions, and selecting
Obtaining permission to use Elsevier material

Notice
No responsibility is assumed by the publisher for any injury and/or damage to persons
or property as a matter of products liability, negligence or otherwise, or from any use or
operation of any methods, products, instructions or ideas contained in the material herein.
Because of rapid advances in the medical sciences, in particular, independent verification
of diagnoses and drug dosages should be made

British Library Cataloguing in Publication Data
A catalogue record for this book is available from the British Library

ISBN: 978-0-08-044629-5

For information on all Elsevier publications
visit our web site at books.elsevier.com

Typeset by Charon Tec Ltd (A Macmillan Company); Chennai, India
www.charontec.com

Printed and bound in Great Britain

Working together to grow
libraries in developing countries

www.elsevier.com | www.bookaid.org | www.sabre.org

ELSEVIER   BOOK AID International   Sabre Foundation

METHODS FOR PHASE DIAGRAM DETERMINATION

Table of Contents

Preface		vii
1	Introduction to Phase Diagrams	1
	John F. Smith	
2	The Role of Phase Transformation Kinetics in Phase Diagram Determination and Assessment	22
	J.-C. Zhao	
3	Correct and Incorrect Phase Diagram Features	51
	Hiroaki Okamoto and Thaddeus B. Massalski	
4	Determination of Phase Diagrams Using Equilibrated Alloys	108
	Yong Zhong Zhan, Yong Du and Ying Hong Zhuang	
5	DTA and Heat-Flux DSC Measurements of Alloy Melting and Freezing	151
	William J. Boettinger, Ursula R. Kattner, Kil-Won Moon and John H. Perepezko	
6	Application of Diffusion Couples in Phase Diagram Determination	222
	A.A. Kodentsov, Guillaume F. Bastin and Frans J.J. van Loo	
7	Phase Diagram Determination Using Diffusion Multiples	246
	J.-C. Zhao	
8	Application of Computational Thermodynamics to Rapidly Determine Multicomponent Phase Diagrams	273
	Y. Austin Chang and Ying Yang	
9	Determination of Phase Diagrams with Reactive or Volatile Elements	292
	Cezary Guminski	
10	Phase Diagram Determination of Ceramic Systems	341
	Doreen Edwards	

| 11 | Determination of Phase Diagrams Involving Order–Disorder Transitions | 361 |

Ryosuke Kainuma, Ikuo Ohnuma and Kiyohito Ishida

| 12 | Determination of Phase Diagrams Involving Magnetic Transitions | 383 |

Ichiro Takeuchi and Samuel E. Lofland

| 13 | Determination of Pressure-Dependent Phase Diagrams | 412 |

Surendra K. Saxena and Yanbin Wang

| 14 | The Determination of Phase Diagrams for Slag Systems | 442 |

David R. Gaskell

| 15 | Determination of Phase Diagrams for Hydrogen-Containing Systems | 459 |

Ted B. Flanagan and Weifang Luo

| 16 | Miscellaneous Topics on Phase Diagrams | 483 |

J.-C. Zhao and Jack H. Westbrook

Preface

When I was approached by Elsevier to put together a book on phase diagram determination, I immediately responded with a positive answer and a book plan. Such a book, in my opinion, is long overdue. There have been several books in recent years on thermodynamic calculations of phase diagrams using the CALculation of PHAse Diagrams (CALPHAD) approach, but only a thin conference proceedings volume on *Experimental Methods of Phase Diagram Determination* edited by J.E. Morral, R.S. Schiffman, and S.M. Merchant and published by the Minerals, Metals and Materials Society (TMS) in 1994. This new book will help offset such imbalance, and emphasize the importance of experimental determination of phase diagrams. Since the Gibbs energy functions in CALPHAD assessments are optimized from experimental phase diagrams and a few thermodynamic measurements (when available), accurate experimental phase diagrams are essential for reliable CALPHAD modeling that is one of the foundations for computational design of multicomponent materials.

Phase diagram determination is both a scientific study and an art in a certain sense, because experience and good practice can play an important role in the quality of the results. The feeling that the "art" part is gradually fading away has been bothering me for quite some time. Hopefully, this book will archive some of the art.

Another important consideration in putting this book together is the need for a comprehensive reference book on the subject, which surprisingly did not exist. I was very encouraged by the positive comments from anonymous reviewers on the book plan. Their comments include words such as "As for the subject matter of the proposed book, indeed it is high time to have an up-to-date book on that subject, which is very central to MSE [materials science and engineering]". An effort has been made in creating this reference book to allow researchers to quickly find various methods in a single volume and to select specific methods for their own research. This volume also serves as a state-of-the-art overview of the subject. Each chapter tries to provide descriptions of the techniques, their underlying principles, their pros and cons, interpretation and reliability of the results, and a number of real examples. The book also covers determination of phase diagrams containing order–disorder transitions, magnetic transitions, as well as phase diagrams in non-metallic systems such as ceramic systems, slags, and hydrides. Traditional well-known techniques such as the equilibrated alloy method and recent developments such as diffusion multiples are covered. The editor hopes this book will serve as an introduction for researchers new to the field and as a refresher for the experienced researchers.

In Chapter 1, Smith provides an introduction to phase diagrams and to this book. He starts with a brief history of the establishment and early application of the phase rule from Gibbs' thermodynamics, and the developments in the twentieth century applying optical and electron microscopy as well as X-ray diffraction to phase diagram determination. The development of the CALPHAD method and its demand for experimental inputs are then discussed to emphasize the importance of reliable experimental data as expressed in the classic cliché, *garbage in equals garbage out*. The phase rule and its importance in phase diagram construction are then discussed along with basic rules for generating reliable experimental data. Also discussed in this chapter are some past efforts in compilation of phase diagram data.

Chapter 2 stresses the importance of understanding the basic phase transformation kinetics in determination and assessment of phase diagrams. The cooling process always shifts the phase transformation temperature away from the equilibrium phase boundary toward lower temperatures, and the opposite is true for the heating process. Real examples are used to show that the shift can be very pronounced for solid–solid phase equilibrium transitions at low temperatures; and the heating data are no better than the cooling data if equilibrium is not reached at low temperatures before heating. The difficulty associated with slow decomposition of some high-temperature phases, especially intermetallic compounds, is also discussed with real examples. This problem has not been adequately discussed before. The slow decomposition kinetics is often manifested as a widening of the composition region of a stoichiometric phase or line compound in a phase diagram. This chapter emphasizes that caution must be used in the determination of phase diagrams at relatively low temperatures.

Okamoto and Massalski, two of the most experienced experts in the world on phase diagram assessments and compilations, illustrate the many correct and incorrect phase diagram features they have seen when performing phase diagram assessments. Based on previous publications of theirs, Okamoto and Massalski use many real examples in Chapter 3 to explain the correct phase diagram topology based on thermodynamics. It is important to understand these features in order to correctly construct phase diagrams from experimental observations and to avoid mistakes that can be easily made when drawing phase boundary lines across experimental data points.

Zhan, Du and Zhuang in Chapter 4 explain the most widely used method – the equilibrated alloy method – for phase diagram determination. They use very simple examples to illustrate the step-by-step process, including alloy preparation, homogenization heat treatment, and determination of phases and phase equilibria via isothermal (static) measurements or cooling/heating (dynamic) experiments. Ample detail is provided to ensure high-quality alloy preparation through induction melting, arc melting, or a powder metallurgy route. The heat treatment process has also been discussed in detail, especially regarding the necessity for long-term heat treatment for low-temperature solid–solid phase equilibria, and the quenching process that is often used to freeze the high-temperature equilibria for analysis at ambient temperature. The determinations of the Cu–Nd binary system, the Al–Be–Si and the Al–Mn–Si ternary systems, and the Al–C–Si–Ti quaternary system are then shown as real examples. Common issues and solutions associated with the equilibrated alloy method are discussed at the end of the chapter to ensure high-quality data.

Chapter 5 is essentially the NIST (National Institute of Standards and Technology) guide to DTA and heat-flux DSC measurements of alloy melting and freezing. Several years in drafting, this chapter provides the most comprehensive guide to date on the recommended practices in measuring the liquidus and the solidus of phase diagrams. The chapter along with its appendices provides valuable information and insight into the DTA and DSC processes, especially the fundamental thermodynamics and kinetics governing these processes. This chapter is a "must read" for those who use or intend to use DTA and DSC measurements to construct phase diagrams.

Kodentsov, Bastin and van Loo in Chapter 6 describe the application of diffusion couples in phase diagram determination. They first explain the principles of the diffusion couple method, the process of making diffusion couples, and basics of the electron probe microanalysis (EPMA). Real examples are then used to illustrate two types of diffusion couples that are used for phase diagram determination, namely the semi-infinite ones and "sandwich" diffusion couples with a thin foil in the middle. The latter configuration is not widely used, but has high potential and advantages in both efficiency and simple sample fabrication. They discuss in detail the various error sources and methods to correct them, again using many real examples encountered in their laboratory. Their examples of missing phases and impurity stabilized extra phases are especially worth noting. Diffusion couples are a powerful way to determine phase diagrams, especially high-temperature isothermal sections. This chapter is a "must read" for those considering using this methodology.

In Chapter 7, a novel diffusion-multiple approach is introduced with real examples. Using a hot isostatic pressing (HIP) process, one can put several pieces of metals together to form several diffusion couples and triples in a single "diffusion multiple" to determine isothermal sections of several systems. The key considerations and steps in making good diffusion multiples are discussed first to stress the importance of sample preparation. The analysis steps involving imaging and phase identification, EPMA, and extraction of tie lines are then discussed. It is shown that two plots, one composition profile (composition–distance plot) and one composition path (composition–composition plot), are essential for reliable interpretation of the experimental results obtained from very concentrated areas of diffusion multiples. The practices used to reduce the chances of errors are discussed at the end of the chapter. By following the many recommendations specified in this chapter, one can determine phase diagrams with high efficiency while maintaining high quality.

Chang and Yang in Chapter 8 describe a novel approach in accelerated phase diagram determination/assessment by coupling CALPHAD modeling with selected experiments. This is a very important and promising direction for multicomponent phase diagram research. Taking advantage of available thermodynamic assessments of many binary (and some ternary) systems, one can perform a preliminary assessment of a ternary (or a higher order) system by assuming there are no ternary (or higher order) compounds and that the solubilities of the elements in the binary (or the higher order) compounds are negligible. Such an approximate assessment is used to predict a phase diagram of interest which guides the selection of key alloy compositions to maximize the significance of the input data points. The phase equilibrium information obtained from the selected alloys is then used to improve the thermodynamic model parameters for better description of the ternary (or the higher order) system. The authors use real examples such as Mg–Al–Sc and Mo–Si–B–Ti systems to illustrate how the process works. Some systems may need multiple iterations to develop a reliable thermodynamic description. This methodology has the advantage of higher efficiency in obtaining a thermodynamic description of a multicomponent system with limited alloys. The users are cautioned that only the composition space that has experimental data is considered to be validated – other composition space still needs to be validated in the future for reliable prediction.

In Chapter 9, Gumiński describes various methods used in the determination of phase diagrams with reactive or volatile elements – some of the most challenging systems to study experimentally. He provides a comprehensive collection of methods used to characterize such systems, including thermal analysis, zone melting, microstructural phase identification, XRD, densitometry, dilatometry, interdiffusion (diffusion couples), hardness, calorimetry, electrical resistivity, electromotive force, coulometric titration, galvanic polarization, magnetic susceptibility, vapor pressure, gas thermal extraction, and thermogravimetry. Using his vast experience, Gumiński provides valuable insight by delineating the precision, accuracy, and issues associated with each method. Some of these methods are used more often by chemists than materials scientists, and they complement Chapter 5 and other chapters in this book to provide a complete set of methodologies. Chapter 9 covers the determination of solid–solid, solid–liquid, liquid–liquid, solid–vapor, liquid–vapor, and liquid–fluidal equilibria with many real examples.

Edwards in Chapter 10 provides an overview of experimental techniques commonly used for ceramic phase diagram determination, with an emphasis on oxide systems. She first discusses in detail the sample preparation and equilibration processes for ceramics that are quite different from those of metals, especially concerning high temperatures and various partial pressure conditions. The partial pressure – defect chemistry type phase diagrams are unique for ceramics, and are becoming very important for many technological applications such as sensors. This chapter also discusses the applicability and issues associated with each technique and how to obtain high-quality data for ceramic phase diagrams.

Kainuma, Ohnuma and Ishida in Chapter 11 describe the determination of phase diagrams involving order–disorder transitions. They first introduce the widely used methods involving thermal analysis such as DSC, electrical resistivity, and XRD. These methods measure the order–disorder transition temperature by monitoring a thermal, physical, or structural change during heating and cooling of an alloy. They then focus the chapter on two alternative methods that extract the order–disorder transition concentration from diffusion couples. The so-called singularity point method and the concentration gradient method developed in recent years can determine the critical ordering concentration and the solubility line of coherent ordering at a fixed temperature, thus allowing order–disorder transitions to be accurately defined on isothermal sections.

Takeuchi and Lofland in Chapter 12 describe various ways to determine phase diagrams involving magnetic transitions. They begin with an introduction to measurements of magnetic properties; then, use examples to illustrate the interaction of ordering and magnetism. In the second half of this chapter, Takeuchi and Lofland introduce the combinatorial/high-throughput approach which is becoming an important tool for accelerated experimentation for materials science. This new approach can greatly accelerate phase diagram studies, especially for understanding highly correlated systems involving complex interplay among magnetism, ordering, and electronic transitions. These complex systems are becoming increasingly important in technological applications.

In Chapter 13, Saxena and Wang provide an overview of various instrumentation and methods available in the world for the determination of high-pressure phase diagrams. Their focus is on ultrahigh pressure research using both diamond anvil cells and large-volume presses. They describe various ways to measure pressure, to reach high pressure and high temperature, and to determine crystal structures. This chapter is an excellent resource for those who want to get involved/started in high-pressure phase diagram research.

Gaskell in Chapter 14 goes through the history of phase diagram determination for slag systems. In a concise manner, he describes the experimental procedures developed by early pioneers in reaching very high temperatures and performing measurements for ternary oxide systems. Some creative methods are described, including the use of a thermocouple for both heating and measurement of a small amount of material to 1880°C. Phase diagrams of slag systems have played an important role in the history of iron and steel making. The art of phase diagram measurements for slag systems is gradually disappearing from common practice. This chapter serves as an important archive of the methods developed by the pioneers in this technology.

Flanagan and Luo in Chapter 15 describe the determination of metal–hydrogen (M–H) phase diagrams. The uniquely fast diffusion kinetics of hydrogen even at ambient temperature makes it possible for the M–H system to reach equilibrium easily at relatively low temperatures. The authors introduce various ways to present the M–H phase equilibria, the useful rules relating the phase diagrams to the reaction enthalpy of the M–H systems, and the difference between local equilibrium and complete equilibrium in multicomponent systems. Various techniques for M–H phase diagram determinations in both open and closed systems are then discussed in detail.

Zhao and Westbrook in Chapter 16 try to tie up a few loose ends in this book by briefly discussing several topics, including the ever-increasing interplay of modeling and experiment, phase diagrams for functional materials, high-throughput phase diagram determination, and the low-temperature phase diagrams. A recommended list of further reading materials is also provided along with a list of recent compilations of phase diagrams.

It is impractical in a single volume to cover every method for phase diagram determination. The editor and the chapter authors try to strike a balance between breadth and depth of coverage in this book. While this book strikes a good balance, there will certainly be some methods that are not covered.

It is worth noting that no attempt was made in this book to cover the experimental measurement of thermodynamic quantities such as enthalpy of formation and specific heat capacity. Two recent books *Measurement of the Thermodynamic Properties of Single Phases* edited by A. Goodwin, K.N. Marsh, and W.A. Wakeham (Elsevier, 2003) and *Measurement of the Thermodynamic Properties of Multiple Phases* edited by R.D. Weir and T.W. de Loos (Elsevier, 2005) have provided comprehensive coverage on the subject. In this regard, the current book complements these two books; and together they provide complete coverage of the experimental determination of phase diagrams and thermodynamic quantities; both of which are essential inputs to computational thermodynamics.

I want to thank my wife Felicia (Fang) for her understanding and tolerance during the many evenings and weekends of writing and editing. Sincere appreciation is extended to the chapter authors who have devoted much time and effort to this project. I also want to thank Jack Westbrook and Mel Jackson for proof reading several chapters of this book. Thanks are also due to Elsevier for initiating the project and for all their efforts in editing and typsetting. Please don't hesitate to contact me if you find errors in this book and/or have suggestions for further improvements. I can be reached at jczhaoemail@gmail.com.

J.-C. Zhao
Latham, NY

CHAPTER ONE

INTRODUCTION TO PHASE DIAGRAMS

John F. Smith

Contents

1 Introduction	1
2 J. Willard Gibbs and the First Equilibrium Diagram	3
3 Twentieth Century Developments	5
4 The CALPHAD Method and Its Need for Data	8
5 Thermodynamic Constraints on Phase Diagrams	10
6 Experimental Considerations	17

1 INTRODUCTION

An interweaving of culture, materials, and technology has been characteristic of man's passage through time. This is illustrated by the fact that anthropologists and others who study the development of man characterize the periods of development as the Paleolithic Stone Age, Neolithic Stone Age, Bronze Age, Iron Age, and the Modern Age. This modern age has been called many things such as Nuclear Age, Space Age, Age of Science, etc. but it remains for the future to provide a lasting name. Materials are important because anything that is used must be made of some form of material. The relationship between materials and technology is self-evident. For instance, the way one shapes a stone into a useful instrument must be quite different from the way one shapes a piece of bronze into an equivalent instrument because the properties of stone are quite different from the properties of bronze. The effect of materials and technology on culture depends upon the relative utility of the things that are made and the time requisite to produce them. For instance the tools available in a modern butcher shop can convert a side of beef into edible portions is a small fraction of the time that it would take a primitive man to do the same job with a sharpened stone. Time is important because there is little room for progress if all of a man's time is consumed in the essentials of simply staying alive, but with a surplus of time man may pursue other activities.

This book is concerned with materials in the context of equilibrium phase diagrams. Various aspects of phase equilibria are of use in the industrial processing of

Prof. Emeritus, Dept. of Materials Science and Engineering, Iowa State Unversity, Ames, IA 50012 materials that are intended to function in useful applications. A quite simple but extremely important example can be found in the electronics industry, and, with the current ubiquitous use of computers, the importance of this contribution is self-evident. In the mid-twentieth century Bardeen, Brattain, and Schockley utilized theory to predict the possibility of a functional transistor. The prediction was soon verified with a point contact transistor. However, to produce large numbers of transistors, it was necessary to refine semiconductor materials to very high purity.

Pfann, a metallurgist at the Bell Laboratories in New Jersey, recognized that the terminal solubilities of impurities in a material differ between the liquid and solid states. In a phase diagram this would be indicated by the separation between the liquidus and solidus lines. Pfann therefore suggested purification by zone refining wherein successive passages of a liquid zone through the solid material utilizes the solubility differences between the liquid and solid to sweep the impurities toward one or the other end of the specimen being purified. Those impurity elements which are more soluble in the liquid than the solid will concentrate at the end toward which the liquid zone moves, and those impurity elements which are more soluble in the solid than the liquid will be concentrated in the other end. Obviously the technique has worked extremely well, and it is now possible to use ion implantation to re-introduce desired impurities into very high-purity semiconductors to produce a multiplicity of transistor junctions in one small chip.

Generally, industrial use of phase diagram information is rather more sophisticated than the above example. The fact that industry believes in the importance of phase diagrams is proven by the fact that industrial funding made possible the Alloy Phase Diagram Program. The continuing industrial need for phase equilibria data is attested by people who supply such data (see e.g., editorials by Agren [1] and Zhao [2]). The invention of the transistor has led to the availability of the ubiquitous computers. This ready availability has led to widespread application of the technique known as the *CALPHAD* method [3] for computer extrapolation of data for the phase equilibria of known systems to predict equilibrium data for more complex unknown systems.

Certainly calculation is more economical and less time consuming than experimental investigation. However, the classic cliché, *garbage in equals garbage out*, means that the calculating procedure must start with reliable input data. Therefore one of the reasons for this book is to describe the means by which an investigator may experimentally generate data in those cases where requisite input data are either of questionable reliability or are not available. A second reason for this book is that the passage of time always results in change, and there are new experimental approaches that need to be made generally known. An example is the diffusion–multiple approach for looking at multi-component systems. This and other examples may be found in the chapters of this book. A third reason is that the shift to computing from experimental investigation has been so extensive that the knowledge and techniques for experimental measurements that were developed during the latter half of the twentieth century are in danger of being lost because most of the people with that "know how" are no longer available.

With regard to the last reason, it seems particularly important at this time to put into the literature a book on experimental techniques to avoid such a loss of knowledge. Generally, experimental techniques, while simple in scientific concept, have an element of art in actual practice. For instance it is very simple and obvious to say that there are only three basic approaches for the determination of the Gibbs energy of formation of a phase, and these are by equilibration with phases of known chemical activity, by vapor pressure measurements, or by measurements with electrochemical cells. However, each of these approaches allows manifold choices, and, though one can set up a number of conceptual experiments to operate for the determination of a particular unknown quantity, construction and operation of an actual working apparatus, applicable to a particular system and temperature regime, requires appreciable "know how".

For instance, the electromotive force of an electrochemical cell can be measured with quite high precision and be highly reproducible and yet give completely erroneous data if one has inadvertently constructed a cell with an unforeseen and undetected side reaction. The heart of this problem is that many of the people who have developed expertise in one or another experimental technique have or soon will become inactive because of retirement or demise. These are the "journeymen" who should pass on their "know how" to the apprentices. Hopefully, some of the materials in this book will help to serve that transfer process.

2 J. WILLARD GIBBS AND THE FIRST EQUILIBRIUM DIAGRAM

In 1876 and 1878 J. Willard Gibbs [4] published work that provides the thermodynamic basis for understanding the equilibrium state of any arbitrary combination of substances under specified conditions. These publications were in the *Transactions of the Connecticut Academy of Sciences* and used rather abstruse mathematics so these were initially neither widely known nor understood, most certainly not in Europe. However, Gibbs had been in communication with James Clerk Maxwell, and Maxwell knew of and understood Gibbs' work. Though Maxwell died only a very few years after the publication of Gibbs' work, Maxwell had made known the work to Van der Waals who in turn made it known to H.W. Bakhuis Roozeboom. Bakhuis Roozeboom successfully incorporated Gibbs' concepts into his own work. This is relevant to the description that follows concerning experimental efforts by Roberts-Austen at the British Mint. The constraints imposed by Gibbs' work upon an equilibrium phase diagram will be discussed in more detail later in this chapter. However, the following description of Roberts-Austen's initial determination of a phase diagram stands in marked contrast to present day techniques for establishing equilibrium phase diagrams.

Sir William Chandler Roberts-Austen* made several significant contributions that pertain to phase diagrams and phase equilibria. Much of the following information can be found in a paper by Kayser and Patterson [5]. Roberts-Austen's career began in 1861 when he became a student at the Royal School of Mines in London and terminated with his death in 1902. During almost his entire career he held concurrent appointments at the Royal School of Mines and at the British Mint. What is believed to be the first temperature-composition plot was made by Roberts-Austen and was published [6] in 1875. The system that was measured was the Ag–Cu system. The experimental technique involved a type of calorimetry wherein an iron ball was in

---

*He was born William Chandler Roberts and his early publications are under the Roberts name. The Austen was added in 1885 in honor of a relative.

contact with the molten alloy. Upon first indication of solidification of the alloy, the Fe ball was removed and immersed in a known amount of water. With the heat capacities of water and iron having previously been measured within the temperature range of ordinary glass thermometers and with the heat capacity of iron at higher temperatures being estimated by extrapolation, the temperature rise of the water determined the heat transfer and hence the temperature of solidification of the initial alloy. Such a cumbersome technique was necessary because of the limits of the then available thermometry. Improvement awaited the development of a stable thermocouple, and a reliable Pt–Pt10%Rh thermocouple was developed by Le Chatelier in 1877. Today, of course, a variety of thermocouples are available with differing sensitivities and for different temperature regimes. For very high temperature measurements, optical pyrometers are available; these are based upon Max Planck's interpretation of black body radiation. For very low temperatures, the He vapor pressure thermometer, platinum resistance thermometer, and solid state thermometers are now available. Thus today's thermometry, which is basic to thermodynamic measurements, is significantly improved when compared to nineteenth century thermometry.

Roberts-Austen utilized the availability of the Le Chatelier thermocouple to design, construct, and then use a photographic strip chart recorder. This allowed direct graphic presentation of thermal arrests or slope discontinuities that are interpretable as phase transitions or phase boundaries. This is believed to be the first such recorder and the principles of its design are basically the same as those of present day pen and ink or digital recorders. Such recorders are powerful scientific tools and have been applied in many ways other than for thermal analysis. In passing, it may be noted that Sorby was contemporaneous with Roberts-Austen and he named the phases which he observed in his metallographic studies of iron–carbon alloys as follows: ferrite for the terminal bcc-Fe solution; austenite for the fcc-Fe solution; pearlite for the Fe–Fe_3C eutectic; cementite for the Fe_3C phase; and martensite for the hardening precipitate phase. Sorby chose the name austenite in honor of Roberts-Austen.

While the British Mint was necessarily interested in noble metals and their alloys, there was also interest in Fe–C alloys for die and stamping materials. In 1897, Roberts-Austen published [7] an Fe–Fe_3C diagram which added freezing point curves to the diagram published by Sauveur [8] the previous year. This 1897 Roberts-Austen diagram has been called "the first phase diagram". However, the temperature-composition diagrams of that period were plotted from dynamic measurements and the Roberts-Austen diagram was not an "equilibrium" diagram since it lacked compatibility with the equilibrium reversibility inherent in the thermodynamic constraints codified by Gibbs. Roberts-Austen did additional work to refine his temperature-composition diagram and published a refined version [9] in 1899. As previously noted, the work of Gibbs was not widely recognized in Europe during this period of time, but Roberts Austen was aware of the work through Bakhuis Roozeboom and invited Bakhuis Roozeboom to modify the 1899 temperature-composition diagram for compatibility with the Gibbs phase rule. The result was the publication [10] of a thermodynamically valid equilibrium Fe–Fe_3C diagram in 1900. Some further discussion of the constraints upon a phase diagram by Gibbs' work will follow later.

A final word concerning the contributions of Roberts-Austen is relevant to current developments in the use and applications of phase diagrams. Since many materials are used in "frozen in" non-equilibrium states and phase diagrams provide a means of showing where the material started and where equilibrium will be reached, the path that the material should follow in transforming from initial state to equilibrium is cogent to understanding the state where it might be in a non-equilibrium "freeze". Because of this interest, programs for the calculation of equilibrium phase diagrams are now being extended to include diffusion data. This interest has led to the recent extension of the title of the *Journal of Phase Equilibria* to the *Journal of Phase Equilibria and Diffusion*. The most probable path from state A to state B is of course the path that will maximize the time rate of production of entropy. This is equivalent to saying that the most likely path will follow the steepest energy gradient (e.g., a ball will roll down a hill in the direction of the steepest elevation contour). What Roberts-Austen contributed is the discovery of solid state diffusion. This happened when two Johannson blocks were clamped together during some work at the British Mint. Johannson blocks have very highly polished flat surfaces. At the conclusion of the work the clamp was forgotten and was left in place for a prolonged period of the order of years. When the blocks were again needed and the clamp was removed. It was found that the blocks had become joined by a cold weld. Roberts-Austen concluded that this meant that atoms were in motion within the solid, that is, there was diffusion.

3 TWENTIETH CENTURY DEVELOPMENTS

The Fe–Fe$_3$C diagram has been shown to be a metastable phase diagram, but has been found to be quite useful in guiding developments in the steel industry. The Fe–C diagram without the metastable Fe$_3$C has also been proven useful in cast iron technology. The utility of these phase diagrams has prompted the experimental determination of a large number of other phase diagrams with the number of such determinations increasing with time through most of the twentieth century.

A broad variety of experimental techniques have been used in these determinations but some of the more common serve as examples. Metallography is widely used and is traceable to Widmanstätten. Widmanstätten was a curator of a museum in Vienna in which a large iron–nickel meteorite was on display. He thought that the polished surface of this meteorite was beautiful, and, since he had at one time worked with etching, he etched the surface and made prints of the pattern and distributed them. Sorby became aware of the Widmanstätten prints and used an optical microscope with polished and etched specimens to make the aforementioned study of the microstructures of iron–carbon alloys. Metallography has an advantage over measurements that involve changes in physical properties such as dilatometry, thermal analysis, etc. in that the microstructures can offer clues concerning invariant reactions. For example, a lamellar structure is indicative of the presence in the system of a eutectic reaction, and rimming is indicative of a peritectic reaction. Quantitative metallography can be used with the Lever Law to determine phase boundaries. Optical metallography is still widely used but the general topic of metallography has now been expanded to include results from electron microscopy. Electron microscopy offers the capability of greater magnification and in addition allows quantitative determination of the compositions of individual grains provided that the microscope is equipped with suitable attachments for composition analysis.

Another powerful tool for the determination of phase diagrams is diffraction. Initially this was X-ray diffraction, but during the past century electron and neutron diffraction have been developed. Each of the three has its own advantages and limitations. X-ray diffraction is the least costly and can be used in a variety of ways in the establishment of phase diagrams. A simple X-ray diffraction pattern can be used for phase identification. X-ray diffraction can also be used for the establishment of the loci of phase boundaries. For instance, in a two-component system, lattice parameters may vary with composition within a single-phase region but are invariant in a two-phase region. Additionally, with measurement of precision lattice parameters, X-ray diffraction can be used to determine gram atomic volumes.

Electron diffraction requires operation in vacuo and can do the things that X-ray diffraction does. Its advantage is that it can focus on a small area so that it can be used to examine individual grains within a microstructure and, as noted above, with the proper attachments can quantitatively determine grain compositions. While X-rays and electrons are scattered by the electron distribution in a solid, neutrons are scattered by nuclei and by magnetic dipoles. Thus neutrons can be used to determine magnetic structures and can distinguish other structures which are difficult or impossible to see with either X-ray or electron diffraction. For instance with X-ray or electron diffraction patterns from systems involving very light elements with heavy elements (e.g., H in Nb), the scattering power of H atoms with one electron is so much less than that of Nb with 41 electrons that the Nb masks the H contribution in the diffraction pattern. This is not true for neutron diffraction patterns, particularly not when D (deuterium) is substituted for H (protium). A similar difficulty arises when the components of a system are closely comparable in atomic number (e.g., Tl and Pb) so that it is difficult to distinguish one species from the other with X-ray or electron diffraction but not with neutron diffraction.

During the twentieth century a great many phase diagrams, largely binary diagrams, have been determined. The stimulus was the increasing diversity of materials that were used. Work in the atomic energy program, in the aerospace industry, and in electronic applications has led to the utilization in one way or another of the vast majority of the elements in the periodic chart. This diversity has led to a need for appropriate phase diagrams to guide the processing of these materials. This in turn has led to the development of more powerful research tools like the electron microprobe, the electron microscope, and other sophisticated electronic devices for monitoring and control of experiments. The first compilation of binary phase diagrams by Max Hansen was published in German in 1936 and was translated, expanded, and published [11] in English in 1958. This compilation contained more than 1300 complete or partial phase diagrams. Two supplements [12,13] followed, one in 1965 and the other in 1969.

After 1969 compilation of phase diagrams languished, though, experimental determinations continued. The information from those experiments was widely scattered through the technical literature and was accessible only with the expenditure of significant effort. Industrial need for ready access to phase diagram information prompted the US National Academy of Sciences to form an ad hoc Panel on Phase Diagrams for Alloys to look into needs for critically evaluated alloy phase diagram data. Because of the large number of possible phase diagrams (~4000 binary systems, ~100,000 ternary systems, etc.) and with experimental work being done in laboratories around the world, it was readily evident that such a program needed worldwide participation. By early 1978, ASM International and NIST (National Institute of Standards and Technology) were chosen to act in concert as the entity to organize this critical compilation. Eleven organizations from around the world participated and 40 category editors from a variety of countries were selected to do the critical evaluation. Funds were raised from industry during a year of extensive travel by Ed Langer and George Bodeen, then respectively Director and President of ASM International. The industrial importance of phase diagrams is proven by the fact that these two gentlemen were able to solicit from 48 industrial contributors funding at a level totaling four million dollars even though the economy at that time was in a state of stress. This money was used for an Alloy Phase Diagram Program for critical evaluation of published phase diagram data. To date, evaluation of data available prior to the mid-1980s for the binary systems has been evaluated and a first edition was published in 1986 with a second edition being published in 1990 in a three-volume compilation [14]; this effectively updates the earlier compilations. Post-1990 publications are regularly noted by Hiro Okamoto in the Supplemental Literature Review section of the *Journal of Phase Equilibria and Diffusion*.

Evaluation of ternary data continues with the major effort being by Effenberg and associates coordinated through the Max Planck Institute in Stuttgart, Germany, and by Raghavan in Kolkata (current Indian name for Calcutta), India. While the overall effort has been quite successful and useful, the undertaking has shown that data for a large number of systems are quite meager while for others data are conflicting with little basis for choosing between the differences. This is indicative of definite deficiencies in the availability of experimental data for a large number of systems. Thus it is quite evident that there is a need for additional experimental work.

The CALPHAD method, which is outlined below, uses both phase diagram data and thermodynamic data to develop a self-consistent equilibrium phase diagram. Representative compilation activity for phase diagram data has just been noted so it seems appropriate to list also representative thermodynamic compilations. An early compilation was published in 1943 by Weibke and Kubaschewski. Kubaschewski continued the compilation activity expanding the available data through six editions of *Materials Thermochemistry* with the most recent edition [15] appearing in 1993. The compilation of Knacke and Barin converted the Kubaschewski data from a 0K reference base to a 298K reference base with the intent of simplifying computations for industrial usage. An extensive amount of compiling of various thermodynamic quantities was done at the US Bureau of Mines, and a number [16] of monographs, each on a specific thermodynamic quantity were published. Ralph Hultgren at the University of California at Berkeley headed a compilation team that published [17] evaluated data for the elements and for a number of binary systems. A significant amount of compiling of thermodynamic data has also been done at the US Bureau of Standards (now NIST). An early result of this team effort was Circular 500 published in 1952. This evaluation was combined with data for the elements edited by Stull and Sinke, published by the American Chemical Society in 1956, to result in the JANAF Tables whose second edition was published [18] in 1971. These tables contain a detailed description of the methods that were used in evaluating the thermodynamic data.

4 THE CALPHAD METHOD AND ITS NEED FOR DATA

The thermodynamic base for establishing equilibrium conditions has been noted as has the applications of the Gibbs constraints to the evaluation of an actual phase diagrams by Bakhuis Roozeboom. This prompted Larry Kaufman to apply the Gibbs energy minimization in a new direction. This was for the prediction of phase equilibria in systems not yet experimentally investigated. The process involves extrapolation from the equilibrium behavior of known systems to estimate the equilibrium behavior of as yet unmeasured systems. The procedure is known as the CALPHAD method and was described in 1970 in a book by Kaufman and Bernstein [3].

The logic of the approach is roughly described by the following. Any chemist will attest that the predominate interaction contributing to the bonding energy of a substance arises from the near neighbor interactions with interaction energies decreasing rapidly with increasing distance. Thus the primary terms in describing the Gibbs energy of any arbitrary system should consist of the Gibbs energies necessary: (a) to convert the components from the form of the chosen reference states to the form of a phase of interest, (b) to sum the interactions between all atom pairs within the phase as a function of composition, and (c) to minimize the Gibbs energy of the system being evaluated for all competitive phases.

In some cases it may be necessary to consider ternary interactions (i.e., the effect of a third unlike near neighbor to a pair of atoms) or even include fourth-order interactions but just as the terms in a Taylor mathematical series decrease in value, the effect of higher order interactions are expected to decrease rapidly. This view is supported by crystal geometry where the number of nearest neighbors is limited by spatial considerations and is a maximum of 12 for atoms of equal size. In this geometry three atoms of different species may be nearest neighbors in the form of a triangle and, with some distortion, four different species may be nearest neighbors in the form a tetrahedron. However, it is well known that atoms prefer like or unlike neighbors so there is a "built in" bias toward one or another configuration with random aggregates being of low probability. Thus geometric and chemical considerations support the mathematical expectation that higher order interactions should decrease rapidly in importance with increasing order of interaction.

In practice, the CALPHAD method is an iterative method of adjusting the parameters that describe the Gibbs energies of various phases in a system to construct a phase diagram which is the best fit to available experimental thermodynamic and phase diagram data for the system. The person doing the calculation is free to weight the input data in the proportions that he thinks justified, and one of the reasons for this book is to give such people a better idea of the relative accuracies and precisions that are attainable by various experimental techniques. For ternary, quaternary, quinary, and other higher order systems, the CALPHAD process should begin with reliable assessments of the constituent binary systems that are to be combined to evaluate any higher order system. Again this is a reason for the publication of this volume because, if certain data are lacking and need to be determined experimentally, an investigator needs to know and choose among the competitive ways of acquiring the desired information.

Experience to date has shown that the binary interactions are of major importance. It is therefore crucial that a database be built that corresponds closely with reliable experimental data for binary systems. In those cases where experimental data are available for a ternary system, use of the CALPHAD method has confirmed that the dominant terms for describing the ternary equilibria are indeed the binary interaction parameters and, in those cases where ternary interactions need be considered, the contributions from the ternary interactions are of secondary and lesser significance. If (a) the CALPHAD method is to be used for the calculation of the expected equilibria of a selected system of arbitrary higher order and (b) the binary interactions are the primary contributions to the resultant equilibria, it seems obvious that the validity of the calculated results hinges critically upon the validity of the data for the binary systems.

Since it has already been indicated that the Alloy Phase Diagram Program has found many binary systems where appreciable data are questionable or lacking, there is evident an obvious need for additional experimental information. In the first three-quarters of the twentieth century experimental investigations were far more common than computational simulations. Now past the end of the century, the balance has been reversed. This reversal results from a combination of economics and time plus the near universal availability of the electronic computer. Experimental work requires time and expense while calculation by computer is comparatively fast and economical. It is now quite apparent that the effort devoted to computer calculation compared to that devoted to experimental evaluation is out of balance; this has been emphasized in a recent editorial by Roberto de Avillez [19]. There are two types of experimental investigations to be considered. One of these is commonly denoted as phase equilibria data and involves measurements of characteristics of the phase equilibria in a system such as invariant temperatures, phase boundaries, tie lines, etc. The other is denoted as thermodynamic data and involves measurements of thermodynamic functions which characterize a phase in a system. The subsequent chapters in this book cover various methods for experimental determination of one or the other type of data.

It seems particularly important at this time to put into the literature a book on experimental techniques to avoid loss of knowledge. Generally, experimental techniques, while simple in scientific concept, have an element of art in actual practice. For instance it has been noted (a) that thermometry is extremely important for all thermodynamic and phase diagram measurements because results are temperature dependent and (b) that the temperature being measured must represent the sample temperature. Heat energy flows to or from a body at any time there is a temperature gradient, so apparatus must be designed to minimize any unwanted heat flow to or from the test specimen. The ways of achieving this goal constitute the "art" of the experimentalist. A litany of problem areas in experimental thermodynamic measurements could be recited, but it is best to let the chapters speak for themselves. Suffice to say, the accumulation of thermodynamic data is highly relevant to CALPHAD calculations, but thermodynamic measurements suffer from nature's tendency to equilibrate temperature. The tendency toward thermal equilibration is characteristic of all substances, and this makes precise thermodynamic measurements among the more difficult of all scientific tasks. Since heat transfer is by conduction, convection, and/or radiation, even a vacuum does not suffice as a thermal barrier because heat can leak through a vacuum by radiation. Thus the most practical way of achieving a constant temperature of a sample is to guard it with a heat shield that is maintained at the same temperature as the test specimen. Hopefully, some of the material in this book will serve to transfer knowledge relevant securing reliable data from a variety of techniques for experimental measurement.

5 THERMODYNAMIC CONSTRAINTS ON PHASE DIAGRAMS

In this section there is a short discussion of the phase rule and the thermodynamic constraints that affect the details of a phase diagram. For more extensive discussions of phase diagrams, there are a number of publications available. A good review of phase diagrams may be found [20] in the Metals Handbook, Vol. 3, while a Readable Introduction is a book by Gordon [21]. Treatments by Rhines [22] and by Prince [23] are thorough discussions of the subject including complex multi-component systems. The present treatment is meant only to highlight the factors that need to be taken into account when constructing or examining a phase diagram.

For any system the number of arbitrarily alterable variables constitute the degrees of freedom, f. For a system in which $P\,dV$ work is the sole work contribution, the first law of thermodynamics in differential form is

$$dE = T\,dS - P\,dV \tag{1}$$

and T and P are system intensive variables. These two plus the number of components, c, yield the total number of variables, n. The number of phases, p, represent the number of ways in which the variables are related, m. Thus the number of degrees of freedom can be represented as

$$f = n - m = c - p + 2 \tag{2}$$

This is the phase rule, and this in combination with thermodynamic considerations determines the constraints that are placed upon a phase diagram.

The nature of phase diagrams may be approached by first considering a one component (unary) phase diagram (see Fig. 1.1). In a single-phase region of such a diagram $f = 1 - 1 + 2 = 2$, so temperature and pressure can be varied independently and still remain in the single-phase region. However, along the lines defining two-phase equilibria, there is only one degree of freedom, so T and P cannot be varied independently but must be varied in combination. This means that, when T is varied along a transition line, P must vary by the defined amount that is required to stay on the line. Finally, if there are three phases, there are no degrees of freedom, and the three phases are in equilibrium at a singular temperature and pressure combination, the triple point. The phase transitions in Fig. 1.1 can readily be shown to be first-order transitions. The liquid–gas transition line can be used as an example. Along the line defining the liquid–gas equilibrium, the Gibbs energies of the two phases must be equal, so:

$$\Delta_t G = \Delta_t H - T\Delta_t S = 0 \quad \text{which leads to} \quad \Delta_t H/T = \Delta_t S \tag{3}$$

At any given pressure at all temperatures below the transition temperature, $T < T_t$, the liquid is the stable phase so the Gibbs energy of the liquid must be lower than

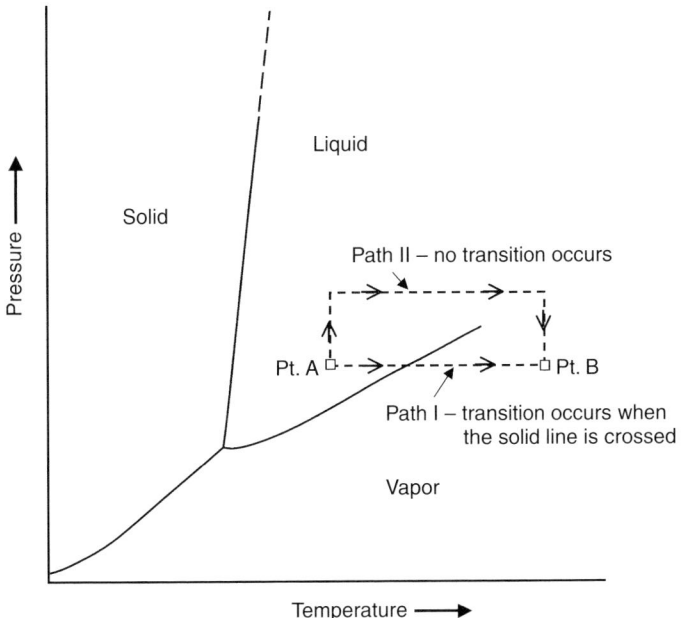

Figure 1.1 A generalized unary phase diagram.

the Gibbs energy of the gas, $G_l < G_g$. The converse is true for $T > T_t$. Thus there must be a slope discontinuity at T_t with dG_l/dT being more shallow than dG_g/dT. Such a slope discontinuity defines a first-order transition. It may also be noted that $dG/dT = -S$ so the slope discontinuity requires $\Delta_t S$, and hence $\Delta_t H$, to be finite at the transition point.

An interesting feature of the liquid–gas equilibrium line in Fig. 1.1 is that the line terminates at a point. This is a point at which $\Delta_t H$ and $\Delta_t S$ vanish and it is called a critical point. Beyond this point there is no distinction between liquid and gas. For any temperature at and above the critical temperature, the Gibbs energy as a function of pressure or temperature must be continuous without slope discontinuity so $\Delta_t H$ and $\Delta_t S$ vanish at the critical point. This must mean that they shrink with increasing temperature along the two-phase line to approach zero at the critical point. The existence of this critical point indicates that, from one point of view, a liquid is simply a highly compressed gas. Alternatively, it seems likely that there is also a critical point at very high pressures as a terminus of the solid–liquid transition line. Pressures sufficient to reach such a critical point might be found in the cores of some planets. For instance the core of the earth will not support a shear wave, and that is the characteristic of a liquid. However, a viscous core could result from high temperature as well as high pressure so the postulate is unproven. The existence of a solid–liquid critical point would indicate that a liquid is a distortion of a solid. Unfortunately, the liquid state of matter is, in contrast to gases and solids, very poorly understood. A detailed understanding remains a problem for the future.

Extending the phase rule to binary, ternary, and higher order systems is relatively straightforward, but displaying the information on a two-dimensional display becomes more difficult. For a binary system, a common way to display the information is a temperature-composition plot at constant pressure. Most often this is for normal atmospheric pressure but can be any pressure for which information is available. For such a plot the constant pressure reduces the constant in the phase rule from 2 to 1 so when the number of components is 2, the phase rule is:

$$f = c - p + 1 = 3 - p \qquad (4)$$

This relation shows that in a single-phase region there are two degrees of freedom. In this case temperature and composition may be varied independently. In a two-phase region temperature may be varied, but at any chosen temperature the compositions of the coexisting two phases are fixed. Changing the system composition at fixed temperature changes only the relative amounts of the two phases but the phase compositions remain unchanged. The relative amounts of the two phases can be determined by the lever rule. Let X_0 represent the system composition of interest, X_2 represent the value of the composition at the boundary between the two-phase region and phase two, and X_1 represent the value of the compositions at the boundary between phase one and the two-phase region. The fraction of phase one in the two-phase region at X_0 is $(X_2 - X_0)/(X_2 - X_1)$ while the fraction of phase 2 at the same composition is $(X_0 - X_1)/(X_2 - X_1)$; this is the lever law.

If three phases are present in equilibrium, $p = 3$ and the temperature and phase compositions are invariant. The four temperature-composition diagrams in Figs. 1.2(a)–(d) indicate a variety of ways in which such three-phase invariant reactions can exist. The number of such reactions is eight; these are designated as eutectic, peritectic, monotectic, eutectoid, peritectoid, monotectoid, metatectic, and syntectic reactions. The nature of the reactions are specified and schematically illustrated in Table 1.1. Examples of most of these reactions are illustrated in Fig. 1.2. In the Fe–Pd system of Fig. 1.2(a) there are: a peritectic reaction, $\gamma \leftrightarrow \delta + L$, near ~1500°C; a monotectoid reaction, $\gamma Fe_1 \leftrightarrow \gamma Fe_2 + \alpha Fe$ at 815°C with a critical point at 900°C; two eutectoid reactions, $\gamma Fe \leftrightarrow \alpha Fe + FePd$ at 605°C and $\gamma Fe \leftrightarrow FePd + FePd_2$ near 740°C; and three congruent transitions $\gamma \leftrightarrow L$ at 1304°C, $FePd \leftrightarrow \gamma$ at 790°C, and $FePd_2 \leftrightarrow \gamma$ at 820°C. In the Fe–Sc system of Fig. 1.2(b) there are: a metatectic reaction, $\delta Fe \leftrightarrow \gamma Fe + L$ at 1360°C; two eutectic reactions, $L \leftrightarrow \alpha Fe + \beta Fe_2 Sc$ at 1200°C and $L \leftrightarrow \alpha Fe_2 Sc + \beta Sc$ at 910°C; two peritectic reactions, $\gamma Fe + \alpha Fe_2 Sc \leftrightarrow \alpha Fe$ at 925°C and $\beta Sc + \alpha Fe_2 Sc \leftrightarrow FeSc_3$ at 800°C; and a eutectoid reaction, $\beta Sc \leftrightarrow FeSc_3 + \alpha Sc$ at 705°C. In Fig. 1.2(c), the Ga–I system illustrates a syntactic reaction and three eutectic reactions. The syntactic reaction is $L_1 + L_2 \leftrightarrow GaI$. Thus among the three systems of Figs. 1.2(a)–(c) there are examples of seven of the eight types of invariant reactions. The only type that is missing is the monotectic, but the diagram in Fig. 1.3(b) shows an example of a monotectic reaction with liquid at the point L dissociating to δ and a second liquid composition. A critical point occurs at the immiscibility maximum below the letter M.

Other constraints result from Gibbs work. An example of such a constraint is that the slope of the phase boundary of a single phase at the point of contact with a three-phase invariant reaction must extrapolate into a two-phase region. A second example is the requirement that, at a congruent transition, dX_t/dT for both phases must be zero. This may be hard to graph for phases with negligible ranges of homogeneity where the radius of curvature approaches an infinitesimal, but it is a

Table 1.1 The types of invariant reactions in a binary equilibrium phase diagram at constant pressure.

Name of reaction	Phase equilibrium	Schematic representation
Eutectic	$L \leftrightarrow s_1 + s_2$	$s_1 > \dfrac{s_1 + L \quad \overset{L}{\vee} \quad L + s_2}{s_1 + s_2} < s_2$
Peritectic	$s_1 + L \leftrightarrow s_2$	$s_1 > \dfrac{s_1 + L}{s_1 + s_2 \;\overset{\wedge}{s_2}\; s_2 + L} < L$
Monotectic	$L_1 \leftrightarrow s_1 + L_2$	$s_1 > \dfrac{s_1 + L \quad \overset{L_1}{\vee} \quad L_1 + L_2}{L + s_2} < L_2$
Eutectoid	$s_1 \leftrightarrow s_2 + s_3$	$s_2 > \dfrac{s_2 + s_1 \quad \overset{s_1}{\vee} \quad s_1 + s_3}{s_2 + s_3} < s_3$
Peritectoid	$s_1 + s_2 \leftrightarrow s_3$	$s_1 > \dfrac{s_1 + s_2}{s_1 + s_3 \;\overset{\wedge}{s_3}\; s_3 + s_2} < s_2$
Monotectoid	$s_{1a} \leftrightarrow s_{1b} + s_2$	$s_{1b} > \dfrac{s_{1b} + s_{1a} \quad \overset{s_{1a}}{\vee} \quad s_{1a} + s_2}{s_{1b} + s_2} < s_2$
Metatectic	$s_1 \leftrightarrow s_2 + L$	$s_2 > \dfrac{s_2 + s_1 \quad \overset{s_1}{\vee} \quad s_1 + L}{s_2 + L} < L$
Syntectic	$L_1 + L_2 \leftrightarrow s$	$L_1 > \dfrac{L_1 + L_2}{L_1 + s \;\hat{s}\; s + L_2} < L_2$

requirement of the Gibbs–Konovalov rule [24]. In any properly drawn binary equilibrium temperature-composition diagram, one will find that, at any selected temperature other than the temperature of an invariant reaction, the phase fields will alternate from single-phase field → binary phase field → single-phase field, repetitively.

Figure 1.2(d) shows a phase diagram for the Co–V system. This system is included to show that a second-order transition can result in a diagram with unusual boundaries. Most magnetic transitions are second order. In this case there is an inflection in the Gibbs energy as a function of a system variable so that the Gibbs energy and its first derivative with respect to the system variable are smooth without a slope discontinuity but the second derivative is zero, that is, an inflection point. Lev Landau [25] utilized group theory to define the conditions that allow a second-order transition

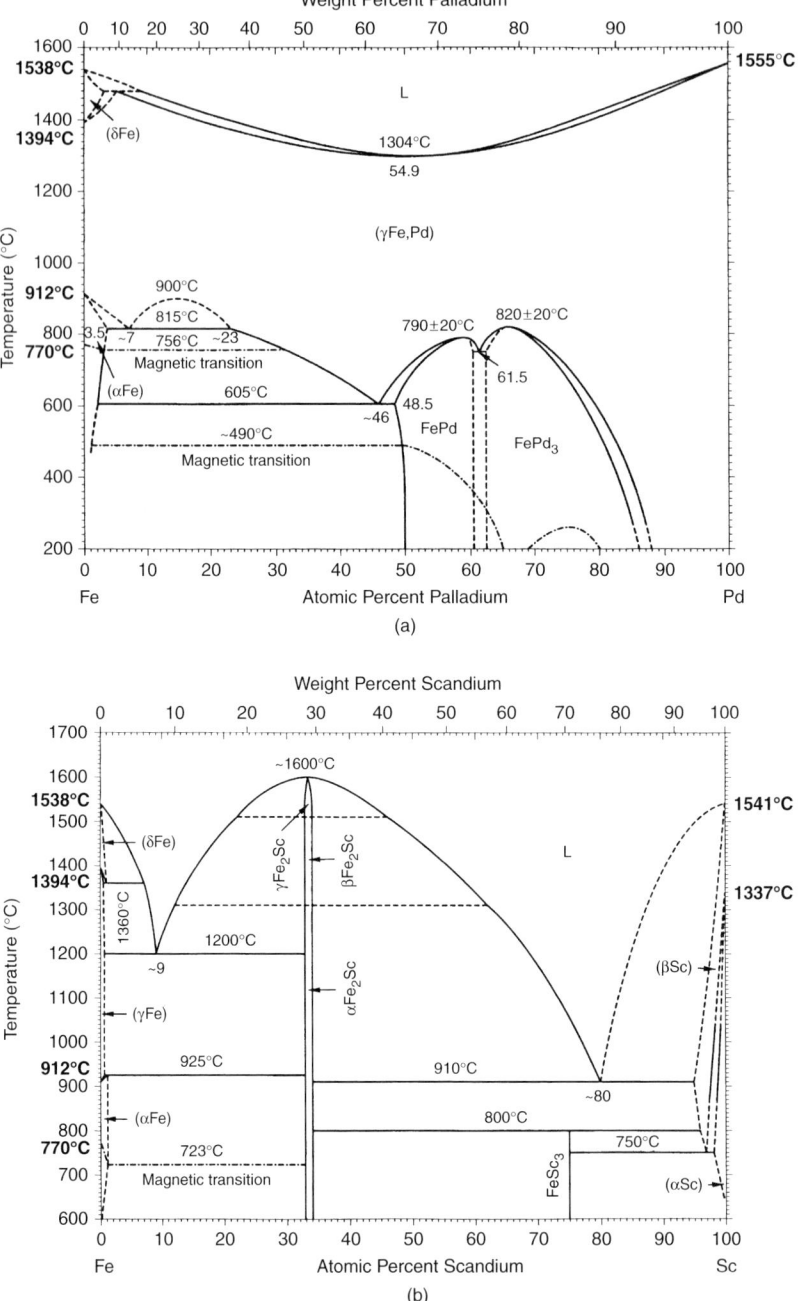

Figure 1.2 Representative phase diagrams [14] illustrating a variety of invariant reactions, congruent transitions, phase boundaries, and ranges of homogeneity: (a) the Fe–Pd system; (b) the Fe–Sc system.

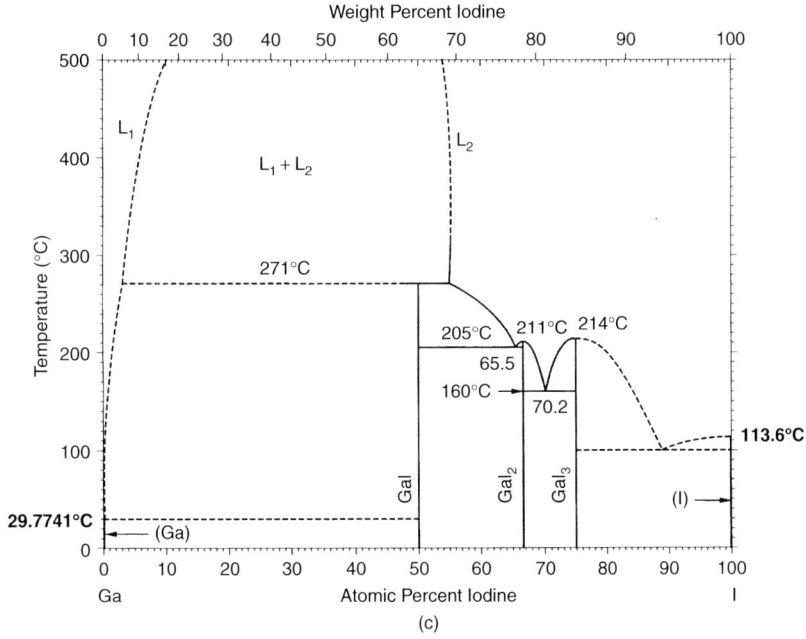

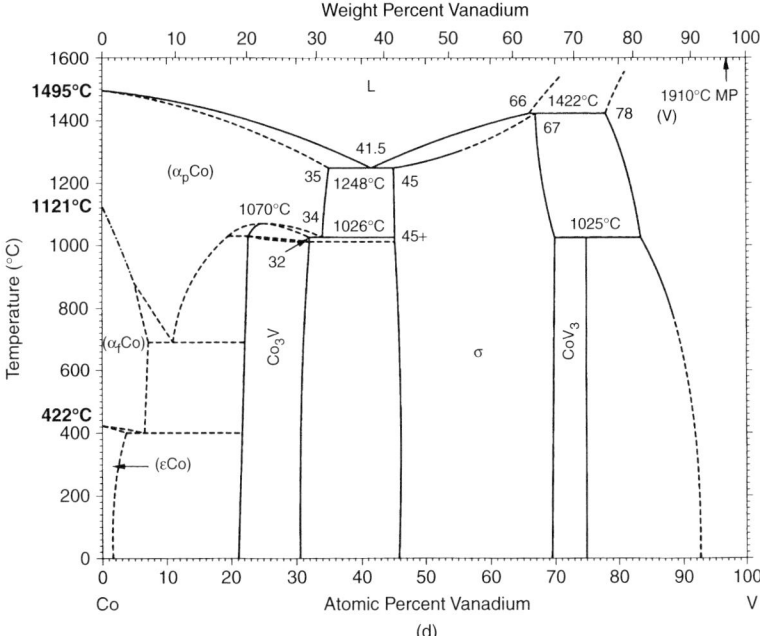

Figure 1.2 (Continued) (c) the Ga–I system; and (d) the Co–V system. Note that all these diagrams show that at any given temperature, the repetitive sequence is single-phase, two-phase, single-phase fields.

in solids. There are four conditions [26] that must be fulfilled to allow a second-order transition:

1. Two structures are involved in a second-order transition and all symmetry elements of the low-symmetry structure must be contained in the high-symmetry structure.
2. At constant composition, the loci of the particles (atoms or dipoles) in the two structures must be definably related.
3. There must be no third order or other odd-order terms in defining the Gibbs energy difference between the structures of the two symmetries.
4. It must be possible for the transformation between the two structures to occur via a continuous path.

These are necessary conditions for a second-order transition to occur, but meeting these conditions does not necessitate that a transformation will be second order, and an example will be noted where a second-order transition changes to a first-order transition There are three types of second-order transitions: order–disorder, displacive, and a combination of the first two.

Though a transition between an ordered magnetic state to a disordered state is most often a second-order transition, magnetic interactions can affect the geometry of an equilibrium diagram. Such effects have been discussed [27,28] in some detail, and magnetic interactions most commonly affect the geometry of phase boundaries. An unusual transition occurs in the Co–V system. In the Co-rich region of the Co–V system in Fig. 1.2(d), there is a ferromagnetic to paramagnetic transition, $\alpha_f Co \leftrightarrow \alpha_p Co$. This transition occurs in pure αCo at a Curie temperature of 1121°C. V additions to Co initially decrease the transition temperature in proportion to the amount of V added, and the transition is second order to a mole fraction of $X_V = \sim 0.03$. At that composition the transition changes from second order to first order with a two-phase region of $\alpha_f Co + \alpha_p Co$ intervening between the ferromagnetic solvus and the paramagnetic solvus. It was Inden [29] who first reported this behavior in the Co–V system. This phenomenon leads to a peculiar geometry in the phase diagram of the system if only first-order transitions and solvus boundaries are plotted; in that case the $\alpha_f Co$ solvus and the $\alpha_p Co$ solvus would converge at a singular point. Thus second-order transitions are normally plotted by lines that are discernibly different from the lines for first-order transitions. This was not done in Fig. 1.2(d) because of uncertainties in the locations of the boundaries in the Co-rich region; however, the existence of a two-phase $\alpha_f Co + \alpha_p Co$ seems well established.

The second-order magnetic transition is classed as a combination of order–disorder and displacive reactions because the magnetic ordering of magnetic dipoles can be determined by neutron diffraction and a displacive effect is shown by magnetostriction which is a structural distortion that accompanies magnetic ordering. The magnetostrictive effect is quite small in Fe, but in Ni it is sufficiently large that Ni can be used as a transducer in ultrasonic cleaning tanks such as those used for cleaning surgical instruments after a surgical operation. A similar type of second-order transition should describe ferroelectric transitions. For instance there has been shown to be a very small atomic displacement when barium titanate, $BaTiO_3$, orders ferroelectrically, but there is a definite electrostriction which is sufficient to allow $BaTiO_3$ to be used as an ultrasonic transducer in non-destructive testing.

Okamoto and Massalski [30] have drawn on their experience in examining the phase diagrams in the Massalski compilation [14] to discuss questionable constructions in phase diagrams. They constructed a phase diagram that purposely shows features that illustrate 20 kinds of detail that are questionable. The diagram and a more detailed discussion can be found in Chapter 3 of this book.

6 EXPERIMENTAL CONSIDERATIONS

This chapter will conclude with a few observations that are relevant to the acquisition of reliable data pertaining to both phase equilibria and thermodynamic data. For phase equilibria data, an earlier publication [31] has indicated that there are three primary criteria for generating reliable data. These are: (1) use of an experimental technique with adequate resolution, (2) establishment and retention of equilibrium for the regime of interest, and (3) adequate characterization of materials to ensure that the determination of equilibria is representative of the system of interest. With regard to the first of these criteria, *adequate resolution*, the original publication illustrated the importance of this criterion with results from the studies of the phase equilibria in the iron-rich region of the Fe–Ni system. Until the study by Romig and Goldstein [32] in 1981, the $\alpha/(\alpha + \gamma)$ and $(\alpha + \gamma)/\gamma$ boundaries between the bcc-Fe(α) and fcc-Fe(γ) phases were uncertain below ~500°C. One of the factors causing this was the fine-grained structure of the alloys in this temperature region. This difficulty was overcome by Romig and Goldstein's use of a scanning-transmission electronic microscope (STEM) with a resolution of $\sim 2 \times 10^{-8}$ m which is about one hundred times finer than the resolution of an electron probe microanalyzer (EPMA) at $\sim 2 \times 10^{-6}$ m. With this difference in capability for analyzing the fine-grained structures of quench-annealed microstructures when annealed below 500°C, the two boundaries were extended with reliability down to room temperature.

An example of an investigation [33] in which the second criterion, *establishment and retention of equilibrium for the regime of interest*, was not adequately considered is a CALPHAD assessment of the Cu–Nb system. Full responsibility for the interpretative failure in that case is due to the author of this chapter who was the research advisor for that investigation. The investigation involved measurements of the vapor pressure of copper over a two-phase field of liquid plus bcc-Nb at temperatures in the range 1400–2000K. These data allowed evaluation of the temperature dependence of the partial molar Gibbs energies of copper in the two-phase region. These experimental results were utilized with Gibbs energy values for the other phases in the system, slightly modified from those used in an earlier assessment by Kaufman [34] (assessment 1, dash-dot lines in Fig. 1.3(a)), to compute a phase diagram that was a good fit to the liquidus points of Allibert et al. [35] (assessment 2). This is the diagram shown by the dashed lines in Fig. 1.3(a). It may be seen that this initial fit closely reproduces the experimental liquidus points of Allibert et al. but indicates a eutectic reaction with liquid dissociating to the Cu-rich and Fe-rich terminal solutions with a eutectic temperature 4°C below the melting temperature of pure copper.

This eutectic disagreed with the then generally accepted phase diagram that indicated a peritectic melting point of Cu terminal solution above the melting

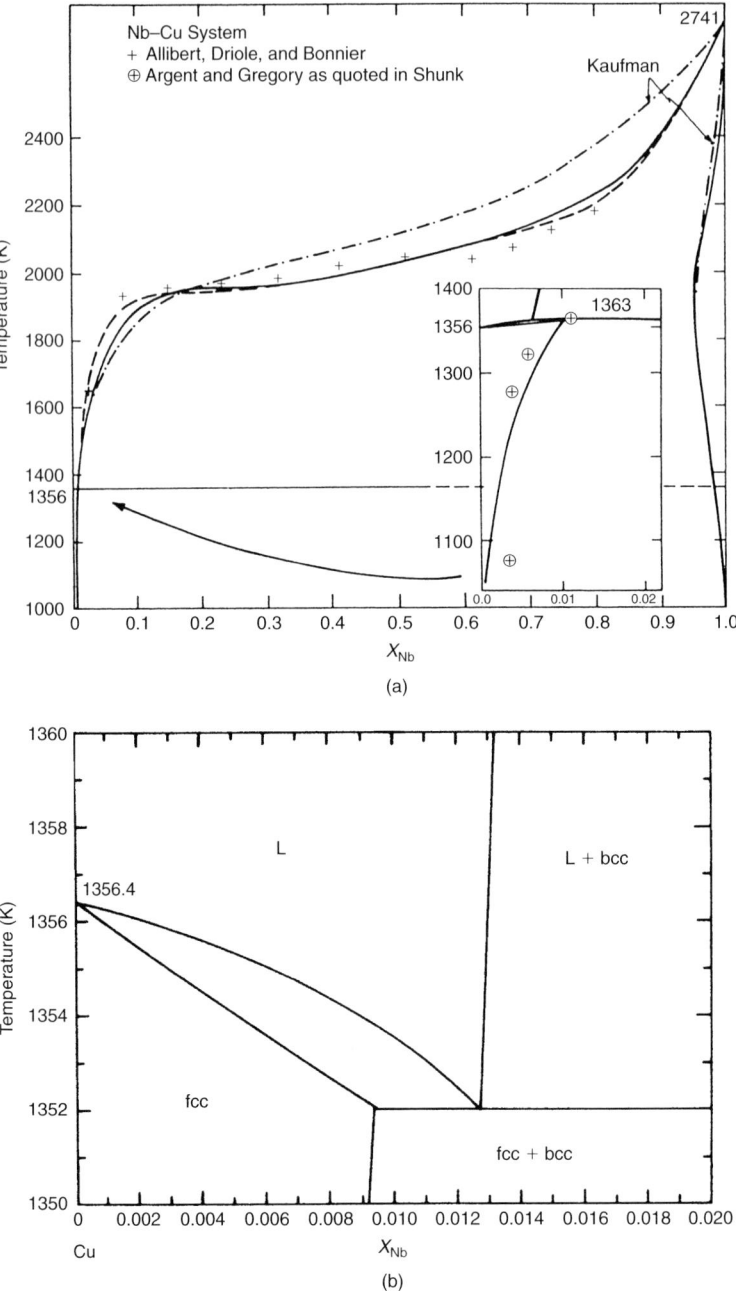

Figure 1.3 Phase diagram of the Cu–Nb binary system: (a) various version of the phase diagram with the dash-dot lines representing the original calculated approximation by Kaufman [34] before thermodynamic data for the liquid and Fe two-phase region were available, the cross (+) representing the experimental liquidus by Allibert et al. [35], the dashed lines representing the 1st CALPHAD fitting by Bailey et al. [33] using the liquidus data (equally weighted in the fitting) from Allibert et al. and some vapor pressure data from the Ames Laboratory, and the solid lines representing the second fitting by Bailey et al. with reduced weighting for the most Cu-rich points above 1900°C, predicting the then accepted peritectic reaction of the insert; and (b) the eutectic reaction predicted by the 1st fitting by Bailey et al. [33]. Experimental data postdating the last assessment is strongly in favor of the equilibrium diagram having a eutectic reaction at a temperature near the melting point of Cu.

point of pure copper. Accordingly, a second fit was made (assessment 3) with a reduction in the weighting of a Cu-rich liquidus point near $X_{Cu} = 0.08$ and 1950K. With this reduced weighting, a peritectic invariant ~5°C above the melting point of pure copper was produced in accord with the then available phase diagrams. This re-calculated diagram is shown by the solid lines in Fig. 1.3(a) with the peritectic reaction enlarged in the inset. This was accepted as correct and was published. Subsequently, more experimental work was done and the system was re-evaluated by Chakrabarti and Laughlin [36]. Their re-evaluation leaves open the question of whether the most Cu-rich invariant reaction is peritectic or eutectic. However, Verhoeven [37] has found eutectic microstructures in solid samples in this Cu-rich region indicative of a eutectic invariant. In the publication of the vapor pressure work, the present author overlooked the dictum of Oswald Kubaschewski – *high temperature data albeit of lesser precision are more likely to represent equilibrium than low temperature data even though of higher precision*. In view of all evidence, the second assessment of the system is most probably the best, and an enlarged view of the eutectic reaction in the Cu-rich region predicted by that assessment is shown in Fig. 1.3(b). Undoubtedly some of the confusion about this system originates in the sensitivity to atmospheric contamination. This is relevant to point three concerning acquisition of reliable data.

With regard to point three, *in any experiment the test specimens should be well characterized with respect to composition*, so the experimentalist should certainly be confident that any impurities in his/her test specimens are not in quantities sufficient to modify meaningful results. It needs to be emphasized that very minor contamination can sometimes make a significant difference to phase equilibria. In the Cu–Nb system just discussed, relatively small amounts of oxygen contamination can result in a flattening of the liquidus to produce a monotectic reaction with an attendant liquid immiscibility. Another example is the allotropic behavior of high-purity calcium metal. Pure calcium metal has been found to have only two allotropes, an fcc form at lower temperatures and a bcc form at higher temperatures. However, intentional contamination [38] with minor amounts of hydrogen causes the appearance of an hcp form at intermediate temperatures. This effect from hydrogen addition has been confirmed by a study of the Ca–H system by Peterson and Fattore [39]. They found a hexagonal phase that was stabilized by hydrogen to exist over the temperature range formerly reported to be the temperature range of stability of a hexagonal allotrope of calcium metal. In considering contamination, it should be noted that, when purities are quoted in mass ppm, the purity may look very good. However, when the atomic weights are drastically different as in the case of hydrogen with either calcium or strontium, conversion of mass ppm to atomic ppm makes the level of contamination more meaningful in the sense that the number of contaminant atoms per constituent atom becomes much larger. A statement quoting 99.99 mass % purity leaves open a contaminant amount of 100 mass ppm, and this should be multiplied by ratio of the gram atomic mass of the solvent species to the gram atomic mass of the impurity species to determine the significance in terms of the atom ratios. Thus if the mass ratio is large, the contaminant level could be much more significant than the quoted purity, for example, for H in Ca, 0.1 mass % becomes 4 at. %.

Finally, the accumulation of thermodynamic data is highly relevant to evaluation of phase equilibria, but thermodynamic measurements suffer from nature's tendency to equilibrate temperature. Thus thermal isolation of a body is extremely difficult because all things transfer heat – even a vacuum by radiation. The transfer is by conduction, convection, or radiation. Metallic solids are primarily electronic conductors while a general solid conducts mostly by smoothing and randomizing vibrational gradients. In addition to these two modes of heat transfer, semiconductors add infrared radiational transfer in the interband region. Liquids transfer heat primarily by convection, and gases transfer heat by convection and collisional exchange. Thus experimental techniques for the determination of thermodynamic quantities are replete with designs to obviate unwanted heat losses or gains and to assure that samples are not in a contaminating environment. It is hoped that the reader will find some answers in the subsequent chapters that are useful for serving his or her own particular needs.

REFERENCES

1. J. Agren, *J. Phase Equilib.*, 19 (1998) 2.
2. J.-C. Zhao, *J. Phase Equilib. Diff.*, 26 (2005) 106.
3. L. Kaufman and H. Bernstein, *Computer Calculation of Phase Diagrams*, Academic Press, New York, 1970.
4. J.W. Gibbs, *Transactions of the Connecticut Academy of Sciences*, Part I (1876); Part II (1878); *Scientific Papers of J. Willard Gibbs*, Ox Bow Press, Woodbridge, CT, 1994.
5. F.X. Kayser and J.W. Patterson, *J. Phase Equilib.*, 19 (1998) 11.
6. W.C. Roberts, *Proc. Roy. Soc. London* 23 (1875) 481.
7. W.C. Roberts-Austen, Report 4, *Proc. Inst. Mech. Eng.* 70 (1897) 33.
8. A. Sauveur, *Trans. Amer. Inst. Mining Eng.* 24 (1896) 863.
9. W.C. Roberts-Austen, Report 5, *Proc. Inst. Mech. Eng.* 72 (1899) 35.
10. H.W. Bakhuis Roozeboom, *Metallographist*, 3 (1900) 293.
11. M. Hansen and K. Anderko, *Constitution of Binary Alloys*, McGraw-Hill, New York, 1958.
12. R.P. Elliott, *Constitution of Binary Alloys*, First Supplement, McGraw-Hill, New York, 1965.
13. F.A. Shunk, *Constitution of Binary Alloys*, Second Supplement, McGraw-Hill, New York, 1969.
14. T.B. Massalski, H. Okamoto, P.R. Subramanian and L. Kacprzak (eds.), *Binary Alloy Phase Diagrams*, 2nd edn., Vols. 1, 2 and 3, ASM International, Materials Park, OH, 1990.
15. O. Kubaschewski, C.B. Alcock and P.J. Spencer, *Materials Thermochemistry*, 6th edn., Pergamon Press, New York, 1993.
16. K.K. Kelley, Contributions to the data on theoretical metallurgy, *U.S. Bur. Mines Bull.*, 383 (1935); 393 (1936); 476 (1949); 477 (1950); 584 (1960); 592 (1961).
17. R.R. Hultgren, P.D. Desai, P.D. Hawkins, M. Gleiser, K.K. Kelley and D.D. Wagman, *Selected Values of Thermodynamic Properties of the Elements*, American Society for Metals, Metals Park, OH, 1973; R.R. Hultgren, R.L. Orr, P.D. Anderson and K.K. Kelley, *Selected Values of Thermodynamic Properties of Metals and Alloys*, Wiley, New York, 1963.
18. D.R. Stull and H. Prophet, Project Directors, *JANAF Thermochemical Tables*, 2nd edn., US Government Printing Office, Washington, D.C, 1971.
19. R.R. de Avillez, *J. Phase Equilib. Diff.*, 26 (2005) 206.
20. H. Baker, Introduction to alloy phase diagrams, in *ASM Handbook Vol. 3: Alloy Phase Diagrams*, ASM International, Materials Park, OH, 1992, p. 1.
21. P. Gordon, *Principles of Phase Diagrams in Materials Systems*, McGraw-Hill, New York, 1968.
22. F.N. Rhines, *Phase Diagrams in Metallurgy*, McGraw-Hill, New York, 1956.
23. A. Prince, *Alloy Phase Diagrams*, Elsevier, New York, 1966.
24. D.A. Goodman, J.W. Cahn and L.H. Bennett, *Bull. Alloy Phase Diagr.*, 2 (1981) 29.
25. L. Landau and L. Lifshitz, *Statistical Physics*, Pergamon Press, New York, 1950, p. 30.
26. J.F. Smith, *J. Phase Equilib. Diff.*, 26 (2005) 5.
27. A.P. Miodownik, *Bull. Alloy Phase Diagr.*, 2 (1982) 406.

28. G. Inden, *Bull. Alloy Phase Diagr.*, 2 (1982) 412.
29. G. Inden, *Physica*, 103B (1981) 82.
30. H. Okamoto and T.B. Massalski, *J. Phase Equilib.*, 12 (1991) 148.
31. J.F. Smith, *Mater. Sci. Eng.*, 48 (1981) P1.
32. A.D. Romig Jr. and J.I. Goldstein, *Metall Trans. A*, 11 (1980) 1151.
33. D.M. Bailey, G.R. Luecke, A.V. Hariharan and J.F. Smith, *J. Less-Common Met.*, 78 (1981) 197.
34. L. Kaufman, *CALPHAD*, 2 (1978) 117.
35. C. Allibert, J. Driole and E. Bonnier, *Comptes Rendus Acad. Sci. Ser. C*, 268 (1975) 1597; C. Allibert and J. Driole, *J. Less-Common Met.*, 51 (1977) 25.
36. D.J. Chakrabarti and D.E. Laughlin, *Bull. Alloy Phase Diagr.*, 2 (1982) 455.
37. J.D. Verhoeven, Iowa State University, Private communication, 1982.
38. B.T. Bernstein and J.F. Smith, *J. Electrochem. Soc.*, 106 (1959) 448.
39. D.T. Peterson and V.G. Fattore, *J. Phys. Chem.*, 65 (1961) 2062.

CHAPTER TWO

The Role of Phase Transformation Kinetics in Phase Diagram Determination and Assessment

J.-C. Zhao

Contents

1 Introduction 22
2 Phase Transformation Kinetics During Cooling and Heating as Related to Phase Diagram Determination 24
  2.1 Shifting of Transformation-Start Temperature with Cooling Rate 26
  2.2 Formation of Metastable Phases During Cooling 27
  2.3 Shifting of Transformation-Start Temperature with Heating Rate 30
  2.4 Analyses of Examples from Cooling and Heating Experiments 35
3 Isothermal Phase Transformation Kinetics as Related to Phase Diagram Determination 41
  3.1 Precipitation of Phases from Quenched Alloys 41
  3.2 Kinetics and Phase Formation in Diffusion Couples/Multiples 46
4 Concluding Remarks 48

1 Introduction

There are two general categories of methods for the determination of phase diagrams [1]. The first category of methods detects the phase boundaries (transuses) by means of thermal, physical, chemical, mechanical, or other property changes during cooling and/or heating of an alloy, and therefore studies the approach of a system to the equilibrium state, as shown in Fig. 2.1. When an alloy with composition X_2 is slowly cooled to T_s, α phase starts to form in the alloy. Because the thermal, physical, chemical, mechanical, and/or other properties of the α phase are usually different from those of the γ phase, by monitoring the variation of a specific property (e.g., electrical resistance) with temperature, or by measurement of the temperature as a function of time (thermal analysis), the transformation-start temperature (T_s) and transformation-finish temperature (T_f) can be evaluated by locating the temperatures at which the property vs. temperature or temperature vs. time curves

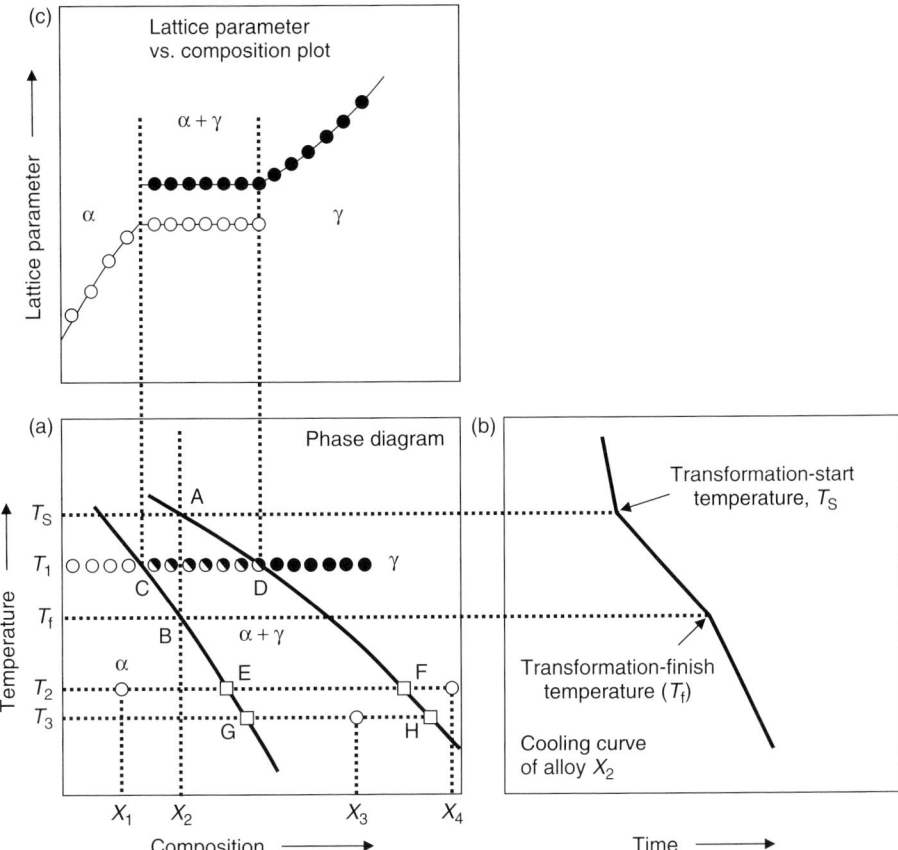

Figure 2.1 Schematic diagram showing two methods for the determination of phase diagrams: (a) schematic phase diagram; (b) the schematic cooling curve of a specific alloy X_2; and (c) schematic plot of lattice parameters vs. composition.

display a change of slope, as shown in Fig. 2.1(b). Theoretically, the T_s and T_f temperatures from several alloys can be used to define the related equilibrium phase boundaries. However, this ideal situation can only be approached when the cooling rate is infinitesimally small and the phase transformation from γ to α starts exactly at T_s and completes exactly at T_f. It is well established that a certain undercooling is necessary to nucleate a particle of a new phase with a size above some critical value. Thus, the real T_s is normally shifted toward lower temperatures (to produce a certain undercooling) during the cooling process. It is important to understand the phase transformation kinetics during cooling and heating, especially how the T_s shifts as a function of cooling rate, to correctly determine and assess equilibrium phase diagrams [1].

The other category of methods employs equilibrated alloys or diffusion couple specimens, and therefore studies the behavior of a system in an equilibrium or local equilibrium state. A series of alloys with different compositions can be annealed at a specific temperature, say T_1, as shown in Fig. 2.1. Some property is measured as a function of composition to test for differences in the state of equilibrium. For example, X-ray diffraction (XRD) may be used for lattice parameter measurement, and the phase boundaries (point C and D in Fig. 2.1(a)) can be located by sharp changes in the lattice parameter vs. composition plot. The lattice parameters become constant in two-phase alloys as schematically shown in Fig. 2.1(c). An alternative method, which takes advantage of the local equilibrium at phase interfaces, locates phase boundaries by means of diffusion couples. For instance, a diffusion couple made up of alloys X_1 and X_4 can be heat treated at temperature T_2 for an extended time period. By studying the compositions as a function of location using techniques such as electron probe microanalysis (EPMA) and extending the compositions to the α/γ phase interface, the equilibrium phase boundaries (points E and F in Fig. 2.1(a)) can be determined by assuming local equilibrium at the phase interface. (Detailed description of this methodology can be found in Chapter 6 of this book). Another way to determine the phase boundaries is to make an alloy such as alloy X_3 and anneal it at a temperature (i.e., T_3 in Fig. 2.1 where it is in a two-phase region) for an extended period of time to reach equilibrium. EPMA can provide the tie-line compositions (points G and H in Fig. 2.1). The alloys used for phase diagram determination are usually melted and then homogenized at a high temperature, and subsequently annealed at a lower temperature such as T_1 or T_3. During the low-temperature annealing, phase precipitation takes place in the two-phase region. It is very important to understand the phase transformation kinetics in order to know whether the annealing time is long enough to reach equilibrium. In addition, the alloys are usually quenched from the test temperature to room temperature for XRD, EPMA, or other analysis. It is important to know whether the equilibrium phases are retained to room temperature or whether a phase transformation takes place during quenching. Thus, for this category of methods, an understanding of the phase transformation kinetics is also very important for the correct determination and assessment of phase diagrams.

This chapter will first explain some basic phase transformation kinetics during cooling and how it affects the measured phase boundaries. Emphasis is placed on how the cooling process shifts the transformation-start temperature away from that for the equilibrium state, often toward that for metastable states [1]. Similarly, heating transformation kinetics will also be briefly introduced and the complication of using heating data to assess equilibrium phase diagrams will be discussed. Subsequently, isothermal transformation kinetics will be briefly introduced and its implication to phase diagram determination will be explained with practical examples. Brief concluding remarks will be provided at the end of this chapter.

2 PHASE TRANSFORMATION KINETICS DURING COOLING AND HEATING AS RELATED TO PHASE DIAGRAM DETERMINATION

Equilibrium states and metastable states are differentiated by their Gibbs energy. An equilibrium state (phase) is the lowest Gibbs energy state (phase) a system can attain under specific external constraints (i.e., temperature, pressure, etc.). A metastable state (phase) has a higher Gibbs energy than the equilibrium state, but it is still stable against fluctuation (i.e., $dG = 0$) [2].

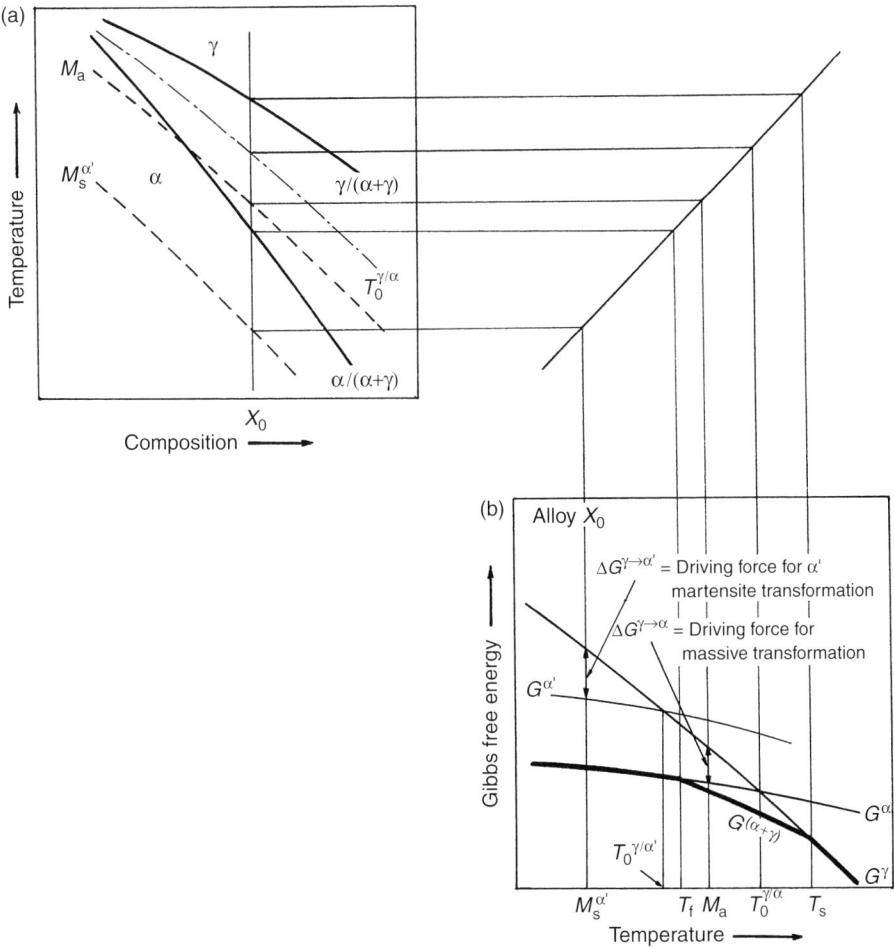

Figure 2.2 Schematic diagram illustrating the variation of Gibbs energy of various stable and metastable phases with temperature. The heavy lines represents the equilibrium (stable) phase(s) at each temperature [1]. See the text for detail.

To understand the formation of equilibrium and metastable phases during cooling, it is worthwhile to examine the temperature dependence of their Gibbs energy, as shown in Fig. 2.2 [1]. The equilibrium phase of alloy X_0 at each temperature is indicated by the heavy line (with the lowest Gibbs energy). At temperatures higher than T_s, the Gibbs energy of the γ phase, G^γ, is lower than the Gibbs energy of the α phase, G^α, so the γ phase is the stable phase. At a temperature between T_s and $T_0^{\gamma/\alpha}$ (the latter defined as the temperature at which $G^\gamma = G^\alpha$), although G^γ is still lower than G^α, the Gibbs energy of the $(\alpha + \gamma)$ two-phase mixture, $G^{(\alpha+\gamma)}$, is lower yet, so the $(\alpha + \gamma)$ two-phase mixture is the equilibrium state. However, the $\gamma \to (\alpha + \gamma)$ transformation can only be realized by long-range diffusion, and thus could take a significantly long time to take place. When the temperature is lower than

$T_0^{\gamma/\alpha}$, G^α is lower than G^γ, and therefore direct $\gamma \rightarrow \alpha$ massive transformation (which produces a massive α phase with composition exactly the same as the γ phase, and involves only local diffusion at the transformation front) becomes thermodynamically possible. However, this transformation can only be realized at a temperature, M_a, at which the driving force, $\Delta G^{\gamma \rightarrow \alpha}$, is large enough to initiate a massive α nucleus. At temperatures below T_f, G^α is the lowest, thus the α phase becomes the stable phase. Suppose a metastable phase, α' martensite, exists in this system; then its Gibbs energy, $G^{\alpha'}$, should be above G^α, and correspondingly, $T_0^{\gamma/\alpha'}$ (as defined by $G^\gamma = G^{\alpha'}$) should be lower than $T_0^{\gamma/\alpha}$, as shown in Fig. 2.2 But the α' martensite transformation can only take place at a temperature, $M_s^{\alpha'}$ (which is below $T_0^{\gamma/\alpha'}$), at which the driving force for the α' martensite transformation, $\Delta G^{\gamma \rightarrow \alpha'}$, is large enough to initiate a martensite nucleus. If other metastable phases exist and their Gibbs energies are higher than $G^{\alpha'}$, they can only be formed at temperatures below $M_s^{\alpha'}$. In this connection, the number of possible metastable phases increases with lower temperatures [1].

2.1 Shifting of Transformation-Start Temperature with Cooling Rate

As mentioned before, only when an infinitesimally small cooling rate (thus an infinitely long cooling time) is used, can the true equilibrium phase boundaries be obtained. In reality, phase transformation kinetics plays an important role. A certain undercooling is necessary to form critical nuclei of a new phase to allow the phase transformation to be detected by a certain method, (e.g., dilatometry or differential thermal analysis (DTA)). In other words, the transformation-start temperature is normally shifted away from an equilibrium phase boundary towards lower temperatures (to produce a certain undercooling) during the cooling process. With increasing cooling rate, the formation of critical nuclei becomes more difficult, and so does the diffusion process; thus greater undercooling is necessary. An example is shown in Fig. 2.3(a) for a Ti–17.5 wt.% Ag (8.6 at.%) alloy [3]; it is clear that the higher the cooling rate, the lower is T_s. By connecting the T_s points of a series of cooling curves, as shown by dashed line in Fig. 2.3(a), a so-called "continuous cooling transformation" (CCT) diagram is obtained. An alternative way to describe the cooling behavior is to directly plot the transformation temperature (T_t) vs. the cooling rate ($\dot{T}$), as shown in Fig. 2.3(b), again for Ti–Ag alloys [3]. The CCT and T_t–$\dot{T}$ diagrams are equivalent because both record the relationship between the transformation temperature and the cooling rate [1,3,4]. For the remainder of this paper, the T_t–$\dot{T}$ diagram will be employed to describe the cooling and heating transformation kinetics.

The lowering of the T_s with increasing $\dot{T}$ is a general behavior that holds for every alloy and lots of systematic investigations confirm it [1,4]. For steels, many of the results are collected in compendia of CCT diagrams [5]. Two examples for nonferrous alloys are shown Fig. 2.4 [6,7].

One may argue that the shifting from equilibrium is not very high at a usual cooling rate of 5°C/min (~0.1°C/s) for the examples shown in Figs. 2.3 and 2.4. It will be shown that for other alloys even at such a slow cooling rate, the shifting can be >500°C.

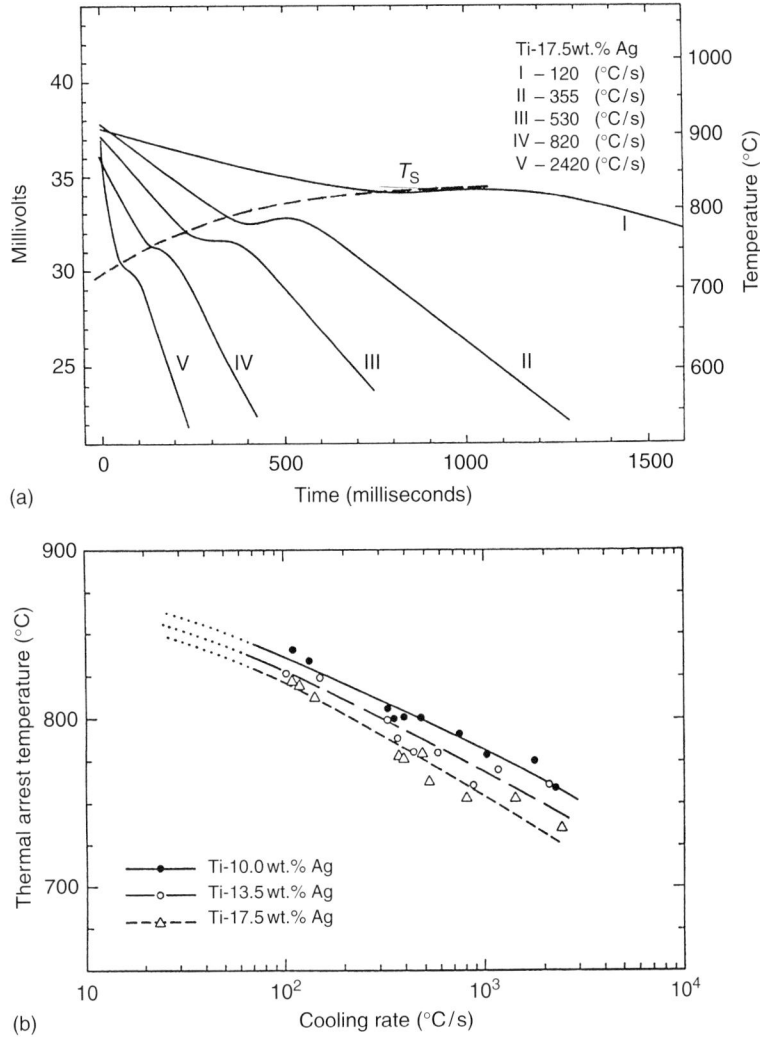

Figure 2.3 Cooling transformation kinetics of three Ti–Ag alloy: (a) a series of cooling curves of the Ti–17.5 wt.% Ag alloy showing the variation of the β → α transformation-start temperatures, T_S, with the cooling rate; and (b) transformation temperature vs. cooling rate $(T_t–\dot{T})$ diagram of three Ti–Ag alloys. From Plichta et al [3].

2.2 Formation of Metastable Phases During Cooling

Above a certain critical cooling rate (CCR), >200°C/s for several Ti–Cu alloys as shown in Fig. 2.5 [8], the transformation-start temperature becomes independent of the cooling rate, (i.e., a plateau forms in the $T_t–\dot{T}$ diagram for each alloy). This plateau corresponds to the formation of a metastable massive transformation [9] or a martensitic transformation product. As mentioned previously, massive transformation is generally believed to involve only local diffusion at the transformation interface. The corresponding massive transformation-start temperature is conventionally

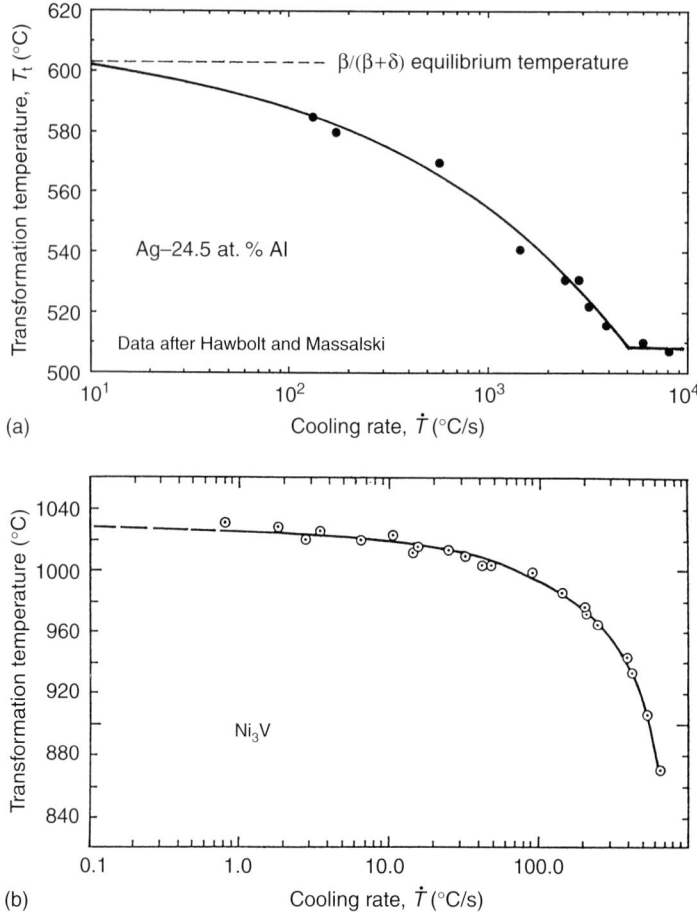

Figure 2.4 Transformation temperature vs. cooling rate (T_t–$\dot{T}$) diagram for the Ag–24.5 at.%Al alloy (a) (data from Hawbolt and Massalski [6]) and Ni_3V (b) (from Perepezko [7]).

designated as M_a. The fact that M_a is independent of the cooling rate implies that long-range diffusion no longer plays a dominant role in this type of transformation, because long-range diffusion is time-dependent and thus should be cooling rate ($\dot{T}$) dependent as well. It is believed that interface movement is the controlling factor of massive transformations [9]. Martensitic transformation also displays a plateau since it involves no long-range diffusion, and thus the martensitic transformation-start temperature, M_S, is independent of cooling rate.

At still higher cooling rates, other metastable phases may be formed, thus displaying multiple plateaus in a T_t–$\dot{T}$ diagram. Two examples are shown in Figs. 2.6 and 2.7 for an Fe–4.23Cr–0.03C alloy [10] and pure cobalt [11], respectively. It is clear that each metastable phase has its own T_s that is independent of $\dot{T}$. Correspondingly the T_t–$\dot{T}$ diagrams display staged kinetics, that is several plateaus, each for a specific metastable phase. This staged kinetics is found to be a general behavior of the cooling transformations, although different alloys may have a different metastable phase

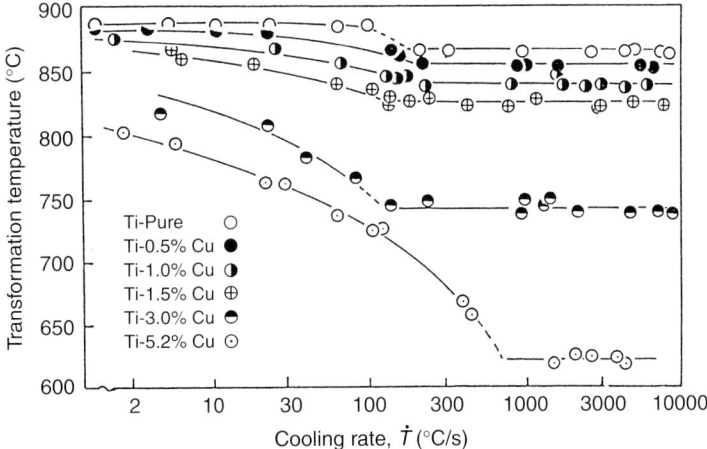

Figure 2.5 Transformation temperature vs. cooling rate $(T_t-\dot{T})$ diagram of Ti and Ti–Cu alloys. The plateaus signify the formation of metastable phases. From Srivastava and Parr [8]. Composition in atomic percent.

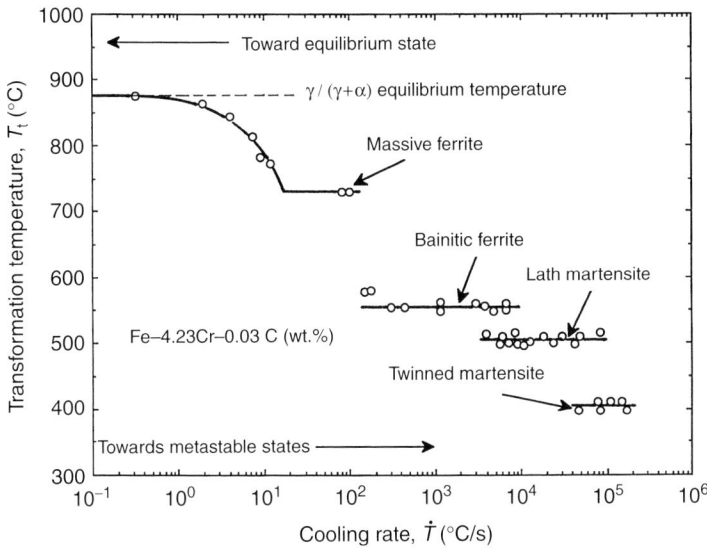

Figure 2.6 Transformation temperature vs. cooling rate $(T_t-\dot{T})$ diagram of the Fe–4.23Cr–0.03C alloy, indicating the formation of four metastable phases. Data from Mirzayev et al.[10].

sequence [1,4,10]. For instance multiple plateaus were observed in pure metals Co, Fe, Ti and Zr [12], iron and iron alloys [4,10,13], Cu–Ga alloys [12,14], and Cu–Al alloys [15].

The fact that each metastable phase possesses its own T_s during cooling is crucial for the correct assessment of metastable phase information. Detailed discussion on metastable phases is beyond the scope of this chapter and can be found elsewhere [1].

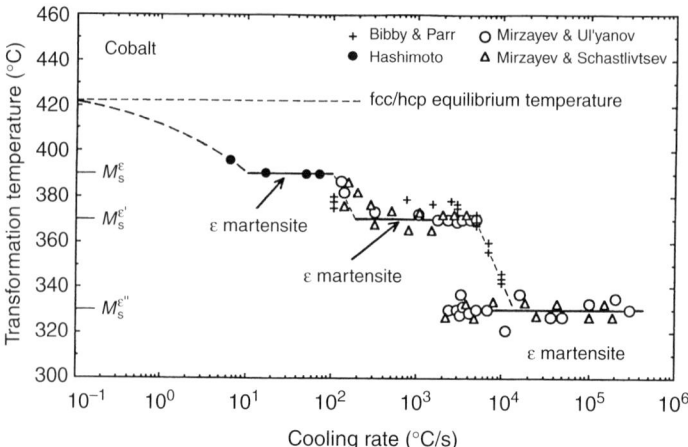

Figure 2.7 Transformation temperature vs. cooling rate $(T_t\text{-}\dot{T})$ diagram of pure cobalt, indicating the formation metastable phases [11].

2.3 Shifting of Transformation-Start Temperature with Heating Rate

Because transformations that occur during heating are similar to these that occur during cooling, the heating transformation also requires a certain superheating to form critical nuclei of the high temperature phase. The heating transformation-start temperature (T_S^h) shifts to higher temperatures with increasing heating rate, as shown in Fig. 2.8 for pure Ti and Zr [16].

As higher heating rates are reached, a massive-type transformation takes place, and its transformation-start temperature is independent of the heating rate, quite similar to what is observed in the cooling process. It should be noted that even though an alloy has a two-phase structure before heating, the T_S^h is still raised with increasing heating rate, and reaches a plateau similar to the massive transformation of the type from single phase to single phase, as shown in Fig. 2.9(a) for a Ti–5.25 at.% Mo alloy $[(\alpha + \beta) \rightarrow \beta]$ [17] and Fig. 2.9(b) for a Ti–4.6 at.% Cr alloy $[(\alpha + \gamma) \rightarrow (\alpha + \beta)]$ [18]. Similar results for the shifting of phase boundaries with heating rate are also found for iron and Fe-base alloys [19,20], Ti-base and Zr-base alloys [16,21], and also Cu-base alloys [22].

Note that when the cooling transformation is such that it has already produced metastable structures (e.g., martensite), the subsequent heating transformation-start temperature is that of the metastable phase to an equilibrium phase and has little relation to the position of the equilibrium phase boundaries. This can be explained with the help of Fig. 2.10. For the situation shown in Fig. 2.10(a), if alloy X_1 is quenched from T_1 to T_3, martensite is produced. When the specimen is heated, the transformation-start temperature is that of the martensite to β phase (often designated as A_s for steels) and has no direct relation with the $\beta/(\alpha + \beta)$ transus. To determine the $\beta/(\alpha + \beta)$ transus, it is necessary to first anneal the specimen at T_3 (or other temperatures) for a sufficiently long time so that the martensite transforms completely to the equilibrium α phase. Then very slow heating can approach the transus (point A). In this connection the A_s temperatures of steels seldom have direct relation to the

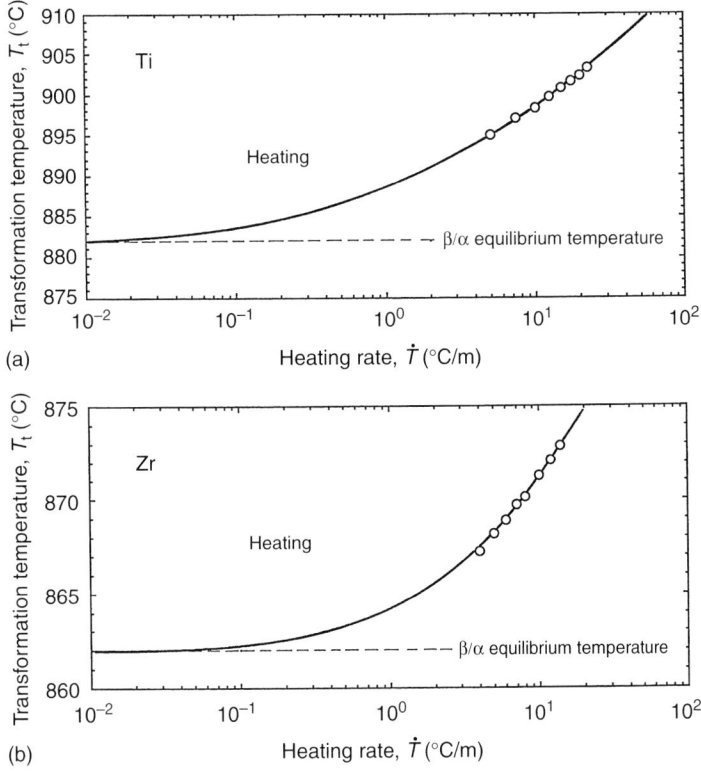

Figure 2.8 Transformation temperature vs. heating rate (T_t–$\dot{T}$) diagram for (a) pure Ti and (b) pure Zr, showing the rising of the α → β transformation-start temperature with increasing heating rate. Data from Zhu and Devletian [16].

γ/(α + γ) transus. For the situation schematically shown in Fig. 2.10(b), if alloy X_2 is quenched from T_4 to T_6, martensite is produced. When the specimen is heated, the transformation-start temperature is again that of the martensite to β phase and has no direct relation with the β/(α + β) transus. Even when the martensite transforms completely to the equilibrium α and β phases with compositions at points C and D after a long-term annealing at T_6, it would still be very difficult to obtain the equilibrium β/(α + β) transus temperature for this alloy (point F) from an heating experiment. This is because only when an infinitesimally small heating rate is used and the compositions of the α phase and the β phase follow the respective phase boundaries from point C to point E and point D to point F, will the equilibrium transus temperature (point F) be obtained. Under normal heating rates, and especially if T_6 is low relative to the solidus temperature of the system, it would be difficult for the β phase to follow the β/(α + β) transus from point D to point F during the heating process; thus, it would be difficult to obtain the equilibrium transus temperature, point F.

For a eutectoid transformation as shown in Fig. 2.10(c), a hypereutectoid alloy X_3 on cooling from T_7 to T_9 may by-pass both points G and H and forms a martensite phase. If no additional annealing is performed, subsequent heating experiment would

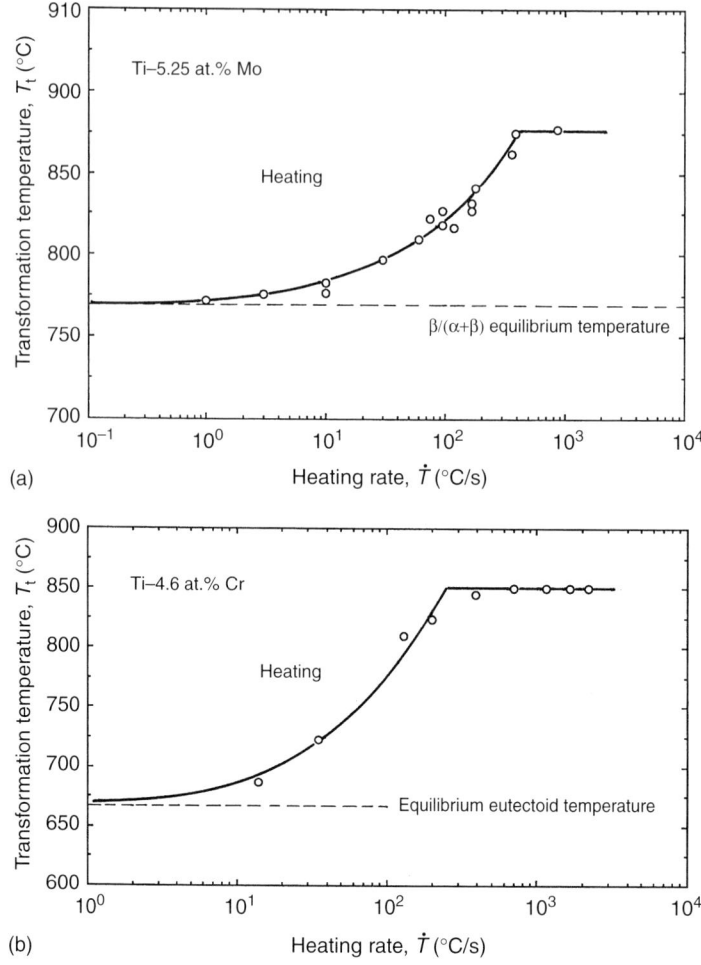

Figure 2.9 Transformation temperature vs. heating rate (T_t–$\dot{T}$) diagrams of a Ti–5.25 at.% Mo alloy (a) showing the shifting of the β/(α + β) transus with increasing heating rate and of a Ti–4.6 at.% Cr alloy (b) showing the shifting of the eutectoid temperature with increasing heating rate. Data from Gridnev et al. [17,18].

give a martensite to β transformation temperature, which has nothing to do with the equilibrium phase boundaries. If the alloy is heated at T_8 for an extended period of time to transform the martensite to the equilibrium α + γ phases, then a slow heating experiment can measure the equilibrium eutectoid temperature, point H.

For a eutectic reaction shown in Fig. 2.10(d), the transformations from a liquid phase to solid phases and vice versa are usually much easier since the temperature is high and the diffusion (especially in the liquid) is fast. In such cases, the undercooling and superheating can be reasonably small; thus, the thermal analysis such as DTA and differential scanning calorimetry (DSC) can be used quite reliably to obtain the liquidus (point I) and invariant reaction temperatures (point J). A more detailed discussion about such measurements is provided by Boettinger et al. in Chapter 5 of this book.

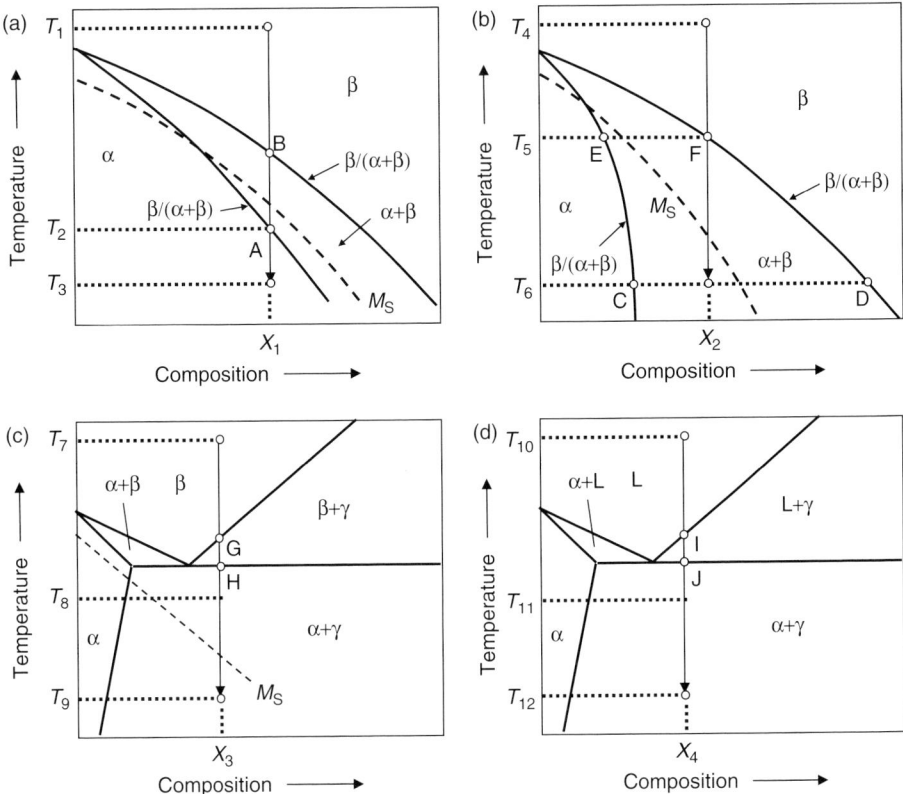

Figure 2.10 Schematic diagram showing the formation of a metastable phase and the subsequent heating process.

For solid–solid phase equilibria, especially at temperatures below half of the homologous solidus temperature, diffusion and phase transformation kinetics can become quite sluggish. In such cases, the undercooling and superheating can be quite pronounced, shifting the transformation-start temperature far away from the equilibrium transuses. Therefore, one needs to avoid using heating and cooling experiments to measure solid–solid equilibria at relatively low temperatures. Big hysteresis of cooling and heating transformation temperatures should be considered as a warning sign that both these temperatures may be far away from the equilibrium temperatures.

The cooling and heating results may be superimposed as shown schematically in Fig. 2.11 so that the shifting of the transformation temperatures with both heating and cooling can be analyzed at the same time. For the situation shown in Figs. 2.11(a) and (b), the transformation-start temperatures of cooling and of heating are associated with the $\beta/(\alpha + \beta)$ and $\alpha/(\alpha + \beta)$ transus respectively, so that at very slow cooling and heating rates, the transformation temperature of cooling should be higher than that of heating. As higher and higher cooling rates are used to determine the $\beta/(\alpha + \beta)$ transus, the measured transformation-start temperature can be shifted all the way from the equilibrium $\beta/(\alpha + \beta)$ transus temperature to the M_a (or M_S) temperature, that is any temperature between the $\beta/(\alpha + \beta)$ transus and the M_a/M_S point

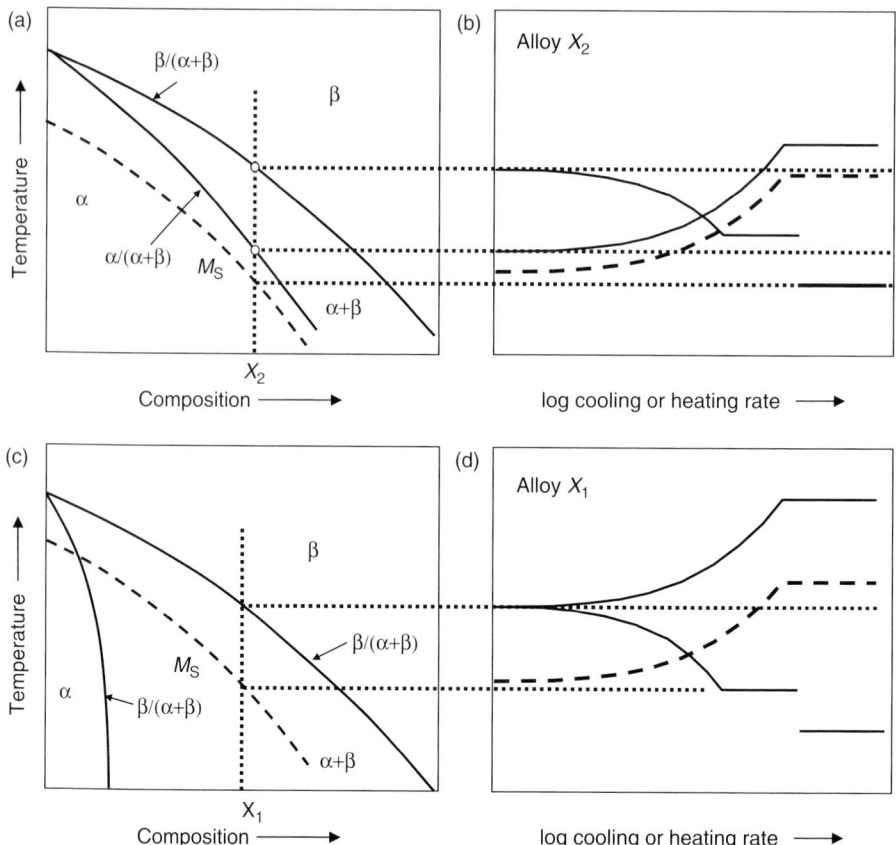

Figure 2.11 Schematic diagram showing the formation of a metastable phase and the subsequent heating process.

is possible, as schematically shown in Fig. 2.11(b). Similar behavior should also be observed for the $\alpha/(\alpha + \beta)$ transus during the heating process. At very high cooling and heating rates, the transformation temperature on cooling can become lower than that on heating. If this behavior is observed, it signifies that the results are already far from equilibrium.

For the situation shown in Figs. 2.11(c) and (d), the transformation-start temperatures on cooling and on heating are both associated with the $\beta/(\alpha + \beta)$ transus if equilibrium phases are reached before heating or cooling. The measured cooling transformation-start temperature will always be lower than that of the heating transformation-start temperature. At high cooling and heating rates, the hysteresis between the cooling and heating transformation temperatures can become quite large, and indicating the system is far from equilibrium. The dashed lines in Figs. 2.11(b) and (d) show the heating transformation-start temperature when a martensite phase is formed during cooling, and the heating experiment is performed without an extended annealing to convert the martensite to the low-temperature equilibrium phases. In such cases, the heating transformation-start temperature has little relationship with the equilibrium transus.

The specific values of "high" cooling and heating rates that cause significant deviation from equilibrium depend on the alloy, its composition, grain size, phase structure, and more importantly the temperature of the transus (relative to the solidus) of the alloy. Unfortunately it is not presently possible to predict which cooling or heating rate would be high enough for a specific alloy to cause significant deviation and how far the value will be shifted away from equilibrium. However, as mentioned before if the high temperature phase is liquid, the shifting of the temperature during slow cooling and heating is less pronounced than that of solid–solid phase transformations because it is much easier to nucleate a critical nucleus from a liquid than from a solid phase. In this connection, the normal thermal analysis method (at very slow cooling and heating) is a good approach to determine equilibrium liquidii. As shown below for high nickel Fe–Ni alloys, for solid–solid transuses, especially at low temperatures, sometimes even a few degrees per minute is a high cooling rate because diffusion is difficult. Under these circumstances, it is recommended to make full use of the results of equilibrated alloys or diffusion couples (the second category of methods as discussed in Section 1) to assess phase boundaries.

It is also worthwhile mentioning that the transformation-finish temperatures for both cooling and heating (M_f, A_f, etc.) have little connection with the equilibrium transuses; because the transformation-start temperatures are already shifted away from equilibrium, the finish temperatures are then shifted even further away from equilibrium. In addition, the cooling transformation is often incomplete at a practical cooling rate, thus the cooling transformation-finish temperature has little to do with a corresponding equilibrium phase boundary. Therefore, no more description of the transformation-finish temperatures is included in this chapter.

Since cooling and heating experiments for phase diagram determination rely on the detection of a phase transformation, it is important to think about transformation kinetics and the transformation temperature relative to the melting points of the elements and compounds. If the transformation temperature is below half of the homologous melting points, one needs to pay special attention to potential problems associated with kinetics of phase transformation and its effect on shifting the transformation temperatures away from the equilibrium phase boundaries.

2.4 Analyses of Examples from Cooling and Heating Experiments

2.4.1 Cooling and Heating Transformation Temperature vs. the $\gamma/(\alpha + \gamma)$ Transus in Fe–Ni

Kaufman and Cohen [23] measured the transformation temperatures of twelve Fe–Ni alloys using a slow cooling rate and heating rate of 5°C/m. Their results are compared to the well-assessed $\gamma/(\alpha + \gamma)$ transus [24] in Fig. 2.12. The transformation-finish temperatures are not included in Fig. 2.12 because they have little relation to the equilibria, as discussed above. It is clear that the cooling T_S temperatures have been depressed from the $\gamma/(\alpha + \gamma)$ equilibrium transus; in high nickel alloys, the shift is especially drastic. For example, the transformation temperature for an Fe–30.7 at.% Ni alloy is depressed ~570°C even at a cooling rate of 5°C/min, as shown by the vertical arrow in Fig. 2.12. One may argue that Kaufman and Cohen intended to measure the M_S and A_S temperatures, not the $\gamma/(\alpha + \gamma)$ equilibrium transus. That's true. However, let's imagine a scenario where we know nothing about the $\gamma/(\alpha + \gamma)$

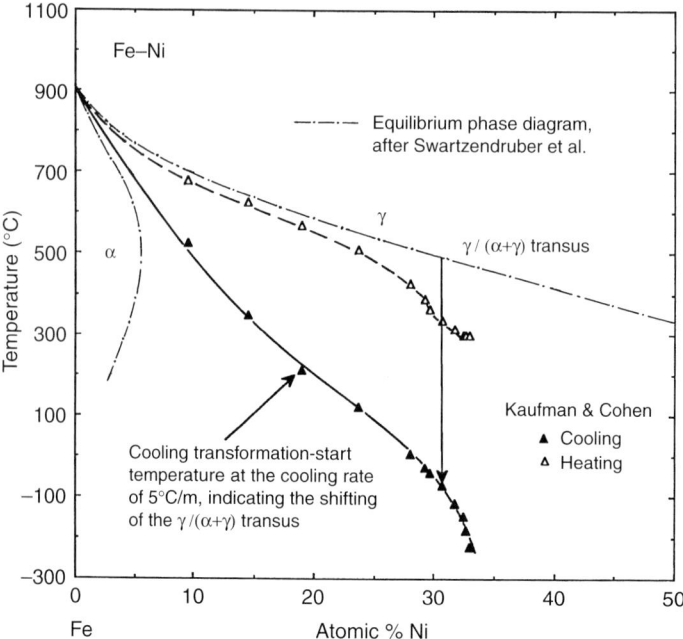

Figure 2.12 Comparison of the cooling and heating transformation-start temperatures with the equilibrium $\gamma/(\alpha + \gamma)$ transus in Fe–Ni, illustrating the shifting of the cooling transformation-start temperature from equilibrium [1,23].

equilibrium transus of the Fe–Ni system and we want to measure it using cooling and heating experiments. A dozen or so alloys are made and then measured using a slow cooling and heating rate of 5°C/min. We would obtain essentially the same results as those of Kaufman and Cohen. If phase equilibrium information from diffusion couples and equilibrated alloys were not available and that only the cooling and heating transformation data were used to assess the $\gamma/(\alpha + \gamma)$ equilibrium transus, the obtained transus would be erroneous. It is therefore very dangerous to use only the cooling and heating data to assess phase boundaries at low temperatures. Only at high temperatures when the diffusion is fast and especially when a liquid phase is involved, can reliable phase equilibria information be obtained from cooling and heating measurements.

The heating transformation-start temperature (A_s) should be above the $\gamma/(\alpha + \gamma)$ transus because heating always shifts the phase boundary to higher temperatures. The fact that in Fig. 2.12 the A_s's (open triangles) on heating are below the transus indicates that a metastable structure has been produced during the cooling process, and therefore the A_s's are for the metastable phase to γ phase transformation and have no direct relation to the equilibrium transus.

2.4.2 The fcc ↔ hcp Phase Equilibria in Co–Cr

The experimental information on the fcc ↔ hcp ($\gamma \leftrightarrow \varepsilon$) transformation during cooling and heating in the Co–Cr system is summarized in Fig. 2.13(a) [25]. Using

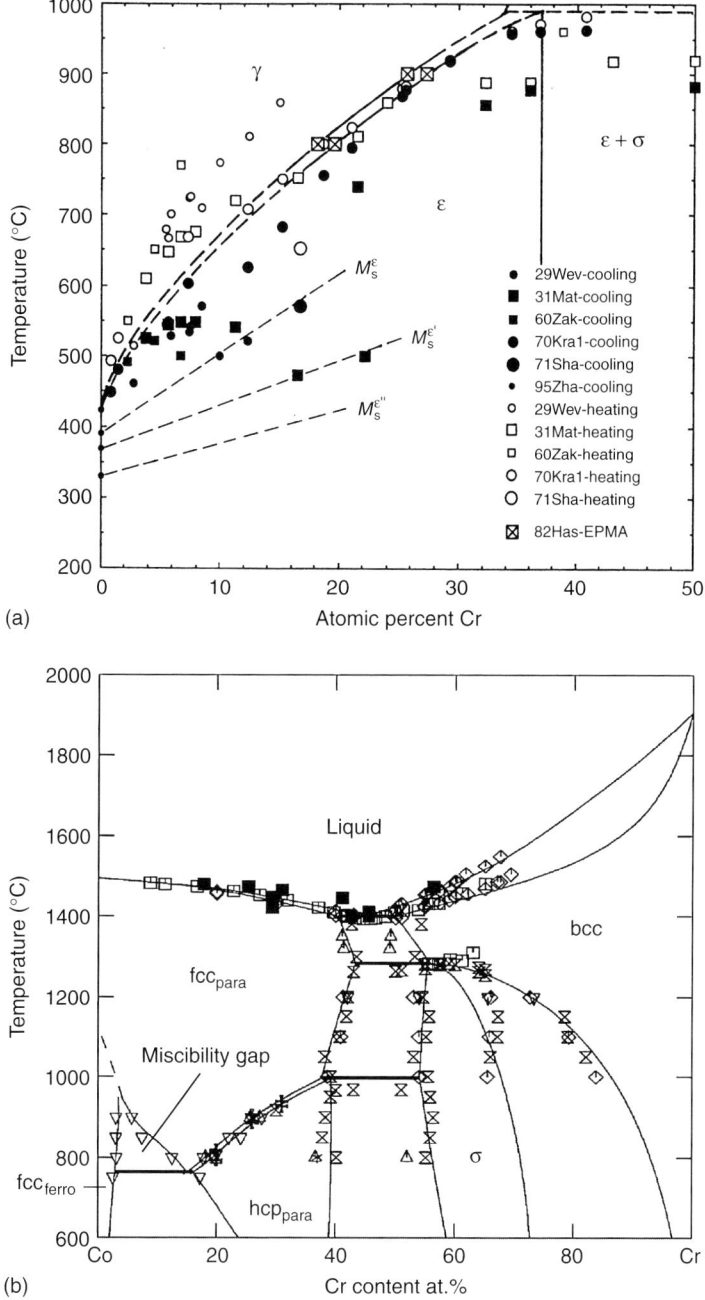

Figure 2.13 The fcc–hcp phase equilibrium in the Co–Cr binary system: (a) an earlier assessment of the fcc–hcp equilibrium to show the results of cooling and heating data in comparison with results from diffusion couples [25]; and (b) an assessment based on more recent experimental results from careful diffusion couple work, revealing a miscibility gap (two-phase region (from approximately 3 to 17 at.%) of fcc_{ferro} and fcc_{para}) induced by the magnetic transition [30] (see text for detail and cited Ref. 25 and 30 for origin of data points shown).

the EPMA results of Hasebe et al. [26], it is possible to make a reasonable assessment of both the $\gamma/(\gamma + \varepsilon)$ and the $\varepsilon/(\gamma + \varepsilon)$ phase boundaries (especially at high temperatures) essentially without invoking any cooling and heating data. Note that all the cooling data (filled symbols in Fig. 2.13(a)) are below the $\gamma/(\gamma + \varepsilon)$ transus, which is consistent with the cooling transformation kinetics, that is the cooling process always depresses the transformation temperatures to lower values. Note also that the cooling data of Krajewski et al. [Ref. 27 – designated as 70Kra1] at high Cr concentrations (>22 at.%) have been depressed away from the equilibrium $\gamma/(\gamma + \varepsilon)$ transus by only ~25–30°C when the temperatures (>~800°C) are still relatively high, as is the diffusion rate. At compositions ranging from 10 to 20 at.% Cr, the cooling transformation temperatures become lower (thus slower diffusion rates); thus the cooling data of Krajewski [Ref. 27 – designated as 70Kra1] have been depressed further away from the equilibrium $\gamma/(\gamma + \varepsilon)$ transus by up to 85°C. It is interesting to note that at low concentrations of Cr (<~5 at.%) the cooling data of Krajewski et al. [Ref. 27 – designated as 70Kra1] seem only slightly shifted away from the equilibrium $\gamma/(\gamma + \varepsilon)$ transus even though the transformation temperatures are very low (~450–600°C, thus even slower diffusion rates), as shown in Fig. 2.13(a). This seemingly "strange" behavior can be explained with the help of the general characteristics of cooling transformation kinetics, and it is related to the variation of CCR with concentration. A more detailed discussion of such transformation kinetics is given elsewhere [4,11]. This behavior of a dip in cooling transformation temperatures at medium concentrations is not uncommon, for example, the cooling data of [Ref. 28–31Mat] for Co–Cr and [Ref. 29–73Kos] for Co–Si. Other cooling data for the Co–Si, Co–Cr and Co–Ge also show this trend [25].

Three lines of M_S temperatures, $M_S^{\bar{\varepsilon}}$, $M_S^{\varepsilon'}$ and $M_S^{\varepsilon''}$ are tentatively drawn as dashed lines in Fig. 2.13(a). There are not enough data points to define these M_S lines which are the author's best guess. The M_S lines become more convincing with more data shown for other systems [25].

It can be seen from Fig. 2.13(a) that all the cooling data are below the $\gamma/(\gamma + \varepsilon)$ transus. The heating data are scattered both above and below the $\varepsilon/(\gamma + \varepsilon)$ transus, thus they cannot be used to assess the phase boundaries.

It turns out that the cooling and heating data shown in Fig. 2.13(a) completely missed a miscibility gap induced by a magnetic transition as shown Fig. 2.13(b) [30]. This miscibility gap (a so-called Nishizawa horn) was recently revealed by careful diffusion couple experiments [30]. The cooling and heating experiments did not detect/"feel" this gap probably because the phase separation kinetics are very sluggish. Interestingly, such a Nishizawa horn was predicted from CALPHAD (calculation of phase diagrams) for several Co-base binary systems long before experimental confirmation. As a matter of fact, earlier experiments with equilibrated alloys also had a hard time finding this miscibility gap probably due to slow phase separation. Diffusion couples have an advantage in this case since they form the equilibrium phases directly from thermal interdiffusion without the need of decomposing a high temperature phase. More discussion on this topic will be provided later in this chapter.

2.4.3 The fcc ↔ hcp Phase Equilibria in Co–Mo and Co–V

An assessment of the Co-rich part of the Co–Mo binary phase diagram is shown in Fig. 2.14 [25]. There are enough experimental results from XRD and diffusion

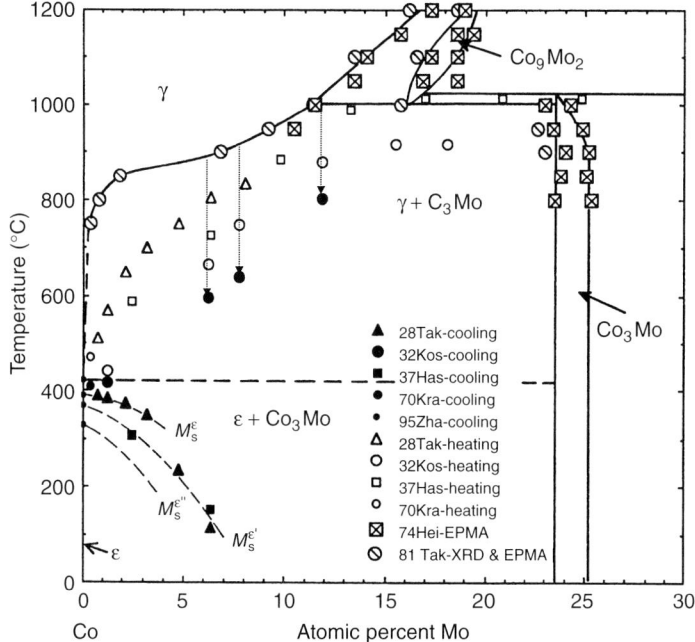

Figure 2.14 Tentative assessment of the Co-rich part of the Co–Mo binary phase diagram comparing cooling and heating data with results from diffusion couples [25].

couples to make a reasonable assessment of the phase boundaries without invoking any cooling and heating data. It can be seen that the cooling data of Koster and Tonn [Ref. 31 designated as 32Kos] have been shifted more than 200°C away from the equilibrium $\gamma/(\gamma + Co_3Mo)$ transus even though the transus for the three alloys with >5 at.% Mo are pretty high (>850°C). If the results from XRD and diffusion couples were unavailable and we had to rely only on the cooling and heating data, it would be very easy to make an erroneous assessment for the phase diagram.

The heating process always shifts the transformation temperature higher than the equilibrium transus. The fact that all the heating data are below the equilibrium $\gamma/(\gamma + Co_3Mo)$ transus indicates that the alloys were not in an equilibrium state before they were heated to high temperatures. This result also shows that the heating transformation-start temperature cannot be used to assess the phase diagram unless equilibrium is reached at low temperature before the alloys are heated.

A new assessment of the Co-rich portion of the Co–V system along with an assessment of the M_S temperatures is shown in Fig. 2.15 [25]. Aoki et al. [Ref. 32 designated as 79Aok in Fig. 2.15] employed a very good way to determine the $\gamma/(\gamma + Co_3V)$ phase boundary. They started by annealing alloy samples at low temperatures (600–1050°C) for 12–1008 h (some samples were annealed stepwise, for example, 650°C for 120 h and 600°C for 240 h, then cooled to the room temperature) so that the samples were in the equilibrium (γ + Co$_3$V) two-phase state before electrical resistance measurements were made during heating and cooling. By using very slow heating and cooling rates (<1°C/min), they were able to accurately measure the $\gamma/(\gamma$ + Co$_3$V) phase boundary. Their cooling and heating data were consistent with XRD results of annealed and quenched samples. Their way of surveying the phase boundary was exemplary, but unfortunately very few investigators have used it. Note that at temperatures above 800°C, the cooling and heating data were very close, whereas at temperatures below 800°C, the data started to deviate because of slower diffusion rates.

When only heating and cooling data were available for the $\gamma \leftrightarrow \varepsilon$ phase equilibrium, it was often assumed that the heating data were more reliable for phase boundary assessment than the cooling data. For instance, in most assessments of cobalt-based binary systems made by Hansen and Anderko [33], the heating data were used to define the $\gamma \leftrightarrow \varepsilon$ phase boundaries. The same practice has been adopted by others [e.g., 34–36]. Actually, it has been shown that heating data are not very reliable for equilibrium $\gamma \leftrightarrow \varepsilon$ phase boundary assessment. Readers are referred to the situations discussed in Figs. 2.10 and 2.11.

It should be mentioned that it is very likely there is a Nishizawa horn in the Co–V system induced by a magnetic transition similar to that in the Co–Cr system. The slow kinetics of phase separation might have prevented the detection of the miscibility gap during the normal heating and cooling experiments.

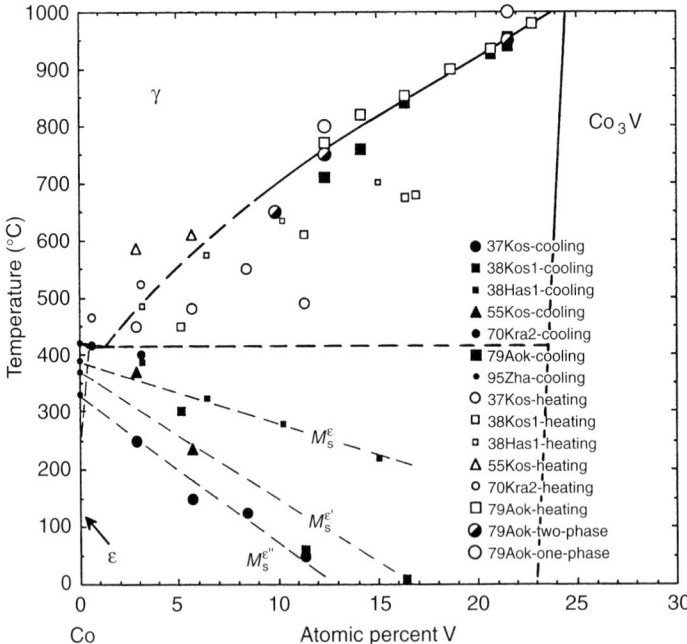

Figure 2.15 Tentative assessment of the Co-rich part of the Co–V binary phase diagram comparing cooling and heating data with results from XRD [25].

3 Isothermal Phase Transformation Kinetics as Related to Phase Diagram Determination

3.1 Precipitation of Phases from Quenched Alloys

The isothermal phase transformation kinetics is usually described using time–temperature–transformation (TTT) diagrams. These diagrams depict the precipitation start (and also finish) time during isothermal annealing at different temperatures. Usually, C-shaped curves (so-called C-curves) emerge when the starting times of precipitation at different temperatures are linked together. An example is shown in Fig. 2.16(b) for a Cu–15Ni–8Sn (wt.%) alloy [37]. A vertical section at 15 wt.% Ni of the Cu–Ni–Sn ternary phase diagram is shown in Fig. 2.16(a) in order to illustrate the relationship between the phase precipitation kinetics and the phase diagram. According to the phase diagram, the equilibrium solvus temperature for the Cu–15Ni–8Sn alloy is ~800°C. By fast quenching from above 800°C, the fcc phase can be retained to room temperature. Isothermal annealing of the quenched specimens at temperatures <800°C leads to the formation of different precipitates depending on the temperature as shown in Fig. 2.16(b). This particular alloy has relatively fast kinetics, thus it is fairly easy to form the equilibrium $\gamma(DO_3)$ phase at temperatures above 500°C. Below this temperature, a metastable spinodal structure, DO_{22} ordering, and $L1_2$ ordering take place before the equilibrium $\gamma(DO_3)$ phase precipitation, thus longer time is required to form the equilibrium phase for phase diagram

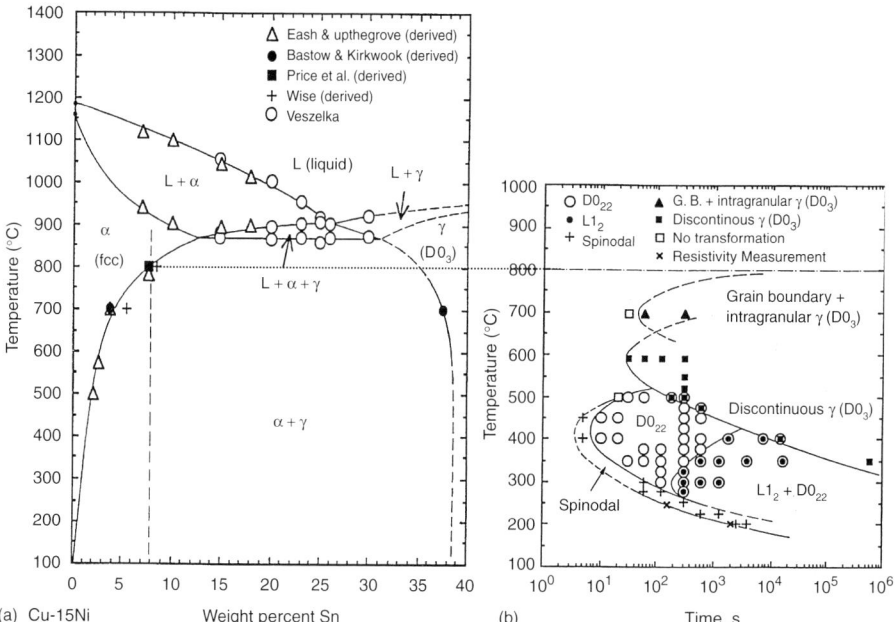

Figure 2.16 Isothermal decomposition kinetics of Cu–15Ni–8Sn alloy and its relationship with the phase diagram: (a) isopleth of the Cu–Ni–Sn ternary phase diagram at 15 wt.% Sn and (b) TTT diagram of the Cu–15Ni–8Sn alloy [37].

determination. For instance, at 300°C, it would take more than 300 h to start to form the equilibrium γ(D0₃) phase. For such low temperatures, the best way to determine the equilibrium phase diagram is to use stepwise annealing. For instance, to determine the phase equilibrium at 300°C, it is much better to first anneal a quenched specimen at 500°C for several hours to promote the γ(D0₃) phase formation and then lower the temperature to 300°C for extended period of time to reach equilibrium at 300°C. Alternatively, one can perform cold deformation (if the specimen is ductile enough) to create microstructural defects such as dislocations and subboundaries to promote the formation of the equilibrium phase.

The Cu–15Ni–8Sn alloy example shows that one needs to understand and pay attention to the isothermal precipitation kinetics in order to reliably determine phase diagrams, especially phase equilibria at low temperature when the diffusion and nucleation kinetics are very slow.

A more problematic situation is the decomposition of a high temperature intermetallic phase. An example is the eutectoid decomposition of Nb₃Si in the Nb–Si binary system. The Nb-rich part of the Nb–Si phase diagram [38] and a TTT diagram [39] of a Nb–19 at.% Si alloy are shown in Fig. 2.17 (a) and (b). Even though the temperature is very high (equilibrium eutectoid temperature for breakdown of Nb₃Si is estimated to be ~1680°C), the decomposition kinetics of the Nb₃Si phase is sluggish. For instance, at 1400°C, it took about 500 h to completely decompose the Nb₃Si phase, Fig. 2.17(b). Even at the nose temperature (~1500°C) of the C-curve where the transformation kinetics is the fastest, it still took about 100 h to complete the eutectoid decomposition. Note that at temperatures close to the equilibrium

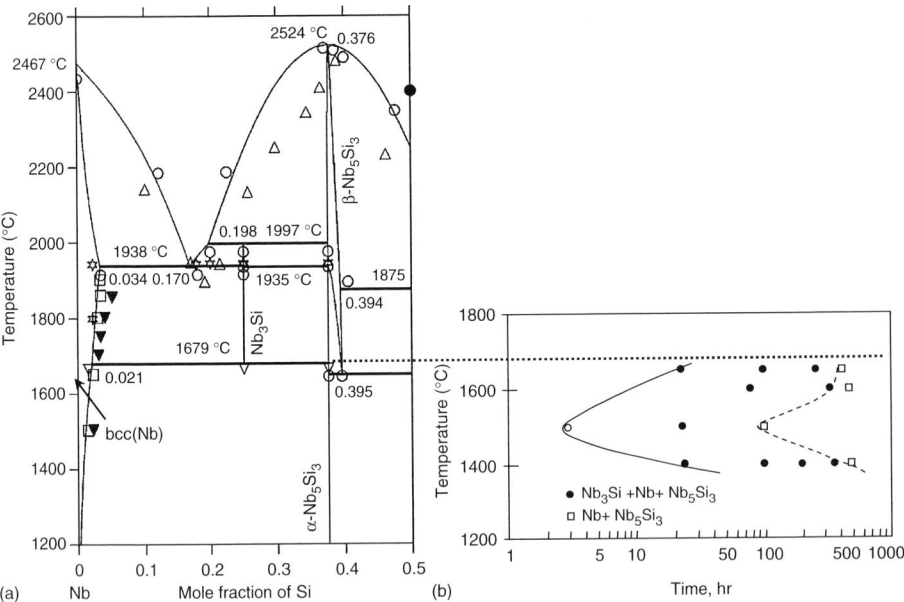

Figure 2.17 Isothermal decomposition kinetics of Nb₃Si to Nb₅Si₃ and Nb and its relationship with the Nb–Si phase diagram: (a) Nb-rich part of the Nb–Si phase diagram [38] and (b) the TTT diagram for the eutectoid decomposition of Nb₃Si [39].

eutectoid temperature of ~1680°C, the kinetics was not very fast due to the C-curve nature of the kinetics. One might think 1680°C is a very high temperature, and therefore the transformation kinetic should be very fast. The result of Fig. 2.17 would thus be surprising. One needs to consider the kinetics in the C-curve behavior and the diffusion kinetics which unfortunately is often unknown a priori.

The problem associated with slow decomposition kinetics of high temperature intermetallic phases is quite common. Two ternary examples are discussed here to further illustrate the problem. The example of the Nb–Cr–Si system is shown in Fig. 2.18. Figure 2.18(a) is the 1000°C isothermal section reported by Goldschmidt and Brand [40] based on 220 alloys that were arc-melted. Figure 2.18(b) [41] is based on experimental result from a diffusion multiple (the methodology will be discussed in

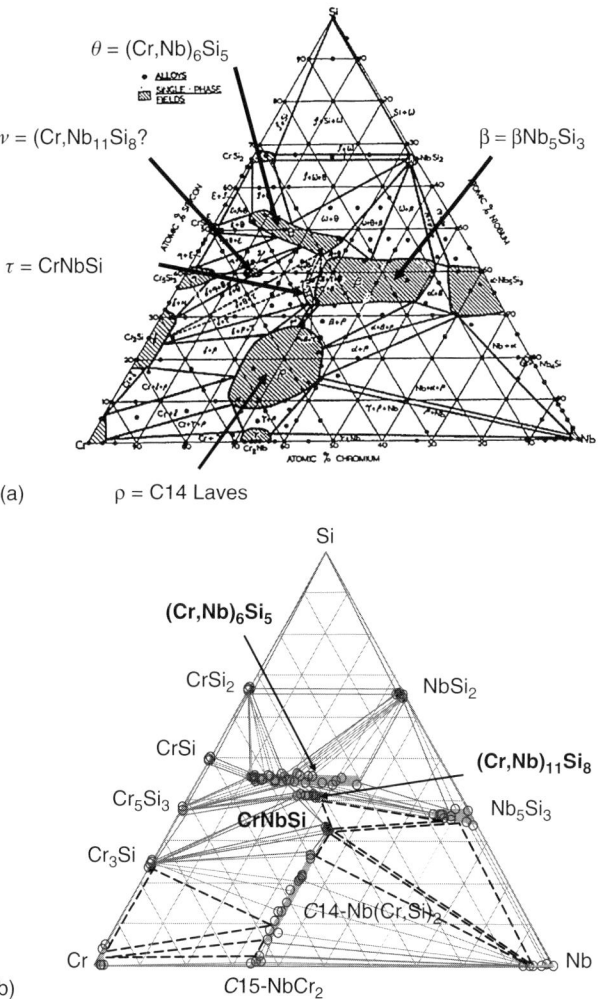

Figure 2.18 Comparison of the 1000°C isothermal sections of the Nb–Cr–Si ternary system from two separate investigations: (a) isothermal section reported by Goldschmidt and Brand [40] based on 220 alloys and (b) isothermal section from a diffusion multiple [41].

more detail in Chapter 7 of this book). The diffusion multiple was annealed at 1000°C for 4000 h and all the phases were formed by thermal interdiffusion reaction without going through the high temperature melting and solidification process. Thus, there is no issue concerning the kinetics of decomposition of a high temperature phase at low temperatures. In this regard, the results (Fig. 2.18(b)) from the diffusion multiple can be regarded as the equilibrium phase diagram. (A more detailed discussion can be found elsewhere [41].) The 220 alloys used by Goldschmidt and Brand were annealed at 1000°C for 336 h (2 weeks) after arc-melting. It is very likely that most of the alloys did not reach equilibrium after the heat treatment. Many Nb-base alloys, such as Nb–Ti–Si alloys studied by Bewlay et al. [42,43] and Nb–Hf–Si alloys studied by Zhao et al. [44] possess very slow kinetics, and they would not be at equilibrium after being annealed at 1000°C for only 336 h. For instance, even after annealing at 1500°C for 100 h several Nb–Hf–Si alloys still had not reached full equilibrium [44]. It is very likely that the Nb–Cr–Si alloys would possess similar slow kinetics as well, thus the data reported in the phase diagram of Goldschmidt and Brand may not represent equilibrium conditions. For instance, the βNb_5Si_3 which was formed directly from solidification (from its primary liquidus surface) and which is not an equilibrium phase at 1000°C for the Nb–Cr–Si ternary system, did not decompose to the equilibrium phases after annealing at 1000°C for only 336 h, Fig. 2.18(a). One common feature of slow decomposition of a high temperature phase is that a stoichiometric phase or line compound (or a phase with a very narrow concentration range) often displays a wide composition region (widening of the composition region of the phase in a phase diagram). For instance, Nb_5Si_3 and Cr_3Si phases have very narrow composition range of Si in the equilibrium binary phase diagrams, whereas in the phase diagram presented by Goldschmidt and Brand, these phases show extensive composition ranges. This indicates that the decomposition process of the off-stoichiometric compositions formed during solidification was incomplete and equilibrium at 1000°C was not reached. Similarly, the C14-Laves phase is essentially a line compound with a very narrow variation of Nb concentration as shown in Fig. 2.18(b). However, it displayed a wide variation of Nb concentration in the non-equilibrium phase diagram, Fig. 2.18(a) (Goldschmidt and Brand called the C14-Laves phase the ρ phase). Again the off-stoichiometric compositions of the C14-Laves phase that were formed during melting and solidification did not decompose to the equilibrium phases at 1000°C after only 336 h of heat treatment.

Another excellent example is shown in Fig. 2.19 that compares the 1000°C isothermal section of the Nb–Cr–Al ternary system determined using melting and heat treatment (Fig. 2.19(a) by Hunt and Raman [45]), solid state reaction of fine powders without melting (Fig. 2.19(b) by Mahdouk and Gachon [46]), and a diffusion multiple (Fig. 2.19(c) by Zhao et al. [47]). The diffusion-multiple result (Fig. 2.19(c)) is consistent with the powder-metallurgy result of Mahdouk and Gachon (Fig. 2.19(b)), and these findings together suggest that the C14-Laves phase is almost a ternary line-compound in the form of $Nb(Cr,Al)_2$ with very little variation of Nb concentration and that the very wide (Nb variation) phase region of the C14-Laves reported by Hunt and Raman (Fig. 2.19(a)) is not real. The apparent widening of the C14-Laves phase region in Fig. 2.19(a) was probably due to sluggish decomposition kinetics: the C14-Laves phase with composition away from $Nb(Cr,Al)_2$ formed during solidification did not have sufficient time to decompose. Interestingly, both Hunt and

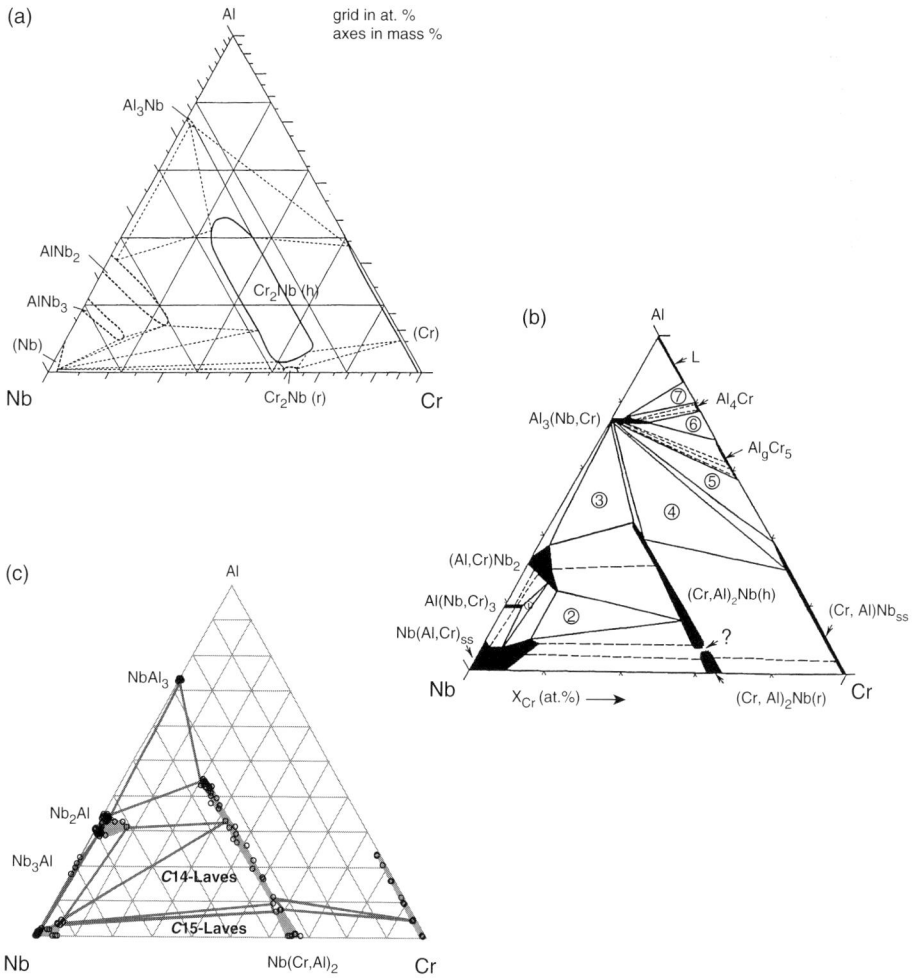

Figure 2.19 Comparison of the 1000 °C isothermal section of the Nb–Cr–Al ternary system: (a) an approximate phase diagram reported by Hunt and Raman [45] based on results from melted and annealed alloys (phase boundaries and tie-triangles were not well defined) (b) results reported by Mahdouk and Gachon [46] based on solid-state reaction of fine powders and (c) results obtained from a diffusion multiple [47].

Raman [45] and Mahdouk and Gachon [46] annealed their samples exactly the same way: 1000 °C for 168 h (1 week). The alloys of Mahdouk and Gachon reached equilibrium through solid-state (powder-metallurgy) reaction, whereas Hunt and Raman did not reach equilibrium for the C14-Laves phase using melting and annealing due to the slow kinetics of decomposition of the C14-Laves phase. Both the diffusion multiple and the fine powder solid-state reactions avoided the problem associated with slow decomposition of a high temperature intermetallic compound.

Even though the examples shown in Figs. 2.17–2.19 are all for high temperature intermetallics, the decomposition of a high temperature solid solution phase may also be kinetically sluggish. A good example is shown in Fig. 2.13 for the miscibility gap (Nishizawa horn) between fcc_{ferro} and fcc_{para}. Such a Nishizawa horn was predicted by thermodynamic modeling before experimental confirmation. Experimental attempts using solution annealed alloys and subsequent phase separation heat treatment failed to produce convincing evidence for the existence of such a miscibility gap. Only careful isothermal diffusion couple experiments [30] revealed the right nature of the phase diagram. As mentioned earlier, cooling and heating experiments completely missed this sluggish phase separation from a solid solution.

3.2 Kinetics and Phase Formation in Diffusion Couples/Multiples

Diffusion couples and diffusion multiples are becoming increasingly important tools for phase diagram determination (see Chapters 6 and 7). Their biggest advantages are: (1) the formation of all the phases at the temperature of annealing/interest without going through the melting and solidification, thus completely avoiding the problem of sluggish decomposition of a high temperature phase and (2) formation of multiple phases in a single sample, thus allowing multiple tie-lines to be extracted from the local equilibrium at the phase interfaces. The biggest disadvantages are: (1) long-term annealing required to grow the phases to sufficient thickness to allow quantitative composition analysis and crystal structure identification and (2) occasional (very rarely) missing phases from diffusion couples/multiples casting doubts about attainment of true equilibrium at the phase interfaces. Practically, the diffusion couples and multiples are recommended only for temperatures higher than half of the homologous temperatures of the system under investigation. Lower temperatures increase: (1) the chance of missing phases, (2) the difficulty in reaching true equilibrium, and (3) the requirement of very long annealing time.

The few rare cases of missing phases had caught a lot of attention. The fact is that essentially all of them occurred at a temperature below half of homologous temperature – probably due to the nucleation difficulty, and the occurrence is very rare. As long as one pays attention and carefully checks both the corresponding binary phase diagrams and the reported multicomponent intermetallic phases, the diffusion couples and multiples can be very reliably used to determine equilibrium phase diagrams, especially for temperatures higher than half of the homologous temperature. It would be the best, if one can afford it, to take advantage of the higher efficiency from the diffusion couples and diffusion multiples and then use a few selected alloys to check areas in a phase diagram where a suspicion of missing phases may arise.

Some investigators doubt whether stoichiometric phases can be formed in diffusion couples/multiples. The example in Fig. 2.20 [48–50] may help to put such doubt to rest. All the intermetallic compounds, most of them essentially stoichiometric, have formed in the diffusion couples/multiples. The formation of all five intermetallic compounds at 1100°C and only four at 1150°C indicates that the bcc + $Ti_5Si_3 \leftrightarrow Ti_3Si$ invariant reaction should be 1125 ± 25°C, which is significantly lower than the previous experimental observation (~1170°C) based on cooling and heating experiments. The result of the isothermal diffusion couple experiments should be much more reliable based on the shifting of cooling and heating temperatures discussed in this chapter.

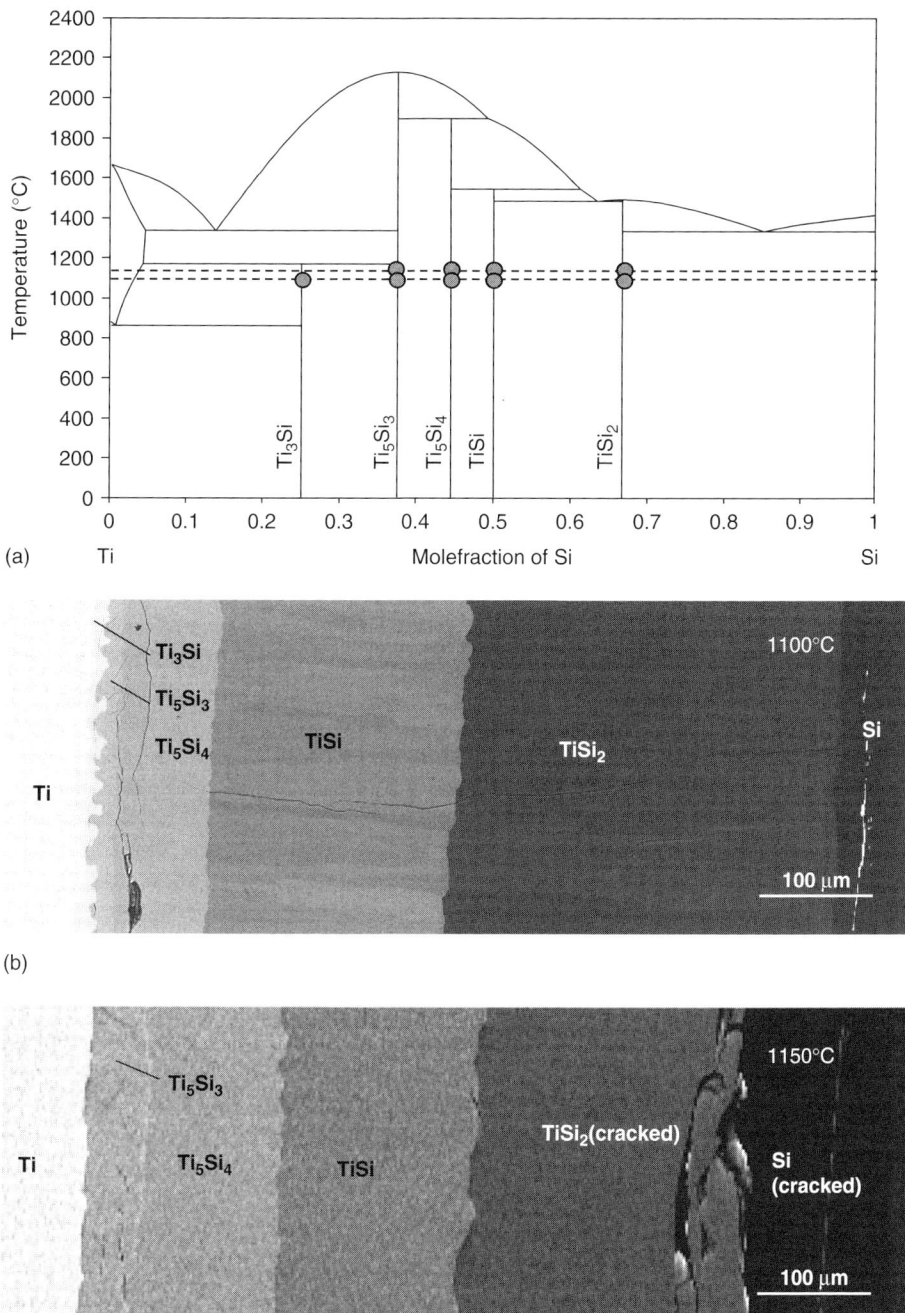

Figure 2.20 Formation of intermetallic compounds in the Ti–Si binary system: (a) phase diagram [48] (b) scanning electron microscopy (SEM) backscattered electron image of a Ti–Si diffusion couple annealed at 1100°C for 4000 h and (c) SEM backscattered electron image of a Ti–Si diffusion couple annealed at 1150°C for 2000 h [49,50].

4 Concluding Remarks

The examples in this chapter show the importance of understanding phase transformation kinetics in reliable determination and assessment of phase diagrams. For both the cooling/heating methods and the isothermal methods, the slow phase transformation kinetics can result in significant deviation from equilibrium, especially for the solid–solid phase equilibria at relative low temperatures when the diffusion and nucleation kinetics are slow. The examples shown in this chapter are somewhat "extreme" cases, not meant to scare researchers away from using these methods for phase diagram determination, but intended to forewarn potential problems associated with slow kinetics such that these problems can be avoided by carefully designed experiments.

There have been sporadic reports of *room temperature* phase diagrams for high melting point elements. The practice started several decades ago [e.g., Ref. 51 for Co–Fe–W], but unfortunately it still persisted nowadays [e.g., Ref. 52 for Co–La–Ti]. Individual alloys were usually melted and homogenized at a very high temperature, which is a good practice. (They were subsequently step-wise heat treated to or simply heat treated at an intermediate temperature, for example, 650–800°C for 45 days for the Co–La–Ti alloys, then slowly furnace cooled to room temperature.) The XRD or other techniques were used to identify the phases in the alloys to establish the phase equilibria. Phase diagrams determined this way should not be called room temperature phase diagrams; instead, they should be regarded as phase diagrams at the temperatures of the final heat treatments. In addition, it is always a better practice to quench the samples from the temperatures of final heat treatments to room temperature to avoid the complication of further phase transformations during the slow furnace cooling. All the "room temperature phase diagrams" reported in the literature for elements with melting points higher than 1000°C should be taken with suspicion and should actually be regarded as phase diagrams at the temperatures of the respective final heat treatments. Slow diffusion and phase transformation kinetics would never allow alloys made up of high melting point elements (e.g., Co–Fe–W) to reach equilibrium at ambient temperature. (For Co–Fe–W alloys, equilibria at 500°C are already extremely difficult to achieve in bulk samples, not even mentioning at ambient temperature).

There may be a way to reach low-temperature phase equilibria for high melting point alloys. The formation of nano-scale intermetallic compounds and solid solutions at ambient temperature using the so-called "metallurgy in a beaker" [53,54] methodology is very encouraging. A benchmark study using this methodology for low-temperature phase diagram determinations will be very valuable to see whether reliable results can be obtained. The methodology may provide critical data for low-temperature phase stability to the CALPHAD modeling.

Acknowledgments

Part of this chapter on cooling and heating kinetics was modified from two previous papers by the author published in *Journal of Phase Equilibria* (JPE) and Z.Metallkunde. The author acknowledges the contribution of Dr. Michael Notis to the JPE paper. He also thanks Prof. Zhanpeng Jin for exciting the author's interest in phase diagrams and Dr. Jack H. Westbrook for proof-reading the manuscript.

REFERENCES

1. J.-C. Zhao and M.R. Notis, *J. Phase Equilib.*, 14 (1993) 303.
2. B. Feuerbacher, *Mater. Sci. Rept.*, 4 (1989) 1.
3. M.R. Plichta, H.I. Aaronson and J.H. Perepezko, *Acta Metall.*, 26 (1978) 1293.
4. J.-C. Zhao and M.R. Notis, *Mater. Sci. Eng. R.*, 15 (1995) 135.
5. G. Vander Voort, *Atlas of Time-Temperature Diagrams for Irons and Steels*, ASM International, Metals Park, OH, 1991.
6. E.B. Hawbolt and T.B. Massalski, *Metall. Trans.*, 2 (1971) 1771.
7. J.H. Perepezko, *Proceedings of the International Conference Solid–Solid Phase Transformations*, H.I.Aaronson, D.E. Laughlin, R.F. Sekerka and C.M. Wayman, (eds.), TMS-AIME, 1982, pp. 231–235.
8. L.P. Srivastava and J.G. Parr, *Trans. TMS-AIME*, 224 (1962) 1295.
9. T.B. Massalski, *Phase Transformations*, ASM International, Metals Park, OH, 1970, pp. 433–483.
10. D.A. Mirzayev, S.Ye. Karzunov, V.M. Schastlivtsev, I.L. Yakovleva and Ye.V. Kharitonova, *Fiz. Metal. Metalloved.*, 61 (1986) 331.
11. J.-C. Zhao and M.R. Notis, *Scripta Metall. Mater.*, 32 (1995) 1671.
12. D.A. Mirzayev and V.M. Schastlivtsev, *Proceedings of the International Conference on Martensitic Transformations (ICOMAT'86)*, Japan Institute of Metals, 1986, pp. 282–287.
13. J.-C. Zhao, *Mater. Sci. Technol.*, 8 (1992) 997.
14. J.E. Kittl and C. Rodriguez, *Acta Metall.*, 17 (1969) 925.
15. J. Jellison and E.P. Klier, *Trans. TMS-AIME*, 233 (1965) 1694.
16. Y.T. Zhu and J.H. Devletian, *Metall. Trans. A*, 22A (1991) 1993.
17. V.N. Gridnev, A.F. Zhuravlev, B.F. Zhuravlev, O.M. Ivasishin and S.P. Oshkaderov, *Fiz. Metal. Metalloved.*, 56 (1983) 985.
18. V.N. Gridnev, Yu.Ya. Meshkov, and N.F. Chernenko, *Sb. Nauchn. Tr. Inst. Metallofiz.*, *Akad. Nauk Ukr. SSR*, No. 17 (1963) 143.
19. W.L. Haworth and J.G. Parr, *Trans. ASM*, 58 (1965) 476.
20. V.N. Gridnev, V.G. Gavrilyuk and S.P. Oshkaderov, *Neue Huette*, 28 (1983) 361.
21. V.N. Gridnev, O.M. Ivasishin and P.E. Markovskii, *Metal Sci. Heat Treat.*, 27 (1985) 43.
22. V.N.Gridnev, Martensitnye Prevrashch, *Dokl. Mezhdunar Konf "ICOMAT-77"*, 1977 (pub.1978), V.N. Gridnev (ed.), Izd. Naukova Dumka, Kiev, USSR, (1978) 246–252.
23. L. Kaufman and M. Cohen, *Trans. AIME*, 206 (1956) 1393.
24. L.J. Swartzendruber, V.P. Itkin and C.B. Alcock, *J. Phase Equilib.*, 12 (1991) 288.
25. J.-C. Zhao, *Z. Metallk.*, 90 (1999) 223.
26. M. Hasebe, K. Oikawa and. T. Nishizawa, *Nippon Kinzoku Gakkai-Si*, 46 (1982) 577.
27. W. Krajewski, J. Kruger and H. Winterhager, *Metall.*, 24 (1970) 480.
28. Y. Matsunaga, *Kinzoku-no-Kenkyu*, 8 (1931) 549.
29. W. Köster, H. Warlimont and T. Godecke, *Z. Metallk.*, 64 (1973) 399.
30. K. Oikawa, G.-W. Qin, T. Ikeshoji, R. Kainuma and K. Ishida, *Acta Mater.*, 50 (2002) 2223.
31. W. Köster and T. Tonn, *Z. Metallk.*, 24 (1932) 296.
32. Y. Aoki, Y. Obi and H. Komatsu, *Z. Metallk.*, 70 (1979) 436.
33. M. Hansen and K. Anderko, *Constitution of Binary Alloys*, McGraw-Hill, New York (1958).
34. U. Haschimoto, *Nippon Kinzoku Gakkai-Si*, 1 (1937) 135.
35. U. Haschimoto, *Nippon Kinzoku Gakkai-Si*, 1 (1937) 177.
36. L. Brewer, *Molybdenum: Physico-Chemical Properties of its Compounds and Alloys*, O. Kubaschewski (ed.), Atomic Energy Review Special Issue No. 7, International Atomic Energy Agency, Vienna, 1980.
37. J.-C. Zhao and M.R. Notis, *Acta Mater.*, 46 (1998) 4203.
38. P.B. Fernandes, G.C. Coelho, F. Ferreira, C.A. Nunes and B. Sundman, *Intermetallics*, 10 (2002) 993.
39. M.G. Mendiratta and D.M. Dimiduk, *Scripta Metall. Mater.*, 25 (1991) 237.
40. H.J. Goldschmidt and J.A. Brand, *J. Less-Common Met.*, 3 (1961) 34.
41. J.-C. Zhao, M.R. Jackson and L.A. Peluso, *Acta Mater.*, 51 (2003) 6395.

42. B.P. Bewlay, R.R. Bishop and M.R. Jackson, *Z. Metallk.*, 90 (1999) 413.
43. B.P. Bewlay, M.R. Jackson and R.R. Bishop, *J. Phase Equilib.*, 19 (1998) 577.
44. J.-C. Zhao, B.P. Bewlay and M.R. Jackson, *Intermetallics*, 9 (2001) 681.
45. C.R. Hunt and A. Raman, *Z. Metallk.*, 59 (1968) 701.
46. K. Mahdouk and J.C. Gachon, *J. Alloy. Compd.*, 321 (2001) 232.
47. J.-C. Zhao, M.R. Jackson and L.A. Peluso, *J. Phase Equilib. Diff.*, 25 (2004) 152.
48. C. Vahlas, P.Y. Chevalier and E. Blanquet, *CALPHAD* 13 (1989) 273.
49. J.-C. Zhao, M.R. Jackson and L.A. Peluso, *Mater. Sci. Eng. A.*, 372 (2004) 21.
50. J.-C. Zhao, unpublished work.
51. W. Köster, *Arch. Eisen.*, 5 (1932) 431.
52. Q.L. Liu, J.J. Nan, J.K. Liang, F. Huang, Y. Chen, G.H. Rao and X.L. Chen, *J. Alloy. Compd.*, 307 (2000) 212.
53. R.E. Schaak, A.K. Sra, B.M. Leonard, R.E. Cable, J.C. Bauer, Y.-F. Han, J. Means, W. Teizer, Y. Vasquez and E.S. Funck, *J. Am. Chem. Soc.*, 127 (2005) 3507.
54. B.M. Leonard, N.S.P. Bhuvanesh and R.E. Schaak, *J. Am. Chem. Soc.*, 127 (2005) 7326.

CHAPTER THREE

Correct and Incorrect Phase Diagram Features

Hiroaki Okamoto[1] and Thaddeus B. Massalski[2]

Contents

1	Introduction	52
2	Phase Rule Violations	53
	2.1 Typical Phase Rule Violations	53
3	Guidelines for Finding Less Obvious Phase Diagram Errors	56
	3.1 Problems Connected with Phase Boundary Curvatures	56
	3.2 Summary of Improbable Phase Diagram Situations	57
	3.3 Van't Hoff Relationship and Charles' Law	59
	3.4 Form of the Liquidus of a Compound Near a Pure Element Side	61
	3.5 Sharpness of the Liquidus of a Compound and its Relation to a Possible Eutectoid Temperature	63
	3.6 Two Compounds with Very Similar Compositions Are Unlikely to Coexist in a Wide Temperature Range	65
	3.7 Asymmetry of the Liquidus Around the Melting Point of a Compound	65
	3.8 Narrowing of the Width of a Two-Phase Field When the Boundaries Are Extrapolated Toward Higher Temperatures	68
	3.9 An Abrupt Change of Slope	70
	3.10 An Excessive Slope Change Associated with a Polymorphic Transformation	72
	3.11 Shape of a Miscibility Gap	73
	3.12 Displaced Miscibility Gap	74
	3.13 An Inverse Miscibility Gap	76
	3.14 Almost Symmetric Syntectic Reaction	76
	3.15 Metastable Melting Point of a Pure Element	77
4	Examples of Phase Diagrams with Improbable Features	79
	4.1 Ag–Pr	80
	4.2 Al–B	80
	4.3 Al–Mn	82
	4.4 Al–Pu	82
	4.5 Al–Se	83

[1] Department of Information Management, Asahi University, Gifu, Japan
[2] Department of Materials Science and Engineering, Carnegie-Mellon University, Pittsburgh, PA, USA

4.6	Au–La	83
4.7	Au–U	84
4.8	B–Bi	86
4.9	B–W	86
4.10	C–Ge	87
4.11	Ca–Eu	87
4.12	Ce–Pr	87
4.13	Cr–Ni	89
4.14	Cu–Hf	90
4.15	Ir–Rh	92
4.16	Lu–Th	92
4.17	Mg–Sb	92
4.18	Mn–Y	94
4.19	Mo–Ru	94
4.20	Nd–Pu	96
4.21	Ni–Sr	96
4.22	Np–Zr	97
4.23	Pd–Zr	97
4.24	Pt–Zr	99
5	Unusual but Correct Phase Diagrams	100
5.1	Apparent Four-Phase Equilibrium	100
5.2	Apparent Five-Phase Equilibrium	100
5.3	Pointed Liquidus	102
5.4	A Straight Line Liquidus	102
5.5	Off-Stoichiometric Melting	104
6	Phase Diagram Below 0°C	105
7	Conclusion	105

1 INTRODUCTION

Recent compilations of assessed phase diagrams, particularly the binary phase diagrams, have placed an increasing emphasis on the fact that all phase diagrams are a consequence of thermodynamics. Hence the experimentally determined, or thermodynamically assessed, phase boundaries must not violate thermodynamic laws, especially the First and Second Laws. This means that the different approaches used by researchers to arrive at the best judgment of the correct phase diagram features and phase boundaries, in any system, should result in a reasonable agreement. The three typical approaches to phase diagram determination are through experimental determination of the actual phase boundaries, through the use of measured thermodynamic quantities followed by computations, or through thermodynamic modeling using carefully assessed thermodynamic parameters.

Extensive efforts have been made in recent years to compile alloy phase diagrams of all binary systems [1–6]. Among these, *Binary Alloy Phase Diagrams* [1] and *Pauling File Binaries Edition* [5] are particularly informative because the former shows the best phase diagrams, crystal structures, and concise texts based on critical evaluations done by category editors of the APDIC (Alloy Phase Diagram International Commission) and the latter contains almost all (likely and unlikely) binary phase diagrams published in the world before 1995. In the course of editing phase diagrams in *Binary Alloy Phase Diagrams* [1], the authors discovered numerous phase diagrams that showed very unlikely phase boundaries in various respects although they did not explicitly violate phase rules. These findings were discussed by Okamoto and Massalski in four articles [7–10].

These unlikely phase diagram features covered by the four articles and some additional features are discussed here. Discovering unlikely features in phase diagrams is helpful not only as a caution for not depending completely on the compiled phase diagrams, such as those mentioned above, but also for making judgments about unlikely thermodynamic models, for critically examining experimentally determined phase diagrams, and even for discovering a new phenomenon when the unlikely feature is confirmed to be real.

In this chapter, we first discuss briefly the explicit violations of the phase rules. This is followed by implicit cases of possible phase rule violations and some more subtle phase boundary features that may come under question when constructing phase diagrams. The primary objective is to provide guidelines for judging whether or not a phase diagram is a reasonable construction. (It must be noted though that the diagram is not necessarily correct even if it looks reasonable, or vice versa. We should also emphasize that essentially identical graphical phase diagram representations can be obtained by using very different thermodynamic data and their temperature or composition dependence.) Because of the nature of the discussion here, curvatures of phase boundaries are referred to in somewhat qualitative terms, such as "sharp", "flat", "steep", "abrupt", etc. The meaning of these words is, of course, relative, as is easily perceived from the fact that the curvature of a phase boundary will vary depending on the scale selected for the phase diagram graphics. Thus, a "sharp" liquidus in our context only implies that the form of the liquidus around a certain compound-like phase appears to be more pointed at the melting point in a temperature–composition (T–x) plot than that of a neighboring phase in the same phase diagram, or more pointed than the form expected from simple thermodynamic considerations. Other examples of curvature terminology used in this text are illustrated graphically in Fig. 3.1.

2 PHASE RULE VIOLATIONS

In this section, thermodynamically incorrect phase diagrams are shown. However, they are explained here only briefly. It is universally recognized that violations of phase rules are of course violations of the laws of thermodynamics, which are the basis of the Gibbs phase rule.

2.1 Typical Phase Rule Violations

In examining a phase diagram, the first thing to do is to check if the Gibbs phase rule and other basic rules are satisfied. The hypothetical phase diagram in Fig. 3.2 shows

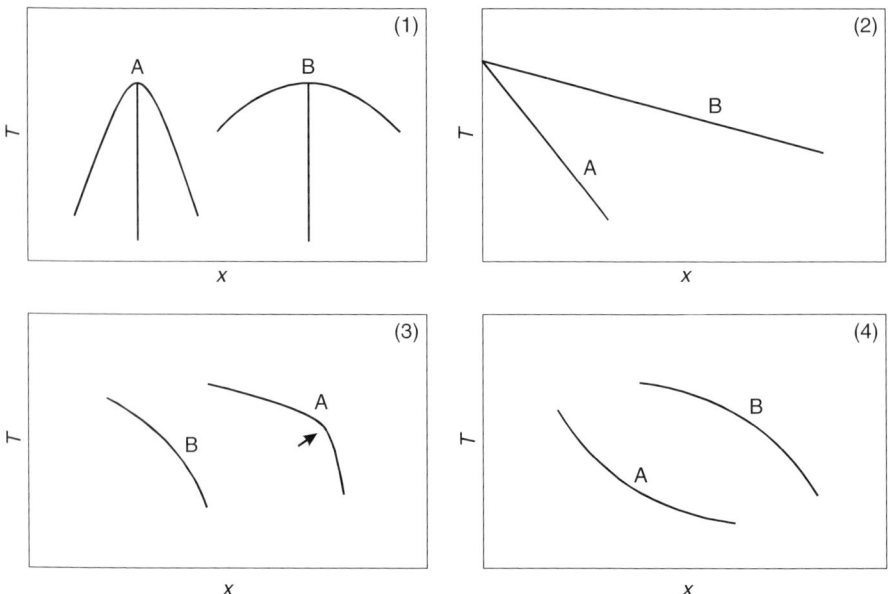

Figure 3.1 Definitions of quantitative words: (1) A has a sharper peak than B; (2) A is steeper than B; (3) A shows a more abrupt change of slope than B; and (4) A is concave and B is convex.

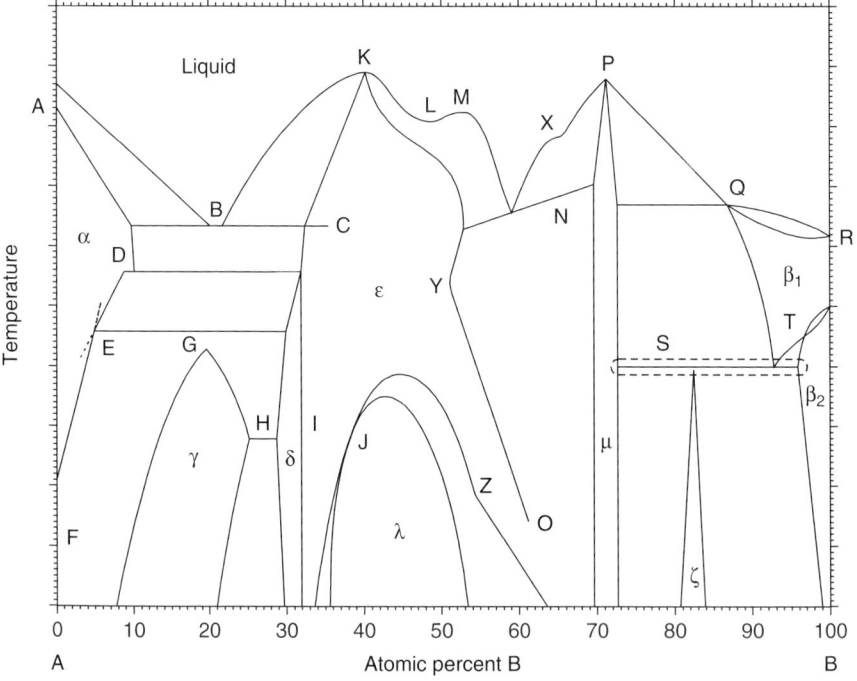

Figure 3.2 Hypothetical binary phase diagram showing phase boundaries that are inconsistent with basic thermodynamic rules (A–T).

problems A to T that are contradictory to the basic rules. When these rules are not met, the diagrams are likely to be wrong, as explained below. However, there are cases when a diagram is valid even though it may look wrong in view of these rules. Such cases are discussed in Section 5.

Problems marked X, Y, and Z in Fig. 3.2 are not violations of phase rules. These points are discussed in Section 3:

- A: A two-phase field cannot be extended to become part of a pure element side of a phase diagram at zero solute. In example A, the liquidus and the solidus must meet at the melting point of the pure element.
- B: Two liquidus curves must meet at one composition at a eutectic temperature.
- C: A tie line must terminate at a phase boundary.
- D: Two solvus boundaries (or two liquidus, or two solidus, or a solidus and a solvus) of the same phase must meet (i.e., intersect) at one composition at an invariant temperature. (There should not be two solubility values for a phase boundary at one temperature.)
- E: A phase boundary must extrapolate into a two-phase field after crossing an invariant point. The validity of this feature, and similar features related to invariant temperatures, is easily demonstrated by constructing hypothetical free energy diagrams slightly below and slightly above the invariant temperature and by observing the relative positions of the relevant tangent points to the free energy curves. After intersection, such boundaries can also be extrapolated into metastable regions of thephase diagram. Such extrapolations are sometimes indicated by dashed or dotted lines.
- F: A phase boundary cannot cross the 0 at.% line except at a phase transition (melting, allotropic transformation, etc.) point of the element A.
- G: The highest point of the γ phase must touch the invariant reaction line.
- H: In a binary system, an invariant temperature line should involve equilibrium among three phases.
- I: There should be a two-phase field between two single-phase fields. (Two single phases cannot touch except at a point. However, second-order and higher-order transformations may be exceptions to this rule.)
- J: When two-phase boundaries touch at a point, they should touch at an extremity of temperature.
- K: A touching liquidus and solidus (or any two touching boundaries) must have a horizontal common tangent at the congruent point (Gibbs–Konovalov rule [11]). In this case, the solidus at the melting point is too "sharp" and appears to be discontinuous.
- L: A local minimum point in the lower part of a single-phase field (in this case the liquid) cannot be drawn without an additional boundary in contact with it. (In this case, a horizontal monotectic line is most likely missing.)
- M: A local maximum point in the lower part of a single-phase field cannot be drawn without a monotectic, monotectoid, syntectic, or syntectoid reaction occurring below it at a lower temperature. Alternatively, a solidus curve must be drawn to touch the liquidus at point M.
- N: The temperature of an invariant reaction in a binary system must be constant. (The reaction line must be horizontal.)

O: A phase boundary cannot terminate within a phase field. (Termination due to lack of data is, of course, often shown in phase diagrams, but this is recognized to be artificial.)
P: The liquidus should not have a sharp peak with discontinuous slope at the melting point of a compound. (This rule is not applicable if the liquid retains the molecular state of the compound, i.e., in case of a strong association.)
Q: The compositions of all three phases at an invariant reaction must be different.
R: Initial slopes of the liquidus and solidus must have the same sign (both up or down).
S: A four-phase equilibrium is not allowed in a binary system.
T: Two separate phase boundaries that create a two-phase field between two phases in equilibrium should not cross one another.

3 Guidelines for Finding Less Obvious Phase Diagram Errors

Besides the explicit errors shown in Section 2, many questionable phase diagram regions exist in proposed phase diagrams, which, although they seemingly do not violate any of the above rules, are nevertheless very unlikely to be true phase representations. Some of these questionable situations become evident only when the respective phase boundaries are extrapolated to the metastable equilibrium region. In performing the extrapolation, it often becomes evident that only through a strange or abrupt (and hardly justified) change of slope could the proposed phase diagram be reconciled in the metastable regions without committing any of the errors A to T.

In this section, guidelines for judging the validity of phase diagrams are shown.

3.1 Problems Connected with Phase Boundary Curvatures

Although phase rules are not violated in these circumstances, three additional unusual situations (X, Y, and Z) have also been included in Fig. 3.2. In each case, a more subtle thermodynamic problem may exist related to the situation. The problems with each of these situations involve an indicated abrupt change of slope of a phase boundary. If X-, Y-, and Z-type situations are to be associated with realistic thermodynamics, the temperature (or the composition) dependence of the thermodynamic functions of the phase (or phases) involved would be expected to show corresponding abrupt and unrealistic variations in the temperature or composition regions where such abrupt phase boundary changes are proposed, without any clear reason for them:

X: Two inflection points are located too closely to one another.
Y: An abrupt reversal of the boundary direction (more abrupt than a typical smooth "retrograde"). This particular change can occur only if there is an accompanying abrupt change in the temperature dependence of the thermodynamic properties of either of the two phases involved (in this case ε or μ in

relation to the boundary). The boundary turn at Y is very unlikely to be explained by any realistic change in the composition dependence of the Gibbs energy functions.

Z: An abrupt change in the slope of a single-phase boundary. This particular change can occur by an unrealistic abrupt change in the composition dependence of the thermodynamic properties of the single phase involved (in this case the ε phase). If the abrupt change of slope is caused by an abrupt change in the temperature dependence of the thermodynamic properties of the ε phase, it is likely that the boundary of the λ phase shows a curvature change at the same temperature. This problem is discussed in Section 3.9 more quantitatively.

3.2 Summary of Improbable Phase Diagram Situations

Some additional improbable phase diagram situations discussed in this section are summarized in Fig. 3.3. Specific examples are explained in Section 4, including those not shown in this figure:

a: The initial opening angle of the G + L two-phase field at 0 at.% B (similar at 100 at.% B), should be much greater than that of the L + S two-phase field because the heat of vaporization of an element is much greater than the heat of fusion. This and other opening angle problems can be examined by using the Van't Hoff relationship, as explained in Section 3.3.

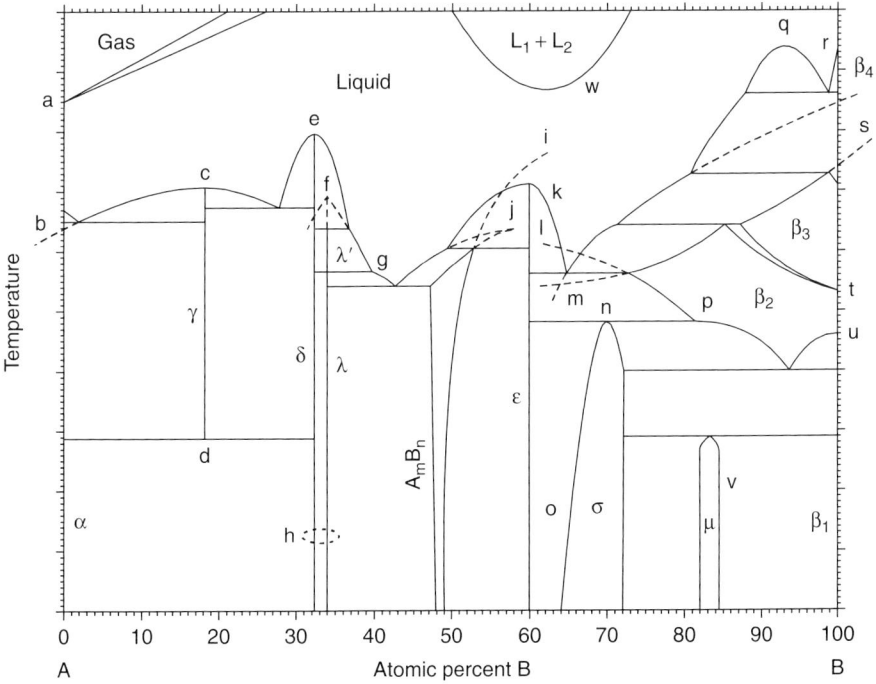

Figure 3.3 Hypothetical binary phase diagram showing phase boundaries that are consistent with basic thermodynamic rules but unlikely (a–w).

b: Extrapolation of the L/L + γ liquidus crosses the 0 at.% line (problem F in Fig. 3.2 in the metastable state). The liquidus cannot be extrapolated below 0 at.% even in the metastable state, as explained in Section 3.4. If an attempt is made to avoid this problem in this case, the liquidus would have to be curved sharply in the metastable region. This is the problem of Z in Fig. 3.2.
c: The peak is too flat in comparison to the peak at point e. The entropy of fusion of most elements is similar (Charles' law), and the trend is the same for intermediate phases. When multiple intermediate phases exist in a system, the sharpness of the liquidus peaks must be similar. This is applicable to peaks in the metastable region also (this explanation is omitted hereafter).
d: The liquidus of a phase that becomes unstable at low temperatures must be more pointed at the melting point. The entropy of fusion of a phase that is unstable at low temperatures is small, and the liquidus would be more pointed as a result (explained in Section 3.5).
e: Opposite of c. Because the peak is pointed, this phase would decompose eutectoidally at a low temperature.
f: When the L/L + λ_2 liquidus is extrapolated, the peak should appear at around the composition of λ_2. A rapid change of curvature is then required near the peak, and the peak would be extremely pointed also (problem P in Fig. 3.2).
g: The enthalpy of transformation is generally smaller than the enthalpy of fusion. The change of slope of the liquidus curves associated with the phase transformation should be small.
h: Two phases, δ and λ_1 with similar compositions are unlikely to coexist in stable forms in a wide temperature range (explained in Section 3.6). If both phases are stable, a very delicate balance is needed between the Gibbs energies of the two phases. (c.f. If the compositions of the two phases are the same, only one phase is stable at any temperature.)
i: In the metastable state, ε transforms into A_mB_n. A stoichiometric phase is unlikely to transform into another phase with different stoichiometry.
j: An intermetallic compound with a specific stoichiometry is most stable at that composition. Therefore, the melting of the phase should occur at around the stoichiometric composition. (Usually, this is only a problem of naming A_mB_n).
k: The liquidus cannot be too asymmetric around the melting point of the phase. Empirically, a liquidus is suspicious if the ratio of widths to the left and right is greater than 2:3, as explained in Section 3.7.
l: When the ε/ε + β_2 phase boundary is extrapolated, its temperature must be higher than the melting point of ε at the composition of ε. Otherwise, β_2 should exist as a stable high-temperature phase.
m: The liquidus and solidus of β_2 should not cross one another when extrapolated (problem T in Fig. 3.2).
n: This problem of solvus lines smoothly touching the invariant line is unlikely, and is often drawn to avoid the problem of K in Fig. 3.2 by mistake.
o: A solid–solid two-phase field must become narrower as the temperature increases. A solid–liquid two-phase field naturally becomes narrower as the temperature increases because the solid phase cannot exist at sufficiently high temperatures. This problem is discussed in Section 3.8.

p: When the σ + β$_2$ two-phase field is extrapolated to the high-temperature side, two boundaries must touch at the maximum temperature (problem J in Fig. 3.2).
q: The miscibility gap of a liquid phase appears in the middle of a phase diagram (e.g., 25–75 at.%). A miscibility gap can form only when an unnatural thermodynamic model is used. This problem is explained in Section 3.12.
r: When the solid solubility of the β$_4$ phase is negligible, the liquids at 100 at.% B should have a slope pointing at 0K, −10 to −20 at.% B. (Derived from Charles' law, as explained in Section 3.3. When B has allotropic forms, as in this case, the slope becomes steeper.)
s: The liquidus and solidus of the β$_3$ phase must cross on the 100 at.% B line at the metastable melting temperature of β$_3$ (problem A in Fig. 3.2). The metastable melting temperature can be estimated from the heat of transformation and the transformation temperature between β$_4$ and β$_3$ at 100 at.% B, as explained in Section 3.15.
t: The opening angle of the β$_2$ + β$_3$ two-phase field at 100 at.% B may be too small (determined by the enthalpy of transformation).
u: The opening angle of the β$_1$ + β$_2$ two-phase field at 100 at.% B may be too large (determined by the enthalpy of transformation).
v: A compound with a constant width over a wide temperature range is unlikely because an abrupt change of slope is needed near the peritectoid temperature. In addition, the almost symmetric shape of μ near the peritectoid temperature is unlikely because the shape on the A-rich side of μ is determined by the thermodynamic relationship between σ and μ and the B-rich side is determined by the relationship between μ and β$_1$, independently.
w: An inverse miscibility gap is unlikely because the ideal entropy of mixing increases (more negative $\Delta_{mix}G$) as the temperature becomes higher. This is explained in Section 3.13.

Some of the problems described above are explained in more detail below. In addition, there are other types of unlikely situations that could not be included in Fig. 3.3.

3.3 Van't Hoff Relationship and Charles' Law (Related to Points a, r, t, and u in Fig. 3.3)

The Van't Hoff relationship [12] is rigorously applicable to any phase transition. Here we choose a liquid to solid transition as an example.

If the liquidus and solidus slopes of a terminal solid solution are wide apart (e.g., a nearly horizontal liquidus and a nearly vertical solidus), or if their slopes are almost the same, the situation should be examined further for possible difficulties. The Van't Hoff relationship for the initial slopes of the liquidus and solidus

$$(dx/dT)_{solidus} - (dx/dT)_{liquidus} = \Delta_{fus}H/RT_m^2 \tag{1}$$

where x is the mole fraction solute, requires the difference of the inverse values of the initial slopes (not angles) must be separated by a certain amount determined by the enthalpy of fusion, $\Delta_{fus}H$, and the melting point, T_m(K), of the pure element itself. Because the variances in $\Delta_{fus}H$ and T_m are not very large among metallic elements, the initial slopes of liquidus and solidus must be "properly" apart.

First $(dx/dT)_{solidus} = 0$ (no solid solubility) is assumed. In this case, Okamoto and Massalski [7] found that the extrapolation of the initial slope of the liquidus calculated from $\Delta_{fus}H$ and T_m using Eq. (1) roughly passes through a fixed point around (130 ± 10 at.%, −500 ± 50°C) for most elements. (Exceptions are semimetals B, Si, Ge, Sb, Te, Se, Bi, and Sn and the group Al, Zn, Cd, Hg, and Ga. They have different conversion points [7].) Paul [13] proved that the convergence point should be 0K, 110 at.% by assuming that the empirical Charles' law (the entropies of fusion for most elements are approximately $1.1R$) is valid (Fig. 3.4(a)). For visual checking of phase diagrams, this convergence point may be taken at 0°C, 100 at.% if the

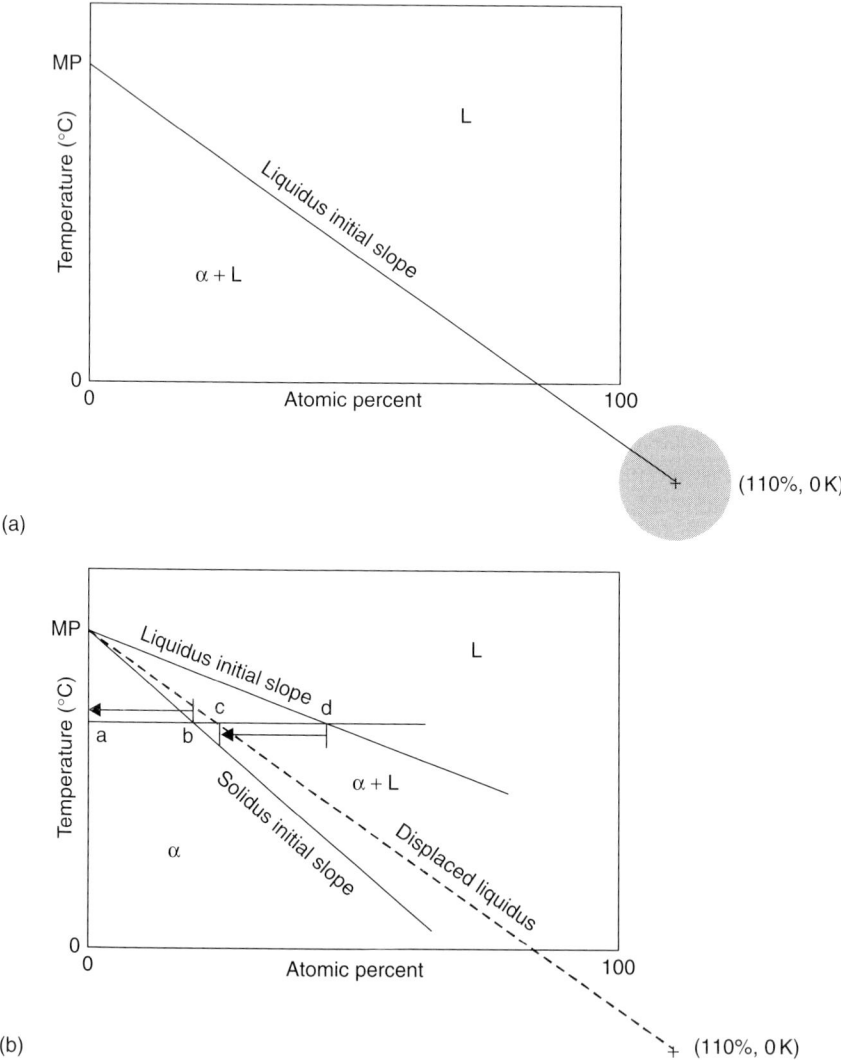

Figure 3.4 Initial slope of the liquidus boundary of a terminal phase converges to a point around (110%, 0K): (a) when the terminal phase has no solid solubility and (b) when the terminal phase has some solid solubility.

melting point of the element is high enough (>~500°C). When the solid solubility is not negligible, the method shown in Fig. 3.4(b) derived from the Van't Hoff relationship (Eq. (1)) can be used. This is helpful for checking whether the opening angle between the liquidus and the solidus is appropriate. First, draw a horizontal line (constant temperature line) to cross both the liquidus and the solidus at any temperature. Measure the distance ab. Determine point c away from point d by the distance ab to the 0 at.% side. Draw a line connecting the melting point and the point c, and check if the line passes around the point at 0K, 110 at.% (or 0°C, 100 at.%).

If the element in question undergoes allotropic transformations, this rule applies to the lowest-temperature modification. For higher-temperature modifications, the initial slope becomes a little steeper due to the smaller enthalpy of fusion than the value expected from the Charles' law.

3.4 Form of the Liquidus of a Compound Near a Pure Element Side
(Point b in Fig. 3.3)

In this section, the form of a liquidus associated with a compound-like phase in the composition range near a pure element end (0 or 100 at.% solute) is considered. Both ends of the liquidus of an intermediate phase must eventually terminate at invariant reactions, such as a eutectic or a peritectic reaction. The termination of the equilibrium portion of the liquidus of an intermediate phase by an invariant reaction is of course expected from the relationship between the liquid phase and the compound phase (i.e., the invariant reaction reflects the thermodynamic competition between the neighboring phases and is not connected with the form of the liquidus of the compound under consideration). It follows from this that the overall validity of a liquidus trend must be judged by taking into account also its projected metastable ranges.

In order to examine the liquidus of a compound including the metastable range, it is convenient to assume that all neighboring solid phases, including the terminal phase, can be suppressed. Then, the liquidus becomes visible in its entire composition range from 0 to 100 at.%. (Only the 0 at.% side is discussed hereafter to eliminate redundancy.) Here we recall that the extrapolated liquidus cannot cross the 0 at.% solute line because the slope of the Gibbs energy function of the liquid phase always (except at 0K) has a negative infinity value at 0 at.% due to the $RTx \log x$ term, which results from the contribution of the ideal entropy of mixing (Fig. 3.5), even when the enthalpy of mixing is strongly positive.

Possible general forms of the liquidus associated with an A_3B-type compound were illustrated in Ref. [7], assuming that the liquid was an ideal solution. Several more general situations for a liquidus are considered below for an AB-type (50 at.%) compound melting at 1000°C. The liquid is now assumed to be a regular solution. Of course, a further deviation from a regular solution behavior would cause additional changes in the form of the liquidus, but the rate of change may be expected almost always to be a slowly changing function in any one phase. Hence, the use of a regular solution approximation may be adequate for the present purpose. (*Note*: The present aim is to show that a too abrupt change of slope in a liquidus near the 0 at.% line is difficult to justify thermodynamically.)

In order to examine possible variations of the curvature of a given liquidus, the curves shown in Fig. 3.6 were calculated for rather extreme situations. Curves $a1$ and $a2$ correspond to a situation where the interaction parameter is positive and large, that is, $G^{ex}(L) = 18\,000x(1-x)$ J/mol. This value cannot be raised much higher because a miscibility gap then develops in the liquidus [7]. Curves $c1$ and $c2$

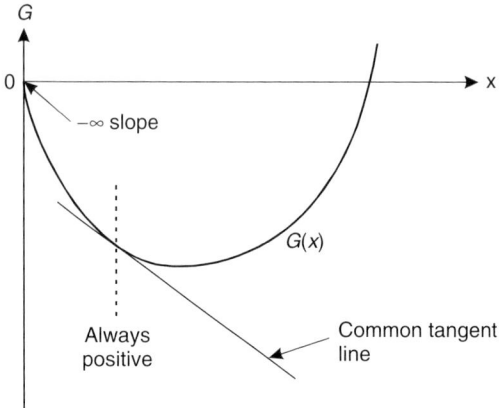

Figure 3.5 The slope of Gibbs energy of mixing is $-\infty$ at 0 at.%.

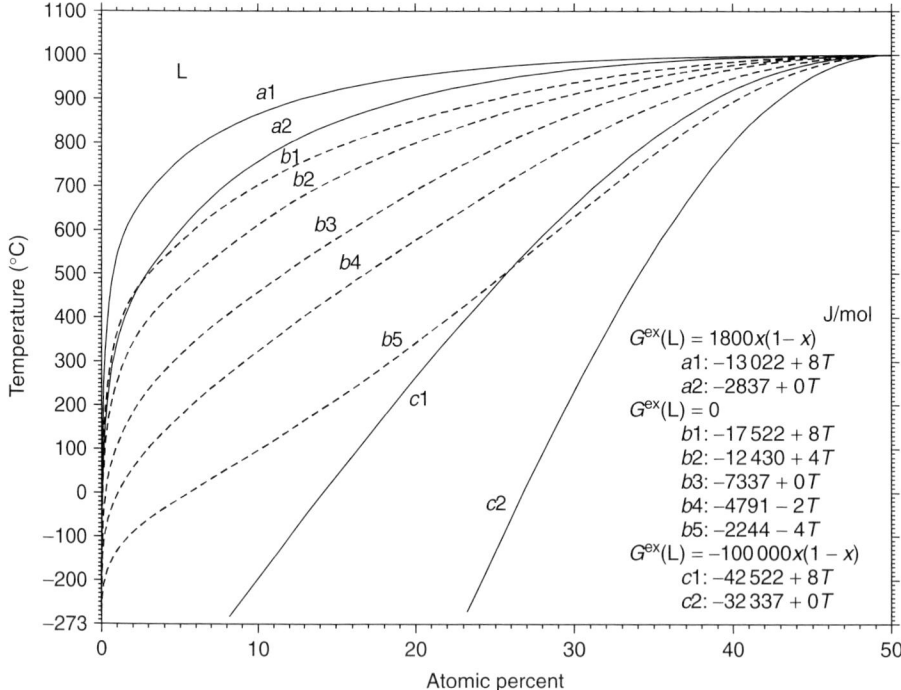

Figure 3.6 Trends of the liquidus of an equiatomic compound melting at 1000°C for typical Gibbs energy of mixing for the liquid phase and lattice stability for the compound phase.

correspond to another extreme situation where the interaction parameter is now negative and large, that is, $G^{ex}(L) = -100\,000x(1-x)$ J/mol, which is extremely negative according to Ref. [7]. In this case, the liquidus does not reach the 0 at.% solute line at all. For comparison, an ideal solution case is also included in Fig. 3.6 ($b1$–$b5$).

Among these curves, $b4$ and $b5$ give the apparent appearance of boundaries projecting to cross the 0% line without a downward turn, especially if the rapidly turning portion of each curve is in an actual diagram obscured from view by the liquidus of another phase, such as the terminal solid solution.

However, these models correspond to rather artificial situations because the coefficients of the T term (-2 or -4) for curves $b4$ and $b5$ may be too negative for a real case, as discussed in Section 3.5. Accordingly, it may be concluded that, in a general case, a liquidus requiring an abrupt change of slope to avoid "crossing" the 0 at.% line when extrapolated should be regarded as suspect.

3.5 Sharpness of the Liquidus of a Compound and its Relation to a Possible Eutectoid Temperature (Points c and d in Fig. 3.3)

A useful notion here is that the "sharpness" of a liquidus of a compound is strongly related to the "stability" of this compound at low temperatures. To illustrate this point, Fig. 3.7 shows various forms of the liquidus of an equiatomic compound

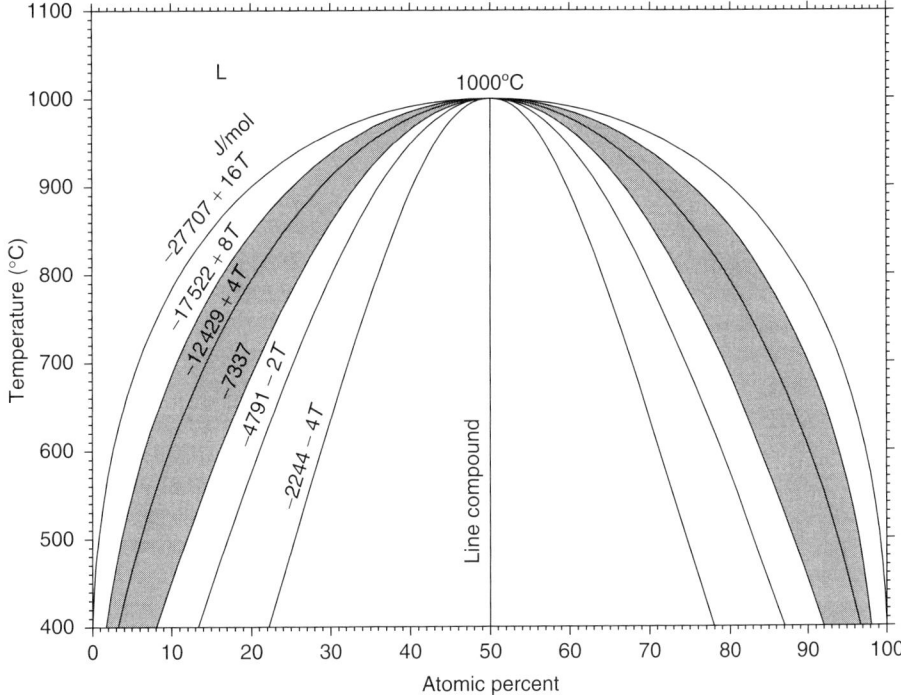

Figure 3.7 Roundness of liquidus of an equiatomic compound melting at 1000°C. The shaded area is the most likely range.

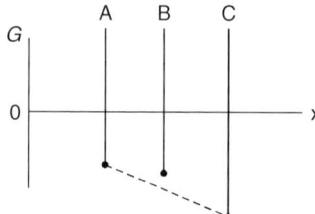

Figure 3.8 Gibbs energy of three line compounds. B is not a stable phase relative to A and C because G(B) is less negative than the average of G(A) and G(C).

melting at 1000°C. The liquid phase is assumed for simplicity to be an ideal solution. The Gibbs energy function of the compound is assumed to be of the form $G = a + bT$, where the reference states are the liquid elements. It is clear from Fig. 3.7 that the liquidus becomes increasingly more flat as the constant term a is made smaller, or the coefficient b is made larger.

As understood from the fact that a is the value of G at 0K, a reflects the relative stability of the compound at low temperatures. Because the stability of a compound increases when its Gibbs energy is lowered, a compound with a flat liquidus must be more stable than one with a sharp liquidus. Hence, when more than one compound exists in a system, a compound with a higher (less negative) Gibbs energy compared with the neighboring compounds cannot "survive" to low temperatures because at low temperatures the Gibbs energy of a two-phase mixture composed of the neighboring compounds becomes even lower (Fig. 3.8). Therefore, a compound with a sharp liquidus is destined to decompose eutectoidally into two neighboring phases. This simple guide can save much time when realistic phase diagram calculations are being attempted.

The smallest possible value of b is limited by the fact that a solid phase at sufficiently low temperatures must become more stable than the liquid phase of the same composition. In the case of an ideal liquid, for example, the Gibbs energy of the liquid phase at equiatomic composition varies with temperature at a rate of -5.763 J/mol·K that is, the value of $R[x \ln x + (1 - x)\ln(1 - x)]$ at $x = 0.5$. Thus, this is the minimum possible value of b in the expression of $G = a + bT$ for the equiatomic compound. (If b is less than -5.763, G would become more negative at high temperatures than the Gibbs energy of the liquid phase, causing an inverse melting.)

For a compound existing at an off-equiatomic composition ($x \neq 0.5$), the limit of b will be even less negative than -5.763. If the compound is moved to positions where $x = 0$ or 1 (i.e., the terminal elements), b becomes equivalent to the entropy of fusion of the elements. Empirically, it is known that the values of b of most elements are approximately constant (the Charles' law) in the range of ~10 (± 2) J/mol·K (~1.1R) with only a few exceptions for nonmetallic elements (c.f. Section 3.3). Because the entropy of fusion of a compound does not differ markedly from the entropies of the terminal elements, the value of b of a compound must be approximately $10 - R[x \ln x + (1 - x)\ln(1 - x)]$ J/mol·K. The b values for various compounds in the same system therefore include a superficial contribution from the $-R[x \ln x + (1 - x)\ln(1 - x)]$ J/mol·K term. Thus, the value of b of most equiatomic compounds will be ~4 (=10 − 5.763) J/mol·K (note that the melting point is not involved), and hence the resulting liquidus of most normal compounds will fall in the shaded range in Fig. 3.7 within a tolerable margin.

The discussion above was based on the assumption that the liquid is an ideal solution. When the enthalpy of mixing cannot be neglected, the sharpness of the liquidus may vary substantially (but systematically), as already discussed in Section 3.4. When the excess entropy of mixing of the liquid phase, $\Delta_{mix}S^{ex}(L)$, is not zero, the "sharpness" of the liquidus will change even for the same b value. However, the magnitude of $\Delta_{mix}S^{ex}(L)$ is usually small (often assumed to be zero), and its influence is also small.

The relationship between the sharpness of the liquidus and the eutectoid decomposition temperature discussed above can be looked at from another viewpoint. That is, when two or more line compounds exist and they are not decomposing eutectoidally, the sharpness of their respective liquidus peaks is expected to be similar. In view of an asymmetric liquidus discussed in Section 3.7, peaks with a width ratio of 1:2 (at the same temperature below the peak temperature) are already too dissimilar even when the fact that the peaks are occurring at different compositions is taken into account.

3.6 Two Compounds with Very Similar Compositions Are Unlikely to Coexist in a Wide Temperature Range (Point h in Fig. 3.3)

It is very unlikely that two compounds with very similar compositions should be able to coexist over a wide temperature range, as shown by h in Fig. 3.3. In order for both δ and λ to remain stable in a wide temperature range, the Gibbs energies, G, of both compounds must vary with temperature in almost exactly the same way because even a very small difference in G makes one of the compounds unstable. This is especially true when the (stable or metastable) congruent melting temperatures of the two compounds are not similar. The compound with a lower melting temperature will be in general more stable at lower temperatures (this is because, in such situations, the temperature dependencies of G of this compound are greater than that of the compound with a higher melting temperature). Accordingly, the high-melting compound will become unstable with respect to its neighbors on cooling sooner or later, and only one compound will remain stable at low temperatures.

More intuitively, this trend can be explained by the fact that the coexisting temperature range is zero when the compositions of the two compounds are the same (polymorphic transformation).

It is very unlikely (but not impossible) that one of the two neighboring compounds exists in a limited temperature range by forming peritectically and decomposing catatectically on cooling because an even more delicate relationship is required between the temperature dependence in G for the two compounds.

3.7 Asymmetry of the Liquidus Around the Melting Point of a Compound (Point k in Fig. 3.3)

A strikingly asymmetric liquidus around a congruently melting compound in a phase diagram is unlikely. A definition of the asymmetry is of course somewhat ambiguous, but from a glance at the La–Au phase diagram (Fig. 3.9), for example,

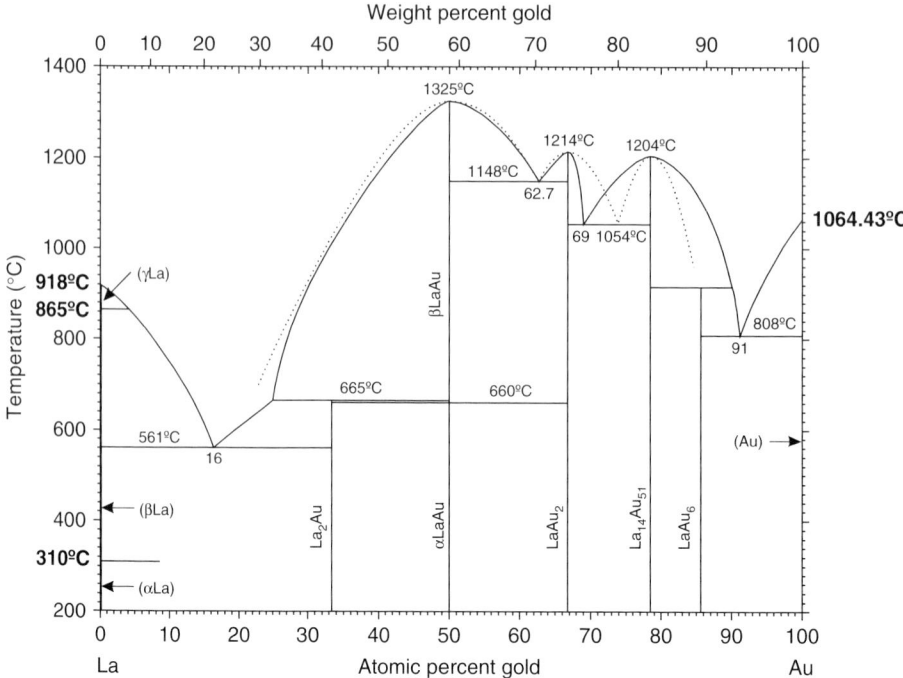

Figure 3.9 La–Au phase diagram. Asymmetry in the LaAu$_2$ liquidus (solid line) is unlikely.

Table 3.1 La–Au thermodynamic data (J/mol).

G(LaAu$_2$)	$= -38\,929 + 0.533\,T$
$\Delta_{mix}G^{ex}(L)$	$= x(1-x)(-97\,488 - 58\,085x)$

it is obvious that the liquidus of βLaAu is more symmetric than that of LaAu$_2$. An approximate judgment can be made by comparing the ratio of the width of the two-phase fields on each side of a compound at the same temperature. If the width on one side is either one half or double that on the other side, the asymmetry may be considered clearly unusual. In our earlier calculation of the La–Au phase diagram [14], we derived a set of almost perfectly symmetric liquidii for all observed compounds (dotted lines). The modeling is illustrated in Table 3.1 by the Gibbs energy functions used for the calculation of the form of the LaAu$_2$ liquidus.

In order to examine here the effect of further changes in this selected Gibbs energy function on the resulting asymmetry, the liquidus on the right-hand side of LaAu$_2$ was recalculated for various modified $\Delta_{mix}G^{ex}(L)$ terms assumed for the liquid phase. In this calculation, the Gibbs energy function of LaAu$_2$ and its melting point are, of course, fixed. Admittedly, the transition between the two branches of the LaAu$_2$ liquidus, on the left- and right-hand sides of stoichiometry, is incompatible with the Gibbs–Konovalov rule [11] because the d^2G/dx^2 value is discontinuous at the congruent point. Hence, the present modeling should be regarded as corresponding to the situation where the zero slope requirement at the congruent point may be

Table 3.2 Calculated LaAu$_2$ liquidus composition at 1148°C and 1054°C.

$\Delta_{mix}G^{ex}(L)$ (J/mol)	Liquidus, at.% Au	
	1148°C	1054°C
$x(1-x)(-200\,000 + 95\,700x)$	72.30	75.68
$x(1-x)(-100\,000 - 54\,300x)^a$	71.21	73.70
$x(1-x)(-97\,488 - 58\,085x)$	71.23	73.69
$x(1-x)(0 - 204\,300x)$	70.61	72.71
$x(1-x)(100\,000 - 354\,300x)^b$	70.20	72.07
$x(1-x)(200\,000 - 504\,300x)^b$	69.91	71.61

[a] Almost identical to the reference curve and not shown in Fig. 3.7.
[b] A very unlikely situation because the partial molar enthalpy for the dilute solution is strongly positive at 0% and strongly negative at 100 at.%.

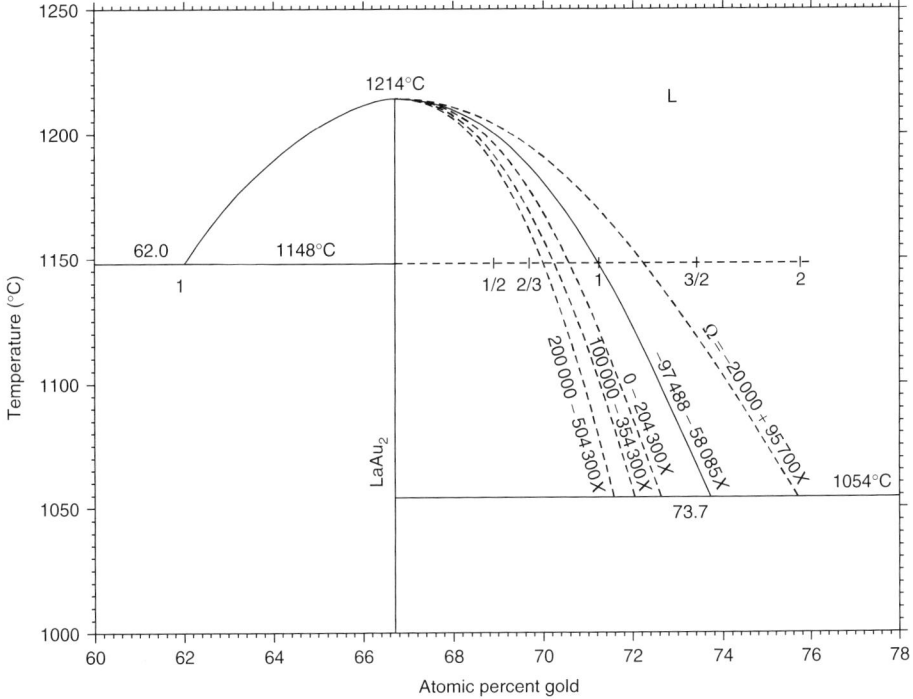

Figure 3.10 Asymmetry introduced in the LaAu$_2$ liquidus by changing the magnitude of the interaction parameter on the Au-rich side of LaAu$_2$.

regarded as a further subtlety that is not material to the more general examination of the liquidus behavior caused by changes in the thermodynamic properties of the liquidus as composition changes. The calculated results (Table 3.2) indicate that it is very difficult to generate asymmetry (Fig. 3.10) even when a drastic change in the thermodynamic behavior of the liquid phase is introduced, as shown by the various forms of the $\Delta_{\text{mix}} G^{\text{ex}}(\text{L})$ function in Table 3.2.

It appears that if the width ratio is 1:1.5 (note the scale at 1148°C in Fig. 3.10), the liquidus is already fairly asymmetric. If the ratio is 1:2, the liquidus is strikingly asymmetric in the sense that an unlikely transition in the thermodynamic properties of the liquid phase would be involved in that composition range.

When checking liquidus asymmetries around compounds in a phase diagram, the above rough measure of the width ratio should be applied at several temperatures slightly away from the melting point because drawing errors often cause an appearance of asymmetry very near the melting point.

Note: (1) If the liquidus is indisputably asymmetric, it is usually found that the corresponding solidus is also asymmetric, which means that the compound is actually a phase with a considerable width. (2) A relatively normal liquidus associated with a compound existing near a pure component may appear asymmetric at low temperatures because one side of it is limited by the edge of the phase diagram (see e.g., the Ba–Be phase diagram in *Binary Alloy Phase Diagrams* [1]).

3.8 Narrowing of the Width of a Two-Phase Field When the Boundaries Are Extrapolated Toward Higher Temperatures
(Point o in Fig. 3.3)

When the high-temperature ends of two boundaries delineating a two-phase field in a given phase diagram are terminated by an invariant reaction or experimental data to guide the form of the boundaries are scarce, extrapolation of these boundaries to higher temperatures may give a clue as to how they should be drawn to satisfy thermodynamic conditions.

Figure 3.11 shows possible relationships between two such boundaries of a two-phase field (in absence of other phases). Obviously, Figs. 3.11(a)–(c) are the most common situations and are not considered here. When at least one of the two phases involved is a liquid, one of these situations is inevitable because a solid must eventually melt. Figures 3.11(d) and (e) are possible when the two phases are solids. In both situations, the two boundaries may remain separated to infinitely high temperatures. This is possible when one type of crystal structure is always more stable when rich in A and the other type when rich in B. Figure 3.23 shows a possible schematic relationship of the Gibbs energies of the two phases. A question may be asked: Which of the two situations, Fig. 3.11(d) or (e), is more plausible? When the temperature is increased, the solid solution of any crystal structure will behave more like an ideal mixture. The contribution of ideal mixing to the Gibbs energy is of the form $RT[x \ln x + (1 - x) \ln(1 - x)]$. This term decreases the Gibbs energy at a rate of $-5.76\,T\,\text{J/mol}$ at $x = 0.5$. On the other hand, the entropy of a possible transformation between two solid phases with different crystal structures is generally much smaller than $5.76\,\text{J/mol·K}$. Accordingly, if the magnitude of the Gibbs energy were drawn with the same scale on a graph, the lattice stability terms (a and b in Fig. 3.12) will tend to become smaller as the temperature increases, that is, the two Gibbs energy curves will approach one another. Accordingly, the two boundaries of a two-phase field should tend to converge.

Hence, Fig. 3.11(d) is more likely to occur than Fig. 3.11(e). When the crystal structures of two phases are the same, the boundary must converge smoothly, as

Correct and Incorrect Phase Diagram Features

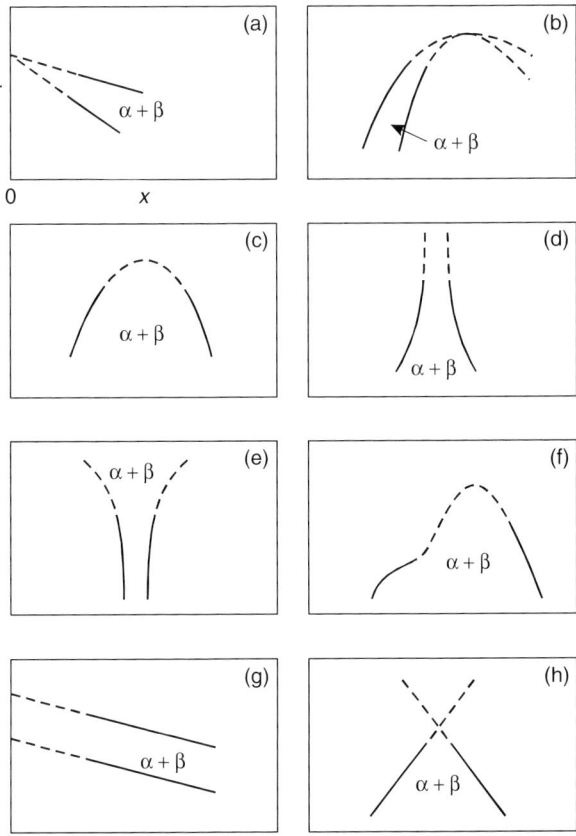

Figure 3.11 Trends of two boundaries of a two-phase field: (a)–(d): normal; (e) and (f): unlikely; and (g) and (h): impossible.

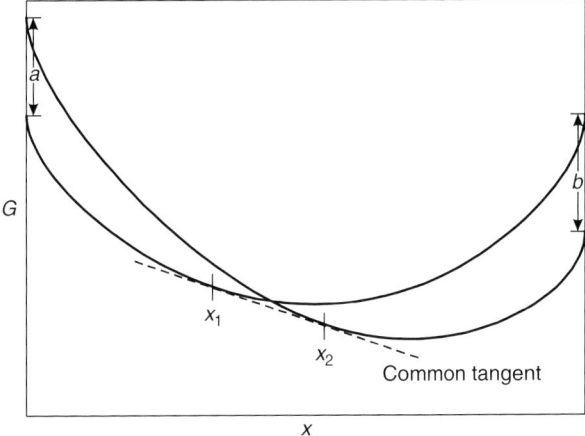

Figure 3.12 The distance between x_1 and x_2 decreases as the temperature increases because the two Gibbs energy curves become similar.

shown in Fig. 3.11(c). Therefore, Fig. 3.11(f) with an abrupt change of slope is unlikely to occur. Figures 3.11(g) and (h) are obviously wrong (errors A and T in Fig. 3.2). It follows that if a two-phase region is drawn as if it were widening at higher temperatures, the two boundaries appear to cross one another when extended, or if the two boundaries appear to cross the pure element line at different temperatures, further experimental work may be needed to check the projected trend.

As is well known, the bcc phase always "wins" at high temperatures due to the large vibrational entropy contribution in the $-TS$ term (Zener effect [15,16]). As a result, the phase relationships between a bcc phase and phases with other structures tend to be of the Fig. 3.11(a) type at the high-temperature end of the two-phase field (when the liquid phase is suppressed).

More intuitively, this trend can be understood by the fact that the solubility of an element in a solid phase increases as the temperature increases. As a result, the remaining two-phase field between two solid phases becomes narrower at higher temperatures.

3.9 An Abrupt Change of Slope (Point Z in Fig. 3.2)

A phase boundary with an abrupt change, as shown at point Z in Fig. 3.2, is not found very often in the equilibrium portions of phase diagrams. However, numerous examples can be found where such an abrupt change is projected to occur (by extrapolation) in the metastable regions of existing phase diagrams. Sometimes such phase boundaries are drawn to avoid other phase rule violations [7], but they then introduce a projected liquidus abruptness that may be unacceptable.

A liquidus is considered to estimate a permissible abrupt change of slope. An abrupt change of slope in the liquidus is caused either by an abrupt change in the thermodynamic properties of the liquid phase or an abrupt change in the thermodynamic properties of the solid phase (in this case an equiatomic compound is assumed) in equilibrium with the liquid. In Section 3.7, it was shown that the slope of a liquidus does not change substantially even when the thermodynamic behavior of the liquid phase changes substantially. We now examine the influence of the change in the thermodynamic behavior of the compound on the slope of the liquidus.

Table 3.3 lists various Gibbs energy functions of an equiatomic compound AB. The one melting at 1000°C is used as the reference state, which was selected from the situation $b2$ in Fig. 3.6. The corresponding liquidus is shown with a solid line in Fig. 3.13. The liquid in this case is an ideal solution. Other curves in Fig. 3.13 were drawn assuming that the thermodynamic behavior of the equiatomic compound changes discontinuously at 800°C as the temperature is lowered. It is then possible to project metastable extensions of the low-temperature segments of the different liquidus forms. In Fig. 3.8, these are drawn to correspond to apparent melting points arbitrarily set at 1600°C, 1300°C, 1100°C, 900°C, and 850°C. As may be seen, the required temperature coefficient of the Gibbs energy (lattice stability) associated with these curves (Table 3.3) varies from -3.322 to $13.765 \, \text{J/mol·K}$. In spite of this drastic change in the temperature coefficient (c.f. Section 3.5), the change of slope of the different liquidii at 800°C is relatively small. As a rough measure, the abruptness of the slope change may be judged by the apparent congruent melting points corresponding to the liquidus portions above and below the temperature at which

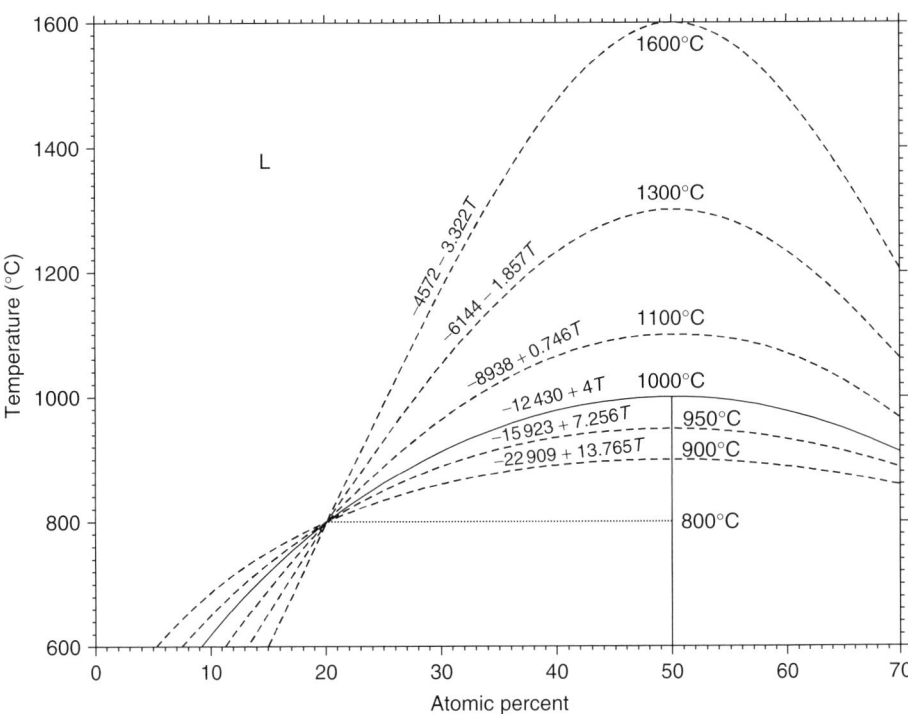

Figure 3.13 Change of slope introduced by different lattice stability parameters.

Table 3.3 Relationship between the Gibbs energy and the melting point of an equiatomic compound (liquid is an ideal solution).

Melting point (°C)	Gibbs energy (J/mol)	Comment
1600	$-4572 - 3.322T$	–
1300	$-6144 - 1.857T$	–
1100	$-8938 + 0.746T$	–
1000	$-12\,430 + 4.000T$	Reference state
950	$-15\,923 + 7.256T$	–
900	$-22\,909 + 13.765T$	–

the slope change occurs. In the present example, a change of only ±10% in the apparent melting point (in K) already seems to require too much change in the curvature. This problem is illustrated with a schematic drawing in Fig. 3.14 for the case of convex curvature change. In order for an abrupt change of slope to happen, the temperature coefficient must change from one level to another within the range where the curvature of the liquidus is changing rapidly. Since no phase transformation

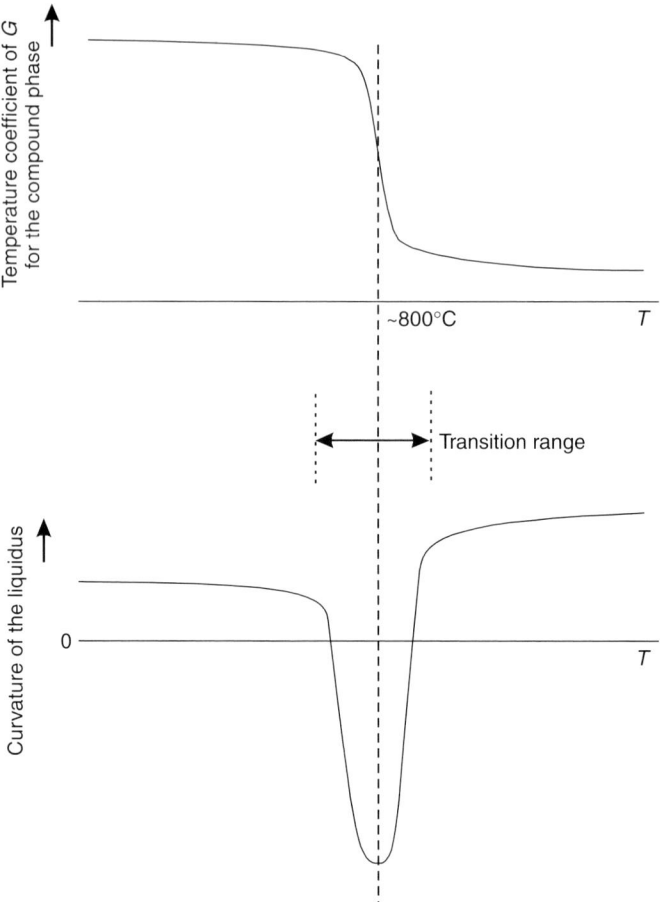

Figure 3.14 Change of the temperature coefficient of G for the compound phase and curvature of the liquidus associated with an abrupt change of slope. Width of transition range is zero for a polymorphic transition.

is assumed to occur in the compound, the change in the temperature coefficient near 800°C as shown in the figure is hardly justifiable. Even if there existed a true phase transformation at 800°C (see Section 3.10), it would be extremely difficult to explain an abrupt change of slope (concave or convex) of a liquidus boundary in terms of an abrupt change in the thermodynamic behavior of the compound.

3.10 An Excessive Slope Change Associated with a Polymorphic Transformation (Point g in Fig. 3.3)

As already shown in Section 3.9, a substantial change in the Gibbs energy function of a compound does not seem to cause much change in the slope of the liquid (or solid) phase boundary in equilibrium with the compound. Similarly, a phase diagram with a marked change in the slope associated with a polymorphic (or allotropic in

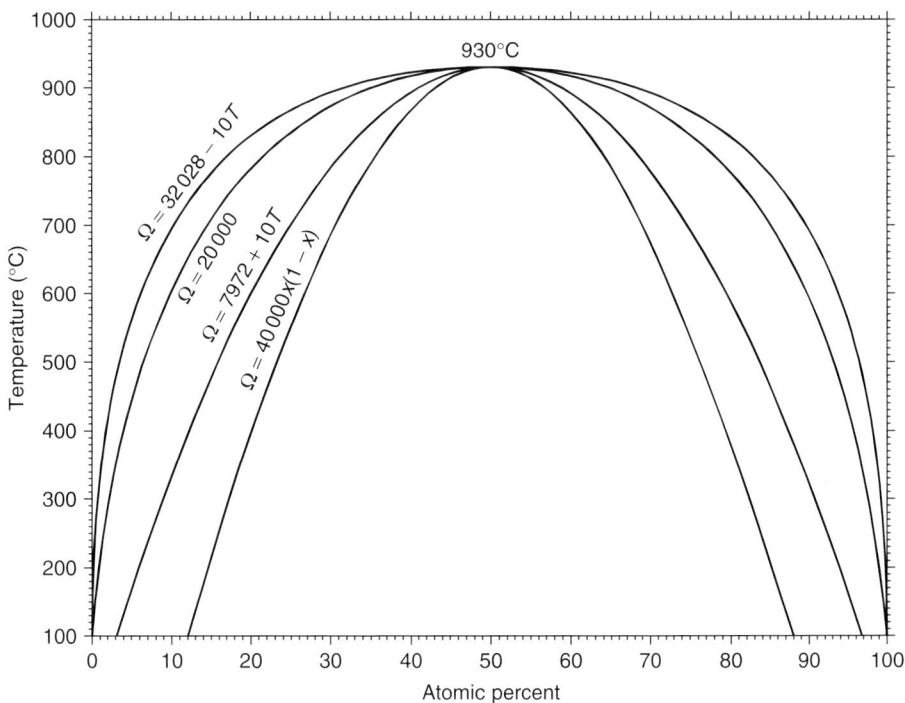

Figure 3.15 Shape of a miscibility gap of a regular solution.

the case of a terminal phase) transformation calls for careful handling because the change in the temperature coefficient of the Gibbs energy function accompanying a polymorphic transformation would be only of the order of $-1\,\text{J/mol·K}$.

3.11 Shape of a Miscibility Gap

Miscibility gaps are found in both the liquid and the solid states. Wilson [12] proved that the shape of the gap near its critical point is uniformly round. An attempt is made here to find possible limits of the curvature of a miscibility gap near its critical point. For a reference miscibility gap, the excess Gibbs energy of a phase (it does not matter whether it is a liquid or a solid) was taken to be of the form $G^{\text{ex}} = 20\,000x(1-x)\,\text{J/mol}$, which means that the interaction parameter $\Omega = 20\,000\,\text{J/mol}$. This yields a critical point at 50 at.% B (of the A–B system) and 930°C. The shape of the calculated gap is quite round, as shown in Fig. 3.15. In order to make the form of this gap sharper, a temperature dependence and composition dependence were introduced to Ω, keeping the critical point at the same location. When the temperature dependence cT is added to Ω, the miscibility gap becomes sharper as c increases (because the critical temperature is fixed, the constant term must be reduced accordingly). Because the usual limit of c is approximately $10\,\text{J/mol·K}$, a miscibility gap for $\Omega = 7972 + 10T\,\text{J/mol}$ is shown in Fig. 3.15 as an example.

When a composition dependence is also added to Ω in the form of $\alpha x(1-x)$, the miscibility gap becomes sharper as the value of α increases. The constant term in

Ω must be reduced as α is increased to maintain the critical temperature unaltered. The maximum possible value of α is then 40 000 J/mol to keep Ω positive for all x. When α exceeds 40 000, a negative constant term must be introduced in Ω, and Ω then becomes negative near $x = 0$ and 1. Under these modeling assumptions, the interaction between the two starting elements becomes too artificial. From this brief analysis, the miscibility gap cannot be structured to have a pointed peak in either case.

It is equally difficult to make a very flat miscibility gap. Even when a temperature dependence amounting to as much as $-10T$ is introduced in Ω, the form of the miscibility gap does not change markedly (Fig. 3.15). On the other hand, when the composition term is modified excessively, two peaks appear in the miscibility gap [17]. The possible existence of a two-peak miscibility gap was discovered first by Borelius [18], as quoted by Hillert [19]. Although Okamoto [17] suggested possible candidates, this type of miscibility gap has not been found yet in actual binary systems.

3.12 Displaced Miscibility Gap (Point q in Fig. 3.3)

The composition of the critical point of a liquid miscibility gap is usually found to occur in the center range of a phase diagram [7]. A binary phase diagram having a liquid miscibility gap that is significantly displaced toward either end of the diagram may be considered to be rather suspect. The question then arises as to how much displacement from a near-central position is tolerable.

To establish the magnitude of the displacement of the critical point, the excess Gibbs energy for the liquid phase of a hypothetical system A–B may be expressed as a function of the B-component concentration x in the usual form given in Eq. (2):

$$\Delta_{\text{mix}} G^{\text{ex}} = \Omega(x) x (1 - x) \tag{2}$$

where $\Omega(x)$ is the interaction parameter, and the excess entropy term is assumed to be negligible. When $\Omega(x)$ is constant (and positive), the composition of the critical point of a miscibility gap (x_c) is, of course, at 50 at.%. Table 3.4 shows the calculated displacement of x for various asymmetric forms of $\Omega(x)$. The magnitude of $\Omega(1)$ has been set arbitrarily at 40 kJ/mol to permit also the calculation of the critical temperature (T_c) corresponding to x_c.

The selected composition-dependence expressions for $\Omega(x)$ are shown graphically in Fig. 3.16. In the first-approximation "quasichemical" model, the solubility of A in B, or vice versa, in the liquid state is considered to be determined mostly by the relative magnitudes of the nearest-neighbor bonds between A and A, B and B, and A and B (Bragg–Williams approximation). Here, the deviation from the relationship $\Omega(x) = \text{constant}$ (Case 1 in Table 3.4) is expected to be small because the relative strengths of the bonds are expected not to vary significantly with composition. Case 2 corresponds to a limit of a normal situation where A is mostly indifferent to the introduction of B, but B strongly repels A in its own environment. Considering the omission of the entropy term and the rather crude nearest-neighbor approximation in the present estimate, it may be said that the location of a miscibility gap in the liquid phase could be viewed as normal if x_c occurred between approximately 25 and 75 at.% (Cases 1 and 2).

As the form of $\Omega(x)$ is made to become more distorted (Cases 3–5), x_c moves away from the center of the phase diagram toward the B side. However, even when

Table 3.4 Effect of asymmetry of the interaction parameter on the critical composition of a miscibility gap.

Case	$\Omega(x)$ (kJ/mol)	x_c (at.%)	T_c (°C)
1	40	50	2132
2	$40x$	74	1986
3	$40x^2$	82	1963
4	$40x^3$	86	1955
5	$40x^4$	89	1952
6	$40(-1 + 2x)$	79	2505

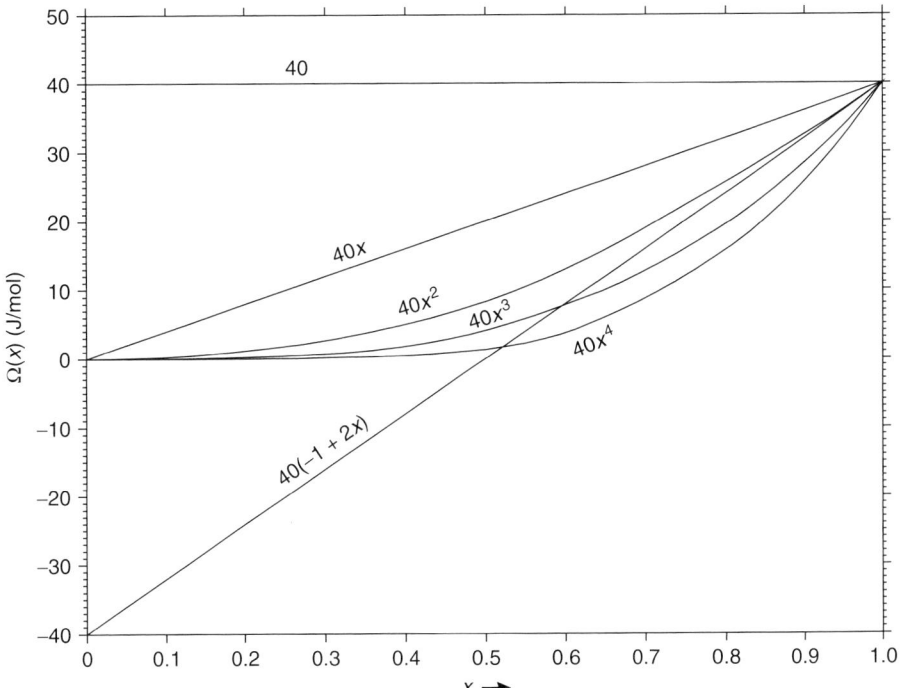

Figure 3.16 Interaction parameters to displace the peak position of a miscibility gap.

$\Omega(x)$ is assumed to have a fourth power dependence on x (Case 5), $x_c = 89$ at.%. Case 6 corresponds to an unusual situation whereby A favorably dissolves B (negative heat of formation), but B rejects A (positive heat of formation). This is of course an unlikely situation because the interaction between A and B should not be in general so drastically affected by the environment even when the next-nearest-neighbor, or higher-order effects, are taken into account. (*Note*: We are considering a liquid phase situation. If water does not dissolve oil, oil does not dissolve water either.) Based mainly on these propositions, a liquid miscibility gap having its critical composition outside the range of 20–80 at.% is very unlikely. If a liquid miscibility gap did actually occur near a pure element, this would signal a most unusual (and unique) situation, and it would be most challenging to determine some existing special reason for it (see Ref. [9]).

Note: (1) If the liquid phase develops associations (existence of molecules in the liquid state), the molecules behave like atoms. The "center range" for the purpose of considering the phase diagram situations should then be taken as falling between the molecular composition and the pure element. The Hg–S system provides an example of this behavior. (2) A miscibility gap in the solid phase can of course occur near a pure element side (e.g., ε and (Cd) in the Ag–Cd phase diagram) because there may be strong crystal lattice effects.

3.13 An Inverse Miscibility Gap (Point w in Fig. 3.3)

A thermodynamic model that might result in the development of an inverse miscibility gap with a minimum-temperature critical point is conceivable. However, the existence of such a miscibility gap in a real system would certainly be unusual because any solution in nature is expected to behave more ideally as the temperature is increased. Hence, a thermodynamic model demanding an inverse miscibility gap with a wider gap at a higher temperature may be considered to be unrealistic by its own nature. An inverse miscibility gap will occur when the contribution from the excess entropy of mixing term overcomes the contribution from the ideal entropy of mixing (maximum $-5.763\,\text{J/mol·K}$); that is, an inverse miscibility gap develops if the magnitude of the excess entropy is greater than $\sim 23x(1-x)\,\text{J/mol}$. A situation corresponding to this large value of 23 ($=5.763 \times 4$) may already be too unnatural. A factor of 10 was assumed to be the limit in Section 3.12.

3.14 Almost Symmetric Syntectic Reaction

In *Binary Alloy Phase Diagrams* [1], syntectic reactions ($L_1 + L_2 \rightarrow s$) are found in only a relatively few binary systems, as listed in Table 3.5. Figure 3.17 shows a typical schematic Gibbs energy functions situation for the liquid and solid phases at the syntectic temperature; immiscibility in the liquid phase and compound formation in a solid phase occur at the same time. This is an unlikely situation unless the nature of the interactions between the elements constituting the binary system is very different in the liquid when compared with the solid state (elements A and B show affinity in the solid state, but they repel each other in the liquid state). Thus, for a syntectic reaction to occur, involvement of two different types of interactions appears to be necessary. Not surprisingly, most systems in Table 3.5 involve a strongly ionic element (alkali, halogen, chalcogen). Moreover, the critical composition of the miscibility gap and the composition at which compound formation occurs are usually substantially different, as, for example, in the Hg–Cl system (Fig. 3.18). Apparently, the tendency of separation into two liquids at one composition is overcome by an even stronger (and opposite) interaction for compound formation at another composition.

Correct and Incorrect Phase Diagram Features

Table 3.5 Systems including a syntectic reaction.

Bi–Br	Bi–Cl	Bi–U	Cd–P	Cd–Rb	Cl–Hg
Cl–In	Cs–In	Ga–I	Ga–K	Ga–Rb	I–Tl
In–Rb	Mn–Y	Ni–S	Sn–P	Pb–U	Pd–Sc

From *Binary Alloy Phase Diagrams* [1].

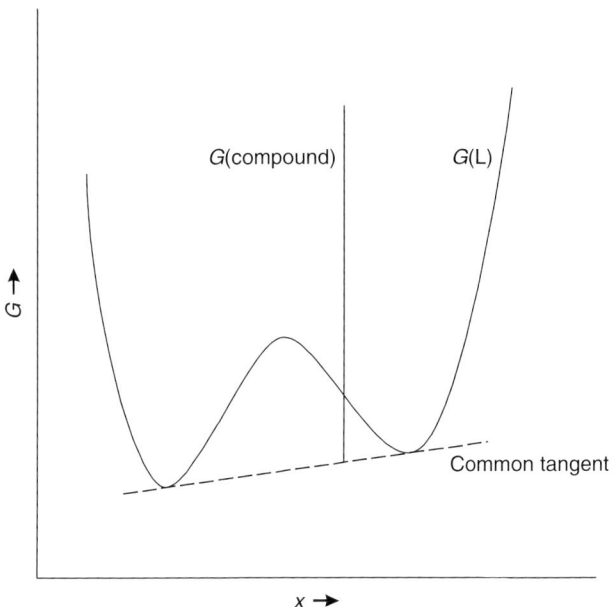

Figure 3.17 Gibbs energy relationship at a syntectic reaction.

3.15 Metastable Melting Point of a Pure Element

When an element exists in two allotropic forms, consistency regarding the metastable melting point of the low-temperature form must be considered. If Δc_p between the two allotropic forms is assumed to be zero, the metastable melting point of the low-temperature phase is given by Eq. (3).

$$T = \frac{\Delta H_m + \Delta H_t}{\dfrac{\Delta H_m}{T_m} + \dfrac{\Delta H_t}{T_t}} \qquad (3)$$

where T_m is the melting point of the high-temperature phase, T_t is the allotropic transformation temperature, ΔH_m is the enthalpy of fusion of the high-temperature phase, and ΔH_t is the enthalpy of allotropic transformation. Table 3.6 shows the melting point of the low-temperature phases of elements with more than two allotropic forms calculated by substituting in the equation above the melting points

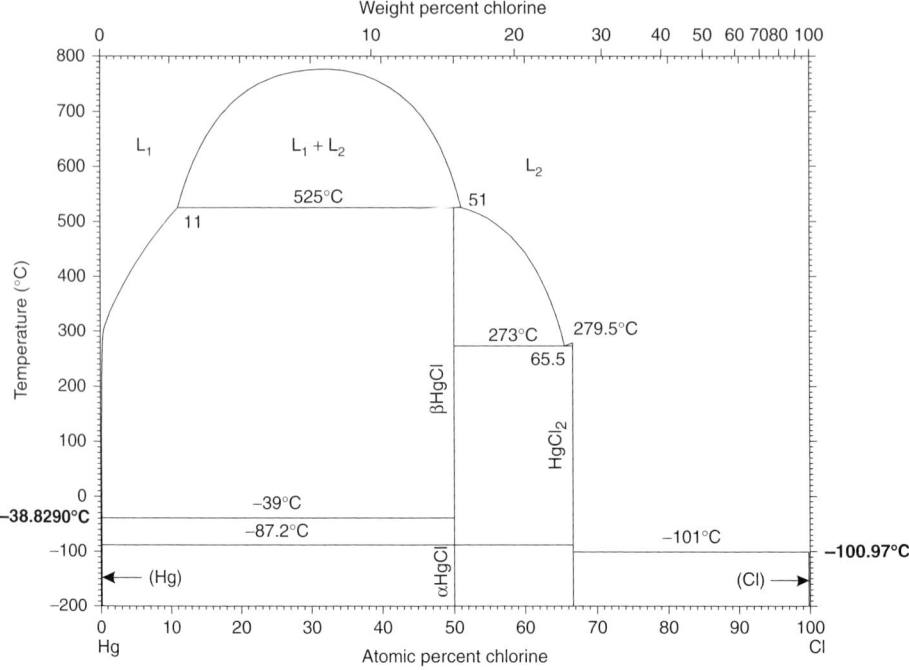

Figure 3.18 Hg–Cl phase diagram with a syntectic reaction.

Table 3.6 Metastable melting point of a low-temperature allotrope.

Element	Normal melting point (°C)	Melting point of low-temperature allotrope (°C)	Difference (°C)
Am	1176	1146	30
Be	1289	1286	3
Ca	842	788	54
Ce	798	771	27
Cm	1345	1332	13
Co	1495	1424	71
Dy	1412	1403	9
Fe	1538	1529	9
Gd	1313	1290	23
Hf	2231	2133	98

(*Continued*)

Table 3.6 (*Continued*)

Element	Normal melting point (°C)	Melting point of low-temperature allotrope (°C)	Difference (°C)
La	918	900	18
Mn	1246	1230	16
Nd	1021	970	51
Np	639	606	33
Pa	1572	1408	164
Pr	931	885	46
Pu	640	571	69
Sc	1541	1491	50
Sm	1074	1030	44
Sr	769	741	28
Tb	1356	1334	22
Th	1755	1658	97
Ti	1670	1408	262
Tl	304	297	7
U	1135	987	148
Y	1522	1508	14
Yb	819	814	5
Zr	1855	1593	262

and the transformation temperatures [1], and the values of the enthalpies of fusion and transformation [20]. When Δc_p is not zero, the estimated metastable melting point must be modified, but the change is small because Δc_p is usually small.

This problem is noticed when the liquidus of the low-temperature allotropic form is extrapolated to the pure element line.

4 Examples of Phase Diagrams with Improbable Features

Errors A to T in Fig. 3.2 are often encountered in explicit forms in numerous published binary phase diagrams. Some errors, such as E, J, K, and Q, are quite common, and are sometimes introduced inadvertently in the drafting stage. In addition, there are less obvious phase diagram problems, as described in Section 3.

These problems occur when experimental data are inaccurate, experimental data are supplemented by speculation, the thermodynamic model is inappropriate, there are errors in the calculation, etc. When the measured experimental data are somewhat suspicious because of the existence of some of the problems described above, it may be helpful to construct a thermodynamic model that approximately reproduces the experimental data. Although the thermodynamic model may be inaccurate, the calculated phase diagram is thermodynamically consistent [8]. The problematic part of the experimental data may then be detected by comparing the experimental and calculated phase diagrams.

Even when experimental data can be reproduced reasonably well by a thermodynamic model, the model must be examined carefully. For example, when Gibbs energy ($\Delta G = \Delta H - T\Delta S$) of a line compound is obtained by optimization, the ΔG value may happen to be appropriate because both ΔH and ΔS are too large. The magnitudes of ΔH and ΔS must be checked [21]. An empirical rule between the entropy of mixing and the enthalpy of mixing [22] and an empirical rule on the magnitude of the enthalpy of mixing [23] for liquid phases may be helpful in identifying suspicious thermodynamic models.

In this section, some specific examples are introduced. In each situation, a possible way of alleviating the difficulty is shown, but there may be alternative interpretations (i.e., corrections) of the indicated error and how to remove it. It might be interesting to think first about the problem in each diagram and then read the explanation.

4.1 Ag–Pr

A "liquidus-projected" melting point of a low-temperature allotropic form in an element often disagrees with the metastable melting point estimated from the enthalpy of transformation data (Section 3.15). The Ag–Pr phase diagram (Fig. 3.19) [24] presents an example of this problem. According to this diagram, the melting point of αPr may be estimated to be ~700°C by extrapolation of the liquidus to 100 at.% Pr. However, the reported enthalpy of transformation of Pr suggests that the melting temperature of αPr must be ~885°C (Table 3.6). This requirement can be satisfied only by introducing unusual inflections in the (αPr) liquidus in the metastable region. Most likely, the shape of the (αPr) liquidus in the stable range is incorrect.

In addition, the AgPr liquidus may be too asymmetric according to the criterion given in Section 3.7.

4.2 Al–B

Figure 3.20 shows the Al–B phase diagram proposed by Carlson [25]. The liquidus of βAlB_{12} consists of two segments (2150–1850°C and 1660–1550°C) on the Al-rich side. These upper and lower segments must be continuous in the metastable range because there is no reported phase transformation in AlB_{12} between 1660°C and 1850°C. However, a connection through extrapolation of the two liquidus segments causes a Z-type problem (requiring an abrupt change in the Gibbs energy of AlB_{12}).

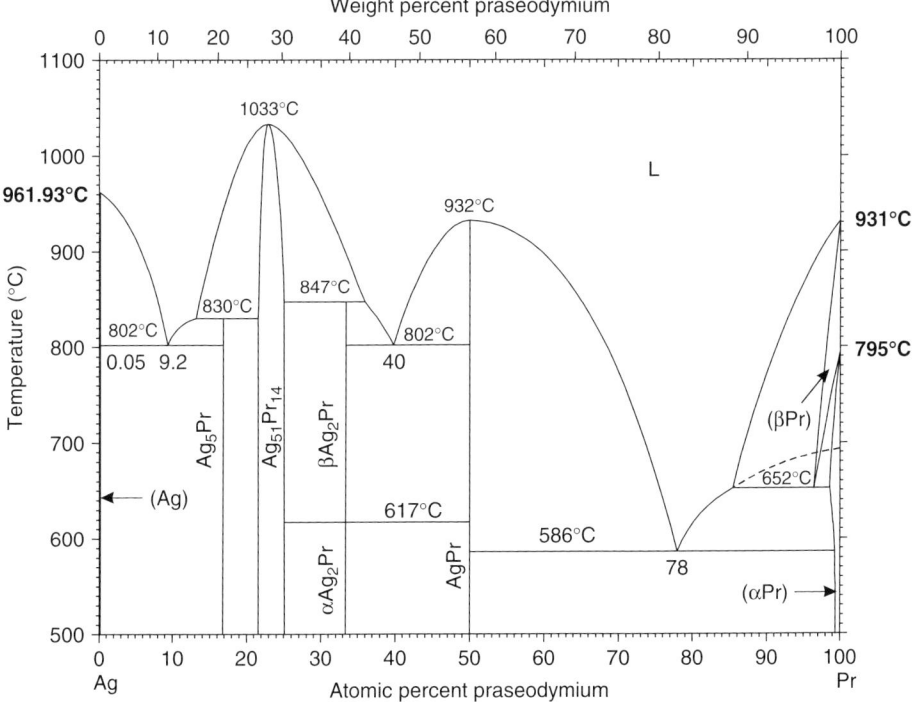

Figure 3.19 Ag–Pr phase diagram [24] (see Section 4.1).

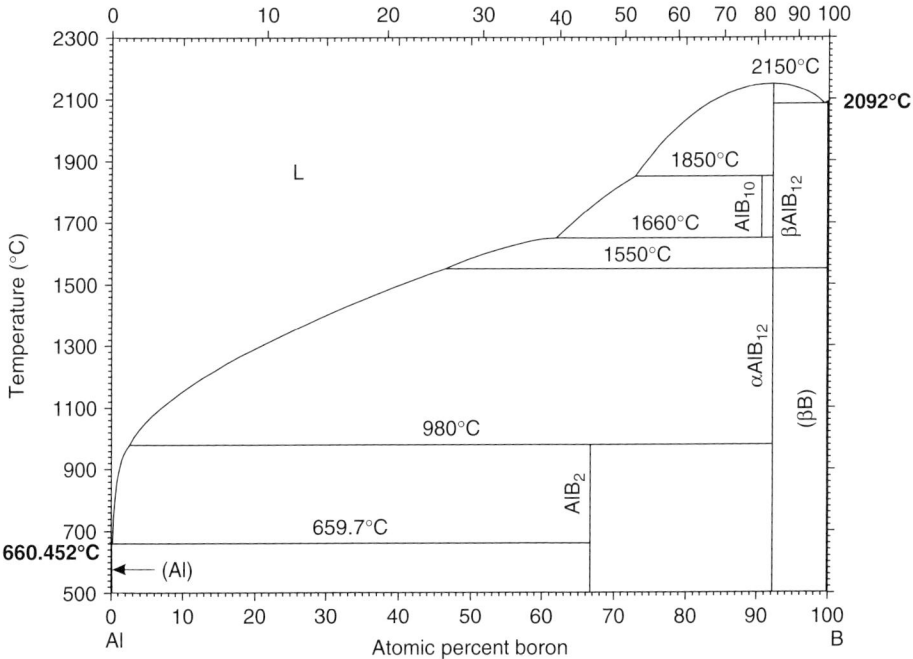

Figure 3.20 Al–B phase diagram [25] (see Section 4.2).

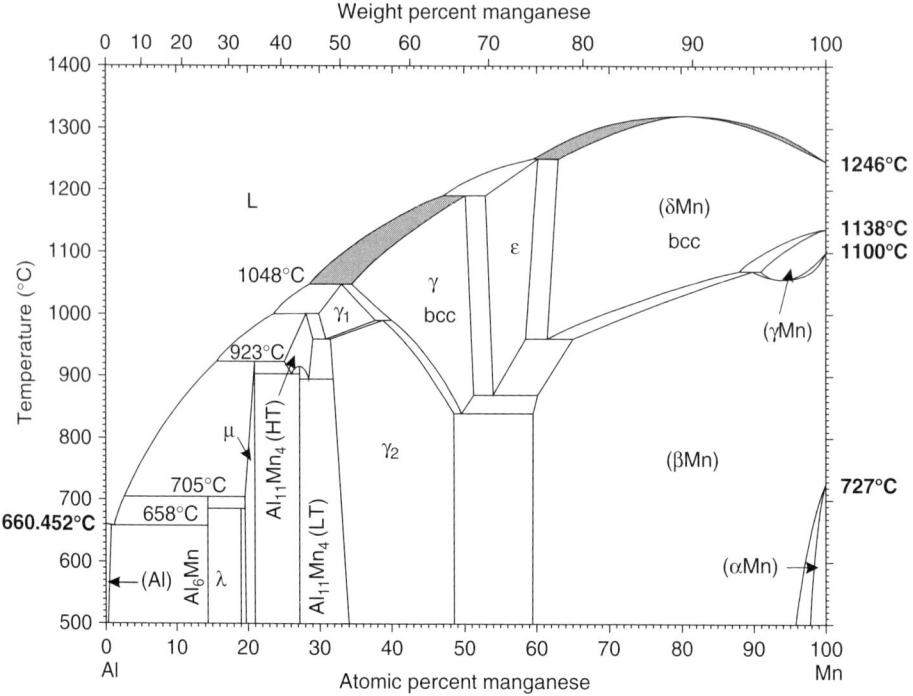

Figure 3.21 Al–Mn phase diagram [26]. Shaded areas must be smoothly continuous (see Section 4.3).

In addition, the existence of AlB$_{10}$ in a limited temperature range in the form shown in this figure is unlikely, as discussed in Section 3.6.

4.3 Al–Mn

Sometimes, one phase appears in more than one place in a binary phase diagram with an appearance of two or more phases. A well-known example is the case of (δFe) and (αFe) phases. When these phases are continuous, a γ loop is observed in the phase diagram. Even when one phase is separated into pieces, the boundaries of these segments must be smoothly continuous in the metastable range. For example, the (δMn) and γ phases in the Al–Mn phase diagram (Fig. 3.21) [26] belong to the same W-type bcc structure. This continuity is supported by isothermal sections of the ternary phase diagrams Al–Co–Mn and Al–Cr–Mn [27]. Therefore, the liquidus and solidus (shaded areas in Fig. 3.21) of these two phases must be smoothly continuous. However, continuous lines cannot be drawn without introducing some inflections.

4.4 Al–Pu

Figure 3.22 shows a partial Pu–Al phase diagram [28]. The two boundaries of the (εPu) + (δPu) two-phase region must converge at 0 at.% Al when extrapolated (problem s in Fig. 3.3). This would be difficult to do without introducing an unusual change of slope in the (εPu) solvus.

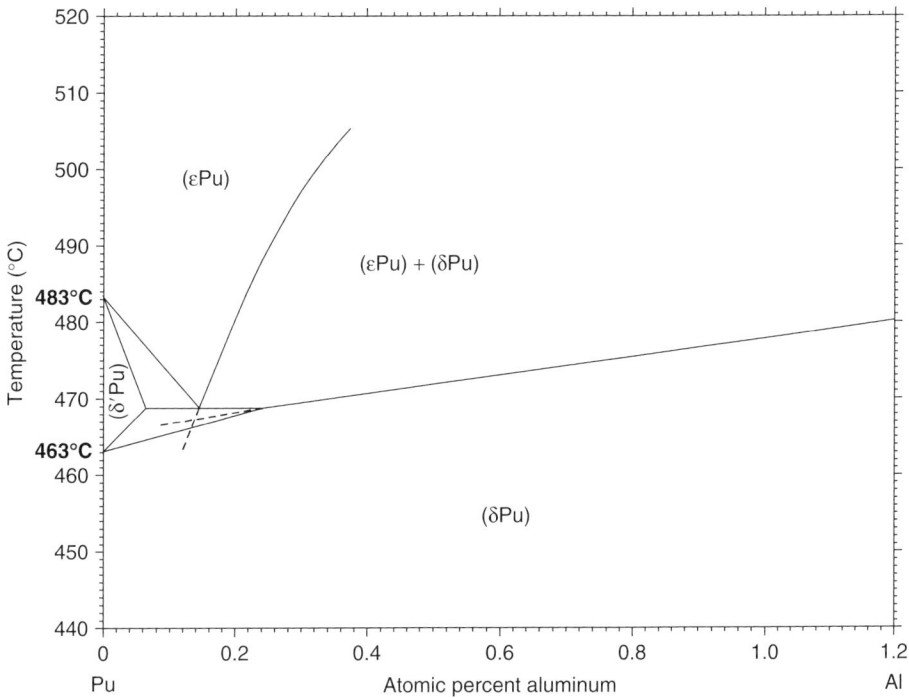

Figure 3.22 Partial Pu–Al phase diagram [28] (see Section 4.4).

4.5 Al–Se

Figure 3.23 shows the Al–Se phase diagram proposed by Howe [29]. If extrapolated smoothly to the Al side beyond the L → (Al) + Al$_2$Se$_3$ eutectic, the L/(L+Al$_2$Se$_3$) liquidus is projected to cross the 0 at.% Se line. However, this liquidus (which is with respect to the Al$_2$Se$_3$ phase) must not be extrapolated to cross the 0 at.% Se line (problem *b* in Fig. 3.3). A slightly changed position of the eutectic could probably improve the situation. The same problem occurs on the Se side of this phase diagram. The above error occurs in phase diagrams in which a eutectic or a peritectic is very close to the 0 or 100 at.% line for each element. If the proposed liquidus is drawn improperly, no room exists between the eutectic or peritectic point and the respective pure element line to allow a change in the course of the liquidus without causing the Z-type problem.

Binary Alloy Phase Diagrams [1] avoided this problem in the Howe diagram [29] by modifying the liquidus slope to become nearly vertical at around 0 and 100 at.% Se. Although this problem was solved at least in appearance, the diagram is still questionable because the Al$_2$Se$_3$ liquidus is too asymmetric in view of the criterion given in Section 3.7.

4.6 Au–La

As discussed in Section 3.7, the problem of an asymmetric liquidus in the La–Au phase diagram (Fig. 3.9) was solved in the calculated diagram. In this figure, the liquidus of

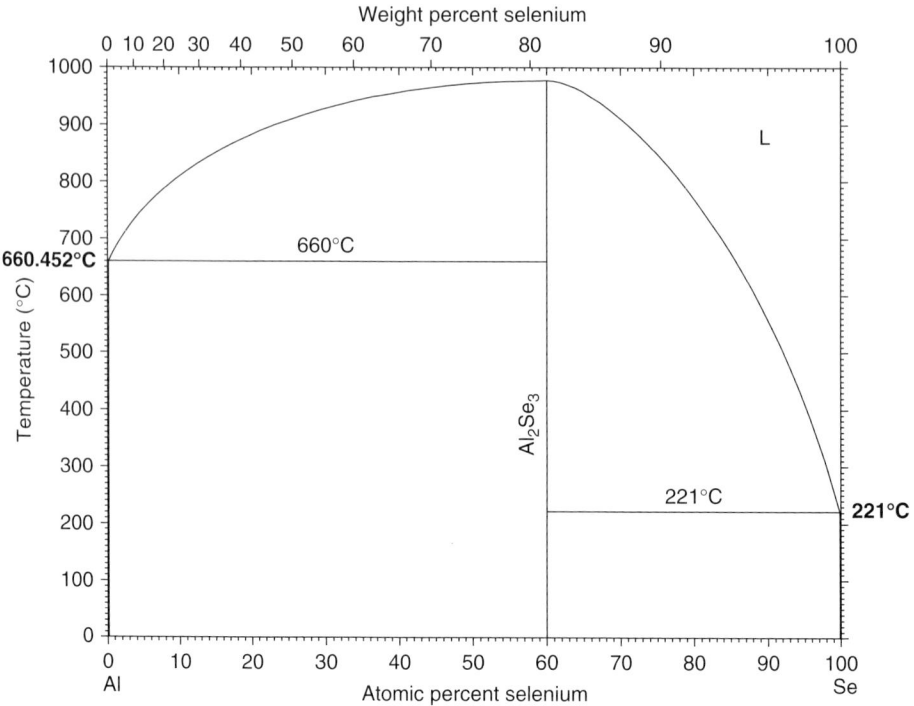

Figure 3.23 Al–Se phase diagram [29] (see Section 4.5).

La$_{14}$Au$_{51}$ was determined based on the calculated eutectic point at 1054°C on the La side. When calculated liquidus peaks of the three compounds βLaAu, LaAu$_2$, and La$_{14}$Au$_{51}$ are compared, the peak of La$_{14}$Au$_{51}$ is too sharp according to the criterion in Section 3.5. In addition, the calculated liquidus of La$_{14}$Au$_{51}$ does not agree with the experimentally determined liquidus. These problems can be solved if the L → LaAu$_2$ + La$_{14}$Au$_{51}$ eutectic temperature is raised approximately 50°C.

4.7 Au–U

Figure 3.24(a) shows the Au–U phase diagram [30]. The proposed liquidus of Au$_3$U is strongly asymmetric (problem k in Fig. 3.3) and a rather abrupt change of slope is required at around 1216°C to avoid the E-type error in Fig. 3.2. Accordingly, the validity of this liquidus was questioned and other possible diagrams were proposed by Okamoto and Massalski [30]. Figure 3.24(b) shows the Au–U diagram revised by Palenzona and Cirafici [31], in which identifications of intermetallic compounds have been corrected from Au$_3$U to Au$_{51}$U$_{14}$. The problem of the E-type error was resolved. However, the problem of asymmetry of the liquidus between Au$_{51}$U$_{14}$ and Au$_2$U remains. If the L → Au$_{51}$U$_{14}$ + Au$_2$U eutectic composition actually is located a little closer to Au$_{51}$U$_{14}$, the asymmetry problem is solved.

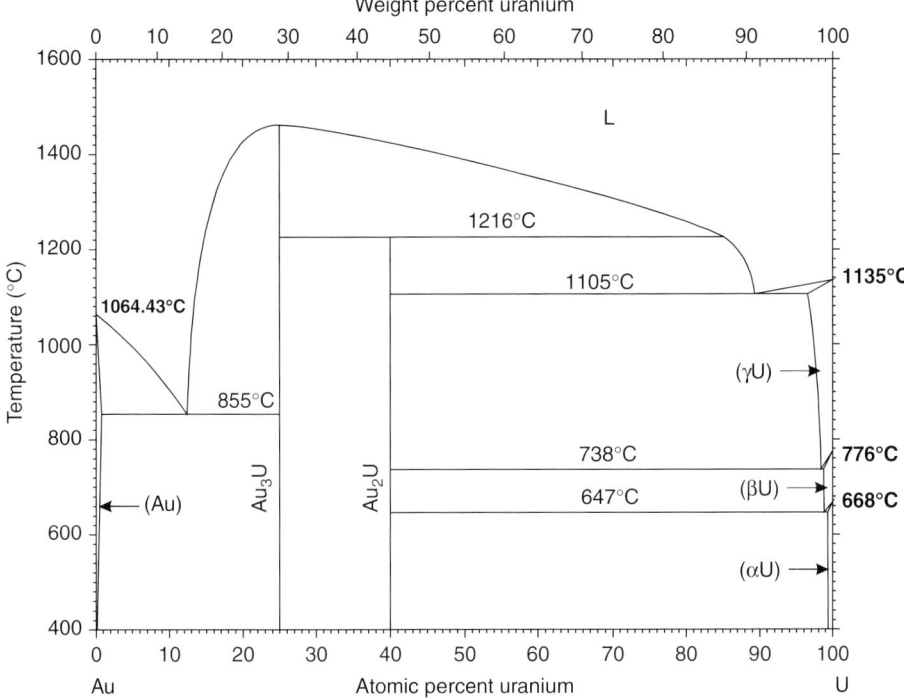

Figure 3.24(a) Assessed Au–U phase diagram [30] (see Section 4.7).

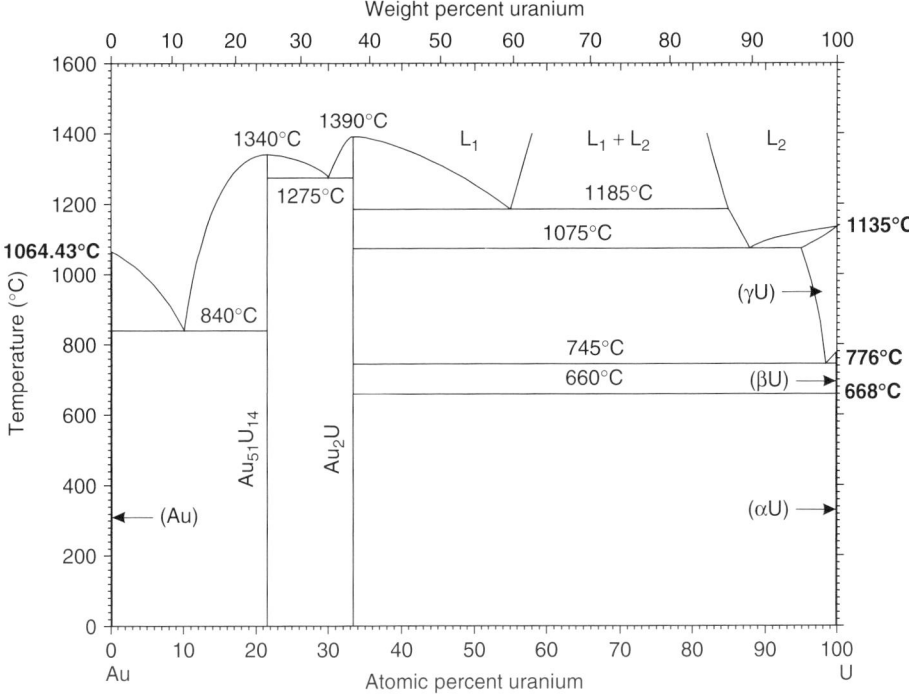

Figure 3.24(b) Revised Au–U phase diagram [31] (see Section 4.7).

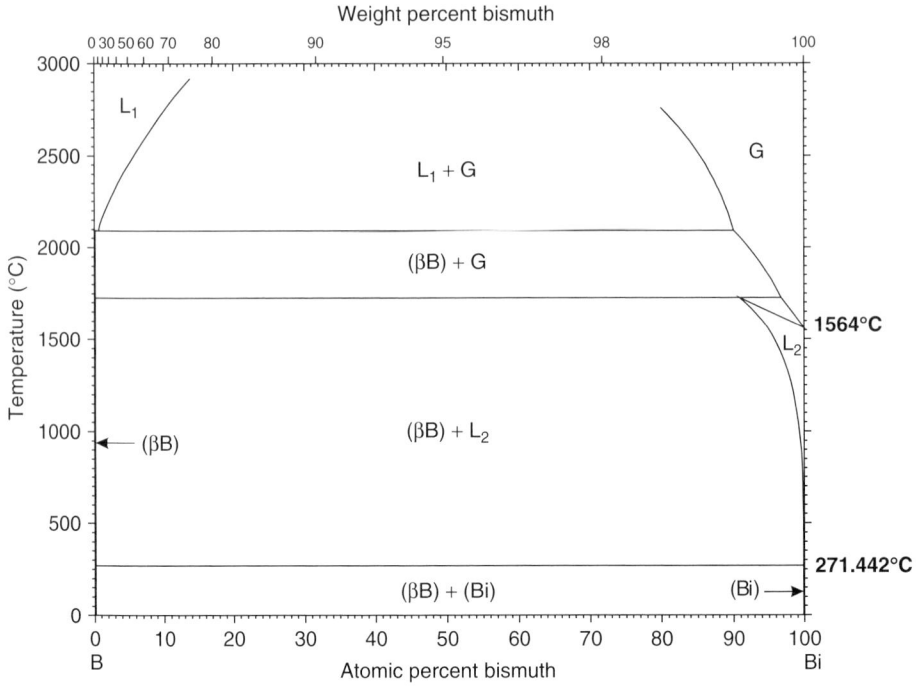

Figure 3.25 B–Bi phase diagram [32] (see Section 4.8).

4.8 B–Bi

Figure 3.25 shows the B–Bi phase diagram compiled by Moffatt [32]. When a liquid phase is in equilibrium with a gas phase, the liquid generally dissolves a diminishing amount of the gas phase with increasing temperatures until the dissolved amount correctly becomes 0 at.% at the boiling point. The convex L/(L + G) boundary must become concave at higher temperatures because it must be connected to the boiling point of B. This requires a strange change of curvature in the phase boundary. Probably, the $L_1/(L_1 + G)$ boundary should be almost vertical.

4.9 B–W

Binary Alloy Phase Diagrams [33] showed the B–W phase diagram as shown with solid lines in Fig. 3.26 (redrawn from Moffatt [32], which is based on the diagram of Rudy [34]). The initial slope of the (W) liquidus is apparently too steep, that is, problem r in Fig. 3.3. (It was not so steep in the diagrams of Rudy [34] and Moffatt [32], but there was an unusual change of slope at ~99 at.% W.) The initial slope was revised in the diagram of Nagender Naidu and Rama Rao [35], as shown with a dashed line. However, the unique overall shape of the (W) liquidus with a change from concave to convex in a narrow composition range is clearly subject to further investigation.

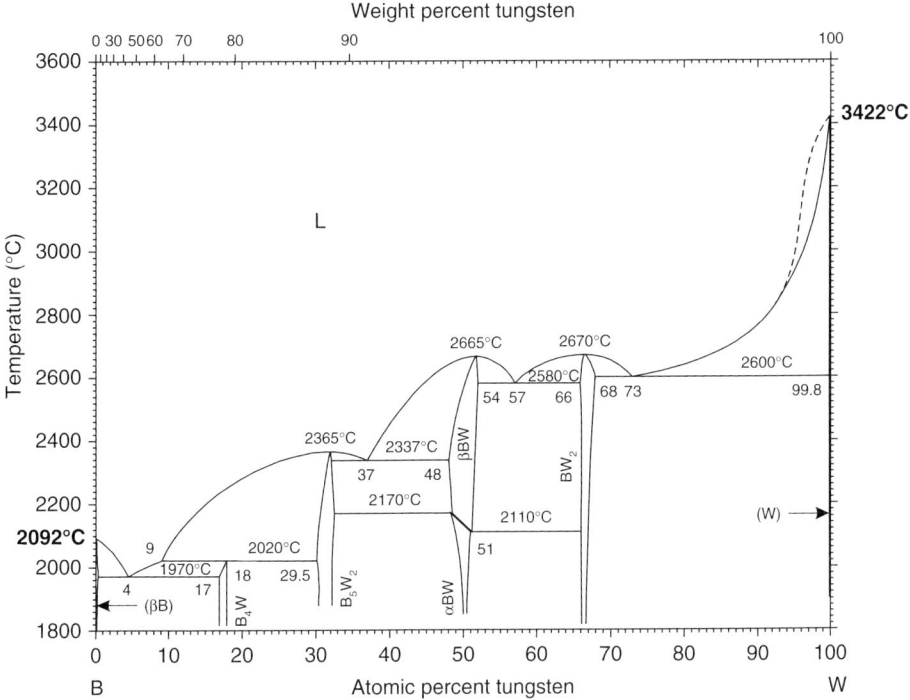

Figure 3.26 B–W phase diagram [33] (see Section 4.9).

4.10 C–Ge

Figure 3.27 shows the C–Ge phase diagram [36]. It is likely that the G + L two-phase field on the Ge end was added to indicate the existence of the G phase. However, the opening angle is too small (problem a in Fig. 3.3) in view of a very large enthalpy of vaporization (~330 kJ/mol) of Ge.

4.11 Ca–Eu

The Ca–Eu phase diagram (Fig. 3.28) [37] is an example of the problem u in Fig. 3.3. The dotted line shows the initial slope of the (βCa,Eu) + (αCa)/(αCa) boundary calculated from the (βCa,Eu)/(βCa,Eu) + (αCa) boundary. Apparently, the initial opening angle of the (βCa,Eu) + (αCa) two-phase field at 0 at.% Eu as proposed is unrealistically large.

4.12 Ce–Pr

Figure 3.29 shows the Ce–Pr phase diagram [38]. If the trend of the [(γCe) + (βCe, αPr)]/(βCe,αPr) solvus is correct, its extrapolation to the Pr-rich side appears to cross the 100 at.% Pr line at ~800°C. Because no two-phase fields should exist above this temperature at 100 at.% Pr in the metastable state, an extrapolation of the solvus should also cross the 100 at.% Pr line at the same temperature. However, this requires

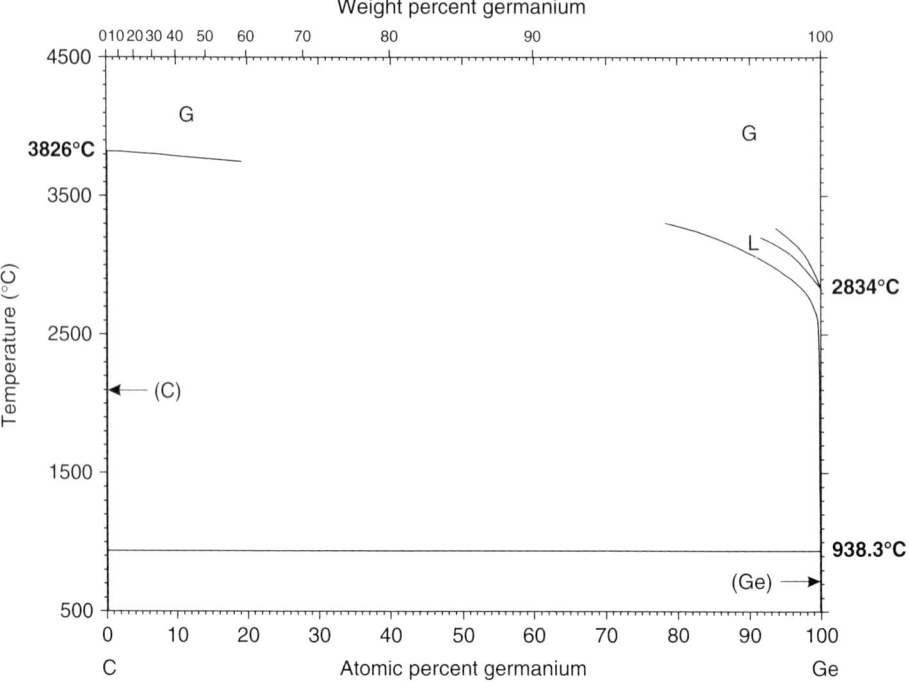

Figure 3.27 C–Ge phase diagram [36] (see Section 4.10).

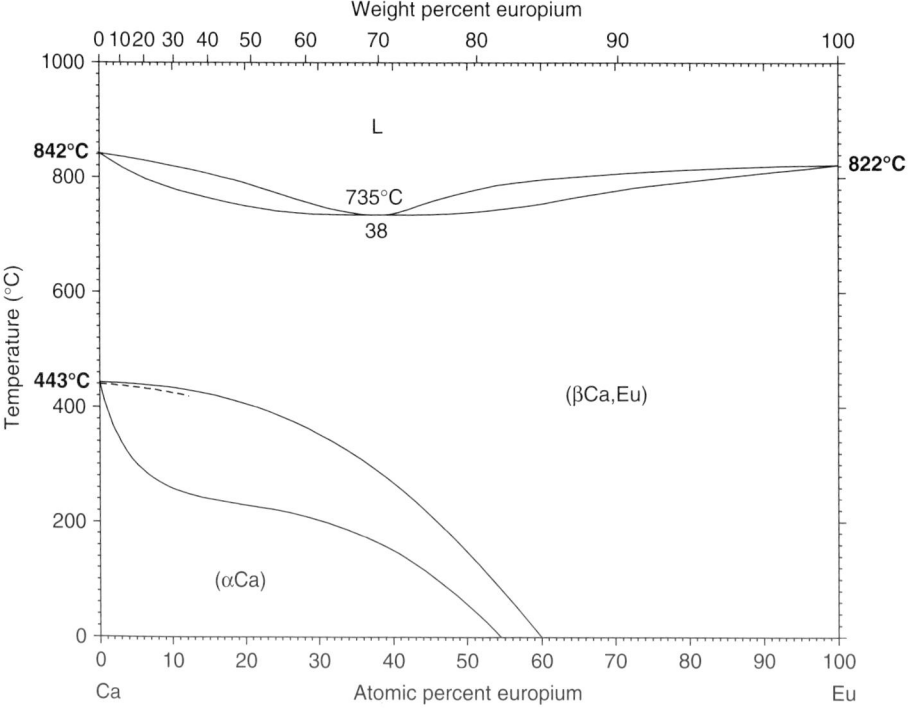

Figure 3.28 Ca–Eu phase diagram [37] (see Section 4.11).

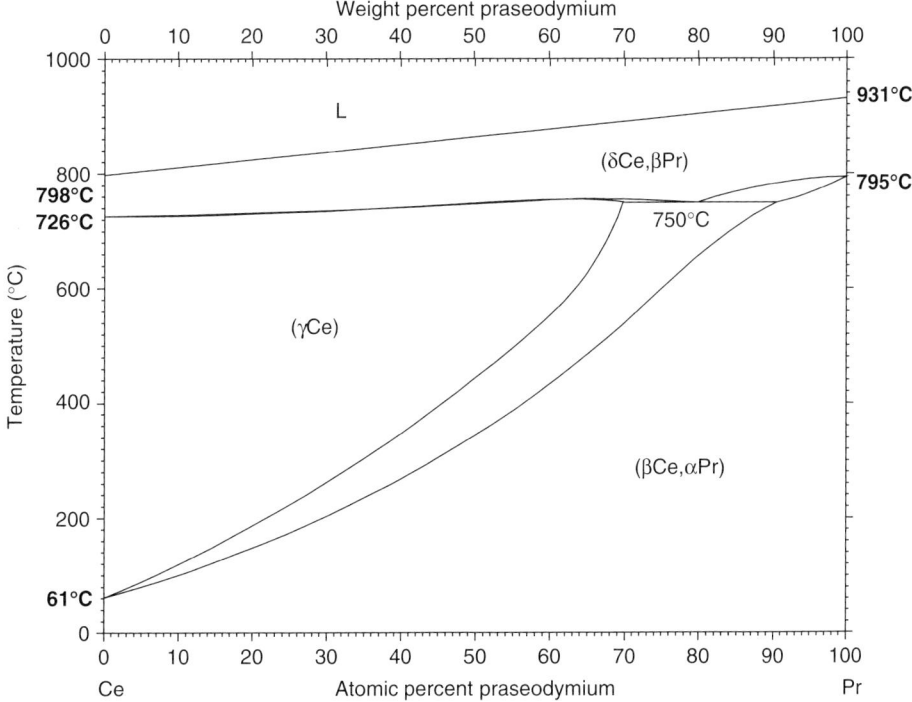

Figure 3.29 Ce–Pr phase diagram [38] (see Section 4.12).

a rather unlikely and abrupt change of slope in the latter solvus. Alternatively, a miscibility gap is needed in the (γCe) phase with its critical point at around 85 at.% Pr, which may also be unlikely for a solid solution phase according to Section 3.12.

If the (γCe)/[(γCe) + (βCe,αPr)] solvus is correct, the [(γCe) + (βCe,αPr)]/(βCe,αPr) solvus must be modified so that it does not cross the 100 at.% Pr line at all when extrapolated smoothly to higher temperatures. This requires unlikely twisting of the phase boundary line. At any rate, at least either one or the other of the solvus boundaries appears to need revision. *Note*: The liquidus and the solidus are so close that they are degenerated nearly into a single boundary.

4.13 Cr–Ni

Figure 3.30 shows the Cr–Ni phase diagram evaluated by Nash [39]. According to Okamoto [40], the existence of the γ′ phase is questionable, at least in the form given in this figure. If the γ′ phase really exists and the (Ni) + γ′ two-phase field is as narrow as shown in the figure, the narrow two-phase field indicates that the (Ni) and γ′ phases should have very similar thermodynamic properties, that is, the temperature and composition dependence of Gibbs energy functions of the (Ni) and γ′ phase is nearly identical. Therefore, the relationship between (Cr) and (Ni) and the relationship between (Cr) and γ′ must be similarly nearly identical. Then the γ′/γ′ + (Cr) boundary should show a trend very similar to that of the (Ni)/(Ni) + (Cr) boundary.

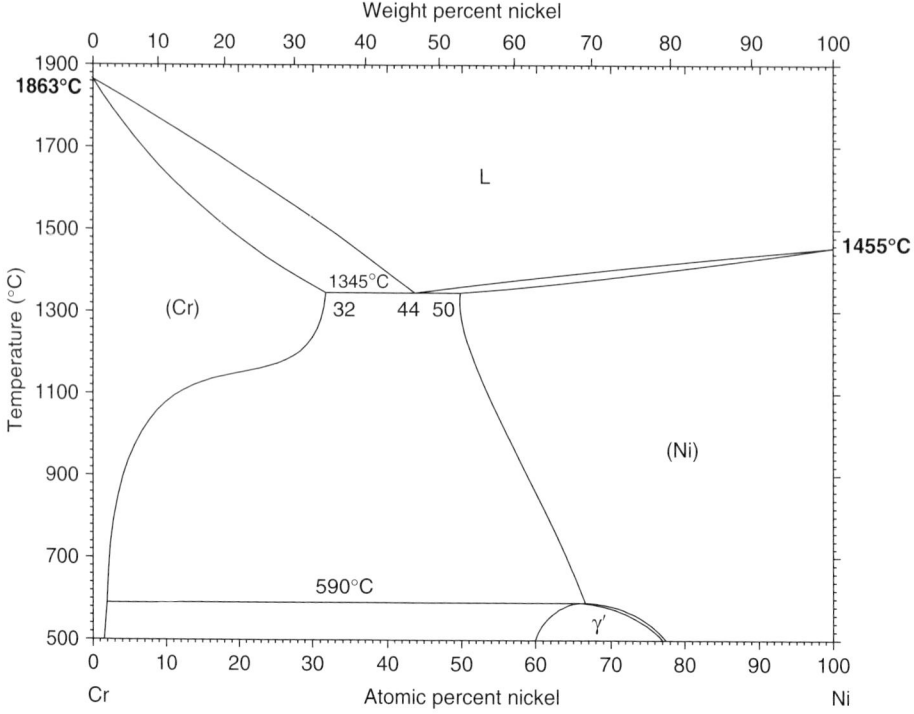

Figure 3.30 Cr–Ni phase diagram [39] (see Section 4.13).

4.14 Cu–Hf

Figure 3.31(a) shows the Cu–Hf phase diagram calculated by Subramanian and Laughlin [41]. Apparently, the liquidus of $Cu_{10}Hf_7$ falls off more steeply than that of the neighboring Cu_8Hf_3. A compound with such pointed liquidus becomes unstable at low temperatures and should decompose eutectoidally (c and d in Fig. 3.3). Therefore, the thermodynamic model employed by Subramanian and Laughlin [41] (Table 3.7) is open to some questions. Inspection of Table 3.7 immediately confirms that Cu_8Hf_3 is predicted to be extremely stable at low temperatures (i.e., at $T = 0K$, the Gibbs energy of Cu_8Hf_3 with respect to liquid standard states is $-210\,207\,J/mol$). Because this value dominates the system, all the other compounds do not remain as stable compounds at low temperatures. In addition, the Gibbs energy of mixing of the liquid is also unusual in their model [41]. The enthalpy term is typical of a regular solution expression, whereas the entropy term is strongly composition dependent. These problems aside, the Cu–Hf phase diagram was recalculated using the values given in Table 3.7. The result is shown in Fig. 3.31(b). As expected, the liquidus boundaries are almost the same as shown by Subramanian and Laughlin [41], but also as expected from the sharpness of the liquidus and from the Gibbs energy functions, all compounds except Cu_8Hf_3 became unstable at low temperatures. There are no reports in the literature that this is in fact the correct picture, so the thermodynamic modeling is only a guide here to what might be expected.

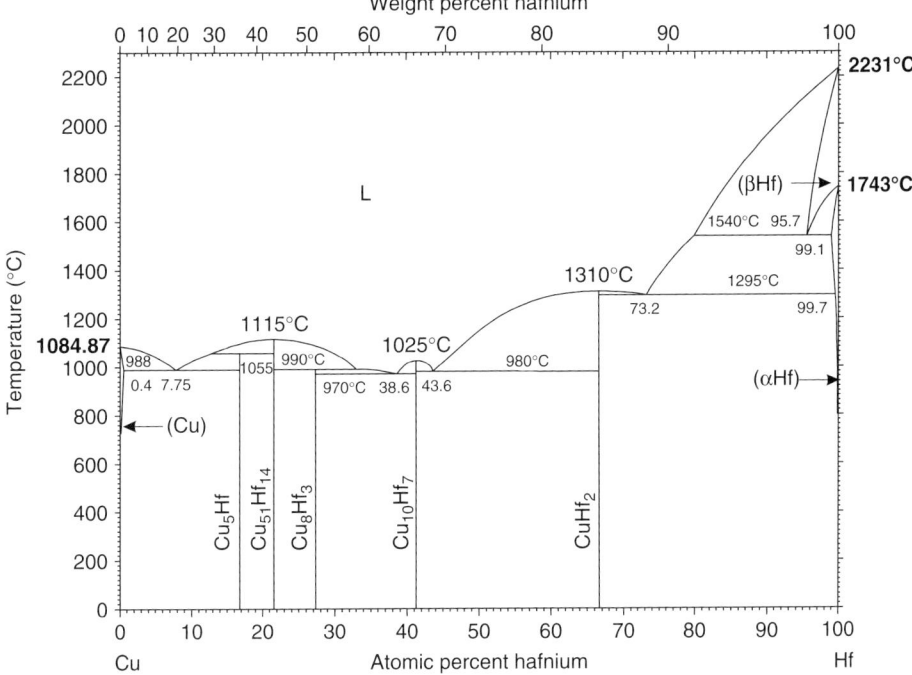

Figure 3.31(a) Cu–Hf phase diagram assessed by Subramanian and Laughlin [41] (see Section 4.14).

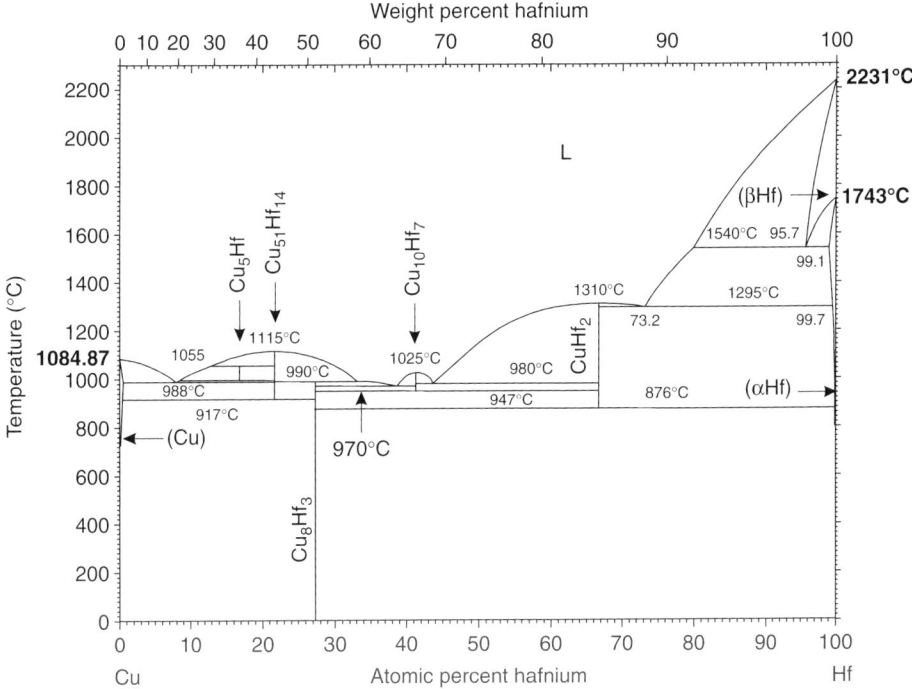

Figure 3.31(b) Cu–Hf phase diagram calculated by using thermodynamic parameters used in Subramanian and Laughlin [41].

Table 3.7 Cu–Hf thermodynamic properties.

Lattice stability parameters for Cu (J/mol)
$G^0(Cu,L) = 0$
$G^0(Cu,fcc) = -13054 + 9.613T$

Lattice stability parameters for Hf (J/mol)
$G^0(Hf,L) = 0$
$G^0(Hf,bcc) = -23390 + 9.341T$
$G^0(Hf,cph) = -29300 + 12.272T$

Excess Gibbs energy of mixing (J/mol)
$\Delta_{mix}G^{ex}(L) = x(1-x)[-77800 - T(83.009 + 248.426x - 596.531x^2 + 328.495x^3)]$
$G(Cu_5Hf) = -33512 - 2.137T$
$G(Cu_{51}Hf_{14}) = -69089 + 11.983T$
$G(Cu_8Hf_3) = -210207 + 126.04T$
$G(Cu_{10}Hf_7) = -23890 - 27.682T$
$G(CuHf_2) = -80005 + 16.393T$

From Subramanian and Laughlin [41].

4.15 Ir–Rh

Figure 3.32 shows the Ir–Rh phase diagram [42]. The shape of the L + (Ir,Rh) two-phase field suggests complete miscibility between (Ir) and (Rh). This is contradictory to the existence of the (Ir) + (Rh) miscibility gap at lower temperatures. If the miscibility gap is real, the influence should be reflected in the L + (Ir,Rh) two-phase field, that is, the two-phase field should become concaved.

4.16 Lu–Th

At first glance, the solidus of the Lu–Th phase diagram (Fig. 3.33) [33] renders no problem. However, the labeling of the (βTh,Lu) phase is incorrect because the bcc βTh and cph Lu cannot form a continuous phase field between them. Hence, either a high-temperature bcc form of Lu exists or some vital portions of the phase diagram are missing. The above example suggests that when a clear conflict with thermodynamical or crystal structure considerations arises, uncertain portions should be either drawn with dashed lines, or omitted altogether in order to emphasize the existing uncertainty.

4.17 Mg–Sb

Figure 3.34 shows the Mg–Sb phase diagram [43]. When the liquidus boundaries of αMg_3Sb_2 are extrapolated to higher temperatures, the metastable melting point of αMg_3Sb_2 is estimated to be at ~900°C from the Mg-rich liquidus and ~950°C from the Sb-rich liquidus. Because there must be only one projected melting point, the extrapolation of the Mg-rich liquidus above the peritectic temperature must be drawn to meet the extrapolation from the Sb-rich side with its peak at the composition of

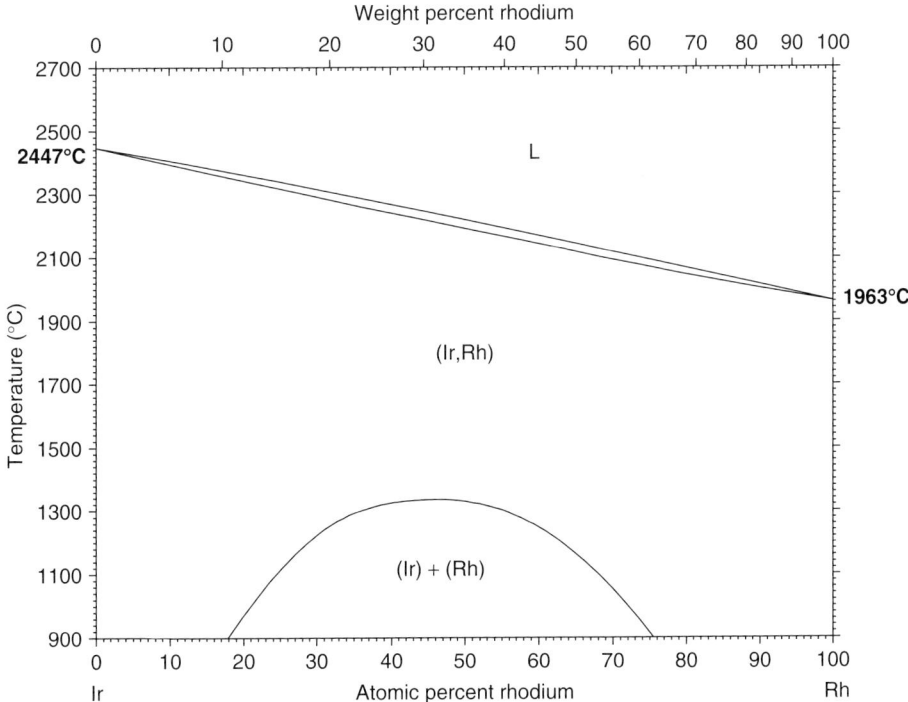

Figure 3.32 Ir–Rh phase diagram [42] (see Section 4.15).

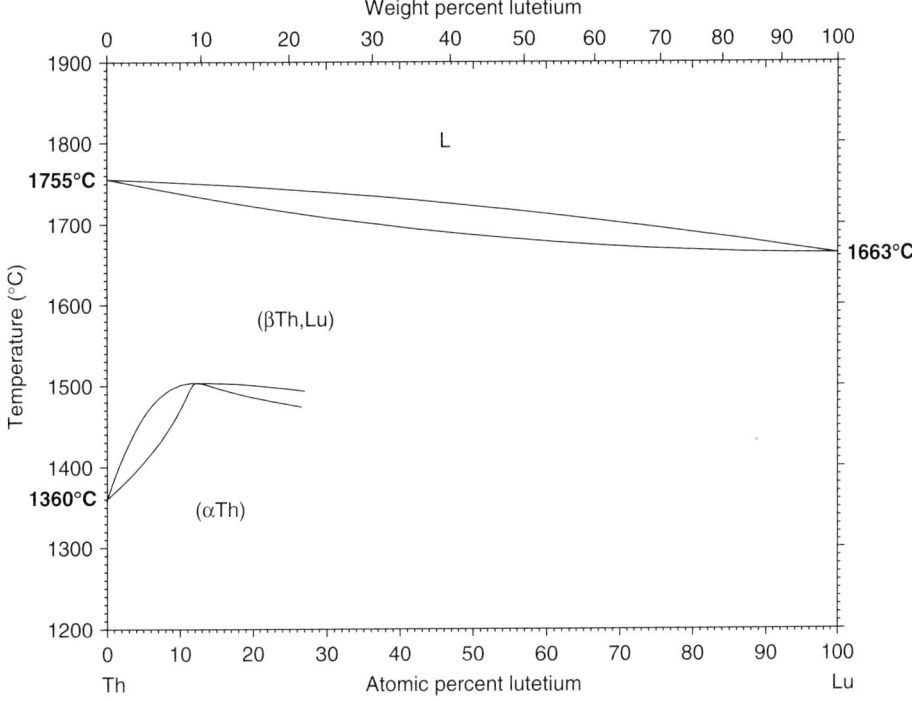

Figure 3.33 Lu–Th phase diagram [33] (see Section 4.16).

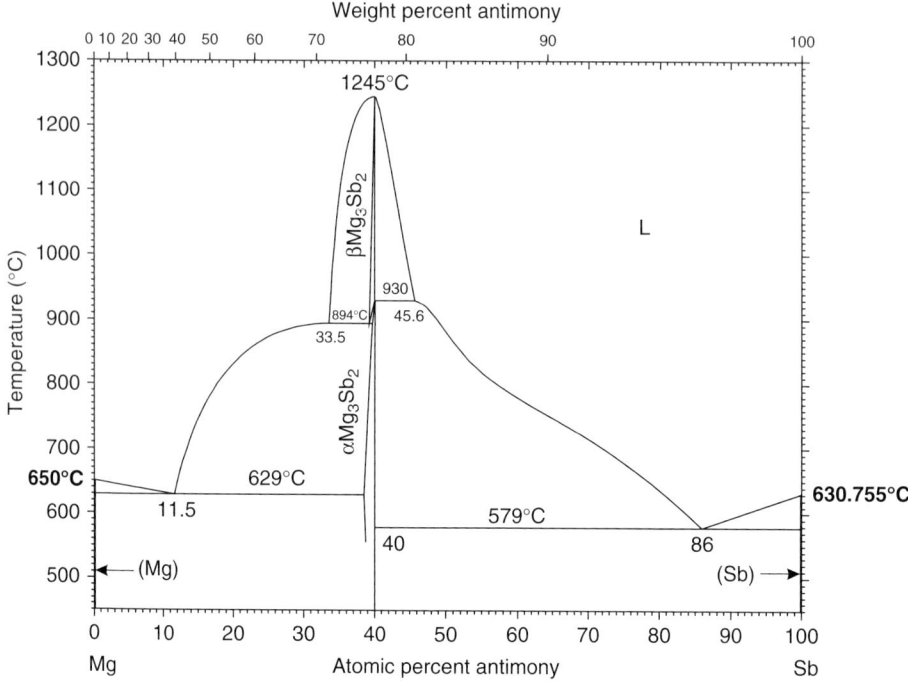

Figure 3.34 Mg–Sb phase diagram [43] (see Section 4.17).

Mg$_3$Sb$_2$. However, an X-type problem in Fig. 3.2 then becomes unavoidable. Most likely, the form of the liquidus on the Mg-rich side in the range of 33 at.% Sb should not be as flat as proposed.

4.18 Mn–Y

There is a strong possibility that the syntectic reaction in the Mn–Y system (Fig. 3.35) [44] is not real, according to the discussion in Section 3.14. If Mn–Y is excluded, Bi–U and Pb–U are the only systems in Table 3.5 not involving strongly ionic elements. This may be an interesting situation whereby the affinities between Bi and U or Pb and U are very high in the solid state, but very low in the liquid state.

4.19 Mo–Ru

In the Mo–Ru phase diagram (Fig. 3.36) [1], the (Mo) + (Ru) two-phase field appears above 1915 °C and below 1143 °C. The boundaries must be smoothly continuous and almost symmetric. However, a strong change of slope is needed to draw continuous lines, as shown with dotted lines. A substantial modification to the phase diagram may be needed because of a new report by Gürler [45] that, for example, the solubility of Mo in (Ru) is approximately 10 at.% at 800 °C (see Ref. [46]).

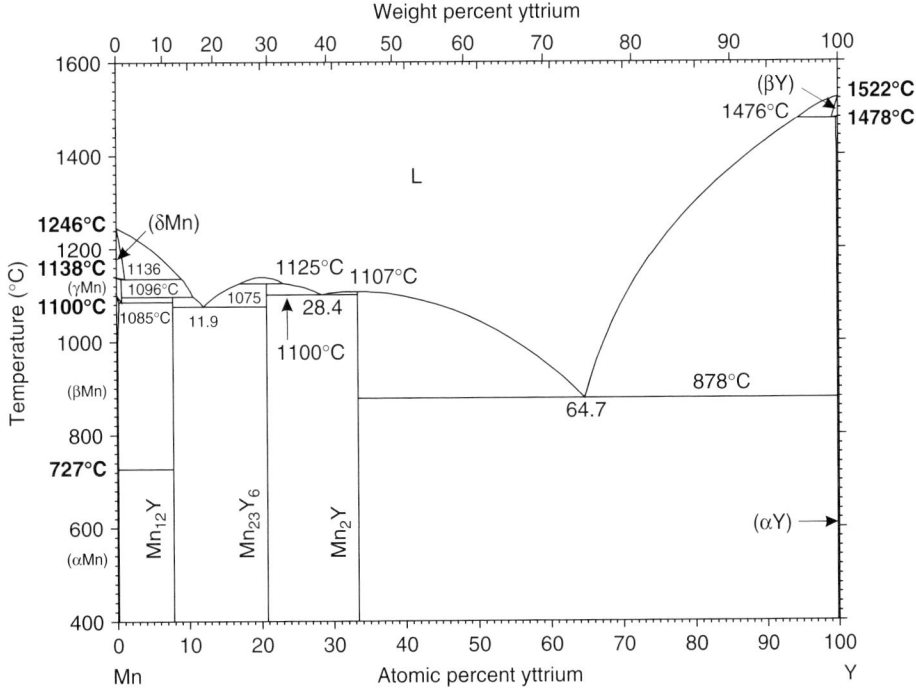

Figure 3.35 Mn–Y phase diagram [44] (see Section 4.18).

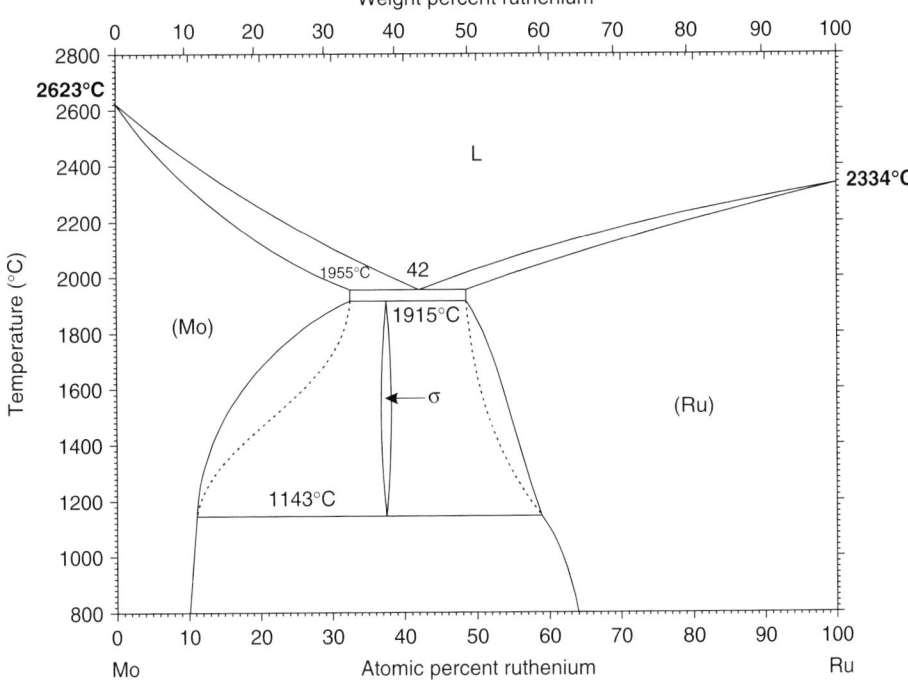

Figure 3.36 Mo–Ru phase diagram [1] (see Section 4.19).

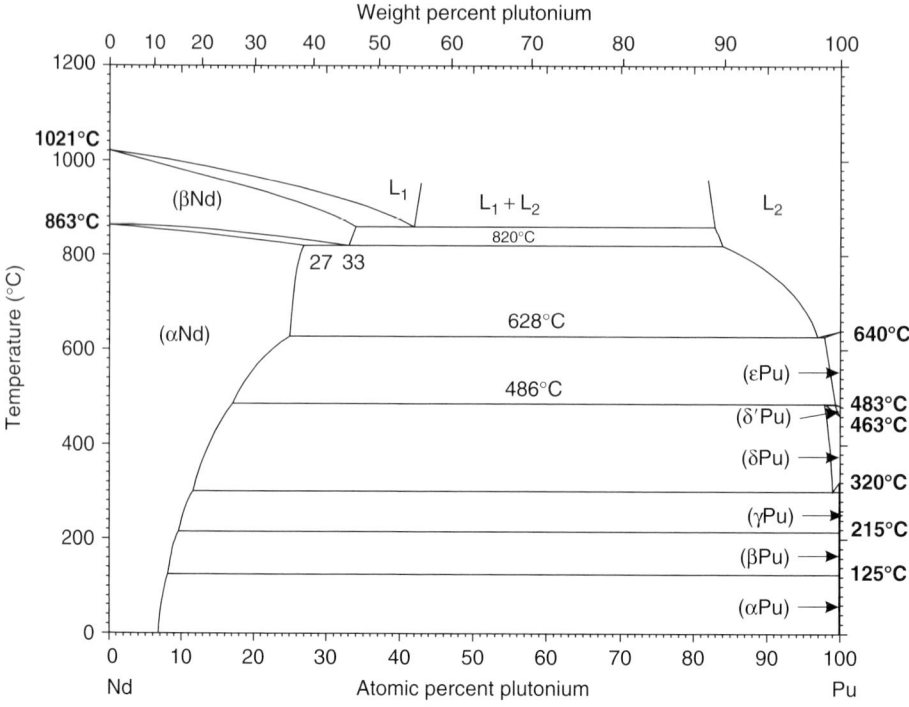

Figure 3.37 Nd–Pu phase diagram [32] (see Section 4.20).

4.20 Nd–Pu

Figure 3.37 shows the Nd–Pu phase diagram [32]. When the two proposed boundaries of the (αNd) + L two-phase field are extrapolated smoothly to higher temperatures, they appear to cross at about 1100°C, causing an obvious T-type problem in Fig. 3.2. This problem may be alleviated by drawing a retrograde boundary on the (αNd) side above 820°C so that the two boundaries cross one another at 0 at.% Pu by smooth extrapolation. However, unless a Y-type problem is introduced, this hypothetical melting of (αNd) becomes higher than that of (βNd), which is unacceptable. Accordingly, the proposed Nd–Pu phase diagram needs substantial modifications. In *Binary Alloy Phase Diagrams* [1], the authors proposed introducing a miscibility gap in the liquid phase to avoid the observed difficulties. Clearly, the proposed diagram must be further explored experimentally.

4.21 Ni–Sr

Figure 3.38 shows the Ni–Sr phase diagram [32]. Because the (Ni) phase shows no solid solubility, the initial slope of the (Ni) liquidus should go through approximately 100 at.%, 0°C when extrapolated, as shown in Section 3.3. However, this is clearly not the case. The Van't Hoff relationship requires the initial slope of the liquidus to be −1421°C/100%, as shown with a dashed line in Fig. 3.38. It is likely that a miscibility gap exists in the liquid state, as in the rather similar Ni–Ba system (see *Binary Alloy Phase Diagrams* [1]).

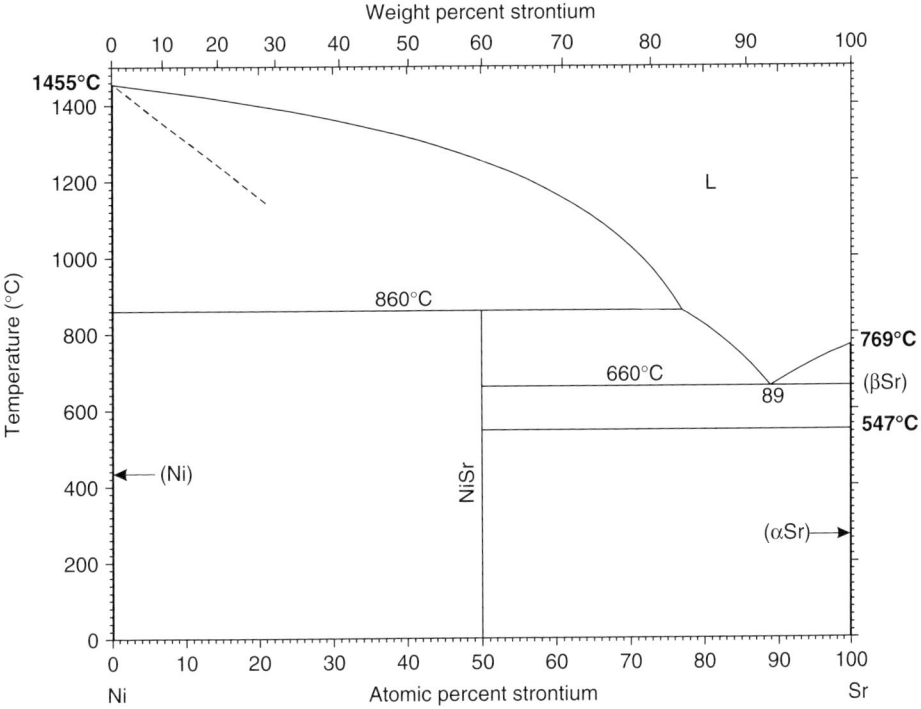

Figure 3.38 Ni–Sr phase diagram [32] (see Section 4.21).

4.22 Np–Zr

Figure 3.39 shows the Np–Zr phase diagram [47]. The Np$_4$Zr and NpZr$_2$ phases show the v-type problem in Fig. 3.3. When a compound with a broad and nearly constant width over a wide composition range decomposes into two neighboring compounds on heating by a peritectoidal reaction, the problem of type o in Fig. 3.3 just below the peritectoid temperature becomes inevitable even if the problem of type n is avoided by adjusting the slope of the phase boundaries.

4.23 Pd–Zr

A thermodynamic model proposed by Stolen and Matsui [48] for the Pd–Zr system may serve as a typical example of inverse miscibility gap in the liquid phase (problem w in Fig. 3.3). In order to reproduce the Pd–Zr phase diagram in *Binary Alloy Phase Diagrams* [1] (Fig. 3.40), Stolen and Matsui [48] obtained an expression for the excess Gibbs energy of mixing of the liquid phase in the form:

$$\Delta_{mix}G^{ex}(L) = x(1-x)[-476\,537 + 241\,068x + 292\,144x^2 \\ + (123.89 + 16.22x - 310.32x^2)T]\,\text{J/mol}$$

where x is the atomic fraction of Zr. The entropy term here is positive and large on the Pd-rich side, and an inverse miscibility gap (overlooked in Ref. [48]) does

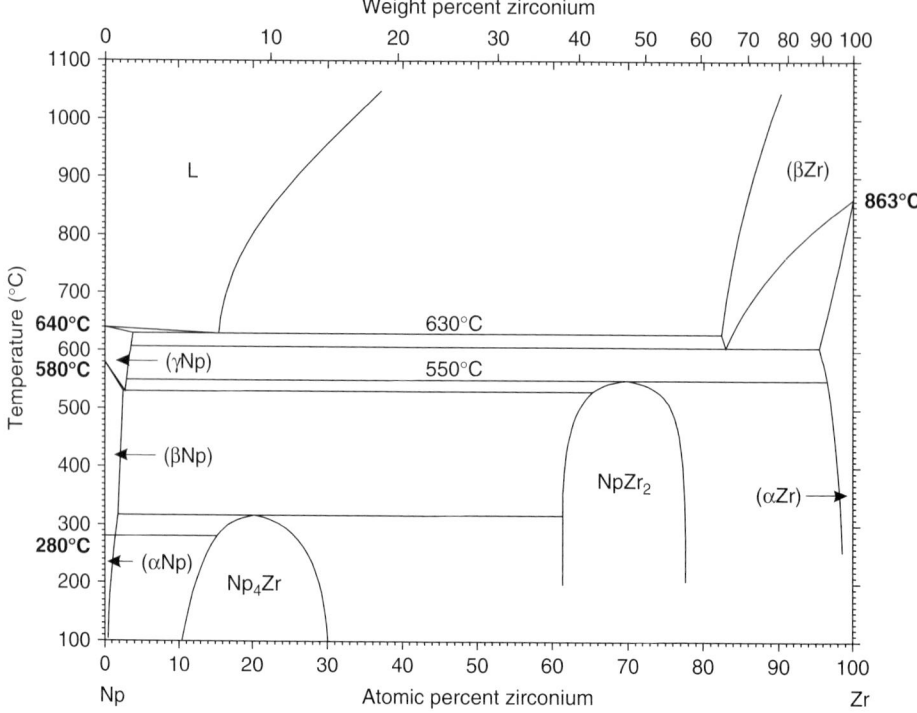

Figure 3.39 Np–Zr phase diagram [47] (see Section 4.22).

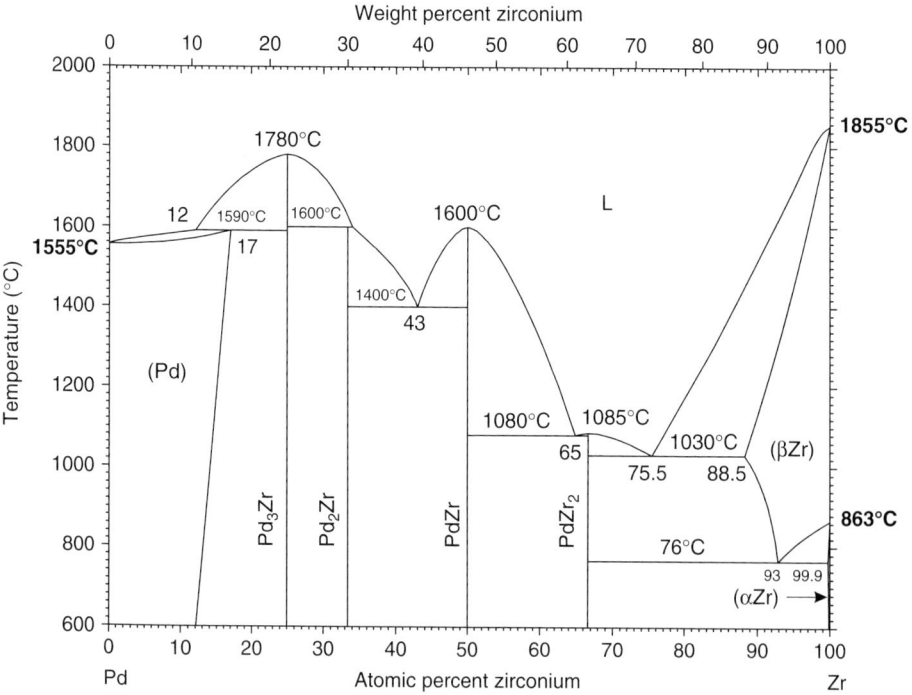

Figure 3.40 Pd–Zr phase diagram [1] (see Section 4.23).

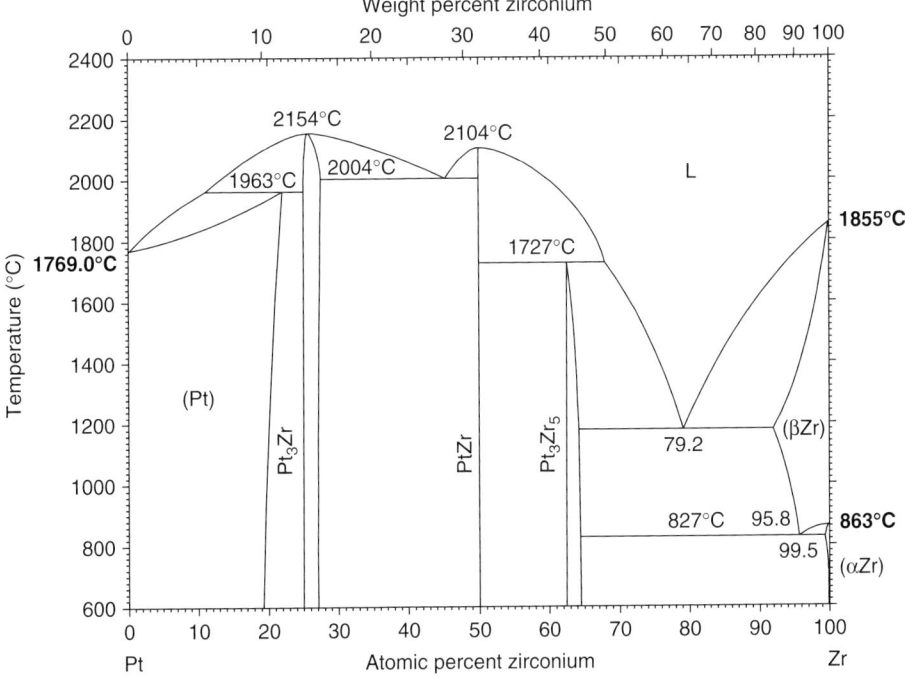

Figure 3.41 Pt–Zr phase diagram [1] (see Section 4.24).

develop with the critical point at $x = 0.387°C$ and 2428°C [8]. Apparently, the model was inappropriate. As a matter of fact, the (Pd) liquidus and solidus in the original phase diagram [1] used for the modeling is itself rather unlikely.

According to Chen et al. [49], similar inverse miscibility gap problems occur unnoticed in many thermodynamic models (Al–Mo [50], Al–Nd [51], Al–W [51], Co–Mo [52], Fe–Si [31], Si–Ta [53], Si–W [53], Sn–Zr [51]).

4.24 Pt–Zr

The Pt–Zr phase diagram (Fig. 3.41) [1] is an example of the problem l in Fig. 3.3. Unless an unlikely sharp bend is introduced in the solidus, the (Pt) phase appears to be extended into the liquid phase. Thus the (Pt) phase is likely to be stabilized at high temperatures. According to Chen et al. [49], some thermodynamic models (Al–Co [54], C–Fe [55], C–Mo [56], C–Nb [57], C–Ni [58], Co–Si [59], Cr–Ta [60], Ni–Ti [61], and Sn–Ti [51]) lead to inversed solidification of low-temperature phases at high temperatures. An example is shown in Fig. 3.42 for the Co–Si system, which was calculated by Chen et al. [49] using thermodynamic parameters proposed by Choi [59]. In this case, the (αCo) and (εCo) phases are extended inevitably into the liquid phase, and the stability of these phases at high temperatures is predictable, as in the case for the Pt–Zr system.

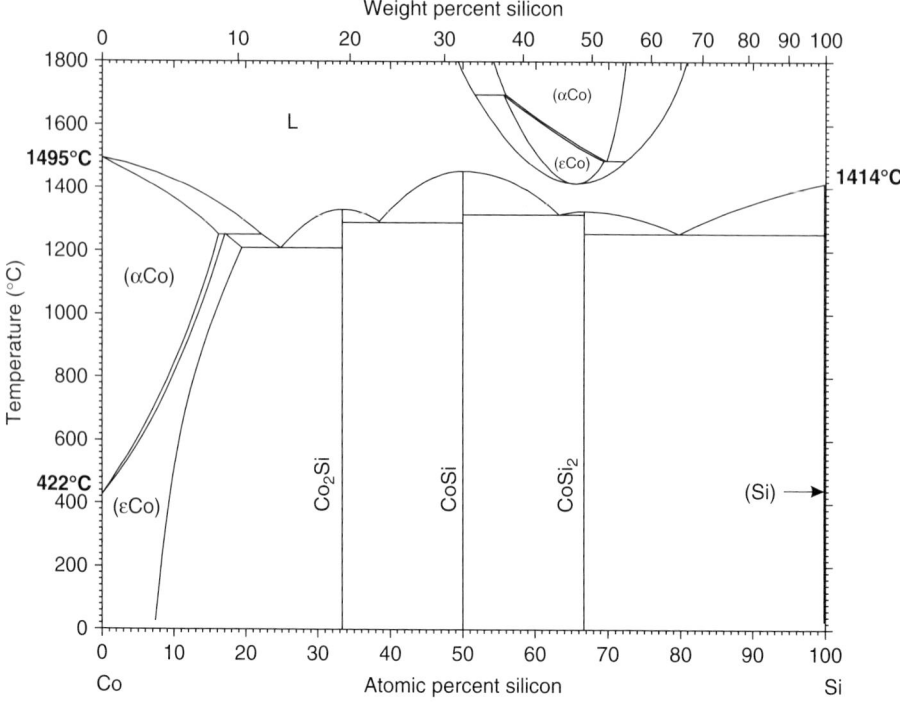

Figure 3.42 Co–Si phase diagram calculated by Chen et al. [49] using thermodynamic parameters proposed by Choi [59] (see Section 4.24).

5 Unusual but Correct Phase Diagrams

In some cases, phase diagrams appear to be violating basic rules or not satisfying the guidelines described in Sections 2 and 3 due to coincidence or other reasons.

5.1 Apparent Four-Phase Equilibrium

If temperature labels are not given, the Gd–Zr phase diagram (Fig. 3.43) [62] might be thought to be erroneous because four phases L, (αGd), (βGd), and (βZr) appear to be in equilibrium. In this case, the (βGd) → (αGd) + L catatectic reaction and the L → (αGd) + (βZr) eutectic reaction are occurring at temperatures 2°C apart. The difference is so small that the two invariant reaction lines appear to be one line.

5.2 Apparent Five-Phase Equilibrium

In the P–Pd phase diagram (Fig. 3.44) [63], five phases, PPd_3, L, $PPd_{4.8}$, L, and PPd_6 appear to be in equilibrium. This is because two eutectic temperatures on either side of $PPd_{4.8}$ have practically the same values, causing an appearance of five-phase equilibrium. Hence, within the resolution scale of the graphics employed and the present experimental information, nothing is wrong with this diagram.

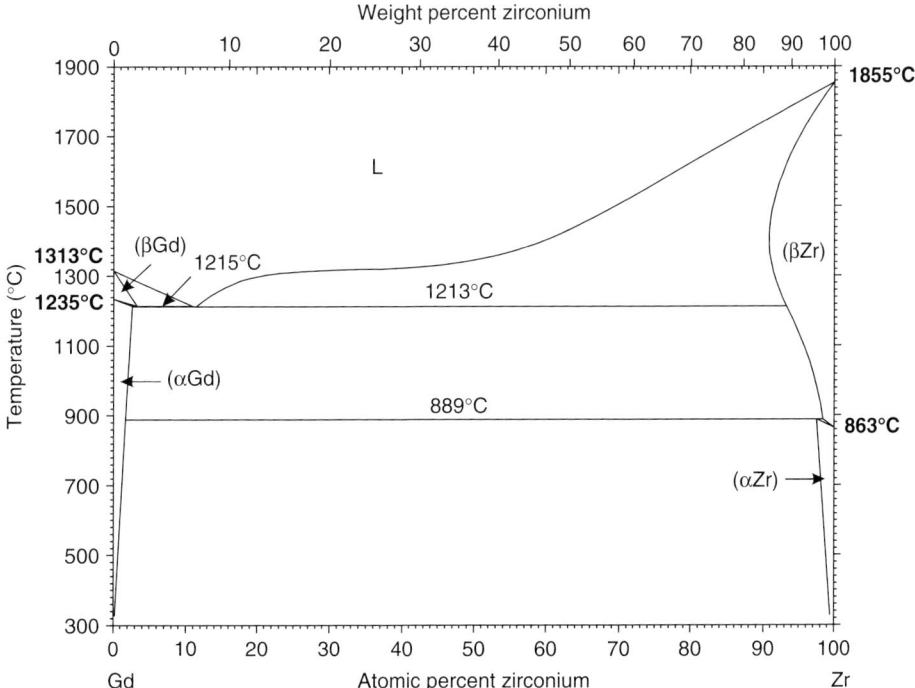

Figure 3.43 Gd–Zr phase diagram [62] (see Section 5.1).

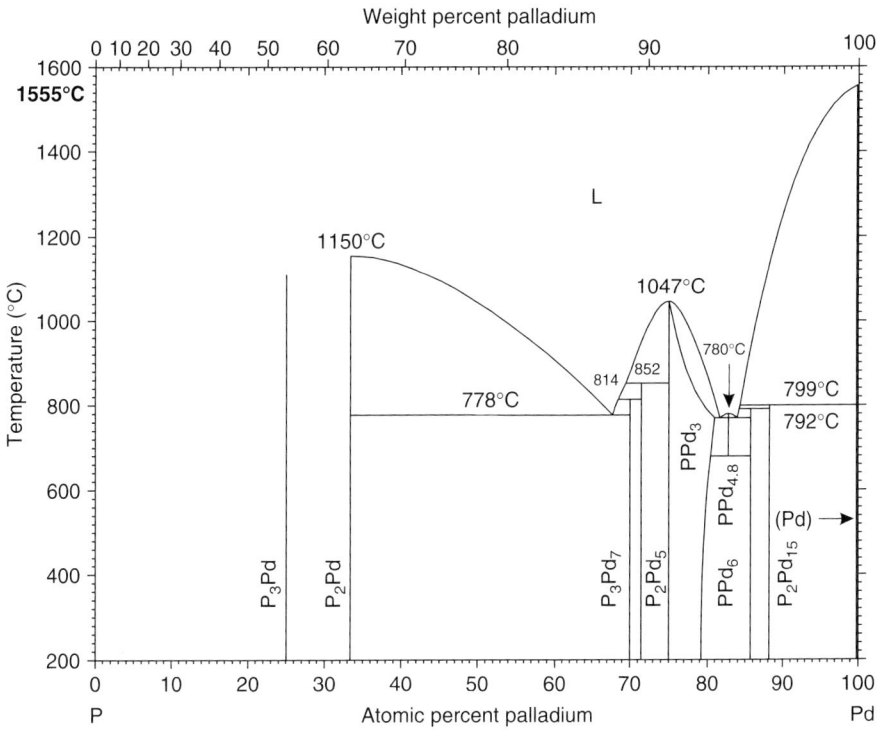

Figure 3.44 P–Pd phase diagram [63] (see Section 5.2).

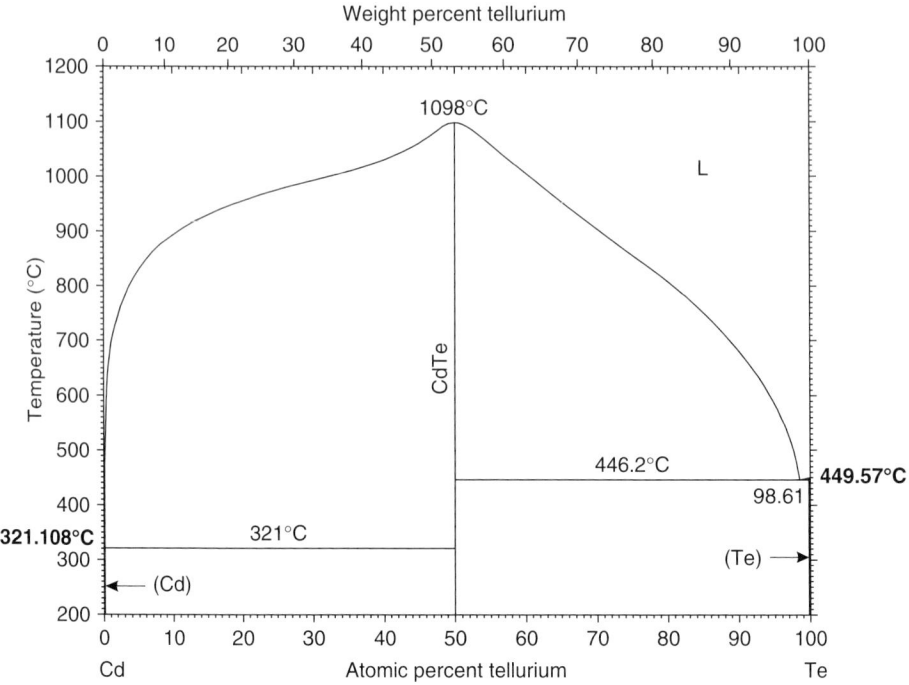

Figure 3.45 Cd–Te phase diagram [64] (see Section 5.3).

5.3 Pointed Liquidus

Some exceptions concerning the Q-type problem are mentioned here. For example, in the Cd–Te phase diagram (Fig. 3.45) [63], the liquidus at the melting point of CdTe appears to be more pointed than ordinary congruent melting points of binary metal–metal systems. These types of pointed melting points are very common in halogen- and chalcogen-based systems, in which strong ionic bonds retain the non-dissociated molecular form of intermediate phase in the liquid state. The diagram appears to consist of Cd–CdTe and CdTe–Te subsystems, with CdTe behaving as an element. If the association of molecules in the liquid state is 100%, the initial slope of the CdTe end can be derived using Eq. (1) and the enthalpy of fusion of CdTe. However, because the association cannot be 100% above 0K, the slope of liquidus at the melting point must be horizontal [64]. However, for viewing the overall trend of the liquidus, the compound should be regarded as an element, allowing a pointed liquidus at the melting point.

Ref. [65] showed the relationship between the sharpness at the melting point of a compound and its heat of formation.

5.4 A Straight Line Liquidus

If the number of experimental data is not enough, the liquidus is sometimes drawn with a straight line, for example B–Pd [66] (Fig. 3.46). Generally, these are doubtful, but it cannot be said categorically that the liquidus is doubtful because it is straight. Figure 3.47 shows the influence of the magnitude of excess Gibbs energy

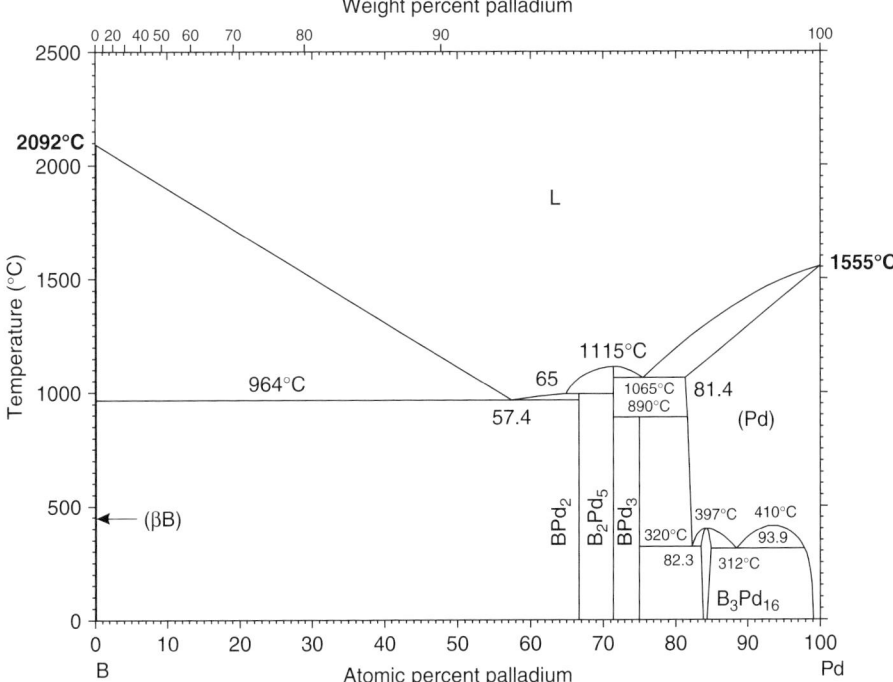

Figure 3.46 B–Pd phase diagram [66] (see Section 5.4).

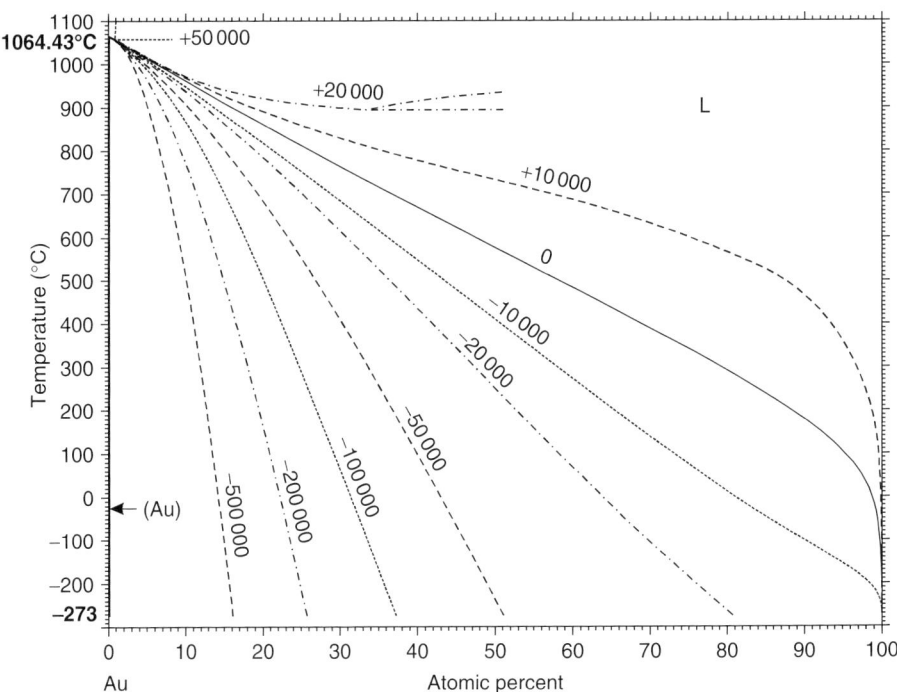

Figure 3.47 The influence of the magnitude of excess Gibbs energy on the form of the liquidus boundary. In this case, the liquidus is almost a straight line up to 90 at.% when $\Omega = -100\,00$ J/mol.

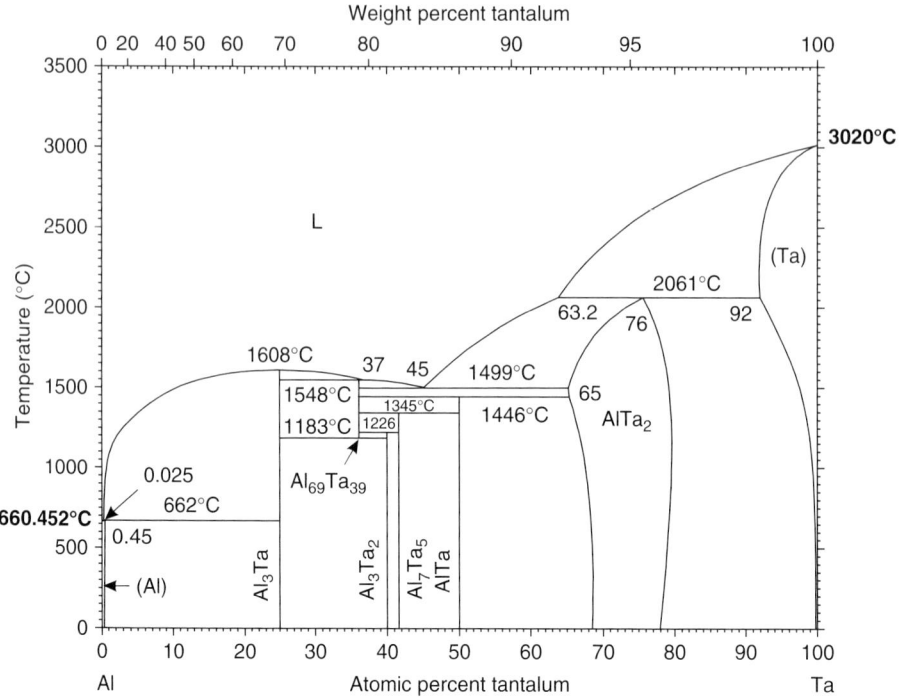

Figure 3.48 Al–Ta phase diagram calculated by Du and Schmid-Fetzer [67] (see Section 5.5).

on the form of the liquidus boundary for the case of Au-based binary systems. The (Au) solid phase was assumed to involve no solid solubility. The lattice stability parameter of Au was derived by assuming that the enthalpy of fusion is 13 000 J/mol [20]. Liquidus curves shown in Fig. 3.47 have been calculated assuming a regular solution behavior for the liquid phase, that is, $\Delta_{mix}G^{ex}(L) = \Omega x(1-x)$, taking Ω to be 50 000, 20 000, 10 000, 0, −10 000, −20 000, −50 000, −100 000, −200 000, and −500 000 J/mol. (x is atomic fraction of solute.) When Ω is large and positive (>19 100 J/mol), a miscibility gap develops in the liquid phase. Therefore, only a part of the liquidus is shown for $\Omega = 50 000$ and 20 000 J/mol. According to this figure, the liquidus is almost straight up to ~90 at.% when $\Omega = 10 000$ J/mol. If this liquidus was determined experimentally, no surprise could be expected if it was drawn as a straight line.

5.5 Off-Stoichiometric Melting

When the liquidus and solidus of AlTa$_2$ phase in the Al–Ta phase diagram [67] (Fig. 3.48) are extrapolated to higher temperatures, congruent melting appears to occur at around 90 at.% Ta, which is far away from the stoichiometry of AlTa$_2$. This is the problem of j in Fig. 3.3. In this case, assignment of "AlTa$_2$" to this phase may be misleading because the crystal structure of AlTa$_2$ is CrFe type σ phase.

6 Phase Diagram Below 0°C

Another feature of interest when considering uncertainties in the phase diagrams is the realization in recent years that the majority of the phase diagrams that have been reported in assessed form are depicted only down to room temperature, or perhaps to 0°C. This is because below 0°C the approach to equilibrium may be considered to be effectively frozen (i.e., in a state of constrained equilibrium) in most systems for practical engineering purposes. The true equilibrium (i.e., unconstrained equilibrium) may be unattainable. Hence, only rarely is an attempt made to assess boundaries as they continue toward absolute zero of temperature (hereafter 0K), but the extrapolated trends of many phase boundaries established above 0°C indicate ranges of solubility at 0K. According to Abriata and Laughlin [68], this could be judged to be in disagreement with the Third Law of thermodynamics which postulates that the entropies of all aspects of a given system reduce to the same value at 0K, often taken to be zero. From this it follows that no solid solubility range exists at 0K. This conclusion contradicts also the idea of any conventional solution model when the enthalpy of mixing of any phase is negative even for a limited composition range. A further exploration is needed in this area.

7 Conclusion

We have attempted here to draw attention of students, researchers, and users of phase diagrams that, despite the vast amount of work done in this area, many of the reported, or assessed phase diagram features can be shown to be thermodynamically impossible, improbable, or unrealistic. Hence, before phase diagrams are used for practical purposes, due caution is always a good approach, and published phase diagram compilations must not be considered as unquestionably reliable. Hopefully, some of the typical errors discussed above, and the several examples of selected binary systems we have shown, may provide a useful guide for spotting potential problems. We have only limited our discussion here to binary systems. Clearly, numerous problems also appear in published or assessed diagrams of ternary and higher-order systems that represent an equally rich area for exercising caution and judgments of reliability.

REFERENCES

1. T.B. Massalski, H. Okamoto, P.R. Subramanian and L. Kacprzak, *Binary Alloy Phase Diagrams*, 2nd edn, ASM International, Materials Park, OH, 1990.
2. H. Okamoto, *Binary Alloy Phase Diagrams*, 2nd edn and Updates, CD-ROM, ASM International, Materials Park, OH, 1996.
3. H. Okamoto, *Desk Handbook: Phase Diagrams for Binary Alloys*, ASM International, Materials Park, OH, 2000.
4. S. Nagasaki and M. Hirabayashi, *Binary Alloy Phase Diagrams*, AGNE Gijutsu Center, Tokyo, Japan, 2001.
5. P. Villars, K. Cenzual, J.L.C. Daams, F. Hulliger, T.B. Massalski, H. Okamoto, K. Osaki and A. Prince, CD-ROM, *Pauling File Binaries Edition*, ASM International, Materials Park, OH, 2002.

6. B. Predel, *Phase Equilibria of Binary Alloys*, CD-ROM, Springer-Verlag, Berlin, Germany, 2003.
7. H. Okamoto and T.B. Massalski, *J. Phase Equilib.*, 12 (1991) 148.
8. H. Okamoto, *J. Phase Equili.*, 12 (1991) 623.
9. H. Okamoto and T.B. Massalski, *J. Phase Equilib.*, 14 (1993) 316.
10. H. Okamoto and T.B. Massalski, *J. Phase Equilib.*, 15 (1994) 500.
11. D.A Goodman, J.W. Cahn and L.H. Bennett, *Bull. Alloy Phase Diagr.*, 2 (1981) 29.
12. A.J.C. Wilson, *J. Inst. Met.*, 70 (1944) 543.
13. E. Paul, *J. Phase Equilib.*, 14 (1993) 135.
14. K.A. Gschneidner Jr., R.W. Calderwood, T.B. Massalski and H. Okamoto, in H. Okamoto, T.B. Massalski (eds.), *Phase Diagrams of Binary Gold Alloys*, ASM International, Metals Park, OH, 1987, p. 157.
15. T.B. Massalski and H.W. King, *Prog. Mater. Sci.*, 10 (1961) 1–78.
16. C. Zenner, in P.S. Rudman, J. Stringer and R. Jaffee (eds.), *Phase Stability in Metals and Alloys*, McGraw Hill Publ. Co., New York, 1966.
17. H. Okamoto, *J. Phase Equilib.*, 14 (1993) 336.
18. G. Borelius, *Ann. Phys.*, 28 (1937) 507.
19. M. Hillert, *J. Phase Equilib.*, 15 (1994) 35.
20. M.W. Chase, *Bull. Alloy Phase Diagr.*, 4 (1983) 123.
21. A.K. Niessen, F.R. de Boer, P.F. de Chatel, W.C. Mattens and S.R. Miedema, *CALPHAD*, 7 (1983) 51.
22. T. Tanaka, N.A. Gokcen and Z. Morita, *Z. Metallk.*, 81 (1990) 49.
23. A.R. Miedema, P.F. de Chatel and F.R. de Boer, *Physica B*, 100 (1980) 1.
24. K.A. Gschneidner Jr. and F.W. Calderwood, *Bull Alloy Phase Diagr.*, 6 (1985) 23.
25. O.N. Carlson, *Bull. Alloy Phase Diagr.*, 11 (1990) 560.
26. A.J. McAlister and J.L. Murray, *Bull. Alloy Phase Diagr.*, 8 (1987) 438.
27. P. Villars, A. Prince and H. Okamoto, *Handbook of Ternary Alloy Phase Diagrams*, ASM International, Materials Park, OH, 1995.
28. M.E. Kassner, D.E. Peterson, in M.E. Kassner and D.E. Peterson (eds.), *Phase Diagrams of Binary Actinide Alloys*, ASM International, Materials Park, OH, 1995, pp. 282–288.
29. J.M. Howe, *Bull. Alloy Phase Diagr.*, 10 (1989) 650.
30. H. Okamoto and T.B. Massalski, *Bull. Alloy Phase Diagr.*, 7 (1986) 532.
31. A. Palenzona and S. Cirafici, *J. Less-Common Met.*, 143 (1988) 167.
32. W.G. Moffatt, *Handbook of Binary Phase Diagrams*, Genium Publishing Corporation, Schenectady, New York, 1978 and Supplements.
33. T.B. Massalski, J.L. Murray, L.H. Bennett and H. Baker, *Binary Alloy Phase Diagrams*, 1st edn, American Society for Metals, Metals Park, OH, 1986.
34. E. Rudy, *Ternary Phase Equilibria in Transition Metal–Boron–Carbon–Silicon Systems, Part V Compendium of Phase Diagram Data*, Air Force Materials Laboratory, Air Force Systems, Wright Patterson Air Force Base, Dayton, OH 1969.
35. S.V. Nagender Naidu and P. Rama Rao, in S.V. Nagender Naidu and P. Rama Rao (eds.), *Phase Diagrams of Binary Tungsten Alloys*, Indian Institute of Metals, Calcutta, India, 1991, p. 21.
36. R.W. Olesinski and G.J. Abbaschian, *Bull. Alloy Phase Diagr.*, 5 (1984) 484.
37. K.A. Gschneidner Jr. and F.W. Calderwood, *Bull. Alloy Phase Diagr.*, 8 (1987) 513.
38. K.A. Gschneidner Jr. and F.W. Calderwood, *Bull. Alloy Phase Diagr.*, 3 (1982) 187.
39. P. Nash, in P. Nash (ed.), *Phase Diagrams of Binary Nickel Alloys*, ASM International, Materials Park, OH, 1991, p. 75.
40. H. Okamoto, *J. Phase Equilib.*, 18 (1997) 221.
41. P.R. Subramanian and D.E. Laughlin, *Bull. Alloy Phase Diagr.*, 9 (1988) 51.
42. N. Tripathi, S.R. Bharadwaj and M.S. Chandrasekharaiah, *J. Phase Equilib.*, 12 (1991) 506.
43. A.A. Nayeb-Hashemi and J.B. Clark, *Bull. Alloy Phase Diagr.*, 5 (1984) 579.
44. A. Palenzona and S. Cirafici, *J. Phase Equilib.*, 12 (1991) 474.
45. R. Gürler, *J. Alloys Compd.*, 285 (1999) 133.
46. H. Okamoto, *J. Phase Equilib.*, 21 (2000) 572.
47. J.K. Gibson, R.G. Haire, M.M. Gensini and T. Ogawa, *J. Alloys Compd.*, 213/214 (1994) 106.
48. S. Stolen and T. Matsui, *J. Nuc., Mater.*, 186 (1992) 242.
49. S.L. Chen, S. Daniel, F. Zhang, Y.A. Chang, W.A. Oates and R. Schmid-Fetzer, *J. Phase Equilib.*, 22 (2001) 373.

50. N. Saunders, *J. Phase Equilib.*, 18 (1997) 370.
51. I. Ansara, A.T. Dinsdale and M.H. Rand (eds.), COST 507, *Thermochemical Database for Light Metal Communities*, Luxembourg, Office for Official Publication of the European Community 1998.
52. A. Davydov and U.R. Kattner, *J. Phase Equilib.*, 20 (1999) 5.
53. C. Vahlas, P.Y. Chevalier, E. Blanquet, *CALPHAD*, 13 (1989) 273–292.
54. N. Dupin and I. Ansara, *Rev. Metall.*, 95 (1998) 1121.
55. P. Gustafson, *Scand. J. Metall.*, 14 (1985) 259.
56. J.O. Andersson, *CALPHAD*, 12 (1988) 1.
57. W. Huang and M. Selleby, *Z. Metallk.*, 88 (1997) 55.
58. A. Gabriel, P. Gustafson, I. Ansara, *CALPHAD*, 11 (1987) 203.
59. S.D. Choi, *CALPHAD*, 16 (1992) 151.
60. N. Dupin and I. Ansara, *J. Phase Equilib.*, 14 (1993) 451.
61. P. Bellen, K.C. Hari-Kumar and P. Wollants, *Z. Metallk.*, 87 (1996) 972.
62. M. Zinkevich, N. Mattern and H.J. Seifert, *J. Phase Equilib.*, 22 (2001) 43.
63. H. Okamoto, *J. Phase Equilib.*, 15 (1994) 58.
64. R.C. Sharma and Y.A. Chang, *Bull. Alloy Phase Diagr.*, 10 (1989) 334.
65. M. Selleby and M. Hillert, *J. Phase Equilib.*, 20 (1999) 288.
66. P.K. Liao, K.E. Spear and M.E. Schlesinger, *J. Phase Equilib.*, 17 (1996) 340.
67. Y. Du and R. Schmid-Fetzer, *J. Phase Equilib.*, 17 (1996) 311.
68. J.P. Abriata and D.E. Laughlin, *Prog. Mater. Sci.*, 49 (2004) 367.

CHAPTER FOUR

DETERMINATION OF PHASE DIAGRAMS USING EQUILIBRATED ALLOYS

Yong Zhong Zhan,[1] Yong Du[2] and Ying Hong Zhuang[1]

Contents

1	Introduction	108
2	Alloy Preparation	109
	2.1 High-Temperature Melting of Alloys	109
	2.2 Arc Melting	110
	2.3 Induction Melting	111
	2.4 Powder Metallurgy Method	112
3	Homogenization Heat Treatment	113
4	Determination of Phase Equilibria: Isothermal Experiments vs. Cooling/Heating Experiments	114
	4.1 Analysis of Quenched Samples to Construct Isothermal Sections (Static Method)	115
	4.2 Analysis of Samples by Heating and Cooling Experiments to Construct Vertical Sections and Liquid Projections (Dynamic Method)	122
5	Examples of Phase Diagram Determination Using Equilibrated Alloys	129
	5.1 The Cu–Nd Binary System	130
	5.2 The Al–Be–Si Ternary System	131
	5.3 The Al–Mn–Si Ternary System	135
	5.4 Multicomponent Phase Diagram and the Al–C–Si–Ti System	137
6	Crystal Structure Identification of New Phases	140
7	Pitfalls	145
	7.1 Verification of the Establishment of True Equilibrium	145
	7.2 Inconsistency Between the Result from DTA Measurement and that from XRD and Microscopy Observation	147
	7.3 Identification of Degenerated Phase Equilibrium	148

1 INTRODUCTION

Phase diagram determination using equilibrated alloys is a traditional, important and widely used method. Employing this method, one can provide phase relationships

---

[1] Institute of Materials Science, Guangxi University, Nanning, Guangxi, China
[2] State Key Laboratory of Powder Metallurgy, Central South University, Changsha, Hunan, China

of alloys under different conditions. The obtained phase equilibria are important experimental data for the optimization of thermodynamic parameters, which in turn can be utilized for calculation of phase diagrams (CALPHAD). A phase diagram is constructed by preparing alloys of required constituents, heat treating at high temperatures to reach equilibrium states, and then identifying the phases, so as to determine liquidus temperatures, solidus temperatures, solubility lines, and other phase transition lines [1,2]. Relying on equilibrated alloys, several techniques are utilized to determine phase diagrams. They include thermal analysis (TA), metallography, X-ray diffraction (XRD), dilatometry, electrical conductivity measurement and magnetic analysis methods, and among others [3,4]. All these methods are based on the principle that when a phase transition occurs in an alloy, its physical and chemical properties, phase composition, and/or structure will vary. By analyzing the temperature, composition, and property changes associated with phase transitions, one can construct phase boundaries according to the phase rule. This chapter deals with this important method for measuring phase diagrams using equilibrated alloys. Several procedures to prepare equilibrated alloys are discussed first. The principles of various experimental methods to measure phase diagrams using equilibrated alloys are then described and followed by real examples. Finally, a few commonly encountered issues and proposed solutions are discussed.

2 ALLOY PREPARATION

Preparation of alloys is a key step for phase diagram determination using equilibrated alloys. For a system that is completely unknown, sample chemistries should be reasonably uniformly distributed throughout the phase diagram. For a partially investigated system, the compositions of the prepared alloys may be selected according to reported experimental data with a focus on measuring unknown regions. To establish a ternary phase diagram, information on the related binary phase diagrams and the melting behavior of ternary compound(s) is of fundamental importance. The same is true of for determination of multicomponent phase diagrams.

It is very important to prepare specimens with homogeneous and credible compositions. Macro-inhomogeneity needs to be avoided in the preparation process since the inhomogeneity may be difficult to remove in subsequent treatments. XRD patterns of inhomogeneous samples may display extra peaks, which may complicate phase identifications. Oxidization or melting loss of the prepared specimens can result in deviations from the nominal compositions, giving rise to inaccurate results. The initial compositions of the samples should be maintained throughout the whole experiments. It is recommended to use the highest purity starting materials available to prepare alloys. In some systems, as little as 0.001% of the extraneous component is enough to result in a significant effect, while in other systems a large quantity of impurities is tolerable. Contamination should be avoided throughout the experimental process, from raw materials to samples for analysis.

2.1 High-Temperature Melting of Alloys

Raw materials in the desired ratios are usually melted in high-temperature furnaces, in which the temperature should be higher than melting points of most of the starting components. To reduce chemical segregation, the alloys can be quenched below their solidus temperatures. Since the electrical conductivities of metals are usually high, methods like high-frequency induction melting and electric arc melting can be employed to effectively prepare alloy samples.

Alloy melting is usually done in a protective atmosphere of inert gas or vacuum to avoid contamination. Melting in vacuum can avoid oxidization and also further purify the metals. Gases that initially exist in the starting materials and trace impurities such as Mg, S, Mn, and Ca volatilize in the vacuum melting process. For example, the partial pressure of O_2 in molten Cu is so high that it can be removed from pure Cu, unless stable oxides are formed. Through the vacuum melting process, H_2 also can be reduced to avoid influencing the properties of metals. However, composition deviation resulting from component volatilization may be a problem for vacuum induction melting, especially when the vapor pressures of the elements are significantly different.

An inert gas (usually Ar or He) with a certain partial pressure is often introduced to an evacuated furnace chamber before arc melting in order to alleviate component volatilization. The partial pressure of these gases is usually selected by balancing the requirements for volatilization reduction and impurity removal. High pressure is favored if degassing is not needed for the starting metals and the vapor pressure of the components (or at least one component) is high. However, in this case the problem of thermal losses induced by conduction and gas convection may arise. The maximum temperature that can be reached in an inert gas atmosphere is much lower than that in vacuum since heat dissipation is much greater in a gas atmosphere than that in vacuum. From this point of view, Ar is better than He due to its lower thermal conductivity.

2.2 Arc Melting

Both consumable and non-consumable electric arc furnaces are suitable for preparation of high melting point alloys. The non-consumable furnace is usually utilized to prepare alloys with low volatility components. The upper electrode is a water-cooled copper pole inlaid with tungsten head, which acts as the cathode. The lower electrode (the anode) is the metal to be melted. A water-cooled copper platform is not only a crucible for the melted alloy, but also part of the positive electrode. Under inert gas (i.e., Ar) shielding or vacuum, an electric arc forms an ion plasma. The electric current passing through the electrode can reach several hundred amperes, melting the anode metals. The power supplied to the metal bath can be adjusted by monitoring the current intensity. Usually arc melting is more capable than induction melting in preparing alloys with very high melting points. Alloys are usually melted at least twice to ensure compositional homogeneity.

Non-consumable electric arc furnaces can be used to melt alloys with a high efficiency. It can prevent harmful gases such as O_2, N_2, and H_2 from contaminating the metals. Moreover, the purity of the molten metals is preserved due to the absence of refractory linings and crucibles as would be the case for induction melting. However, when samples containing volatile components such as Mn, Sb, and Pb, are arc melted, large melting loss would lead to noticeable compositional deviations. For this case, other ways such as high-frequency induction melting or powder metallurgy methods may be used. For phase diagram determination involving extremely reactive and volatile elements, please refer to Chapter 9 of this book by Guminski.

2.3 Induction Melting

Another widely used process for high-quality alloys, high-frequency induction furnace melting has several advantages. Firstly, since the heating body only includes metals, the heating process is short and a high temperature may be obtained rapidly. Secondly, the natural stirring effect of the metallic molten bath originating from the induction current helps to homogenize alloys. And, finally, the melting process can be easily carried out in an inert gas or vacuum atmosphere.

A coil is the main unit of a high-frequency induction furnace. It is hollow, heavy duty, and highly conductive copper water-cooled tubing wound into a helical coil. The coil is contained within a steel shell and magnetic shielding is used to prevent heating of the supporting shell. The frequency of coreless induction melting can be as high as 10 000 cycles per second and it is an important parameter that influences the melting quality. The higher the operating frequency, the greater the maximum amount of power that can be applied to a furnace of a given capacity and the lower the amount of turbulence induced. The required frequency is high when the samples to be melted have small size or high electrical conductivity.

Since high-frequency induction melting is carried out in lined crucibles, one should select only lining materials that do not react with the molten metal at the operating temperature. For alloy samples, refractory oxides such as Al_2O_3 crucibles are usually used. If no proper crucible is available, one solution is to coat the internal surface of the crucible with inert materials. Graphite is the best material for induction heating. As a result, if some elements (i.e., copper) are difficult to be induction heated, they could be melted in a graphite crucible. If the alloys react with graphite, refractory oxide crucibles can be packed with a graphite layer. To reduce melting losses, the crucible should be tall and narrow since the evaporation rate of liquid metals is proportional to its exposed area.

High-frequency induction melting is suitable for making a wide variety of metals with various melting points. Compared with arc melting, this method can be successfully used to prepare alloy samples that contain volatile components such as Mn, Sb, and Pb [5–7]. If the content of volatile components is high, melting loss should be taken into account. The amount of additional starting materials added to make up for the melting loss is based on preliminary experiments in which samples with similar compositions are melted and weighed to determine the loss percentage. Usually 1–3% additional amount of the volatile element compensates for the melting loss.

Another way for solving this problem is to put a mixture of starting metals in a quartz ampoule, which is then evacuated and encapsulated. The sealed ampoule is then placed in a high-frequency induction furnace for melting. If the components possess quite different boiling temperatures (T_b) (i.e., the Al–Cd–Cu system), the component with low T_b (Cd) will have boiled and volatilized before the component with high T_b (Cu) is melted. Because the upper part of the quartz ampoule is much cooler than the lower part, the Cd vapor condenses on the inner wall of the upper part, resulting in deviation from the nominal composition. In order to avoid such a deviation, the Cd may be put at the bottom of the ampoule and a graphite cap used to cover the upper part of the ampoule, as illustrated in Fig. 4.1. The induction coil is firstly lifted to preheat the graphite cap and then lowered to melt

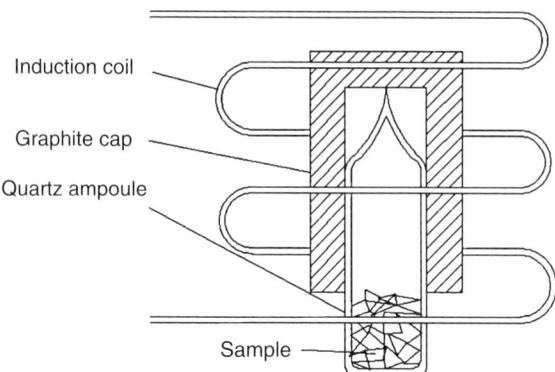

Figure 4.1 Schematic diagram of an induction melting unit to cope with alloys containing elements of very diverse boiling points.

the metals. After the metals have been completely melted, the coil is lifted again to heat the upper part so that its temperature is higher than that of the sample. Several ternary phase diagrams with quite different T_b have been determined using this method to prepare samples [8,9].

For induction melting, arc melting as well as the powder metallurgy method (to be discussed subsequently), master alloys may be useful in the preparation of homogeneous alloys with well-controlled compositions. For instance, to add precise amounts of boron into Ni-base alloys, a Ni–40 at.% B master alloy (or other compositions) can be made by arc melting a compact of Ni and boron powder mixture. The master alloy so obtained can be analyzed for its composition (in case there is preferential oxidation or evaporation of boron). Instead of adding pure boron into Ni, the Ni–40 at.% B master alloy can be used to make Ni–B alloys by using different ratios of Ni and the master alloy. There are two advantages in this case compared to the use of pure boron. Boron has a melting point about 2092°C and it is not a good electrical conductor, thus it cannot be induction melted. If one forces it to be induction melted by melting Ni first, the low-density boron will float, thus making it difficult to homogenize the alloy composition and to produce the exact boron content. The Ni–40 at.% B master alloy, melting at around 1020°C, can be induction melted. The density difference between the master alloy and Ni is much less than that between pure boron and Ni, making alloy homogenization much easier. Moreover, the master alloy has far less easily oxidized than pure boron. Similar ideas of master alloys can be used to solve difficult problems in making alloys with elements of disparate melting points, volatile elements, and so on.

2.4 Powder Metallurgy Method

For alloys that have large quantities of oxidizable or volatile components, or that may undergo severe compositional segregation during solidification, high-temperature melting methods may not be ideal. These alloys may be synthesized by the powder metallurgy method. An example is the experimental determination of rare earth (RE) – magnesium (Mg) alloy phase diagram. When melted, Mg and light RE elements (i.e., Ce, La) are easy to volatilize. In a lesser vacuum, Mg may be severely oxidized. Zhou et al. [10] prepared the RE–Mg–Ni alloys by mixing the powders of pure components and master alloys within a glove box with ball milling and cold pressing. The compacts were then encapsulated in quartz tubes and sintered. Since melting does not occur in this process, melting loss and oxidation are greatly reduced. Consequently, accurate compositions of the prepared samples are obtained.

It should be pointed out, however, that non-equilibrium condition could occur during the sintering process. As a result, liquid phase can appear at a temperature lower than the solidus temperature or the eutectic temperature. This transient liquid phase leads to overgrowth of crystals and compositional segregation. Since the sintered samples are usually in a non-equilibrium state with multiphase coexistence, further homogenization treatment is required.

3 Homogenization Heat Treatment

To accurately establish a phase diagram, it is important to ensure that the samples reach the equilibrium state. As the melted alloys cool through the liquidus and solidus and gradually solidify, non-equilibrium structures are easily generated from solidification segregation. Even for powder metallurgy samples, homogenization treatment is required since the sintering process is usually short and it usually does not react the powders fully into a homogeneous composition without additional high-temperature treatment.

Homogenization heat treatment involves putting samples into a furnace at a high temperature below the solidus temperature of the alloys for an extended period of time. The samples are then quenched from the elevated temperatures to room temperature. The quenching process is trying to freeze the phases at the homogenization temperature to room temperature for analysis. Alternatively, the sample may be slowly cooled to room temperature. This process generates less thermal stress in the samples, thus making the lattice parameter measurements more accurate.

Homogenization is achieved through diffusion, thus annealing time and temperature are two important factors. Since the diffusion coefficient increases greatly with temperature, the heat treatment temperature should be raised as high as possible to accelerate the homogenization process. The homogenization temperature can be as high as about 50°C below the solidus temperature. If there is an allotropic transformation, the samples could be kept at a temperature slightly lower than the transformation point for adequate time, to ensure the completion of the phase change.

If there is a phase transition at low temperatures, very long heat treatment time is required to attain equilibrium. For example, even after samples have been heat treated at 170°C (corresponding to the order–disorder transition temperature) for 1 year, true equilibrium could not be reached in the Cu–Au binary system [11]. With increasing annealing time, the two-phase structure gradually disappeared while the superstructure of $CuAu_3$ could be clearly detected. For systems in which there are compounds formed via peritectic reactions, the samples should be annealed at low temperatures for extended periods of time to ensure thorough reaction. An example is the peritectic reaction, $L + Al_3Ni_2 \rightarrow Al_3Ni$ at 854°C, in the Al-rich side of the Al–Ni system [12]. Due to the slow peritectic reaction, when the sample is cooled at a typical rate to lower temperatures, three phases exist in a metastable equilibrium state. Three-phase coexistence is an equilibrium state in binary systems only at the exact three-phase equilibrium temperature. During the solidification, the composition locus of the liquid phase changes along the liquidus line until it transforms into the eutectic mixture of (Al) and Al_3Ni. Three phases (Al), Al_3Ni, and Al_3Ni_2, are observed by using XRD. However, if an alloy with composition Al_3Ni is annealed at a temperature slightly lower than the peritectic temperature for an extended period of time (830°C for 3 days), only the equilibrium phase Al_3Ni is observed.

For the isothermal decomposition of a solid phase, especially via an eutectoid reaction, the time and temperature of heat treatment are even more important. In the Co–Gd binary system, the Co_5Gd decomposes into $Co_{17}Gd_2$ and Co_7Gd_2 at about 850°C. To determine this phase transition, the sample needs to be annealed at a temperature slightly lower than 850°C for sufficient time to ensure complete decomposition. If annealing temperature is much lower than 850°C, it is too sluggish for Co_5Gd to decompose into equilibrium phases $Co_{17}Gd_2$ and Co_7Gd_2. Discussion on the difficulty of decomposing high-temperature phases can also be found in Chapter 2 of this book.

The annealing process should be carried out in vacuum or an inert gas in order to avoid/reduce oxidization and unwanted environmental interaction. The annealing furnaces are usually heated through electrical resistance or an induction heater. The prepared samples are packaged with refractory materials such as Ta and put in a quartz tube. The tube is then evacuated and encapsulated. The samples in the quartz tube are slowly heated to the required temperature. Since the solid-state reaction rate increases exponentially with temperature, the accuracy, and uniformity of the constant temperature region should be strictly ensured. If room temperature lattice parameters need to be determined precisely, the alloys should be slowly cooled after the homogenization heat treatment, usually less than 10°C/h.

For some alloy systems, elemental segregation is very heavy during the melting process. Such segregation may be hard to remove completely by a homogenization treatment. A combination of cold work and homogenization heat treatment is much more effective. This is only possible if the sample has sufficient ductility to survive cold deformation. One problem of the heating and cooling method is that selective oxidization or selective volatilization of a certain component may occur in the remelting or heat treatment process, so the experimental condition must be strictly controlled.

4 Determination of Phase Equilibria: Isothermal Experiments vs. Cooling/Heating Experiments

Determination of a phase diagram implies the construction of phase boundary curves including liquidus, solidus and invariant reaction lines, and so on. Since the phase transition rates of various systems differ from one other, it takes different time to reach equilibrium. To determine a phase diagram with equilibrated alloys, two kinds of methods, static and dynamic, are used.

As a matter of fact, true equilibrium is difficult to reach under most experimental conditions, as discussed in more detail in Chapter 2. However, equilibrium may be approached from two directions: slow cooling and heating or long-term isothermal heat treatment. Overall, the longer the time to approach equilibrium, the closer to the equilibrium one can get. For the cooling and heating process, the liquidus temperature measured from cooling using differential TA (DTA) is usually lower than the equilibrium liquidus line. When the heating and cooling rates are reduced, the degree of overheating and overcooling could be reduced, thus approaching equilibrium more closely; however, the DTA peak-to-background ratio is degraded. There are optimal cooling and heating rates, and Chapter 5 discussed the TA in greater detail.

To investigate phase equilibria at elevated temperatures, two approaches could be utilized. The first one is to study the phase equilibria directly at the temperature of interest using a hot-stage microscope or high-temperature in-situ XRD. The method has the advantages of: (1) having no complications from quenching; (2) allowing more straightforward interpretation of the results; and (3) allowing a study in a continuous temperature range, which is impractical for quenching experiments. Its disadvantages include less availability of such equipment, more difficult experiments, and potentially greater preferential evaporation, oxidation, or other environmental interactions. Sometimes, in-situ high-temperature experiments are essential to study complex phase equilibria in narrow temperature and composition ranges.

The other approach is to retain the high-temperature phase equilibria to room temperature by quenching. This requires that the quenching/cooling rate is high enough to preserve the high-temperature phases and their structures to room temperature. Liquid nitrogen, water, iced-water, salt-water solution (brine), and oil are common quenching media. Occasionally, some phase transitions take place very fast; it is not possible to retain the high-temperature phase to room temperature even by a fast quench. One example is the γ-Fe (fcc) to α-Fe (bcc) transition in pure iron and dilute Fe alloys. In such cases, care needs to be exercised in interpreting the result.

4.1 Analysis of Quenched Samples to Construct Isothermal Sections (Static Method)

The success of the quenching method depends on the preservation of high-temperature equilibrated phases to room temperature. Phase transition speed and the quenching efficiency are two factors to be considered. The method is appropriate for investigating systems in which the phase transition is sluggish and long-term homogenization is thus needed. One limitation of this method is the difficulty to determine phase transition temperatures accurately. Another problem associated with the method is that the quenched sample may not represent the equilibrium state at elevated temperatures, thus care needs to be exercised when interpreting the experimental observations.

4.1.1 Metallography, Scanning Electron Microscopy/Energy Dispersive X-ray Microanalysis, and Electron Probe Microanalysis

Metallographic analysis is one of the key tools for determining phase diagrams. Microstructure examinations are routinely used to investigate the number of phases (single phase or multiphase) and invariant reaction types (eutectic or peritectic, or others). In particular, the characteristics of each phase such as composition, size, shape, distribution, color, orientation, and hardness may be examined. Metallographic analysis is carried out based on the assumption that the observed microstructure can represent the true structure of the samples. This tool has been successfully used to determine even multicomponent phase diagrams, such as the Al–Cu–Mg–Si quaternary phase diagram, since the 1920s [13].

Prior to metallographic observation, the surface of a specimen should be cleaned and polished. Small samples are usually inlaid in polymer resin or other mounting materials, mechanically ground and finally subjected to polishing with emery paste of Al_2O_3, Cr_2O_3, Fe_2O_3, MgO, or carborundum. Sometimes, electrolytic polishing and electro-spark polishing are used. In order to clearly identify the microstructure, etching methods are adopted to reveal different phases. For example, different phases may be selectively colored with different corrosive agents. Other etching methods include electrochemical etching and magnetic field added etching. The later etching technique can display the difference of magnetic phases and non-magnetic phases.

Optical microscopy (OM) is a classical technique for phase identification. There is tremendous value in examining a sample optically since many phases can be easily differentiated in OM. A simple examination of a sample from low magnifications to high magnifications can reveal approximate volume fractions, microstructure type, homogeneity, and potential surface contamination such as oxidation. The contrast mechanism is complex in OM and some phases can be more easily differentiated in OM than in electron microscopy. The major limitations of OM include: (1) limited magnifications (usually ≤2000x) prevent its use for observations of potential fine microstructure features and (2) essential absence of composition and crystal structure information of the phases. Many early phase diagrams were determined using OM alone.

Scanning electron microscopy (SEM) is another widely used tool for phase diagram determination. In SEM, a focused and collimated electron beam impinges upon the surface of a sample, creating backscattered electrons, secondary electrons, characteristic X-rays, and Auger electrons, among other signals. By rastering the electron beam around a small rectangular area of the sample, and putting a signal such as secondary electron intensity on a screen in a raster mode, a corresponding image is obtained. Two predominate imaging modes are backscattered electron imaging and secondary electron imaging. The main contrast of backscattered electron images comes from the difference in the average atomic number at each point in the sample, thus it carries some compositional information. Contrast mechanism for secondary electron imaging is predominately the surface topology of the sample with small contrast contribution from backscattered electrons.

Scanning electron microscopes are often coupled with energy dispersive X-ray microanalysis (EDX). EDX is a microanalytical technique that uses the characteristic spectrum of X-rays emitted by the different elements in a specimen after excitation by high-energy electrons to obtain quantitative or qualitative compositional information about the samples. When employed to measure a phase diagram, EDX can provide information on compositions of individual phases and the distribution of alloying elements and micro-inhomogeneity.

Sometimes the volume fraction of phases in a multiphase alloy needs to be measured in order to establish a phase diagram. The fraction can be obtained by metallographic method or XRD. For the metallographic method, it is extremely important to examine the whole sample carefully to make sure the microstructure is uniform, even though heterogeneous, and free from macro-segregation. Only under such a condition, can reliable volume fraction information be extracted. Images representative of the whole alloy are taken and usually processed with imaging software to extract the volume fraction data from the area proportions. With known densities and formula molar weight of the phases, the weight fraction and mole fraction of the phases can be computed. In general, trace amounts of phases can also be identified with this method. The imaging analysis software is prone to error and should be carefully checked with other methods. One commonly used method for trace phase analysis is to preferentially etch away the matrix phase and analyze the residual amount carefully.

Very often the metallographic method is more sensitive to small volume fractions of phases than is XRD. Although limited by the necessity of obtaining a well-polished or etched surface, metallography reveals the presence of quenched liquid by its characteristic microstructure much better than does XRD. In addition, the microstructure gives other important information about the order of occurrence of the phases present and allows estimation of the quantities of each phase more accurately. The sequence of occurrence contains very valuable information on the location of the alloy composition in the projected liquidus surface and also information on the reaction type (eutectic, peritectic, etc.). In contrast, XRD spectrum can only identify the phases but not invariant reaction types. Through microscopic analysis, the phase transition products such as eutectics or eutectoids can be clearly identified. Moreover, the compositions of hypoeutectic, hypereutectic, hypoeutectoid, and hypereutectoid can be correctly determined.

The major shortcoming of the metallographic method in phase diagram determination is its difficulty in detecting fine precipitates. Many terminal solid solutions have decreasing solubility with decreasing temperature. After an alloy is homogenized, annealed at a lower temperature, and subsequently quenched, precipitates may be so small that even SEM cannot detect them. Only careful transmission electron microscopy (TEM) examination can clearly identify them. Ample annealing time should be given to make sure the precipitation takes place.

For samples containing brittle phases, care needs to be exercised when performing microscopic observations, especially when the matrix phase is brittle. The second phase may fall out during sample preparation, leading to erroneous volume fraction analysis. The brittleness could also yield intergranular cracking during the quenching process, thus reducing the chance to detect a quenched liquid or other particles at grain boundaries. For samples with a large amount of brittle phase(s), the phase boundaries are better determined by using the XRD method.

Currently the metallographic method is not just a static technique for determination of phase diagrams. Dynamic observation of phase change can be realized by installing a heating and cooling stage on OM or SEM. By in-situ investigation, information on phase relationships can be obtained for solid-state phase transition, melting, and crystallization of alloy.

Electron probe microanalysis (EPMA) is another important technique for phase diagram determination. It is essentially a dedicated SEM with wavelength dispersive spectrometers (WDS) attached. As an elemental analysis technique, it uses a focused beam of high-energy electrons (5–30 KeV) in the SEM to impinge on a sample to induce emission of characteristic X-rays from each element. Its spatial resolution for X-ray microanalysis depends mainly on the accelerating voltage of the electron beam and the average atomic weight of a phase in a specimen, and usually ranges from one to several microns.

Quantitative matrix correction procedures have been well developed over the last several decades. These quantitative procedures have been demonstrated to produce error distributions characterized by a standard deviation of less than 3% relative when the samples are in the ideal form of a metallographically polished bulk solid. Standards utilized in EPMA are in the form of pure elements or simple compounds (e.g., MgO or GaP). This analytical approach provides great versatility in the analysis of multi-element unknowns of virtually any composition with the exception of very light elements (atomic numbers less than 6). Detection limits are of the order of 100 ppm. Spatial distribution of elemental constituents can be visualized quantitatively by digital composition maps and displayed in a gray scale or false colors.

The EPMA technique is mainly used to measure the position of two-phase and three-phase regions, and then establish the boundaries of single-phase regions. To determine the triangle of the three-phase equilibria with EPMA, only one alloy within the three-phase region is needed. The phase composition is measured when the electron beam hits the grains of each of the three phases. Other techniques such as XRD would require many more alloys to determine the phase boundaries through the lattice constant vs. composition curves within single- and two-phase regions. The EPMA measurement is far less time-consuming. For a two-phase equilibrium, grains of each phase (in a range of several microns) are easily selected and analyzed by EPMA. The measurements yield tie lines. When the compositions of two phases are very close to each other, care should be taken in using EPMA for tie-line determination. A combination of EPMA with XRD and metallography should be used to determine the phase boundaries accurately.

4.1.2 XRD

XRD is widely used to determine the presence of different phases in a sample and thus place the alloy in the right phase region with known composition of the alloy either from the nominal compositions, chemical analysis, or EPMA. The other essential application of XRD is the determination of the crystal structure of a new phase. One of the most important tasks in the determination of ternary and higher-order phase diagrams is to find out whether there are ternary or quaternary compounds. In this regard, XRD is indispensable for phase diagram research. Quantitative determination of the amounts of different phases in multiphase mixtures can be done by peak-ratio calculations [14].

Currently, XRD data are automatically collected and analyzed. The preparation of XRD specimens is usually the most critical factor influencing the quality of the result. The ideal specimen for X-ray powder diffraction is a statistically significant amount of randomly oriented powders with crystallite size less than $10\,\mu m$, mounted in a manner without preferred crystallite orientation. The ideal size of a XRD specimen depends on the crystal symmetry. For high symmetry crystals such as cubic and hexagonal, the specimen need not be very fine. For low symmetry crystals such as monoclinic and triclinic, finer specimens are favored.

The preparation method of XRD powder specimens depends on the properties of the alloy samples. Brittle samples can be easily fragmented and then ground in agate mortars without cold working induced internal stress. The prepared powder can be directly analyzed by XRD. High ductility metallic samples can be made into powder by stainless files or diamond wheel grinding. If the sample is ferromagnetic, non-magnetic grinding wheel or stainless files should be chosen. The ground powder specimens can then be successfully separated from the abrasive materials using a magnet. For an alloy with good ductility, a high internal stress may be induced by cold working, resulting in undesirable XRD peak broadening. Annealing treatment (so-called relief annealing) is required for this kind of specimen to remove the stress. The temperature for relief annealing is usually low (about 400°C), and the annealed specimens should be slowly cooled to room temperature.

There are many cases where the phases have different brittleness and hardness. As a result, the sizes of the particles may differ significantly. After sieving, the relative content of each phase in the powder will change. This kind of sample, therefore, should be ground until all the particles can pass through the required screen mesh. One needs to avoid reaction of certain components of the alloy with oxygen or nitrogen in the air or in the following annealing treatment. Oxygen or nitrogen may form interstitial solid solutions or even new phases, thus greatly influencing the accuracy of a phase diagram determination. Another process leading to composition variation is preferential oxidation of some components in the alloy powders. In this case, the preparation of powders should be carried out in vacuum or an inert gas atmosphere such as inside a glove box.

Phase identification is usually accomplished by comparing the peaks and relative intensities from the investigated specimen with those taken from a very large set of "standard" data provided by the International Center for Diffraction Data (ICDD) or other databases. The current ICDD release contains well over 150 000 XRD patterns, both experimental (about 94 000) and calculated (about 59 000), from known inorganic and many organic crystalline substances. Professional software such as Jade (from Materials Data, Inc., USA) can be used to easily access this and other massive (and continually growing) databases. Jade includes an automated search-match function that compares the sample pattern with that taken from the ICDD database. With good data from a single-phase sample, Jade's automated search-match program will usually identify the phase successfully. For most two-phase samples the identification of the dominant phase is usually successful with more hunting for the identification of the second phase. As the number of the phase increases, some knowledge of the likely constituents will be required to successfully identify all phases. Fortunately the ability to visually compare the sample pattern to a large number of possible phases is a manageable task.

Based on the XRD results, two methods, peak intensity method and lattice parameter method, are usually used for the determination of phase boundaries.

The peak intensity method is widely used in phase diagram determination. Different phases have different crystal structures that are characterized by their distinctive XRD peaks. The phases in an alloy can be determined by its XRD patterns to have either a single phase or multiple phases. In a single-phase region, there is only one type of XRD spectrum observed. If the crystal structure is known, the diffraction intensity of each peak can be calibrated by its position. In this case, the position of diffraction peaks may vary with alloying composition, but additional peaks should not appear. Two situations may cause the appearance of additional peaks in a supposedly single-phase alloy. Firstly, either complete equilibrium has not been reached in the sample or extraneous components are incorporated. Secondly, the alloy composition has deviated from the nominal one, and the altered composition is located in a two-phase region.

In a two-phase region where two kinds of crystal structures coexist, the XRD pattern consists of peaks from both phases. As the composition changes, the position of each diffraction peak does not shift, but the relative intensity varies accordingly. The Ag–Cu binary phase diagram shown in Fig. 4.2 is discussed here as an example. At 650°C, the alloy with composition P on line xy should be in the $\alpha + \beta$ phase region. When point P gradually approaches point x, then the XRD intensity of the α phase increases and that of the β phase decreases. When P reaches point x, the XRD pattern should be completely that of the α phase, and the β phase has vanished. To determine the phase boundaries of $\alpha/(\alpha + \beta)$ and $\beta/(\alpha + \beta)$, several alloys in the $(\alpha + \beta)$ two-phase field should be prepared and subjected to XRD analysis. By measuring the diffraction intensity of one sharp peak from each phase, an intensity ratio vs. composition curve can be plotted. When the curve is extrapolated to the point of $I_\beta/I_\alpha = 0$, the corresponding composition is determined to be the phase boundary between α and $(\alpha + \beta)$. If it is extrapolated to $I_\alpha/I_\beta = 0$, then the phase boundary between β and $(\alpha + \beta)$ at this temperature.

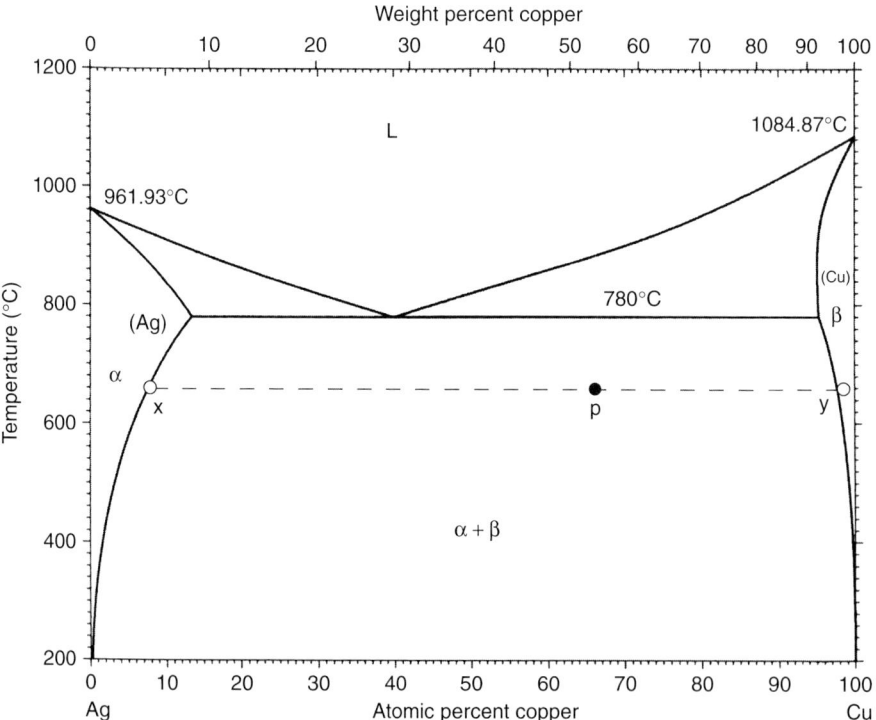

Figure 4.2 Ag–Cu phase diagram illustrated for the peak intensity method [12].

The lattice parameter method for the determination of the phase boundary of a solid solution is very straightforward: plot the lattice parameter against the compositions of the alloys and find the composition at which the lattice parameter first becomes constant. When solute element B is added to solvent A, the lattice parameter of A phase may decrease or increase due to the difference in their atomic sizes. In the single-phase region, lattice parameter changes continuously with the composition of B. However, in a two-phase field, since the composition of each phase remains constant at a given temperature, the lattice parameter does not vary with the alloying composition. Thus, one can plot the lattice parameter against composition. The composition at which the lattice parameter first becomes constant corresponds to the solid solubility at this temperature. If the solid solubility at different temperatures is measured by this method, then the phase boundary can be established.

It should be pointed out that the lattice parameter vs. composition curve usually deviates from Vegard's law. As a result, more samples in the single-phase region should be prepared to determine a phase diagram accurately. Three types of lattice parameter vs. composition curves are shown in Fig. 4.3 [4]. In Fig. 4.4(a), the lattice parameter increases continuously with the solute atoms, and the reverse case is shown in Fig. 4.4(b). An exceptional case is shown in Fig. 4.4(c) in which the variation of the lattice parameter with composition for fcc(Ni) phase of the Fe–Ni binary phase diagram is illustrated. For the fcc(Ni) phase, its lattice parameter does not vary monotonically with the solute concentration. For this kind of extreme case, more samples could be prepared to determine the boundaries. The (Ag) solid solution in the Al–Ag system is another special case in which the lattice parameter is nearly constant within the composition range of 0–10 at.% Al. In this case, it is difficult to apply the lattice parameter method to determine the phase boundary. The peak intensity method could be used.

Generally the lattice parameter of single-phase alloys does not change noticeably with the quenching temperature. Therefore, to determine a phase boundary using this method, one can quench alloys in the two-phase region from different

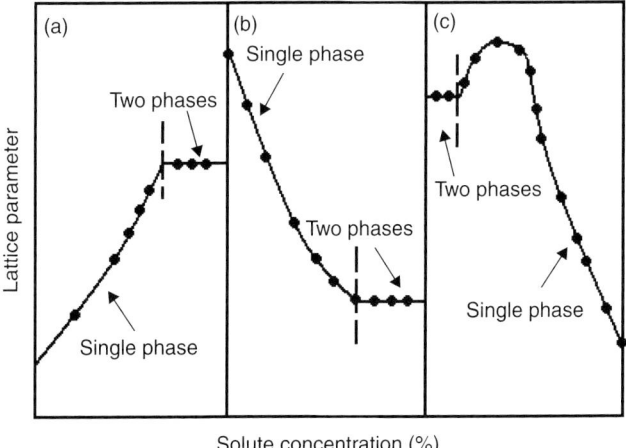

Figure 4.3 Different trends of lattice parameter vs. solute concentration curves [4].

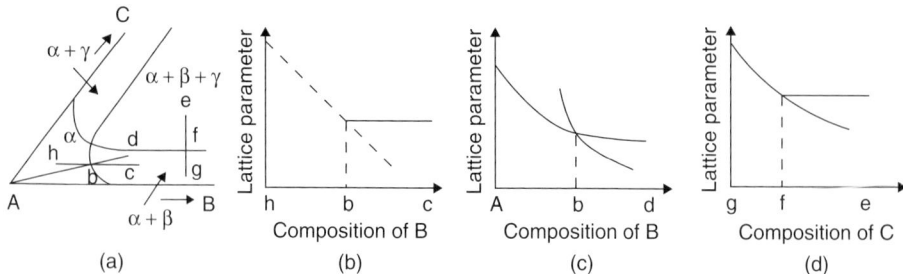

Figure 4.4 Determination of ternary alloy phase diagram using lattice parameter method: (a) isothermal section of the A-rich part in a ternary system; (b) variation of lattice parameter with composition for α along the tie line; (c) variation of lattice parameter with composition for α not along the tie line; (d) variation of lattice parameter with composition for α along a line being across a two-phase and three-phase regions [2].

temperatures and then measure the corresponding lattice parameters. Bradley and Lu [15] had successfully used this method to determine the α phase boundary of the Cr–Al binary system.

The determination of a phase boundary in a ternary system using the lattice parameter method is illustrated with the help of Fig. 4.4. If the alloy compositions are on a tie line or its extension line, that is, line hbc in Fig. 4.4(a), one can determine the phase boundary using the same procedure applied to binary systems. Figure 4.4(b) indicates schematically the variation of the lattice parameter against composition in the case of alloy compositions along the tie line. If the compositions are not on a tie line but on another straight line such as Abd in Fig. 4.4(a), then the lattice parameters in a two-phase region may vary with the composition. The phase boundary is the intersection of the two curves, as shown in Fig. 4.4(c). To determine the boundary between two-phase and three-phase regions, one can prepare alloys with compositions along a straight line (such as line efg in Fig. 4.4(a)) which crosses the two regions. The lattice parameter varies regularly with composition in a two-phase region. Within a three-phase region, the lattice parameter remains constant. Consequently, the boundary between two-phase and three-phase regions corresponds to the intersection, as indicated in Fig. 4.4(d). If the crystal structures of the phases are available, one can measure the interplanar distance (d) or 2θ value of a high-angle sharp peak and then locate the phase boundary by plotting the d- or 2θ-composition curves.

4.2 Analysis of Samples by Heating and Cooling Experiments to Construct Vertical Sections and Liquid Projections (Dynamic Method)

In the dynamic method, phase equilibrium is determined when an alloy is heated and/or cooled while a certain property such as heat flow or electrical conductivity is monitored continuously to detect the temperature at which a phase transition takes place. There are two advantages over the static methods discussed before. Firstly, it is possible to quickly explore the limits of phase stability since quenching is not involved. Secondly, the possibility of a phase change occurring during quenching is avoided. The dynamic method, however, is not suitable for investigation of systems with slow phase transitions. Any physical or chemical property that changes with composition and temperature may be used as a parameter for the dynamic method to determine a phase diagram. Commonly adopted analyses include high-temperature (hot stage) metallography, high-temperature in-situ XRD analysis, dilatometry, electrical conductivity, and TA including both DTA [16] and differential scanning calorimetry (DSC) [17]. Both the DTA and heat-flux DSC methods are discussed in great detail in Chapter 5 of this book for measurements of solidus and liquidus lines; therefore, only a brief introduction is given here.

4.2.1 DTA

A phase transition usually involves an enthalpy change (evolution or absorption of heat); thus, thermal properties are commonly monitored to detect phase changes. When a specimen is heated or cooled under uniform conditions, a structural change will be identified with a temperature anomaly by plotting time vs. temperature.

Two major techniques to detect phase transitions are developed based on enthalpy change [2,18]. The first one is TA where temperature vs. time curve shows a thermal arrest at a phase transition point. The second technique is DTA in which a test sample and an inert reference sample are heated and cooled under identical conditions and a temperature difference between the test sample and reference sample are recorded. Since the signal is differential, it can be amplified with a suitable diffusion couple (DC) amplifier to increase sensitivity. As a result, DTA is more sensitive than TA. The differential temperature is then plotted either against time or against temperature. When a phase transition takes place in a sample that involves release of heat, the test sample temperature rises temporarily above that of the reference sample, resulting in an exothermic peak. Conversely, a transition accompanied by absorption of heat reduces the temperature of the test sample compared to that of the reference sample, leading to an endothermic peak. For example, in the case of a binary system, if the overall specimen composition does not vary during a heating process, a phase transition results in a slope change on the TA curve and a thermal spike on the corresponding DTA curve.

Experimental parameters that affect the DTA results include heating and cooling rate and the environment surrounding the sample. In general, the effect of lowering the heating rate is the same as that of lowering the sample weight: sharper peaks and increased resolution, but reduced sensitivity. For most applications, a heating rate of 10°C/min is adopted, since this appears to offer a compromise between quality of resolution and time taken per experiment. However, in cases where an accurate phase transition temperature is required, such as a degenerated equilibrium, it may be necessary to investigate the effect of heating (and/or cooling) rate on the DTA trace. Heating rates as low as 2°C/min are not uncommon in this type of study. For the decomposition process where two or more concurrent reactions are taking place, changing the heating rate may lead to additional information. If the kinetics of a reaction is to be investigated, it is particularly important to minimize the temperature difference across the sample. This can be achieved either by decreasing the sample weight or by lowering the heating rate. It is not uncommon to find that a change of the sample mass also affects the observed kinetics. It is, therefore, generally preferable to reduce the heating rate instead.

The atmosphere around the sample is another important factor that deserves attention. It may be either static or flowing, and may be either inert or reactive, depending on the compatibility with the process under investigation. For the determination of alloy phase diagrams, inert gas is usually chosen in order to avoid contamination, evaporation, and reaction with the environment such as oxidation. DTA is particularly useful for phase equilibria at higher temperatures (>1000°C) or in aggressive environments, where heat-flux DSC instruments may not be able to operate.

4.2.2 DSC

DSC is a TA technique that measures the energy absorbed or emitted by a sample as a function of temperature or time. When a phase transition occurs, DSC provides a direct calorimetric measurement of the transition energy at the transition temperature by subjecting the sample and an inert reference material to identical temperature regimes in an environment heated or cooled at a controlled rate. DSC equipment can be utilized not only to determine the liquidus line, solidus line, and other phase transition points on a phase diagram, but also to measure some thermodynamic parameters such as enthalpy, entropy, and specific heat, which are important for investigation of second-order phase transitions. One deficiency of DSC is that its usage temperature is usually in the range of −175–1100°C, much lower than that of DTA.

There are two types of DSC systems in common usage (Fig. 4.5) [16]. In the power-compensation DSC (Fig. 4.5(a)), the temperatures of the sample and reference are controlled independently using separate but identical furnaces. The temperatures of the sample and the reference are made identical by varying the power input to the two furnaces. The energy required to do so is a measure of the enthalpy or heat capacity changes in the sample compared to the reference.

In the heat-flux DSC, the sample and reference are connected with a low-resistance heat-flow path (a metal disk). The assembly is enclosed in a single furnace, as illustrated in Fig. 4.5(b). Enthalpy or heat capacity changes in the sample cause a difference in its temperature relative to the reference. The resulting heat flow is smaller compared to that in DTA because the sample and the reference are in good thermal contact. The temperature difference is recorded and related to enthalpy change in the sample using calibration experiments.

The heat-flux DSC is a subtle modification of DTA, differing only by the fact that the sample and reference crucibles are linked by a good heat-flow path. The sample and the reference are enclosed in the same furnace. The difference in energy required to maintain them at a nearly identical temperature is provided by the heat changes in the sample. Any excess energy is conducted between the sample and the reference

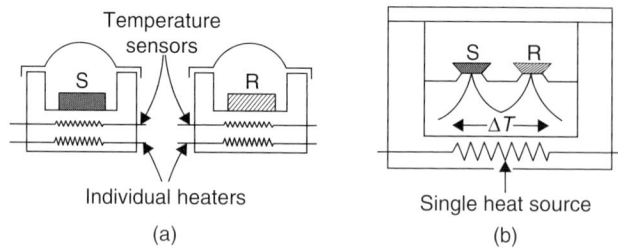

Figure 4.5 Two types of DSC equipments: (a) power-compensation DSC and (b) heat-flux DSC [16]. S: sample; R: reference material.

through the connecting metallic disk, a feature absent in DTA. As in modern DTA equipment, thermocouples are not embedded in either the specimen or reference material. The small temperature difference that may develop between the sample and the inert reference (usually an empty sample pan and lid) is proportional to the heat flow between them. The small temperature difference is important to ensure that both containers are exposed to essentially the same temperature program.

TA techniques have the advantage that only a small amount of material is needed for the measurement. This ensures uniform temperature distribution and high resolution. The sample can be encapsulated in an inert atmosphere to prevent oxidation, and low heating rates lead to higher accuracy. The reproducibility of a phase transition temperature can be checked with repeated heating and cooling through a critical temperature range.

During a first-order phase transformation, a latent heat is evolved and the transformation obeys the classical Clausius–Clapeyron equation. Second-order phase transitions do not have accompanying latent heats, but abrupt variations in compressibility, heat capacity, and coefficient of thermal expansion can be detected.

Because of the pressure sensitivity of liquid–vapor transitions, additional precautions are called for when measuring boiling points or enthalpy changes. The ambient pressure is required, and the peak area does not correspond to the latent heat of vaporization in any simple way. The transition temperature T' is related to the pressure P by the Clausius–Clapeyron equation:

$$\mathrm{Ln}(P) = \frac{L}{RT'} + C \qquad (1)$$

where L is the molar heat of vaporization and C is an integration constant. L can be obtained using the above equation and a set of measured P and T' values, assuming: (1) L is independent of temperature, (2) the volume of the vapor phase far exceeds that of the liquid, and (3) the vapor behaves like an ideal gas.

Greater care is needed when using DSC to study solid–solid phase transitions where the enthalpy changes are much smaller than those associated with vaporization. Stored energy in the form of elastic strains and defects can contribute to the energy balance. Thus, the physical state of the initial solid and the final state of the product become important. This stored energy reduces the observed enthalpy change.

4.2.3 Dilatometry

Dilatometry is another technique used to study phase transitions in alloys. This technique utilizes the change in volume associated with nearly all transitions and measures the change of length of a specimen as it is heated and cooled at a fixed rate [4].

The relationships of length vs. time and temperature vs. time are measured simultaneously, so as to plot the length vs. temperature curve. The dilatometry curve is similar to the cooling curve in the TA, and the phase transition temperature can be determined. The dilatometry curve of a carbon steel during heating and cooling is shown in Fig. 4.6. It is clear that during the heating process, the volume shrinks as the sample transform from the α phase (bcc) to the γ phase (fcc), and the volume increases when cooled.

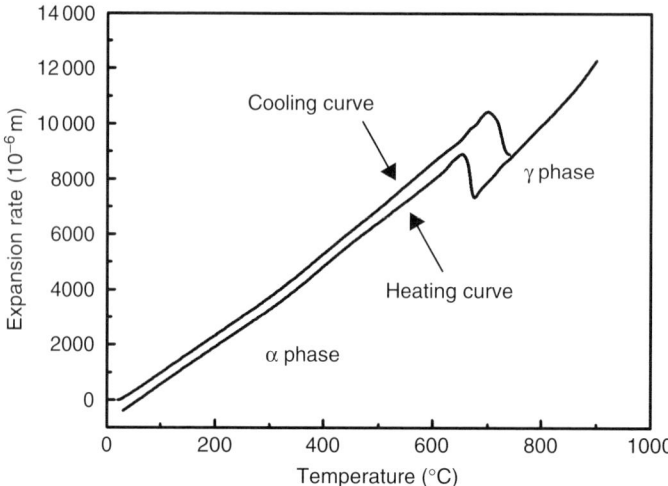

Figure 4.6 Variation of expansion rate as a function of temperature for a carbon steel (C: 0.59%, P: 0.024%, Mn: 0.92%, S: 0.033%, Si: 0.25%; in wt.%).

A dilatometer consists of a sensing element such as transducer, variable capacitor, and dial gauge which is activated by the specimen positioned in the furnace. Changes in length are transmitted to the sensor by means of a push rod. The dilatometer for automatically recording thermal expansion studies has been described elsewhere [17]. The specimen is enclosed in a thermal mantle, which is heated or cooled by passage of gas. Temperatures are measured by a thermocouple and plotted on one axis of an X–Y recorder, and changes in length are recorded simultaneously.

The dilatometry method is an important dynamic technique with high sensitivity. It has been used to investigate several phase transitions that are difficult to be observed by TA. When studying solid–state phase changes with TA, the hysteresis effect may result in a difference of phase transition temperatures (T_p) of the two processes, for example, T_p in heating process is much higher than that in cooling process. This problem can be easily solved by using dilatometry method, in which the samples may be kept at any temperature for a time long enough to reach phase equilibrium. Since the heating or cooling rate of this method can be much lower than that of the TA methods, the hysteresis effect is significantly reduced. This advantage is more obvious in the determination of solid-state reactions. It should be pointed out that the samples for dilatometry measurements should have a uniform composition and accurate dimensions.

4.2.4 Electrical Conductivity

Measurement of electrical conductivity is a useful auxiliary technique for the determination of alloy phase diagrams. The basic procedure to determine phase regions is to plot the electrical conductivity against composition or to plot the electrical conductivity against temperature for a fixed composition. To investigate subsolidus equilibria, this method appears to have advantages over DTA. The measurements do not require spontaneous heat effects and can therefore be observed during arbitrarily slow rates of temperature change or upon allowance of sufficient time for

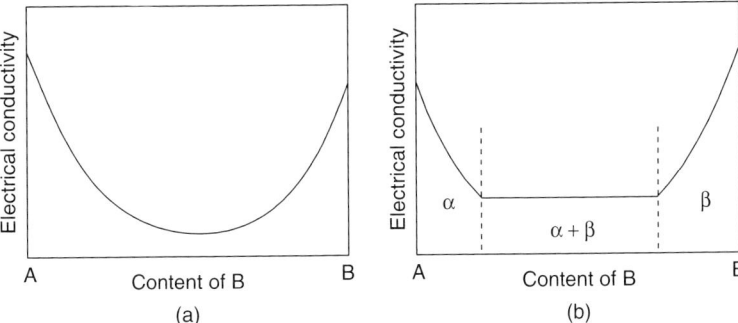

Figure 4.7 Variation of electrical conductivity with composition for (a) a continuous solid solution and (b) a binary eutectic system.

equilibrium. At subsolidus temperatures, formation of a new phase will, in most cases, be evidenced by a change of slope in the conductivity vs. temperature curve. In cases of polymorphic transitions, a number of effects are possible depending upon the type of transition.

For solid solutions, electrical conductivity often decreases markedly due to the addition of a small amount of solute element. For systems that form a continuous solid solution, the electrical conductivity vs. composition curve usually exhibits a "U" shape, as shown in Fig. 4.7(a). The minimum electrical conductivity of solid solution is much lower than that of the pure metals. For an alloy system where there are two terminal solid solutions, the electrical conductivity vs. composition curve exhibits a shape similar to that of Fig. 4.7(b). The decreasing parts represent solid solutions while the near flat part is the two-phase region.

In binary alloy systems, the electrical conductivity of an intermetallic phase may be a little lower than one or both the pure metals. If the intermetallic phase is stoichiometric, its stoichiometry should be marked by the intersection of two approximate lines, as illustrated in Fig. 4.8(a) [4]. On both sides of the intermetallic phase are two two-phase regions. If the intermetallic phase has a wide homogeneity range and it is an ordered structure, a cuspidal point close to the stoichiometric composition may occur on the electrical conductivity vs. composition curve, as shown in Fig. 4.8(b). If the entire single-phase region is away from a stoichiometric composition, then the electrical conductivity may change monotonically (Fig. 4.8(c)). However, one cannot conclude that there must exist an intermetallic phase when one of the above relationships is observed. When a superstructure is formed in a solid solution at low temperatures, the electrical conductivity may also increase dramatically due to less electron scattering.

To measure electrical conductivity, the dimension, and shape of a sample should be considered. It is known that the measurement result is greatly influenced by defects such as shrinkage cavities, bubbles, and inclusions, so it is difficult to prepare satisfactory samples for a phase diagram determination. Temperature coefficient of electrical conductivity (TCEC) is usually measured to avoid these problems. This method is developed based on the Matthiessen's rule on the electrical resistance of solid solutions. Experimental results have shown that the TCEC vs. composition

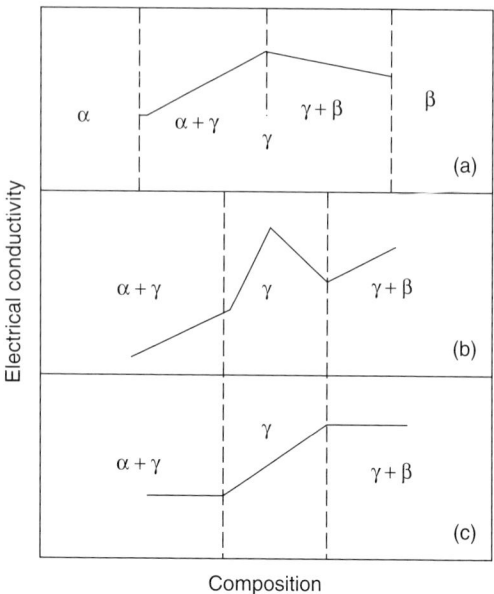

Figure 4.8 Different kinds of electrical conductivity vs. composition curve for binary systems with intermetallic phase γ [4]: (a) stoichiometric compound (γ), (b) ordered compound γ with wide homogeneity range, and (c) disordered compound with wide homogeneity range.

curve is very similar to the electrical conductivity vs. composition curve, so it can be used to determine a phase transition.

Conductivity measurements can be made in a number of ways. Since absolute values of conductivity are not necessary, specimens can be recorded in any manner that assures an ohmic contact and constant geometry throughout the temperature range to be investigated. In the case of metals, wires, or rods can be fabricated and welded to noble-metal leads. A variety of measurement techniques are possible utilizing either direct or alternating current. In most cases, however, a simple measurement of voltage drops across a resistor in series with the sample and power source is adequate. The voltage drop can be recorded simultaneously with temperature using an X–Y recorder. Temperatures are cycled slowly through the region of interest generally at 0.5°C/min or less. It is also possible to heat or cool at a faster rate to perform DTA and conductivity measurements simultaneously.

At high pressures where the use of dynamic technique is a matter of necessity rather than convenience, resistance measurements have found wide applications. Like other measurements at high temperatures, the effect of pressure on resistance is by no means simple. Resistance changes exhibited by metals at phase transitions induced by pressure have been used as a pressure calibration tool.

4.2.5 Magnetic Measurement

Magnetic analysis is another technique for the determination of an alloy phase diagram. Curie temperature (T_c) and saturation magnetization (σ) are two important parameters usually used for this method. Since both T_c and σ are only affected by the

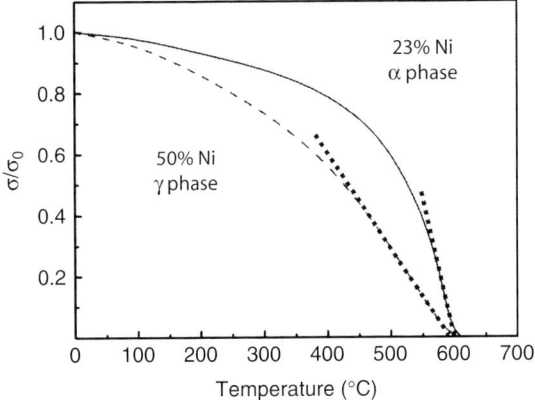

Figure 4.9 The σ–T curve of two Fe–Ni alloys.

atomic arrangement but not by the imperfection of crystal lattice, their values depend essentially only on an alloy composition and the corresponding phase compositions. By measuring these magnetic properties, phase relationships of alloy systems that include ferromagnetic components or compounds can be determined. A more detailed description of magnetic measurements is given in Chapter 12 of this book.

The fundamental assumption of this method is that the saturation magnetization is a function of temperature, and the total value of σ is the linear superposition of each single phase. For a single-phase alloy, there is only one value of T_c, which can be determined by extrapolating the steepest tangent to the temperature axis, as shown in Fig. 4.9. The σ–T curve of an alloy that consists of two ferromagnetic phases is a superposed effect of the two phases. The mass ratio of these two phases is equal to the ratio of their σ values.

The magnetic analysis method is based on three fundamental conditions. Firstly, the composition of the sample is known. Secondly, the σ(T) and T_c values of all the phases in the sample should be known. Thirdly, the phase distribution in the sample does not change during the measurement. Thus, for an alloy system with n phases among which m ($m \leq n-1$) phases are ferromagnetic, the volume fractions of these phases can be accurately determined from σ(T) and T_c.

The shape of σ–T curve is often used for phase detection, as shown in Fig. 4.9. The two σ–T curves correspond to 23% Ni (bcc) and 50% Ni (fcc) alloys in the Fe–Ni system, respectively. It is clear that their Curie temperatures are approximately the same, but the shapes of the σ–T curves are quite different. Thus, the characteristics of the phases can be identified from these curves.

5 EXAMPLES OF PHASE DIAGRAM DETERMINATION USING EQUILIBRATED ALLOYS

In this section, several phase diagrams are used as examples to show step by step how they were determined. We start with the Cu–Nd binary example, followed by a simple ternary Al–Be–Si system and a very complex ternary system

5.1 The Cu–Nd Binary System

The first example is the Cu–Nd system (Fig. 4.10). The crystal structures of the binary compounds (NdCu, $NdCu_2$, $NdCu_4$, $NdCu_5$, and $NdCu_6$) were determined by several groups [18–21]. XRD, DTA, and metallography were combined to establish this binary phase diagram.

The starting materials were 99.9% purity Nd and 99.999% purity Cu chips. They were cleaned in acetone, flushed with water and then dried. The Cu chips were further immersed in dilute sulfuric acid to remove the oxides on the surface of the chips. The binary alloys were prepared in a high-frequency induction furnace under argon shield with pressure slightly higher than one atmosphere. Aluminum oxide crucibles with caps were used as the container. For the Nd-rich samples (⩽50 wt.% Cu), the melting time was about 4 min, and for the other alloys 4–15 min were required, depending on their compositions. The melting loss was less than 0.6% for all alloys. The as-cast alloys were sealed in evacuated quartz tubes and then subjected to homogenization heat treatments. The samples with 0–51 at.%

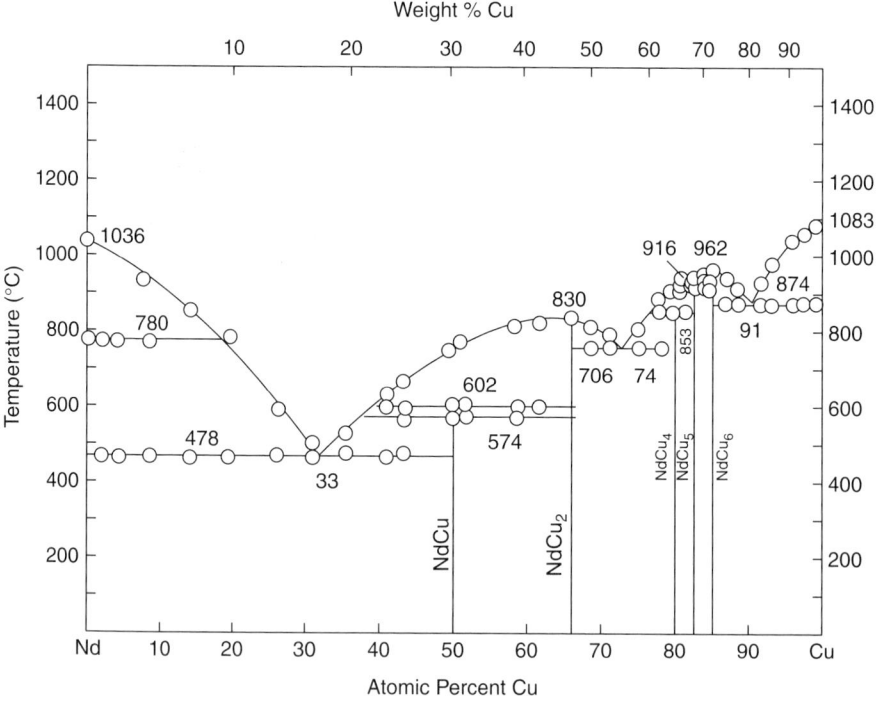

Figure 4.10 Cu–Nd phase diagram based on results from XRD, DTA, and metallography [22].

Cu were annealed at 410°C for 42 days, while those with 51–67 at.% Cu were annealed at 500°C for 50 days. The other samples were homogenized at 720°C for 42 days. The annealed samples were slowly cooled to room temperature. Chemical analysis was performed on the homogenized alloys to determine the deviation from nominal compositions.

The surfaces of the annealed samples were slightly ground and polished before being ground into powders. The powders were sieved with 325 meshes and then encapsulated in evacuated glass tubes for residual stress relief treatment which was performed at 410°C for 5 days for alloys with Cu <51 at.% and 500°C for 3 days for the other alloys. The annealed powders were cooled to room temperature at a rate of 10°C/h.

In order to obtain phase distribution data at elevated temperatures, bulk samples of alloys with composition between Nd and NdCu were encapsulated with Ta foils, sealed in evacuated quartz tubes and then annealed at elevated temperatures for 20 h, followed by water quenching. They were then ground into powders and subjected to the same treatment as described above for XRD analysis.

One rule for selecting XRD radiation source is that the characteristic X-ray mass absorption coefficient of the sample should be as low as possible in order to increase the peak to background ratio. For the alloys with Cu content above 50 at.%, XRD was performed using CuKα radiation with 26 kV and 22 mA. For the other alloys, better XRD patterns were obtained by using FeKα radiation. The phase boundaries were determined with the phase disappearance method. XRD results show that the peak positions of XRD patterns of two-phase alloys coincide with those of the adjacent single phases, that is, no peak shift was observed. Thus all the binary compounds are essentially stoichiometric ones. The lattice parameter measurements also indicated that the mutual solid solubility between Cu and Nd was negligible.

Phase transition temperatures were measured using DTA with Al_2O_3 crucibles as container under argon atmosphere. Cu, Ag, Sb, Al, and Zn were used for temperature calibration. The detail for the DTA measurements have been described elsewhere [22], but the results are summarized in Table 4.1 together with the XRD results.

Microstructure examination of several as-cast samples (especially those having nominal composition close to eutectic or peritectic point) was performed. The results indicate that there are three eutectic reactions and three peritectic reactions. The OM observations were consistent with those of DTA measurements.

Based on the experimental results from XRD, DTA, and metallography, the Cu–Nd phase diagram is established as shown in Fig. 4.10.

5.2 The Al–Be–Si Ternary System

Both the Al–Si and Al–Be binary systems are well established in the literature [12]. The Al–Si is of a simple eutectic type with very little solubility of Al in Si, but some solubility of Si in Al. The Al–Be is also of a simple eutectic type with essentially no solubility between the elements. The Be–Si binary system was recently determined by Pan et al. (Fig. 4.11(c)) [23]. They arc melted six alloys from high-purity Be and Si pieces, homogenized them at 1050 ± 1°C for 7 days, and then quenched them into water or heavy oil. OM, XRD, SEM/EDX, and DTA were employed to

Table 4.1 Summary of DTA and XRD measurements in the Cu–Nd binary system.

Composition at.% Cu	Phase at room temperature	Phase transition temperature from DTA (°C)								
		1	2	3	4	5	6	7	8	9
2.24	Nd + NdCu	470			778					
4.43		471			782					
8.64	Nd + NdCu	480			778					920
14.59		475								864
20.14	Nd + NdCu	469								790
26.98		482								600
31.73		474								510
36.20	Nd + NdCu	486								534
41.75		477		600						640
44.36	Nd + NdCu	485	568	602						670
50.00	NdCu		572	604						754
51.64	NdCu + NdCu$_2$		576	614						777
59.20			574	607						813
62.17	NdCu + NdCu$_2$			602						818
66.67	NdCu$_2$									830
69.41	NdCu$_2$ + NdCu$_4$					758				813
71.90						764				800
75.81	NdCu$_2$ + NdCu$_4$					759				810
78.74	NdCu$_2$ + NdCu$_4$					762	846			890
80.00	NdCu$_4$						853			900
81.50	NdCu$_4$ + NdCu$_5$						853		908	928
83.33	NdCu$_5$								916	936
84.12	NdCu$_5$ + NdCu$_6$								920	940
84.75	NdCu$_5$ + NdCu$_6$								918	953
85.71	NdCu$_6$									962

(*Continued*)

Table 4.1 (Continued)

Composition at.% Cu	Phase at room temperture	Phase transition temperature from DTA (°C)								
		1	2	3	4	5	6	7	8	9
87.78	$NdCu_6$ + Cu							876		944
88.95	$NdCu_6$ + Cu							872		918
92.26								872		932
93.82	$NdCu_6$ + Cu							874		980
96.79								876		1044
98.20								874		1068
99.56	$NdCu_6$ + Cu							880		1082

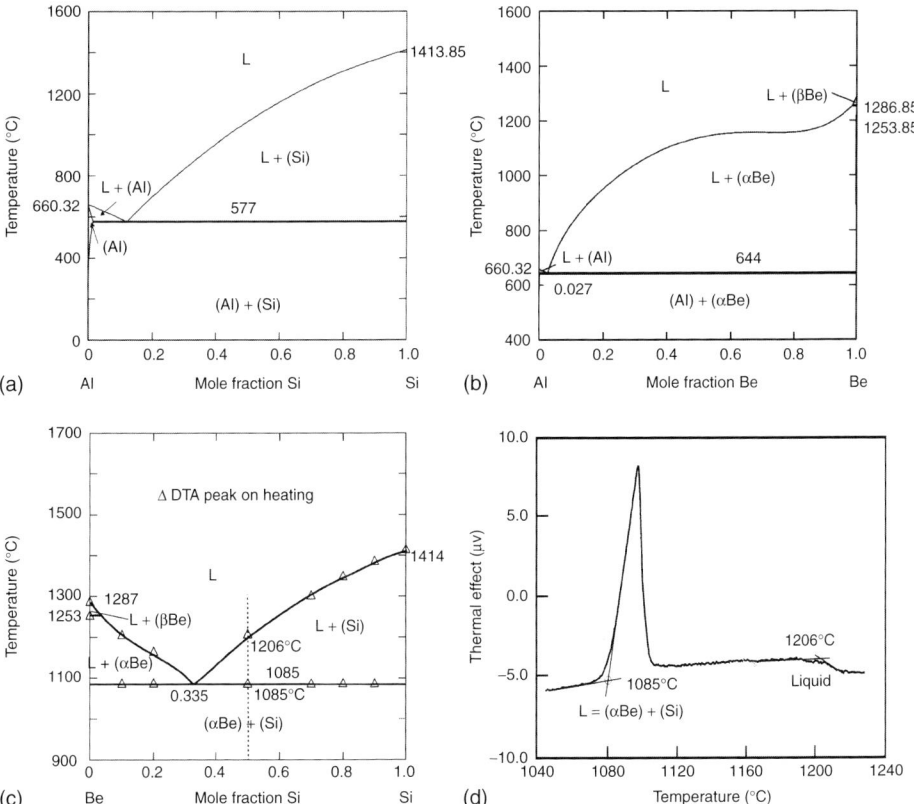

Figure 4.11 Phase diagram of the (a) Al–Si, (b) Al–Be, and (c) Be–Si binary system. The DTA heating curve (heating rate: 5K/min) of a $Be_{50}Si_{50}$ alloy (at.%) annealed at 1050°C for 7 days is shown in Fig. 4.11(c).

characterize the six alloys. The XRD and SEM analyses showed no new phases and the lattice parameters of Be and Si in all alloys were essentially the same as those of pure Be and pure Si, indicating essentially no solubility between Be and Si. The microstructure from SEM examination clearly indicated that the Be–Si binary system is of a simple eutectic type. About 0.1 g of each alloy was taken from the samples homogenized at 1050 ± 1°C for 7 days to perform DTA analysis. The homogenized samples instead of the as-cast ones are used for DTA measurement since the former yields the equilibrium state. The measurement was carried out in argon atmosphere between room temperature and 1500°C with a heating and cooling rate of 5°C/min. The temperature was measured with Pt–Pt/Rh thermocouples and calibrated to the melting temperatures of a few standards, Al (660.32°C), Au (1064.18°C), and Si (1413.85°C). These standards were chosen to be widely distributed in the temperature range of interest. The accuracy of the temperature measurement was estimated to be ±2°C based on the calibration curve. Figure 4.11(d) shows the DTA curve for the alloy $Be_{50}Si_{50}$ (at.%). The phase transition temperatures are marked in Fig. 4.11(c). The onset of the first DTA peak corresponds to the invariant reaction temperature, and the liquidus temperature was taken to be the peak maximum of the last DTA peak. The phase transition temperatures were taken from the heating curves since reactions between samples and Al_2O_3 crucible material could not be completely excluded and supercooling phenomenon usually accompanies solidification/cooling process. Thus, the heating

Table 4.2 Summary of the phases and DTA peaks for the Al–Be–Si alloys annealed at 530°C for 8 days.

No	at.% Al	at.% Be	at.% Si	Phase[a]	DTA signal (°C)[b]
1	50	40	10	(Al) + (αBe) + (Si)	571, 634, 1104
2	50	35	15	(Al) + (αBe) + (Si)	571, 706, 1071
3	50	30	20	(Al) + (αBe) + (Si)	571, 768, 1028
4	50	16	34	(Al) + (αBe) + (Si)	570, 870
5	50	10	40	(Al) + (αBe) + (Si)	571, 817, 944
6	74.69	22.87	2.44	(Al) + (αBe) + (Si)	571, 633, 1014
7	78.76	16.08	5.16	(Al) + (αBe) + (Si)	571, 608, 911
8	87.62	10.32	2.06	(Al) + (αBe) + (Si)	572, 638, 817
9	89.48	7.41	3.11	(Al) + (αBe) + (Si)	571, 631, 784
10	90.53	5.77	3.7	(Al) + (αBe) + (Si)	571, 620
11	86.81	5.78	7.41	(Al) + (αBe) + (Si)	571, 605

[a] Identified with XRD, metallography, and SEM/EDX methods.
[b] Obtained on heating with a heating rate of 5K/min.

data are usually more reliable. The results from the six alloys were compiled into the phase diagram shown in Fig. 4.11(c), clearly showing a simple eutectic type phase diagram with a degenerated feature among liquid, (αBe) and (βBe) phases.

With all the three binary phase diagrams established, the next step in determining the Al–Be–Si ternary system is to construct isothermal sections, isopleths (vertical section), and to investigate whether there are any ternary compounds/phases [24]. An isothermal section at 530°C was established by studying 11 ternary alloys listed in Table 4.2 using XRD and SEM/EDX. XRD examinations of both as-cast and annealed samples showed that there is no ternary compound in the Al–Be–Si system, since all XRD patterns can be indexed to (Al), (αBe), and Si. The lattice parameters measured from the ternary alloys agreed well with those of the unary and binary (only in the case of Al-based fcc solid solution with Si) phases, thus, the mutual solubility among the pure elements in the ternary system is essentially negligible at 530°C. Figure 4.12(a) shows the experimental isothermal section at 530°C in which the dominating phase equilibrium is the three-phase region: (Al) + (Si) + (αBe).

For ternary systems, it is usually necessary to establish typical isopleths in order to gain better understanding of the phase relationship, especially the invariant reaction temperatures. A vertical section at 50 at.% Al of the simple Al–Be–Si ternary system was determined using DTA supplemented with XRD and SEM/EDX analysis (Fig. 4.12(b)).

Based on three binary phase diagrams, the isothermal section at 530°C, and the vertical section at 50 at.% Al, a reaction scheme for the Al–Be–Si system is constructed in Fig. 4.12(c). Such a reaction scheme is a very useful tool to describe phase equilibria in multicomponent systems. For example, one can draw a draft isothermal section at any temperature by just looking at the reaction scheme.

5.3 The Al–Mn–Si Ternary System

The measurement of this complex system is very difficult mainly due to the existence of 10 ternary compounds, the effects of metastability, incomplete reactions, as well as substantial undercooling. Krendelsberger et al. [25] made major contributions to the understanding of the various phase equilibria in this ternary system. Their procedure for sample preparation, phase identification and phase transition temperature measurement is similar to that employed for the Be–Si system. Figure 4.13 shows the measured isothermal section at 550°C. At this temperature, nine ternary phases and several extended solid solutions of binary phases occur. The phases are identified using XRD technique. Phase compositions are measured by means of EPMA.

The liquidus surface was determined by metallographic inspection of the as-cast alloys combined with DTA data of the alloys equilibrated at 550°C. In the first step, the primary phase in the as-solidified microstructure was identified with EPMA or SEM/EDX technique. These analyses yield a preliminary map of the areas for primary crystallization of the ternary phases as well as the binary phases. In the second step, the thermal arrests in the DTA heating curves of the alloys equilibrated at 550°C were recorded. The first onsets of these arrests are interpreted to correspond to invariant reactions among three solid phases observed in the annealed alloy plus either liquid

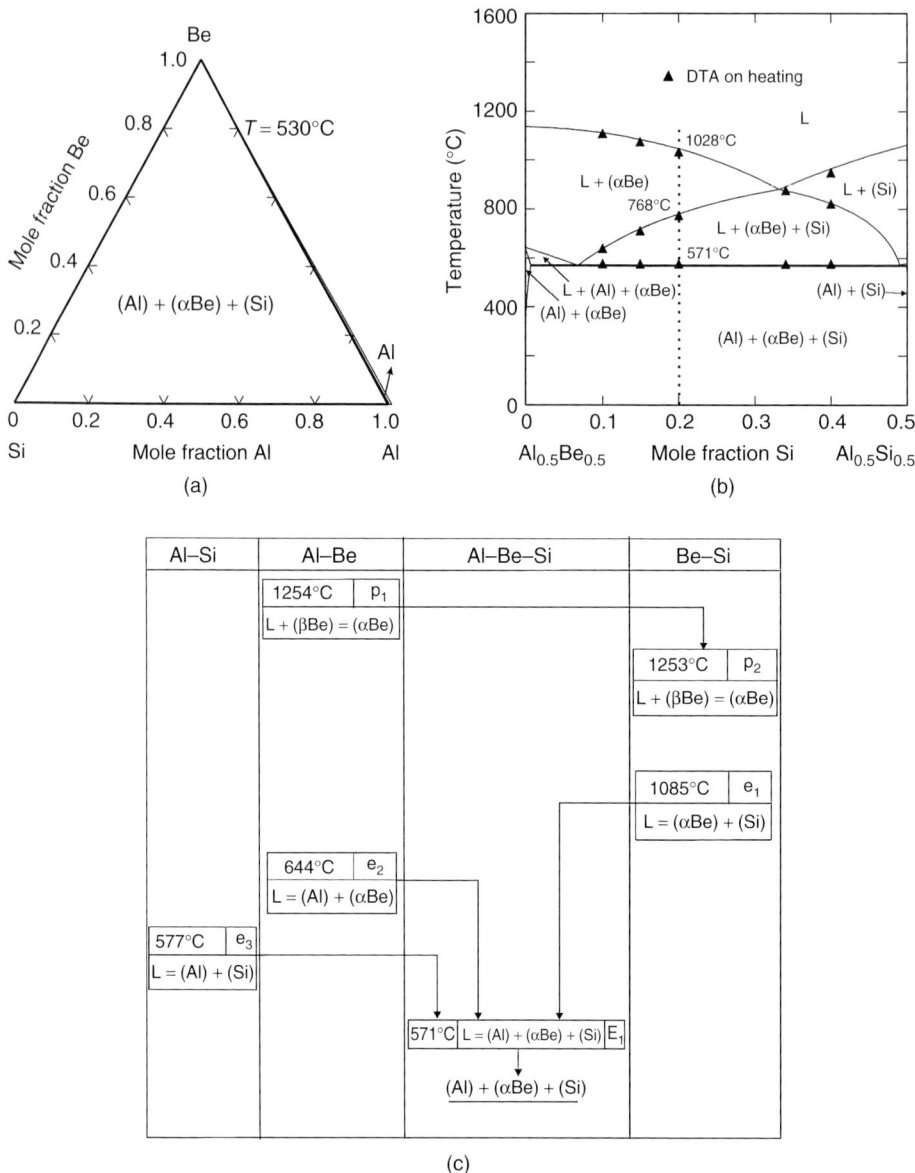

Figure 4.12 The Al–Be–Si system: (a) 530°C isothermal section; (b) vertical section at 50 at.% Al (the phase transition temperatures corresponding to the alloy $Al_{0.5}Be_{0.3}Si_{0.2}$ are indicated in the diagram); and (c) reaction scheme.

phase or an additional solid phase. In general, the thermal effect resulting from an invariant reaction in solid state is much smaller than that from an invariant reaction involving the liquid phase. The peak maximum of the last thermal arrest could be taken to be the liquidus temperature. It is usually time-consuming to assign the other thermal effects to particular phase boundaries since quenching above and

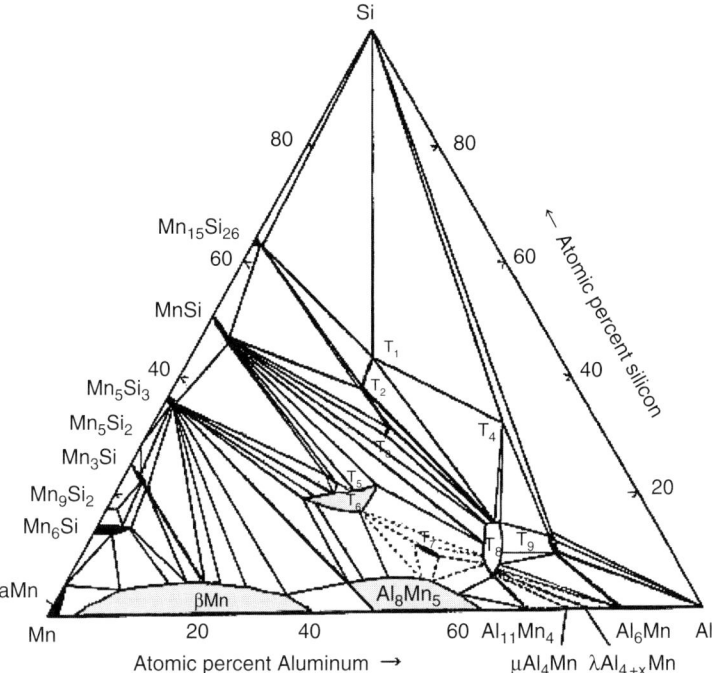

Figure 4.13 Isothermal section at 550°C in the Al–Mn–Si system.

below these thermal effects is needed to locate these boundaries accurately. In the third step, individual analysis and presentation of each invariant reaction observed resulted in refinement of the compositions of the liquid phase participating in these reactions. Figure 4.14 presents the experimental liquidus projection of the Al–Mn–Si system. Interested readers are referred to the original work of Krendelsberger et al. [25] for details.

5.4 Multicomponent Phase Diagram and the Al–C–Si–Ti System

From the viewpoint of practical applications, multicomponent phase diagrams are very important since nearly all current commercial alloys contain multiple elements; many exceeding 10 elements. Due to the complex topology of phase relations, up to now few quaternary systems have been completely investigated. To show the composition of a quaternary alloy system, the compositional percentage of three components must be plotted. When pressure remains constant, four-dimensional (4-D) diagrams are needed to represent the temperature–composition equilibrium diagram of a quaternary system. Since it is difficult to image such 4-D diagrams, three-dimensional (3-D) diagrams such as isothermal and isobaric sections are usually used. In these iso-sections, temperature and one or two compositional variables are shown in two-dimensional or 3-D space.

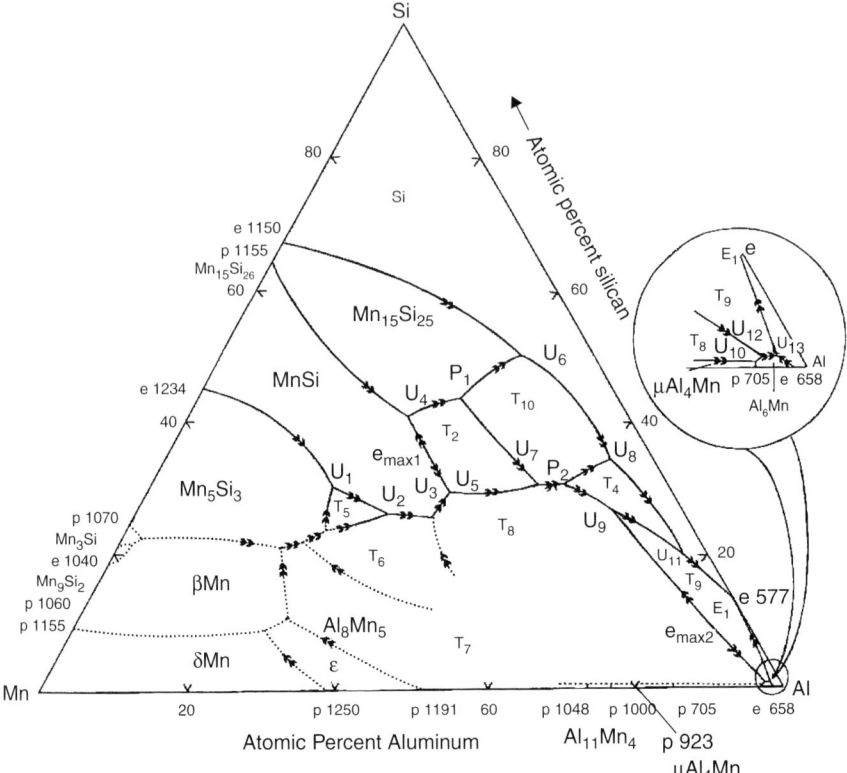

Figure 4.14 Liquidus projection of the Al–Mn–Si system.

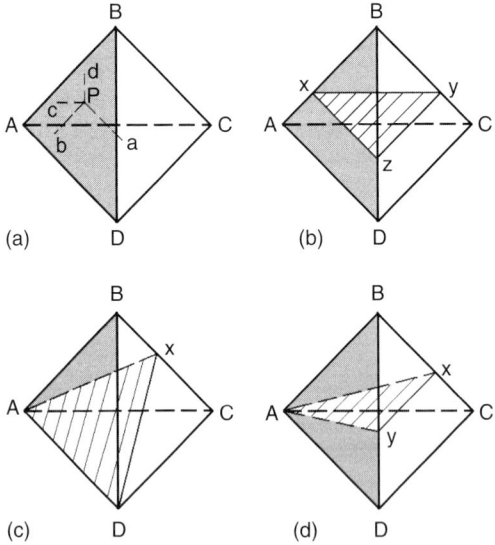

Figure 4.15 Schematic representation of a quaternary phase diagram: (a) equilateral tetrahedron illustrating the representation of compositions in a quaternary system; (b) a section with a constant concentration of B; (c) a section with a constant composition ratio of B and C; and (d) a section with a constant composition ratio of B with both C and D.

In a 3-D diagram, an equilateral tetrahedron can be used to represent three compositional variables. When four lines are drawn across an arbitrary point in the tetrahedron, they are intercepted by the opposite triangle plane. It is clear that the total length of these lines equals to the side length of the regular tetrahedron. As illustrated in Fig. 4.15(a), four lines run across point "P" in the regular tetrahedron ABCD. Line "Pa" parallel to "AD" and is intercepted by plane "BCD", "Pb" parallel to "AB" and is intercepted by plane "ACD", "Pc" parallel to "AC" and is intercepted by plane "ABD", "Pd" parallel to "BD" and intercepted by plane "ABC". The overall length of these lines equals to each side of the tetrahedron. Therefore, if the length of side is 100%, then the composition of point "P" is: %A = Pa, %B = Pb, %C = Pc, and %D = Pd. The pure components, binary, ternary, and quaternary compositions are represented by four apexes, six edges, four triangle planes, and points in the regular tetrahedron, respectively.

To investigate alloys with a specific composition, temperature is considered as variable and a series of compositional sections are obtained. As shown in Fig. 4.15(b), the plane triangle "xyz" is a section with a constant content of component B. If it is the basal face of the triangular prism, the coordinate system used for ternary systems can be adopted in this case.

Another kind of section is to fix the constituent ratio of two components. This is shown as plane ADx in Fig. 4.15(c), where the compositional ratio of B and C is constant. This kind of section represents the pseudo-ternary equilibrium that consists of an intermetallic phase (x) and two pure components (A and D). The plane Axy in Fig. 4.15(d) indicates the case where compositional ratio of one component with the other two remains constant. A pseudo-ternary system is composed of two intermetallic phases and one pure component in such a section.

Determination of a quaternary phase diagram is similar to that of binary and ternary systems. The basic methods still consist of TA, metallography, XRD, and other physical–chemical analysis. An example discussed here is the determination of the 1000°C isothermal section of the Al–C–Si–Ti quaternary system [26]. The first step was to prepare the equilibrated alloys with a powder metallurgy method. Powder mixtures with compositions lying mainly in the two-dimensional sections Al_3Ti–Si–C and Ti_3Al–Si–C were prepared from commercial powders of Al (purity > 99 wt.%, particle size $d < 150\,\mu m$), carbon (spectrographic grade graphite, $d \approx 1\,\mu m$), Si (purity > 99 wt.%, $d < 50\,\mu m$), α-hexagonal SiC (purity > 99.8 wt.%, $d < 50\,\mu m$), and cubic TiC with a C:Ti atomic ratio of 0.98 (purity > 99 wt.%, $d < 20\,\mu m$). The mixtures were ball mixed for 10 min in a tungsten carbide mortar, cold-pressed under 240 MPa into small parallel rods weighing about 2.5 g (30 × 4 × 6 mm), placed on alumina boats, and loaded into silica tubes. The silica tubes were sealed under approximately 0.3 atm of purified argon and heated firstly at 700 ± 5°C for 24 h and then at 1000 ± 5°C for 400 h using a conventional horizontal tube furnace. At the end of the isothermal treatment at 1000°C, the silica tube was pushed out of the furnace and rapidly cooled in air. Preheating for 24 h at 700°C was necessary to avoid the selective vaporization of Al during the early stage of the reaction. Moreover, a heating time of 400 h appeared to be long enough for the completion of the reactions without excessive Al vaporization. Finally, although the cooling rate was not very fast, it was sufficient to freeze equilibria established during isothermal annealing at 1000°C.

Different techniques were used to characterize the homogenized alloys. In the first step, one face of the treated sample was roughly polished and characterized using XRD. Some samples were diamond polished to a finish of 1 μm and examined by OM and SEM. Further characterization of the observed phases was performed using EPMA. For each phase, at least 10 different points were analyzed and averaged for Al, Si, and Ti concentrations with an estimated accuracy within ±1 at.%. After having verified that no foreign impurities were detectable in the fluorescence spectra of the analyzed sample, the carbon content was determined by difference with an estimated accuracy of ±3 at.%. Analyses of the Al–Si alloys resulting from the solidification of the liquid phase were performed using either EPMA or XRD. For hypoeutectic or near-eutectic alloys (less than 15 at.% Si), the composition was determined using EPMA; and for hypereutectic alloys the Si content was estimated by comparing the intensities of the XRD peaks characteristic for Al and Si in the samples with those in the cast Al–Si standards. Table 4.3 shows the phases and their compositions determined with XRD and EPMA methods for some of the samples.

The last step was to construct the 1000°C isotherm using the classical equilateral tetrahedron representation. Representative two-dimensional sections through this tetrahedron are shown in Fig. 4.16.

6 CRYSTAL STRUCTURE IDENTIFICATION OF NEW PHASES

One of the key functions of the experimental determination of phase diagrams is to find and identify new phases. The CALPHAD approach works reasonably well to predict ternary and higher-order equilibria when the following conditions are met: (1) all the related binary systems are well assessed; (2) there are no ternary or higher-order phases present; and (3) the solubility of a third or fourth element in binary compounds are limited. The process breaks down when new phases appear. It is up to the experimental work (and potentially combined with ab initio calculations such as density functional theory (DFT) calculations) to find and identify new phases. Identification of a new phase involves the study of: (1) the stoichiometry or composition range of the phase, (2) the exact space group of the crystal structure, (3) site occupancy of the elements in the crystal structure, and (4) whether the phase belongs to one of the binary phases or an ordered variant of a binary phase. In this section, we will use an example to describe the common procedure to identify crystal structures using XRD.

When the Cu–Ho–Sb ternary phase diagram was systematically investigated, a ternary compound Cu_5Ho_7Sb was found. Its crystal structure was determined by means of X-ray powder diffraction [27]. The sample investigated was prepared by arc melting pieces of the constituent metals in a high-purity argon atmosphere. The purities of the starting elements were 99.999% Cu, 99.9% Ho, and 99.9% Sb. To ensure homogeneity, the sample was melted three times using a low electric current to minimize the vaporization losses of Sb. The weight loss of the sample during melting was less than 0.6%, thus no chemical analysis was carried out after melting. The sample was enclosed with Ta foil, sealed in an evacuated quartz tube and then annealed at 700°C for 30 days. The sample was subsequently cooled to 500°C with a cooling rate of 10°C/h

Table 4.3 Phases and their compositions for the samples annealed at 1000°C for 400 h in the Al_3Ti–Si–C triangle.

No.	Initial mixture composition (at.%)				Mass loss during preparation	Phases (by XRD)	Phase composition (at.%) (by EPMA)			
	0.75Al–0.25Ti	0.5SiC	Si	C			C*	Al	Si	Ti
17	82	10	0	8		(Al)	ND	97.7	≈0.8	≈1.5
18	84	6.4	0	9.6	1.5	Ti_3SiC_2	31.0	6.5	12.3	50.2
						TiC	51	ND	ND	49
						Al_3Ti (blocks)	ND	70.8	4.4	24.8
19	78	10	12	0	0.7	L = Al + Si	ND	90.5	8	≈1.5
20	79	10	11	0	1.0	Ti_3SiC_2	32.2	1.5	16.6	49.7
						Al_3Ti (blocks)	ND	63.2	12.4	24.4
						$Ti_5Si_3C_x$	10.3	7.7	34.5	47.5
21	74	10	16	0	0.9	L = Al + Si	ND	88.5	10.5	≈1.5
						$Ti(Al,Si)_2$	ND	14.5	52.4	33.1
						Ti_3SiC_2	36.7	ND	15.8	47.5
						$Ti_5Si_3C_x$	10.3	7.7	34.5	47.5
22	78	10	0	12	1	L = Al + Si	ND	96.5	≈2	≈1.5
						TiC	47.9	ND	ND	52.1

(*Continued*)

Table 4.3 Phases and their compositions for the samples annealed at 1000 °C for 400 h in the Al$_3$Ti–Si–C triangle.

No.	Initial mixture composition (at.%)			Mass loss during preparation	Phases (by XRD)	Phase composition (at.%) (by EPMA)			
	0.75Al–0.25Ti	0.5SiC	Si	C			C*	Al	Si	Ti
						Ti$_3$SiC$_2$	31.2	2.7	15.5	50.6
						Al$_3$Ti (platelets)	ND	66.2	8.6	25.2
23	96	4	0	0	0.4	(Al)	ND	ND	ND	ND
24	88	0	2	10	2.6	Al$_3$Ti (blocks)	ND	ND	ND	ND
						TiC	ND	ND	ND	ND
25	80	20	0	0	1.5	(Al)	ND	96.2	≈2	≈1.5
						Ti$_3$SiC$_2$	29.8	3.6	16.2	50.4
						Al$_3$Ti (blocks)	ND	64.8	9.9	25.3
26	70	15	15	0		(Al)	ND	85.5	13	≈1.5
						Ti(Al,Si)$_2$	ND	13.1	54.2	32.7
						Ti$_3$SiC$_2$	33.6	1.0	16.8	48.6

ND: not determined
★ Estimated accuracy of ±3 at.%

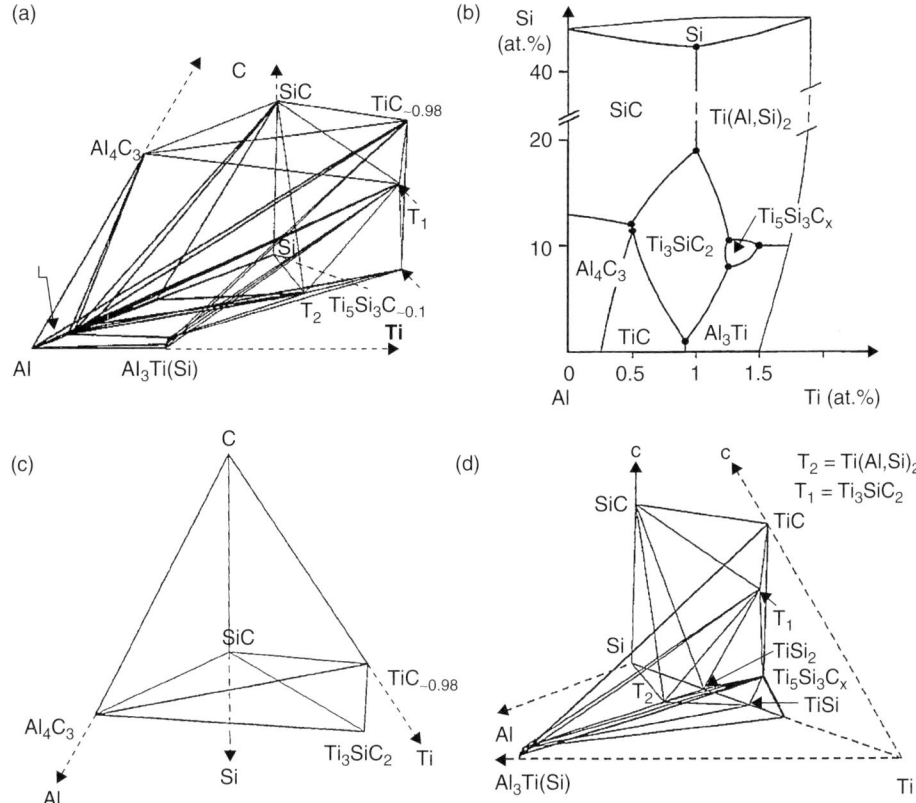

Figure 4.16 Partial view in true perspective of the 3-D 1000°C Al–C–Si–Ti isotherm: (a) solid–liquid phase equilibria; (b) characteristic lines of the liquidus surface: conic projection from the C apex onto the Al–Ti–Si plane; (c) solid–solid phase equilibria in the C-rich corner; and (d) solid–solid phase equilibria in the Ti-rich corner.

and annealed at this temperature for 1 week, followed by water quenching. The X-ray powder diffraction data used for the crystal structure determination was collected on a powder diffractometer using CuKα radiation with 40 kV and 200 mA. The 2θ scan range was from 15° to 100° with a step size of 0.02° and a count time of 2 s per step.

All reflections for Cu_5Ho_7Sb are well indexed using the JADE 5.0 program [28] in a tetragonal unit cell with the lattice parameters $a = 0.62008(3)$ nm, $c = 0.87807(4)$ nm. Each unit cell contains 13 atoms based on the lattice parameters and the density (7.81 g/cm^3) was measured by the specific gravity bottle method. The space group was determined to be P4/nbm (No.125) according to the reflection conditions (no reflection: $hk0$: $h + k = 2n$ and $0kl$: $k = 2n$). The refinement of the structure was performed using DBWS 9411 program [29]. The pseudo-Voight function was used for the simulation of the peak shapes. The DMPLOT plot view program [30] for Rietveld refinement was used to demonstrate the refinement results. A total of 23 parameters were adjusted during the crystal structure measurement of Cu_5Ho_7Sb. Details of the refinement are summarized in Table 4.4. Table 4.5 presents the atomic coordinates, occupancy and thermal parameters. Beq is the isotropic atom displacement

Table 4.4 Rietveld refinement data of Cu_5Ho_7Sb.

Fomula	Cu_5Ho_7Sb
Space group	P4/nbm
Radiation wavelength CuKα (nm)	0.154178
Lattice parameters (nm)	
a	0.62008(3)
c	0.87807(4)
Unit cell volume (nm³)	0.33762
Calculated density (g/cm³)	7.839
Formula units per unit cell	$Z = 1$
Scan range	$15° \leqslant 2\theta \leqslant 100°$
Residual values	
R_p	0.088
R_{wp}	0.118
$R_{expected}$	0.033

Table 4.5 Atomic and thermal parameters for the Cu_5Ho_7Sb compound.

Atom	Position	x	y	z	Occupation	Beq(Å²)*
Cu1	2a	0.25	0.25	0	0.5 (50% Cu + 50% vacancy)	2.04(3) 2.13(2)
Cu2	4e	0.00	0.00	0.00	1	1.34(1)
Ho1	4g	0.25	0.25	0.3450(4)	1	2.22(3)
Ho2	4h	0.75	0.25	0.3271(4)	0.75 (75% Ho + 25% vacancy)	2.11(5)
Sb	4h	0.75	0.25	0.3271(4)	0.25 (25% Sb + 75% vacancy)	

parameter for the atoms. The observed and calculated diffraction patterns were in excellent agreement.

The basic process of crystal structure identification is roughly as follows. Firstly, XRD patterns are collected. Single-phase diffraction patterns are much easier to index than multiple-phase samples. When only multiple-phase samples are available, it is be very helpful to separate out the peaks belonging to the known phases. Secondly, the diffraction patterns are matched with all known crystal structures in a database. When a structure match is not found, the basic lattice group is then estimated and used to match the peak positions and intensities. Thirdly, when the pattern is indexed, fundamental parameters such as lattice parameter, ideal molecular formula, and space group are then determined. Finally, Rietveld refinement to modify the crystal structure and calculation of credence factor are performed.

In the case of ionic crystals, rationality of the structure may be further checked with the bond valence theory.

7 PITFALLS

Even though the method of equilibrated alloys is a standard way for phase diagram determination, it is not immune from pitfalls. The following cases are used to illustrate a few common issues and solutions.

7.1 Verification of the Establishment of True Equilibrium

One of the most tedious issues in phase diagram determination is to judge if a particular alloy has achieved equilibrium. One usually anneals the samples at elevated temperatures for extended periods of time and then quenches them in order to retain the phase equilibrium at elevated temperatures. The following strategies can be used to verify if true equilibrium is reached.

The first strategy is to subject representative samples to several periods of annealing [31]. Using this strategy, the quenched sample is ground and then returned to the furnace for a second or third period of equilibration. The variation of phase distribution and their lattice parameters against annealing time may prove helpful. Figure 4.17 shows the measured isothermal section at 1400°C for the ZrO_2–CeO_2–$LaO_{1.5}$ system. Establishing accurate ZrO_2-based phase diagrams has proven to be difficult because of the sluggish kinetics at low temperatures, the formation of metastable phases, evaporation at high temperatures, and so on. In order to guarantee that real equilibrium was reached, a representative sample (20 mol% ZrO_2–10 mol% CeO_2–70 mol% $LaO_{1.5}$) prepared with two different experimental procedures was annealed at 1400°C for 400, 500, 750, and 820 h, respectively. The observed phase distribution and lattice parameters are shown in Fig. 4.18 as a function of the annealing time, and they were clearly identical for different annealing times and for both co-precipitation samples and arc-melted ones. These results indicate that 400 h annealing was enough for the attainment of equilibrium (one needs to be careful that a lack of change can also indicate not sufficient diffusion, thus not in equilibrium).

Another strategy is to approach equilibrium from different directions. For example, a specimen was found to consist of a liquid and a solid phase when quenched from a certain condition. To check whether this is the true equilibrium state, one would prepare a sample of quenched liquid phase by quenching it from a temperature above the liquidus, and a second sample with the same composition by firing at subsolidus temperatures. These two samples would then be equilibrated at the temperature of interest and should yield identical phase state at equilibrium.

A third and fast approach is to check the microstructure of a specimen. A uniform distribution of phases throughout the sample is a necessary indication that equilibrium may have been achieved. This is not a sufficient criterion since the metastable phases may also distribute uniformly in the microstructure even at high temperatures for extended periods of time. Often even a simple check with the phase rule is a good start to separate a non-equilibrium state from an equilibrium state.

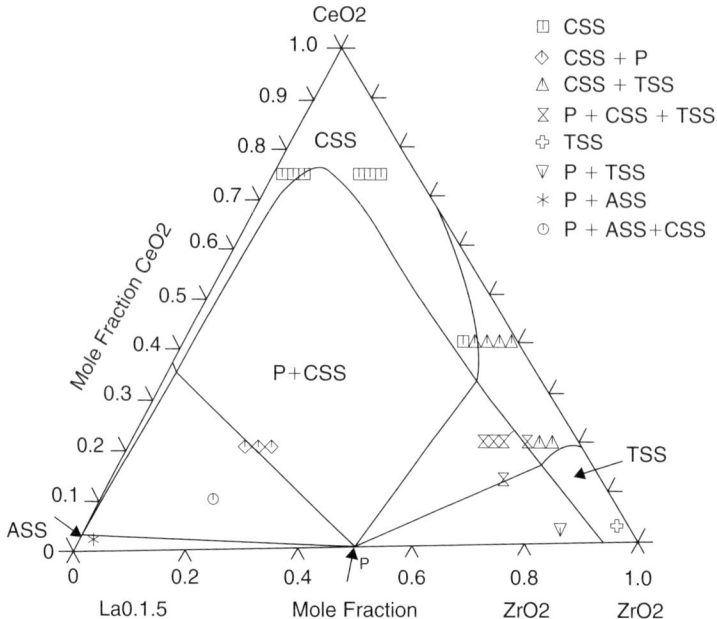

Figure 4.17 Isothermal section at 1400°C in the ZrO_2–CeO_2–$LaO_{1.5}$ system. P: $Zr_2La_2O_7$; Css: Cubic ZrO_2 solid solution; Tss: tetragonal ZrO_2 solid solution; and Ass: solid solution based on low-temperature hexagonal $LaO_{1.5}$.

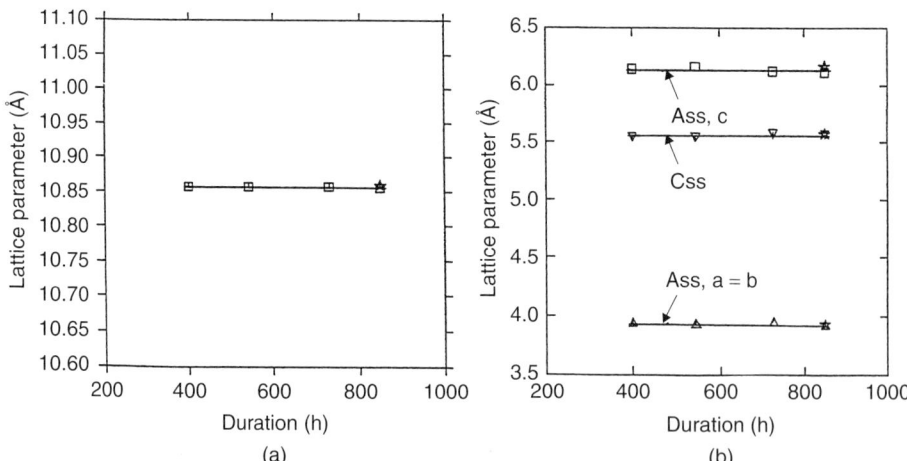

Figure 4.18 Lattice parameters of the phases for the sample (20 mol% ZrO_2–10 mol% CeO_2–70 mol% $LaO_{1.5}$) annealed at 1400°C for different times. The star symbol corresponds to the arc-melted samples, and the other symbols to the co-precipitated sample. (a) P phase and (b) Ass and Css phases. P: $Zr_2La_2O_7$; Ass: solid solution based on low-temperature hexagonal $LaO_{1.5}$; Css: cubic ZrO_2 solid solution.

For instance, the phase rule dictates that a three-phase equilibrium can only exist at a specific temperature for a binary system. If the three phases exist at multiple temperatures, it indicates that true equilibrium has not been reached at some of the temperatures.

7.2 Inconsistency Between the Result from DTA Measurement and that from XRD and Microscopy Observation

Although DTA is an effective approach to measure phase transition temperatures, it could yield erroneous results due to slow kinetics (such as nucleation barrier) of reactions during measurements [32]. Figure 4.19(a) shows the phase relation in Al-rich side of the Al–Mn system derived from XRD analysis and microstructure observation of the equilibrated and quenched alloys. The DTA measurement of the same alloys yields the phase relation shown in Fig. 4.19(b). As shown in Fig. 4.19, the two sets of phase relations are inconsistent. XRD and microstructure observation indicate the existence of a phase field L + λAl_4Mn, while the DTA data suggest no such a phase field. Thus, one of the experimental methods must have yielded unreliable data. In order to resolve the issue, a thorough check of the obtained data was performed. Several possible problems and their proof check in XRD and microstructure observation of the equilibrated and quenched alloys are as follows. Firstly, it is possible to have wrong phase identification. This was not the case since XRD and SEM/EDX analyses confirm each other. Secondly, the annealing temperature(s) was not what it was assumed. This can be ruled out since for each annealing temperature at least three different furnaces were used and it was unlikely that the temperature controllers of all three furnaces were wrong in exactly the same way. Thirdly, λAl_4Mn forms only as a metastable phase upon cooling the alloys from the phase field L + μAl_4Mn. This was not the case either because

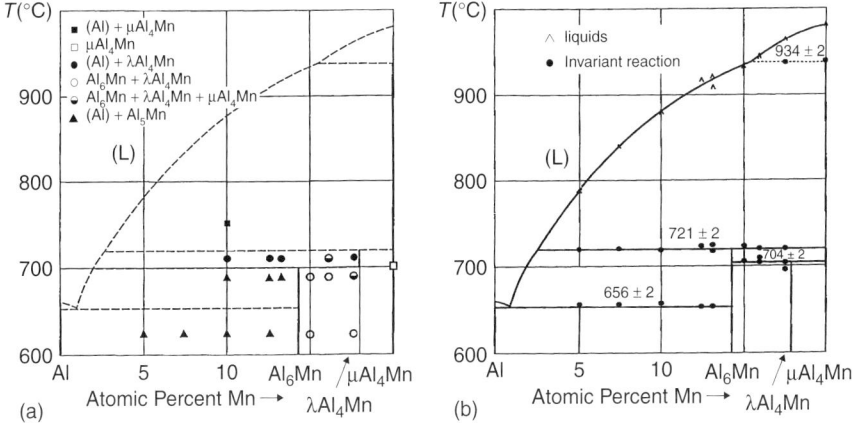

Figure 4.19 Phase relations in the Al-rich side of the Al–Mn system: (a) derived from equilibration and quench experiments and (b) derived from DTA measurement [32]. The DTA results are believed to be non-equilibrium due to slow kinetics in the melting of Al_6Mn.

λAl₄Mn formation is observed upon heating L + Al₆Mn containing alloys to 715°C (4 days) and reacts back to Al₆Mn + liquid (which solidifies to Al) at 690°C. Fourthly, the quenched alloys did not represent the high-temperature equilibria because λAl₄Mn formed upon cooling from the melt of the phase field L + Al₆Mn. In this case such alloys must contain (Al) + Al₆Mn + λAl₄Mn upon solidification. However, for the annealed and quenched alloys containing less than 14.3 at.% Mn, which were free of Al₆Mn, only (Al) and λAl₄Mn are observed. Lastly, the occurrence of λAl₄Mn is due to Si-contamination from the starting materials or during annealing. This was unlikely because starting materials and experimental conditions were selected specifically to minimize this risk. In order to identify any effects of contamination by traces of Si, representative specimens were placed in Y_2O_3 stabilized ZrO_2 crucibles and sealed in Ta cylinders. Thus, there is no reason to doubt the results based on XRD and microstructure observation.

What then could be possible reasons for the unreliable DTA result? Firstly, there was potential contamination due to Si. As discussed above, this was unlikely to play a role. Secondly, there was a possibility of contamination with oxygen. During the DTA runs there was slight oxidation presence. Due to the extremely low oxygen solubility in Al-rich compounds, this potentially led to Al_2O_3 formation but consequently should not (or only marginally) affect the phase relationships in Al-rich side of the Al–Mn system. Thirdly, the kinetics of the reaction Al₆Mn → L + λAl₄Mn was too slow. In order to investigate the kinetics associated with the formation of λAl₄Mn, time-dependent annealing and quench experiments were performed for the alloy Al–12.5 at.%Mn annealed at 712°C for 5, 50, 100, and 500 min, followed by water quenching. XRD analysis showed no trace of λAl₄Mn within the first 50 min and after 500 min the transformation to λAl₄Mn was observed but not complete. At a heating rate of 5K/min, it took only 12 min to heat the alloy from 690°C (where Al₆Mn was found even after prolonged anneal) to 750°C (where L + μAl₄Mn was observed to coexist). Reducing the heating rate to 2K/min and assuming the unlikely case that the phase field L + λAl₄Mn ranges from 690°C to 750°C, the phase λAl₄Mn has only 30 min to form or will not be detected by DTA. This suggests that the formation of λAl₄Mn during incongruent melting of Al₆Mn is a slow nucleation-controlled process. This nucleation barrier is responsible for the erroneous conclusions from DTA data. Care should be exercised to check the general consistency of the phase diagram data obtained from different experimental methods.

7.3 Identification of Degenerated Phase Equilibrium

For some binary systems in which one element has much higher melting point than the other, the invariant equilibrium at the low melting point element side is of the degenerated nature. This means that the invariant reaction temperature is very close to the melting point of the low melting element. In order to elucidate the degenerated nature, precise experiment is required. The invariant equilibrium in the extreme Al side of the Al–Cr system is indicative of the degenerated feature. In the literature some reported a eutectic reaction type (L ↔ (Al) + Al₄₅Cr₇), whereas some others suggested a peritectic one, L + Al₄₅Cr₇ ↔ (Al). The degenerated feature associated with the invariant equilibrium among liquid, Al₄₅Cr₇, and (Al) can be unambiguously clarified by means of DSC measurements with a few different

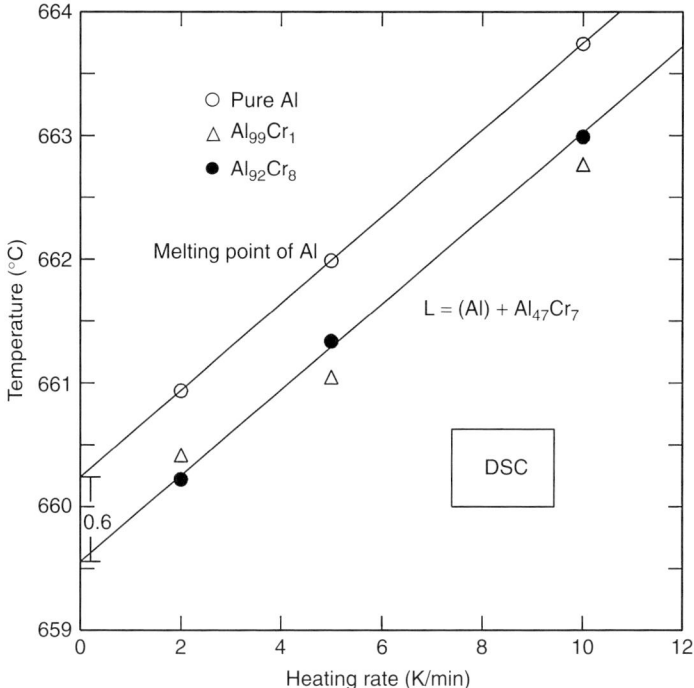

Figure 4.20 Invariant reaction temperature as a function of heating rate for the Al–Cr alloys annealed at 630°C for 31 days [33]. The composition is in atomic percent.

heating rates together with XRD and EPMA techniques [33]. Figure 4.20 presents the invariant reaction temperatures measured with DSC against the heating rate. It clearly indicates the degenerated feature for the invariant equilibrium among liquid, (Al), and $Al_{45}Cr_7$: the eutectic reaction temperature is only 0.6°C lower than the melting point of Al. Such a small temperature difference cannot be resolved by means of DTA technique since the experimental uncertainty resulting from DTA measurements is about 2°C. The degenerated nature of the invariant equilibrium cannot be resolved with microstructure observation either because the eutectic composition is very close to the composition of (Al) phase. The DSC approach could be employed to identify the degenerated equilibrium in ternary and multicomponent systems also [33].

REFERENCES

1. F. Levy, *Phys. Kondens. Materie*, 10 (1969) 85.
2. J.K. Liang, *Phase Diagram and Crystal Structure*, Science Press, Beijing, China, 1993 (In Chinese).
3. C.K. Kuo, Z.X. Lin and T.S. Yen, *High Temperature Phase Equilibria and Phase Diagrams*, Shanghai Scientific and Technical Publishers, Shanghai, China, 1987 (In Chinese).
4. X. Lu, *Phase Diagrams and Phase Transitions*, University of Science and Technology China Press, Hefei, China, 1990.
5. J. Liu, F. Jiao, Y. Zhuang, F. Ma and Z. Sun, *J. Alloys Comp.*, 384 (2004) 192.
6. L. Zeng and G. Lin, *J. Alloys Comp.*, 395 (2005) 101.

7. G. Cai and L. Zeng, *J. Alloys Comp.*, 403 (2005) 127.
8. J. Zheng, Y. Gan and D. Li, *Acta Physica Sinica*, 21 (1965) 1487.
9. J. Zheng, R. Chen and D. Li, *Acta Physica Sinica*, 24 (1975) 174.
10. H. Zhou, Y. Wang and Q. Rao, *J. Alloys Comp.*, 407 (2006) 129.
11. X. Lu and J. Liang, *Acta Physica Sinica*, 22 (1966) 669.
12. T.B. Massalski, H. Okamoto, P.R. Subramanian and L. Kacprzak, *Binary Alloy Phase Diagrams*, 2nd edn, ASM International, Materials Park, OH, 1990.
13. M.L.V. Gayler, *J. Inst. Met.*, 30 (1923) 139.
14. F. Teng, Y. Wang and X. Jiang, *XRD Structure Analysis and Characterization of Materials Properties*, Science Press, Beijing, 1997, p. 1.
15. A.J. Bradley and S.S. Lu, *J. Inst. Met.*, 60 (1937) 319.
16. M.I. Pope and M.D. Judd, *Differential Thermal Analysis*, Heyden & Son Ltd., London, 1977, p. 1.
17. J.B. MacChesney and P.E. Rosenberg, in A.M. Alper (ed.), *Phase Diagrams: Materials Science and Technology*, Academic Press, New York and London, 1970, p. 156.
18. K.A. Gschneider Jr. (ed.), *Handbook on the Physics and Chemistry of Rare Earths*, Vol. 2, North-Holland Publishing Company, Amsterdam, Netherland 1979, p. 29.
19. J.H. Wernick and S. Geller, *Acta Cryst.*, 12 (1959) 662.
20. K.H.J. Buschow and A.S. Van der Goot, *Acta Cryst. B*, 27 (1971) 1085.
21. K.H.J. Buschow and A.S. Van der Goot, *J. Less-Common Met.*, 20 (1970) 309.
22. J.X. Zheng and L.Q. Nong, *Chinese Phys. Lett.*, 32 (1983) 1449.
23. Z. Pan, Y. Du, B.Y. Huang, H.H. Xu, Y. Liu and R. Schmid-Fetzer, *J. Mater. Sci.*, 41 (2006) 2525.
24. Z. Pan, Y. Du and B.Y. Huang, *Z. Metallk.*, 96 (2005) 1301.
25. N. Krendelsberger, F. Weitzer and J.C. Schuster, *Metall. Mater. Trans.*, A33 (2002) 3311.
26. J.C. Viala, N. Peillon, F. Bosselet and J. Bouix, *Mater. Sci. Eng.*, A229 (1997) 95.
27. J. Li, L. Zeng and H. Ning, *J. Alloys Comp.*, 353 (2003) L5.
28. Jade 5.0, *XRD Pattern Processing*, Materials Data Inc., Livermore, CA., USA. 1999.
29. R.A. Yong, A. Sakthivel, T.S. Moss, C.O. Paive-Santos, *J. Appl. Crystallogr.*, 28 (1995) 366.
30. H. Marciniak and R. Diduszko, *DMPLOT-Plot View Program for Rietveld Refinement Method*, Version 3.38, 1997.
31. Y. Du, M. Yashima, T. Koura, M. Kakihana and M. Yoshimura, *CALPHAD* 20 (1996) 95.
32. J.C. Schuster, Y. Du and F. Weitzer, *International Conference on Alloy Thermodynamics*, Vienna, Austria, 2004.
33. Y. Du, J.C. Schuster and Y.A. Chang, *J. Mater. Sci.*, 40 (2005) 1023.

CHAPTER FIVE

DTA and Heat-Flux DSC Measurements of Alloy Melting and Freezing

William J. Boettinger,[1] Ursula R. Kattner,[1] Kil-Won Moon[1] and John H. Perepezko[2]

Contents

1 Introduction		152
1.1 Focus of this Chapter		152
1.2 Information Sought from DTA/Heat-Flux DSC Measurements		153
1.3 Relevant Standards		154
1.4 Major Points		155
2 Instruments and Operation		155
2.1 Variations Among Instruments		155
2.2 Samples		159
2.3 Reference Materials		164
2.4 Calibration and DTA Signal from Pure Metals		164
2.5 Major Points		169
3 Analysis of DTA Data for Binary Alloys		170
3.1 General Behavior for a Binary Eutectic System: Example Ag–Cu Alloy Melting		171
3.2 Problems with Solidus Determination on Heating		176
3.3 Problems with Liquidus Determination on Heating		180
3.4 Supercooling Problem with Liquidus Determination on Cooling		186
3.5 Eutectic Reactions vs. Peritectic Reactions		191
3.6 Major Points		192
4 Analysis of DTA Data for Ternary Alloys		194
4.1 Al-Rich Corner of Al–Cu–Fe Phase Diagram		194
4.2 Al–20% Cu–0.5% Fe		195
4.3 Al–6% Cu–0.5% Fe		197
4.4 Major Points		198
5 Concluding Remarks		198
Appendix A Glossary		201
Appendix B Recommended Reading		204
Appendix C Model for Simulating DTA Response for Melting and Solidification of Materials with Known or Assumed Enthalpy vs. Temperature Relations. Also Method for Determining Thermal Lag Time Constants of DTA/DSC Instruments		205

[1] Metallurgy Division, National Institute of Standards and Technology (NIST), Gaithersburg, MD, USA.
[2] Department of Materials Science and Engineering, University of Wisconsin-Madison, Madison, WI, USA.

Appendix D Expressions for the Rate Dependence of Melting Onset Temperatures
for a Pure Metal 205
Appendix E Enthalpy vs. Temperature Relations for Dilute Binary
Solid Solution Alloy 211
Appendix F Binary Phase Diagrams and DTA Response 213
Appendix G Tutorial on Melting and Freezing of Multicomponent Alloys 213
G.1 Aluminum Alloy 2219 213
G.2 Udimet 700 218

1 Introduction

1.1 Focus of this Chapter

This chapter is focused on **differential thermal analysis (DTA)**[1] and heat-flux differential **scanning calorimetry (HF-DSC)**[2] of metals and alloys. A thermal analysis guide focused only on metals and alloys is appropriate because they behave quite differently from molecular materials such as polymers and organics. Firstly, metals are good conductors of heat so that temperature differences at any instant can often be ignored within the small (200 mg) samples typical of DTA. Secondly, liquid metals and alloys are typically monatomic. Freezing and melting occur rapidly in response to changes in temperature compared to other materials. Melting and freezing transformations, once initiated, take place within, at most, a degree of **local thermodynamic equilibrium**.[3] Therefore the guide also focuses on melting and solidification behavior because special methods can be employed that are not necessarily useful for a broader class of materials and processes. The metallurgist should expect a high level of precision when proper attention is paid to technique and interpretation of data.

Thermal analysis is a widely used experimental measurement technique with a long history [1–11]. In contemporary application, most work is performed with instruments that are commercially available and under computer control. Under these conditions even the novice user can obtain data and software generated analysis of measurements in a relatively short time. However the expert may want to consider specific sample characteristics in order to judge whether the traditional measurement protocols are satisfactory or if modifications are necessary. The specific sample characteristics also impact the interpretation and analysis of the thermal analysis measurements and the manner in which the software might be used. Best practice may also require a departure from standard practice to include, for example, thermal cycling schedules and holding treatments in order to address specific sample characteristics. Similarly, the proper interpretation of the thermal analysis signals should include consideration of the multicomponent phase equilibrium as well as analysis of the phase evolution during solidification and melting that again is specific to a given sample. It is the intention of this guide to provide the thermal analysis user with the considerations

---

[1] Items in bold grey are defined in the glossary, Appendix A.
[2] The Guide does not treat power-compensating scanning calorimetry (PC-DSC).
[3] However, supercooling of a liquid is often required to initiate freezing (nucleation supercooling).

that are necessary for proper sample preparation and to illustrate how the sample characteristics influence the proper interpretation and analysis of measurements. Since the specific characteristics of a given sample determine the best practice in measurement and analysis, it is not possible to include all possible advice. Instead, the cornerstone of best practice should be the flexibility to alter procedures in response to specific sample characteristics and types of information sought.

In the remainder of the introduction, we describe the different types of information usually sought from DTA/HF-DSC during the melting and freezing of alloys. In Section 2, the details of instruments, operation and calibration are described. The goal is to describe the thermal lags between the sample and sample thermocouple that must be understood to enable good analysis of data. In Sections 3 and 4, the response of the DTA to binary and ternary alloys, respectively, is detailed. Here a strong use is made of the enthalpy vs. temperature relation for melting and freezing of alloys with different phase diagram features when one assumes full diffusional equilibrium and when diffusion in the solid is so slow as to be ignored. Section 5 makes some concluding remarks. The appendices include a glossary of terms that are printed bold grey in the text, recommended reading, a heat flow model for the DTA and a description of DTA response for alloys with more than three components.

1.2 Information Sought from DTA/Heat-Flux DSC Measurements

DTA and HF-DSC thermal measurement instruments attempt to measure the difference in temperature between a sample and an inert standard during programmed heating or cooling. As such, the measurements are sensitive to the difference between the enthalpy vs. temperature relation of a sample and the **enthalpy** vs. temperature relation of a standard. The enthalpy vs. temperature relations of pure metals are well known and output information from such pure metals from a DTA/HF-DSC is used primarily for calibration. Figure 5.1 shows the general shape of the DTA output for the melting and freezing of a pure metal under ideal conditions. On heating, melting

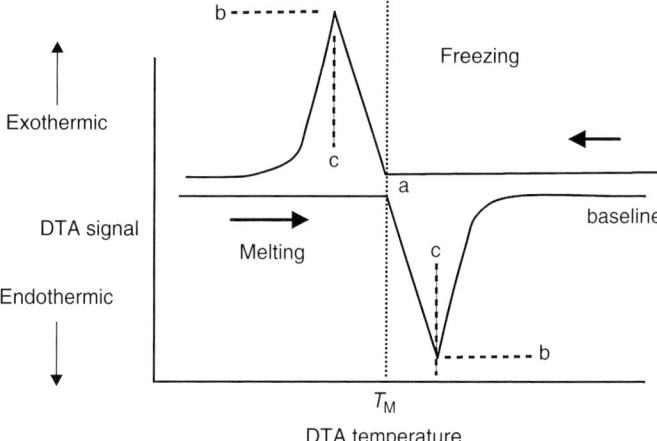

Figure 5.1 DTA responses to melting and freezing of a pure material under ideal conditions. a: onset temperatures (taken here as equal to melting point, T_M); b: peak signals; c: peak temperatures.

requires an input of heat and the downward peak is endothermic. On cooling, freezing releases heat and the upward peak is exothermic. Peak shapes have a linear portion up to the maximum deflection from the horizontal followed by an exponential return to the horizontal. The initial detection of melting or freezing is indicated by the beginning of the linear portion of the peak and is used for temperature calibration. The area of the peak is used for calibration of the heat flow. The details of the ideal DTA shape shown in Fig. 5.1 as well as offsets from the true melting temperature and rounding of various curve features are described in Section 2.4.

Once the instrument is calibrated, various types of information for melting and solidification of alloys using a DTA/HF-DSC are sought. The most common measurement involves liquidus and solidus temperature determination. These terms will be used here to refer to thermodynamic quantities that depend only on the alloy composition (at fixed pressure). For phase diagram research one might also want to determine other thermodynamic temperatures associated with melting and freezing; for example, in binary alloys, invariant reaction temperatures and their character; viz., eutectic and peritectic. In ternary and higher order systems, information is sought regarding additional thermodynamic transitions with one or more degrees of freedom.

Related measurements of interest can include the amount of supercooling possible prior to solidification and the microstructurally and solid diffusion rate sensitive final freezing temperature/incipient melting point. One might want to know the supercooling tendency of a particular alloy in contact with a certain type of crucible or in the presence of a grain refiner. These temperatures are not purely thermodynamic in origin, but are needed for the modeling of castings, for the determination of maximum heat treatment temperature, etc. DTA/HF-DSC may at times be used for the identification of alloys from within a small class of alloys as a quality control method; that is, the DTA plot could be used as a "finger print" of the alloy.

A more difficult measurement is the determination of the enthalpy vs. temperature relation for the sample. One may want the thermodynamic (equilibrium) enthalpy–temperature relation, or for practical considerations, one may want the enthalpy vs. temperature relation as it would apply under certain casting conditions where freezing would end at a final temperature generally lower than the solidus.

DTA and DSC are sometimes used in a non-scanning mode where the temperature is held constant and evidence for isothermal transformation is detected. Such an approach is sometimes used for solidification studies involving measurement of nucleation kinetics and diffusion bonding.

1.3 Relevant Standards

ASTM Subcommittee E37.01 is responsible for thermal analysis test methods. A summary of relevant ASTM Standards is given below. Many of the methods suggested in these documents have been developed with a broad array of materials in mind. With a focus on metal and alloy melting and freezing, variations of these methods are sometimes more appropriate as we will describe in this document. Various other international standards are available. In particular we note many relevant standards of the Deutsche Institut für Normung (DIN), Association Française de Normalisation (AFNOR) and the British Standard Institution (BSI). Many of these are available from ANSI web store (http://webstore.ansi.org).

1.3.1 Terms and Definitions

ASTM E473, "Standard Terminology Relating to Thermal Analysis," is a compilation of definitions of terms used in other ASTM documents on all thermal analysis methods including techniques besides DTA and DSC.

ASTM E1142, "Terminology Relating to Thermophysical Properties," is a compilation of definitions of terms used in other ASTM documents that involve the measurement of thermophysical properties in general.

1.3.2 Calibration and Sensitivity

ASTM E967, "Practice for Temperature Calibration of DSC and DTA," presents simple recipes for calibration for fixed mass and heating rate using two pure materials to obtain a linear correction for conversion of measured temperature to actual temperature. The *onset temperature* extracted from the melting peak is determined by the extrapolation method (see Section 2.4.3). For some materials the standard suggests using the peak for calibration, a method not recommended for metals.

ASTM E968, "Standard Practice for Heat Flow Calibration of DSC," uses sapphire as heat capacity standard. The method is described in Section 2.4.4.

ASTM E2253, "Standard Method for Enthalpy Measurement Validation of Differential Scanning Calorimeters," presents a method using three small masses to determine the detection limit of DTA/DSC.

1.3.3 Analysis of Data

ASTM E928, "Standard Test Method for Determining Purity by DSC," employs comparison of the shape of the melting peak of an impure sample to the shape for a high purity sample to determine the concentration of the impurity. The method uses the "1/F plot" which examines the down slope of the melting peak.

ASTM E794, "Standard Test Method for Melting and Crystallization Temperatures by Thermal Analysis," employs the extrapolated onset determination method.

ASTM E793, "Standard Test Method for Enthalpies of Fusion and Crystallization by DSC," uses area on signal vs. time plot for comparison to known heats of fusion of pure materials.

ASTM E1269, "Standard Test Method for Determining Specific Heat Capacity by DSC," uses sapphire or aluminum as a standard.

1.4 Major Points

- This guide focuses on melting and solidification of metals and alloys.
- Different types of information are sought by different users.
- Some deviation from standard practice may be useful for specific measurements.

2 INSTRUMENTS AND OPERATION

2.1 Variations among Instruments

For precision work, it is important for users to understand the operation of their particular DTA/HF-DSC, as different instruments employ different designs and control features and employ different data analysis and presentation methods.

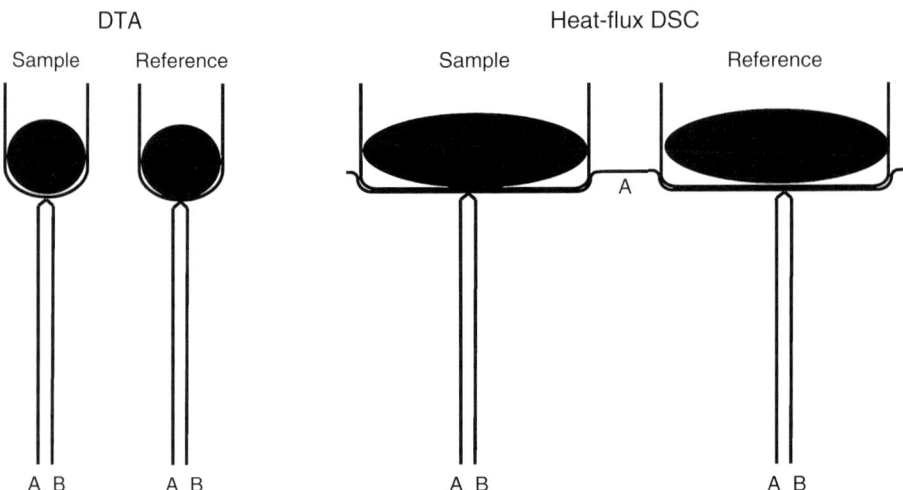

Figure 5.2 Schematic of DTA and HF-DSC geometries (not to scale). A and B denote the different legs of thermocouples.

2.1.1 Heat Transfer between System Components

A DTA/HF-DSC instrument consists of a single furnace and two crucibles with thermocouples (Fig. 5.2). One crucible is for the sample being tested and the other is for a reference material, often alumina powder. The sample or reference thermocouple temperatures are given by the voltage difference between thermocouple legs A and B for either cup with suitable cold junction compensation. In the DTA, the difference signal is obtained by shorting the B legs and measuring the voltage between the A legs. In the heat-flux DSC, the connecting metal strip is often used as an active sensing element to obtain the difference signal by measuring the voltage between legs B and B.

Depending on the geometrical arrangement of these parts, the atmosphere, and the temperature range of operation, heat flow will have different contributions from radiation, convection and conduction processes. The relative contributions will change the size and heating rate dependence of the temperature differences and thermal lags during heating and cooling between the sample, the reference and the thermocouples. The manner in which these lags influence the interpretation of DTA data will be a central focus of this document.

Appendix C presents a model for the heat transfer within a DTA instrument and an experimental method to determine three time constants for the instrument. With these time constants, the sample mass and an enthalpy vs. temperature relationship for the sample, one can calculate DTA curves. This procedure permits one to understand an important difficulty of DTA analysis as demonstrated by the fact that the sample thermocouple in a DTA does not remain isothermal during the melting of a pure metal. This is not due to the kinetics of the melting process, but is due to the difference between the sample temperature and the sample thermocouple temperature [12–14].[4]

---

[4] It is to be noted that cooling and heating curves obtained with direct immersion of a thinly protected thermocouple in the melt are often superior to conventional DTA methods because of greatly reduced thermal lags. Such experiments require custom equipment designed for the specific alloy of interest to provide suitable atmosphere and refractory coatings (see e.g., Ref. [12–14]). Because the DTA is a common desktop instrument available in many laboratories, its capabilities and limitations are nevertheless worth consideration.

The model will be used to clarify calibration procedures in Section 2.4.1 and aid in the interpretation of DTA signals from alloys in Sections 3 and 4 and Appendix G.

2.1.2 DTA/Heat-Flux DSC vs. Power-Compensating (True) DSC

This document is limited to a description of devices with one heater. The term differential scanning calorimeter was originally applied to instruments with separate heaters for the reference and the sample. The differential signal of a true DSC is the power difference required to keep the temperatures of the reference and sample identical during a defined temperature ramp. Thus, the signal is in units of watts per unit sample mass. This type of instrument is today called a power-compensating DSC. Many instruments with a single heater are called a DSC, but are more properly called a heat-flux DSC. The basic measurement, just as for a DTA, is a difference in voltage between the output of the sample thermocouple and the reference thermocouple. Although quantitative measurements of heats of transformation can be evaluated by either DTA or heat-flux DSC, there are differences in sensitivity and accuracy. In a heat-flux DSC, the heat flow path between the sample and the reference is provided by a metal strip whose thermal conductivity is high and well defined. This minimizes the temperature difference between the sample and reference and also makes the temperature difference directly proportional to the heat flux between the two. For both, a calibration procedure can be used to relate the difference signal integrated over time to an enthalpy (see Section 2.4.4). Such instruments are not capable of the same accuracy as a power-compensating DSC, but are usually built to operate at the higher temperatures ($>1000°C$) necessary for metallurgical work.

2.1.3 How Does the Instrument Control the Heating Rate?

In its simplest configuration, the control thermocouple for a DTA[5] is in the furnace. The controller applies power to maintain a linear heating or cooling rate for this thermocouple. In this case, there is usually a thermal transient between the furnace thermocouple and the reference and sample thermocouples, which is particularly noticeable at the beginning and end of a DTA scan.

A second method uses the reference thermocouple to control the application of power. This has the advantage of preventing the initial (and final) transients. Finally some instruments use the sample thermocouple for control in an attempt to maintain a constant heating rate in the sample itself. However while control of a sample thermocouple to a constant heating rate may be possible for phase changes with sluggish kinetics or very small enthalpy of transformation, it is practically impossible to maintain a constant heating rate during the melting or freezing of metals and alloys. This control method is not recommended for melting and solidification studies. Some instruments provide user selectable choices of control method. Modulated DTA/DSC is an instrument that adds a small sinusiodally varying temperature oscillation on the usual temperature program. This last subject, as well as power-compensating DSC, is beyond the scope of the present document.

---

[5] In the remainder of this document, we will usually drop the distinction between DTA and HF-DTA for brevity.

2.1.4 What Is the Signal, Millivolts or Kelvin?

As the name implies, the signal for a DTA is a difference. This is the y-axis of standard DTA plots (see Fig. 5.1). It is the difference in either the voltage or the derived temperature difference between the sample thermocouple and the reference thermocouple. When the signal is given as temperature difference, a reference table or equation has been used by the instrument to obtain temperature difference from a voltage difference. The choice of one or the other by individual manufacturers is mostly historical or an attempt to provide numbers with which the user is comfortable. Unfortunately, both types of output are not usually available on a single instrument. In some cases the millivolt signal is divided by the sample mass for presentation. The signal can either be endothermic or exothermic; that is, indicating heat into or out of the sample. For transformations of materials that remain close to equilibrium, heating scans lead to endothermic signals and cooling scans lead to exothermic signals. Unfortunately, there is no common practice for whether a positive y-axis is endothermic or exothermic. In this guide we will always assume that an endothermic signal is negative on the y-axis. Some instruments use a thermopile between the sample and the reference. A thermopile is a number of thermocouples connected in series. Connecting many thermocouples in series produces a higher differential voltage signal.

2.1.5 Plotting Signal vs. Temperature or Time

Figure 5.3(a) shows experimental data for the sample and the reference thermocouple temperatures plotted vs. time for the melting of pure Ag. The difference between these temperatures can be plotted vs. time, vs. reference thermocouple temperature or vs. sample thermocouple temperature as shown in Fig. 5.3(b). The output table or graphics display of a particular instrument may provide some or all of these choices as the x-axis of DTA plots. Significant differences in the shapes of the curves are evident depending on the x-axis choice. A display vs. temperature is useful for directly reading temperatures of thermal events. A display vs. time is required when peak area (Kelvin·second or millivolt·second) is calculated from the data for quantitative measures of total enthalpy change associated with a thermal event. For display or numerical output, some instruments choose the temperature as the reference thermocouple temperature, while others choose the sample thermocouple temperature. For precision work the latter is preferable because the changes in sample thermocouple temperature are more closely related to events in the material of interest. Most importantly, it is necessary for the user to know which temperature is being used by the instrument software. Then replotting of data can be performed using external manipulations if desired.

Sections 3 and 4 of this document will delve in considerable detail into the expected shape and interpretation of DTA curves for alloys. For the melting and solidification of a pure metal, DTA curves are simple and one can define several features of the curve as seen in Fig. 5.1. The **baseline** is the graphical feature corresponding to the ΔT signal over temperature ranges where no transformation of the sample takes place. The baseline is horizontal if the signal is a constant. The onset temperature is the beginning of the deviation from the baseline measured by some quantitative measure as described in Section 2.4.3. The **peak signal** is the maximum deviation of the signal (Δy-value) from the baseline and the **peak temperature** is the temperature (x-value) corresponding to the peak signal. The term **peak** will be used regardless of whether it is negative or positive. The temperature at which the signal

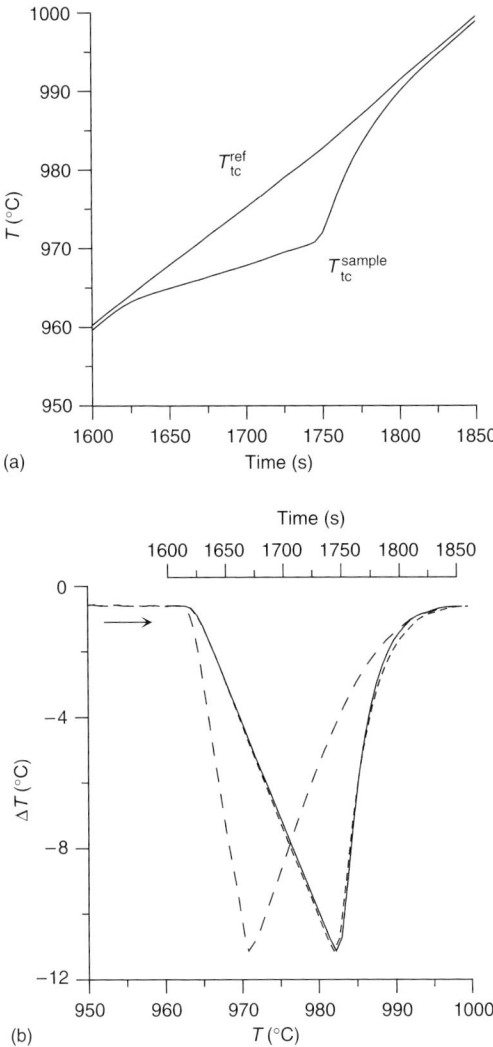

Figure 5.3 DTA melting of a 211.6 mg of pure Ag at 10K/min: (a) reference thermocouple temperature T_{tc}^{ref} and sample thermocouple temperature T_{tc}^{sample} vs. time and (b) differential signal $\Delta T = T_{tc}^{sample} - T_{tc}^{ref}$ vs. time (solid), ΔT vs. sample temperature (long dashed) and ΔT vs. reference temperature (short dashed). For this data, the reference temperature remains quite linear in time as the sample melts, so that a linear scaling of the time axis makes the plots with x-axes of time and reference temperature practically identical.

returns to the baseline has no significance with respect to thermal events in the sample and reflects heat transfer characteristics within the instrument (see Appendix C).

2.2 Samples

2.2.1 Mass

Figure 5.4 compares the signal during melting of different masses of pure Ag with a fixed mass of the reference. A large sample produces a larger peak signal (deflection

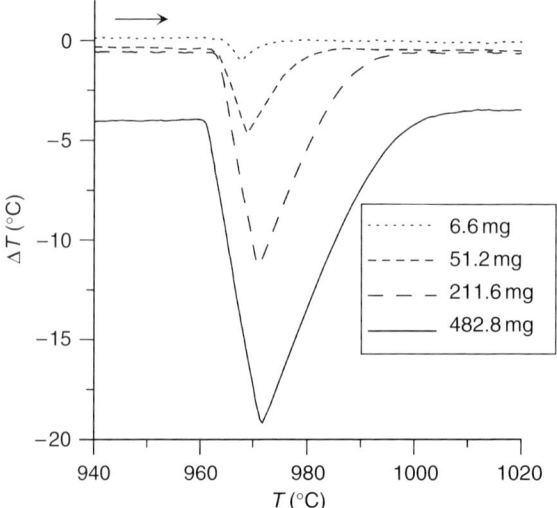

Figure 5.4 Effect of sample mass on DTA signal for pure Ag. The reference mass was held constant. Heating rate is 10K/min.

from the baseline). However, the larger signal delays the temperature at which the signal returns to the baseline, making detection of closely spaced thermal events more difficult. On the other hand, large masses decrease the response time between the sample and the sample cup. For careful work, sample mass vs. signal output should be examined for optimal performance. These items are discussed further in Appendix D.

2.2.2 Shape
Sample shape is a significant factor if the user wants to extract information on the first melting cycle. The shape will typically not conform to the shape of the sample cup. The thermal contact area between the sample and the cup will change during the melting process. This change can lead to a DTA peak shape that is difficult or impossible to interpret. Sometimes the effect of changing sample shape on first melt is more subtle. For the melting of pure Ag and for Ag–Cu eutectic, Fig. 5.5 shows a shift of the baseline in the y-direction after the first melting that is not evident for the second melt. A more rounded onset and absence of linearity of the down slope is also evident for the first melt (the expected linearity is described in Section 2.4.2). If data from a first melt is desired, success may be achieved by careful grinding of the sample to conform to the crucible shape. In general, it is preferable to discard data obtained from the first melting run of a sample.

2.2.3 Powder Samples
In general, the use of powder samples of metals and alloys can increase chances of contamination from atmosphere (oxidation), reduce heat flow within the entire sample mass and reduce the area of contact with the sample cup. However supercooling studies are enhanced by the use of powdered samples due to isolation of the most active heterogeneous nucleation sites into a small fraction of the powder population [15].

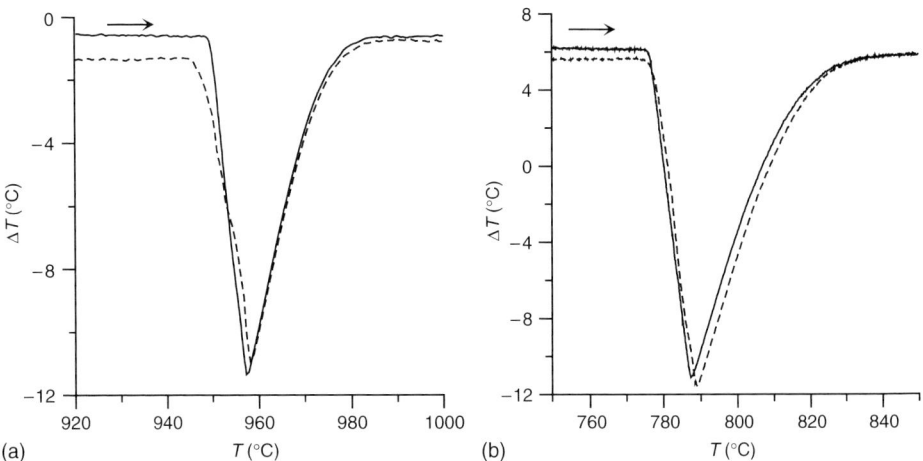

Figure 5.5 DTA signal for melting of (a) pure Ag (211.6 mg; 10K/min) and (b) Ag–Cu eutectic (231.3 mg; 15K/min) comparing first melt (dashed) and second melt (solid). Note the offset of the baselines, the more rounded onsets and the lack of linearity of the downslopes for the first melt compared to the second.

2.2.4 Inert Powder Cover

Some standard practice suggests mixing Al_2O_3 powder with powder samples or covering samples with a layer of Al_2O_3. This is done in an attempt to match the heat capacity and thermal conductivity of the sample and the (alumina) reference and produce a flatter baseline. Such practice must be weighed against the possibility of reaction between the sample and the powder cover. If the wetting angle between the cup and the sample is close to 180°, some authors suggest that use of an inert powder cover increases the thermal conduction to the sample cup.

2.2.5 Lid

A lid is useful for high vapor pressure materials to reduce material loss and contamination of the other parts of the instrument. A lid will also prevent sample radiation loss and help maintain an isothermal sample.

2.2.6 Atmosphere

For precision work, the use of commercial purity inert gas is not adequate. High purity inert gas coupled to a Ti getter furnace is usually adequate for most metals. Helium has a higher thermal conductivity than Ar; the choice can alter the thermal transport rates in the DTA instrument [16]. Calibration should be performed with the same atmosphere as will be used for samples.

2.2.7 Crucible Selection/Reaction

High purity alumina is a standard DTA cup material. Zirconia and yttria cups can be made/purchased for highly reactive metals. The inside of alumina cups can also be coated with zirconia or yttria washes. Coatings of boron nitride may be useful for some materials. Graphite is usually an excellent crucible material for some nonreactive metals, like Au and Ag. For the selection of oxide crucibles, Table 5.1 gives the

Table 5.1 Enthalpies of formation of solid oxides ({ } liquid, ( ) gas).

																	He
{H$_2$O}	286																
{H$_2$O$_2$}	94																
Li$_2$O	595	BeO	610								B$_2$O$_3$	426	(CO$_2$)	197	O	-10	(OF$_2$)	18	Ne
Li$_2$O$_2$	318				2/n M$_m$O$_n$(s) ↔ 2m/n M(s) + O$_2$(g) ΔS = 184±10 J/K								(CO)	111		-15			
					M(s) + O$_2$(g) ↔ MO$_2$(g) ΔS = 15±5 J/K														
					O$_2$ ↔ 2 O ΔH = 498 kJ														
Na$_2$O	415	MgO	601	Ref.:	A.W. Searcy, "Enthalpy and Predictions of Solid-State Reaction Equilibria" in "Chemical and Mechanical Behavior of Inorganic						Al$_2$O$_3$	558	SiO$_2$	455	P$_2$O$_3$	273	SO$_3$	151	(OCl$_2$)	-81	Ar
Na$_2$O$_2$	261				Materials," Eds. A.W. Searcy, A.V. Ragone and U. Colombo, Wiley-Interscience, New York, NY, 1970, pp. 33-55										P$_2$O$_5$	61					
NaO$_2$	130				"O. Kubaschewski, C.B. Alcock and P.J. Spencer, "Materials Thermochemistry," 6th Ed., Pergamon Press, Oxford, UK, 1993										P$_2$O$_3$*	300					
Na$_2$O$_2$*	256																			
K$_2$O	361	CaO	635	Sc$_2$O$_3$	623	Ti$_2$O$_3$	490	V$_2$O	456	Cr$_2$O$_3$	376	MnO	385	Fe$_{0.95}$O	264	CoO	239	NiO	240	Cu$_2$O	171	ZnO	348	Ga$_2$O	355	GeO	212	As$_2$O$_3$	198	SeO$_2$	112	Br
KO$_2$	142					TiO$_2$	472	VO	431	CrO$_2$	293	Mn$_3$O$_4$	346	Fe$_3$O$_4$	280	Co$_3$O$_4$	226			CuO	155			Ga$_2$O$_3$	363	GeO$_2$	363	As$_2$O$_5$	185	Se$_2$O$_3$	82	
K$_2$O$_2$*	142					TiO$_3$	507	V$_2$O$_3$*	406	CrO$_3$*	196	Mn$_2$O$_3$*	319	Fe$_2$O$_3$*	274											GeO$_2$*	290					
						TiO*	543	VO$_2$*	357			MnO$_3$*	270																			
								V$_2$O$_5$*	310																							
Rb$_2$O	330	SrO	604	Y$_2$O$_3$	627	ZrO$_2$	547	NbO	410	MoO$_2$	294	TcO$_2$	216	RuO$_2$	150	RhO	92	PdO	96	Ag$_2$O	30	CdO	258	In$_2$O$_3$	308	SnO	285	Sb$_2$O$_3$	240	TeO$_2$	161	I
Rb$_2$O$_3$	176	SrO$_2$	326					NbO$_2$	399	MoO$_3$	248	TcO$_3$	180	RuO$_4$	46	Rh$_2$O$_3$	119	PdO*	115							SnO$_2$	290	SbO$_2$	290			
								Nb$_2$O$_5$*	380							RhO$_4$	96											SbO$_2$*	227			
Cs$_2$O	318	BaO	581	La$_2$O$_3$	597	HfO$_2$	556	Ta$_2$O$_5$	408	WO$_2$	295	ReO$_2$	213	OsO$_2$*	147	IrO$_2$	111	PtO$_2$	-85	Au$_2$O$_3$	1	HgO	91	Ti$_2$O	166	PbO	217	Bi$_2$O$_3$	191	Po		At		Rn
Cs$_2$O$_3$	188	BaO$_2$	318							W$_3$O$_8$	284	ReO$_3$	205	OsO$_4$	97									TiO$_3$	129	Pb$_3$O$_4$	179						
										WO$_3$*	281	Re$_2$O$_7$*	178													PbO$_2$*	137						
Fr		Ra		Ac																													

Ce$_2$O$_3$	599	Pr$_2$O$_3$	603	Nd$_2$O$_3$	602	Pm		Sm$_2$O$_3$	609	EuO*	590	Gd$_2$O$_3$	607	Tb$_2$O$_3$	622	Dy$_2$O$_3$	621	Ho$_2$O$_3$	627	Er$_2$O$_3$	632	Tm$_2$O$_3$	629	Yb$_2$O$_3$	604	Lu$_2$O$_3$	625
CeO$_2$	594	PrO$_2$	475							Eu$_2$O$_3$	554			TbO$_2$	486												
ThO$_2$	613	Pa		UO$_2$	541	NpO$_2$	514	PuO$_2$	527	Am$_2$O$_3$	585	Cm		Bk		Cf		Es		Fm		Md		No		Lw	
				U$_4$O$_9$	502																						
				U$_3$O$_8$	447																						
				UO$_3$*	408																						

negative of the enthalpy of formation from the elements per mole of O (not O_2) for various oxides of the elements. Use of this table is shown by the following example that examines the suitability of using Al_2O_3 or Y_2O_3 as a crucible for Ti. The numbers taken from the table are: Al_2O_3 ($\Delta H_f = -558$ kJ/mol), Y_2O_3 ($\Delta H_f = -627$ kJ/mol), TiO ($\Delta H_f = -543$ kJ/mol). For the possible reaction of Al_2O_3 with Ti

$$\tfrac{1}{3}(Al_2O_3) + Ti \rightarrow TiO + \tfrac{2}{3} Al$$

$$\Delta H_{reaction} = \Delta H_f^{products} - \Delta H_f^{reactants}$$
$$= (-543 + \tfrac{2}{3} \cdot 0) \text{kJ/mol} - (-558 + 0) \text{kJ/mol} = 15 \text{ kJ/mol}$$

whereas for the possible reaction of Y_2O_3 with Ti

$$\tfrac{1}{3}(Y_2O_3) + Ti \rightarrow TiO + \tfrac{2}{3} Y$$

$$\Delta H_{reaction} = \Delta H_f^{products} - \Delta H_f^{reactants}$$
$$= (-543 + \tfrac{2}{3} \cdot 0) \text{kJ/mol} - (-627 + 0) \text{kJ/mol} = 74 \text{ kJ/mol}$$

The equilibrium constant, K, is given by

$$\ln K = -\frac{\Delta H_{reaction}}{RT}$$

At the melting point of Ti, $RT = 16.2$ kJ/mol, so the equilibrium constant is of order unity, $\exp(-1)$, for the reduction of Al_2O_3, whereas the equilibrium constant is of order $\exp(-4)$ for the reduction of Y_2O_3. Thus a significant amount of Al_2O_3 can be reduced by Ti, whereas negligible amount of Y_2O_3 is reduced by Ti. Yttria is a better crucible for Ti than alumina.

The observed absence of a distinct reaction product between a crucible and a sample can be misleading. Reaction products can sometimes go into liquid or solid solution in the sample, affect the thermal measurement, and not be easily observable; for example, Al and O from Al_2O_3 into Ti alloys.

2.2.8 Evaporation

Excessive evaporation can reduce the sample mass and lead to incorrect measurement of enthalpy. Preferential loss of one or more components of an alloy can also lead to composition change. Mass loss is one of the reasons to employ simultaneous DSC/TGA (thermal gravimetric analysis). In addition, metals and alloys with high vapor pressure will lead to contamination of the furnace interior and thermocouple support assemblies. A good rule of thumb is to avoid materials with vapor pressures higher than approximately 100 Pa at the melting point (e.g., Zn and Cd).

2.2.9 Initial Metallurgical State of Alloy Samples

At heating rates typical of a DTA, one cannot be certain that the microstructure of an alloy sample is in *global equilibrium* prior to or during analysis. A characteristic time required to eliminate spatial concentration gradients within phases is $\lambda^2/4D_S$ where D_S is the solid diffusion coefficient and λ is a distance that characterizes the modulation of the concentration. Near the solidus, D_S is typically $10^{-13}\,\text{m}^2/\text{s}$ for substitutional solutes or $10^{-10}\,\text{m}^2/\text{s}$ for interstitial solutes. For solidified samples cooled at between 5 and 15K/min, typical spacings of dendritic structures are approximately 100–200 μm. Hence, hold times of the order of 10^5 s or 10^2 s may be required for substitutional or interstitial alloys, respectively, to become homogenized. For interstitial alloys, normal thermal equilibration of the instrument prior to scans is usually sufficient to eliminate concerns about the prior homogeneity state of the sample and global equilibrium is likely to be maintained at slow scan rates in the DTA. However, the microstructural state of the substitutional materials must always be considered. If the user wants to determine the thermodynamic solidus, a series of melting runs following isothermal holds at successively higher temperatures below the proposed solidus value should be conducted to ensure that the sample has been annealed into an equilibrium structure (see Section 3.2.2).

2.3 Reference Materials

The ideal reference material is a substance with the same thermal mass as the sample, but with no thermal events over the temperature range of interest. The thermal mass is the product of the mass and the **heat capacity**. This provides the optimum condition to obtain a **horizontal baseline**. When the baseline is not horizontal, various methods (see e.g., Ref. [11]) are used to fit and subtract a non-constant baseline from the signal. For some instruments, such a subtraction is done by the software. A horizontal baseline is not essential for the measurement of transition temperatures, but is important for quantitative enthalpy measurement. Alumina powder obtained from the instrument maker is often used as a convenient reference material for DTA and DSC of metals and alloys, even though the heat capacity differs significantly from that of most metals.

2.4 Calibration and DTA Signal from Pure Metals

2.4.1 Fixed Point (Pure Metal) Enthalpy vs. Temperature Curves and DTA Response

Calibration is performed by melting a pure metal because it exhibits a particularly simple enthalpy vs. temperature relation $H_S(T_S)$ as shown in Fig. 5.6(a). The subscript S is used to denote sample enthalpy and sample temperature as distinguished from, for example, reference material enthalpy, sample thermocouple temperature. Below the melting point, T_M, the slope, $dH_S(T_S)/dT_S$, of the curve is the solid heat capacity. At the melting point, the curve experiences a jump in enthalpy, L, equal to the **latent heat** of fusion. Above the melting point the slope is the liquid heat capacity. Figure 5.6(b) shows the derivative curve, $dH_S(T_S)/dT_S$, where we employ the **delta function** $L\delta(T_S - T_M)$ of strength L located at $T_S = T_M$ to represent the jump of the $H_S(T_S)$ curve at T_M. Figure 5.6(c) shows schematic melting and cooling DTA curves. A useful way to think about the response of a DTA to melting and solidification is to consider an instrument response or transfer function that converts the delta function of Fig. 5.6(b) into the quasi-triangular shaped peaks for melting and freezing seen in Fig. 5.6(c). Fourier methods are often used to represent instrument response functions. However due to the intrinsic non-linearity of DTA response

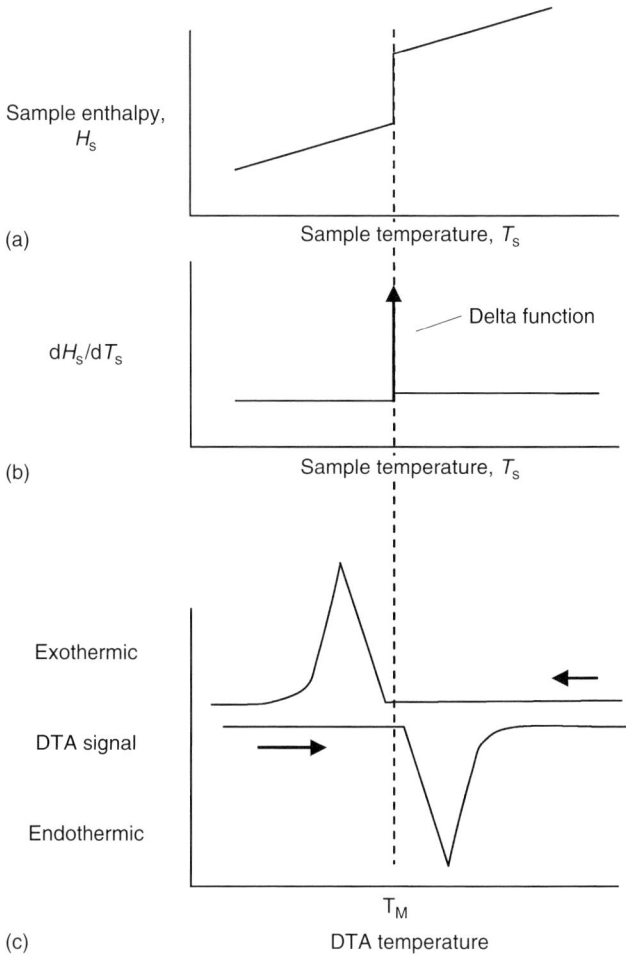

Figure 5.6 Schematic diagram showing (a) enthalpy vs. temperature for a pure metal; (b) corresponding derivative $dH_s(T_S)/dT_s$ curve; and (c) DTA signal for melting (bottom) and freezing (top). The small difference in heat capacity of liquid and solid leads to a small offset of the baseline before and after melting. The onsets in the DTA curves are shown with a small deviation from the melting point, T_M due to heat flow limitation in the DTA. This difference on melting is adjusted to zero by the calibration procedure, at least for one heating rate.

to the enthalpy curves, the heat flow model of Appendix C is preferred. The model confirms the central role that the rate of change of the sample enthalpy with respect to sample temperature, $dH_S(T_S)/dT_S$, plays in the DTA response, Eq. (C3). This derivative is usually referred to as the heat capacity; however, the term is sometimes ambiguous, especially for alloys, in that it may or may not include the heat of fusion. In this document, the derivative $dH_S(T_S)/dT_S$ will be taken to include the heat of fusion. The derivative vs. temperature curve concept will be employed throughout this document as it becomes quite useful to understand and interpret alloy behavior in Sections 3 and 4 and Appendices F and G.

Table 5.2 Fixed point temperatures for calibration [17–19].

Substance	T (°C)	References
In	156.5985	18
Sn	231.928	18
Bi	271.402	18
Cd	321.069	19
Pb	327.462	19
Zn	419.527	18
Sb	630.628	19
Al	660.323	18
Ag–28.1% Cu★	779.63	19
Ag	961.78	18
Au	1064.18	18
Cu	1084.62	18
Ni	1455	17
Co	1494	17
Pd	1554.8	19

★Eutectic composition. In this document, the % symbol will refer to percentage by mass.

2.4.2 Temperature Calibration: Effect of Instrument Thermal Lags on Onset Determination

Temperature calibration of a DTA is typically performed using the melting of a pure metal or series of pure metals. A list of commonly used calibration substances is given in Table 5.2. Metals of 99.99% purity are usually adequate for metallurgical work. Research involving International Temperature Standards (ITS-90) employ higher purity metals and are available as NIST Standard Reference Materials (www.nist.gov).

In order to perform the calibration, the onset of melting must be determined from the melting data. For precision work, one must quantify various subtleties in the detection of melting for a pure metal and the thermal lags inherent in the instrument.

To directly observe the effects of the thermal lags, the melting and freezing of pure Sn was performed in a DTA instrument with an additional thermocouple (type K) immersed directly into the Sn. The sheathed thermocouple was only 250 µm in diameter so that its thermal response is quite rapid. Figure 5.7(a) and (b) show the temperature vs. time and associated DTA plots obtained with the immersed thermocouple and with the ordinary instrument thermocouple located beneath the sample

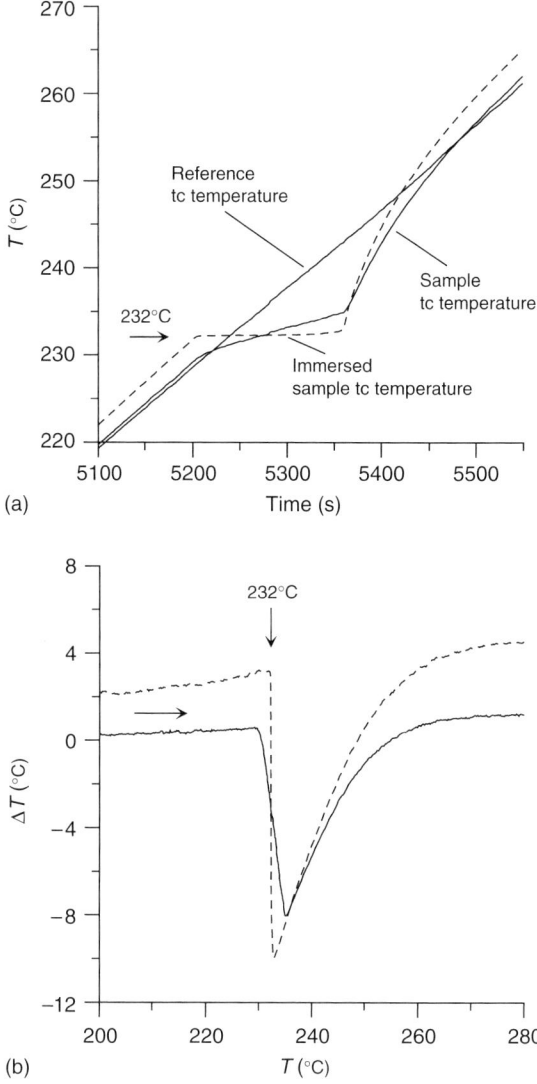

Figure 5.7 (a) Measured temperature vs. time plots and (b) associated DTA plots for melting of pure Sn using instrument thermocouple (solid) and an immersed thermocouple (dashed). Mass = 163.2 mg; heating rate = 5K/min.

cup. One notes the perfectly isothermal plateau at 232°C for the immersed thermocouple and the associated vertical drop in the DTA plot at the Sn melting point. The instrument sample thermocouple, on the other hand, is not isothermal during melting nor is the drop vertical on the associated DTA plot. This difference is due to the heat transfer limitations between the metal sample and the sample thermocouple and gives rise to the linear section of the DTA peak shape. In the section on supercooling effects (Section 3.4.1), we will show the corresponding data during cooling experiments with the immersed thermocouple.

2.4.3 Temperature Calibration: Choice of Onset Temperature

In normal DTA where the thermocouple is not immersed directly in the sample, the thermal lags cause the linear portion of the peak shape as well as a less sharp demarcation of the initiation of melting. This loss of sharpness can lead to difficulties of calibration for highly precise work as described in Appendix D. In Fig. D1(c), the temperature of the first detectable deviation from baseline is labeled T_T^{onset}. The subscript T is used to denote thermocouple temperatures. Due to the inherent trouble of picking this temperature experimentally and due to the noise in the data, an alternative melting onset temperature is commonly employed. This procedure takes advantage of the fact that the thermocouple temperature vs. time, $T_T(t)$, curve quickly becomes linear after the melting onset; the onset temperature, T_T^{extrap} is taken as the intersection of a linear fit to the downward sloping linear section of the melting peak and the extrapolation of the baseline.

Following Eqs. (D3) and (D4), both T_T^{onset} and T_T^{extrap} deviate from T_M by an amount that depends on heating rate and on sample mass through the parameter ρ and the time constants, which depend on the area of contact between sample and cup. Thus calibration should ideally be conducted for a mass and heating rate similar to those that will be used for the test samples.

In addition to this consideration, it is best to choose the first deviation from baseline method for onset determination. We show in Appendix D that picking the first sign of melting actually has a smaller deviation from the true melting point and smaller heating rate dependence than the onset determined by the extrapolation procedure. This is one reason to prefer using the first detectable deviation from baseline for calibration. A second, and more important reason, comes from the principle that the same method for the determination of the onset should be used for the calibration *and* the actual measurement on alloys. Alloys usually have a melting range and, therefore, do not melt with a linear section of the DTA curve. Thus the onset for alloy melting is better taken using the first detectable deviation method. With these points in mind, it makes sense to use the first detectable deviation from baseline method for the calibration.

We note that ASTM E967 and articles on instrument calibration [20,21] use the extrapolation onset temperature, T_T^{extrap}, to determine all onsets, even for alloys. With the use of computers for recording the thermocouple voltages, instrument manufacturers introduced software for the analysis of the DTA curves and determination of the reaction temperatures. In most cases the software uses the extrapolation method for the temperature determination T_T^{extrap} instead of T_T^{onset}.

2.4.4 Quantitative Enthalpy and Heat Capacity Calibration

In order to determine the quantitative heat flux, the difference signal (ΔT) must be converted to one of energy units by a calibration process. The method uses the heat of a specific transformation per unit mass in a standard sample, such as heat of fusion. The instrument sensitivity coefficient, S, at various temperatures is determined from the ratio

$$S = \frac{m_{standard} \Delta H_{standard}}{\int_{t_1}^{t_2} \Delta T(t) dt}$$

obtained from melting a series of standards (usually pure metals). The integral is the area of the deviation from the baseline of the peak determined from a plot of ΔT vs. time. Alternately the original voltage difference from the thermocouples can be used instead of the temperature. The quantity $m_{standard}\Delta H_{standard}$ is the product of the mass of the standard and the heat of fusion per unit mass. For an unknown sample then,

$$m_{sample}\Delta H_{sample} = S\int_{t_1}^{t_2} \Delta T(t)dt$$

where S may be interpolated between the two standards spanning the temperature range of interest. The calibration is often performed by the instrument software and converts the DTA signal area from units of millivolt·seconds or Kelvin·seconds to Joules. In order to obtain accurate heat-flux information, the same heating/cooling rate, gas flow rate, sample/reference cups and temperature range should be employed for standard and unknown.

After conversion of the DTA signal to energy, some instruments provide a second calibration to directly provide heat capacity measurement. For many heat-flux DSC instruments, heat capacity calibration is done by software with data obtained by performing a heating scan with a pair of empty cups, to get the baseline, and performing a second heating scan with the chosen reference material in one cup and an empty second cup. More details are given in ASTM E968.

2.5 Major Points

- Important factors for proper technique:
  - Determination of what temperatures are being recorded and presented by the instrument and instrument software.
  - Calibration.
  - Crucible and atmosphere selection.
  - Consideration of thermal gradients and associated thermal lag times within the instrument.
- Determination of instrument time constants provides quantitative insight into thermal lags important in consideration of calibration procedures and general DTA response.
- For alloys, the microstructure and sample history may influence the measured response to thermal analysis.

3 Analysis of DTA Data for Binary Alloys

The general response of DTA during cooling and heating for a series of alloys in a eutectic system where there is limited solid solubility is given schematically in Fig. 5.8. In such general descriptions it is usually assumed that there is no barrier to nucleation on cooling and that solute diffusion is adequate to maintain the phases uniform in concentration at each temperature during the process. Under these

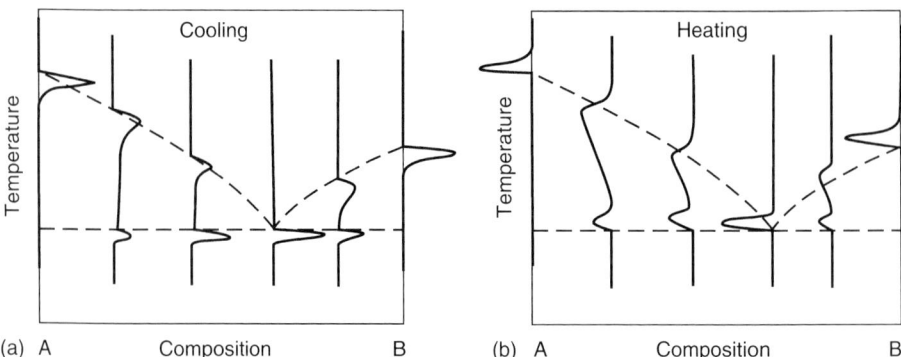

Figure 5.8 Schematic DTA response on cooling and heating of pure A and B and four other alloys superimposed on a simple eutectic phase diagram.

assumptions, the phase fractions at each temperature follow the simple **lever rule** and the general shape of the DTA response is readily predicted, if the phase diagram and heats of fusion are known.

For other binary phase diagram features, such as peritectic reactions, the DTA response is not so easily understood. Thus the model of Appendix C is used to simulate how various phase diagram features are reflected in the DTA signals. It is our belief that such simulations can provide the DTA user with the understanding to enable more accurate interpretation of DTA results from samples with unknown phase diagrams.

For the simulations, we not only use the full equilibrium (lever rule) to obtain enthalpy vs. temperature curves but also another limiting case that is important in solidification. If an alloy has been solidified (even in the DTA), it may contain spatial gradients of composition within the phases. A limiting case of this kind of solidification is given by the **Scheil (Scheil–Gulliver)** approach. This approach, well known in the solidification and casting field, assumes that no diffusion occurs in the solid during freezing, but that the liquid composition is uniform [22]. DTA melting simulations using the lever enthalpy–temperature relation apply to an alloy equilibrated prior to melting and where diffusion was adequate to guarantee spatial concentration uniformity of all phases during melting. DTA melting simulations using the Scheil enthalpy relations apply to a microstructure that was previously solidified and remelted, where there was no solid diffusion during *both* processes. Clearly these are extreme cases; interpreting DTA data from the melting of an equilibrated sample as well as an as-cast sample requires an analysis of solid diffusion for both the freezing process and the melting process that is beyond the scope of this document (Ref. [23] treats diffusion during melting). Enthalpy vs. temperature relations can be computed more simply for the extreme cases without the requirement of any kinetic information or geometry of the microstructure. The enthalpy vs. temperature relations are obtained from two different methods in this document: analytical expressions applicable to straight line phase diagrams (Appendix E) and numerical data applicable to real alloy systems obtained from a calculation of phase diagrams (CALPHAD) assessment of the thermodynamic functions.

3.1 General Behavior for a Binary Eutectic System: Example Ag–Cu Alloy Melting

The Ag–Cu alloy system is used to illustrate various features of DTA analysis.

3.1.1 Enthalpy vs. Temperature Curves

Figure 5.9(a) shows the Ag–Cu phase diagram, and Fig. 5.9(b) shows the enthalpy vs. temperature curves for six Ag-rich Ag–Cu alloys, under equilibrium and Scheil conditions. The thermodynamic data of Ref. [24] were employed to construct the phase diagram and to produce the enthalpy vs. temperature information using code developed by Ref. [25]. Other code, such as Thermo-Calc or Pandat[6] could have been used. The 1% Cu alloy exists as a single-phase solid solution alloy with liquidus and solidus temperatures defining the temperature range where liquid phase and the Ag-rich face centered cubic (FCC) phase coexist. (In this document, the % symbol will refer to percentage by mass.) With increasing temperature, the enthalpy curve has a sharp increase of slope at the solidus for the equilibrium curve but a smooth increase in slope for the Scheil curve.[7] For the 5% Cu alloy, the enthalpy curve has a sharp increase of slope at the solidus for the equilibrium curve but an isothermal jump for the Scheil curve corresponding to the eutectic. The enthalpy curve is concave up between the solidus and the liquidus, combining the effects of changing heat capacity of the solid and liquid phases as well as the heat of fusion associated with the changing fractions of liquid and solid phases. The 9% Cu and 15% Cu compositions are two-phase alloys when solid; the solidus is the eutectic temperature. The enthalpy curves for both Scheil and equilibrium conditions have isothermal (vertical) rises in enthalpy corresponding to the eutectic melting of a mixture of Ag-rich phase and Cu-rich phase. The height of the isothermal rise is larger for Scheil than for equilibrium conditions and is larger for the 15% alloy than for the 9% alloy. Upon completion of eutectic melting, all of the Cu-rich phase has melted and the enthalpy curve rises with heating corresponding to the continued melting of the Ag-rich phase. For the Ag–28.1% Cu alloy, the enthalpy curves has only an isothermal rise in enthalpy corresponding the eutectic melting.

For the Ag–15% Cu and 23% Cu alloys there is little difference between the Scheil and equilibrium calculations.[8] This is because there is so little non-eutectic Ag-rich FCC phase that any microsegregation within this phase has little effect on the heat evolution and fraction of eutectic that forms. For the other alloys, significant differences arise between equilibrium and Scheil enthalpies that are important to the interpretation of DTA curves. Most striking is the presence of an isothermal rise in the enthalpy corresponding to the eutectic in the 5% Cu alloy for the Scheil enthalpy whereas no rise is present for the equilibrium enthalpy. The microsegregation

---

[6]The use of a specific trade name of instruments or software in this document is for illustration purposes only and does not constitute an endorsement by NIST.
[7]Technically, the Scheil approach predicts some eutectic for all alloys. In this case, the mass fraction of eutectic is very small, 0.003 and would not be detectable.
[8]We note the difference in solid heat capacity predicted by the equilibrium and Scheil models. This difference is due to the fact that the equilibrium computation permits the solid state adjustment of the phase fraction and compositions of the Ag-rich and Cu-rich phases as temperature changes, whereas the Scheil calculation assumes that these phases have fixed composition and phase fraction as the temperature changes.

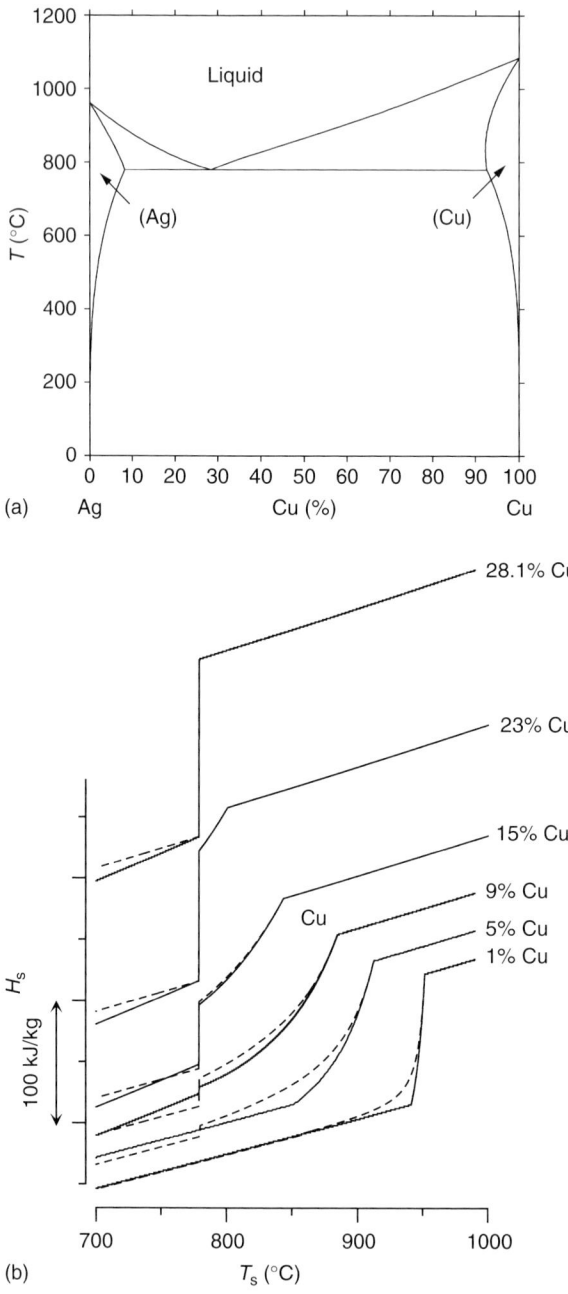

Figure 5.9 (a) Ag–Cu phase diagram; (b) enthalpy vs. temperature curves for five Ag–Cu alloys (solid, equilibrium; dashed, Scheil); the curves have been shifted vertically for clarity, only differences of enthalpy are relevant.

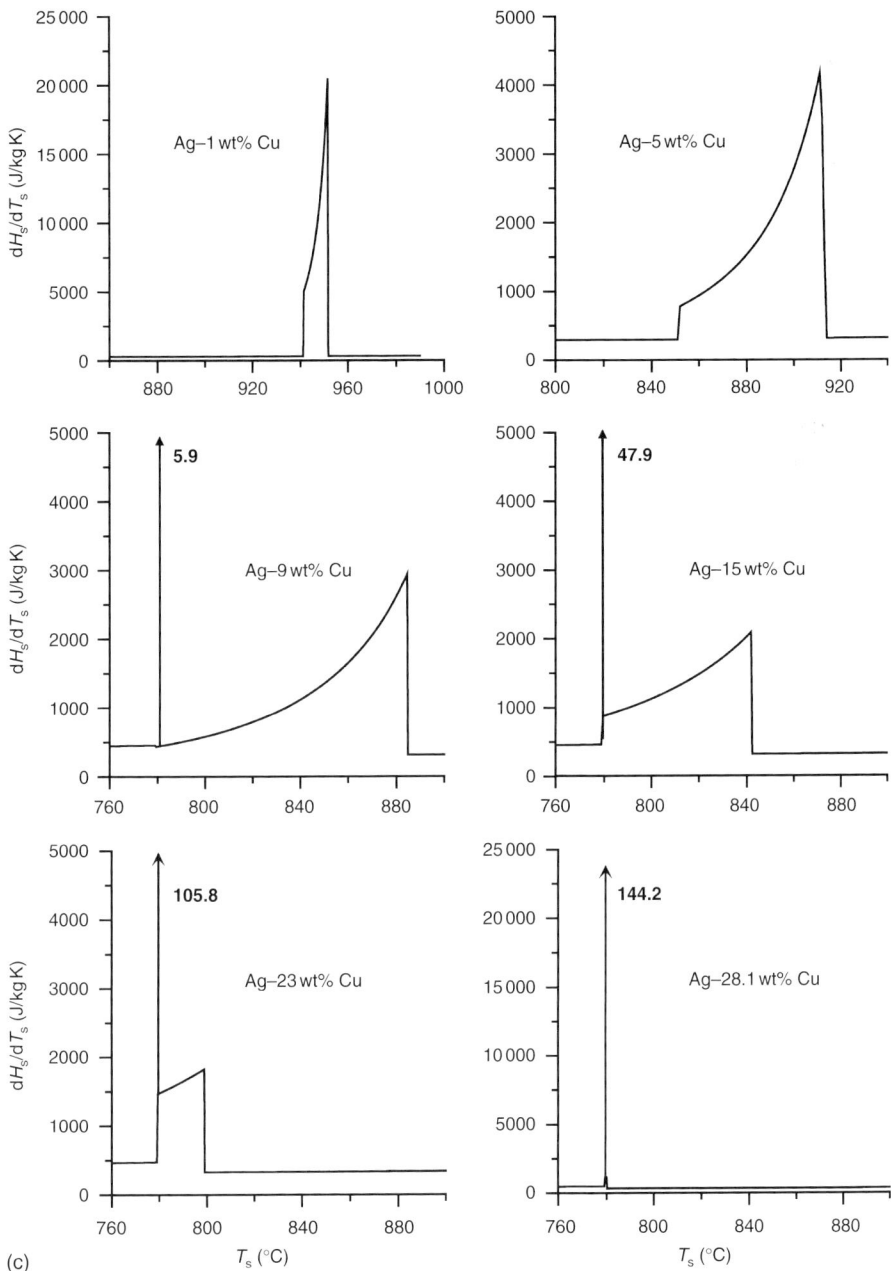

Figure 5.9 (Continued) (c) dH_S/dT_S vs. T_S curves for equilibrium conditions for various Ag–Cu alloys derived from the equilibrium enthalpy curves. The vertical arrows at the eutectic temperature (779°C) represent delta functions of the indicated strength in kJ/kg.

within the Ag-rich FCC phase predicted by the Scheil calculation leads to the formation of eutectic, whereas when solid diffusion is adequate to keep the Ag-rich FCC phase uniform in composition, no eutectic forms.

3.1.2 Derivative of Enthalpy vs. Temperature Curve and their Relation to DTA Curves

As developed in Section 2.4.1, the temperature derivative (slope) of the enthalpy vs. temperature curve determines the DTA response. Figure 5.9(c) shows these derivatives for the Ag–Cu alloys for the equilibrium enthalpy. The delta functions corresponding to the isothermal jumps in enthalpy are labeled with the size of the jump. Just as the DTA response to a delta function for a pure metal is altered by the instrument, so is the response to the alloy dH_S/dT_S curve. This alteration can be quantitatively viewed through the following sequence of operations. First flip a dH_S/dT_S vs. T_S curve (from Fig. 5.9(c)) so that the positive y-axis become negative. Second, broaden this curve in the direction of the x-axis (temperature axis). This shape is a rough approximation to the DTA curve for melting.

The experimentally determined DTA melting responses of a series of Ag–Cu alloys at three heating rates are shown in Fig. 5.10. It can be seen that the general shapes of the DTA curves are indeed modified versions of the derivative curves, dH_S/dT_S vs. T_S. For Ag–1% Cu and Ag–28% Cu only a single peak is observed. For Ag–9% Cu and Ag–15% Cu two peaks are clearly shown. The first peak begins near the eutectic temperature and the maximum signal of the second peak occurs near the liquidus. The reason the maximum signal of the second peak is associated with the liquidus temperature will be discussed in Section 3.3. For Ag–5% Cu at the fastest heating rate, a small signal is evident at the eutectic temperature. At the slowest heating rate, the small signal for the eutectic melting is absent. The two peaks for Ag–9% Cu and Ag–15% Cu are for the completion of eutectic melting and the final melting of the Ag-rich FCC phase. For Ag–23% Cu, while there are not two distinct peaks, there is a slight bulge or shoulder for the final melting of the Ag-rich FCC phase before the return to the baseline that corresponds to the liquidus. Clearly, the peak temperature corresponding to the end of melting of the Ag-rich phase is too close in temperature to the strong signal from the end of melting of the eutectic for the response times of the instrument. Thus no distinct peak is observed for the liquidus.

3.1.3 Comparison to Experiment

Figure 5.11 shows how the measured DTA scans at 5K/min compared to the calculated DTA scans using the equilibrium and Scheil enthalpies. For the 15% Cu, 23% Cu and 28.1% Cu, the model predicts all observed features as well as trends of amplitude and observed temperatures with alloy composition. The general shapes of the experimental curves for the 1% Cu and 5% Cu alloys generally lie between the shapes of the lever calculation and Scheil simulation in the lack of sharpness of the onset. This intermediate behavior indicates that solid diffusion is not sufficient to permit full equilibration to be reached at each temperature during heating. The general agreement between the calculated and measured results validates the use of the model to simulate and assist in the data interpretation of more complex alloys.

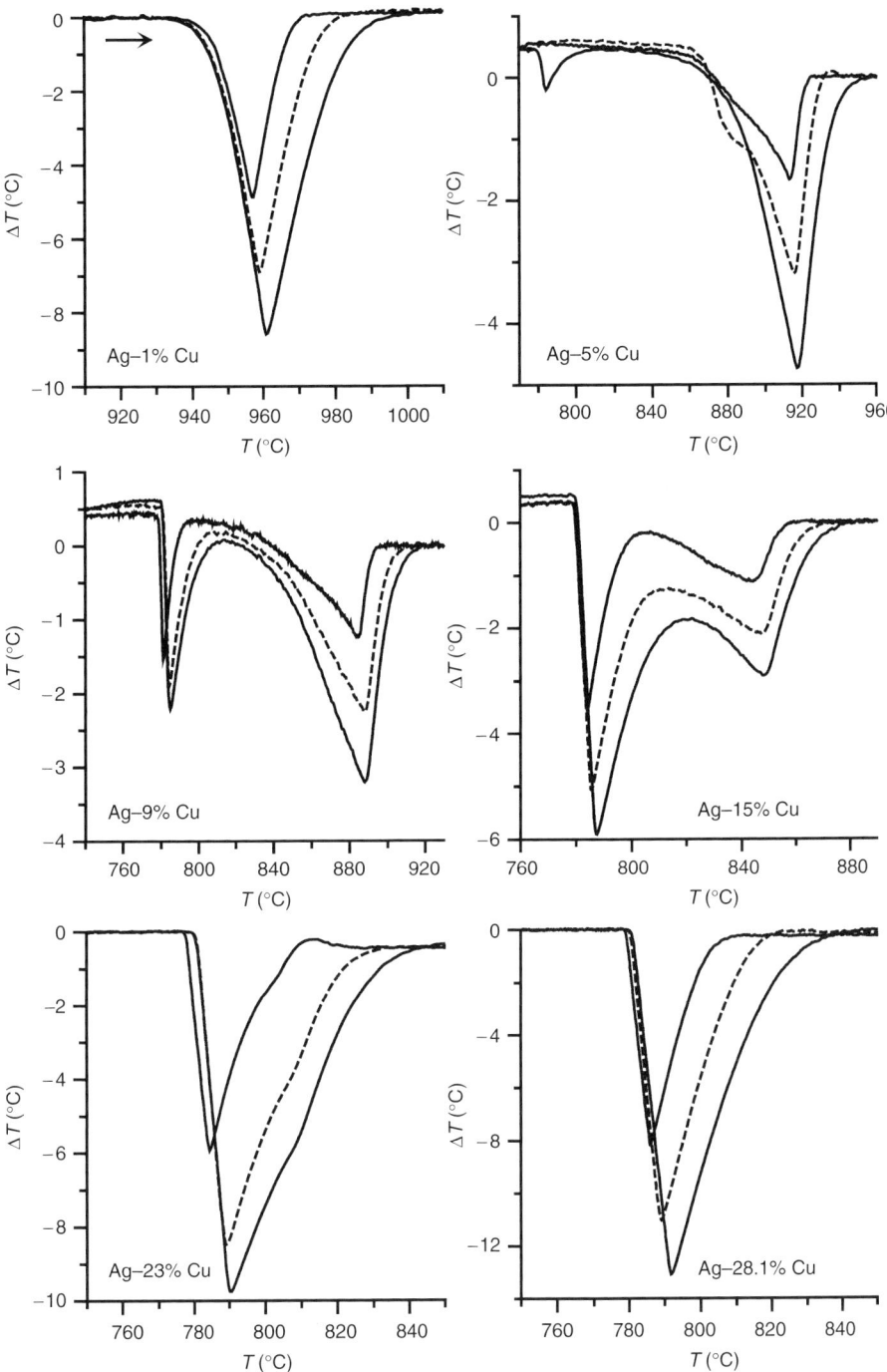

Figure 5.10 Experimental DTA melting scans for a series of Ag–Cu alloys at three different heating rates, 5K/min (solid), 10K/min (dashed) and 15K/min (solid). The peak height is bigger with faster heating. Note that the small signal for the eutectic melting is absent in the Ag–5% Cu alloy at the slower heating rates. Note also the absence of a distinct second peak in Ag–23% Cu. Percentage is by mass.

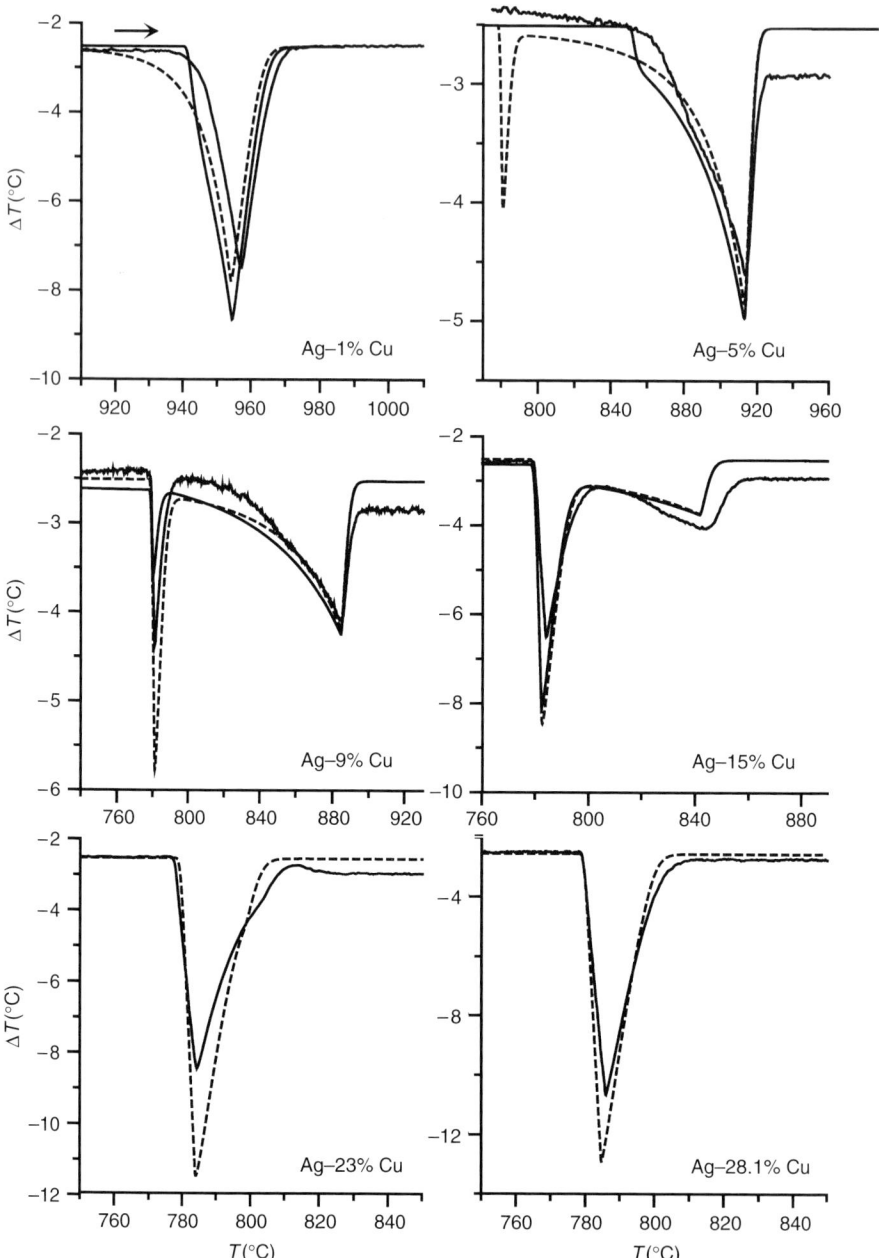

Figure 5.11 Comparison of experimental and calculated DTA curves for a series of Ag–Cu alloys melted at 5K/min. The experimental DTA curves are shown solid and contain noise; the computed DTA curves for lever are solid and for Scheil are dashed. Lever and Scheil curves for the two alloys, Ag–23% Cu and 28.1% Cu, are essentially identical.

3.2 Problems with Solidus Determination on Heating

3.2.1 Incipient Melting Point vs. Solidus

The temperature at which a solid alloy begins to melt depends on the history of the material. The terms *solidus*, *non-equilibrium solidus* and *incipient melting point*, are variously used in the literature to describe the minimum temperature where some liquid forms. As mentioned earlier, this guide only uses the term solidus to refer to the thermodynamic quantity. Cast alloys often begin melting at temperatures below the solidus of the average composition, the incipient melting point, which stems from the existence of composition gradients within individual phases or the presence of extra phases in the alloy microstructure. Whether a DTA sample is taken directly from a casting or if it has been frozen in the DTA, it is uncertain if the solid microstructure is in global equilibrium at heating rates typical of a DTA. Hence, the microstructural state of the material must *always* be considered when performing an alloy melting experiment. This point was previously discussed in Section 2.2.9.

3.2.2 Effect of Hold Time Prior to Melting

If the user wants to determine the thermodynamic solidus rather than the incipient melting point, a series of melting runs following isothermal holds at successively higher temperatures below the proposed solidus value should be conducted to ensure that the sample has been annealed into an equilibrium structure. An example of this situation is shown in Fig. 5.12 for a Ag–5% Cu alloy. After a first melting scan to obtain good thermal contact with the alumina crucible, the sample was cooled at 15K/min

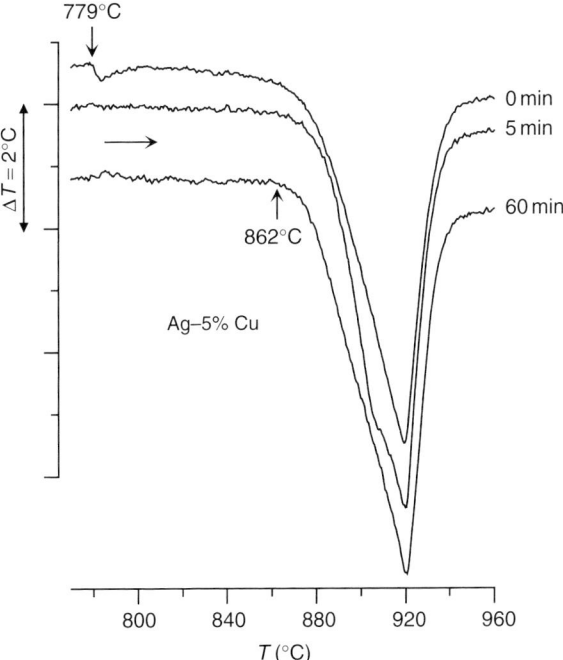

Figure 5.12 DTA signal for melting of Ag–5% Cu showing the effect of hold for the indicated times at 760°C and subsequent heating at 15K/min.

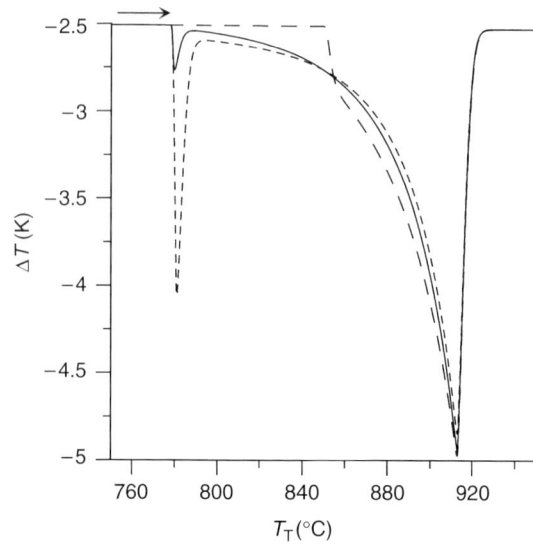

Figure 5.13 Computed DTA curve for an enthalpy vs. temperature relation derived from a back diffusion model of solidification at 5K/min of Ag-5% Cu is shown solid. $D_S = 1 \times 10^{-14} m^2/s$, $\lambda_2 = 3.8 \times 10^{-5}$ m. The equilibrium curve is long dashed and the Scheil is short dashed. The peak eutectic signal at approximately 780°C is intermediate between that for the Scheil and equilibrium cases but is not as small as the measured eutectic signal with no annealing in Fig. 5.13.

to 760°C and then immediately remelted at 15K/min. This cycle was repeated with successively increasing hold times at 760°C. The temperature of the large peak remains relatively fixed, but the onset varies considerably. For the sample with zero hold time, the onset of melting occurs at the eutectic temperature, 779°C. This onset disappears with the 5 min hold, the onset increasing to 862°C, where it remains with further increase of the hold time. The lower onset occurs without hold time because of the metallurgical state of the alloy after freezing. Such a sample contains dendritic coring of the Ag-rich phase (**microsegregation**) and eutectic consisting of Ag- and Cu-rich phases. The amount of time needed to homogenize the phases and dissolve the eutectic depends on the hold temperature, which affects the solid diffusion coefficient, and on the distance between dendrite arms. The required hold time increases as the square of the length scale. The finer the length scale, the shorter the hold time. Therefore, as a corollary, more rapid preceding freezing results in a finer spacing and a shorter hold time. Significantly, this is contrary to normally accepted DTA practice where slower cooling is usually thought to be better, an idea based on the assumption that the slow rates in a DTA are sufficient to guarantee freezing following the lever rule.

Figure 5.13 shows the computed DTA curve for an enthalpy vs. temperature relation when diffusion can occur in the solid as derived from a *back diffusion* model of solidification, Eq. (E11), at 5K/min for a Ag–5% Cu alloy. The predicted strength of the eutectic signal is less than that for the Scheil case, but is not as small as the measured eutectic signal with no annealing in Fig. 5.12. This might indicate that the fraction of eutectic in the DTA cast alloy has undergone further reduction during the short equilibration time required before the heating scan was begun.

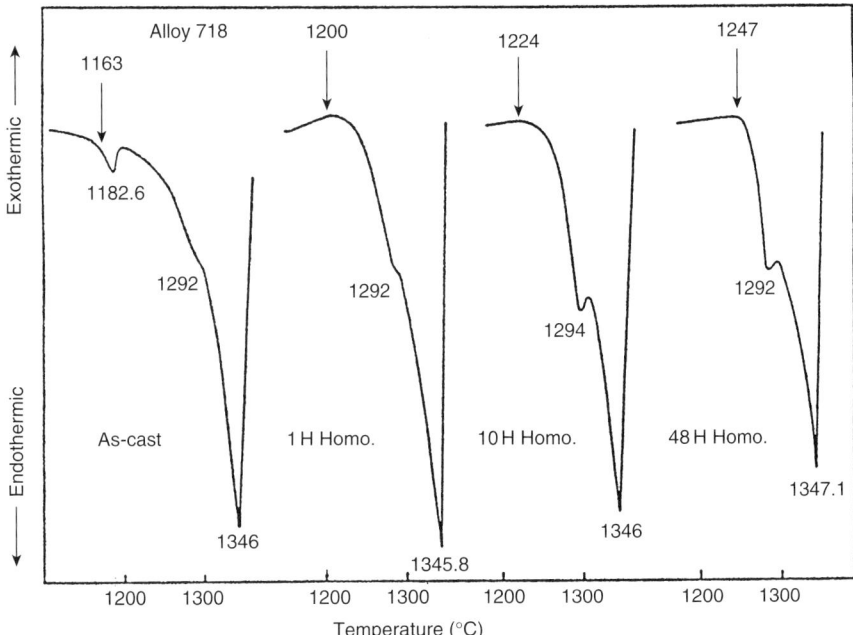

Figure 5.14 DTA signal for Inconnel 718 showing effect of annealing time [26]. With annealing, the $(Fe,Cr)_2Nb$ Laves phase dissolves and the onset of melting increases from 1163°C to 1247°C.

Figure 5.14 shows another example of how annealing time prior to melting changes the DTA signal for Inconnel 718 alloy [26]. With annealing, the $(Fe,Cr)_2Nb$ Laves phase present in the industrially cast sample dissolves and the onset of melting increases from 1163°C to 1247°C to reveal the true solidus temperature. The lower temperature is the incipient melting point of the casting and is important for constructing heat treatment schedules for the casting.

3.2.3 Errors Caused by Using Extrapolated Melting Onset (Tangent Construction)

It has already been noted (Section 2.4.3) that the temperature of the first visible onset of melting is generally preferable over linear extrapolation methods for solidus determination (in spite of the prevalent use of the latter). This is especially true for solid solution alloys. Unlike a pure metal or an alloy containing eutectic, there is no physical basis for a linear section of the DTA curve. There is thus no theoretical basis for using the intersection of the baseline with the linear extrapolation of any portion of the DTA curve to determine the solidus temperature. As soon as a deflection from the baseline occurs, melting is surely taking place. Compare the DTA curves for the melting behavior of pure Ag and eutectic (Fig. 5.5, second melts) to the curves for Ag–1% Cu and Ag–5% Cu (Fig. 5.11); For the pure metal and the eutectic, the onset is quite sharp and the DTA curve is linear after the onset; the DTA curves for the alloys with no eutectic have no linear portion near the onset of melting from which to construct an extrapolated onset. Examples of errors introduced using the

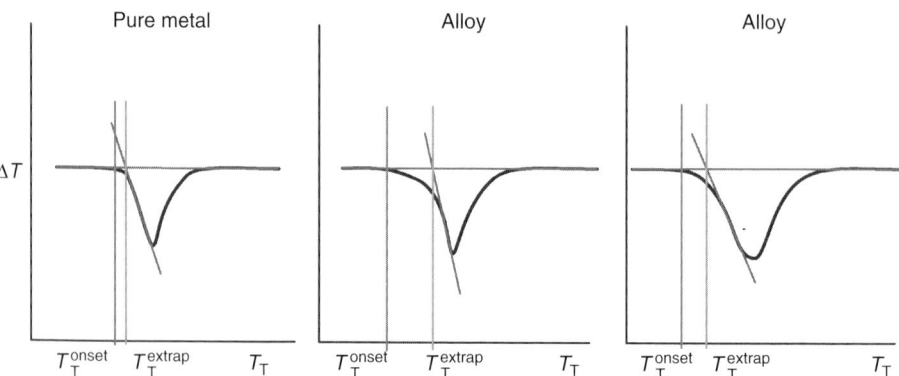

Figure 5.15 Schematic DTA plots showing error introduced by using the extrapolation method for onset determination rather than the first detectable departure from baseline.

extrapolation method of determining onset compared to the first detectable departure from baseline are shown schematically in Fig. 5.15.

3.3 Problems with Liquidus Determination on Heating

It is customary to select the peak temperature of the last thermal event on heating as the liquidus temperature. However in the results for Ag–23% Cu (mass), Fig. 5.10, it was seen that a distinct peak did not exist for the end of melting because the liquidus temperature was too close to the eutectic temperature. Other subtleties of liquidus determination on melting require exposition.

3.3.1 General DTA Curve Analysis

The alloy, Ag–15% Cu, is examined in more detail to indicate which features of its DTA curves are important and which features are unimportant during melting and freezing. This discussion may appear obvious to some but examples abound in the literature of incorrect conclusions drawn from DTA curves. Figure 5.16 repeats the dH_S/dT_S vs. T_S using the equilibrium enthalpy for Ag–15% Cu from Fig. 5.9(c) and also shows the computed DTA melting and freezing scans at 5K/min. As the dH_S/dT_S plot shows, there are only two temperatures that have significance: the vertical jump at 842°C (liquidus temperature) and the delta function at 779°C (eutectic temperature). Features on the DTA scans are labeled with: i: important or n: not important depending on how well they approximate these two temperatures. The significant temperatures i are close to the liquidus and the eutectic temperature. In contrast, the heat transfer details produce other features n on the DTA scans which bear no direct relation to the above two temperatures. Furthermore, note the asymmetry of the interpretation of the melting and freezing scans. For example, the peak on melting near the liquidus temperature is important but the peak just below the liquidus on cooling is unimportant. Similarly the temperature of minimum deflection from baseline between the eutectic and liquidus temperatures on heating has no meaning yet the

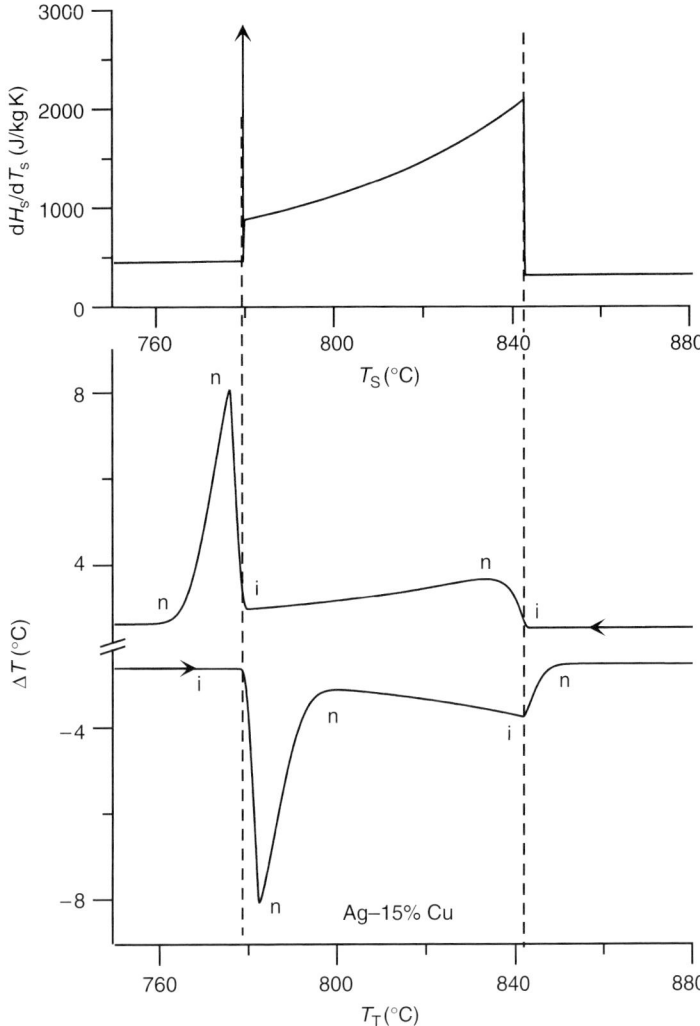

Figure 5.16 dH_S/dT_S vs. T_S (top) and computed DTA scans for melting and freezing at 5K/min (bottom) for Ag–15% Cu using equilibrium enthalpy. Features on the DTA scans are labeled with: i: important or n: not important.

minimum deflection between the liquidus and eutectic temperatures on cooling is important; it is the onset of eutectic freezing.

We also note that after the completion of melting or the completion of freezing, no extra heat beyond the liquid heat capacity is required for continued sample heating or cooling. The sample temperature increases/decreases rapidly after the completion of melting/freezing to catch up with the reference and furnace temperatures; the shape of the return and the temperature of the return of the DTA signal to the baseline are only measures of the efficacy of heat transfer. They have no metallurgical or thermodynamic meaning.

In solidification of binary alloys, one often employs two numbers to describe the liquidus and solidus curves for dilute alloys, the liquidus slope, m, in units of K/% (for example) and the partition coefficient, k, which is the ratio of the solid composition to the liquid composition. So far, the DTA response has been described for an alloy where the liquidus slope, m, is negative and the partition coefficient, k, is less than unity. See Section 3.3.5 for further complicating factors for alloys with $m > 0$ and $k > 1$.

3.3.2 Details of Computed Behavior of an Alloy on Melting

For a solid solution alloy with no eutectic, the equilibrium equation for enthalpy, Eq. (E11), and numerical solution of Eq. (C1) are employed to determine the temperature–time histories and the DTA signal for alloy melting. Figure 5.17 shows the time histories of the temperatures of the various parts of the DTA instrument for a hypothetical Ni-base alloy of composition of alloying addition $C_0 = 2\%$ ($k = 0.5$, $m = -10 K/\%$) with solidus/liquidus values of 1688K/1708K. It is important to note that the value of the thermocouple temperature at the instant when the sample completes melting at the liquidus temperature (Fig. 5.17(a) at ≈290 s) is ≈1720K, 12K higher than the liquidus temperature. Figure 5.17(b) shows that the peak temperature is near 1720K. The figure also shows the DTA plots of a 5% alloy and also pure Ni for comparison. Figure 5.17(c) shows how heating rate affects the computed DTA scan.

3.3.3 Small Liquidus Solidus Separation

Figure 5.18(a) shows calculated DTA peak temperatures superimposed on the phase diagram for $0\% < C_0 < 20\%$. Figure 5.18(b) shows the difference between the peak temperature and the liquidus temperature. One notes that the peak temperature lies above the liquidus temperature with an error increasing as the concentration of the alloying element decreases. In other words, if the freezing range of the alloy is small (of the order of the temperature difference between the onset and peak temperature of a pure metal melting DTA scan), the peak temperature of the DTA can only serve as an upper bound on the liquidus temperature.

3.3.4 Resolution of Difficulties Using Temperature Cycling Near the Liquidus

Cycling experiments can be used to determine the liquidus for small freezing range alloys or for alloys where the fraction of the final phase on melting is small. This approach was demonstrated in Ref. [27] for complex superalloys and is shown in Fig. 5.19. First an upper bound on the liquidus is determined from the peak temperature of an ordinary DTA scan, 1399°C in Fig. 5.19(a). Then a lower bound is determined by the cooling curve, 1358°C in Fig. 5.19(b) as the nucleation temperature. Clearly significant supercooling has occurred. Next the sample is cooled, reheated and held at a temperature below the upper bound temperature, 1380°C in Fig. 5.19(c) for 10 min, and then heating is reinitiated. Signs of an endothermic signal are sought. If absent, a fully liquid state is implied for the hold temperature and the hold temperature is above the liquidus temperature. If present, a partially solid state is implied at the hold temperature and the hold temperature is below the liquidus. In Fig. 5.19(c) with a hold temperature of 1380°C, no endothermic signal

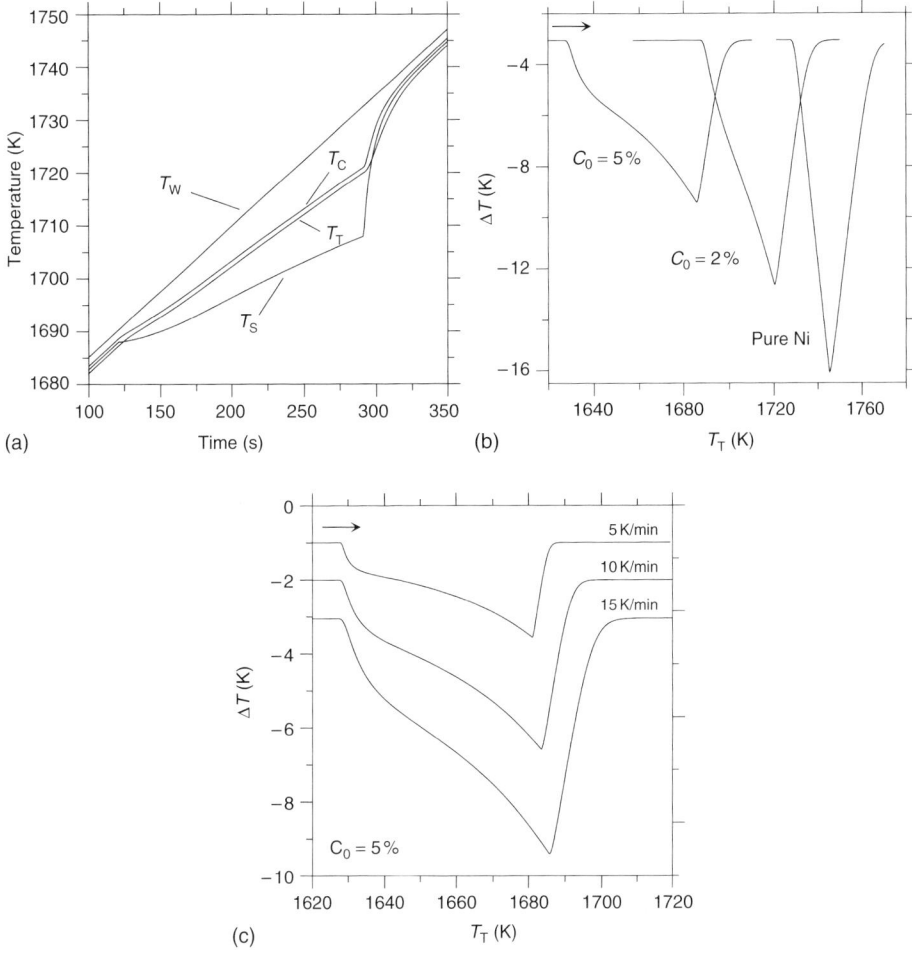

Figure 5.17 (a) Plot of calculated temperatures vs. time for sample (T_S), sample cup (T_C), sample thermocouple (T_T) and furnace wall (T_W) for melting at a heating rate of 15K/min of a 180 mg binary alloy sample that melts following the lever rule enthalpy, with $T_M = 1728$K, $m = -10$K/%, $k = 0.5$, $C_0 = 2$%. The model of Appendix C and the instrument time constants from Table C1 for Ni were employed. (b) Calculated DTA curves for pure Ni and alloys with $C_0 = 2$% and 5% following the lever rule at 15K/min. (c) Calculated DTA curves for alloy with $C_0 = 5$% following the lever rule at heating rates of 5, 10 and 15K/min.

was seen. Thus the cycle was repeated at a lower hold temperature. This process is repeated until an endothermic sign of melting is seen. With this process the liquidus temperature can be bracketed to within a few degrees, 1378–1379°C in the present case.

3.3.5 Alloys with $k < 1$ and $k > 1$: Peak Temperature

A second problem with respect to liquidus temperature determination may be encountered when dealing with an alloy for which the partition coefficient $k > 1$. The $k > 1$ case applies whenever the liquidus/solidus separation (freezing range)

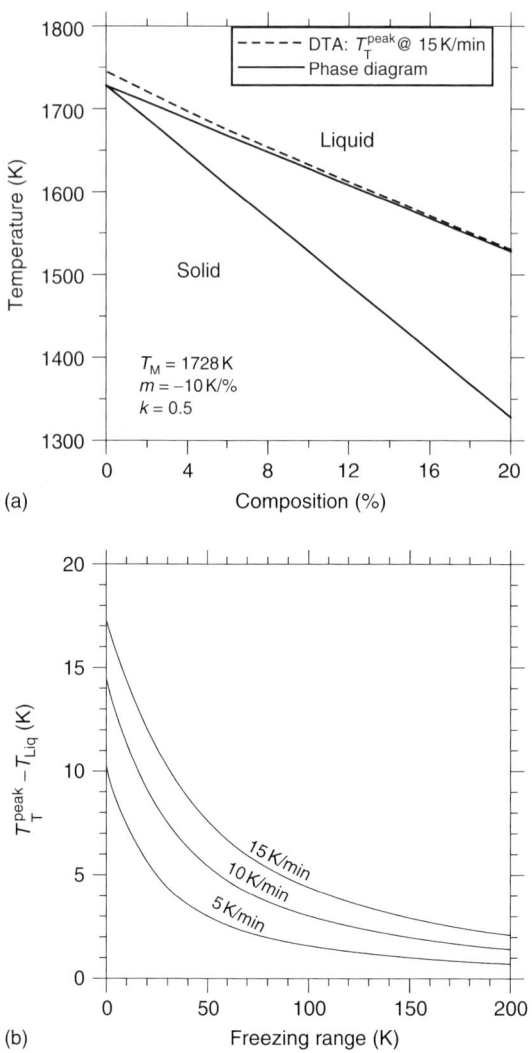

Figure 5.18 (a) DTA peak temperature calculated at a heating rate of 15K/min for different alloy compositions using a lever assumption. Peak temperature is superimposed on the phase diagram ($T_M = 1728$K, $m = -10$K/%, $k = 0.5$). (b) The difference between the peak temperature and the liquidus temperature as a function of freezing range at three heating rates. The freezing range is given by $mC_0[(k-1)/k]$.

decreases with decreasing temperature. The more common situation, $k < 1$, applies whenever the liquidus/solidus separation increases with decreasing temperature (see Fig. 5.20(a)). To demonstrate this effect we compare the computed DTA response of two alloys with identical heats of fusion, and liquidus and solidus temperatures with $k = 2$ and $k = 0.5$. Again we use the straight line liquidus and solidus model, Eq. (E2), and enthalpy given by Eq. (E9), and the computation method of Appendix C. The derived dH_S/dT_S vs. T_S curves are shown in Fig. 5.20(b). The

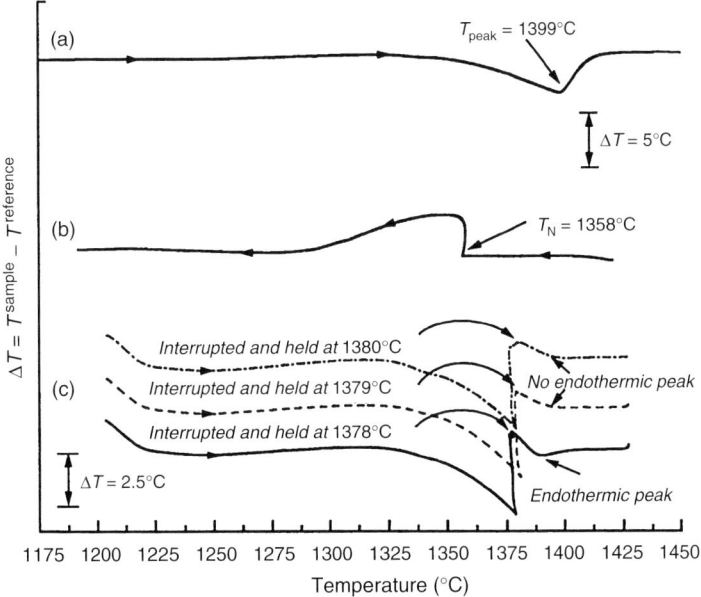

Figure 5.19 (a) Normal DTA scan on heating for a small freezing range Ni-base superalloy. (b) Normal DTA scan on cooling. Supercooling has occurred. (c) Cycling DTA to determine the liquidus temperature [27].

computed DTA scans for three heating rates are shown in Fig. 5.20(c) and (d). For $k < 1$, the DTA curve exhibits a sharp peak at 1681°C (for 5K/min). For $k > 1$, the DTA curve has a broad peak at 1639°C and a shoulder at 1679°C (for 5K/min). The liquidus temperature for both is 1678°C. Hence it is clear that the peak temperature for the $k > 1$ case bears no relationship to the liquidus. For $k < 1$, more of the latent heat is released higher in the freezing range compared to the case for $k > 1$ where more of the latent heat is released lower in the freezing range. In other words, for $k < 1$, the value of dH_S/dT_S increases with temperature, whereas for $k > 1$, the value of dH_S/dT_S decreases with temperature (Fig. 5.20(b)). Because the $k < 1$ case is much more common for alloys, there is a strong association of the peak temperature on the DTA with the liquidus temperature. This indicates that such is not always the case. This point is reinforced by the simulation shown in Fig. 5.21 using real thermodynamic data for Bi-rich and Sb-rich alloys of the Bi–Sb isomorphous system [28].

3.3.6 Failure to Completely Melt

Another difficulty in the determination of the liquidus temperature can occur when the alloy contains only a small fraction of the high temperature phase and the liquidus is very steep. Failure to run the DTA to a sufficient high temperature to reach the liquidus and/or failure to notice a small signal can lead to significant errors. This problem is discussed by Moon et al. [29] for Sn–Cu–Ag alloys where the intermetallic compound Cu_6Sn_5 is the primary phase.

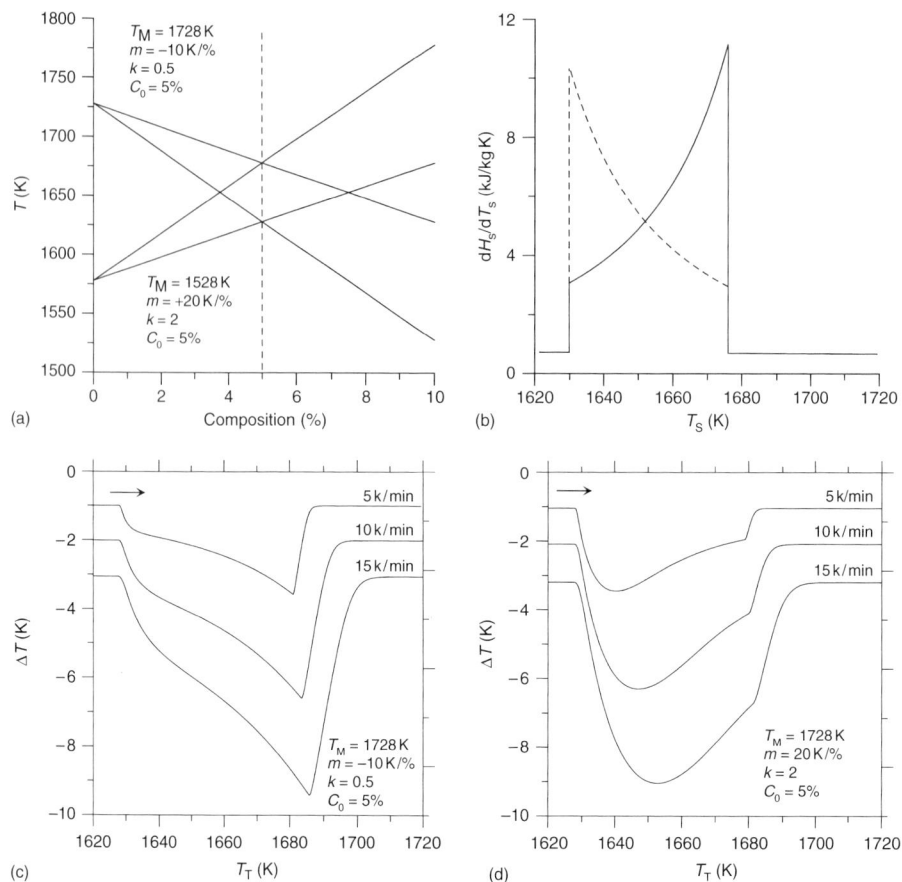

Figure 5.20 Comparison of two hypothetical alloys with identical heat of fusion, liquidus and solidus temperatures but differing having $k < 1$ and $k > 1$. (a) Phase diagram; (b) dH_S/dT_S for $k < 1$ (solid) and $k > 1$ (dashed); (c) DTA for $k < 1$ and (d) DTA for $k > 1$. For $k < 1$, the latent heat is released higher in the freezing range compared to $k > 1$ where the latent heat is released lower in the freezing range.

3.4 Supercooling Problem with Liquidus Determination on Cooling

3.4.1 Onset of Freezing

Many metals and alloys are prone to supercooling prior to the initial nucleation of the solid metal from the melt. In other words, a solid phase does not begin to form during cooling at the temperature given by the phase diagram. When a solid phase does begin to form, it usually nucleates on foreign crystals; examples of possible nucleating sites are the ceramic grains in the crucible, oxide on the melt surfaces or fine particles dispersed throughout the liquid. Nucleation temperatures can vary from a fraction of a degree below the liquidus to several hundred degrees below the liquidus depending on alloy system and other factors. The failure to nucleate with small supercooling complicates the measurement of a liquidus temperature during

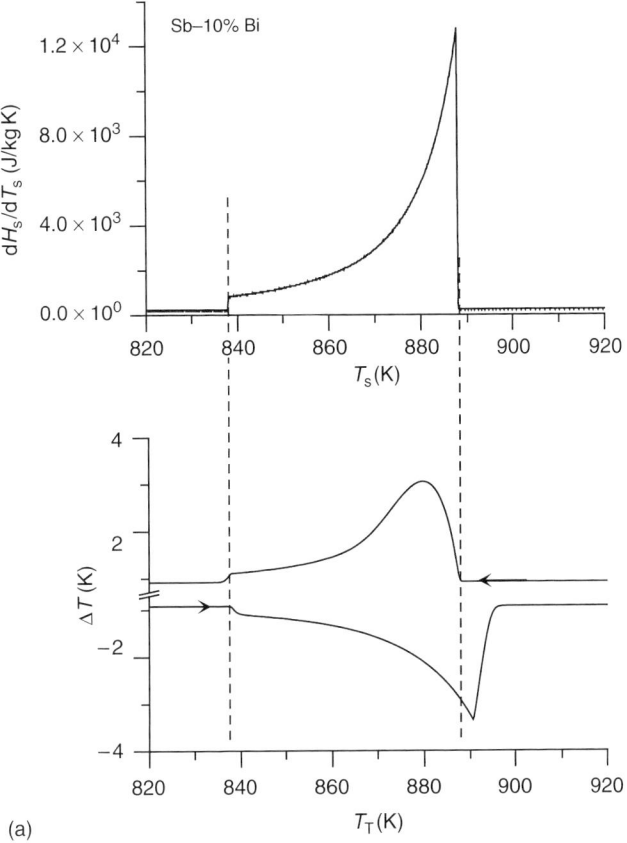

(a)

Figure 5.21 Comparison of Bi–10% Sb with $k > 1$ and Sb–10% Bi with $k < 1$ alloys. The dH_S/dT_S curves computed for lever law using the thermodynamic functions of Ref. [28] and the computed DTA scans for heating and cooling are shown.

cooling.[9] The events occurring during supercooling are demonstrated in Fig. 5.22 showing data obtained with a fine thermocouple immersed in Sn in a DTA cup (same setup as used for Fig. 5.7). Here significant supercooling has occurred. For the immersed thermocouple, the temperature reheats fully up to the melting point of 232°C as the heat of fusion is released rapidly due to the very fast dendritic freezing resulting from the large supercooling. The sample temperature does not drop again until freezing is complete. The use of the direct immersion thermocouple allows the determination of the temperature where solid first forms (nucleation temperature) in this sample as 206°C, 26°C below the melting point of 232°C. Using the instrument thermocouple, the nucleation temperature would have been measured as 203°C with a recalescence only up to 217°C. The presence of supercooling is detected quite easily in the DTA plots using either temperature (Fig. 5.22(b)) by

---

[9] Nucleation of liquid on melting is not difficult due to the large number of microstructural defects (grain boundaries, defects, etc.).

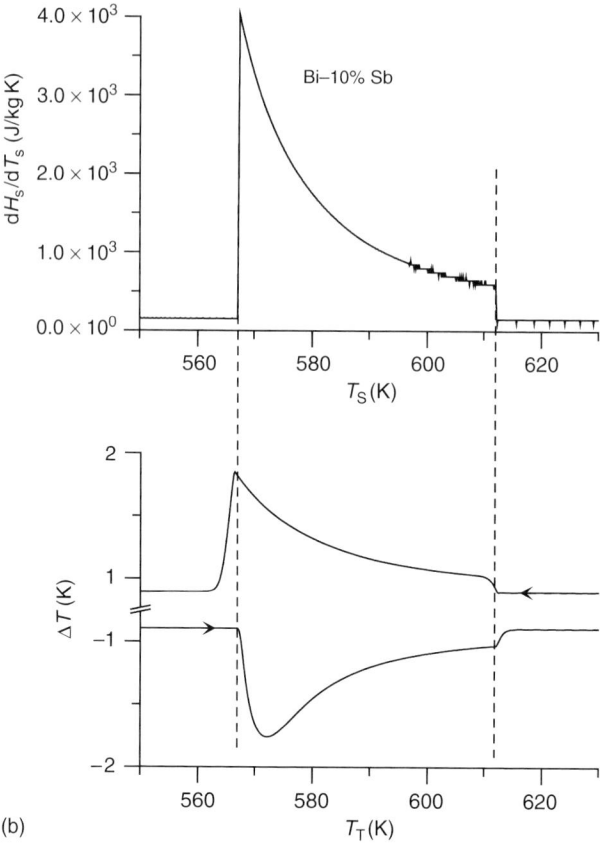

(b)

Figure 5.21 (Continued)

the positive slope of the ΔT vs. T graph. Supercooling provides the most extreme example of where the instrument thermocouple temperature and the sample temperature differ due to the limitations of heat transfer within the DTA.

3.4.2 Slope of DTA Curve on Initial Freezing

Smaller amounts of supercooling in some samples can be harder to detect as they may not lead to a positive sloping DTA curve. In Appendix D, we have noted that the slope of the DTA on melting of a pure material is given by the ratio of two heat flow response times, $-t_{W,C}/t_{S,C}$ (wall to cup/sample to cup). This slope is easily determined experimentally by melting a pure material of similar mass. The rising linear portion of the DTA peak after the initiation of freezing (without supercooling) of a pure metal has the same slope. Thus, if the measured slope after the onset of freezing is more vertical than $-t_{W,C}/t_{S,C}$, one can be certain that supercooling has occurred and the measured onset temperature of freezing would be lower than the melting point. This simple test for the presence of initial supercooling would not necessarily be conclusive for an alloy because the slope during alloy freezing with no supercooling would also be less vertical than $-t_{W,C}/t_{S,C}$ due to slope of the liquidus curve.

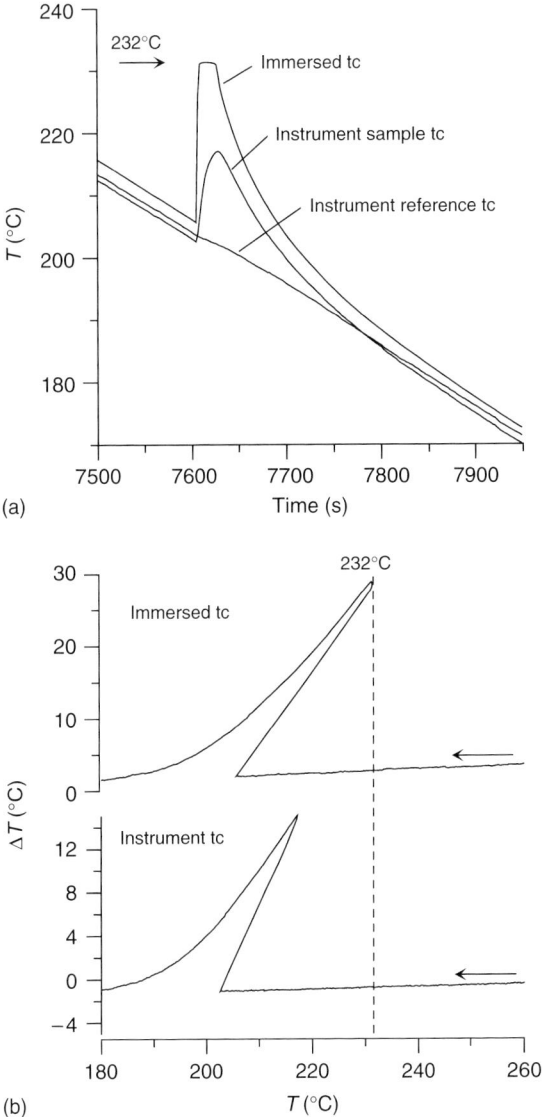

Figure 5.22 (a) Measured temperature vs. time plots for the freezing of pure Sn. The instrument thermocouple readings (sample and reference) and the readings from a thermocouple immersed directly into the Sn in the sample cup are shown. Significant supercooling has occurred prior to freezing inducing the rapid temperature rise upon solidification. (b) Associated DTA plots for both thermocouples; mass = 163.2 mg; heating rate = 5K/min.

3.4.3 Simulation of DTA Response for Alloys with Supercooling

To gain insight into the DTA response when supercooling occurs for an alloy, a model of dendritic growth into a supercooled liquid for a non-dilute alloy has been coupled to the DTA response model [23]. The sample, cup, thermocouple and wall temperature histories are shown for a particular set of materials parameters in Fig. 5.23(a)

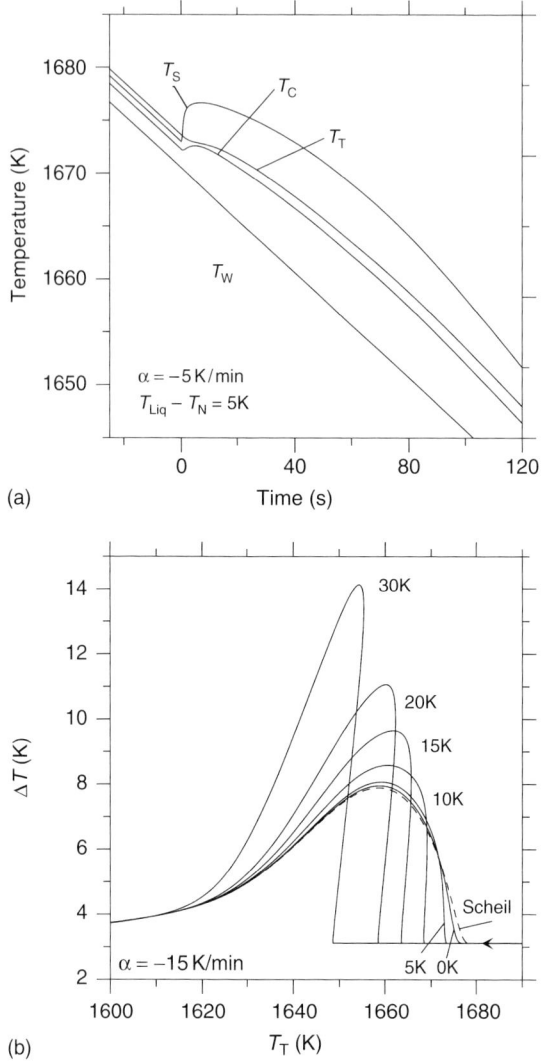

Figure 5.23 (a) Plot of calculated sample, sample cup, sample thermocouple and furnace wall temperatures vs. time curves for dendritic freezing of a 180 mg sample of alloy with $C_0 = 5\%$ at 15K/min with an initial supercooling of 5K. (b) Effect of supercooling on the freezing DTA signal of the same alloy. The curves are for 0K, 5K, 10K, 15K, 20K, 30K supercooling. From Ref. [23].

for an initial supercooling of 5K below the liquidus temperature. Several features can be noted: (1) recalescence of the sample temperature occurs as the latent heat is released during the dendritic network propagation across the sample; (2) after the recalescence temperature maximum, the sample temperature descends as the dendritic network thickens, increasing the fraction solid; (3) the cup temperature exhibits a smaller recalescence due to the heat transfer restrictions of the DTA cell and (4) the sample thermocouple actually exhibits no recalescence.

The DTA curves for initial supercoolings of 0K, 5K, 10K, 15K, 20K and 30K are shown along with the DTA curve for the Scheil path in Fig. 5.23(b). The initial rise of the DTA signal departs from the baseline at a lower temperature and rises more rapidly as the supercooling increases. The latter indicates the recalescence of the sample due to the rapid initial dendritic growth. The rise has a backward (positive) slope for supercoolings of 10K and above. Here the sample recalescence is large enough that the thermocouple temperature also recalesces (Fig. 5.23(a)); clearly, the absence of a backward DTA rise is not proof that supercooling is absent and that a valid liquidus temperature has been determined.

3.5 Eutectic Reactions vs. Peritectic Reactions

3.5.1 Diffusion

From a purely thermodynamic point of view, eutectic reactions, $L \rightarrow \alpha + \beta$, and peritectic reactions, $L + \alpha \rightarrow \beta$, in binary alloys both take place at a fixed temperature and exhibit an isothermal jump in the enthalpy vs. temperature curve at the transition temperature. However, as is well known in the solidification literature [22], they are quite different in their diffusion kinetics. For eutectic solidification, $L \rightarrow \alpha + \beta$, both solid phases form directly from the liquid; that is, locally one has $L \rightarrow \alpha$ and $L \rightarrow \beta$. Thus the necessary solute redistribution occurs in the liquid ahead of the individual interfaces, which are in close proximity. On the other hand, the peritectic reaction, $L + \alpha \rightarrow \beta$, requires the complete disappearance of the α, a process that involves solute diffusion in two solid phases at the peritectic temperature. The kinetics is therefore quite different for eutectic and peritectic alloys because of the extremely different rates of diffusion in liquids and substitutional solids. (If only interstitial solid diffusion is required, the peritectic reaction occurs more easily than but still not as fast as the eutectic reaction.) Very little research has been performed on DTA analysis of systems containing peritectic reactions.

If one assumes as a limiting case that no diffusion occurs in the solid upon cooling, solidification merely switches from the freezing of the high temperature phase, $L \rightarrow \alpha$, to freezing of the low temperature phase, $L \rightarrow \beta$, at the peritectic temperature and below. Then the β phase usually surrounds the α phase in the microstructure resulting in a coarser two-phase solid microstructure than in a eutectic microstructure. More complex arrangements of the two solid phases can occur for both eutectic and peritectic reactions especially if interface attachment kinetics is sluggish (usually encountered for crystals that grow from the liquid with crystallographic facets). Then the two solid phases grow independently from the melt with very little communication of the solute fields in the liquid. This leads to much coarser mixture of the two sold phases.

When the eutectic portion of a microstructure melts, both solid phases melt very close to a common temperature. This is because the phases usually exist as a fine two-phase intermingled microstructure. The melting DTA signal looks like that of a pure material (Fig. 5.5(b)). For melting of an alloy with a phase diagram containing a peritectic reaction, the two solid phases are not intermingled as closely as they would be in a eutectic alloy. The melting response of the two-phase microstructure can occur over a range of temperatures due to the requirements of solid diffusion. The DTA response, as in freezing, again depends of the rate of solid diffusion with lever and Scheil enthalpies representing the extremes of behavior.

3.5.2 DTA Response

To illustrate these points we have chosen an alloy in the Sn–Au system that has both a peritectic reaction (L + Sn_2Au → Sn_4Au) at 252°C and a eutectic reaction (L → Sn_4Au + Sn) at 217°C as shown in Fig. 5.24(a). Figure 5.24(b) shows the enthalpy vs. temperature and dH_s/dT_s vs. temperature curves for Sn–25% Au calculated for equilibrium (dashed) and Scheil (solid) conditions using the thermodynamic description of Ref. [30]. For the equilibrium curves, note the presence of two isothermal jumps in enthalpy with the corresponding delta functions in the dH_s/dT_s curves at the two invariant temperatures. For the Scheil curve the delta function is absent for the peritectic reaction. This absence occurs because the assumption of no solid diffusion in the Scheil model prevents the "peritectic reaction". When traversing the peritectic temperature during solidification, the freezing process merely switches from L → Sn_2Au to L → Sn_4Au and there is no isothermal drop in enthalpy. For melting of an alloy frozen under these conditions, again with no solid diffusion assumed, melting switches from Sn_4Au → L to Sn_2Au → L at the peritectic temperature and there is no isothermal rise in enthalpy. Real solidification and melting will be intermediate to these two cases. The corresponding computed DTA curves for melting and freezing are shown in Fig. 5.24(c) and (d). Note how the peak signal for the "peritectic" reaction near 252°C is much reduced when the Scheil enthalpy is used. The computed DTA response agrees favorably with experimental results for Sn–25% Au alloy shown in Fig. 5.24(e).

DTA responses for the basic cases in binary eutectic and peritectic systems are summarized in Appendix F.

3.6 Major Points

- Melting onset depends on metallurgical state of sample prior to analysis.
- Slow cooling and heating rates do not necessarily guarantee an equilibrated sample at each instant.
- The melting onset during heating should be determined by the first deviation from baseline method. The common practice of using the intersection temperature of a tangent to the DTA curve and the extrapolated baseline can lead to significant error because alloy DTA scans do not in general posses a linear section.
- Annealing of samples in the instrument prior to melting is sometimes required to obtain the thermodynamic solidus.
- Peak temperature on heating for alloys with small freezing ranges may overestimate the liquidus temperature. Cycling experiments can be used to obtain a true liquidus.
- Liquidus temperature determination on heating for alloys with partition coefficient $k > 1$ is more difficult than alloys with $k < 1$.
- Peritectics do not produce as sharp a melting peak as do eutectics.
- Not all temperatures that can be extracted from alloy DTA scans have meaning with regard to the alloy. Some temperatures are merely an indication of thermal lags within the instrument.

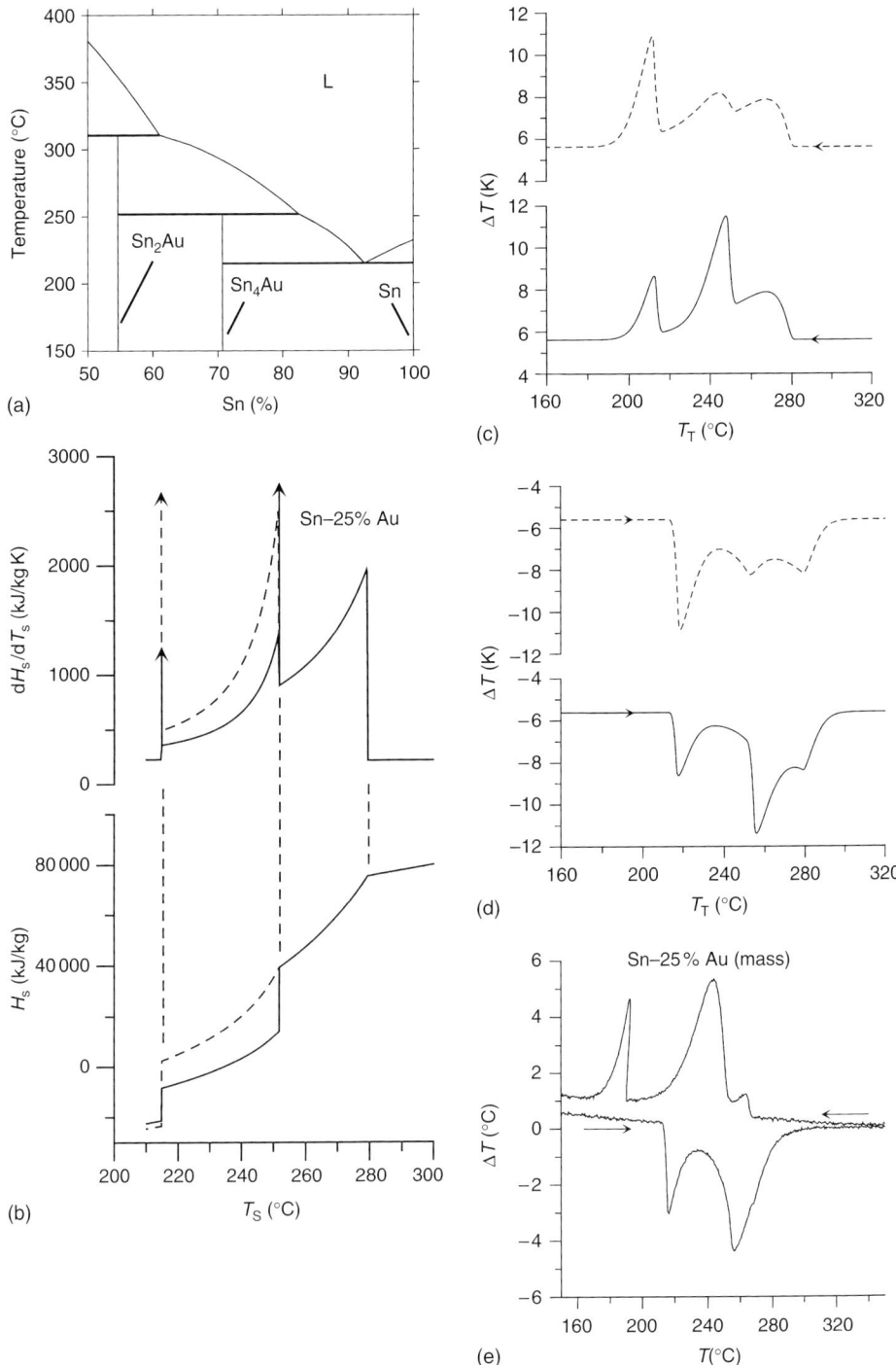

Figure 5.24 (a) Phase diagram for Sn-rich Sn–Au alloys; (b) H_S and dH_S/dT_S curves for Sn–25% Au calculated for equilibrium (solid) and Scheil (dashed) conditions using the thermodynamic description of [30]. Computed DTA curves for (c) freezing and (d) melting. Note how the amplitude of the peak for the "peritectic" reaction near 252°C is much reduced when the Scheil enthalpy is used. (e) Experimental melting and freezing curves at 5K/min.

4 ANALYSIS OF DTA DATA FOR TERNARY ALLOYS

The analysis of DTA curves for ternary and higher order systems is best accomplished with knowledge of the general thermodynamic aspects of multicomponent phase equilibrium. See Appendix B for selected readings. An excellent discussion of the DTA signals obtained during the melting and solidification of ternary alloys is given by Ref. [31]. In this work, Gödecke treats the situation when complete equilibrium is maintained during melting and freezing; that is, that all phases nucleate without superheating or supercooling and that the compositions of all phases are uniform at every temperature and thus follow the predictions of the lever rule. While this is an excellent starting point for understanding alloys, we highlight here some of the different responses obtained when interfacial equilibrium is maintained, but diffusion is not adequate to maintain uniformity of the phases during melting and solidification. We do this by comparing the DTA response to enthalpy temperature relations obtained from a thermodynamic calculation for two ternary aluminum alloys using, as in previous sections, full equilibrium and Scheil conditions. Appendix G gives examples of similar DTA response for alloys with more than three components.

4.1 Al-Rich Corner of Al–Cu–Fe Phase Diagram

First we discuss solidification. Figure 5.25 shows the liquidus surface (projection) of the Al-rich corner of the Al–Cu–Fe ternary system. The diagram is like a topographic map where the temperature takes the place of elevation. The various surfaces represent the range of liquid alloy composition and the temperatures where various phases are first to solidify. The phase is labeled on the surface. In the aluminum-rich corner, the FCC aluminum solid solution forms. At higher Fe content and with no Cu, the Al_3Fe phase is first to form. At high Cu content with no Fe, the θ phase, Al_2Cu, is the first phase to form. Other regions where the α phase, $Al_6(Fe,Cu)$, and the β phase, Al_7Cu_2Fe, form first are indicated. Intersections of these surfaces occur in this diagram along lines (e.g., ab, bd, cd, de) that descend in temperature. Many of these lines correspond to valleys.[10] They represent the **monovariant eutectic** solidification of two solid phases whose liquidus surfaces have intersected. Where two monovariant eutectic valleys join to form a third monovariant valley that continues to descend, such as at (d), an invariant occurs, variously termed a transition reaction, Class II four-phase invariant reaction [32], quasi-peritectic [33] or ternary peritectic [34]. When three descending monovariant valleys intersect, such as at (e), a different type of invariant occurs, termed a ternary eutectic reaction or Class I four-phase invariant reaction.

The solidification path is the entire history of how the liquid composition, whose concentration is assumed uniform at each temperature, changes as the various solids form from the melt. During primary (or first) solidification, the path moves away from the composition of the primary phase, until it intersects one of the valleys. Solidification proceeds down in temperature following the valley and traversing whatever invariant

---

[10] We have not discussed the situation where the liquidus surfaces intersect to form a monovariant peritectic reaction as for the intersection of Al_3Fe and $Al_6(Fe,Cu)$ liquidus surfaces on "bc" in Fig. 5.25. Such an intersection is not a "valley" but a change in slope of the liquidus. Lever solidification follows this line but Scheil solidification crosses it, switching solidification from Al_3Fe to $Al_6(Fe,Cu)$.

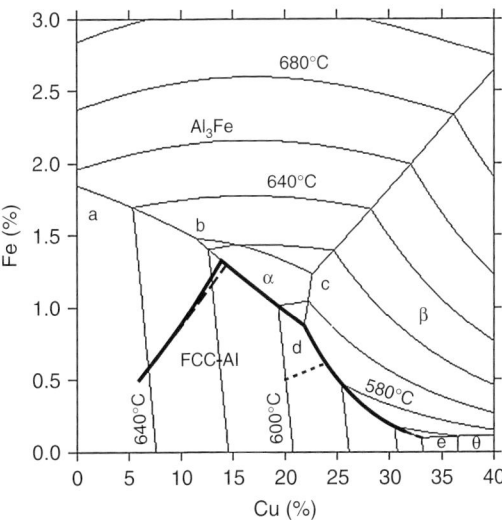

Figure 5.25 Liquidus surface of Al-rich corner of Al–Cu–Fe alloy with lines of constant temperature (labeled), monovariant lines (ab, bc, bd, cd, de, ef and eg) and invariant points (a, b, c, d, e) at 654°C, 623°C, 608°C, 592°C and 548°C, respectively. The diagram is computed from thermodynamic data of Ref. [39]. The calculated paths of the liquid concentration for the two alloys discussed in Figs. 5.26–5.28 are indicated: Al–6% Cu–0.5% Fe equilibrium is solid (ends along line de), Scheil is long dashed (ends at point e) and Al–20 wt% Cu–0.5% Fe equilibrium and Scheil are short dashed (both end at point e and are essentially identical).

reaction is encountered. The approximate paths of liquid concentration superimposed on the liquidus surface for the two alloys discussed below, Al–20% Cu–0.5% Fe and Al–6% Cu–0.5% Fe for the equilibrium calculation are shown in Fig. 5.25.

For the illustration of the DTA response, we employ the model of Appendix C, using an alloy mass of 200 mg, heating and cooling rates of 5 K/min and instrument time constants obtained by linear interpolation to 600°C between the results obtained experimentally for Sn and Ag in Appendix D; that is, $t_{S,C} = 9.8\,s$, $t_{W,C} = 23.6\,s$, $t_{T,C} = 20.9\,s$.

4.2 Al–20% Cu–0.5% Fe

With the brief preceding description of the phase diagram, we proceed to describe the DTA response. The alloy Al–20% Cu–0.5% Fe has been chosen for exposition because it exhibits the most common behavior for a ternary alloy; that is, a three stage process on cooling that progresses from freezing of one, then two and finally three solid phases from the melt. Figure 5.26 shows the DTA and dH_s/dT_s plots under equilibrium conditions. The solidification sequence is

$$L \rightarrow FCC$$

$$L \rightarrow FCC + Al_7Cu_2Fe$$

$$L \rightarrow FCC + Al_7Cu_2Fe + Al_2Cu$$

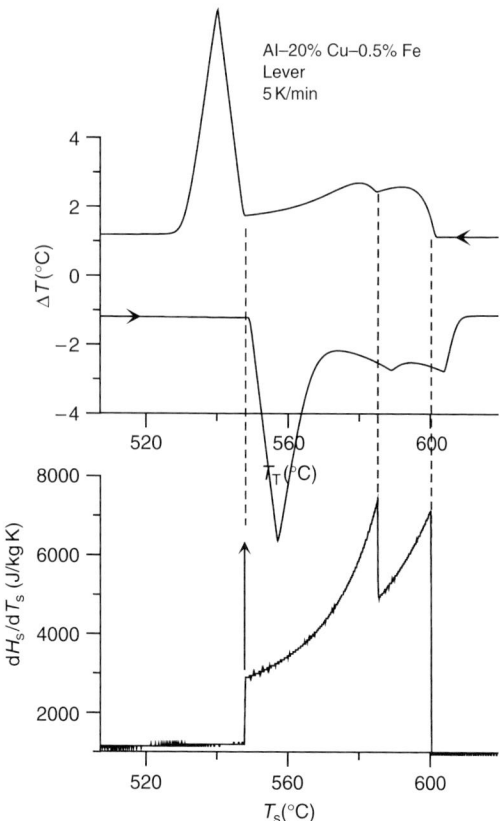

Figure 5.26 DTA and dH_S/dT_S plots for Al–20% Cu–0.5% Fe under equilibrium conditions. The aluminum alloy thermodynamic database [39] was employed. The solidification sequence is: L → FCC, L → FCC + Al$_7$Cu$_2$Fe, L → FCC + Al$_7$Cu$_2$Fe + Al$_2$Cu.

With decreasing temperature, the dH_S/dT_S plot first encounters a jump in value at the liquidus temperature, ≈598°C. The derivative value then slowly decays until another jump occurs when the monovariant valley (de) is encountered at ≈584°C. The value of dH_S/dT_S then slowly decays again until the delta function associated with the ternary eutectic temperature (e) is encountered at ≈548°C. The corresponding DTA plot on cooling shows an onset associated with the liquidus, a second onset associated with encountering the monovariant valley and a third onset associated with the ternary eutectic. Supercooling may be required to enable the nucleation of each new solid phase. Thus, problems can occur in identification of onsets during cooling not only for the first phase to form from the melt but also for each new phase in the various later stages of solidification.

On heating, an onset is encountered that is associated with the ternary eutectic melting. All three solid phases are melting, but only one, Al$_2$Cu, is completely consumed by this process. The other solid phases, FCC and Al$_7$Cu$_2$Fe, although reduced in amount, still remain. The temperature associated with the maximum deflection does correspond to the *time* when Al$_2$Cu is gone, but the temperature of the sample thermocouple at this time has no significance due to the thermal lags of the instrument; this case is analogous to the peak shape during melting of a pure metal in that the peak temperature after an invariant reaction has no thermodynamic meaning.

On continued heating, according to the dH_S/dT_S plot, the rate at which the remaining FCC and Al_7Cu_2Fe melt increases with increasing temperature until a sudden drop in the dH_S/dT_S value occurs. This drop indicates that all of the Al_7Cu_2Fe has melted. This drop in the dH_S/dT_S value produces the maximum deflection in the DTA plot associated with the temperature where all of the Al_7Cu_2Fe has melted. According to the dH_S/dT_S plot, the rate at which the remaining FCC melts increases with increasing temperature until a sudden drop in the dH_S/dT_S value occurs. This drop indicates that all of the FCC has melted. The final peak temperature in the DTA scan is thus associated with the temperature when all of the FCC has melted. We have used the term "associated with" to indicate that the measured onset and peak temperatures suffer from same thermal lags from the temperatures where the events are truly taking place in the instrument; this deviation appears in Fig. 5.26 as the offset of the DTA peaks from the dashed vertical lines indicating the thermodynamic temperatures. We note finally that, for this alloy, there is little difference between the lever and Scheil enthalpies because the fraction of the FCC phase (in which dendritic coring can exist) occupies a relatively small amount of the total alloy.

4.3 Al–6% Cu–0.5% Fe

This alloy has been chosen for exposition because it exhibits the second most common behavior for a ternary alloy, that is, a four (five for Scheil) stage process on cooling that progresses from the freezing of one solid phase, then two solid phases, interrupted by an invariant reaction, continued freezing of two solid phases and finally three solid phases. The solidification sequence is

$$L \rightarrow FCC$$

$$L \rightarrow FCC + Al_6(Fe,Cu)$$

$$L + Al_6(Fe,Cu) \rightarrow Al_7Cu_2Fe + FCC \text{ (invariant); lever path only}$$

$$L \rightarrow FCC + Al_7Cu_2Fe$$

$$L \rightarrow FCC + Al_7Cu_2Fe + Al_2Cu \text{ (invariant); Scheil path only}$$

Solidification first involves the formation of the FCC phase from the melt. The liquid is enriched in Cu and Fe as the temperature drops until the concentration reaches the monovariant valley (bd), $L \rightarrow FCC + Al_6(Fe,Cu)$, at 616°C. Then simultaneous liquid freezing to FCC and $Al_6(Fe,Cu)$ phases occurs with decreasing temperature until the valley encounters a second valley that is between the FCC and Al_7Cu_2Fe liquidus surfaces (de). Under the equilibrium assumption, all of the $Al_6(Fe,Cu)$ dissolves at the invariant temperature (d) at 592°C, in the process forming more FCC and the new phase, Al_7Cu_2Fe. Subsequently, solidification continues with decreasing temperature (along de) forming more FCC and Al_7Cu_2Fe. Solidification is completed at 556°C on (de) between (d) and (e) at 556°C. In contrast, under Scheil assumptions, the invariant temperature (d) is merely the temperature (592°C) where solidification

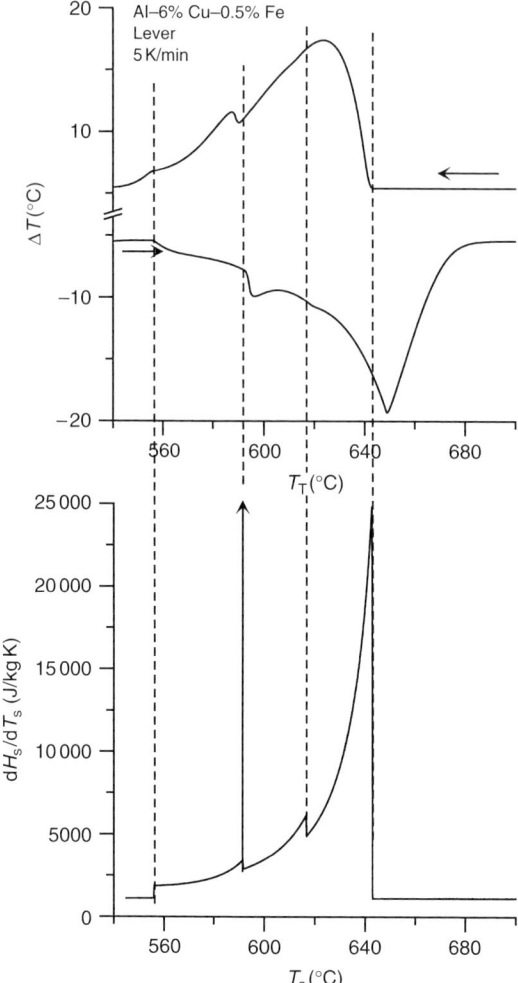

Figure 5.27 DTA and dH_S/dT_S plots for Al–6% Cu–0.5% Fe under equilibrium conditions. The aluminum alloy thermodynamic database [39] was employed. The solidification sequence is: L → FCC, L → FCC + Al$_6$(Fe,Cu), L + Al$_6$(Fe,Cu) → FCC + Al$_7$Cu$_2$Fe, L → FCC + Al$_7$Cu$_2$Fe. Note the delta function at 592°C corresponding to the invariant reaction, L + Al$_6$(Fe,Cu) → FCC + Al$_7$Cu$_2$Fe.

switches from L → FCC + Al$_6$(Fe,Cu) to L → FCC + Al$_7$Cu$_2$Fe as the temperature drops. For Scheil conditions, the curve (de) is followed as FCC and Al$_7$Cu$_2$Fe solids form, the temperature descending and the liquid concentrations increasing to the end of (de) until the ternary eutectic temperature at (548°C) and liquid concentration are reached at (e). Here all liquid disappears via the ternary invariant eutectic, L → FCC + Al$_7$Cu$_2$Fe + Al$_2$Cu.

Figure 5.27 shows dH_S/dT_S and melting and freezing DTA plots under equilibrium conditions for this alloy. Figure 5.28 shows dH_S/dT_S and melting and freezing

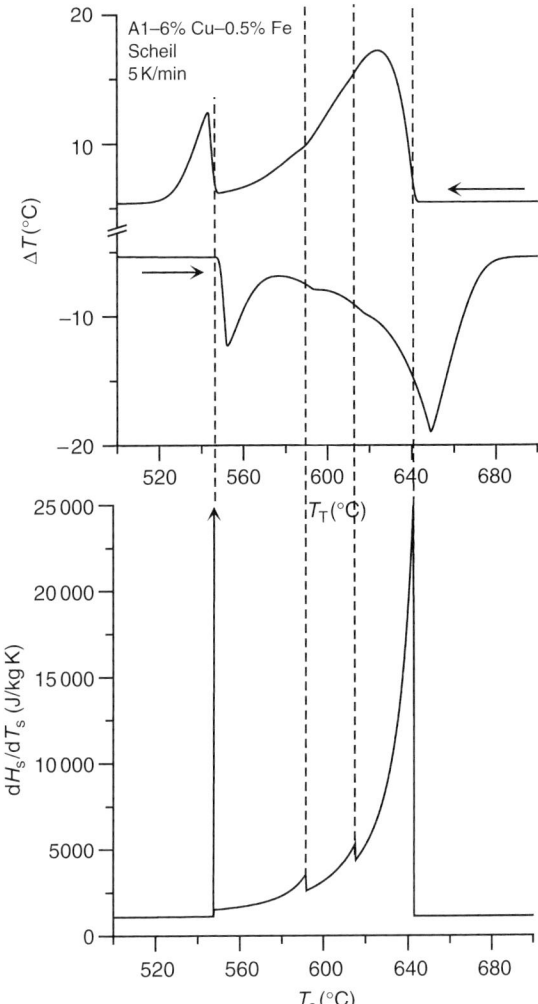

Figure 5.28 DTA and dH_S/dT_S plots for Al–6% Cu–0.5% Fe under Scheil conditions. The aluminum alloy thermodynamic database [39] was employed. L → FCC, L → FCC + Al_6(Fe,Cu), L → FCC + Al_7Cu_2Fe, L → FCC + Al_7Cu_2Fe + Al_2Cu. Note that the delta function in the dH_S/dT_S plot at 592°C in Fig. 5.27 corresponding to the invariant reaction, L + Al_6(Fe,Cu) → FCC + Al_7Cu_2Fe, is gone.

DTA plots under Scheil conditions. Note the presence of the delta function at 592°C in the dH_S/dT_S plot corresponding to the invariant reaction, L + Al_6(Fe,Cu) → FCC + Al_7Cu_2Fe, in the equilibrium case but not for the Scheil case. Also note the presence of the delta function corresponding to the ternary eutectic at 547°C in the Scheil case and its absence in the equilibrium case. These differences are due to the assumptions of complete diffusional equilibrium in the first case and no solid diffusion in the second case, again representing extremes of behavior. Just like a peritectic reaction in a binary alloy (Section 3.5), the reaction L + Al_6(Fe,Cu) →

FCC + Al_7Cu_2Fe requires solid state diffusion and thus does not occur under the Scheil assumption. On solidification, one merely switches at 592°C from the formation of FCC + Al_6(Fe,Cu) to the formation of FCC + Al_7Cu_2Fe. Al_6(Fe,Cu) is not consumed on continued cooling but remains in the alloy's microstructure. These differences in the dH_S/dT_S curves are then reflected in the DTA scans.

4.4 Major Points

- Knowledge of ternary alloy phase equilibrium is necessary to interpret data for ternary and higher order systems.
- Ternary eutectic invariant peritectic reactions may have quite different DTA signals compared to ternary invariant eutectic reactions.
- Failure to nucleate during cooling DTA scans can occur with solid phases that form through secondary solidification processes in addition to primary solidification.

5 Concluding Remarks

We have attempted to highlight two important effects that must be carefully considered in the measurement of alloy melting and freezing using a DTA/HF-DSC. The first is concerned with the heat flow and the small but important temperature differences that exist within the different parts of the instrument. Thermocouple signals are the only information available and the thermocouple, not directly immersed in the metal, is not at the sample temperature. This is most easily seen in the shape of the DTA curves for the melting of a pure material, where one might naively think that the sample temperature ranges from the onset temperature to the peak temperature during melting due to an erroneous concept of kinetics. This is clearly not the case. Thus the heat transfer between sample, through the sample container and finally to the thermocouple must be considered. A model is presented to quantify these effects and to determine heat transfer characteristic of any DTA/HF-DSC instrument.

The second important effect is that deviations from full equilibrium can and do occur during the melting and freezing of alloy samples at the rates encountered in DTA. Indeed the small sample in the DTA cup is best viewed as a small casting with all of the complications that can occur in castings. However, the loss of full equilibrium is due mostly to the slow rate of solute diffusion in alloys and is thus analyzable without recourse to complex models that involve so-called reaction kinetics. This concept is widely used in the successful modeling of castings and can be applied to DTA response. This idea has been developed in the current document in the broad usage of the comparison of full equilibrium and Scheil approaches to place bounds on expected DTA response.

The heavy use of simulations of the DTA response in this document should not be taken to imply that measurements are unnecessary. The simulations were performed using thermodynamic data that remains under constant revision by the CALPHAD community. Indeed these thermodynamic assessments are constructed from data sometimes obtained from incorrect interpretations of transformation temperatures using the DTA. In addition many complications of diffusion kinetics and microstructure variations can occur that are not considered here.

The emphasis of this document on problems with DTA analysis may also have led the reader to the impression that accurate temperature determination with the DTA is nearly impossible. This is not true and it is not the intention of this guide to discourage DTA measurements. The intention is to increase user awareness of the factors that influence the DTA signal by providing physical explanations that enable thorough and meaningful analysis of the results and temperature determinations.

APPENDIX A GLOSSARY

Back (solid) diffusion: Diffusion in the solid phase(s) to attain equilibrium composition throughout the solid phase.

Baseline: The signal recorded without a thermal event such as melting taking place.

CALPHAD: **Cal**culation of **pha**se **d**iagrams. The Gibbs energy of a phase is represented as a function of composition, temperature and pressure. The adjustable parameters of the functions for the individual phases are derived from experimental phase equilibria and thermochemical data. The descriptions of constituent subsystems can be combined for the calculation of a multicomponent system.

Degrees of freedom: Number of variables that can be changed without changing the identity of the phases in equilibrium. The number of degrees of freedom, F, can be obtained from the phase rule, $F = N - P + 2$, where N is the number of components and P is the number of phases in equilibrium. For condensed phase systems the pressure is usually dropped as a variable and the phase rule is modified to $F = N - P + 1$.

Delta function: A function $\delta(x)$ that has the value of infinity for $x = 0$ and the value zero elsewhere. The integral from minus infinity to plus infinity of a delta function is 1.

Differential scanning calorimetry (DSC): See *Heat-flux differential scanning calorimetry (HF-DSC)* or *Power-compensating differential scanning calorimetry (PC-DSC)*.

Differential thermal analysis (DTA): Thermal analysis using a reference. Sample and reference are heated in one furnace. The difference of sample temperature and reference temperature is recorded during programmed heating and cooling cycles.

Endothermic: Heat is consumed during a reaction or transformation, that is, the enthalpy change is positive. In a heating experiment, melting is endothermic.

Enthalpy (H): Heat content of a phase.

Equilibrium: State of the lowest free energy when no further change of a system is possible. Phases in equilibrium at a given composition, temperature and pressure have no driving free energy to react or transform. See *Global equilibrium* and *Local equilibrium*.

Eutectic: Thermodynamic situation where a liquid is simultaneously saturated with respect to several solid phases and cooling decreases the amount of liquid and increases the amount of all solids. The situation may have zero or more degrees of freedom depending on the number of phases and components.

Exothermic: Heat is released during a reaction or transformation, that is, the enthalpy change is negative. In a cooling experiment, solidification is exothermic.

Fraction of solid vs. temperature: Amount of solid phase(s) formed during the solidification of an alloy.

Gibbs–Thomson effect: The contribution of the interface energy to the Gibbs free energy due to the effect of the shape (curvature) of the interface between phases. Phases with high curvature interfaces exhibit altered melting temperatures and solubilities.

Global equilibrium: A complete equilibrium where there are no gradients of temperature or pressure and the compositions of coexisting phases are uniform.

Heat capacity (Cp): Rate of enthalpy (heat content) change of a phase with respect to temperature at constant pressure.

Heat-flux differential scanning calorimetry (HF-DSC): Sample and reference are heated in one furnace. The difference of sample temperature and reference temperature is proportional to the heat flux between the sample and the reference. Contrast with *Power-compensating differential scanning calorimetry*.

Incipient melting: Initial melting of an alloy. This may occur below the solidus temperature of the average alloy composition due to microsegregation from prior solidification.

Interface undercooling: The departure of the interface temperature (and interface phase concentrations) from that given by the phase diagram. The interface undercooling that drives the kinetic processes involved in solidification and melting is typically quite small ($<1K$) for metals and alloys.

Interstitial solutes: Elements that occupy positions between the solvent element atoms in a crystal lattice. Diffusion of interstitial solutes is rapid because the large number of vacant sites.

Invariant reaction: A reaction with zero degrees of freedom. Also called invariant equilibrium. Examples are eutectic and peritectic reactions in a condensed phase system consisting of two components.

Latent heat: Heat change during a reaction or transformation. This quantity is only exactly defined for invariant reactions or transformations.

Lever rule: An expression of conservation of matter in which the relative phase amounts are determined from the overall alloy composition and the coexisting phase compositions, assumed to be in *global equilibrium* at each temperature.

Lever rule solidification: One limiting case of solidification where complete diffusion for the liquid and the solid phases are assumed. The phases are in thermodynamic equilibrium and phase amounts and compositions are linked by the lever rule at each temperature during cooling.

Liquidus: Boundary in a phase diagram between phase field regions containing *only* the liquid phase and phase fields regions containing liquid and solid phase(s). This boundary represents the limit where the fraction of solid is zero.

Local equilibrium: Equilibrium conditions are only fulfilled at the interface between phases.

Metastable equilibrium: Equilibrium when one or more of the stable phases are absent.

Microsegregation: Concentration gradients in the solid phase formed during solidification because of insufficient diffusion in the solid phase.

Nucleation supercooling: Extent of supercooling needed for nucleation to occur. See *Supercooling*.

Onset temperature: The temperature where the signal first departs from the baseline and which corresponds to the leading portion of a peak. Two onset temperatures are described in this document: one obtained by an extrapolation method and the other obtained as the first detectable deviation from baseline.

Peak: The entirety of the deviation and return of the DTA signal from and to the baseline.

Peak signal: The amplitude of the maximum deviation of the signal from the baseline. Can be exothermic or endothermic.

Peak temperature: The temperature corresponding to the peak signal.

Peritectic: Thermodynamic situation where a liquid is simultaneously saturated with respect to multiple solid phases and cooling decreases the amount of liquid and at least one of the solid phases while increasing the amount of the other solid phase(s). The situation may have zero or more degrees of freedom depending on the number of phases and components.

Phase: A portion of a material that is distinguishable by its state, composition and/or crystal structure. A phase may exhibit smooth variations in composition.

Power-compensating differential scanning calorimeter (PC-DSC): Sample and reference are heated independently in two furnaces so that their temperature difference stays zero. The difference in power used to heat the sample and reference is a measure of the enthalpy of the sample. Contrast with: *Heat-flux differential scanning calorimeter*.

Recalescence: The rapid increase in temperature due to release of the latent heat due to the solidification from a supercooled condition.

Scheil–Gulliver solidification: One limiting case of solidification where complete diffusion for the liquid phase and no diffusion for the solid phases are assumed. This type of solidification produces the worst case of microsegregation. Also called Scheil solidification.

Solidification path: The change in liquid composition and the sequence of solid phases that form as temperature decreases.

Solidus: Boundary in a phase diagram between phase field regions containing *only* solid phase(s) and phase fields regions containing solid and liquid phase(s). This boundary represents the limit where the fraction of liquid is zero.

Supercooling: Failure of a solid phase to nucleate at the temperature given by the phase diagram (thermodynamic equilibrium). See: *Nucleation supercooling*.

Substitutional solutes: Elements that replace the solvent element randomly in its crystal lattice. Diffusion is usually occurs by a vacancy mechanism.

Thermal conductivity: The time rate of heat flow, under steady conditions, through unit area, per unit temperature gradient in the direction normal to the area.

Tie line: An isothermal, isobaric line connecting the compositions of a pair of phases in equilibrium.

Transition reaction: Invariant reaction in a multicomponent system where a number of phases react to form a number of different phases. This reaction is neither a eutectic nor a peritectic reaction. However, since this reaction is somewhat similar to a peritectic reaction it is also called quasi-peritectic. Another term for this reaction is Class II reaction.

Variant eutectic reaction: A multiphase equilibrium with liquid and at least one degree of freedom with eutectic character. A monovariant (one degree of freedom) eutectic is also called a eutectic valley.

Appendix B Recommended Reading

Monographs on ternary phase diagrams:

- F.N. Rhines, *Phase Diagrams in Metallurgy*, McGraw-Hill Book Comp., 1956.
- G. Massing and B.A. Rodgers, *Ternary Systems*, Dover Publications, NY, 1960.
- A. Prince, *Alloy Phase Equilibria*, Elsevier, Amsterdam, 1966.
- D.R.F. West, *Ternary Equilibrium Diagrams*, 2nd edn, Chapman & Hall, London, UK, 1982.
- D.R.F. West and N. Saunders, *Ternary Phase Diagrams in Materials Science*, Maney-Institute of Materials, London, UK, 2002.

A classic in the field, general focus on a wide range of materials, not specific to alloys:

- P.D. Garn, *Thermoanalytical Methods of Investigation*, Academic Press, NY, 1965.

A concise summary of techniques and interpretation:

- M.E. Brown, *Introduction to Thermal Analysis; Techniques and Applications*, Chapman & Hall, NY, Chapter 4, 1988.

A general treatment of solidification:

- *Principles of solidification*, Vol. 15; *Casting, Metals Handbook*, 9th edn, ASM, Metals Park, OH, 1988.
- H. Biloni and W.J. Boettinger, Solidification, in P. Haasen and R.W. Cahn (eds.), *Physical Metallurgy*, 4th edn, North Holland, Amsterdam, 1996, p. 669.

An excellent summary of the DTA response to the solidification of ternary alloys under equilibrium (lever law) melting and solidification:

- T. Gödecke, *Z. Metallk.*, 92 (2001) 966 [In German].

A general guide to good technique for phase diagram determination:

- R. Ferro, G. Cacciamani and G. Borzone, *Intermetallics*, 11 (2003) 1081.

A review, focused on Al alloys, on DTA techniques with a major focus on solid state reactions and with some melting work. Effect of prior sample history made clear:

- M.J. Starink, *Inter. Mater. Rev.*, 49 (2004) 191.

APPENDIX C MODEL FOR SIMULATING DTA RESPONSE FOR MELTING AND SOLIDIFICATION OF MATERIALS WITH KNOWN OR ASSUMED ENTHALPY VS. TEMPERATURE RELATIONS. ALSO METHOD FOR DETERMINING THERMAL LAG TIME CONSTANTS OF DTA/DSC INSTRUMENTS

We employ an approach using ordinary differential equations (ODE) in which $T_S(t)$, $T_C(t)$, and $T_T(t)$ are the sample, sample cup and thermocouple temperatures. Thus thermal gradients within these three objects are ignored.[11] As described in Section 2.1.3, different instruments control the heating rate in different ways. For this model, we will assume that the furnace wall temperature, $T_W(t)$, is controlled by the instrument to be a prescribed linear function of time. An additional triplet of variables can be used to describe the temperatures of the reference sample, cup and thermocouple. A system of ODE's that describes these six unknown temperatures is given in Ref. [23] as well as the reduced set that we employ here. We consider only heat flow between: (a) the sample and the sample cup; (b) the sample cup and the furnace wall; and (c) the sample cup and the thermocouple. We neglect heat flow between the sample and the reference cups.[12]

We let $h_{S,C}A_{S,C}$, $h_{W,C}A_{W,C}$ and $h_{T,C}A_{T,C}$ be the products of the heat transfer coefficients $h_{X,Y}$ and areas $A_{X,Y}$ for the heat flow (a), (b) and (c) described above, respectively. We also let m_S and H_S be the sample mass and enthalpy/unit mass, m_C and C_p^C be the crucible mass and heat capacity/mass and m_T, and C_p^T be the thermocouple mass and heat capacity/mass. Defining instrument time constants $t_{S,C}$, $t_{W,C}$ and $t_{T,C}$ by

$$\begin{cases} t_{S,C} = m_C C_p^C / h_{S,C} A_{S,C} \\ t_{W,C} = m_C C_p^C / h_{W,C} A_{W,C} \\ t_{T,C} = m_T C_p^T / h_{T,C} A_{T,C} \end{cases} \quad (C1)$$

---

[11] A similar treatment is found in Ref. [35] and one explicitly dealing with radiative heat transfer is found in Ref. [36]. A finite element treatment that includes thermal gradients is given by Ref. [37].
[12] Examination of experimental temperature–time data for the reference thermocouple during the melting of small (180 mg) samples of pure Ni at 5K/min in a DTA show less than 0.8K variations from linearity. This indicates that very little heat flows between the reference and sample cups in these experiments and validates the use of the simple model.

a heat balance gives

$$\begin{cases} m_S \dot{H}_S = \dfrac{m_C C_p^C}{t_{S,C}} (T_C - T_S) \\ \dot{T}_C = \dfrac{1}{t_{S,C}} (T_S - T_C) + \dfrac{1}{t_{W,C}} (T_W - T_C) \\ \dot{T}_T = \dfrac{1}{t_{T,C}} (T_C - T_T) \end{cases} \quad (C2)$$

where the dot represents time differentiation. For situations where transformation kinetics can be ignored, the enthalpy is only a function of temperature and

$$\dot{H}_S = \frac{dH_S}{dt} = \frac{dH_S}{dT_S} \frac{dT_S}{dt} = \frac{dH_S}{dT_S} \dot{T}_S \quad (C3)$$

Inclusion of kinetic effects produces a time dependent sample enthalpy function that is described in Ref. [23].

The time parameter, $t_{S,C}$, is the characteristic response time for heat flow between the metal *sample* and the crucible *cup*; $t_{W,C}$ is the response time between the furnace *wall* and the *cup*; and $t_{T,C}$ is the response time between the *thermocouple* and the *cup*. Due to the presence of the area of contact between the sample and the cup in Eq. (C1), the response time, $t_{S,C}$ depends on the size and thus the mass of the sample. We will assume that these response times are independent of temperature over the course of melting or solidification. We will further assume that the furnace heating rate, α, is constant such that the interior furnace *wall* temperature is given by

$$T_W(t) = \text{constant} + \alpha t \quad (C4)$$

Analytic solution of these equations is possible for a pure material and is given in Ref. [23], the results of which are given in Appendix D with respect to melting onset analysis. Numerical solution is necessary for alloys and is possible using standard software packages, such as Mathematica or Matlab. Simulation of DTA plots for alloys with known or assumed enthalpy vs. temperature curves are reported throughout this document.

The time constants appropriate to a particular instrument were obtained as follows. Using a least squares procedure, measured DTA data was compared to the computed DTA response for initial guesses of the time constants. Minimization of the error using an iterative procedure led to values for the thermal lags given in Table C1 for a Perkin-Elmer DTA 1600 using the melting of ≈200 mg samples of Sn, Ag and Ni. The decrease in the size of the thermal lags with increasing melting point suggests the increasing importance of radiation in the heat flow. We have also found that the

Table C1 Thermal lag time constants determined for a DTA using the model of Appendix C.

	Sn (232°C)			Ag (961°C)			Ni (1453°C)		
Mass (g)	0.163			0.237			0.180		
Heating rate (K/min)	5	10	15	5	10	15	5	10	15
$t_{S,C}$ (s)	13.5	15.3	15.9	6.2	6.7	6.7	6.5	6.0	5.5
$t_{W,C}$ (s)	34.5	29.2	26.3	13.0	12.7	13.1	5.1	5.0	4.7
$t_{T,C}$ (s)	27.6	18.9	17.0	14.5	12.4	11.7	7.4	6.5	6.7
$t_{W,C}/t_{S,C}$	2.56	1.91	1.65	2.10	1.90	1.96	0.78	0.83	0.85

Table C2 Thermal lag time constants determined for a HF-DSC using the model of Appendix C.

	Ag (961°C)		
Mass (g)	0.214	0.214	0.214
Heating rate (K/min)	5	10	15
$t_{S,C}$ (s)	12.5	12.0	11.7
$t_{W,C}$ (s)	18.2	18.3	18.7
$t_{T,C}$ (s)	0.82	0.37	0.11
$t_{W,C}/t_{S,C}$	1.46	1.53	1.60

parameter, $t_{S,C}$, is approximately inversely proportional to the 1/3 power of the sample mass, rather than a 2/3 power that might be expected from the increased contact area between the sample and the sample cup. This may be due to the high contact angle of the samples with the alumina cup. Parameters for an HF-DSC (Netsch DSC 404C) are given in Table C2. Note the much smaller value of $t_{T,C}$ for this instrument probably due to the direct welding of thermocouple wires to the platinum cup support structure. Please note that the parameters given in these tables are meant only as a guide. Each user should determine a set of lag parameters for the specific instrument and operation conditions being used. We also note that a more complex heat flow model of the DTA can be employed (see e.g., the appendix in Ref. [23]) that can include additional time parameters that characterize the heat transfer between the sample and the reference cups as well as other effects that are likely more important for HF-DSC.

APPENDIX D EXPRESSIONS FOR THE RATE DEPENDENCE OF MELTING ONSET TEMPERATURES FOR A PURE METAL

For a pure metal, analytical solution to the differential equations in Appendix C is possible [23] if one takes an enthalpy function given by

$$H_S(T_S) = \begin{cases} C_p^{S0}(T_S - T_M); & T < T_M \\ C_p^{S0}(T_S - T_M) + L; & T > T_M \end{cases} \quad (D1)$$

where C_p^{S0} is the (constant) heat capacity per unit mass (assumed the same for solid and liquid) and L is the heat of fusion per unit mass. For convenience we define the ratio ρ as

$$\rho = \frac{m_S C_p^{S0}}{m_C C_p^C} \quad (D2)$$

where m_S is the sample mass and m_C and C_p^C are the crucible mass and heat capacity/mass. The solution is broken into three time regimes: before melting, during melting and after melting. For DTA instruments, the "DTA signal" is the difference between the sample and the reference thermocouple temperatures, which is the same in this simplified model as the difference between the sample temperature and the furnace wall temperature.

Figure D1 shows the calculated temperature vs. time behavior and the associated DTA curve, which is a plot of $(T_T - T_W)$ vs. T_T. Prior to melting ($t < 0$), the temperatures are all linear in time, with the same slope α, but with different temporal or temperature offsets. These offsets are established during the initial transient of the instrument after heating is initiated. The sizes of all temperature offsets are directly proportional to the heating rate. During melting, $0 \leq t < t_M$, the sample temperature is assumed constant consistent with negligible temperature differences within the metal sample and the assumption of equilibrium, $T_S(t) = T_M$. During this regime both the cup and the thermocouple temperature–time curves exponentially approach linear behavior with slope $\alpha t_{S,C}/(t_{S,C} + t_{W,C})$, where α is the heating rate as in Appendix C. The corresponding slope on the DTA plot is $-t_{W,C}/t_{S,C}$, which is independent of heating rate. For $t > t_M$, the sample, sample cup and thermocouple return as the sum of three exponential functions of time to the linear temperature behavior obtained during the first regime, $t < t_M$.

The difficulty of picking the melting onset is shown in Fig. D1(b), which is an enlarged view of the temperature histories near the initiation of melting at $t = 0$. It is seen that only a gradual change in slope of the thermocouple temperature begins at $t = 0$ marked T_T^{onset}. The onset temperature of the sample thermocouple, T_T^{onset} differs from melting point of the sample T_M, the two related by

$$T_T^{onset} = T_M + \alpha \, (t_{S,C}\rho - t_{T,C}) \quad (D3)$$

which can either increase or decrease from T_M with increasing heating rate depending on the values of the time constants and the ratio ρ. In principle the deviation from the melting point with heating rate might be zeroed by a proper heat flow arrangement.

An extrapolation procedure is commonly used to determine an alternate melting onset. This procedure takes advantage of the fact that the thermocouple temperature vs. time, $T_T(t)$, curve becomes linear quickly after the melting onset as described

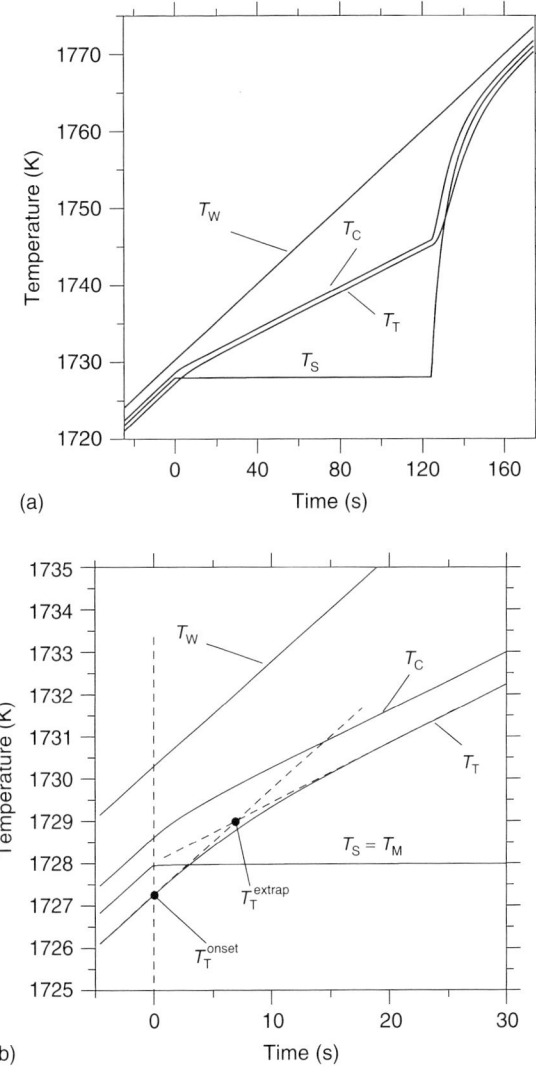

Figure D1 (a) Calculated temperatures within a DTA instrument as functions of time for pure Ni, sample (T_S), sample cup (T_C), thermocouple (T_T) and furnace wall (T_W); (b) enlarged region near $t = 0$;

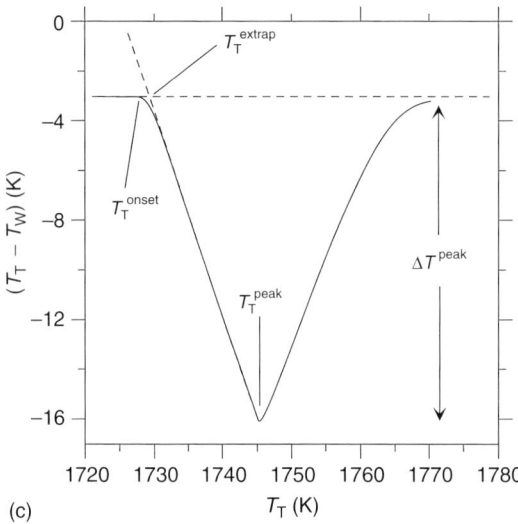

Figure D1 (Continued) (c) associated DTA curve. The figures show the possible sources of error in determining the melting onset temperature for calibration purposes.

above. Then the onset temperature is taken from the DTA curve as the intersection of the extrapolated linear portion with the extrapolated baseline. The melting onset picked in this way is denoted T_T^{extrap}. It is always *above* T_M with a deviation from the melting point that increases with heating rate according to

$$T_T^{extrap} = T_M + \alpha \left(\frac{t_{S,C}(t_{S,C}\rho + t_{W,C}\rho + t_{W,C})}{t_{S,C} + t_{W,C}} \right) \quad (D4)$$

For the values of the time constants given in Table C1 for Ni, and a typical value of $\rho = 0.5$, these two choices for onset temperature can be compared numerically. T_T^{onset} is α times (2.6 s) below T_M, which equals 0.6K for a heating rate of 15K/min. T_T^{extrap} is α times (5.2 s) above T_M, which equals 1.3K at a heating rate of 15K/min. Thus picking the onset (or calibrating the instrument) by the extrapolation method is more heating rate sensitive than using the first sign of deflection from the baseline. One might argue, however, that noise in the baseline makes the latter choice less preferable. In any case, using the same heating rate and method for picking the onsets for calibration and measurements on alloy samples reduces the errors.

Another important piece of information that can be obtained from the analytical solution is the characteristic time it takes for the sample thermocouple temperature behavior to return to a slope of α; or equivalently the temperature change that occurs before the DTA plot returns to the baseline. The smaller this time is, the closer detectable thermal events can be. The return to baseline is the sum of three exponential functions in time [23]. The characteristic times are:

$$t_{T,C} \text{ and } \frac{2t_{S,C}t_{W,C}}{(t_{W,C} + \rho t_{S,C} + \rho t_{W,C}) \pm \sqrt{(t_{W,C} + \rho t_{S,C} + \rho t_{W,C})^2 - 4\rho t_{S,C}t_{W,C}}} \quad (D5)$$

which, for the parameters in Table C1 for Ni and $\rho = 0.5$ are 7.4, 1.8 and 9.0 s, respectively.

If we assume that the longest time above dominates and is denoted by τ, the temperature interval of a DTA scan over which the signal returns from the peak temperature to baseline is approximately $|\Delta T_{peak}| + \alpha\tau$ or $\alpha\tau$ more than the peak height. Thus detecting two events within this temperature interval is very difficult in a DTA instrument. Reduced peak signals are inherent in the heat-flux DSC, and thus provide some advantage for detecting events separated by small temperature intervals.

APPENDIX E ENTHALPY VS. TEMPERATURE RELATIONS FOR DILUTE BINARY SOLID SOLUTION ALLOY

For simulation and understanding of DTA response to simple alloys, it is useful to examine a hypothetical alloy with straight line liquidus and solidus curves, T_{Liq} and T_{Sol}, both terminating at a eutectic temperature T_E. The lines are given by

$$T_{Liq} = T_M + mC_0 \quad (E1)$$
$$T_{Sol} = T_M + (m/k)C_0$$

where C_0 is the composition, T_M is the pure solvent melting point, k is the partition coefficient and m is the liquidus slope taken as negative for a eutectic type diagram.[13] Equivalently, m/k is the slope of the solidus. If C_0 is such that $T_{Sol} > T_E$, the equilibrium freezing range is given by

$$T_{Liq} - T_{Sol} = \frac{m(k-1)C_0}{k} \quad (E2)$$

First we examine different models for the liquid composition, $C_L(f_S)$ as a function of fraction solid, f_S, where $0 < f_S < 1$ and then the corresponding enthalpy curves.

---

[13] For a dilute alloy, m and k are related by the expression, $m = -(1-k)[(RT_M^2)/(L_M)]$, where R is the gas constant and L_M is the heat of fusion per mole of the pure solvent. This reduces to the normal melting point depression law (van't Hoff Law) if $k = 0$.

For an equilibrium liquid–solid mixture (lever law), $C_L(f_S)$ is given by

$$C_L(f_S) = C_0[1 - (1-k)f_S]^{-1} \tag{E3}$$

For Scheil behavior $C_L(f_S)$ is given by

$$C_L(f_S) = C_0[1 - f_S]^{k-1} \tag{E4}$$

For the Ohnaka [38] model for back diffusion

$$C_L(f_S) = C_0\left[1 - f_S\left(1 - \frac{2ak}{1+2a}\right)\right]^{\frac{k-1}{1-\frac{2ak}{1+2a}}} \tag{E5}$$

with $a = D_s t_f/\lambda^2$ and t_f and λ the local freezing time and dendrite spacing, respectively. Here, the local freezing time is taken as the equilibrium freezing range divided by the cooling rate. The Ohnaka form recovers the equations for Scheil and lever as the parameter a spans the range from 0 to ∞, respectively.

It is important to determine if the liquid concentration increases to the eutectic composition when using Eqs. (E3–E5); that is, whether some value of f_S exists where

$$C_L(f_S) = C_E \tag{E6}$$

The value of fraction solid f_S^P if and where this occurs is the fraction of primary phase and the fraction of eutectic will be $(1 - f_S^P)$.

The enthalpy is given by

$$H(T) = C_p^{S0}T + L(1 - f_S(T)) \tag{E7}$$

where L is the heat of fusion and where $f_S(T)$ is obtained by solving the expression

$$T = T_M + mC_L(f_S) \tag{E8}$$

for f_S in terms of T. Eq. (E8) is the liquidus curve combined with one of the forms for $C_L(f_S)$ from Eq. (E3–E5). Then the enthalpy can then be expressed as a function of temperature. For an equilibrium liquid–solid mixture (lever law), the enthalpy is given by

$$H(T) = C_p^{S0}T + L\left[1 - \left(1 - k\left(\frac{T_S - T_{Sol}}{T_{Liq} - T}\right)\right)^{-1}\right] \tag{E9}$$

For Scheil behavior, the enthalpy is given by

$$H(T) = C_p^{S0}T + L\left[1 - \left(\frac{k-1}{k}\right)\left(\frac{T_{Liq} - T}{T_{Liq} - T_{Sol}}\right)^{1/(k-1)}\right] \tag{E10}$$

For the back-diffusion form, the enthalpy is given by

$$H(T) = C_p^{S0}T + L\left[1 - \left(\frac{1}{1 - 2ak/(1+2a)}\right)\left[1 - \left(1 - \left(\frac{k-1}{k}\right)\left(\frac{T_{Liq} - T}{T_{Liq} - T_{Sol}}\right)\right)^{\frac{1 - \frac{2ak}{1+2a}}{k-1}}\right]\right]$$

(E11)

In these equations, T_{sol} is strictly given by Eq. (E1) for the alloy composition of interest, C_0, even if $T_{sol} < T_E$. These enthalpy expressions are only valid for if no eutectic forms. If the eutectic is encountered, the enthalpy suffers a jump discontinuity at T_E of height $L(1 - f_S^P)$.

APPENDIX F BINARY PHASE DIAGRAMS AND DTA RESPONSE

To illustrate the general shape of DTA curves for different cases in binary eutectic and peritectic systems we have chosen hypothetical systems based on Ni. The parameters used for the calculation of the DTA curves are: $\alpha = 15$K/min, $t_{S,C} = 5.67$ s, $t_{W,C} = 4.65$ s, $t_{T,C} = 5.5$ s, $L = 290$ kJ/kg, $C_p = 0.75$ kJ/kg and $m_S = 177$ mg. Three alloy compositions were chosen for each system and the results for the eutectic system are shown in Fig. F1 and the results for the peritectic system are shown in Fig. F2.

APPENDIX G TUTORIAL ON MELTING AND FREEZING OF MULTICOMPONENT ALLOYS

We now explore the DTA melting response of a pair of complex alloys. These were first described in Ref. [23]. As input to the calculation we use enthalpy vs. temperature values obtained for full equilibrium (lever law) and for Scheil freezing assumptions. DTA melting simulations, using the lever enthalpy–temperature relation, would apply to an alloy equilibrated prior to melting and where diffusion was adequate to guarantee spatial concentration uniformity of all phases during melting. DTA melting simulations, using the Scheil enthalpy calculations, would apply to a microstructure that was solidified *and* remelted with no solid diffusion. Clearly these are extreme cases. The melting of an equilibrated alloy as well as an as-cast sample requires an analysis of solid diffusion for both the freezing process and the melting process. The thermodynamic parameters of Ref. [39,40] were used in conjunction with the methods of Ref. [41] to give the enthalpy–temperature relations.

G.1 Aluminum Alloy 2219

Al alloy 2219 has typical mass composition of 6.63% Cu, 0.03% Mn, 0.2% Fe and 0.1% Si. Table G1 gives the sequence of phase formation "reactions" listed in the

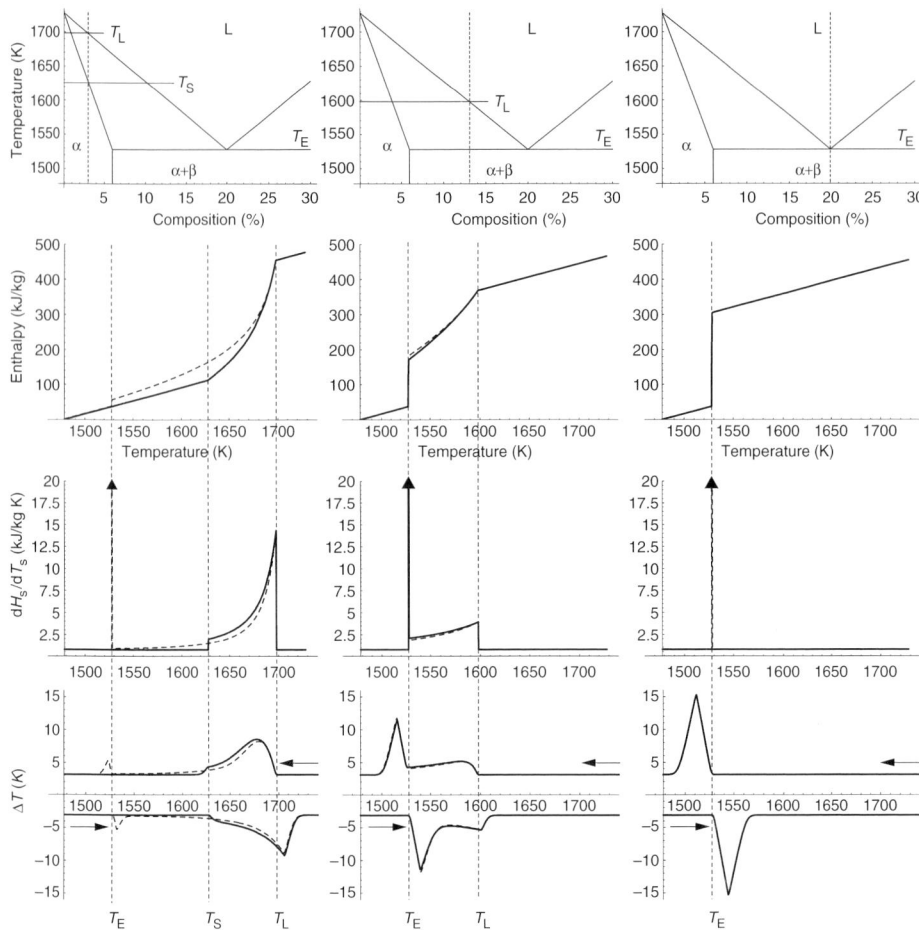

Figure F1 Phase diagram, enthalpy, dH_S/dT_S and computed DTA curves for three different compositions in a hypothetical binary eutectic system. Lever calculations are solid, Scheil are dashed.

order of decreasing temperature. More phases occur, and the final solidification temperature is lower for the Scheil assumption due to the microsegregation. Figure G1 shows the phase fractions as a function of temperature. Off of the scale of Fig. G1 are the α-AlSiFe and Si phases with maximum phase fractions of 8×10^{-4} and 3×10^{-4}, respectively.

Figure G2 shows the values of dH_S/dT_S obtained from the calculated enthalpy–temperature curves and the DTA simulation for 5K/min. Comparison of Fig. G1 with the dH_S/dT_S curves allows one to recognize the cause of the various peaks in the DTA signal. During melting, a peak occurs when all of a particular phase has completely melted (see however problems associated with phase diagram lines that are converging with decreasing temperature, Section 3.3.5). For example for the lever melting, the first peak on heating at ≈820K indicates that all of the Al_2Cu phase has melted. The second peak is barely detectable at ≈840K and

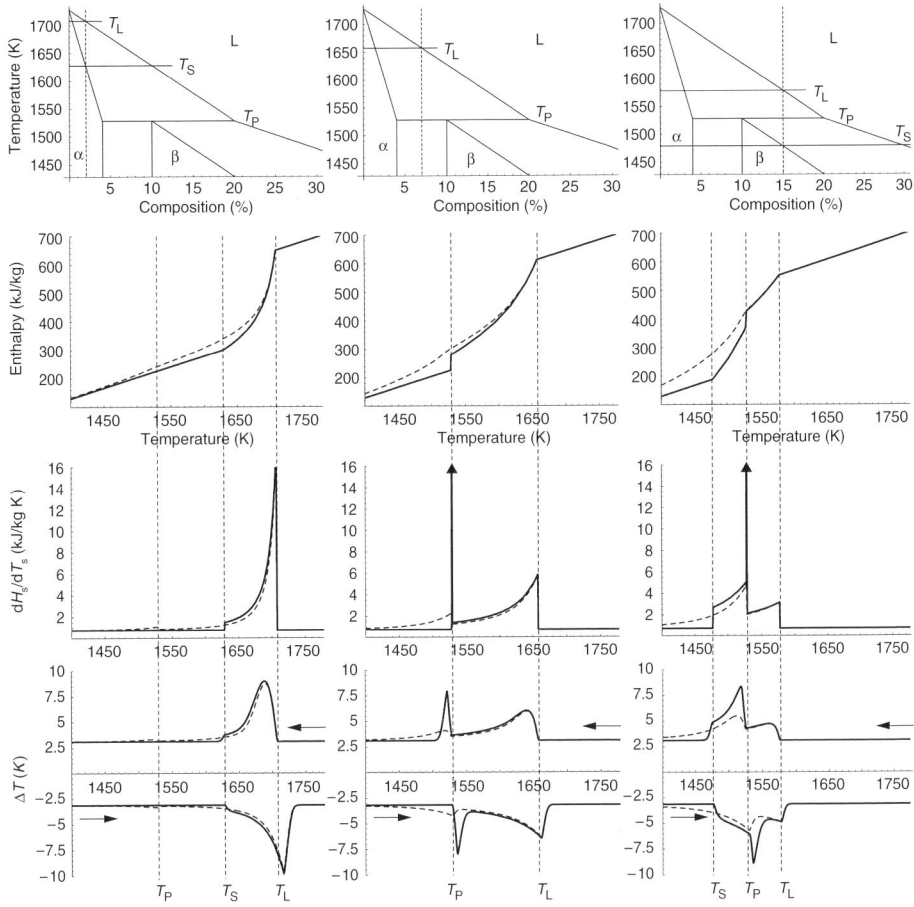

Figure F2 Phase diagram, enthalpy, dH_S/dT_S and computed DTA curves for three different compositions in a hypothetical binary peritectic system. Lever calculations are solid, Scheil are dashed.

Table G1 Sequence of phase formation during lever and Scheil freezing of 2219 Al Alloy.

Lever	Scheil
L → FCC	L → FCC
L → FCC + Al_6Mn	L → FCC + Al_6Mn
L + Al_6Mn → FCC + Al_7Cu_2Fe	L → FCC + Al_7Cu_2Fe
L → FCC + Al_7Cu_2Fe	L → FCC + Al_7Cu_2Fe + Al_2Cu
L → FCC + Al_7Cu_2Fe + $Al_{20}Cu_2Mn_3$	L → FCC + Al_7Cu_2Fe + Al_2Cu + α-AlFeSi
L → FCC + Al_7Cu_2Fe + $Al_{20}Cu_2Mn_3$ + Al_2Cu	L → FCC + Al_7Cu_2Fe + Al_2Cu + α-AlFeSi + Si (Invariant reaction @797 K)

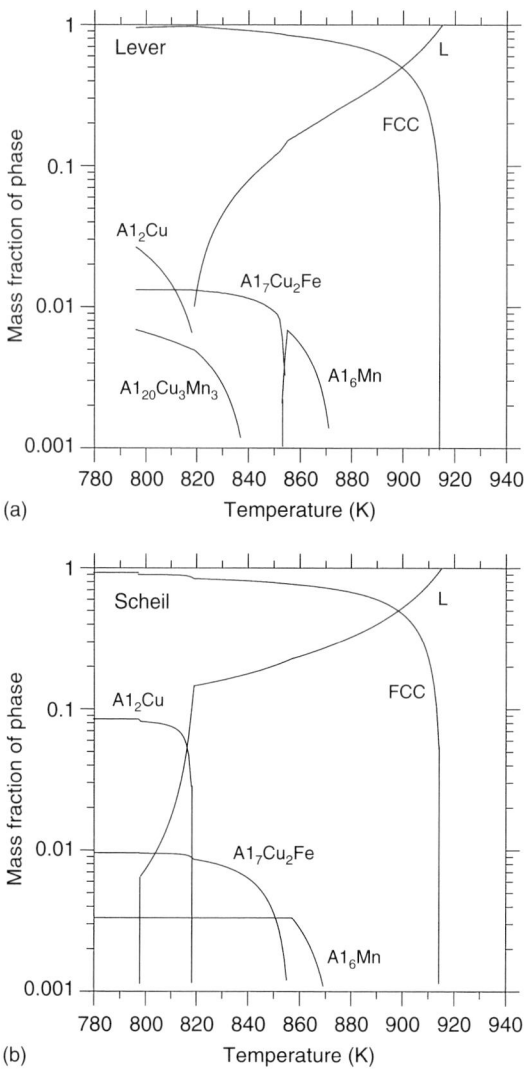

Figure G1 Phase fraction vs. temperature computed using an aluminum alloy thermodynamic database [39] for lever and Scheil conditions for Al 2219 alloy.

indicates that all $Al_{20}Cu_2Mn_3$ has melted. The third peak at ≈855K indicates that all of Al_7Cu_2Fe has melted and similarly for other peaks. Note that for the Scheil DTA calculation, a peak is visible at ≈797K. This is due to the melting of the invariant quinary eutectic at 797K (Table G1) where the Si phase completely disappears and reductions also occur in the phase fractions of the other phases. The

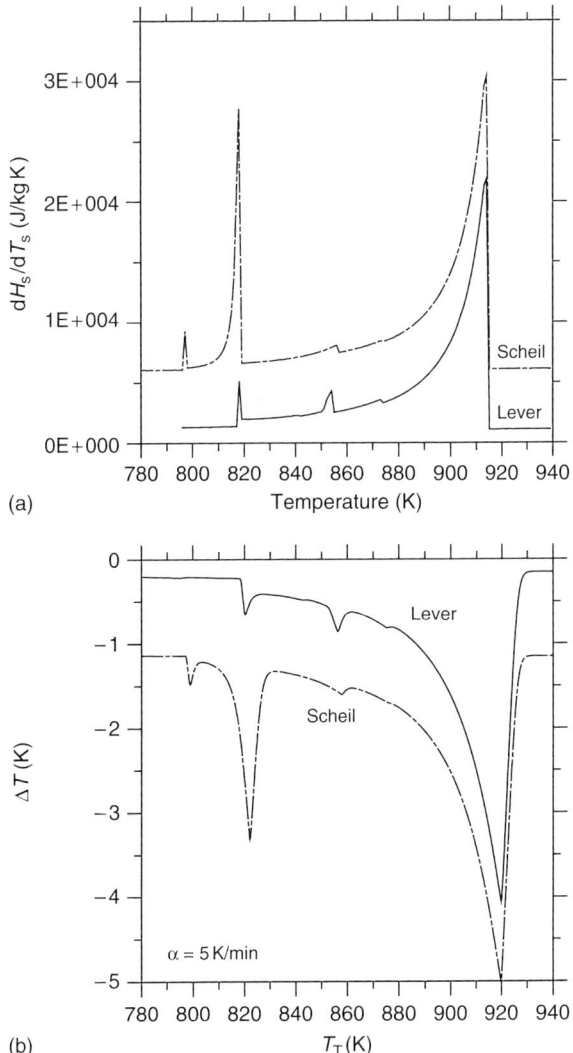

Figure G2 (a) dH_S/dT_S obtained from the enthalpy–temperature predictions for Al 2219 alloy computed using thermodynamic database [39] for lever and Scheil conditions for Al 2219 alloy. The curve for "Scheil" is shifted up by 5×10^3 J/kg K for clarity. (b) Corresponding DTA plots for melting at 5 K/min. The curve for "Lever" is shifted up 1 K for clarity.

alloy under consideration has five components, the reaction involves six phases, and there are zero degrees of freedom. Thus despite the small fraction of Si phase, the signal from the invariant eutectic melting is relatively large.

Simulations of alloy freezing were performed. Figure G3 compares the melting and freezing signals for the Al 2219 alloy at 5 K/min (Scheil enthalpy used for both). The vertical dashed lines are the liquidus temperature and the temperature

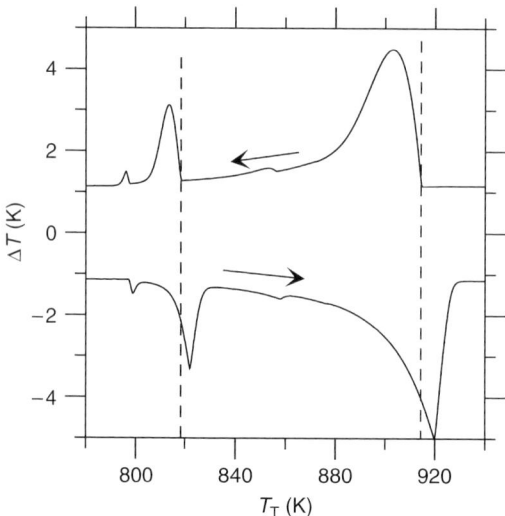

Figure G3 Calculated DTA plots for melting and freezing at 5K/min of Al 2219 alloy using the Scheil enthalpy–temperature relation.

Table G2 Sequence of phase formation during lever and Scheil freezing of Udimet 700.

Lever	Scheil
L	L → FCC
L → FCC	L → FCC + MC
L → FCC + MC	L → FCC + MC + MB_2
L → FCC + MC + MB_2	L → FCC + MC + M_3B_2
	L → FCC + MC + M_3B_2 + σ
	L → FCC + MC + M_3B_2 + σ + γ′
	L → FCC + MC + σ + γ′ + MB_2

where the Al_2Cu phase disappears, or first appears, on melting or solidification, respectively. The peak temperatures are clearly offset from these dashed lines and should not be used to characterize the melting or freezing process.

G.2 Udimet 700

A second example is the Ni alloy, Udimet 700, with mass composition of 15% Cr, 18.5% Co, 5% Mo, 3.5% Ti, 4.4% Al, 0.07% C and 0.025% B. The thermodynamic

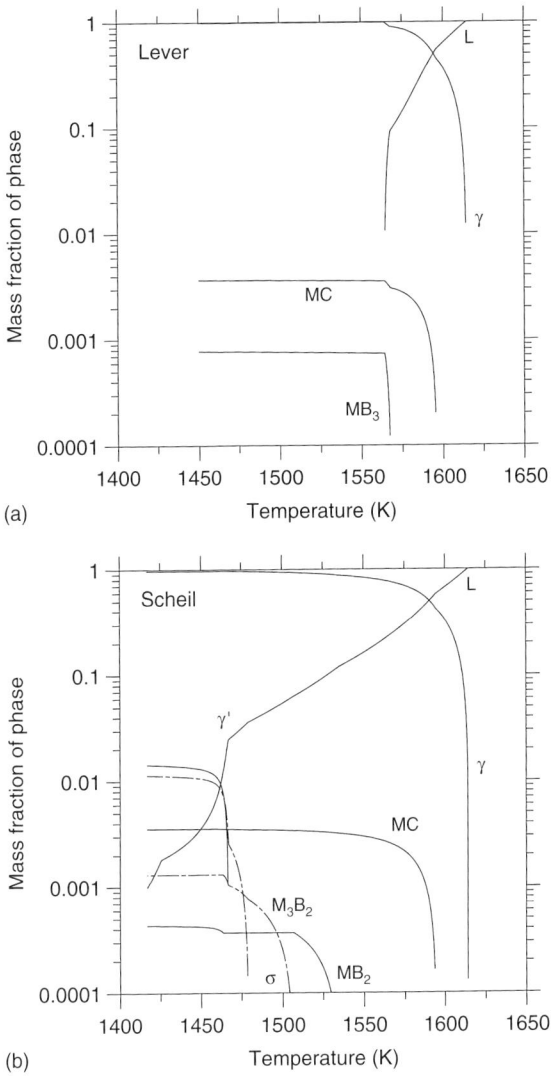

Figure G4 Phase fraction vs. temperature computed using Ni alloy thermodynamic database [40] for lever and Scheil conditions for Udimet 700 alloy.

parameters were taken from Ref. [40]. The phase formation sequences are given in Table G2, the phase fractions in Fig. G4 and dH_S/dT_S and the DTA signals in Fig. G5. To be noted here is the very large difference between the sizes of the freezing ranges of the two cases. The particular peaks and their sizes for real DTA signals for this alloy would be difficult to predict given the large difference between the diffusion rates of the interstitials B and C and the substitutional elements in the FCC phase.

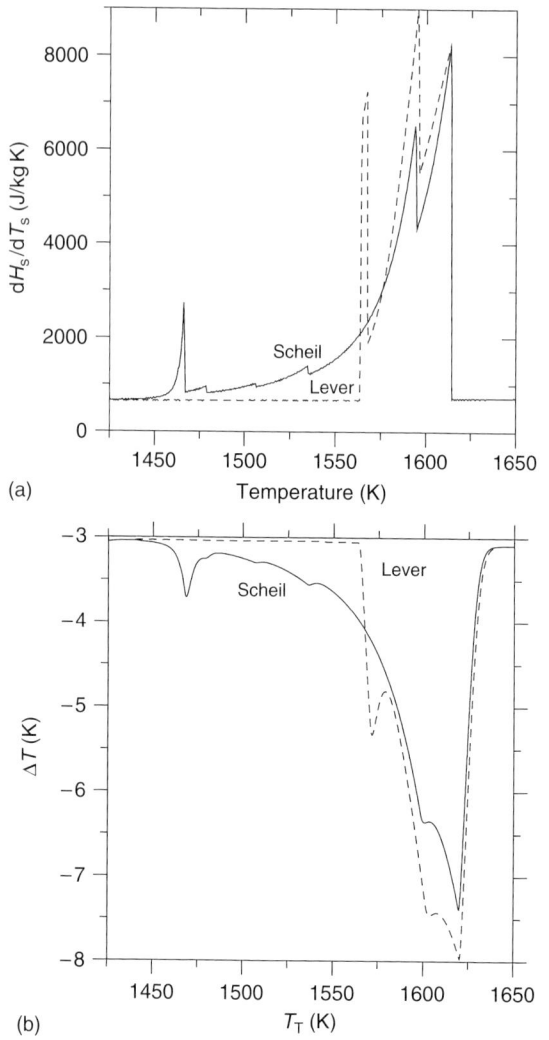

Figure G5 (a) dH_S/dT_S obtained from the enthalpy–temperature predictions for Udimet 700 alloy computed using a thermodynamic database [40] for lever and Scheil conditions for Udimet 700 alloy. (b) Corresponding DTA plots for melting of 180 mg sample at 15K/min.

REFERENCES

1. M.J. Vold, *Analyt. Chem.*, 21 (1949) 683.
2. P.D. Garn, *Thermoanalytical Methods of Investigation*, Academic Press, NY, 1965, Chapters 4 and 5.
3. N.F. Tsang, in W.J. Smothers and Y. Chiang (eds.), *Handbook of Differential Thermal Analysis*, Chem. Publ. Comp., NY, 1966, p. 91.
4. A.P. Gray, in R.S. Porter and J.F. Johnson (eds.), *Analytical Calorimetry*, Vol. 1, Plenum Press, NY, 1968, p. 209.
5. A.D. Cunningham and F.W. Wilburn, in R.C. Mackenzie (ed.), *Differential Thermal Analysis*, Vol. 1, Academic Press, London, UK, 1970, p. 31.
6. J.L. McNaughton and C.T. Mortimer, in H.A. Skinner (ed.), *Thermochemistry and Thermodynamics*, International Review of Science: Physical Chemistry, Series Two, Vol. 10, Butterworths, London, 1975, p. 1.

7. E.M. Barral and R.J. Gritter, in J.H. Richardson and R.V. Peterson (eds.), *Systematic Materials Science*, Vol. IV, Academic Press, NY, 1978, Chapter 39.
8. W. Heyroth, *J. Thermal Analy.*, 31 (1986) 61.
9. J.H. Flynn, *J. Thermal Analy.*, 34 (1988) 367.
10. R.F. Speyer, *Thermal Analysis of Materials*, Marcel Dekker, NY, 1994.
11. J. Opfermann, *ThermoKinetics: A Software Module for the Kinetic Analysis of Thermal Measurements using Multivariate Linear Regression Software*, Netzsch-Gerätebau GmbH, Selb, Germany, 2000.
12. H. Fredriksson and B. Rogberg, *Metal Sci.*, 13 (1979) 685.
13. H. Fredriksson, *ASM Metals Handbook*, 9th edn, Vol. 15, *Casting*, ASM International, Metals Park, OH 1988, p. 182.
14. H.B. Dong, M.R.M. Shin, E.C. Kurum, H.Cama and J.D. Hunt, *Metall. Mater. Trans.*, 34A (2003) 441.
15. J.H. Perepezko, *Mater. Sci. Eng.*, 65 (1984) 125.
16. A. Rabinkin, Solidus–Liquidus, Brazing Course Lecture Notes, ASM/AWS *International Brazing and Soldering Conference 2003*.
17. C.R. Barber, *The International Practical Temperature Scale of 1968, Metrologia*, 5 (1969) 35.
18. H. Preston-Thomas, *The International Temperature Scale of 1990 (ITS-90), Metrologia*, 27 (1990) 3.
19. R.E. Bedford, G. Bonnier, H. Maas and F. Pavese, *Metrologia* 33 (1996) 133.
20. G.W.H. Höhne, H.K. Cammenga, W. Eysel, E. Gmelin and W. Hemminger, *Thermochim. Acta*, 180 (1990) 1.
21. G.W.H. Höhne, *J. Thermal Analy.*, 37 (1991) 1987.
22. M.C. Flemings, *Solidification Processing*, McGraw Hill, NY, 1974, p. 177.
23. W.J. Boettinger and U.R. Kattner, *Metall. Mater. Trans.*, 33A (2002) 1179.
24. F.H. Hayes, H.L. Lukas, G. Effenberg and G. Petzow, *Z. Metallk.*, 77 (1986) 749.
25. U.R. Kattner, W.J.Boettinger and S.R. Coriell, *Z. Metallk.*, 87 (1996) 522.
26. W.D. Cao, R.L. Kennedy and M.P. Willis, in E.A. Loria (ed.), *Superalloys 718, 625 and Various Derivatives*, TMS, Warrendale, PA, 1991, p. 147.
27. R.I. Wu and J.H. Perepezko, *Metall. Mater. Trans.* A, 31A (2000) 497.
28. Y. Feutelais, G. Morgant, J.R. Didry and J. Schnitter, *CALPHAD*, 16 (1992) 111.
29. K.-W. Moon, W.J. Boettinger, U.R. Kattner, F.S. Biancaniello and C.A. Handwerker, *J. Electr. Mater.*, 29 (2000) 1122.
30. H.S. Liu, C.L. Liu, K. Ishida and Z.P. Jin, *J. Electr. Mater.*, 32 (2003) 1290.
31. T. Gödecke, *Z. Metallk.*, 92 (2001) 966 [In German].
32. F.N. Rhines, *Phase Diagrams in Metallurgy*, McGraw-Hill Book Comp., NY, 1956.
33. A. Prince, *Alloy Phase Equilibria*, Elsevier, Amsterdam, 1966, p. 211.
34. G. Massing and B.A. Rodgers, *Ternary Systems*, Dover Publications, NY, 1960, p. 49.
35. R.D. Shull, in R.D. Shull and A. Joshi (eds.), *Thermal Analysis in Metallurgy*, TMS, Warrendale, PA, 1992, p. 95.
36. H.B. Dong and J.D. Hunt, *Mater. Sci. Eng.* A, 413–414 (2005) 470.
37. D.K. Banerjee, W.J. Boettinger, R.J. Schaefer, M.E. Williams, in M. Cross and J. Campbell (eds.), *Modeling of Casting, Welding and Advanced Solidification Processes VII*, TMS, Warrendale, PA, 1995, p. 491.
38. I. Ohnaka, *Trans. Iron Steel Inst. Japan*, 26 (1986) 1045.
39. N. Saunders, *Mater. Sci. Forum*, 217–222 (1996) 667.
40. N. Saunders, in R.D. Kissinger, D.J. Deye, D.L. Anton, A.D. Cetel, M.V. Nathal, T.M. Pollock and D.A. Woodford (eds.), *Superalloys 1996*, TMS, Warrendale, PA, 1996, p. 101.
41. W.J. Boettinger, U.R. Kattner and D.K. Banerjee, in B.G. Thomas and C. Beckermann (eds.), *Modeling of Casting, Welding and Advanced Solidification Processes VIII*, TMS, Warrendale, PA, 1998, p. 159.

CHAPTER SIX

APPLICATION OF DIFFUSION COUPLES IN PHASE DIAGRAM DETERMINATION

A.A. Kodentsov, Guillaume F. Bastin and Frans J.J. van Loo

Contents

1 Introduction	222
2 General Principles of the Diffusion Couple Method	223
3 Experimental Procedures	225
3.1 Preparation of Diffusion Couples	225
3.2 Analytical Techniques and Specimen Preparation	226
3.3 Preparation of Diffusion Couple Specimens for EPMA	227
4 Variations of the Diffusion Couple Technique	228
5 Error Sources Encountered in the Diffusion Couple Experiments	236
6 Concluding Remarks	244

1 Introduction

This chapter will explain the basics of applying the diffusion couple technique in phase diagram determination. The technique came into use in the early 1960s, when a number of seminal papers emerged on multiphase diffusion, solid-state bonding, application of electron probe microanalysis (EPMA) in studying interfacial reactions and the like. These articles elegantly covered the subject and are still being widely cited today.

The simplicity and versatility have undoubtedly contributed to the success of this technique as a "research tool" in solid state and materials science. Nevertheless, new experimental arrangements and procedures were developed to exploit greater potential of diffusion couples in establishing phase relations in multicomponent systems. Most of such information, developed by many investigators, has been published in different journals, proceedings and research reports; but no comprehensive treatment of the subject is available. As a consequence, investigators have to devote time and effort to solve the often basic procedural problems that have already been solved by others. Worse still, the distinct advantages of using the diffusion couple technique for studying phase equilibria in very complex systems are not fully recognized.

Laboratory of Materials and Interface Chemistry, Eindhoven University of Technology, Eindhoven, The Netherlands

This chapter is written with the intention of remedying this situation, although it is not meant to be an exhaustive literature survey or a total overview. We will use practical examples to illustrate the process of using diffusion couples in phase diagram determination. Each example is chosen for its clarity of illustration and pertinence to the particular point of discussion.

We would like to emphasize that this chapter is based mainly on the authors' own experience in this research field; and throughout this chapter, reference will be made to our original publications upon which much of the current text is based. Readers are referred to the original articles to gain a better understanding and clearer insight into the finer nuances of the diffusion couple method.

The systematic layout of this chapter, beginning with the introduction of the principles of the method and ending with the description of different error sources encountered in the diffusion couple experiments, should make a newcomer feel comfortable to decide whether this technique is suitable to his/her particular research project. In the case of a positive answer, one should be able to design diffusion couple experiments best satisfying his/her needs. For an experienced investigator, there should be enough details available on specific experimental arrangements to compare and learn from one another.

Before we consider the diffusion couple method in any details, we shall discuss some fundamental aspects of multiphase interdiffusion which form a natural introduction to our subject.

2 General Principles of the Diffusion Couple Method

The use of diffusion couples in phase diagram studies is based on the assumption of local equilibria at the phase interfaces in the diffusion zone [1–3]. The latter implies that an infinitesimally thick layer adjacent to the interface in such a diffusion zone is effectively in thermodynamic equilibrium with its neighboring layer on the other side of the interface. In other words, the chemical potential (activity) of species varies continuously through the product layers of the reaction zone and has the same value at both sides of an interphase interface.

This concept of local equilibrium being attained at interfaces plays a prominent role in the diffusion theory. It is to be remarked, however, that since diffusion takes place only by virtue of a thermodynamic potential gradient, the total system in which diffusion occurs cannot be at equilibrium. Throughout this discussion, we assume that local equilibrium is established and maintained at the interfaces in the diffusion zone, which means that diffusion is very slow compared with the rate of reaction to form a new phase. An experimental consequence of local equilibrium is the proportionality of the product layer thickness to the square root of the reaction time [4].

A diffusion-controlled interaction in a multiphase binary system which will invariably result in a diffusion zone with single-phase product layers separated by parallel interfaces in a sequence dictated by the corresponding phase diagram. The reason for the development of only straight interfaces with fixed composition gaps follows directly from the phase rule. Three degrees of freedom are required to fix temperature and pressure and to vary the composition. Reaction morphologies consisting of two-phase structures (i.e., precipitates or wavy interfaces) are, therefore, thermodynamically forbidden for binary systems, assuming that only volume-diffusion takes place.

In a ternary system, on the other hand, it is possible to develop two-phase areas in the diffusion zone because of the extra degree of freedom. The diffusion zone morphology, which develops during solid-state interaction in a ternary couple, is defined by type, structure, number, shape and topological arrangement of the newly formed phases. The resulting microstructure of the reaction zone can be visualized with the aid of the so-called *diffusion path*. This is a line on the ternary isotherm, representing the locus of the average composition in planes parallel to the original interface throughout the diffusion zone. Naturally, the diffusion path in a ternary system must fulfill the law of conservation of mass. If no material is lost or created during the interaction, then the diffusion path is impelled to cross the straight line between the end-members of the reaction couple (so-called mass balance line) at least once.

If phases are separated by planar interfaces, the diffusion path crosses the two-phase region on the isotherm parallel to a tie-line and it can be assumed that local equilibrium exists along the whole interface. However, this is not necessarily always the case. Wavy interfaces or isolated precipitates can form at or near a phase interface. In the theoretical framework, the formation of non-planar interface is thought to be related to the formation and release of supersaturation of elements during the diffusion process.

Kirkaldy and Brown [4] formulated a number of rules, which relate the composition of the reaction zone to the phase diagram. These rules were subsequently interpreted by Clark in an elegant manner that most researchers could comprehend [5]. Referring for details especially to the paper by Clark, we will summarize the main ideas here using a hypothetical reaction couple of an A–B–C-ternary system shown in Fig. 6.1.

For a given diffusion couple under conditions of chemical equilibrium, the reaction path involves a time-independent sequence of intermediate layers. The plot gives information about the order of the product layers, their morphology, and their compositions. For example, in the hypothetical system shown in Fig. 6.1, a solid line crossing a single-phase field on the isothermal section (e.g., a–b) denotes an existing layer of that phase in the reaction zone of the couple A/Z. A dashed line parallel to a tie-line in a two-phase field (g–h) represents a straight interface between two single phases. A solid line crossing tie-lines on the isotherm (b–c) represents a locally equilibrated two-phase zone (in fact, a wavy interface) in the couple. A solid line entering a two-phase field and returning to the same phase field (d–e–f) represents a region of isolated precipitates. A dashed line crossing a three-phase field (e.g., i–j or k–l) implies an interface in the diffusion structure with equilibrium among three phases, either a two-phase layer adjacent to a layer consisting of a different phase (e.g., i/j interface) or adjacent two-phase layers with one common phase (e.g., k/l interface).

Before we start to show specific examples of phase diagram determination by means of diffusion couples, it is important to make some general comments concerning sample preparation and analytical techniques used in this method.

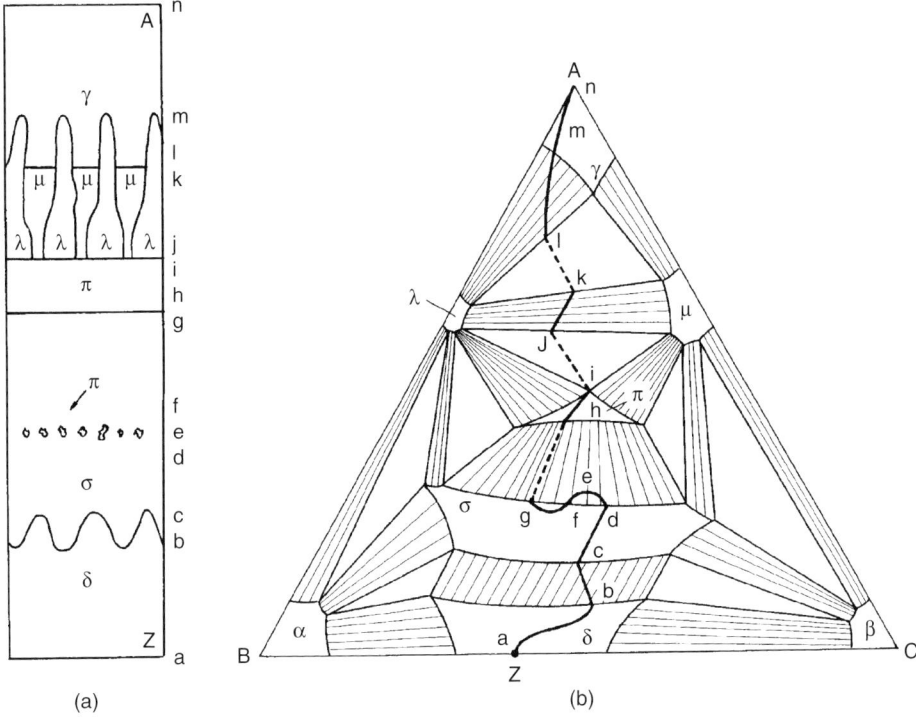

Figure 6.1 A reaction zone structure in a hypothetical couple A/Z of the A–B–C system (a) and the corresponding diffusion path plotted on the isotherm of the ternary diagram (b). The lower-case letters relate the structure to the appropriate composition on the isotherm. (Note all paths in three-phase fields must be denoted by dashed lines, as a three-phase layer cannot form in a ternary diffusion couple.)

3 Experimental Procedures

3.1 Preparation of Diffusion Couples

Several techniques are available to make solid-state diffusion couples, that is to bring two (or more) materials in such intimate contact that one diffuses into the other. Most commonly, the bonding faces of the couple components are ground and polished flat, clamped together and annealed at the temperature of interest. Depending upon the initial materials, various protective atmospheres can be used (e.g., vacuum and inert gas). After the heat treatment, quenching of the sample is desirable in order to freeze the high-temperature equilibrium.

For some metallic systems, different electrolytic and electroless plating techniques can be utilized to fabricate diffusion couples. In this case, one half of the couple is typically a piece of a bulk alloy. The second half of the couple is formed by plating the second alloy onto the bulk alloy substrate.

Other plating techniques include plasma spraying and chemical vapor deposition (CVD). These techniques are suitable for both metals and non-metals. Thermal evaporation, electron beam evaporation or laser evaporation can also be used to deposit the second component layer onto a bulk substrate.

It is also possible to create a multiphase diffusion couple by annealing a substrate material in a reactive gas atmosphere [6]. Such experiments can be carried out in a conventional way (under isothermal conditions) or by imposing a temperature gradient perpendicular to the diffusion direction in the couple. The latter method enables one to observe competing phase reactions "simultaneously" as a function of temperature [7].

3.2 Analytical Techniques and Specimen Preparation

Sometimes phases present in the reaction zone of diffusion couples can be detected simply using an optical microscope. In this respect, a polarized light microscope provides an interesting option. Grains of some crystalline phases exhibiting optical anisotropy show distinctly different colors in white polarized light due to reflection pleochroism. However, in order to obtain information about the phase equilibria existing in a system at a specific temperature, the phase boundary concentrations within the diffusion zone have to be measured.

Different measurement techniques can be used to determine the chemical compositions on both sides of the interfaces. These are Auger electron spectroscopy (AES), secondary ion mass spectrometry (SIMS), Rutherford backscattering spectrometry (RBS), EPMA, and analytical electron microscopy (AEM). In this chapter we shall be concerned with only the last two methods, which are most suitable for the investigation of "bulk" diffusion couples. The others are used for determining composition depth profiles in studies of interdiffusion and reactions in thin-film diffusion couples.

The EPMA has been used since the early 1960s to measure concentration profiles in diffusion couples and has proven itself an indispensable tool in the determination of phase diagrams. There are literally hundreds of studies, which employed this microanalytical technique. By contrast, the AEM has become available for diffusion couple investigations somewhat later. Currently, EPMA is a widely accepted technique for the analysis of phase compositions in multicomponent systems. The complexities in specimen preparation and quantification of the results have prevented AEM from becoming a widely used method for diffusion couple studies. Nevertheless, these two techniques could form a powerful combination for the study of phase stability and reactive diffusion in solids.

In the EPMA, high energy electrons are focused to a fine probe and directed at the point of interest in the diffusion couple. The incident electrons interact with the atoms in the sample and generate, among other signals, characteristic X-rays. These X-rays are detected and identified for qualitative analysis and with the use of suitable standards they can be corrected for matrix effects in order to perform quantitative analysis. EPMA is used to investigate bulk solid samples while AEM is dealing with electron transparent thin films. The procedures used for X-ray quantification are quite different for both techniques.

The principal advantage of EPMA is its ability to measure compositions of very small volumes of a specimen. The spatial resolution for bulk specimens is limited to approximately 1 µm. For thin foils, resolutions better than 50 nm are attainable routinely with AEM. Obviously, in order to obtain the most reliable results, the operating conditions must be optimized. Accelerating voltage, beam current and counting time are, perhaps, the most critical parameters. The optimization of these parameters has been discussed in detail elsewhere [8] and will only be summarized briefly here.

- A ratio of operating accelerating voltage to excitation potential for the measured characteristic X-ray radiation of approximately 2 to 3 is desirable in order to maximize peak to background ratio and minimize the X-ray generation volume.
- The value of beam current has to be chosen such that the X-ray counts are maximized for a statistically meaningful analysis without greatly increasing the electron beam size.
- Counting times must be long enough to allow the accumulation of sufficient X-ray counts for statistically meaningful results (without producing too much carbon contamination).

Presently, two different techniques, wavelength dispersive spectrometry (WDS) and energy dispersive spectrometry (EDS) can be used to collect X-ray spectra from samples being analyzed. Both systems have specific advantages and disadvantages and the choice will depend on the materials studied. In many modern instruments both types of spectrometers can be used simultaneously. Some of the characteristics of these analytical systems, which are important with respect to quantitative analysis of diffusion couples include the following.

For WDS: (1) elemental detection for atomic number $Z \geqslant 4$; (2) X-ray energy resolution as good as 5 eV; (3) a high counting rate and low background which gives a detectability limit on the order of 100 ppm; and (4) higher accuracy in the composition than EDS. The disadvantages of WDS include: (1) sequential detection of the elements thus longer analysis time; (2) much higher equipment cost; and (3) relatively high beam current required.

For EDS: (1) elemental detection for atomic number $Z \geqslant 11$ (in routine applications) or $Z \geqslant 6$ with windowless or ultra-thin window detectors; (2) maximum X-ray energy resolution approximately 140 eV at MnK_α; (3) a low count rate and a low peak to background ratio compared with the WDS system, thus lower accuracy; (4) detectability limit on the order of 1000 ppm; and (5) difficulty in obtaining very accurate compositions. The advantages include simultaneous detection of all elements for faster analysis and much cheaper equipment cost – thus far more widely available.

For a very detailed account of the electron beam-based techniques, the excellent book by Goldstein et al. [8] should be consulted.

3.3 Preparation of Diffusion Couple Specimens for EPMA

As is the case in any microanalytical technique, specimen preparation is crucial. One must be careful not to introduce artifacts connected with the preparation of diffusion couple samples.

Standard metallographic procedures can be used to prepare bulk multiphase couples for EPMA. The critical requirement is to prevent "smearing" across the interphase interfaces. Good metallographic practice and occasionally a final electropolish may prevent contamination of a given phase from an adjacent area of the sample. In addition, special care has to be taken to minimize the height differences among the components and phases in the diffusion couple.

Once a suitable specimen has been prepared it can be analyzed by EPMA, and the measured concentrations can be used to define the tie-lines in the equilibrium phase diagram.

4 VARIATIONS OF THE DIFFUSION COUPLE TECHNIQUE

To stress the salient features of the technique, we will confine the discussion to phase diagram determination in ternary systems.

There are several variations of the diffusion couple method. In the first variant, the sample is a classical semi-infinite diffusion couple, which means that after the diffusion annealing the couple ends still have their original compositions. If volume diffusion in a semi-infinite couple is the rate-limiting step, local equilibrium is supposed to exist, in which case the rules described above can be used to relate the reaction zone morphology, developed during isothermal diffusion, to the phase diagram. The main feature of this variation of the diffusion couple method is that the phase composition of the reaction zone is independent of time and that the diffusion path is fixed. The versatility of this technique in constructing isothermal cross-sections of ternary systems has been demonstrated repeatedly (see e.g., Refs. [9–12]).

We will illustrate the method using a case study of phase relations in the Ni–Cr–Ti system at 850°C [12]. Figure 6.2 shows a typical microstructure of the

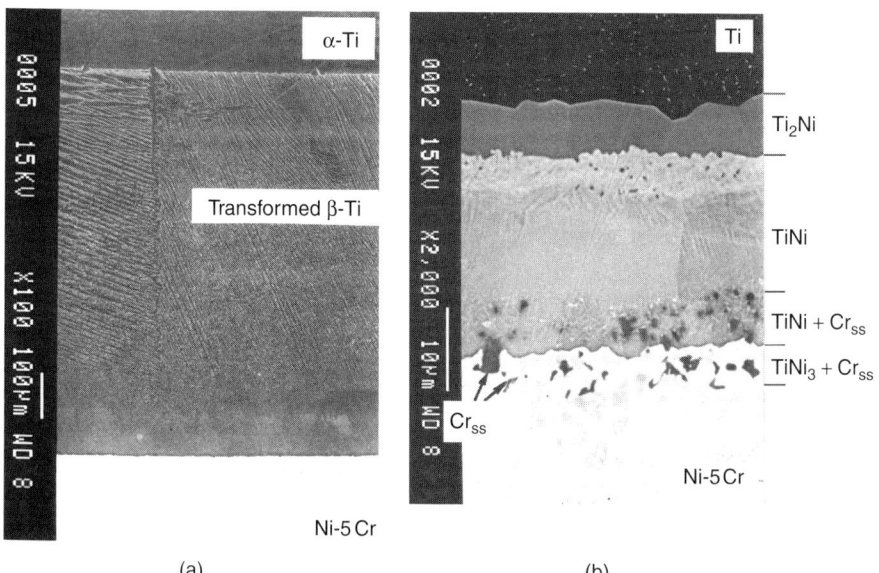

Figure 6.2 Backscattered electron images (BEI) of the microstructure of the reaction zone in the diffusion couple Ti/Ni–5 at.% Cr after annealing at 850°C in vacuum for 100 h: (a) general view and (b) magnified area of the interfacial region.

reaction zone in a Ti/Ni–5 at.% Cr diffusion couple after annealing at 850°C in vacuum for 100 h. Essentially, the microstructure consists of three layers of the binary intermetallic compounds $TiNi_3$, TiNi and Ti_2Ni, and a very thick layer of β-Ti. Concentration of Cr was measured of about 1.2 and 2.3 at.% in the $TiNi_3$ and TiNi phases, respectively, whereas practically no chromium was found with EPMA in the Ti_2Ni reaction product.

Nickel can also be detected in the β-Ti-solid solution beyond the intermetallic layers within a "Widmanstätten"-type structure that is clearly visible on Fig. 6.2. The maximum concentration of Ni in the vicinity of Ti_2Ni/β-Ti-interface was estimated as about 2.5 at.%.

It is worth mentioning that a small amount of Ni or Cr increases the stability of the bcc. β-Ti structure. This can be seen from the binary Ti–Ni and Ti–Cr phase diagrams [13]. The layer of the β-Ti(Ni)-solid solution was formed in the couple at 850°C and, then transformed into the α (and ω)-Ti structure upon cooling.

Close inspection of the transition zone revealed isolated precipitates of the Cr-based solid solution (Cr_{ss}) dispersed inside the $TiNi_3$-layer and also within the TiNi-phase in the vicinity of the $TiNi_3$/TiNi interface (Fig. 6.2(b)). In some areas of the interfacial region the precipitates are rooted to the Ni–5 at.% Cr endmember. This reaction morphology indicates the existence of the three-phase equilibria $TiNi_3 + Cr_{ss} + Ni_{ss}$ and $TiNi_3 + Cr_{ss} + TiNi$ in the Ni–Cr–Ti system at this temperature.

Similar reaction products were also found in the diffusion zone of the couple Ti/Ni–17 at.% Cr after heat treatment at 850°C in vacuum for 100 h (Fig. 6.3). However, the volume fraction of the Cr-based solid solution within the product layer is considerably higher and the Cr-based precipitates are generally coarser.

The concentration of chromium in the $TiNi_3$- and TiNi-matrix layers is also increased (up to approximately 2.7 and 3.5 at.%, respectively), while the Cr-content in the Ti_2Ni remains very low (about 0.5 at.%).

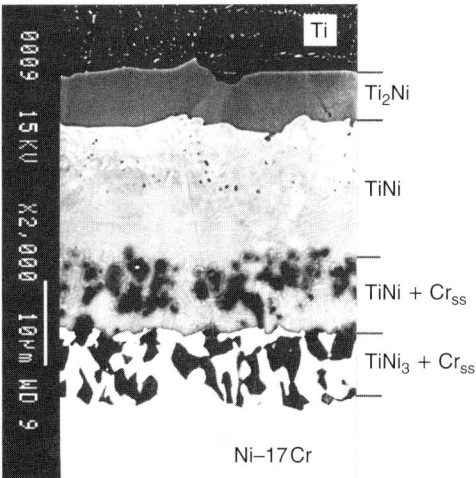

Figure 6.3 BEI of the reaction zone in the couple Ti/Ni–17 at.% Cr after annealing at 850°C in vacuum for 100 h.

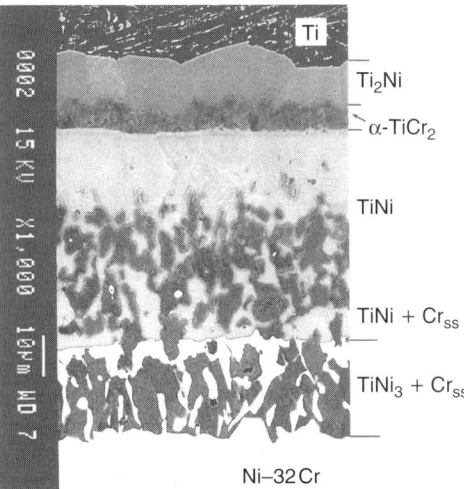

Figure 6.4 BEI of the microstructure of the reaction zone between Ti and Ni–32 at.% Cr-alloy after heat treatment at 850°C in vacuum for 400 h.

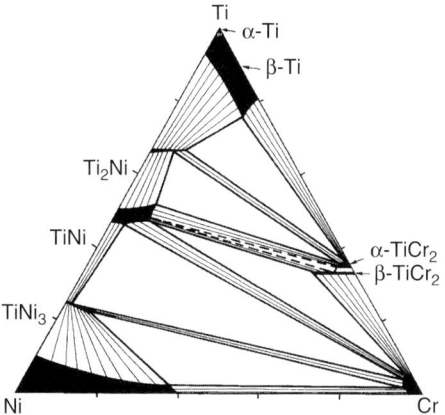

Figure 6.5 The 850°C isothermal section through the Ni–Cr–Ti ternary phase diagram [12].

When a Ni-solid solution containing 32 at.% of Cr was used as end-member of the diffusion couple, the reaction with Ti at 850°C led to the formation of a α-TiCr$_2$ layer between the TiNi and Ti$_2$Ni intermetallics (Fig. 6.4). The concentration of Ni in the α-TiCr$_2$ phase was found to be about 2 at.%.

At the Ni–Cr-side of the couple, an interpenetrating type structure was observed. Domains of the TiNi$_3$-phase (with approximately 6 at.% of Cr) are rooted in the adjoining Ni–Cr substrate, and elongated precipitates of a Cr-based solid solution are also constituents of the next two-phase product layer TiNi + Cr$_{ss}$. The microstructure changes gradually into the TiNi-layer free of the precipitates.

Analysis of the reaction zone morphology evolved in these diffusion couples shows the three-phase equilibria (TiNi$_3$ + Cr$_{ss}$ + Ni$_{ss}$, TiNi$_3$ + Cr$_{ss}$ + TiNi and TiNi + Ti$_2$Ni + α-TiCr$_2$) in the Ni–Cr–Ti system at 850°C (Fig. 6.5).

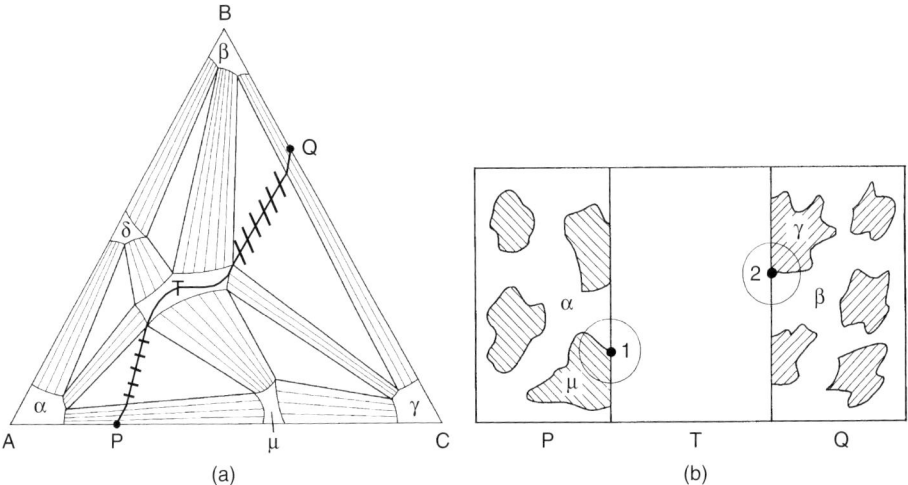

Figure 6.6 Determination of phase equilibria on the isotherm of an A–B–C ternary system using two-phase alloys as end-members: (a) schematic isothermal section and (b) schematic view of a possible reaction zone in a hypothetical diffusion couple P/Q.

An interesting feature of the diffusion zones developed in these samples is a significant enrichment of chromium in the Ni-based solution in the vicinity of the alloy/reaction product-interface. This implies that in the phase region of the Ni-based solid solution on the ternary isotherm, the diffusion paths proceed in the direction of increasing chromium concentration before entering the three-phase region $TiNi_3 + Cr_{ss} + Ni_{ss}$. This is connected with the higher affinity of Ni toward Ti as compared with that for Cr, and with mass balance requirements.

Since a semi-infinite diffusion couple follows a unique diffusion path, very often a large number of couples must be studied experimentally to construct a ternary isotherm. However, the number of samples can be reduced appreciably by using multiphase terminal materials in diffusion couples. In studying a ternary system, the chance "to hit" interfaces at which three phases are in equilibrium is much higher when two-phase alloys are used as end-members. Schematically this procedure is shown in Fig. 6.6. If a diffusion couple between P and Q is assembled, the reaction zone after annealing at a specific temperature might exhibit the morphology indicated in Fig. 6.6(b). In area 1, microprobe measurements will reveal the three-phase equilibrium $\alpha + \mu + T$, whereas from area 2 the equilibrium triangle $\beta + \gamma + T$ existing on the isotherm can be found.

Application of this technique can be demonstrated by studying phase relations in the Ag–Fe–Ti ternary system at 850°C [14]. The microstructure of the reaction zones in the diffusion couples based on Ag and various two-phase binary Fe–Ti alloys is given in Fig. 6.7.

At the interface in the diffusion zone between pure Ag and an equilibrated two-phase alloy with a nominal composition $Fe_{80}Ti_{20}$ after annealing at 850°C for 100 h in vacuum, three phases were found in equilibrium: α-Fe-based solid solution

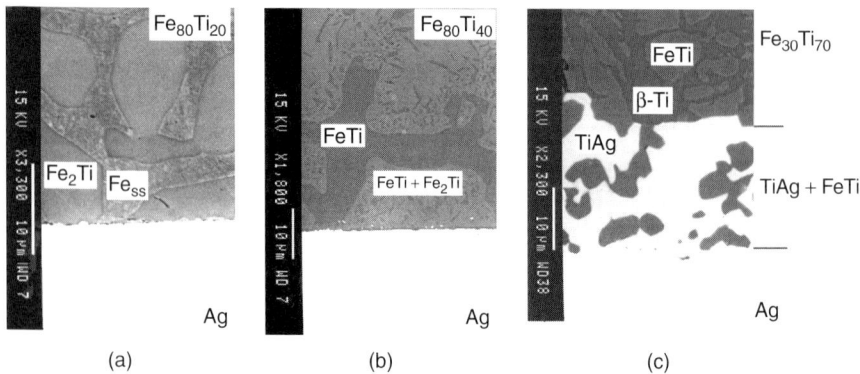

Figure 6.7 Reaction zone morphology developed in the diffusion couples based on binary two-phase Fe–Ti alloys and Ag after annealing at 850°C in vacuum for 100 h: (a) $Fe_{80}Ti_{20}$/Ag (BEI); (b) $Fe_{60}Ti_{40}$/Ag (BEI); and (c) $Fe_{30}Ti_{70}$/Ag (secondary electron image).

(Fe_{ss}), Fe_2Ti and a Ag-based solid solution (Fig. 6.7(a)). Obviously, such a morphology indicates a three-phase equilibrium Fe_{ss} + Fe_2Ti + Ag in the Ag–Fe–Ti system at this temperature. No silver was detected in the Fe-based solid solution or in the Fe_2Ti inside the diffusion zone. Approximately 0.5 at.% of Ti was found in the Ag-based solid solution in equilibrium with Fe_{ss} and Fe_2Ti at the contact interface. The solubility of Fe in Ag is very low and could not be detected reliably by EPMA.

A similar type of morphology was developed in the diffusion zone between Ag and equilibrated alloy $Fe_{60}Ti_{40}$ (Fe_2Ti + FeTi) after heat treatment under the same conditions (Fig. 6.7(b)). The intermetallic compounds Fe_2Ti and FeTi are in equilibrium with Ag within the reaction zone. This implies a three-phase equilibrium Fe_2Ti + FeTi + Ag at the Ag–Fe–Ti isotherm at 850°C. No Fe is dissolved in the Ag-solid solution, whereas the Ti concentration in the solid solution near the interface was estimated to be ~1.0 at.%.

Interaction between Ag and an equilibrated two-phase alloy with a nominal composition $Fe_{30}Ti_{70}$, consisting of FeTi intermetallic and β-Ti (with nearly 18 at.% of Fe) at 850°C leads to the formation of the two-phase layer TiAg + FeTi in the diffusion zone (Fig. 6.7(c)). Two interfaces where three phases exist in equilibrium are found. The TiAg-matrix is in equilibrium with both β-Ti (with ~14.5 at.% of Fe and ~2.0 at.% of Ag) and FeTi, which is a constituent of the starting two-phase alloy. The solubility of Fe in the TiAg-matrix is negligible and the solubility of Ag in the FeTi is estimated at ~0.5 at.%. This two-phase reaction layer also adjoins the Ag-based solid solution. The observed morphology indicates the existence of the three-phase equilibria FeTi + TiAg + Ag and FeTi + TiAg + β-Ti on the isotherm.

When a two-phase $Ti_{80}Ag_{20}$ alloy, containing the Ti_2Ag-intermetallic and α-Ti after equilibration, is used as an end-member of a diffusion couple, the reaction with Fe at this temperature results in the formation of continuous layers of the binary compounds Fe_2Ti and FeTi (Fig. 6.8). The micrograph in Fig. 6.8 shows only a part of the diffusion zone close to the reaction product/Fe-interface. The original end-member, composed of α-Ti and Ti_2Ag, is much further away from the interface (about 500 μm).

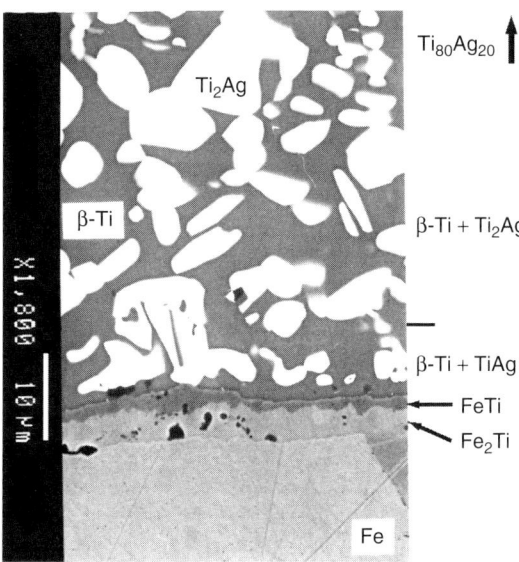

Figure 6.8 BEI of the reaction zone in the diffusion couple $Ti_{80}Ag_{20}/Fe$ after heat treatment in vacuum at 850°C for 100 h. Note that the original (α-Ti + Ti_2Ag) alloy is much further away from the interface (about 500 μm).

Deeper in the alloy, a two-phase layer β-Ti and TiAg was formed. The β-Ti-matrix is also a constituent of the next two-phase layer β-Ti + Ti_2Ag adjacent to the initial two-phase Ti–Ag alloy. (Note: Ag stabilizes the hexagonal α-Ti structure, while the presence of Fe increases the stability of the bcc β-Ti-based solid solution [13].) The composition of the β-Ti-based solid solution in equilibrium with the intermetallic compounds Ti_2Ag and TiAg inside the diffusion zone was determined as $Ti_{84.0}Fe_{13.5}Ag_{2.5}$. Analysis of the morphology developed in this diffusion couple reveals the three-phase equilibria FeTi + TiAg + β-Ti, β-Ti + TiAg + Ti_2Ag and β-Ti + Ti_2Ag + α-Ti in the Ag–Fe–Ti system at 850°C. All equilibria found in the multiphase diffusion couples are consistent with the ternary Ag–Fe–Ti diagram given in Fig. 6.9. The experimentally established phase relations were also confirmed by examining a number of ternary equilibrated alloys and, in addition, they were verified by a thermodynamic assessment of this system.

Further development of this technique for studying phase diagrams is related to changing the "macrostructure" of the classical diffusion couple. A sample is prepared by joining two plane-parallel slices of metal (alloy) through a thin layer of the third metal (alloy) as shown schematically in Fig. 6.10(a). In such a layered system, in which the central part is eventually consumed, the diffusion path is not fixed as in the semi-infinite diffusion couple. The phase composition of the complex diffusion zone is changing continuously with time as a result of the overlapping of two quasi-equilibrated diffusion zones. To relate the morphology and composition of the reaction zone to the phase diagram, rules similar to those explained before still can be used. For instance, in Fig. 6.10 the composition found in the phases α_1 and α_2 at $t = t_3$ and t_4, respectively, are the end points of two tie-lines in the two-phase region.

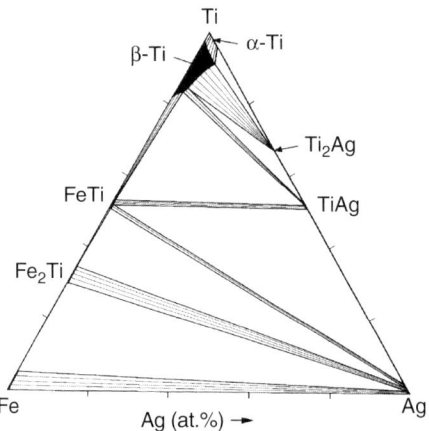

Figure 6.9 The 850°C isothermal section through the Ag–Fe–Ti ternary phase diagram [14].

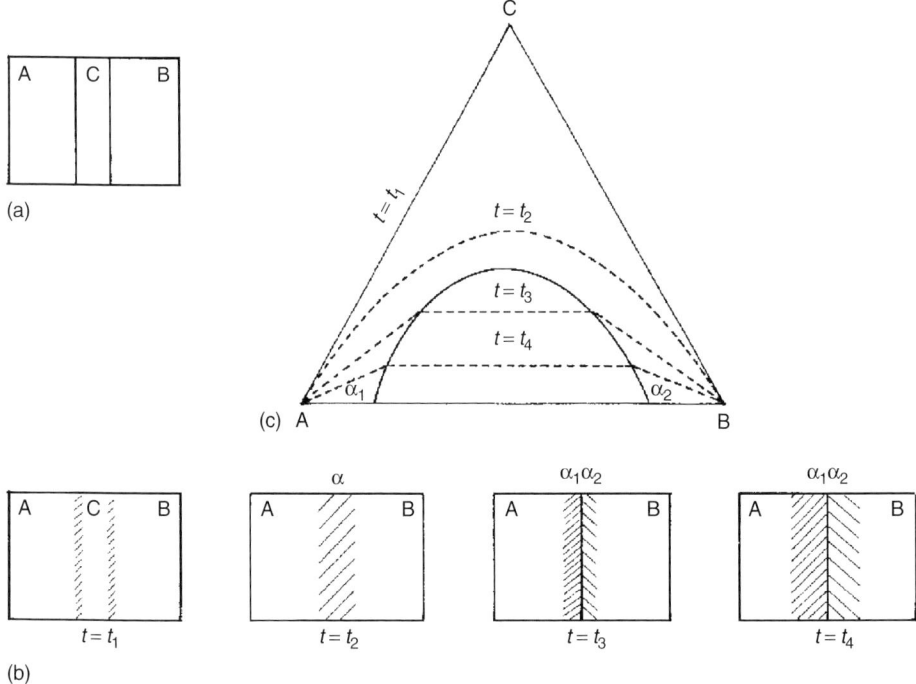

Figure 6.10 Schematic view of the reaction zones and diffusion paths on an isotherm after various annealing times: (a) initial "sandwich" sample; (b) reaction zone morphology for different annealing times; and (c) diffusion paths for various annealing times.

A relatively simple example of constructing an isothermal section of the Ni–Cr–V ternary system at 1150°C is used to illustrate this method. In this case, "sandwich" samples were prepared [15] by joining and subsequent annealing at 1150°C in vacuum of vanadium and chromium discs (about 2 mm thickness) with

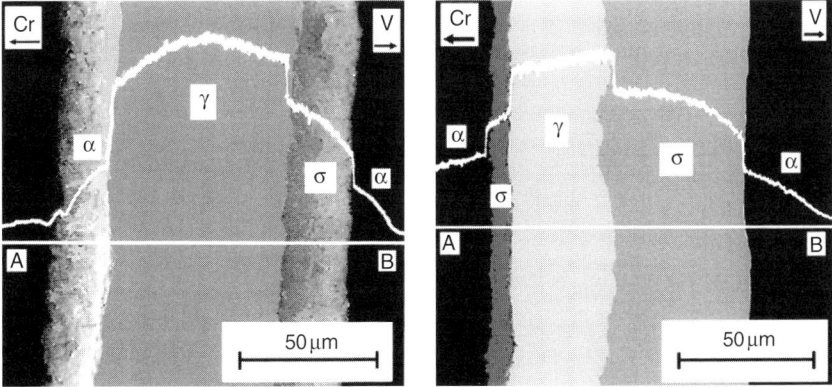

Figure 6.11 Microstructure of the diffusion zone in the finite "sandwich" sample V/Ni foil (100 μm)/Cr after annealing at 1150°C in vacuum for (a) 16 h and (b) 49 h. Backscattered electron images together with the variation of the Ni K_α signal across the zone of interaction (along the line AB). The fcc and bcc solid solutions are denoted as γ and α, respectively.

an intermediate Ni-foil. It is imperative that before the process of interference/overlapping starts (i.e., before V reaches the Cr–Ni interface, or before Cr reaches the Ni–V interface), local equilibria have been established inside the reaction zone in each of the "semi-infinite" couples of which the "sandwich" sample was originally composed. Based on the information about diffusion kinetics in the constituent Ni/V and Cr/Ni binary couples [15], the thickness of the Ni-interlayer was chosen as 100 μm.

Examination of the "sandwich" samples after heat treatment at 1150°C for 4, 16 and 25 h revealed the formation at the V-side of a continuous layer of the σ-"VNi" phase containing some Cr. A typical microstructure of the reaction zone after the annealing is shown in Fig. 6.11(a). The Cr content in the product layer was found to increase with increasing reaction time.

Subsequent annealing of the sample up to 49 h led to the reaction zone morphology shown in Fig. 6.11(b). A characteristic feature of the diffusion zone developed under these conditions is the formation of two layers of the σ-phase, separated by a layer of the Ni(V,Cr)-based solid solution. The latter observation indicates that the diffusion path crosses the σ-phase field on the ternary Ni–Cr–V isotherm twice.

This isotherm was also investigated using "semi-infinite" couples of the NiCr/V type and by means of traditional methods of equilibrated alloys. Figure 6.12 represents the experimentally determined isotherm and the diffusion paths found in the "sandwich" samples after various annealing times. One can see that the results obtained by the different techniques are in good agreement.

Before completing this part of the discussion, it should be mentioned that within the past 15 years, this variation of the diffusion couple technique has proven itself as a valuable tool in phase diagram studies (see, e.g., Refs. [15–19]). The efficiency of this method is very high: by making only one finite "sandwich" sample, re-annealing and investigating after various annealing times, a lot of information can be gained about the whole isotherm.

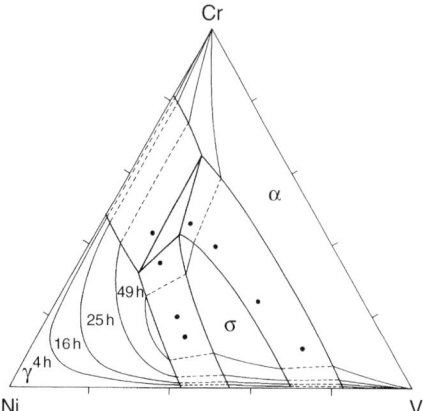

Figure 6.12 Experimentally determined isothermal cross-section through the Ni–Cr–V phase diagram at 1150°C [15]. The compositions of the equilibrated alloys examined are depicted as points.

At a first level, the diffusion couple technique is nothing more than a tool for establishing a correlation between the morphology developed in the diffusion zone of a diffusion couple and a certain type of phase relationship in the system. However, it must be added immediately that like other seemingly simple methods used in materials science, the proper use of diffusion couples for determination of multi-component phase diagrams is by no means a trivial procedure (as it might seem at first sight). A number of error sources may appear when multiphase diffusion experiments are used to establish phase equilibria.

5 Error Sources Encountered in the Diffusion Couple Experiments

The experimental results may contain errors directly attributable to the nature of the sample. One of the possible dangers is a system in which the terminal compositions (end-members) are solids, but a liquid phase exists at the annealing temperature. It is then possible for the diffusion path to wander into the liquid phase region, with disastrous results indeed. Poor adherence at the interfaces in the diffusion zone and accelerated reaction rates due to defects such as cracks and grain boundaries may also complicate the interpretation of diffusion couple experiments.

Another source of error comes from the experimental measurements themselves. There are difficulties associated with accurate determination of phase boundary concentrations in the reaction zone for both semi-infinite and finite diffusion couple techniques. Several issues concerning the electron-beam microanalytical techniques are to be noted here.

Firstly, the determination of a chemical composition with EPMA has an inherent experimental error associated with data counting statistics and data correction procedures.

Secondly, the volume of X-ray generation in the "bulk" samples is determined by the electron scattering in the target, and not by the incident electron beam size.

Because of the scattering effects, it is impossible to generate characteristic X-rays from (and hence, to determine the chemical composition of) a volume much smaller than 1–2 μm in diameter. In other words, explicit measurements of interfacial compositions are not possible with any electron-beam techniques. It is, however, possible to measure concentration gradients in the vicinity of the interfaces in the multiphase diffusion zone. When the phases in question have a reasonable layer width, the composition at the interface can be determined by extrapolating the measured concentration profiles to the interface locations to obtain the interface compositions. This approach, however, is not always very accurate. In most cases, the necessity for extrapolation is the most serious limitation of the EPMA and the diffusion couple method. When steep concentration gradients occur near the phase boundaries in the diffusion-grown layers, the extrapolation may lead to large errors. The same problem arises when the equilibrium compositions at precipitate/matrix interfaces are determined in a two-phase diffusion reaction product.

Thirdly, the X-ray absorption effect may cause problems. This occurs when X-rays produced at one point within the specimen travel through materials of different compositions, and perhaps of different mass absorption coefficients on their way to the spectrometer. To avoid such absorption effects, it is necessary to orient the phase boundary in the diffusion couple perpendicular to the sample surface and the interphase interface parallel to the X-ray path to the spectrometer.

Finally, accurate microprobe analysis near the interfaces is sometimes very difficult owing to fluorescence effects. When dealing with the spatial resolution in quantitative EPMA, of primary concern is the range of electron propagation. This can be defined as the distance the electrons diffuse away from the point of impact of the incident electron beam on the specimen until they have lost so much energy as to be incapable of exciting primary characteristic X-ray radiation. This range is typically on the order of 1–2 μm, yielding a volume of primary excitation of approximately 2–4 μm diameter. In many cases, however, the primary characteristic X-ray radiation can be powerful enough to excite one or more elements present in the sample, thus giving rise to enhanced X-ray production (i.e., fluorescence). In other words, if the sample contains an atom species with a critical excitation energy less than the energy of the primary characteristic X-rays being absorbed, fluorescence will occur. The main trouble here is that this "characteristic" fluorescence usually takes place in a much larger volume, sometimes an order of magnitude greater than the volume of primary excitation due to the fact that X-rays can travel through solids much more easily than electrons. As a consequence, the spatial resolution can be drastically reduced. In fact, in analyzing multiphase diffusion couples the additional X-ray excitation may occur on one side of the phase interface even though the electron beam and the analysis point are on the other side of the interphase interface. This may lead to the erroneous observation that X-rays of a certain element may be detected in locations where the element is not present. In this context, it is relevant to mention that for the case of K–K fluorescence (K_α-radiation of one element exciting K_α-radiation of the other), the excitation of secondary radiation is especially bad in targets containing elements with atomic numbers differing by two (for atomic number $Z > 21$), for example combinations of the elements Fe/Ni and Cu/Co.

A number of schemes have been introduced to correct the fluorescence (and absorption) "uncertainty" [8]. One should, however, bear in mind that in all these correction procedures it is implicitly assumed that the primary and secondary production of X-ray radiation as well as subsequent absorption, all take place in the same homogeneous (single phase) matrix. Apparently, these conditions are violated to an increasing extent as the incident electron beam approaches an interface or when the size of particles decreases below a certain limit. Corrections for this effect are possible but difficult [20]. It is best to determine how large the effect is by using undiffused couples and to correct the data accordingly. In a binary diffusion couple, usually one of the elements does not suffer from the fluorescence effect and one could measure the concentration of that element by microprobe analysis to describe the distribution of elements across the reaction zone. If, for instance, pure Cu and Co are clamped together, without any diffusion taking place, CoK_α characteristic X-ray radiation can apparently be measured at a distance of up to 40 μm from the interface in the pure Cu [20]. About 4.8 at.% of Co appears to be in Cu at the interface. Measuring CoK_α radiation diminishes the error, but still an apparent Cu concentration of 2 at.% is found in cobalt near the interface, whereas Cu X-ray radiation can be detected at about 15 μm from the interface in cobalt.

Obviously, in the case of a ternary system, the situation can be even more complex. The reader interested in greater details concerning the correction procedures is invited to refer to the references cited.

Another group of errors arise from the formation of a quasi-equilibrated diffusion zone. When the diffusion couple techniques are used for phase diagram determination, it is of fundamental importance to be sure that equilibrium values are obtained at the interphase interfaces and that all equilibrium phases pertaining to the system under the conditions of the diffusion couple experiment, are formed in the reaction zone.

Sometimes certain phases seem to be missing in a diffusion couple when investigated by microscopic or microprobe analysis. One of the reasons for the absence of an equilibrium phase might be the presence of a barrier layer at the interface, such as oxide films at the contact surface or the presence of impurities in the starting materials. In the latter case, the segregation of impurities, which may be present only in the ppm level in one of the end-members can cause enrichment in the diffusion zone, making nucleation of a certain phase difficult.

Even though the absence of certain phases might be due primarily to difficulties in nucleation, it is important to point out that the apparent absence of a particular intermediate phase in a diffusion zone cannot automatically be interpreted as the result of nucleation problems. It is possible that the phase is present in such a minute quantity that it cannot be determined easily by the experimental techniques available. It should be noted that a kinetic consideration of product phase growth in itself is not so essential for the present discussion. However, when taking into account the overall context of a diffusion couple experiment, the question of the apparent absence of a certain equilibrium phase in a reaction zone or, conversely, the formation of metastable phases during interaction, becomes an important issue. For instance, in a famous example of a Ti/Al diffusion couple only $TiAl_3$ was found [21]. Yet, several other intermetallic phases should have been formed according to the phase diagram [13]. On the other hand, when incremental $Ti/TiAl_3$ diffusion couples were annealed at 800°C, all the possible intermetallic phases in the equilibrium phase diagram were indeed found in the reaction zone: $Ti/Ti_3Al/TiAl/TiAl_2/TiAl_3$. However, when

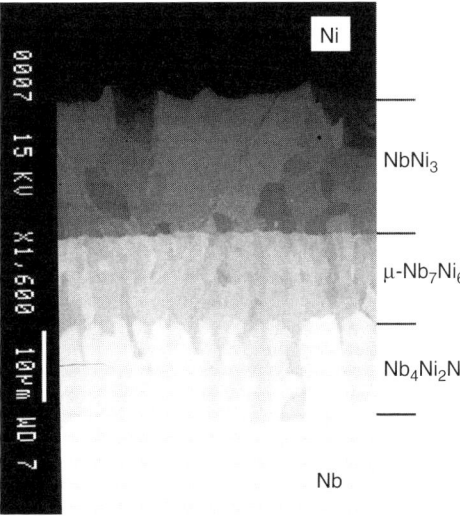

Figure 6.13 Backscattered electron image of an annealed (1100°C; 196 h; vacuum) binary Nb/Ni couple showing the formation of the "impurity-stabilized" phase Nb_4Ni_2N within the reaction zone.

a layer of Al was joined on the outside of the $TiAl_3$ layer of the above couple and annealed at 625°C for 15 h, the compounds $TiAl_3$, TiAl and $TiAl_2$ disappeared, resulting in the original configuration of $Ti/TiAl_3/Al$. Clearly in this case, nucleation of Ti_3Al, TiAl, $TiAl_2$ was not the problem. Rather, the apparent absence of the other phases is due to the relatively slow diffusion in these compounds, that is due to slow growth kinetics.

An additional reason for caution here is the occasional formation of a non-equilibrium phase in a diffusion zone, often stabilized by impurities present in the end-members. Very telling examples are carbides like Mo_5Si_3C [22] or Mo_6Ni_6C [23], which might be confused with the purely binary phases Mo_5Si_3 and MoNi. The same is true for oxides such as Ti_4Ni_2O and nitrides such as Nb_4Ni_2N; they might be confused with the binary compound Ti_2Ni or with an, in fact nonexisting, binary "compound" Nb_2Ni. Further complications may result from the fact that the "impurity-stabilized" phases often grow in series with genuine binary compound layers, as shown in Fig. 6.13 for the Nb/Ni couple annealed at 1100°C. Three distinct product layers are clearly visible on the micrograph. According to the binary Nb–Ni phase diagram, however, only two products are expected to form in the reaction zone of the Nb/Ni couple at this temperature: $NbNi_3$ and the μ-phase Nb_7Ni_6 [13]. EPMA revealed that the reaction layer formed at the Nb-side of the diffusion zone is a ternary compound with a composition close to the Nb_4Ni_2N, which indeed exists in the ternary Ni–Nb–N system at 1100°C (Fig. 6.14) [24]. The nitrogen was present as an impurity in the starting Nb.

An easy way to verify the purely binary character of an equilibrium phase layer is the use of incremental diffusion couples. The terminal compositions are then chosen quite close to the apparent phase in question, in such a way that only this phase

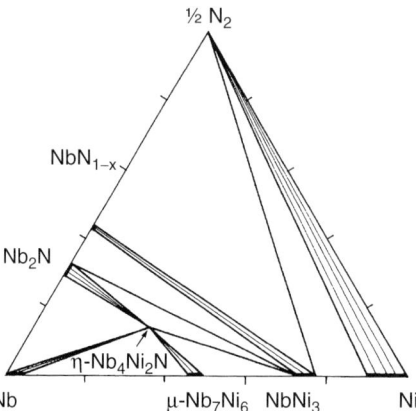

Figure 6.14 The Nb–Ni–N ternary phase diagram at 1100°C and 10^{-8} bar of nitrogen partial pressure [24].

might be formed. The end-members may be single-phase or two-phase alloys. A true equilibrium phase then grows parabolically with time as a relatively thick layer, whereas an impurity-stabilized phase ceases to grow after some time since the impurity from the end-member is totally consumed [25]. Figure 6.15 schematically illustrates this procedure.

It is also worthwhile to show one more example. At first sight, it might not appear to have much to do with the diffusion couple method. However, it is important to realize that as soon as gaseous species such as oxygen and nitrogen, and under some circumstances also sulfur, phosphorous, arsenic or antimony, take part in the interaction, the gas atmosphere may decide which reaction path would occur.

The Mo–Si–N system is an instructive illustration of how the formation of a volatile reaction product affects the diffusion path. The interesting point in this system is the dependence of the reaction path upon the total pressure and, especially, the partial pressure of nitrogen. This can be appreciated by looking at Fig. 6.16, which represents reaction zone morphologies developed between Si_3N_4 and Mo after annealing at 1300°C in vacuum. Two different types of Si_3N_4 were used in these experiments: fully dense hot isostatically pressed ceramics (Fig. 6.16(a)) and 50% porous silicon nitride (Fig. 6.16(b)). Despite the apparent similarity of "chemical" identity of the reactants in these couples, the corresponding diffusion paths were found to be remarkably different. In the reaction couple based on dense Si_3N_4-ceramics, only Mo_3Si was detected in the diffusion zone, whereas layers of $MoSi_2$ and Mo_5Si_3 are formed in the diffusion couple consisting of porous silicon nitride. It can be seen in Fig. 6.16(b) that the whole initial Mo – end-member of the couple has been consumed, and so has the Mo_3Si layer that must have been present at shorter annealing times.

The difference in behavior of these couples indicates that the type of reaction products formed at an elevated temperature in the diffusion zone between Si_3N_4 and any metal (or alloy) depends on the chemical potential (activity) of silicon

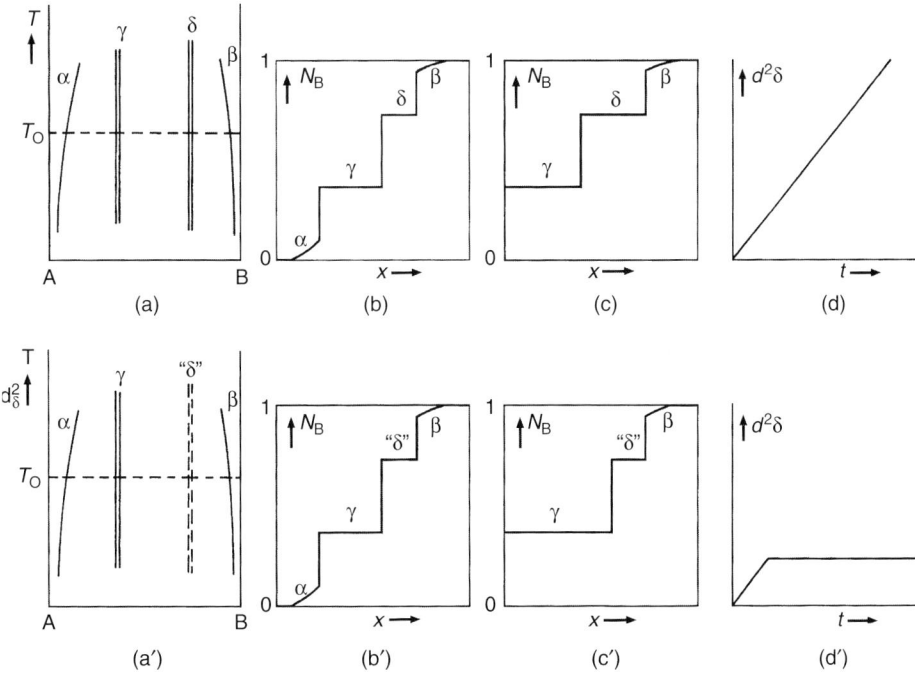

Figure 6.15 Comparison of the growth of an equilibrium phase ((a)–(d)) vs. a non-equilibrium, impurity-stabilized compound ((a′)–(d′)): (a) phase diagram of an A–B system; (b) and (c) the concentration profile for component B in a couple A/B and γ/B (d) plot of square of the layer thickness, d_δ^2 vs. time in the γ/B couple; (a′)–(d′) the analogs when the δ-phase is a non-equilibrium, impurity-stabilized compound. The growth of the δ-layer in a couple γ/B reaches a maximum value soon and stays without further increase.

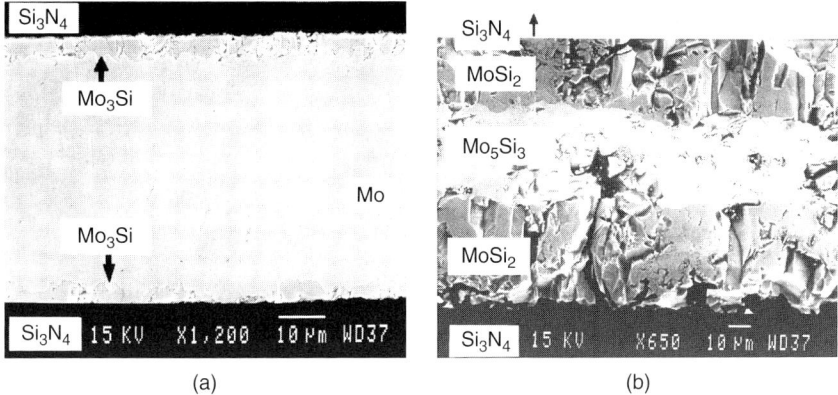

Figure 6.16 (a) Secondary electron image (SEI) of the transition zone between Mo and dense Si_3N_4 after annealing at 1300°C for 32 h in vacuum and (b) SEI of the reaction zone between Mo and 50% porous Si_3N_4 after annealing at 1300°C for 50 h under flowing argon containing about 500 ppm of nitrogen.

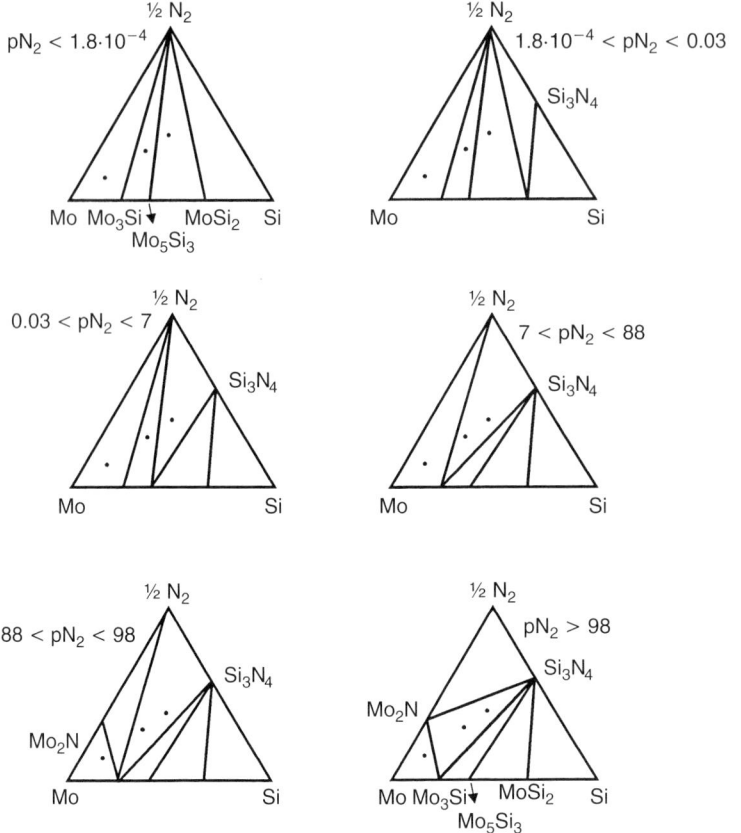

Figure 6.17 Isothermal sections through the Mo–Si–N phase diagram at 1300°C at various partial pressures (bar) of nitrogen. The compositions of experimentally investigated powder mixtures are depicted as points.

and, hence, on the activity (fugacity) of nitrogen at the contact surface. In diffusion couples consisting of dense Si_3N_4-ceramics and non-nitride forming metals, the interior of the couple is not in direct contact with the surrounding atmosphere. Nitrogen, which is formed by the interfacial reaction, cannot escape easily. A nitrogen pressure (fugacity) will be built up at the contact surface. This pressure determines the activity of Si at the metal/ceramic interface. Thus, it dictates the isothermal cross-section through the phase diagram Me–Si–N and, more specifically, the position of the monovariant equilibrium N_2-gas + Si_3N_4 + Me_xSi_y.

Figure 6.17 represents the phase equilibria in the Mo–Si–N system at 1300°C and various partial pressures (fugacities) of nitrogen. These isothermal cross-sections were created using thermodynamic data on pertinent binary systems [26] and the results of X-ray diffraction analysis of the Mo + Si_3N_4 powder compacts equilibrated at 1300°C under various nitrogen pressures [27].

There are, of course, other ways of presenting the same information as shown in Fig. 6.18. This type of construction (a so-called stability diagram) displays condensed

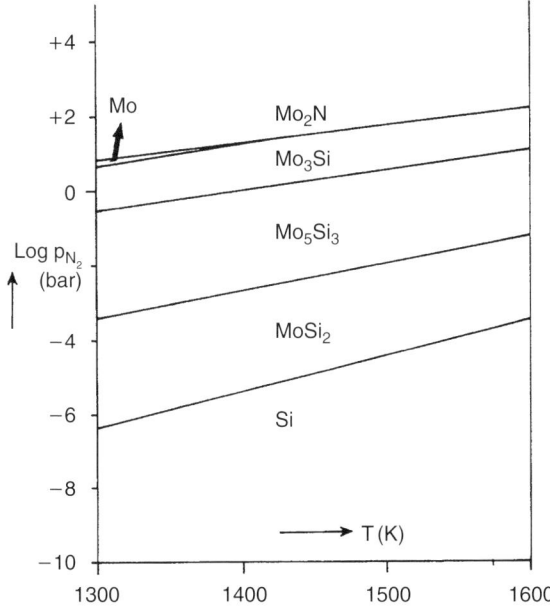

Figure 6.18 Stability diagram showing solid phases in equilibrium with solid Si_3N_4 in the Mo–Si–N system as a function of temperature and nitrogen partial pressure (fugacity).

phases in equilibrium with solid Si_3N_4 as a function of temperature and nitrogen partial pressure (fugacity). The thermodynamic activity of Si at the metal/ceramic interface is related to the N_2-partial pressure through the equilibrium constant of Si_3N_4.

If one looks at Figs. 6.17 and 6.18 and tries to predict the type and sequence of the reaction products in a Si_3N_4/Mo diffusion couple at 1 bar partial pressure of N_2, the following layer sequence Mo | Mo_3Si | Mo_5Si_3 | (N_2) | Si_3N_4 may appear. Experimentally, however, only Mo_3Si was found in the reaction zone after interdiffusion of dense Si_3N_4 with Mo at 1300°C under 1 bar of nitrogen or argon (containing N_2), or in vacuum ($\sim 10^{-9}$ bar), and no Mo_2N had been formed (Fig. 6.16(a)).

These observations suggest that the nitrogen pressure in the gas atmosphere surrounding the couple does not influence the interfacial reaction in the central part of the joint. This becomes quite clear since N_2-gas has to escape somehow from the interior of the transition zone and this is difficult in the fully dense diffusion couple. Therefore, a nitrogen pressure (fugacity) will build up at the contact surface. According to calculations for the Mo–S–N system at 1300°C this pressure is between 7 and 88 bar, because Mo_3Si has been found in equilibrium with Si_3N_4 but Mo_2N has not been formed.

On the other hand, no N_2-pressure can build up at the reaction interface when a 50% porous ceramic is used as an end-member because nitrogen can escape through the open pores in the Si_3N_4. The reaction products in this couple depend entirely on N_2-partial pressure in the surrounding atmosphere since this pressure determines the activity of Si at the Si_3N_4/metal interface.

The message here is very simple. The formation of volatile products during interfacial reactions can profoundly affect the diffusion path which must be taken into consideration in interpreting results of diffusion couple experiments.

6 CONCLUDING REMARKS

The diffusion couple technique is unquestionably a powerful and viable technique in establishing phase relationships. The efficiency of this method is very high. Yet, it is not without its shortcomings. One area is in the spatial resolution of the microanalytical techniques used. Errors attributable to extrapolated interface compositions can then be minimized and steeper concentration gradients can be measured. In this respect, AEM is very promising.

In recent years, the research on phase diagrams involving diffusion methods took on a new life, due primarily to the advance in preparation of so-called diffusion multiples [28,29]. A diffusion multiple is an assembly of three or more different metal (or ceramic) blocks, in intimate interfacial contact, arranged as a triple, quadruple, etc., and subjected to a heat treatment to allow interdiffusion to occur. More detail about this rather complex technique can be found in Chapter 7 of this book. Judicious cautions should be exercised in correlating compositions of the reaction products formed during interdiffusion in such complicated diffusion zones with the corresponding equilibrium phase diagram. Especially when two-phase regions are developed within the diffusion structures (which is seldom the case), one must be careful not to attribute too much precision to the locations of the phase boundaries deduced from EPMA measurements of such two-phase regions.

It should be emphasized that although the diffusion couple technique is a unique and powerful tool in materials science studies, it is most useful in conjunction with other methods. For instance, in order to increase the reliability of the information obtained about an isothermal cross-section of an equilibrium phase diagram, a combination of the diffusion method with an investigation of selected equilibrated alloys (or powder compacts) is desirable. Multiphase diffusion couples can be used to map a constitutional diagram and help to select the compositions of the alloys to be examined. The precise location of the phase field boundaries as well as direction of tie-lines on the isotherm can be obtained from selected equilibrated alloy samples.

REFERENCES

1. J.S. Kirkaldy, *Can. J. Phys.*, 36 (1958) 917.
2. J.B. Clark and F.N. Rhines, *Trans. ASM*, 51 (1959) 199.
3. L.S. Darken, *Trans. AIME*, 175 (1948) 184.
4. J.S. Kirkaldy and L.C. Brown, *Can. Metall. Quart.*, 2 (1963) 89.
5. J.B. Clark, *Trans. Met. Soc., AIME*, 227 (1963) 1250.
6. W. Lengauer, *Acta Mater.*, 39 (1991) 2985.
7. W. Lengauer, *J. Solid State Chem.*, 91 (1991) 279.
8. J.I. Goldstein, D.E. Newbury, P. Echlin, D.C. Joy, A.D. Romig Jr., C.E. Lyman, C. Fiori and E. Lifshin, *Scanning Electron Microscopy and X-ray Microanalysis*, Plenum Press, New York, 1992.
9. A.D. Romig Jr. and J.I. Goldstein, *Metall. Trans. A.*, 9A (1978) 1599.
10. F.J.J. van Loo, J.W.G.A. Vrolijk and G.F. Bastin, *J. Less-Common Met.*, 77 (1981) 121.

11. F.J.J. van Loo and G.F. Bastin, *J. Less-Common Met.*, 81 (1981) 61.
12. J.A. van Beek, A.A. Kodentsov and F.J.J. van Loo, *J. Alloy. Compd.*, 270 (1998) 218.
13. T.B. Massalski, J.I. Murray, L.H. Bennet and H. Baker (eds.), *Binary Alloy Phase Diagrams*, ASM, Metal Park, OH, 1986.
14. J.A. van Beek, A.A. Kodentsov and F.J.J. van Loo, *J. Alloy Comp.* 221 (1995) 108.
15. A.A. Kodentsov, S.F. Dunaev and E.M. Slusarenko, *J. Less-Common Met.*, 135 (1987) 15.
16. A.A. Kodentsov and E.M. Slusarenko, *J. Less-Common Met.*, 141 (1988) 225.
17. A.A. Kodentsov and E.M. Slusarenko, *J. Less-Common Met.*, 153 (1989) 351.
18. J.A. van Beek, A.A. Kodentsov and F.J.J. van Loo, *J. Alloy. Compd.*, 217 (1995) 97.
19. M.R. Rijnders, J.A. van Beek, A.A. Kodentsov and F.J.J. van Loo, *Z. Metallkd.*, 87 (1996) 732.
20. G.F. Bastin, F.J.J. van Loo, P.J.C. Voster and J.W.G.A. Vrolijk, *Scanning*, 5 (1983) 172.
21. F.J.J. van Loo and G.D. Rieck, *Acta Metall.*, 21 (1973) 61.
22. F.J.J. van Loo, F.M. Smet, G.D. Rieck and G. Verspui, *High Temp. High Press.*, 14 (1982) 25.
23. C.P. Heijwegen and G.D. Rieck, *Metall. Trans.*, 4 (1973) 2159.
24. H. Holleck, *Binäre und ternäre Carbid- und Nitridsysteme der Übergangsmetalle*, Gebrüder Borntraeger, Berlin–Stuttgart, 1984.
25. F.J.J. van Loo, *Prog. Solid State Chem.*, 20 (1990) 47.
26. P. Rogl and J.C. Schuster (eds.), *Phase Diagrams of Ternary Boron Nitride and Silicon Nitride Systems*, ASM, Materials Park, OH, 1992.
27. E. Heikinheimo, A. Kodentsov, J.A. van Beek, J.K. Klomp and F.J.J. van Loo, *Acta Metall. Mater.*, 40 (1992) S111.
28. J.-C. Zhao, *Adv. Eng. Mater.*, 3 (2001) 143.
29. J.-C. Zhao, *J. Mater. Res.*, 16 (2001) 1565.

CHAPTER SEVEN

PHASE DIAGRAM DETERMINATION USING DIFFUSION MULTIPLES

J.-C. Zhao

Contents

1 Introduction	246
2 Diffusion-Multiple Fabrication	248
3 Analysis of Diffusion Multiples and Extraction of Phase Diagram Data	256
3.1 Imaging Examination and Phase Analysis	256
3.2 EPMA Profiling	257
3.3 Extraction of Equilibrium Tie Lines	259
4 Sources of Errors	266
5 Concluding Remarks	269

1 INTRODUCTION

A diffusion multiple is an assembly of three or more different metal blocks, in intimate interfacial contact, and subjected to a high temperature to promote thermal interdiffusion to form solid solutions and intermetallic compounds [1–5]. It is an expansion of traditional diffusion couples [e.g., 6,7] and the little known "ternary diffusion couples" [8–11]. For the purpose of phase diagram determination, a diffusion multiple is nothing more than a sample with multiple diffusion couples and diffusion triples in it. An example is schematically shown in Fig. 7.1 which has eight diffusion couples (shown by the dotted lines) and four diffusion triples (dotted circles). The local equilibrium at the phase interfaces serves as the foundation to extract phase equilibrium information from diffusion multiples in the same way as that from diffusion couples (discussed in detail in Chapter 6 of this book).

The biggest advantage of the diffusion-multiple approach in phase diagram determination is its high efficiency in both time and raw materials usage. An entire ternary phase diagram can be obtained from a tri-junction region of a diffusion multiple. By creating several tri-junctions in one sample, isothermal sections of multiple ternary systems can be determined without making dozens or even hundreds of individual alloys, thus saving the usage of raw materials. The diffusion-multiple approach can also save the electron probe microanalysis (EPMA) time since there is no need to exchange

GE Global Research, Niskayuna, NY, USA

Methods for Phase Diagram Determination (Edited by J.-C. Zhao)
9780080446295

Copyright © 2007. Elsevier Ltd.
All rights reserved.

Phase Diagram Determination Using Diffusion Multiples

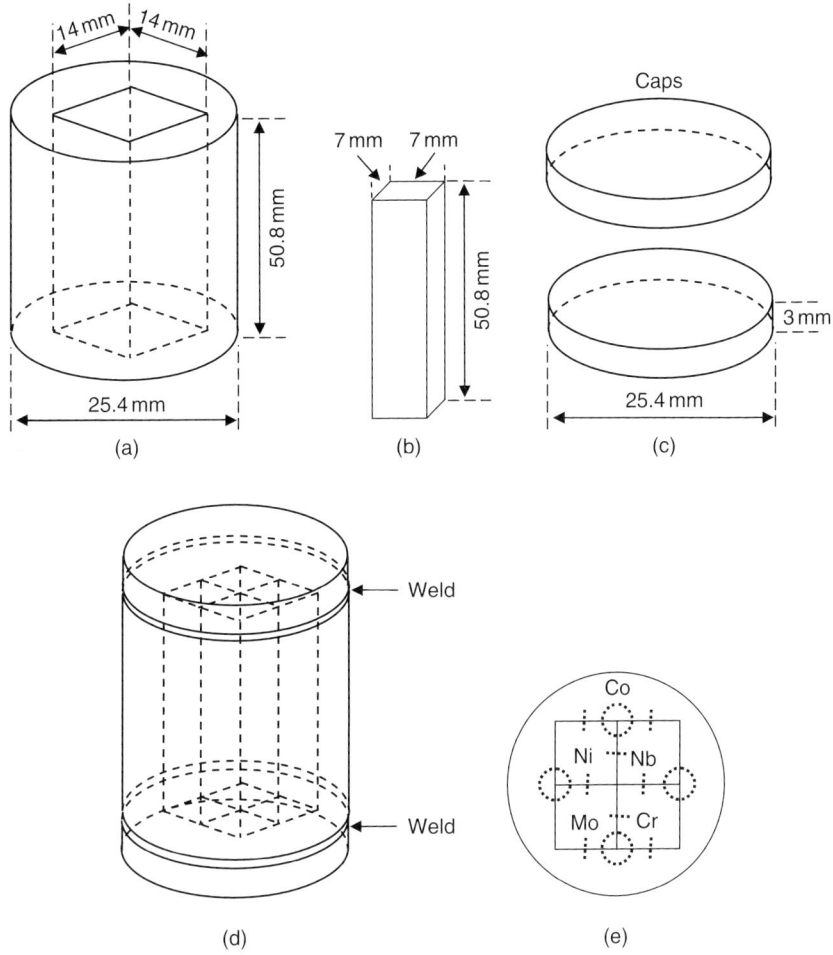

Figure 7.1 Schematic illustration of the components (a)–(c), assembly (d), and cross-sectional view (e) of a Co–Cr–Mo–Nb–Ni diffusion multiple.

many alloy samples in and out of the EPMA system, which is very time-consuming – one needs to wait for a good vacuum to start the analysis each time.

The term "diffusion multiple" was coined to reflect the much expanded capability in recent years in mapping various properties as a function of composition and phases, thus extending the methodology far beyond phase diagram determination. In this sense, a diffusion multiple is primarily used to create compositional variations of both solid solutions and intermetallic compounds to allow their properties to be measured/mapped without synthesizing one composition at a time. Several micron-scale property probes/measurement tools have been developed to enable high throughput measurements of hardness, modulus, thermal conductivity, optical properties, dielectric constants, and other properties. Such property measurements together with the localized composition measurements allow effective construction of composition–structure–property relationships [1–5]. A review of the state-of-the-art of micron-scale property mapping can be found elsewhere [5].

As part of this book on methods for phase diagram determination, this chapter will only address the phase diagram mapping part of the diffusion-multiple approach. The main topics to be discussed are: (1) how to design and make good diffusion multiples; (2) how to perform effective analyses using EPMA and electron backscatter diffraction (EBSD); (3) how to extract phase diagram data from the EPMA results; and (4) how to take advantage of the diffusion-multiple approach while avoiding its limitations.

2 Diffusion-Multiple Fabrication

The most important step in determining phase diagrams using the diffusion-multiple technique is to make a good diffusion-multiple sample. One needs to spend time upfront to design it with several key considerations as discussed herein. Very successful and less successful examples of diffusion multiples will be used to illustrate these key considerations.

The diffusion multiple shown in Fig. 7.1 was assembled from several components: one cylinder of pure Co (25.4 mm in diameter and 50.8 mm in height) with a 14 × 14 mm square opening along the cylindrical axis (Fig. 7.1(a)), one prismatic bar each of pure Cr, Mo, Nb, and Ni of 7 × 7 × 50.8 mm dimensions (Fig. 7.1(b)), and two pure Co disks (caps) of 25.4 mm diameter and 3 mm thickness. All the components were cut to shape using wire electro-discharge machining (EDM). The re-cast layer formed on all cut surfaces during wire EDM was removed using grit (Al_2O_3 particles) blast and subsequent mechanical grinding to a metal finish. To make the pieces close to the final dimensions shown in Fig. 7.1, each surface was given a 25 μm margin for the loss during grit blasting and grinding of the re-cast layer, that is, the bar inserts of Cr, Mo, Nb, and Ni were cut to the dimensions of 7.050 × 7.050 × 50.850 mm. The pieces were not polished – usually grinding with 1200 grit SiC paper is good enough to make a good diffusion multiple for phase diagram determination. The amount ground away is different for each component/piece and a perfect match among them could not realistically be expected. It is good enough at this stage if the four bar inserts could be fitted into the square opening in the cylindrical Co piece, even though a little loose. It is usually the case that the components are not perfectly aligned, thus the supposed quaternary junction in the center of the diffusion multiple only occasionally yield lots of quaternary phase equilibrium information.

All the pieces were ultrasonically cleaned in acetone or alcohol before assembly. The pure Cr, Mo, Nb, and Ni pieces were put inside the opening in the cylindrical Co bar. Two Co caps were put at the ends of the cylindrical Co bar with the inserts inside, as shown in Fig. 7.1(d). Thin Ni strips were spot-welded onto the Co caps and the Co cylinder to keep the pieces from falling apart during transport to electron beam (EB) welding. The EB welding was performed in vacuum and was performed along the outer circular edges of the Co cylinder, as shown in Fig. 7.1(d). At this stage, the prismatic pieces were still loose inside. The EB-welding step was very important since it kept the inside of the assembly in a vacuum state which would allow the subsequent hot isostatic pressing (HIP) to squeeze all the components together from the outside. The HIP run was performed at 1100°C for 4 h at 200 MPa argon pressure (the welded Co cylinder and caps served as an HIP can). At 1100°C, the Co was very soft and deformed plastically under the 200 MPa pressure, eliminating the loose gaps among the components and achieving intimate interfacial contacts. The diffusion multiple was then cooled down from the HIP unit, and subsequently wire EDM cut into three equal-height pieces parallel to the Co caps. One such piece of the diffusion multiple was wrapped in Ta foil and sealed in an evacuated quartz tube, back-filled with pure argon. To absorb any oxygen that might diffuse across the quartz tube, some pure yttrium pieces were wrapped inside a Ta foil and put inside the quartz tube before it was sealed off.

The sealed quartz tube with the diffusion-multiple inside was put into an air furnace and heated to 1100°C and held at that temperature for 1000 h. Upon completion of the diffusion annealing, the quartz tube was quickly taken out of the furnace and smashed into a tank of water. The diffusion multiple was thus water quenched to room temperature. It was then ground and polished for optical microscopy and scanning electron microscopy (SEM) examination, followed by EPMA and EBSD.

SEM images taken from the four tri-junctions of the diffusion multiple are shown in Fig. 7.2. The formation of intermetallic compounds, the wide diffusion

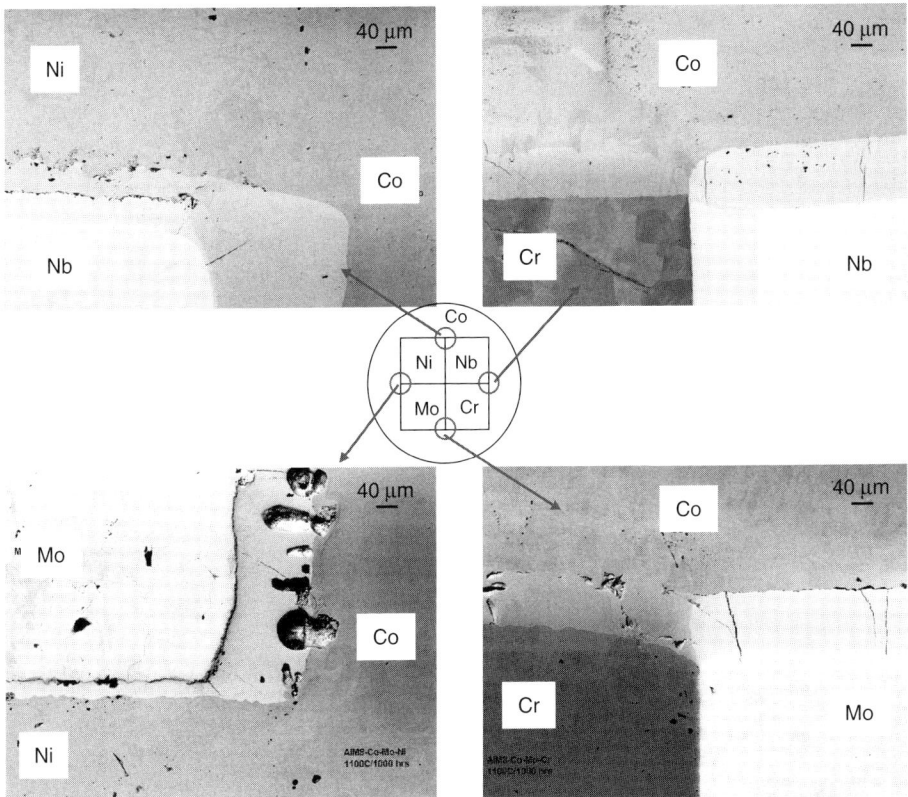

Figure 7.2 SEM images of the four tri-junctions of the 1100°C diffusion multiple shown in Fig. 7.1 showing good interdiffusion and the formation of intermetallic phases between/among the elements. Some of the SEM images were rotated with respect to the schematic diagram shown in the middle.

zones and the good integrity of the sample indicate a very successful fabrication of the diffusion multiple. Detailed analysis of the Co–Cr–Mo tri-junction (lower right in Fig. 7.2) will be described in the next sections of this chapter.

As mentioned earlier, many considerations need to be taken to design and fabricate a successful diffusion multiple. These considerations are discussed here:

• Selection of the components/end members and the annealing temperatures of a diffusion multiple need to be made for the alloy systems of interest. Depending on the temperatures and the compositional regions of interest, one can select pure elements, alloys, or intermetallic compounds as the components of a diffusion multiple. The temperature is an extremely important factor to consider since the diffusion kinetics may be too slow at low temperatures to allow effective interdiffusion to form the phases with sufficient thickness for meaningful analysis. Practically speaking, long-term diffusion annealing should be conducted at a temperature above half of the homologous melting temperature. When a system contains a low melting element or a low eutectic temperature, one can instead select an intermetallic compound or an alloy with high melting point as a component of a diffusion multiple rather than use a pure element as a component. An example is shown in Fig. 7.3 for the Ni–Al–Cr, Ni–Al–Pt, and Ni–Al–Ta systems. Instead of using pure Al as a member of the diffusion multiple, which would limit the annealing temperature to less than 660°C, a Ni–54.5 at.% Al intermetallic compound (B2 NiAl phase) was employed. The use of this intermetallic compound allows the diffusion multiple to be annealed at 1200°C for much faster diffusion. The downside is that only the high-temperature part of the respective phase diagrams can be determined. In this case, this was not a problem since the interest was on the high-temperature Ni-rich regions. Only when pure elements are used, will the entire ternary isothermal sections be obtained.

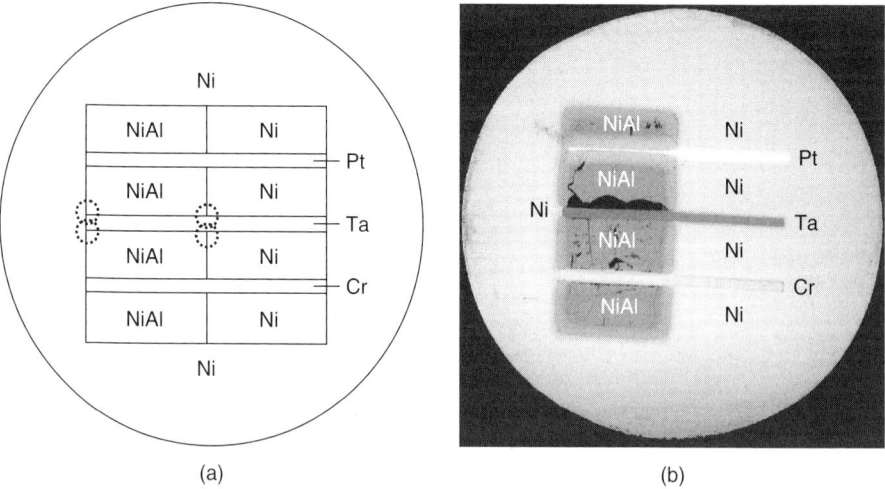

Figure 7.3 The geometry (a) and optical image (b) of a diffusion multiple made up of Ni–NiAl–Pt–Ta–Cr. The diffusion multiple was HIP'ed at 1200°C for 4 h and subsequently annealed at 1200°C for 96 h making the total annealing time at 1200°C for 100 h.

- After selecting the components and the temperatures of annealing, one needs to decide which component will be used as the outer case/matrix of the diffusion multiple (equivalent of an HIP-can). Brittle intermetallic compounds, when used as components of a diffusion multiple, should only comprise the inserts (inner pieces). One advantage of the HIP process in fabricating diffusion multiples is that brittle intermetallics such as NiAl can be easily accommodated as shown in Fig. 7.3. In this case, even though the arc-melted NiAl inserts had a lot of porosity and some small cracks, they were successfully used as components of the diffusion multiple. If no component element is suitable as the outer case, then a separate material can be used as an HIP-can as is the case shown in Fig. 7.4(a) which illustrates a diffusion multiple of Pd, Pt, Rh, Ru inserts within a Cr case/matrix [12,13]. Since the precious metals are expensive, only Cr would be a reasonable choice as the HIP-can. However, Cr is brittle, difficult to weld, and would evaporate rapidly at 1200°C based on vapor pressure data. All these properties prevent Cr from being used as the HIP-can. Therefore, a separate pure Ti HIP-can was used to contain the precious metals and the Cr matrix. Ti is easy to weld using EB, and as an HIP-can it prevents Cr from evaporation (This diffusion multiple was not sliced until after the diffusion annealing to avoid exposing Cr during annealing). One also needs to check the yield strength of the outer shell/HIP-can material to make sure that at the temperature of HIP, the strength is low enough to allow plastic deformation to close the gaps between/among the components inside. This estimation is also important for the design of the outer case/HIP-can dimensions. If the outer case is too thick, it may not deform enough to close the gap at a reasonable HIP time (e.g., 4–8 h) and argon pressure (e.g., 200 MPa). One can increase the HIP temperature to make the materials easier to deform; however, the eutectic temperature or the lowest solidus temperature of the system set the upper temperature limit.
- The next step is to design a geometry for the diffusion multiple. An important consideration is the availability of the raw materials. Some materials are difficult or very expensive to purchase or obtain in large and bulk form; others are readily available. The geometry should be designed to use the smallest amount of the expensive or hard to get materials. For instance, only foils of Pd, Pt, and Rh of 250 μm thickness and a pure Ru piece with two steps on it (500 μm thick on one side and 1000 μm thick on the other) were used to make the diffusion multiple shown in Fig. 7.4(a). Some rough estimation of the diffusion distance using the simple square root Dt (diffusion coefficient D multiplied by time t) calculation is very useful to determine the size (especially thickness) of the components. Again for the case of Fig. 7.4(a), the annealing time was only 40 h at 1200°C due to the very thin (250 μm) foils used to make the diffusion multiple. The diffusion distance during the anneal needs to be constricted to less than the foil thickness. If one does not need the phase equilibrium information of an entire ternary system, one can think of using very thin foils of a very high melting point element within the system as a component of the diffusion multiple. As the high melting point component is consumed by the diffusion process to become alloys with lower melting points, the diffusion may become faster. One needs to pay special attention when two relatively very high melting point elements are side by side, then the diffusion distance may be very small and the bonding between them during

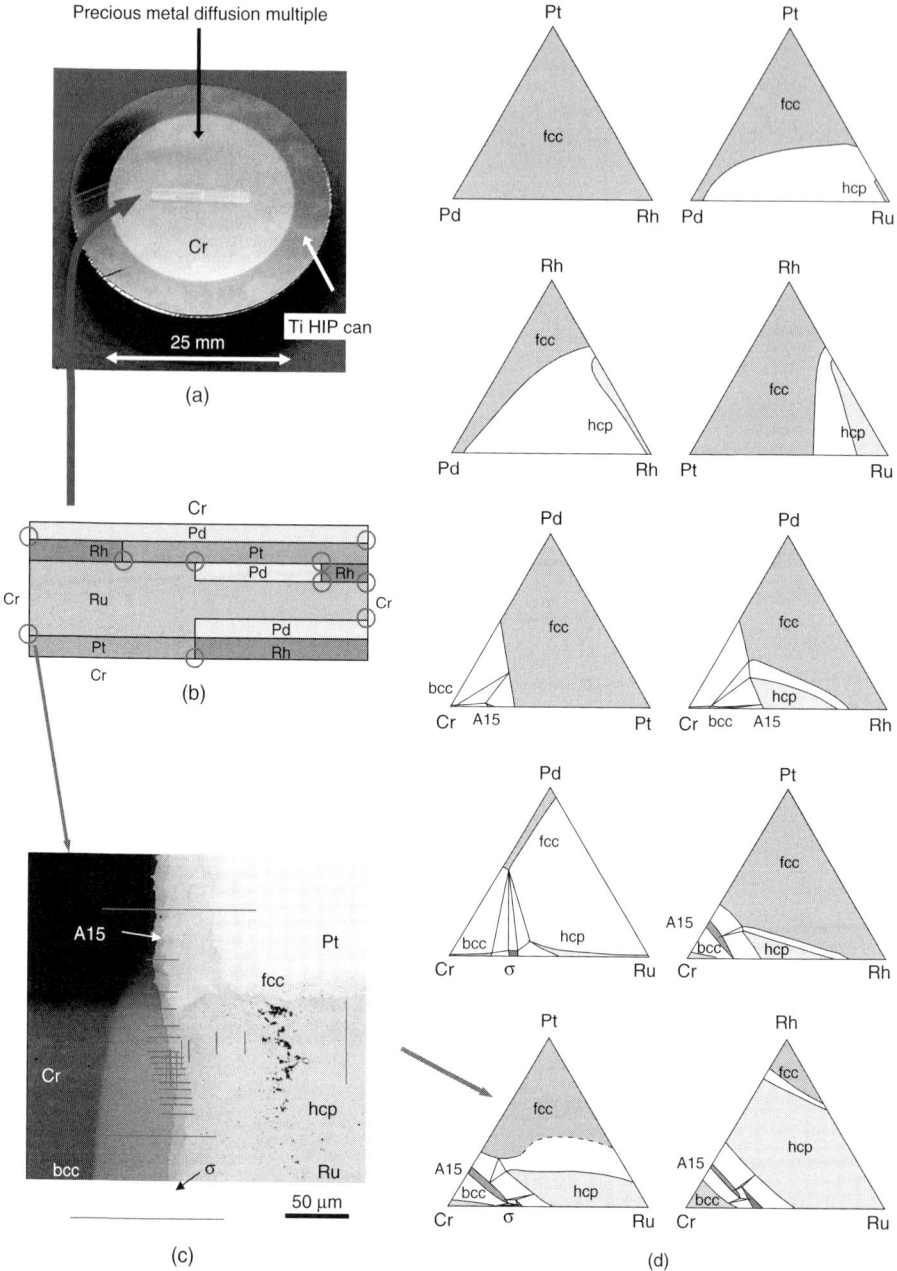

Figure 7.4 A diffusion multiple for rapid mapping of ternary phase diagrams in the Pd–Pt–Rh–Ru–Cr system: (a) optical image of the sample; (b) arrangement of the precious metal foils in the diffusion multiple to create many tri-junctions shown in circles; (c) BSE image of the Cr–Pt–Ru tri-junction showing the formation of the A15 ($Cr_3(Pt,Ru)$) and σ ($Cr_x(Ru,Pt)_y$) phases due to interdiffusion of Pt, Ru, and Cr as well as electron microprobe scan locations (lines); and (d) 10 ternary phase diagrams (isothermal sections at 1200°C) obtained from this single diffusion multiple. The phase diagrams are plotted on atomic percent axes with the scales removed for simplicity [12,13].

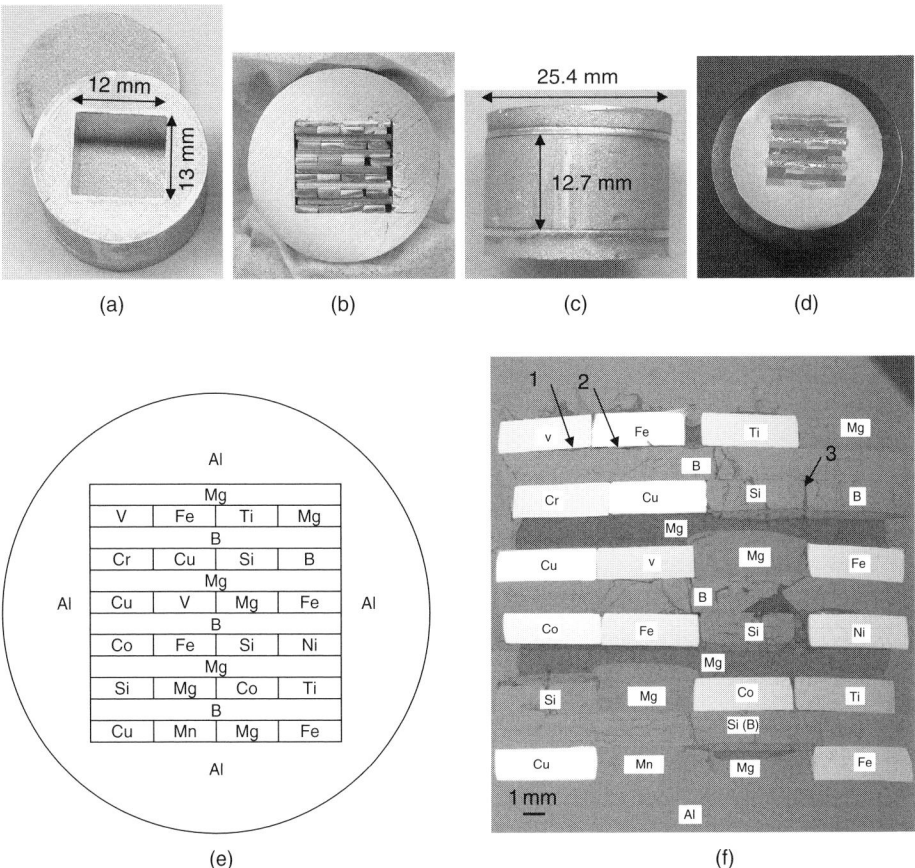

Figure 7.5 A diffusion multiple for Al-based systems: (a) picture of the Al matrix with an 12 mm × 13 mm opening and Al caps; (b) the assembled diffusion multiple before putting on the Al caps; (c) picture of the diffusion multiple after EB welding of the caps; (d) optical image of the diffusion multiple after HIP'ed at 350°C for 4 h, heat treated at 400°C for 500 h, and then sectioned, ground and polished; (e) schematic diagram showing the arrangement of elements; and (f) low-magnification SEM image showing the various components in the diffusion multiple.

HIP may be very weak, thus preventing any meaningful diffusion reaction. An example is shown in Fig. 7.5 for an Al-based diffusion multiple. The smaller pieces of the elements are 1 × 3 × 12.7 mm and the longer pieces are 1 × 12 × 12.7 mm. The outer pure Al case/matrix is 25.4 mm in diameter and 12.7 mm in height with an opening of 12 × 13 mm. After surface preparation, the pieces came in different sizes as shown in Fig. 7.5(b), leaving gaps in various locations. Since many elements are involved, the annealing temperature cannot be much higher than 400°C. In order to give a margin for potential overheating during HIP, the HIP temperature was set at 350°C. At 350°C, Al is soft enough to deform very effectively to close the gaps between/among the elements. However, many elements such as B, Si, V, Fe, Cr, Ni, and Ti have very high melting points relative to the HIP temperature. Not much metallurgical bonding was created between/among these elements during HIP (however, the bonding between pure Al and each of these elements is relatively good). Gaps between V and B, Fe and B, and B and Si did not close or were created during cooling from HIP temperature to room temperature as shown in locations 1–3 in Fig. 7.5(e). One needs to avoid putting two brittle elements/components side by side, especially if either of them deforms at the HIP temperature (e.g., B and Si do not have any plasticity at 350°C). Diffusion multiples can be made in various shapes and forms as long as an HIP-can can be made to deform effectively at the temperature of interest. When brittle intermetallics such as NiAl are used as a component of a diffusion multiple, it is a good idea to arrange for some equivalent tri-junctions in case one had a crack at the junction. For instance, four equivalent Ni–Ta–NiAl tri-junctions as shown by dotted circles in Fig. 7.3(a), were designed for the diffusion multiple in Fig. 7.3. In this case, the crack, occurring at two locations did not prevent the sample from providing very useful ternary data.

- One also needs to think of how to protect a diffusion multiple from elemental evaporation or an unwanted environmental interaction (e.g., oxidation and pesting). When elements in a diffusion multiple are not prone to evaporation or environmental interaction, one can make a diffusion multiple using HIP and then slice it into several pieces for heat treatment at different temperatures. This will save materials and the effort of making several diffusion multiples. However, when one or more elements in the diffusion multiple is prone to evaporation or environmental interaction, it will be necessary to make one diffusion multiple for each heat treatment temperature. The HIP-can can protect the inside elements from the environment. For instance, a Ti HIP-can was used to protect Cr from evaporation in the diffusion multiple shown in Fig. 7.4. The diffusion multiple was not sliced to pieces for different temperatures. A sliced diffusion multiple of Ni–NiAl–Ta–W–R88 (Rene 88) was destroyed, Fig. 7.6, after being annealed at 700°C for 4000 h. Even though the diffusion multiple was wrapped in Ta foil and yttrium pieces to absorb oxygen during heat treatment were placed inside the sealed quartz tube back-filled with pure Ar, apparently enough oxygen diffused across the quartz tube to cause severe pesting reaction of the Ta and W. There were significant forces built up during the pesting to deform the cylindrical sample into the shape pictured in Figs. 7.6(a) and (b). All the metallic Ta and W pieces were gone. If the diffusion multiple had not been sliced, the R88 outer case/HIP-can would have been able to protect the Ta and W from pesting.

- One additional consideration in making a diffusion multiple is whether to quench or furnace cool the sample from the diffusion annealing temperature to room temperature for analysis. A quenched sample is more likely to retain the high-temperature equilibria to room temperature, but it may promote more cracking of the brittle intermetallic phases. Slow furnace cooling may reduce cracking, but it may result in phase precipitation to complicate the analysis. Quenching is recommended unless it is not possible to do so or unless the cracking is very severe. For instance when the diffusion heat treatment is performed in a conventional vacuum furnace, it is difficult to quench the sample into water.

Upfront considerations of the above points can dramatically increase the success rate of any diffusion multiple fabrication. The EB welding and HIP processes are very convenient steps for making diffusion multiples. The HIP process has been effectively

Phase Diagram Determination Using Diffusion Multiples

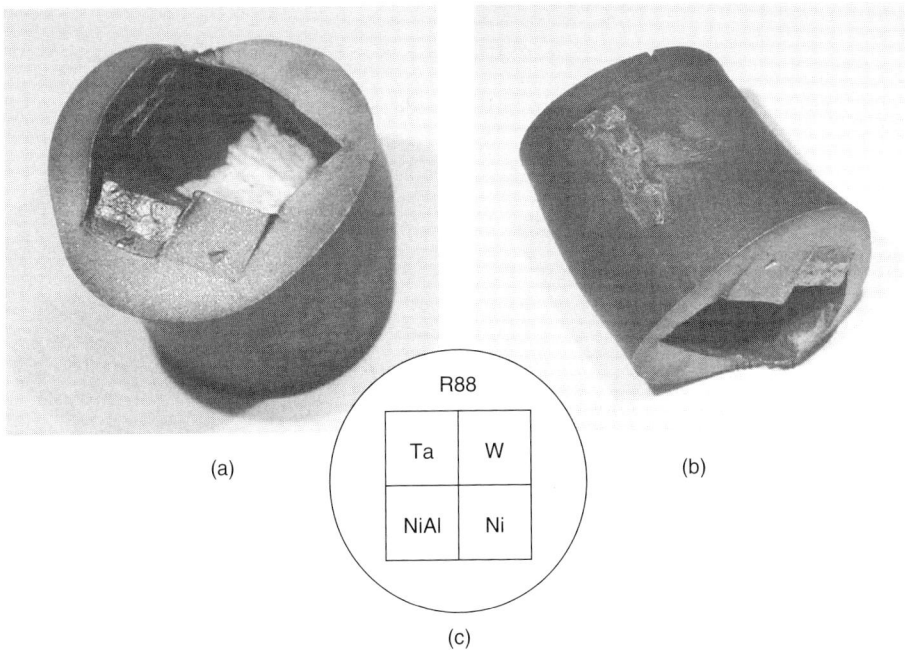

Figure 7.6 An unsuccessful diffusion multiple due to a pesting reaction: (a) and (b) pictures (taken at different angles) of the Ni–NiAl–Ta–W–R88 (Rene 88) diffusion multiple after annealing at 700°C for 4000 h; and (c) schematic cross-sectional view of the diffusion multiple.

used to make many successful diffusion multiples. With a good design, the HIP process makes it possible to include many binary diffusion couples and ternary diffusion triples into a diffusion multiple. The HIP process also helps to crack any surface oxide such as Al_2O_3 to make intimate interfacial contacts for interdiffusion reactions.

Determination of the low-temperature part of a phase diagram containing only high melting point elements, for example, a 500°C isothermal section of a Co–Cr–Mo system, is very difficult using equilibrated alloys; it is even harder to do so using diffusion couples and multiples. It is impractical to promote sufficient interdiffusion at such low temperatures to form the intermetallic compounds and solid solutions to determine equilibrium phase diagrams. A two-step process can be employed to determine phase diagrams at relatively low temperatures. One can first anneal a diffusion multiple at a high temperature (above half the homologous melting point), and then anneal it at low temperatures of interest to examine the phase precipitation. In such studies, the high-temperature annealing serves the purpose of creating the compositional variations/intermetallics, that is, creating many alloys simultaneously. The process also serves, in a sense, as an equivalent to the homogenization annealing of equilibrated alloys. The low-temperature annealing would be the equivalent of heat-treating many alloy compositions at the temperature of interest to examine their phase formation and equilibria. Theoretically, with such a process it should be possible to obtain the isothermal sections at relatively low temperatures. The analysis process is non-trivial for complex systems. A combination of diffusion multiples with a few selected alloys is highly recommended.

3 ANALYSIS OF DIFFUSION MULTIPLES AND EXTRACTION OF PHASE DIAGRAM DATA

3.1 Imaging Examination and Phase Analysis

The Co–Cr–Mo ternary system in the diffusion multiple shown in Fig. 7.1 will be used to illustrate the key processes analyzing a diffusion multiple and extracting phase equilibrium data. Since most diffusion multiples contain brittle intermetallic phases, it is very important to cut, grind, and polish the samples very carefully. Even so, cracking of the intermetallics is often unavoidable. As long as the cracking is not at critical locations, one can get lots of phase equilibrium data from a diffusion multiple.

Simple optical microscopy is usually used first to examine the phase formation and the integrity of a diffusion multiple. Most phases can usually be seen with an optical microscope, although sometimes it is hard to tell which one is which when many phases are present. Optical examination from low magnifications to high magnifications along with information on the existing binary phase diagrams can give clues of the phases present. After optical examination, SEM is often performed to obtain backscattered electron (BSE) images such as those shown in Fig. 7.2. The atomic number contrast along with some energy dispersive spectroscopy (EDS) analysis (which is often available with SEM) can help to further define some or most of the phases. Note that sometimes the contrast from different grain orientations can confound the atomic number contrast. A high quality and high contrast BSE image provides lots of good information about the phases and related equilibria.

EBSD is a very useful tool to perform crystal structure analysis to aid phase identification [14–16]. EBSD can be used to identify crystal structures of micron-size phases in a regularly polished sample (without going through the trouble of making transmission electron microscopy (TEM) thin foil specimens). Commercial EBSD systems are available as an attachment to regular scanning electron microscopes. As a focused EB impinges on a phase, it generates BSEs in addition to secondary electrons, Auger electrons, X-rays and others. The BSEs escaping the sample are further scattered/diffracted by the crystal lattice, thus producing a spatially resolved intensity variation. When a phosphor screen or another type of detector is used to capture the BSEs, a pattern (similar to a Kikuchi map in TEM) is obtained. Sophisticated algorithms have been developed to automatically capture and index the EBSD patterns [16]. The sample is usually tilted at about 60–70° to face the EBSD detector to maximize the collection of diffracted BSEs.

From a phase diagram mapping standpoint, EBSD is important to detect phase boundaries and identify phases, and is critical to efficient experimental planning in the "identification" and "screening" activities. If a list of expected phases in a sample can be generated, then EBSD can be used to rapidly detect the spatial positions of the phase boundaries on the diffusion multiple. The spatial positions of these phase boundaries can be directly related to the quantitative compositions measured by the EPMA, resulting in phase boundary positions on the phase diagram. Phase identification is accomplished by a direct match of the diffraction bands in an experimental EBSD pattern with simulated patterns generated using known structure types and lattice parameters. In this regard, all known crystal structures of a system should be put into the EBSD system software. For instance, the Cr–Pt–Ru ternary system has five phases: bcc, fcc, hcp, A15 (Cr_3Pt), and (Cr–Ru) σ phase (Fig. 7.4(c)). Their crystal structure data (space group, atom positions, and lattice parameters) are provided based on crystal structure information from the three binaries: Cr–Pt, Cr–Ru, and Pt–Ru [17]. When ternary intermetallic compounds are known from crystal structure databases, their crystal structure information should also be provided. In the Cr–Pt–Ru case, no ternary compounds were reported. All the EBSD patterns from the Cr–Pt–Ru system belonged to the five known structures. The EBSD analysis greatly helped locate the interface between the A15 and the σ phases – the SEM image alone cannot differentiate them (Fig.7.4(c)). An EBSD phase map of this same area locates the positions of the phase boundaries (especially the A15/σ phase boundary), which were used for intelligent placement of EPMA scan locations.

The power of EBSD is its capability for effective crystal structure identification with a micron-scale resolution yet without laborious sample preparation. Polished metallographic samples are usually good for analysis with only one additional step. Most metallographic procedures finish polishing with a 1 μm diamond medium. This finishing step leaves a damage layer on the surface on the order of 0.5–1 μm, particularly for metallic samples. This level of surface damage can severely degrade the quality of EBSD patterns due to the shallow depths (~100 nm) of beam interaction. To relieve this surface damage, vibratory polishing with a 0.05 μm silica suspension for several hours is suggested. For diffusion-multiple samples, too many hours' vibratory polishing can induce topographic relief due to the differences in hardness among different components/elements. Such relief is undesirable for EPMA analysis. A trial and error method can be used to select a balanced time. Usually a few hours are a good starting point.

If a completely unknown ternary compound should appear, the EBSD technique may not be the best way to identify it effectively. X-ray or TEM work would be preferred for detailed crystal structure identification. In a sense, all crystal structure identification (including EBSD, X-ray diffraction (XRD), and electron diffraction in TEM) of an unknown phase is more or less a matching game. In other words, one needs first to identify which crystal system it belongs to (fcc, bcc, tetragonal, etc.) and then gradually identify the detailed space group.

If most of the phases can be identified from the diffusion multiple using optical microscopy, SEM and EDS analysis, and EBSD, one can construct a rough topology of the phase diagram without the phase boundaries being accurately defined. Since the usage of all these tools is less expensive than EPMA, as much upfront analysis should be performed using them to define the locations of phases, their interfaces and tri-phase junctions. This information is then used to most effectively place the EPMA line scans onto the sample (or the corresponding BSE image) such as those shown in Figs. 7.4(c) and 7.7.

3.2 EPMA Profiling

The EPMA (or microprobe as it is usually called) is an essential and powerful tool to analyze diffusion multiples. EPMA is a technique capable of chemically analyzing the composition of a solid with high spatial resolution and sensitivity [18–20]. The volume excited under typical EPMA conditions is on the order of a cubic micron. EPMA can detect elements spanning most of the periodic table, from Be to U, with typical detection limits of 0.1% atomic.

EPMA is naturally well suited to achieving efficient and accurate composition profiling with high spatial resolution. The marriage of high-speed computer automation and very precise sample stage and X-ray spectrometer hardware has provided a powerful analytical tool. Modern EPMA systems are usually equipped with a motorized stage that has a positional precision of 0.5 μm in the X-, Y-, and Z-directions, and can move over a range of several centimeters. Such a system allows automated data collection by advance logging of the EPMA line scan positions such as those shown in Figs. 7.4(c) and 7.7 into an executable file that allows the instrument to collect data without the presence of the operator.

Since EPMA machine time is usually expensive, it is very important to achieve very high throughput EPMA while maintaining acceptable accuracy and precision in the resultant compositions measured. In measuring large numbers of phases and intermetallic compounds with compositions varying only slightly among some of them, one needs to carefully select the analysis parameters that would give the most productive results. A large effort should be placed on both qualitative and quantitative examination of diffusion multiples in advance of more thorough quantitative profiling in order to balance the accuracy needed with the total acquisition time during analyses of up to a few thousand data points per ternary system. Usually the EPMA analysis is performed on the order of 1 min per point.

The main procedures for automating the electron microprobe effort can be broken down into four activities: identification, screening, analysis design, and acquisition. The "identification" activity involves a combination of light optical and backscatter electron imaging coupled with some form of qualitative analysis to distinguish meaningful regions from artifact (topographic or crystallographic differences) and so confirm/gauge the information obtained during imaging or phase analysis as described in Section 3.1. Sometimes, more than one section of a specific multiple needs to be analyzed due to porosity, cracking of intermetallic compounds during sample preparation, or other deleterious effects along a particular binary or ternary region in a given metallographic section.

The phases identified by BSE contrast would be compared to adjacent regions using semi-quantitative surveys to determine presence of unique phases or diffused regions that could not be clearly defined in previous imaging efforts – a critical step to avoid spending hours performing unnecessary quantitative analysis. This was often done using X-ray counting integration for specific elements present over a pre-set time, or by performing an EDS spectral acquisition on the areas for comparison. In some cases a quick wavelength dispersive spectroscopy (WDS) count rate meter comparison was used. This "identification" activity allows the locations of all significant phases in the diffusion multiple to be logged into the computer.

The "screening" activity includes using EDS, WDS, or both for semi-quantitative and quantitative analysis. The key is to obtain a matrix of count rates from identified phases for the elements present. Typically screening was performed in an automated mode and analysis was done over a very small subset of the actual analysis matrix to approximate the composition ranges to be measured. This gives vital information to allow the selection of all analysis parameters.

In the "analysis design" activity, the goal is to optimize all the conditions for the impending analyses. The primary EB energy, beam current, choice of standards, spectrometer crystals, detector bias, and pulse height discrimination must be selected from results of the abbreviated "screening" trials. It must be made certain that the statistical accuracy required in crucial regions is met and that a pre-determined number of composition profiles are tested to "map" the diffusion multiple most efficiently. Also important here are X-ray counting times, dead time error, detection limits, and the accuracy required. The "step size" (distance between points along a profile) is an extremely important parameter during EPMA analysis. To increase the step size from 1 μm to 2 μm will cut the EPMA run time in half. However, some regions with very thin phases may be missed during a 2 μm step run. One solution to this is to segment a scan into multiple sections with different step sizes, for example, 1 μm for thin phase regions and larger steps for thick and large phase regions. Since the acquisitions are performed in an automated fashion, one can alter specific sections of the acquisition to accommodate needs of greater sensitivity, counting time on peak and background positions, etc. as a function of the composition ranges.

The "acquisition" activity includes intermittent analyses of the standards as a measure of the "drift" of beam current, stage, spectrometer position, and other instrument stability issues. This can help salvage data by correcting acquired values using known variation from the standard. It also helps identify long-term stability limits.

Due to the large number of data points collected for a given EPMA acquisition, the challenge is to minimize the X-ray counting times at each data point. A few unnecessary seconds spent counting at each data point represent many hours of acquisition time for typical EPMA runs on diffusion multiples.

Having some knowledge of the approximate levels of elemental concentration within a phase and along a gradient to be measured is extremely important in making decisions about the counting statistics and dwell times needed for each measurement for all elements and for both peak and background X-ray integrations.

All the information collected from the preliminary analyses is used to effectively place the EPMA scans onto a diffusion multiple as shown in Fig. 7.7 for the Co–Cr–Mo ternary system and to select the optimum step size(s) for each scan. EPMA analysis can then be performed to obtain compositions of the points along these scans. Compositions of a total of 1557 points were collected for the Co–Cr–Mo ternary system.

3.3 Extraction of Equilibrium Tie Lines

Note there are usually no two-phase mixture regions in diffusion multiples – two-phase mixtures are thermodynamically forbidden for binary diffusion couples. Even though two-phase mixtures are allowed by thermodynamics for ternary systems, they seldom appear in the diffusion annealing process. Two phases reach local equilibrium at an interphase interface; and three phases reach local equilibrium at a tri-junction.

As the EPMA beam travels at a specified step size along a line scan across different phases, the composition of each point is obtained. Since the beam hits single-phase regions far more often than an interphase interface, the population density of data inside a single-phase region should be much higher than that of two-phase regions in the phase diagram. Only when the EPMA point is at or near an interphase interface – while sampling X-ray signals from both phases, will a composition inside a two-phase region in the phase diagram be obtained. Therefore, when the compositions of points in an EPMA scan are plotted onto an isothermal section (similar

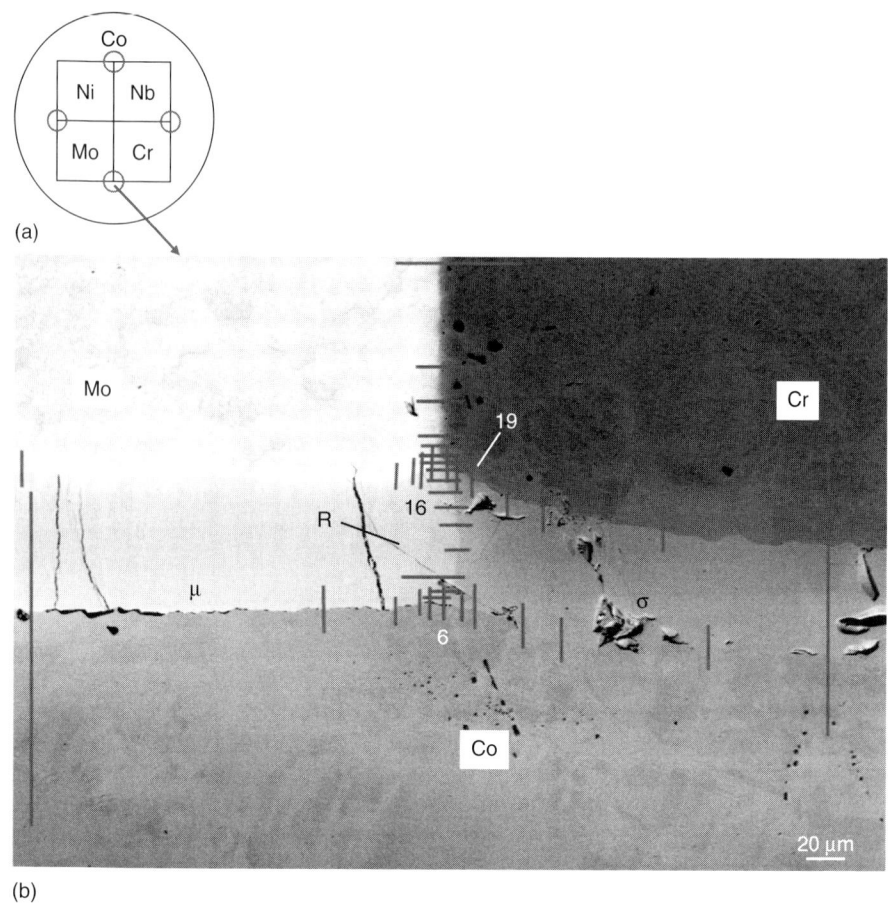

Figure 7.7 Schematic diagram (a) of the diffusion multiple shown in Fig. 7.1 and the EPMA line scan positions marked onto a BSE image (b) of the Co–Cr–Mo tri-junction.

to a diffusion path), the EPMA data are close together in the single-phase regions and scarce in the two-phase regions. In addition, as the EPMA beam moves from one side of an interphase interface to another, the compositions of the two phases should be on a straight line.

By simply plotting the at.% Mo concentration against the at.% Cr concentration of all the 1557 data points obtained from line scans shown in Fig. 7.7(b) in a triangular plot (i.e., on a isothermal section) without any data reduction/processing, one can see a good distribution of data points across the entire ternary isothermal section, Fig. 7.8(a), which indicates that the EPMA line scans are well positioned/placed. By connecting all the data points together for each EPMA scan, one can see the behavior discussed above: densely populated points in single-phase regions and a few data points along a straight line in two-phase regions. Thus, the approximate phase boundaries can be estimated by looking at the density and alignments of data points, Fig. 7.8(b), especially when the binary phase diagram information is used to help bracket the phase regions along the edges of the isothermal section.

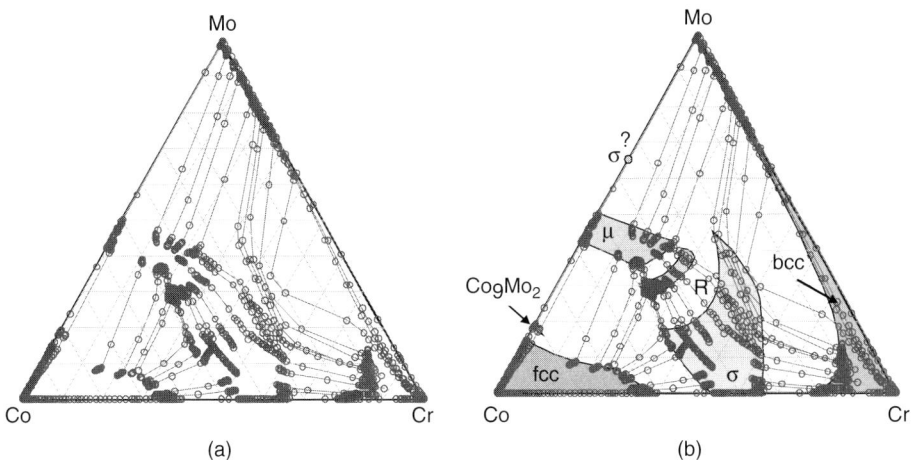

Figure 7.8 Plots of at.% Mo concentration against at.% Cr concentration in a triangular format for all the EPMA data collected for the Co–Cr–Mo ternary system: (a) plot of all data without any data reduction/processing and (b) estimated phase boundary locations based on knowledge of the related three ternary systems and the ternary R phase.

The right-hand edge of Fig. 7.7(b), far from Mo, is essentially a binary diffusion couple of Co and Cr. The solubility of Co in Cr and Cr in Co and the composition range of the σ phase are all consistent with the binary phase diagram [17]. Moving from the right-hand edge gradually toward the center of Fig. 7.7(b), more and more Mo diffuses into the fcc Co phase, the σ phase, and the bcc Cr phase, thus tie-lines with higher and higher Mo concentrations are obtained as shown in Fig. 7.8(b). It can be seen that Mo partitions higher in the σ phase than in the fcc and bcc phases. The top edge of Fig. 7.7(b), far from Co, is essentially a binary diffusion couple of Cr and Mo. Cr and Mo are mutually soluble, forming a continuous bcc solid solution. Moving from the top edge gradually toward the center in Fig. 7.7(b), more Co is diffused into the bcc phase. Similarly, the left-hand side of Fig. 7.7(b), far from Cr, is essentially a diffusion couple of Co and Mo. There is about 15 at.% solubility of Mo in Co and low solubility of Co in Mo, all consistent the existing binary Co–Mo phase diagram [17]. Both the μ (Co_7Mo_6) phase (called ε phase in recent phase diagram compilation [17]) and the Co_9Mo_2 phase were observed. The Co_9Mo_2 phase is hard to see in Fig. 7.7(b), but can be clearly seen in a high contrast BSE image taken from a binary Co–Mo region of the diffusion multiple, Fig. 7.9. There are contradictory reports concerning the stability of the σ phase in the Co–Mo binary system. Quinn and Hume-Rothery observed the eutectoid decomposition of the σ phase at ≤1250°C [21], whereas Heijwegen and Rieck observed the σ phase formation in Co–Mo diffusion couples annealed at 1000°C for 48 h [22]. The absence of the σ phase in the binary Co–Mo diffusion couple region shown in Fig. 7.9 is consistent with the result of Quinn and Hume-Rothery. Moving from the left-hand edge gradually toward the center, more Cr has diffused into the μ, bcc, and fcc phases and a bit into the Co_9Mo_2 phase. At the center of the tri-junction, all three elements interdiffused to form the ternary compositions and a ternary compound, the R phase that can

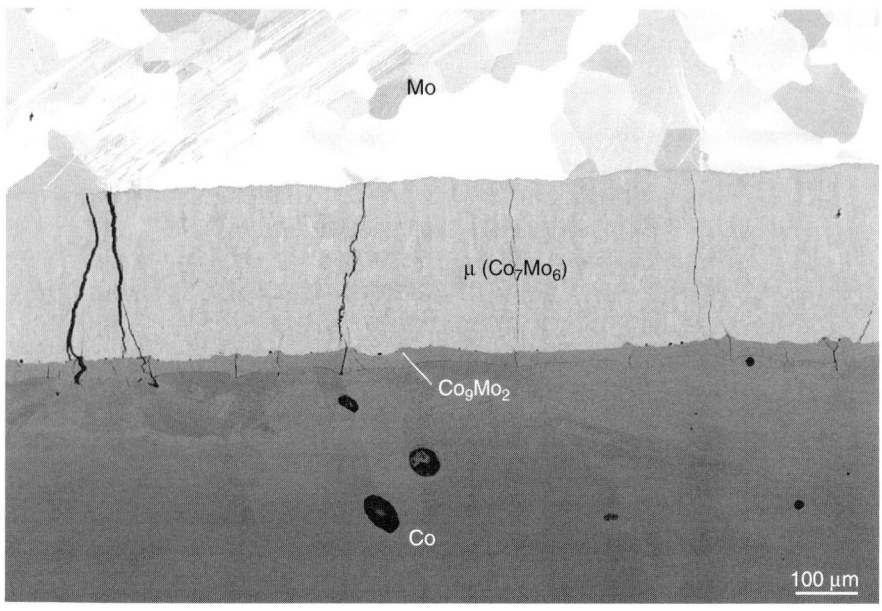

Figure 7.9 SEM BSE image of a Co–Mo binary region of the diffusion multiple shown in Fig. 7.1 annealed at 1100°C for 1000 h, clearly showing the formation of the μ phase and the Co_9Mo_2 phase and the absence of the σ phase. The bcc Mo phase shows a strong orientation contrast.

be seen in a high contrast BSE image shown in Fig. 7.10. The arrangement of the R phase in relation to other phases in Figs. 7.10(a) and (b) created tri-junctions 1–4 that indicates the existence of the corresponding four three-phase triangles: bcc(Mo) + μ + R, bcc(Mo) + R + σ, fcc(Co) + μ + R, and fcc(Co) + R + σ.

During EPMA data reduction to extract equilibrium tie lines, it is very beneficial to make two plots for each EPMA scan, as shown in Fig. 7.11 for a scan (scan #16 in Fig. 7.7(b), from right to left) that started in the σ phase, crossed the R phase, and ended in the μ phase. The first plot, Fig. 7.11(a), shows the compositions of individual points against location (distance in X-direction). This helps to extrapolate to the local equilibrium compositions at the phase interface (Fig. 7.11(a)). The simple straight line extrapolation to the phase interface between the μ phase and the R phase can result in reasonable values for the equilibrium tie line between the two phases (Fig. 7.11(a)). However, since the Cr and Mo concentrations of the σ phase vary quickly with distance, it is hard to extrapolate reliable tie-line compositions, especially since it is not known whether linear extrapolations are still valid. The second graph plots one element (Mo) against another (Cr) (Fig. 7.11(b)). This plot basically shows the compositional path of the scan in the corresponding phase diagram. Since the tie line must be a straight line passing through the two-phase region and also passing through any points along the path inside the two-phase region, this graph can be used to define the tie lines very quickly, as shown by the heavy dotted lines in Fig. 7.11(b). Both plots are very important for the extraction of equilibrium tie lines.

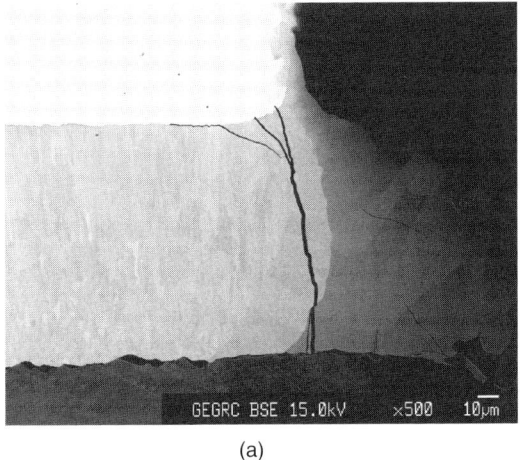

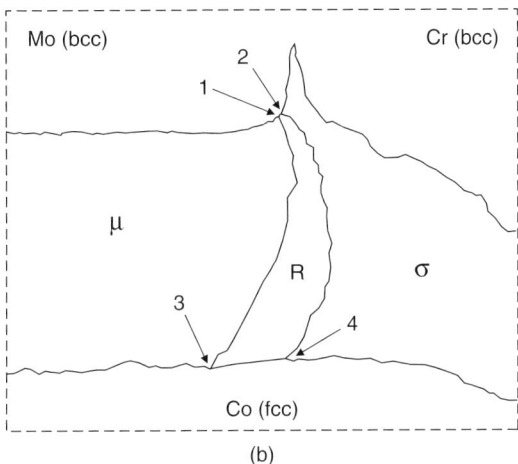

Figure 7.10 SEM BSE image (a) and a corresponding schematic (b) of the Co–Cr–Mo trijunction of the diffusion multiple shown in Fig. 7.1 annealed at 1100°C for 1000 h showing the formation of a ternary R phase. This image was not taken from the sample shown in Fig. 7.7(b), but a sister sample cut from the same diffusion multiple.

The EPMA data from another scan (scan #19 in Fig. 7.7(b), from left to right) are plotted in both composition–distance and composition–composition formats in Fig. 7.12. In this case, the extrapolation of the σ phase compositions using the composition–distance plot is difficult, but relatively straightforward for the composition–composition plot with the heavy dotted lines shown in Fig. 7.12(b).

A final example of tie-line data extraction is shown in Fig. 7.13 for a scan started in the R phase, crossed the σ phase, and ended in the fcc phase (scan #6 in Fig. 7.7(b), from top to bottom). It is difficult to decide whether the data in the first 15 μm of Fig. 7.13(a) are from a single-phase region with smoothly varying Cr and Mo concentrations or two separate phases (R and σ). The composition–composition plot in Fig. 7.13(b) clearly shows the tie-line information.

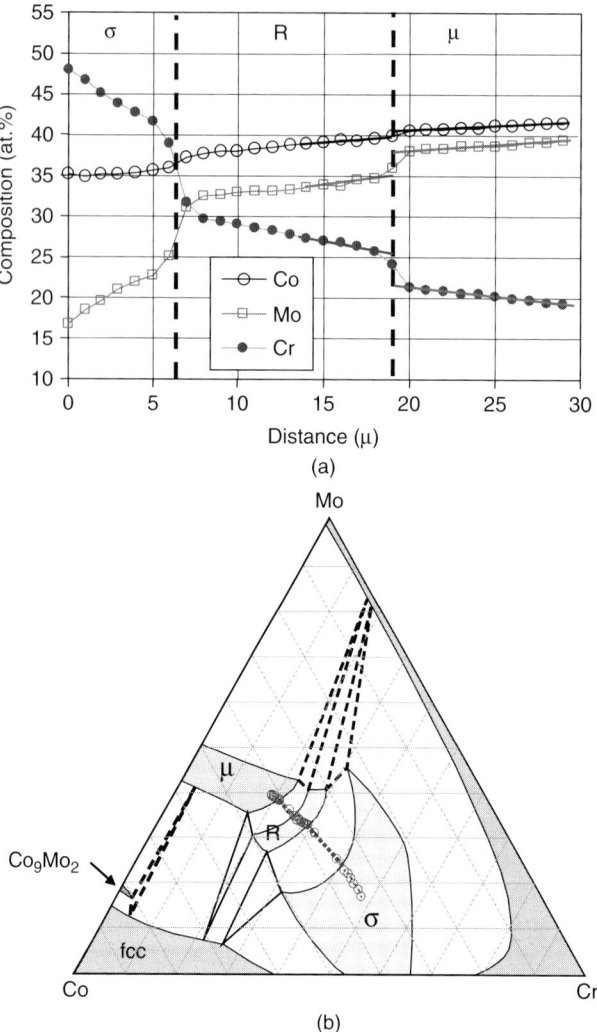

Figure 7.11 Plots used to extract equilibrium tie lines from an EPMA scan that started in the σ phase, crossed the R phase, and ended in the μ phase (scan #16 in Fig. 7.7(b), from right to left): (a) plot of compositions of the elements in at.% against distance and (b) plot of the at.% Mo against at.% Cr onto an isothermal section, showing the tie-line compositions at the end of the heavy dotted lines.

The examples in Figs. 7.11–7.13 clearly show the usefulness of composition–composition plots in tie-line data extraction. Sometimes, one needs the composition–distance plot together with a BSE image to define the tie lines and phase boundaries. This is especially true when the compositions of the two phases are very close.

By performing analyses similar to that shown in Figs. 7.11–7.13, lots of tie lines were obtained to construct the equilibrium isothermal section of the Co–Cr–Mo

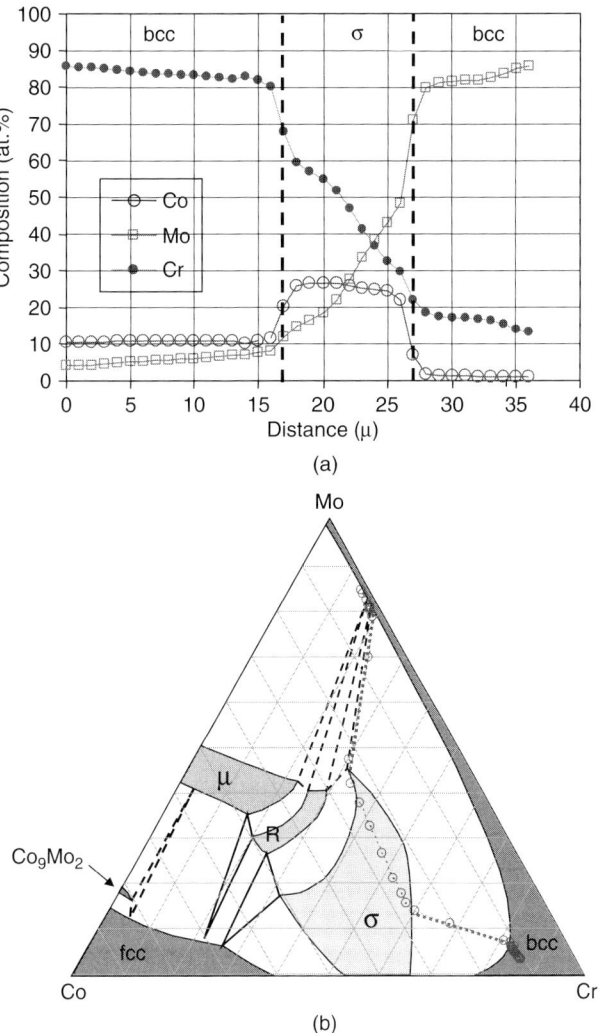

Figure 7.12 Plots used to extract equilibrium tie lines from an EPMA scan that started in the Cr-rich bcc phase, crossed the σ phase, and ended in the Mo-rich bcc phase (scan #19 in Fig. 7.7(b), from left to right): (a) plot of compositions of the elements in at.% against distance and (b) plot of the at.% Mo against at.% Cr onto an isothermal section, showing the tie-line compositions at the end of the heavy dotted lines.

ternary system (Fig. 7.14). The three-phase tie-triangles are obtained by extrapolating the related three two-phase region tie lines.

Since large amounts of EPMA data are generated during the analysis of the diffusion multiple, automated plotting procedures have been developed based on Microsoft Excel Spreadsheet and MatLab software. These programs helped to reduce the data extraction time from days to hours.

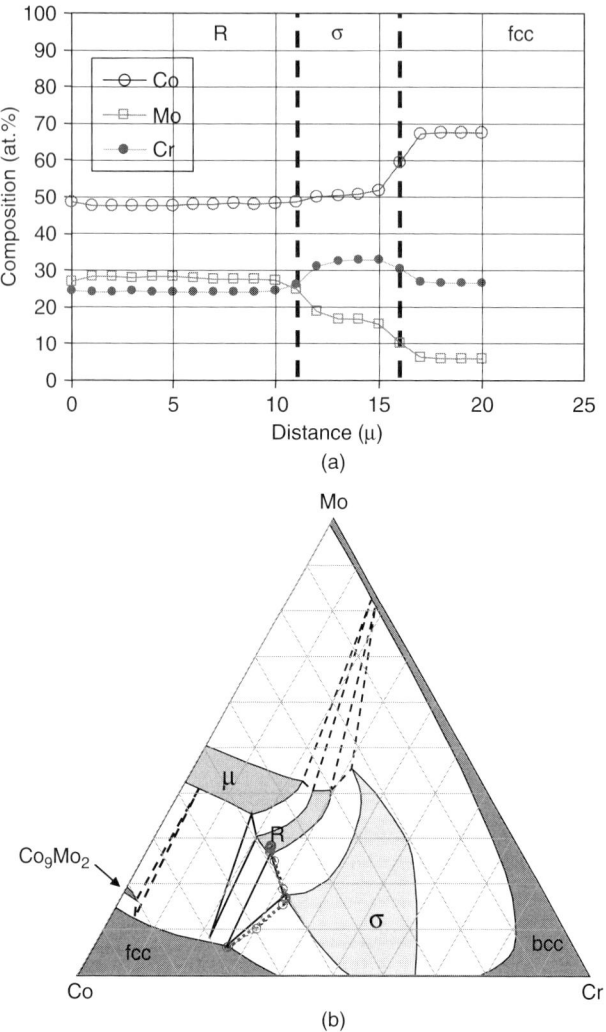

Figure 7.13 Plots used to extract equilibrium tie lines from an EPMA scan that started in the R phase, crossed the σ phase, and ended in the fcc phase (scan #6 in Fig. 7.7(b), from top to bottom): (a) plot of compositions of the elements in at.% against distance and (b) plot of the at.% Mo against at.% Cr onto an isothermal section, showing the tie-line compositions at the end of the heavy dotted lines.

4 SOURCES OF ERRORS

The local equilibrium at the phase interfaces is the basis for using diffusion multiples to map phase diagrams. The existence of such local equilibrium and its reliability in establishing equilibrium tie-line information has been demonstrated for many years in diffusion couples [6,7]. One can refer to the discussion of this topic in Chapter 6 of this book.

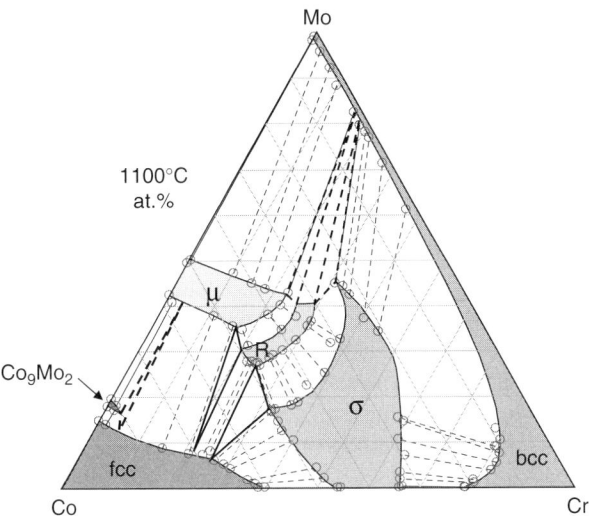

Figure 17.14 The 1100°C isothermal section of the Co–Cr–Mo ternary system determined from the diffusion multiple shown in Fig. 7.1 that was annealed at 1100°C for 1000 h. The phase diagram is plotted in at.% axes with the scales from 0% to 100% for each element removed for simplicity.

For each tie line, we can only obtain one set of data from one polished cross-section of a diffusion multiple. This is different from analysis of individual alloy samples from which several repeats can be made for a single tie line. Fortunately, the consistency of the tie line trends in the diffusion-multiple results (e.g., Fig. 7.14) gives as much confidence as that from repeated results from individual alloys.

The major source of error lies in the interpretation and extraction of equilibrium tie lines from EPMA results. There is lots of information condensed in a very small area of a sample, and one cannot practically perform EPMA analysis at every location in the interdiffusion zone. It is not even practical to use very small steps such as 1 μm for all the scans. One can fail to detect phases that are actually there, simply by performing the EPMA line scans at step sizes that are too large.

It is very convenient for EPMA experiments and also for automated data reduction to arrange the EPMA line scans along either the X or Y directions, as in the examples in Figs. 7.4(c) and 7.7(b). Sometimes, such a scan traverses an interphase interface at an angle not perpendicular to the interface, for example, line scan #19 in Fig. 7.7(b). If the compositional gradient is very steep at that location, the corresponding tie line extracted may not be the real tie line, but at an angle with it. This can be found out by examining the consistency of the tie-line orientation and by examining whether tie-line crossing is observed. For instance, a slight tie-line crossing is observed for the σ + bcc two-phase region in Fig. 7.14. When this happens, the related "tie lines" should be regarded as less reliable data comparing to others. Fortunately, many state-of-the-art EPMA systems allow titled line scans be easily placed nowadays to reduce such a problem.

On rare occasions, one of the phases may not form by interdiffusion reaction in diffusion couples, which makes one wonder whether a similar situation might happen in diffusion multiples. A famous example is the Ti–Al binary system [23]. When

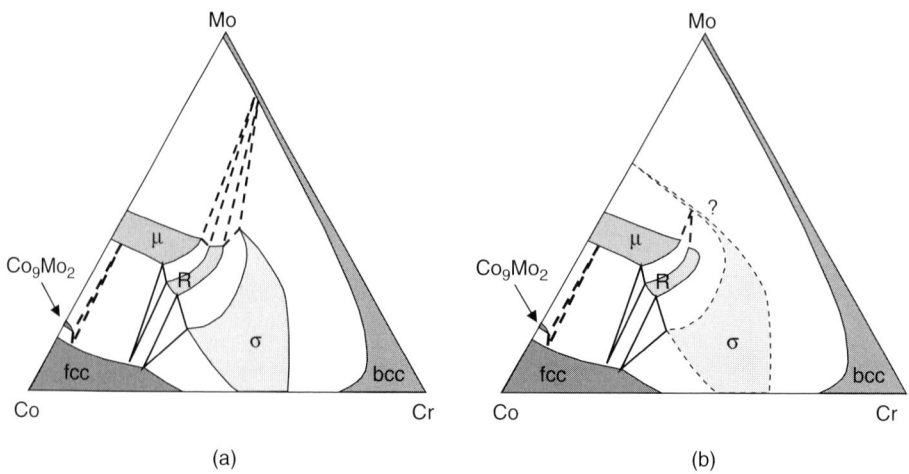

Figure 7.15 Schematic Co–Cr–Mo isothermal sections at 1100°C showing the two possibilities regarding the stability of the σ phase: (a) current version of the phase diagram in which the σ phase in the Co-Mo binary system is unstable at 1100°C and (b) a plausible isothermal section at 1100°C if the σ phase is stable at this temperature. The schematic phase diagrams are plotted in at.% axes with the scales from 0% to 100% for each element removed for simplicity.

Ti/Al diffusion couples were made, only the TiAl$_3$ phase was found, but when Ti/TiAl$_3$ diffusion couples were made, all the compounds appeared [23]. A close examination of the cases reported with missing phases in diffusion couples shows that in most such cases equilibration was attempted at temperatures below half of the homologous melting points, although the exact reason for the absence of phases under these conditions is still not well understood. All the recent phase diagram determination work using diffusion multiples has been concentrated on high temperatures. That may be the reason that we have not seen a missing phase situation. Even though the absence of phases was very rare, when using diffusion couples and diffusion multiples in determining phase diagrams one should always be watchful for the possibility of missing phases (especially at low temperatures). The absence of the σ phase in the Co–Mo binary diffusion couple discussed in Fig. 7.9 is consistent with the result of Quinn and Hume-Rothery [21] who observed the eutectoid decomposition of the σ phase at ≤1250°C, but contradict with the result of Heijwegen and Rieck [22] who reported the formation of the σ phase in a 1000°C annealed diffusion couple. If the result of Heijwegen and Rieck were repeated in the future, then a missing phase situation would have appeared. The 1100°C isothermal section of the Co–Cr–Mo would be look like Fig. 7.15(b). Since the σ phase has already formed in the Co–Cr binary system and the Co–Cr–Mo ternary system in the same diffusion multiple, it is hard to believe that it is a missing phase situation for the Co–Mo binary. Thus, the phase diagram reported in Figs. 7.14 and 7.15(a) are very likely the equilibrium isothermal section at 1100°C of the Co–Cr–Mo system.

One can reduce the chances of error by carefully following these tips:

- Perform the diffusion annealing at a high enough temperature to grow the phases to sufficiently large physical dimensions for reliable phase examination and

EPMA analysis. Thin phases or phases in small dimensions can be missed during analysis.
- Become familiar with the related binary systems and the ternary systems by carefully studying the literature information. It is very important to check all the reported phases and their stability ranges.
- Make sure all the phases are carefully checked with EBSD. It is important to make sure all the phases in the related binary systems are formed in the diffusion multiple, and any ternary phases reported in the literature for the particular ternary system are also present. High quality and high contrast BSE images are very useful in revealing the phases, especially when combined with EDS, optical microscopy and EBSD.
- Examine multiple pieces of the same diffusion multiple to make sure the phases are not inadvertently removed during sample preparation. Occasionally, a brittle phase may fall out of the sample during cutting, grinding, or polishing. One needs to investigate the cracks and porosities in the tri-junction areas to make sure they are not the results of phases that have been removed during specimen preparation. This can be done by examining the same location for different slices of the same diffusion multiple.
- Correlate the EPMA results and BSE images with EBSD results for consistency. If inconsistency is found, one needs to examine whether the interpretation of the EPMA results is correct. Automation in data extraction is a great means to save time, but one needs to be watchful to make sure it does not introduce error.
- Use equilibrated alloys to check regions of the phase diagrams in doubt. This is especially important for complex phase diagrams with multiple phases. In case the funding situation wouldn't allow the making of individual alloys, the regions in doubt should only be reported as tentative results. For instance, the two three-phase regions bcc + μ + R and bcc + R + σ in Fig. 7.14 are reported as tentative results as indicated by the dashed lines.

5 Concluding Remarks

The best way to check the reliability of the diffusion-multiple approach is to compare the phase diagrams obtained from diffusion multiples to those obtained from equilibrated alloys and diffusion couples. A careful comparison for many systems has been performed [6]. The excellent agreements demonstrate that diffusion multiples can be used to determine phase diagrams at orders of magnitude increases in efficiency without sacrificing the quality of the data. A comparison of results for three ternary systems is shown in Figs. 7.16 and 7.17 [1,2,13,24,25]. The good agreement is very apparent.

The combination of the use of diffusion multiples with results on selected equilibrated alloys would be a good safety check against any potential occurrences of metastable or missing phases. Phase diagrams mapped at two or more temperatures are also very useful to check the consistency and reliability of the results.

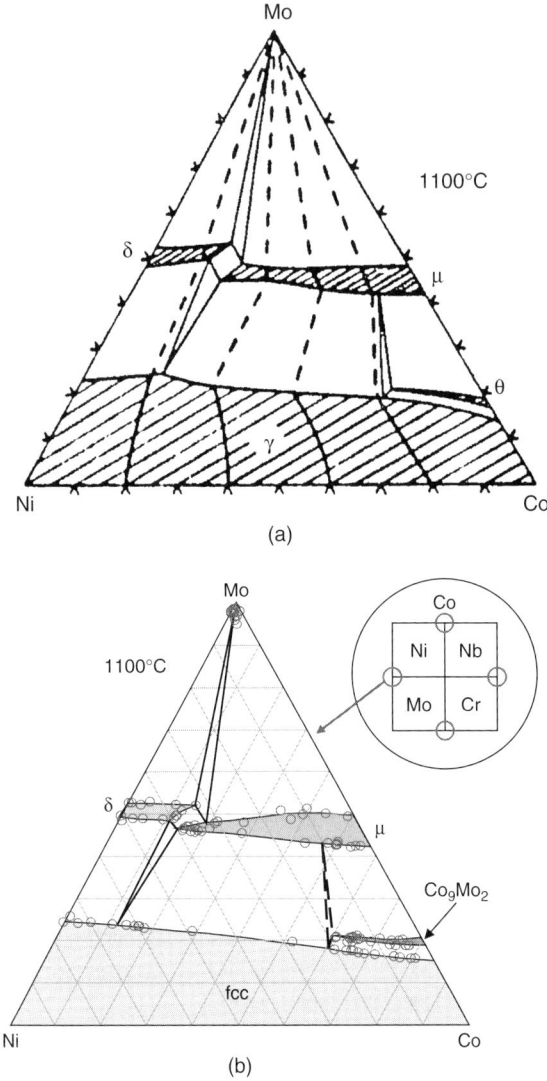

Figure 7.16 Comparison of the 1100°C isothermal section of the Co–Mo–Ni ternary system: (a) results obtained from nine equilibrated alloys and eight diffusion couples [24] and (b) results obtained from the diffusion multiple shown in Fig. 7.1.

No method is foolproof for phase diagram determination. One needs to be watchful for potential pitfalls for each method, including the diffusion-multiple method. The biggest source of error lies in the interpretation of extracted tie-line information. By following the suggestions discussed in this chapter, one can reliably determine phase diagrams using the diffusion-multiple approach, taking full advantage of its efficiency while maintaining the high quality of results.

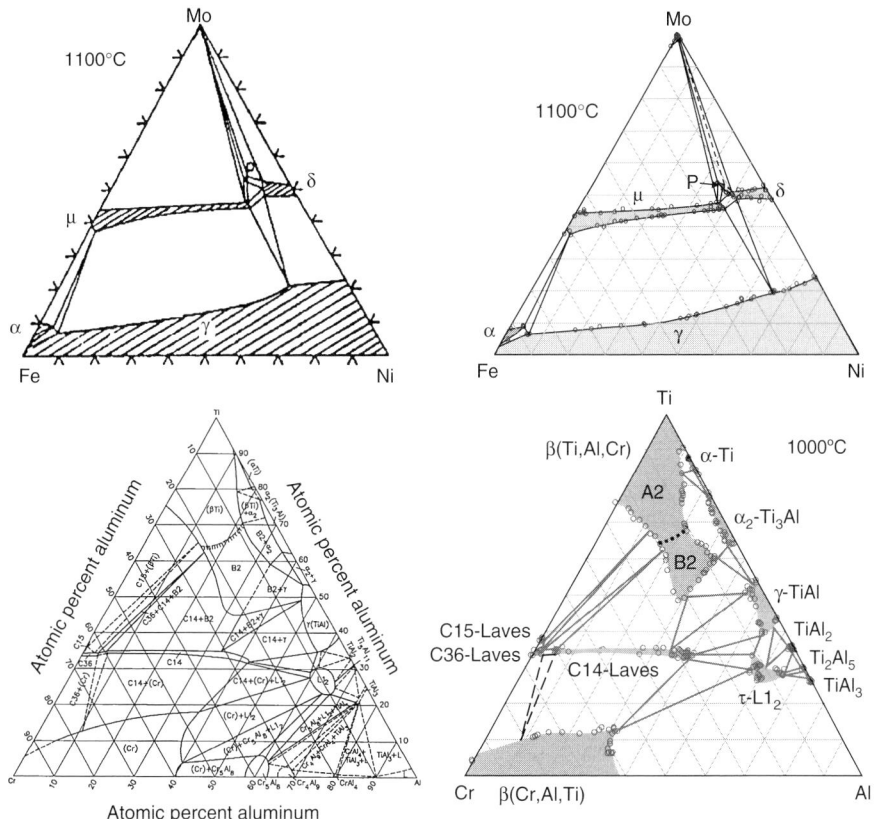

Figure 7.17 Comparison of phase diagrams determined from diffusion multiples with those obtained from equilibrated alloys and diffusion couples: (a) the 1100°C isothermal section of the Fe–Mo–Ni ternary system obtained from equilibrated alloys and diffusion couples [24]; (b) the 1100°C isothermal section of the Fe–Mo–Ni ternary system obtained from a diffusion multiple [1,2]; (c) the 1000°C isothermal section of the Al–Cr–Ti ternary system obtained from more than 100 equilibrated alloys [25]; and (d) the 1000°C isothermal section of the Al–Cr–Ti ternary system obtained from a single diffusion multiple [13].

Acknowledgments

The author is grateful to L.N. Brewer, E.L. Hall, M.F. Henry, M.R. Jackson, L.A. Peluso, A.M. Ritter, and J.H. Westbrook for their help and/or valuable discussions. Special appreciation is extended to L.A. Peluso for an elegant description of EPMA in Section 3.2. Various parts of the work were supported by General Electric (GE) Company internal funding, the US Air Force Office of Scientific Research (AFOSR) under grant number F49620-99-C-0026 with C. Hartley as a program manager; and Defense Advanced Research Projects Agency (DARPA) under the Accelerated Insertion of Materials (AIM) program (Grant number: F33615-00-C-5215) with Dr. L. Christodoulou as the project manager and Dr. R. Dutton as the project monitor. I am especially thankful to my wife, Felicia for her understanding and tolerance during the manuscript preparation.

REFERENCES

1. J.-C. Zhao, *Adv. Eng. Mater.*, 3 (2001) 143.
2. J.-C. Zhao, *J. Mater. Res.*, 16 (2001) 1565.
3. J.-C. Zhao, M.R. Jackson, L.A. Peluso and L. Brewer, *MRS Bull.*, 27 (2002) 324.
4. J.-C. Zhao, *Annu. Rev. Mater. Res.*, 35 (2005) 51.
5. J.-C. Zhao, *Prog. Mater. Sci.*, 51 (2006) 557.
6. A.D. Romig, *Bull. Alloy Phase Diagr.*, 8 (1987) 308.
7. F.J.J. van Loo, *Prog. Solid State Chem.*, 20 (1990) 47.
8. M. Hasebe and T. Nishizawa, in G.C. Carter (ed.), *Application of Phase Diagrams in Metallurgy and Ceramics*, Vol. 2, NBS Special Publications, No. 496, Gaithersburg, MD, 1978, p. 911.
9. Z. Jin, *Scand. J. Metall.*, 10 (1981) 279.
10. J.-C. Zhao and Z. Jin, *Z. Metallkde.*, 81 (1990) 247.
11. Z. Jin and C. Qiu, *Metall. Mater. Trans.*, 24A (1993) 2137.
12. J.-C. Zhao, M.R. Jackson, L.A. Peluso and L.N. Brewer, *JOM*, 54(7) (2002) 42.
13. J.-C. Zhao, *J. Mater. Sci.*, 39 (2004) 3913.
14. F.J. Humphreys, *Scripta Mater.*, 51 (2004) 771.
15. D.J. Dingley and V. Randle, *J. Mater. Sci.*, 27 (1992) 4545.
16. A.J. Schwartz, M. Kumar and B.L. Adams (eds.), *Electron Backscatter Diffraction in Materials Science*, Kluwer Academic/Plenum, New York, 2000.
17. T.B. Massalski (ed.), *Binary Alloy Phase Diagrams*, 2nd edn, ASM International, Materials Park, OH, 1990.
18. V.D. Scot and G. Love, in V.D. Scot and G. Love (eds.), *Quantitative Electron-Probe Microanalysis*, Ellis Horwood Ltd., Great Britain, 1983, p. 32.
19. J.I. Goldstein, D.E. Newbury, P. Echlin, D.C. Joy, A.D. Romig, C.E. Lyman, C. Fiori and E. Lifshin (eds.), *In Scanning Electron Microscopy and X-ray Microanalysis*, Plenum Press, New York, 1981, p. 111.
20. J.T. Armstrong, *Proceedings of the Annual MSA/MAS Meeting*, 1999 p. 561.
21. T.J. Quinn and W. Hume-Rothery, *J. Lesss-Common Met.*, 5 (1963) 314.
22. C.P. Heijwegen and G.D. Rieck, *J. Lesss-Common Met.*, 34 (1974) 309.
23. F.J.J. van Loo and G.D. Rieck, *Acta Metall.*, 21 (1973) 61.
24. F.J.J. van Loo, G.F. Bastin, J.W.Q.A. Vrolijk and J.J.M. Hendriks, *J. Less-Common Met.*, 72 (1980) 225.
25. V. Raghavan, *J. Phase Equili. Diff.*, 26 (2005) 349.

CHAPTER EIGHT

Application of Computational Thermodynamics to Rapidly Determine Multicomponent Phase Diagrams

Y. Austin Chang[1] and Ying Yang[2]

Contents

1 Introduction	273
2 Ternary Mg–Al–Sr System	275
2.1 Experimental Method	276
2.2 Thermodynamic Models	276
2.3 Experimental Results and Discussion	277
3 Quaternary Mo–Si–B–Ti System	282
4 Concluding Remarks	289

1 Introduction

Phase diagrams are the maps for materials and processing development; they have traditionally been determined purely by experimentation that is meticulous, time-consuming, costly, and impractical for obtaining multicomponent phase diagrams over wide ranges of compositions and temperatures. The recent advance made in calculation of phase diagrams (CALPHAD) using the phenomenological or CALPHAD approach has enabled us to rapidly obtain multicomponent phase diagrams with minimum amount of experimental effort. The essence of the CALPHAD route is to obtain a thermodynamic description of a multicomponent system based on the descriptions of its constituent lower order systems, normally binaries and ternaries, via extrapolation [1]. The term "thermodynamic description" refers to a set of thermodynamic models with parameters for all the phases in a system that provide thermodynamic values for these phases. The thermodynamic values can be used to calculate the phase diagrams. Experience has shown that the thermodynamic description so obtained works quite well in many cases [2–5]. The exceptions are when a higher

---

[1] Department of Materials Science and Engineering, University of Wisconsin, Madison, WI, USA
[2] CompuTherm, LLC, Madison, WI, USA

order phase, that is a quaternary phase, appears and/or one of the lower order phases extends its range of homogeneity into the quaternary compositional space. In these cases, a preliminary thermodynamic description of this multicomponent system can be obtained via extrapolation from those of its constituent lower order systems. This description enables us to calculate a phase diagram of the system in question for identifying a limited number of samples for experimentation. The results obtained from these samples are used to obtain an improved description, occasionally a final description. Our experience shows that this strategy can minimize the experimental efforts involved in determining multicomponent phase equilibria [6–9]. This means that the major amounts of experimental efforts involved in determining multicomponent phase diagrams are those primarily for the binaries and the ternaries.

The thermodynamic values calculated from quantum mechanics or first principles are not accurate enough to predict multicomponent phase diagrams, thus for the foreseeable future, we will always need experimentally measured phase equilibria of binaries and ternaries in addition to thermodynamic data of phases obtained either experimentally or calculated theoretically, or both. However, computational thermodynamics can minimize the effort involved to obtain a ternary phase diagram. In other words, we can obtain a preliminary thermodynamic description of a ternary by extrapolation from those of its constituent binaries. Even though this preliminary description is not adequate in most cases due to the rather strong interaction between the component elements, it can serve as an intelligent guide to identify critical samples for experimental investigations. The newly acquired experimental ternary phase equilibria can then be used to develop an improved description of this ternary. In contrast to the traditional approach to study a large number of ternary alloys in order to determine the ternary phase diagram, this approach immediately identifies few alloys for experimentation. The power of this approach is its ability to obtain a description of a multicomponent alloy system such as the potentially useful very high-temperature Nb–Si-based alloys via extrapolation when thermodynamic descriptions of a large number of relevant binaries and ternaries are available. Thus, we have only to perform experimentation to verify the calculated diagrams instead of carrying out a large number of alloys to establish the phase diagrams of the multicomponent system. In addition, the thermodynamic properties of all the phases allow us to calculate the thermodynamic factors for evaluating the diffusion coefficients for a multicomponent system when the mobilities of the diffusing species are known. The Gibbs energy functions of the phases so obtained can also be used to calculate the driving force for studying the kinetics of phase transformation to provide inputs to model the microstructure with known processing parameters. The microstructure of the materials governs the mechanical and other properties, ultimately the performance of the material.

In this chapter, we focus primarily on the use of this approach to rapidly develop thermodynamic descriptions of a model ternary and a quaternary system as examples. It is worth noting that in developing thermodynamic databases of Al-, Fe-, Ti-, and Ni-based alloys for practical applications, Zhang and co-workers have found that whenever there is difficulty in obtaining good thermodynamic databases for commercial alloys, they always trace the sources to inappreciable model parameters of phases in one of the key constituent ternaries [10]. It is understood that in general thermodynamic descriptions of binaries are better established since more experimental data are available. In fact, it is safe to make the statement that unless reasonable descriptions for binaries are available, it becomes nearly impossible to obtain good descriptions of multicomponent systems.

2 TERNARY MG–AL–SR SYSTEM

As part of a program to develop magnesium alloys for future application as automotive power train components, we have been attempting to develop thermodynamic descriptions of key ternary systems such as Mg–Al-based alloys with the addition of either alkaline or rare earth elements. Since magnesium alloys have the lowest density, replacement of the automotive components with these alloys will reduce weight, thus energy consumption. One of the key ternaries is Mg–Al–Sr; yet there were no experimental phase equilibrium data reported in the literature in 2004–2005. We decided to first calculate a tentative isotherm [9] using a preliminary thermodynamic description obtained via extrapolation based on those of the constituent binaries Mg–Al, Mg–Sr, and Al–Sr [11–13]. The tentatively calculated isotherm is shown in Fig. 8.1 at

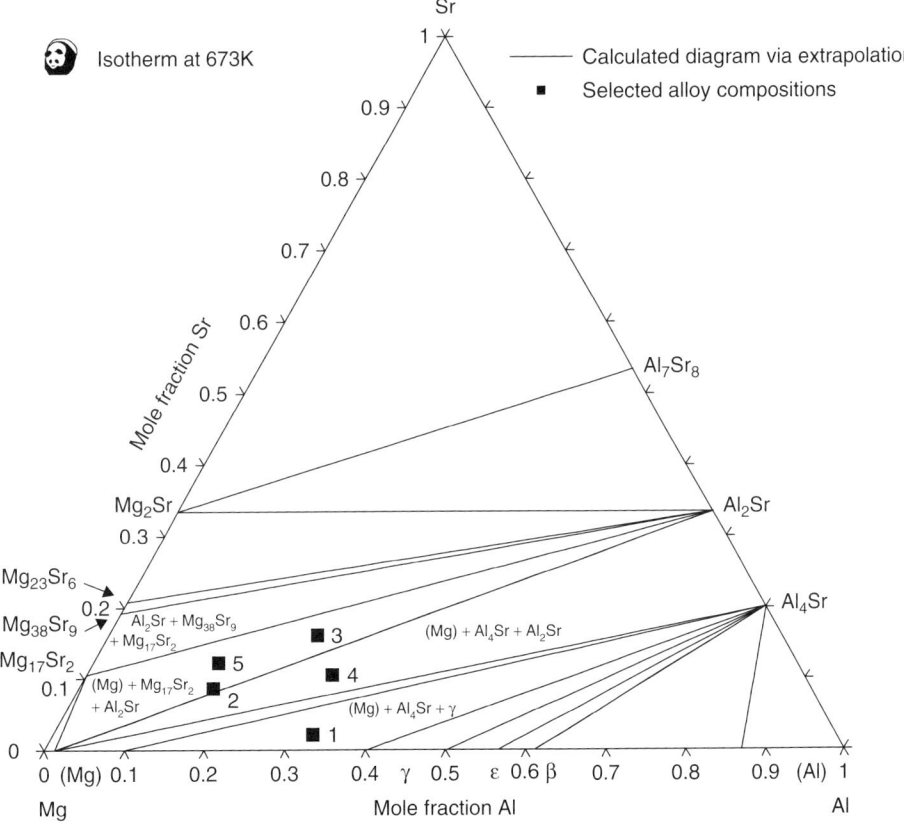

Figure 8.1 Tentative isotherm of Mg–Al–Sr at 400°C calculated using a thermodynamic description obtained via extrapolation based on those of the constituent binaries and the compositions of the five alloys selected for experimentation.

400°C with the assumption that all binary intermetallic phases have no mutual solubilities. This is obviously overly simplified. Since we were only interested in the phase equilibria in the Mg–Al rich region, we identified a total of five samples with Sr concentrations less than 20 mol% for experimental investigation. The composition of sample 1 lies in the calculated three-phase equilibrium of (Mg) + γ-$Mg_{17}Al_{12}$ + Al_4Sr; that of sample 2 in the two-phase equilibrium of (Mg) + Al_2Sr, perhaps in the three-phase equilibrium of (Mg) + Al_2Sr + $Mg_{17}Sr_2$; that of sample 4 in the three-phase equilibrium of (Mg) + Al_4Sr + Al_2Sr; and those of samples 3 and 5 in the three-phase equilibrium of (Mg) + Al_2Sr + $Mg_{17}Sr_2$. These samples were prepared in both the as-cast and 400°C annealed states and subsequently characterized. The experimental results obtained were then used to obtain a thermodynamic description of Mg–Al–Sr as given below. In the following, we give a brief description of the experimental method, next the thermodynamic models, and lastly experimental results and discussion.

2.1 Experimental Method

Alloys were prepared with raw materials of Mg ingot (99.9%), Al lump (99.5%), and Sr ingot (99%) by induction melting in a mild steel crucible with a flowing gas mixture of 99% Ar and 1% SF_6 to protect the alloys from oxidation. The melted alloys were poured into a mild steel crucible. The solidified samples were then remelted with stirring, followed by casting into a cold steel bowl coated with Boron Nitride (BN). Each of the cast samples was cut into two pieces with one of them characterized using scanning electron microscopy (SEM), electron probe microanalysis (EPMA), and X-ray diffraction (XRD) for the primary phase of solidification as well as the compositions and crystal structures of the phases present. The other half was annealed at 400°C for 770 or 835 h in an ultrahigh-purity argon atmosphere (99.998% Ar). Each of the annealed samples was also characterized using SEM, EPMA, and XRD. Unfortunately, two of the five annealed samples 4 and 5, were overly oxidized. The results obtained for these two samples were not used in developing a thermodynamic description of this ternary. The readers are referred to the literature for details [9]. Prior to presenting the experimental results obtained and the developed thermodynamic description, we will first present the thermodynamic models used for the phases in this system including the equation used for extrapolation.

2.2 Thermodynamic Models

There are three types of solutions in this ternary system. The first type includes substitutional solid solutions of the component elements Mg, Al, and Sr. The second one includes pseudo-binary solutions of two stoichiometric binary intermetallic phases. One example is a solid solution $(Mg,Al)_{17}Sr_2$, consisting of binary $Mg_{17}Sr_2$ and its counter binary phase $Al_{17}Sr_2$. The third one is a binary intermetallic compound, whose composition extends into the ternary compositional space but not as a line compound such as γ-$Mg_{17}Al_{12}$. The Gibbs energy of a phase normally consists of three terms, the Gibbs energy of the components in the reference states, the ideal Gibbs energy of mixing, and the excess Gibbs energy. For a substitutional phase, the excess term is represented by the following equation,

$$^{ex}G_m^\varphi = x_{Mg}x_{Al}L_{Mg,Al}^\varphi + x_{Mg}x_{Sr}L_{Mg,Sr}^\varphi + x_{Al}x_{Sr}L_{Al,Sr}^\varphi \\ + x_{Mg}x_{Al}x_{Sr} \tag{1}$$

The term G is the Gibbs energy, with the superscript ex denoting that this G is an excess quantity, x_i's the mole fractions of the component i, and the L^φ the interaction parameters of the binary and the ternary phases, respectively. Values of the binary parameters are obtained from experimental data for the phases and those of the ternary parameters are likewise obtained. However, a preliminary value of the ternary parameter is set to be zero during extrapolation.

For the second type of phases, since the component binary intermetallics have negligible range of homogeneity with respect to Sr or Mg(Al), this phase can be described by a two-sublattice model written as $(Mg,Al)_{1-x}Sr_x$ with Mg and Al occupying one sublattice and only Sr the other. A familiar example in the chemical literature is a solid solution of NaCl and KCl, normally represented as (Na,K)Cl. It is obvious that we can treat the interaction between Mg and Al on one of the two sublattices in $(Mg,Al)_{1-x}Sr_x$ as that between Mg and Al occupying either the hcp solid solution lattice or fcc solid solution lattice. The other sublattice of the intermetallic phase is occupied exclusively by Sr. In other words, we can use the first term in Eq. (1) to describe the excess Gibbs energy of such a ternary phase except we use y_i to represent the mole fractions of the component elements Mg and Al to differentiate them from x_i. While x_i represents the mole fraction of the component i in the alloy, y_i the mole fraction of i on one of the sublattices. For the third type, the γ-$Mg_{17}Al_{12}$ intermetallic phase, a three-sublattice model (**Mg**,Sr)$_{10}$(**Mg**,Al)$_{24}$(**Al**,Mg)$_{24}$ [9,11,14] was used for its simplicity, even though a four-sublattice model is more consistent with the crystal structure. The element in bold denotes that particular element is predominant on the particular sublattice.

2.3 Experimental Results and Discussion

According to the 400°C isothermal section (Fig. 8.1) calculated from the extrapolated thermodynamic description, samples 1, 2, and 3 should be in the phase regions of (Mg) + γ-$Mg_{17}Al_{12}$ + Al_4Sr, (Mg) + Al_2Sr, and (Mg) + Al_2Sr + $Mg_{17}Sr_2$, respectively. However, the EPMA results of these samples given in Table 8.1 show that extensive homogeneity ranges along the direction parallel to the Mg–Al binary exist for all three intermetallics $(Mg,Al)_{1-x}Sr_x$ with $x = 1/5, 9/47$, and $2/9$, respectively, corresponding to Al_4Sr, $Mg_{38}Sr_9$, $Mg_{17}Sr_2$. The solubility of γ-$Mg_{17}Al_{12}$ also extends appreciably into the ternary region. These results suggest that the preliminarily obtained thermodynamic description via extrapolation is not adequate to describe the ternary phase equilibria. On the basis of these results obtained on samples 1–3 in the as-cast and the 400°C annealed states and the published binary descriptions, we developed a thermodynamic description of Mg–Al–Sr.

Comparisons between the calculated phase equilibria from this newly obtained description and experimental observation are shown in Fig. 8.2. For sample 1, the

Table 8.1 EPMA measured compositions for phases in equilibrium at 673K.

Alloy number	Composition (at.%)	Heat treatment	Phases	Mg (at.%)	Al (at.%)	Sr (at.%)	O (at.%)
1	Mg–3.4Sr–32.5Al	Annealed, 673K, 770 h	Al_4Sr	6.08	72.85	18.78	2.29
			$\gamma\text{-}Mg_{17}Al_{12}$	57.13	37.28	3.91	1.68
			(Mg)	92.60	6.36	0.01	1.03
2	Mg–8.5Sr–16.5Al	Annealed, 673K, 835 h	Al_4Sr	13.42	63.93	18.59	4.06
			$Mg_{17}Sr_2$	75.13	13.75	9.26	1.86
			(Mg)★	98.18	0.57	0.38	0.87
3	Mg–16Sr–26Al	Annealed, 673K, 770 h	Al_4Sr	15.60	63.46	19.12	1.82
			$Mg_{38}Sr_9$	61.56	20.43	15.87	2.14
			$Mg_{17}Sr_2$	76.53	11.15	9.71	2.61

★The (Mg) phase is measured from energy dispersive spectrometry (EDS) and the data for (Mg) have wide scatter for this sample.

gross composition, represented by a filled square, lies in the three-phase equilibrium of (Mg) + $\gamma\text{-}Mg_{17}Al_{12}$ + Al_4Sr. The open triangles represent the compositions of the three co-existing phases (Mg), $\gamma\text{-}Mg_{17}Al_{12}$, and Al_4Sr, respectively. Figures 8.3(a) and (b) shows the backscattered electron (BSE) images of sample 1, in the as-cast and 400°C annealed states. The as-cast sample shows that Al_4Sr is the primary phase of solidification followed by the formation of (Mg) and the annealed sample shows the existence of three phases, (Mg) + $\gamma\text{-}Mg_{17}Al_{12}$ + Al_4Sr. The results from samples 2 and 3 are presented in Fig. 8.2 and Table 8.1. Since the calculated phase equilibria agree with experimental observations in the Mg–Al rich region, the calculated phase boundaries for alloys with Sr contents less than 20 at.% are shown as solid lines. In the current modeling, thermodynamic functions of the binary Mg–Sr and Al–Sr intermetallic phases with higher Sr contents were obtained via extrapolation and the calculated phase boundaries involving those phases are shown as dashed lines. The dashed lines indicate that the calculated phase boundaries may be incorrect in view of the appreciable solubilities observed in the Mg–Sr and Al–Sr intermetallics in the Sr-poor region; it is likely that appreciable solubilities also exist in the Sr-rich intermetallics. However, when we focus on the phase equilibria in the Sr-poor region, there is reasonable agreement between the model calculated and experimental measured phase equilibria within the uncertainties of the experimental values, estimated to be about 1–2 at.%. One noticeable exception is the calculated composition of $Mg_{38}Sr_9$ in the three-phase equilibrium of $Mg_{38}Sr_9$ + $Mg_{17}Sr_2$ + Al_4Sr. This is due to the difficulty of the simplified two-sublattice model, $(Mg,Al)_{38}Sr_9$, used for this phase. Had we used a more realistic model such as a three-sublattice model for $\gamma\text{-}Mg_{17}Al_{12}$ (or a four-sublattice model), we would have been able to properly describe the range of homogeneity of this phase with regard to the three component elements. However, because the Mg concentration in most of the commercial alloys is higher than 90%, this phase is unlikely to form from solidification. Therefore, it is not necessary to use a highly complicated thermodynamic model.

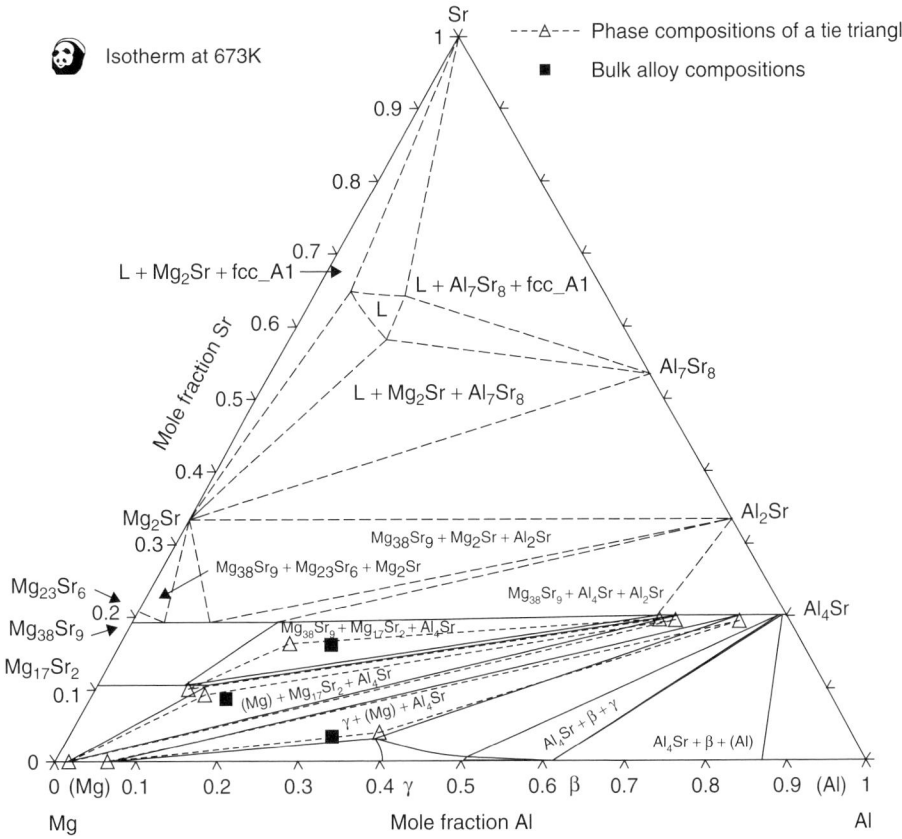

Figure 8.2 Comparison of the calculated isotherm using the description of Mg–Al–Sr obtained based on the results acquired from samples 1–3 (see Fig. 8.1) with experimental data obtained in the present study. Samples 1–3 contain 3.4, 8.5, and 16 at.% Sr, respectively.

The calculated liquidus projection is compared with the primarily solidified phases observed from the five as-cast alloys. As shown in Fig. 8.4, the observed primary phase of solidification for samples 1, 3, and 4, denoted as solid circle, is Al_4Sr, consistent with the calculation. The observed primary phase of solidification from sample 2, denoted as triangle, is $Mg_{17}Sr_2$, also consistent with the calculation. The composition of sample 5 lies on or near the monovariant line $L + Mg_{17}Sr_2 + Mg_{38}Sr_9$. Indeed, the as-cast microstructure of sample 5 shows that both $Mg_{17}Sr_2$ and $Mg_{38}Sr_9$ solidify from the melt, consistent with the calculated monovariant line for the three-phase equilibrium of $L + Mg_{17}Sr_2 + Mg_{38}Sr_9$. Even though the data for samples 4 and 5 were not used in obtaining the thermodynamic description, the phases present in the as-cast microstructure of these two samples are also consistent with the calculated liquidus projection.

After we obtained a thermodynamic description of the Mg–Al–Sr system, an experimental study on the phase equilibria of Mg–Al–Sr by Parvez et al. appeared in the literature [15]. These authors studied a total of 22 alloys prepared by induction melting under an atmosphere of argon with 1% SF_6, followed by slow cooling.

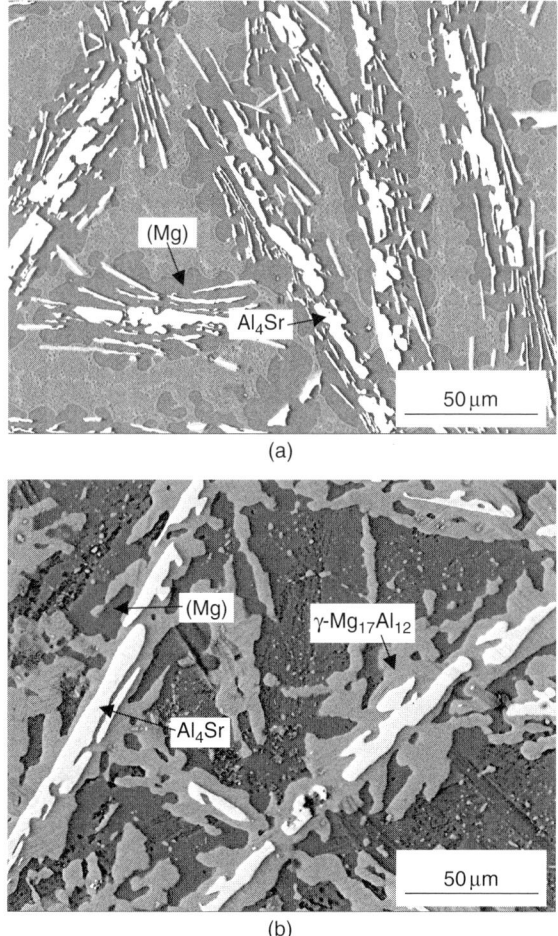

Figure 8.3 BSE images of sample 1 (64.1Mg–3.4Sr–32.5Al) in the as-cast state (a) and 400°C annealed state (b).

These slowly cooled samples were characterized by XRD, metallography, and differential scanning calorimetry (DSC) with slow heating and cooling rates (2–5°C/min). It is likely that they might not have identified all the XRD peaks but suggested possible existence of four new ternary phases. However, on the basis of our XRD analysis [9], it is likely that their diffraction patterns correspond to solid solutions of the binary intermetallics. Although their alloys were not annealed at 400°C, they did state that they cooled their samples slowly. Moreover, the microstructure taken from different locations in as-cast and post-DSC samples were observed to be similar. It is likely that the phases they obtained may not differ significantly from ours at 400°C. Indeed as shown in Fig. 8.5, their experimental results are not inconsistent with our calculated isotherm at 400°C. The symbols △, □, ○, etc. represent the nominal compositions of the alloys. For example, samples 10 and 11 (solid triangles with one corner pointed upward) were reported by

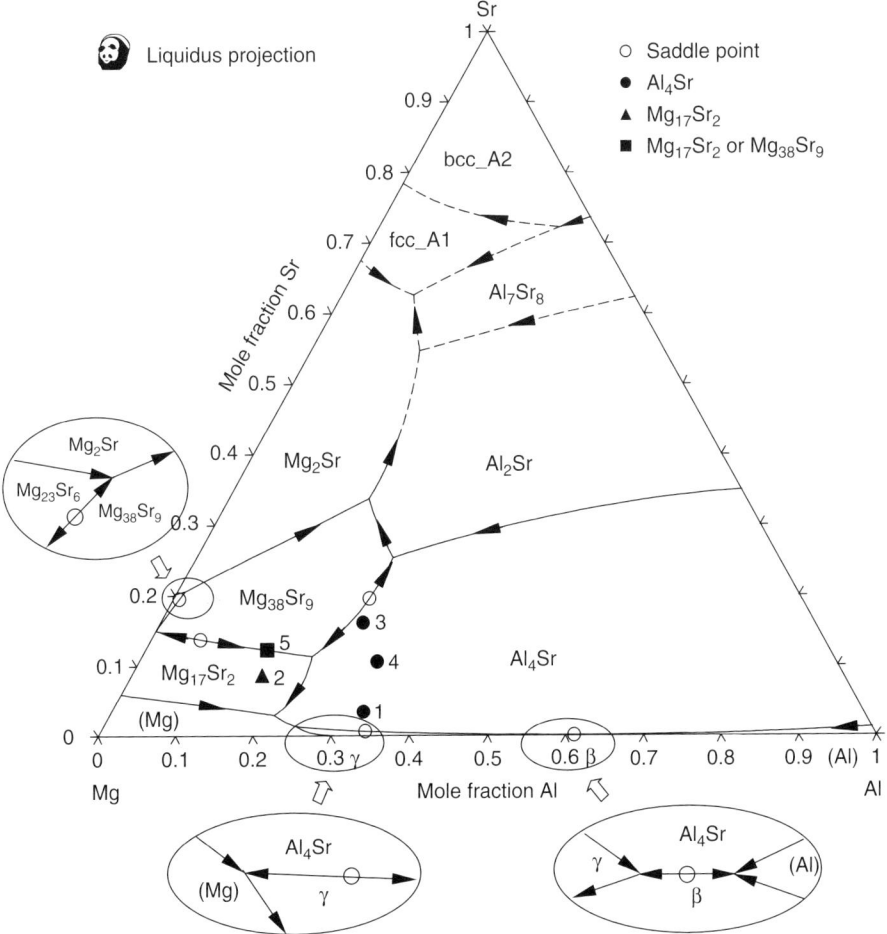

Figure 8.4 Comparison of the calculated liquidus projection using the description of Mg–Al–Sr obtained based on the results obtained for sample 1–3 (see Fig. 8.1) with experimental data obtained in the present study. The results of sample 4 and 5 were not used in obtaining the thermodynamic description.

Parvez et al. [15] to be in the (Mg) + Al_4Sr + τ_2 three-phase field, likely due to their misidentification of the XRD patterns as a result of the existence of $(Mg_{0.2}Al_{0.8})_4Sr$ instead of Al_4Sr. As shown in Fig. 8.6, the primary phases of solidification for 19 out of their 22 alloy samples are consistent with the calculated liquidus projection (the primary phase information for samples 12–14 were not clearly given by Parvez et al. [15]). We thus concluded that the thermodynamic description obtained by Cao et al. [9] provides a reasonable understanding of the phase equilibria in the (Mg,Al)-rich region of the Mg–Al–Sr system. However, as noted earlier, the calculated phase equilibria using this description for alloys in the compositional vicinity of the $Mg_{38}Sr_9$ phase and for compositions higher than 20 at.% Sr must be used with caution. Additional experimental data are needed.

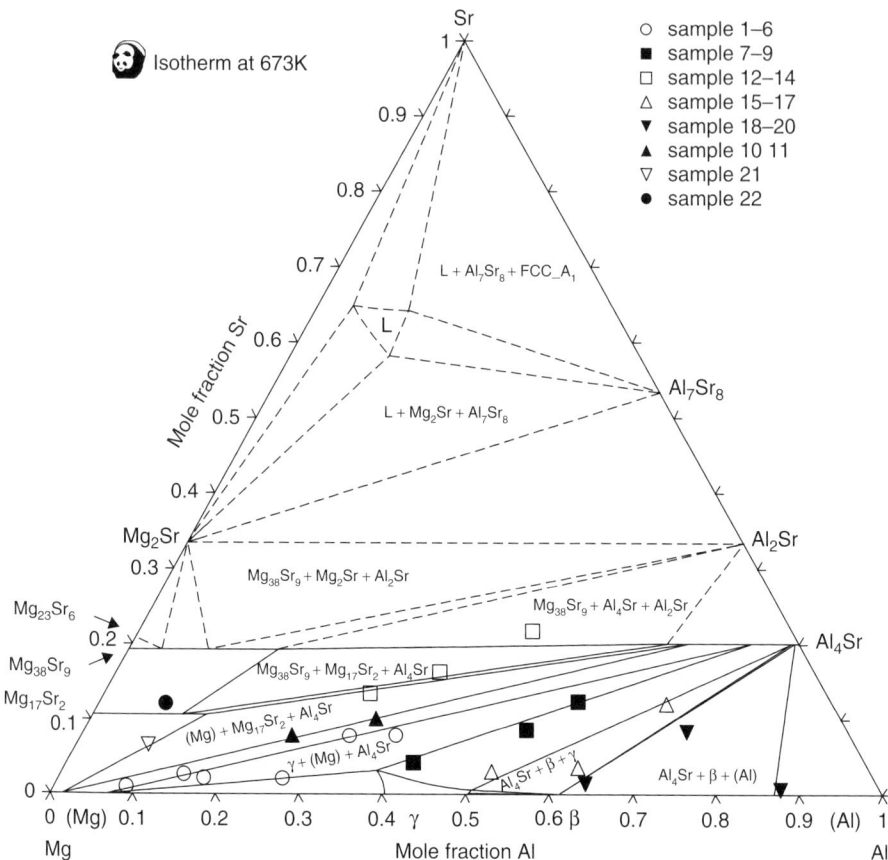

Figure 8.5 Comparison of the calculated isotherm of Mg–Al–Sr at 400°C with the experimental isotherm of Parvez et al. [15].

3 Quaternary Mo–Si–B–Ti System

As noted in the introduction, experience has shown that a reliable thermodynamic description of a quaternary system can be obtained in many cases from those of its constituent lower systems, that is, binaries and ternaries. The exceptions are when a new quaternary phase occurs or the range of homogeneity of a lower order phase extends significantly into the quaternary compositional space. We recently encountered the second case when we attempted to obtain a thermodynamic description of Mo–Si–B–Ti. Since the experimental method and thermodynamic models used are essentially the same as those presented in Section 2, we will give a brief presentation of the background on this quaternary system, and then the thermodynamic description obtained.

Figure 8.7 shows a schematic isothermal tetrahedron of Mo–Si–B–Ti at 1600°C, focusing on the phase equilibria in the (Mo,Ti)-rich region of this quaternary system [6–8,16]. The intermetallic phases in the Mo–Ti rich region of

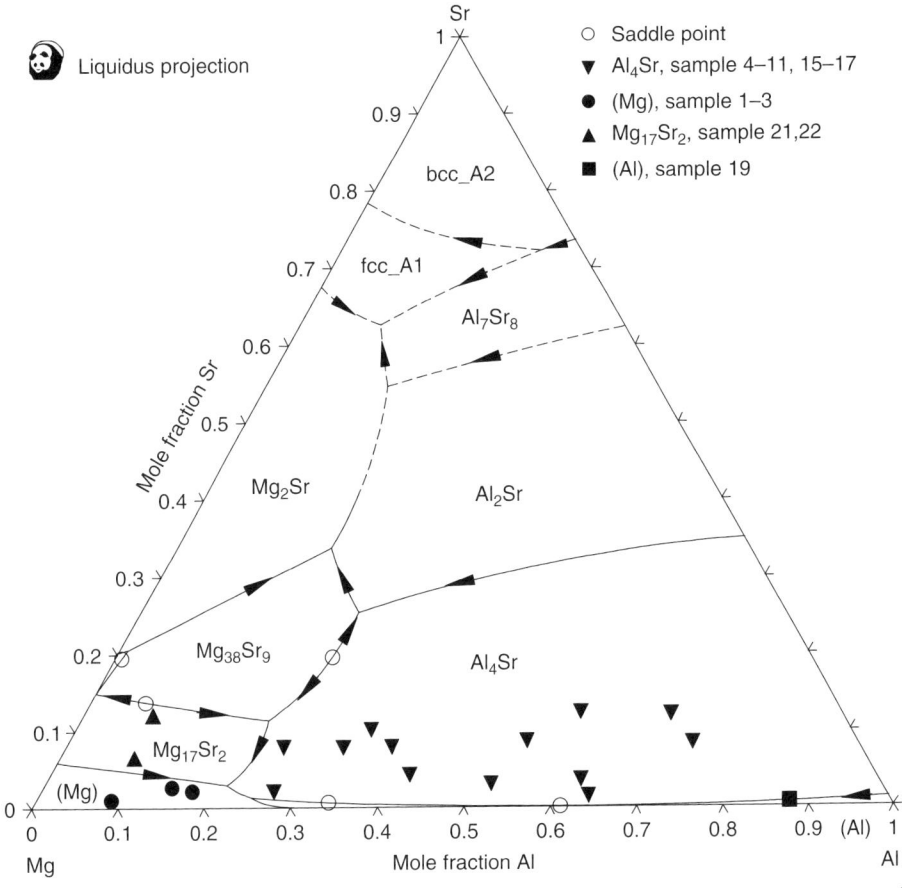

Figure 8.6 Comparison of the calculated liquidus projection of Mg–Al–Sr with the experimental projection of Parvez et al. [15].

Mo–Si–B–Ti are Mo_3Si (A15), Mo_5Si_3 (T1, tI32, $D8_m$), Mo_2B, MoB, Mo_5SiB_2 (T2, tI32, $D8_1$), and Ti_5Si_3 ($D8_8$). It is worth noting that even though both T1 and T2 have the same Pearson symbol, their structures differ. The prototype for T1 is W_5Si_3 and that for T2 is Cr_5B_3. We use bcc to denote the (Mo,Ti) solution phase, A15 the $(Mo,Ti)_3Si$ intermetallic phase, T1 the $(Mo,Ti)_5Si_3$ intermetallic phase, exhibiting the $D8_m$ structure, $D8_8$ the $(Mo,Ti)_5Si_3$ intermetallic phase, exhibiting the $D8_8$ structure, and T2 the $(Mo,Ti)_5SiB_2$ intermetallic phase. The T1 phase is stable in ternary Mo–Si–B but unstable in Ti–Si–B as shown in Fig. 8.7. Yet, T1-Mo_5Si_3 dissolves a considerable amount of its counter phase, "T1-Ti_5Si_3", that is unstable in Ti–Si–B. The symbol " " denotes that the phase is normally unstable. This is also true for $D8_8$-Ti_5Si_3; it also dissolves an appreciable amount of its counter phase, "$D8_8$-Mo_5Si_3" that is unstable in Mo–Si–B. This information suggests that the Mo_5SiB_2 phase (T2, tI32, $D8_1$) must also dissolve a considerable amount of its counter phase, "T2-Ti_5SiB_2", that is also unstable in Ti–Si–B. The question is, how much. The traditional method is to carry out a series of experiments to determine

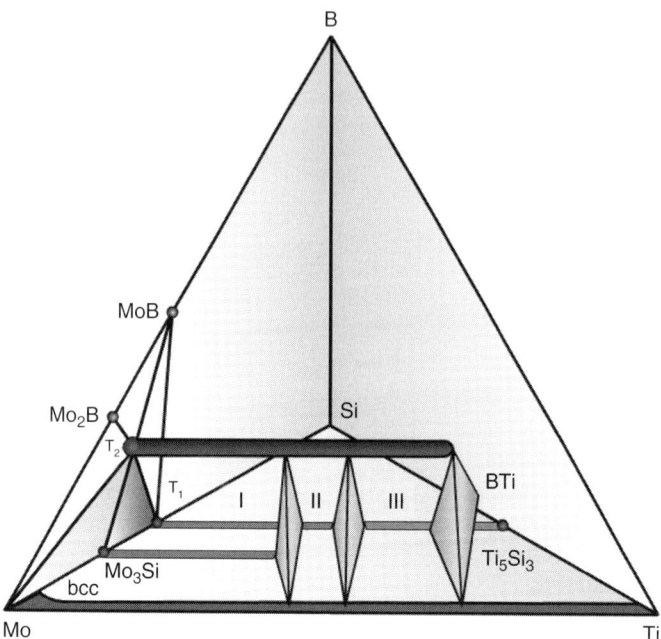

Figure 8.7 A schematic tetrahedron displaying the phase relationship among (Mo,Ti) (bcc), T2, Mo$_3$Si, Mo$_5$Si$_3$ (T1), and Ti$_5$Si$_3$ on the metal-rich side of the Mo–Si–B–Ti quaternary system at 1600°C. This diagram is drawn based on the calculated results from the preliminary thermodynamic modeling [6].

its solubility in the quaternary compositional space as well as various possible heterogeneous equilibria. We used basically the same approach as we did for the Mg–Al–Sr ternary system presented in Section 2. We first estimated a Gibbs energy value for T2-Ti$_5$SiB$_2$ and then obtained a thermodynamic model for the T2-(Mo,Ti)$_5$SiB$_2$ phase assuming ideal mixing of Mo and Ti on their sublattice. Models for all other phases were obtained by extrapolation in the usual manner from those in the lower order systems. We then had a preliminary thermodynamic description for the quaternary Mo–Si–B–Ti system and thus were able to calculate the phase equilibria involving the bcc, A15, T1, D8$_8$, and T2 phases as shown in Fig. 8.7. On the basis of these preliminary calculated phase equilibria, we prepared a total of three samples and subsequently used the experimental data to obtain an improved description. The calculated phase equilibria using this improved description allowed us to identify another set of six samples. The experimental results obtained from the second set of samples led to a final thermodynamic description. We will now present the results in more detail.

The preliminarily calculated phase equilibria, shown in Fig. 8.7, served as a guide to select a minimum number of alloys for experimental determinations. The calculation shows the existence of three four-phase equilibria among these phases, bcc + A15 + T2 + T1 (I), bcc + T2 + T1 + D8$_8$ (II), and bcc + T2 + D8$_8$ + TiB (III), respectively. From the Mo–Si–B ternary to the four-phase equilibria (III), three "windows" exist and are separated by the two four-phase equilibria (I) and (II). In the left-hand window, there are two three-phase equilibria, bcc + A15 + T2 and A15 + T1 + T2. There is one three-phase equilibrium, bcc + T2 + T1, in the middle window, and another one, bcc + T2 + D8$_8$, in the third window. Based on this preliminary calculation, three alloy compositions, Mo–11Si–19B–10Ti, Mo–11Si–19B–20Ti, and Mo–11Si–19B–30Ti, were selected in such a way that the first alloy composition was located in the calculated three-phase bcc + A15 + T2 field, the second in bcc + T1 + T2, and the third in bcc + D8$_8$ + T2. These three alloys were studied experimentally either to verify or to improve this preliminary thermodynamic description; the experimental details were given elsewhere [6]. The experimental results showed the existence of bcc + A15 + T2 in the first sample consistent with the calculation. However, the experimental results also showed the same three-phase equilibrium in the second sample, indicating the calculated four-phase equilibrium (I) should have been richer in Ti–Si–B. The three-phase equilibrium of bcc + T2 + D8$_8$ found in the third sample is consistent with the calculation. An improved description was developed using the newly obtained experimental results focusing primarily on the model parameters of the T2 phase. The newly calculated phase equilibria using the improved description were next used to identify an additional six alloys for further experimental investigations. The compositions of the first two alloys were in the compositional vicinity of the newly calculated four-phase equilibrium (I), those of the next two were in that of the newly calculated four-phase equilibrium (II), and those of the last two were in that of the newly calculated four-phase equilibrium (III). These six alloy compositions were Mo–28.5Ti, Mo–32.5Ti, Mo–35Ti, Mo–37.5Ti, Mo–55Ti, and Mo–57.5Ti with 18 mol% Si and 9 mol% B in each alloy, respectively. The experimental results from the second group of six alloys as well as those in the first group, a total of nine, were then used to develop another improved and final thermodynamic description.

In order to test the predictive capability of the current thermodynamic description, it was decided to compare the calculated phase equilibria with the experimental results obtained from three additional samples. The arbitrarily selected compositions of these samples lay within the narrow compositional region between the two four-phase equilibria (I) and (II). They were Mo–10Si–10B–25Ti, Mo–10Si–10B–27.5Ti, and Mo–10Si–10B–30Ti. It is important to point out that the experimental results obtained from these three alloys were used only for comparisons between calculation and experimentation, not for optimization of the model parameters. If the calculated results are in good agreement with experimental data from the third group of alloys, the current thermodynamic description is believed to be a reliable knowledge base for predicting and understanding the phase equilibria among the bcc, T2, A15, T1, and D8$_8$ phases.

A thorough experimental investigation was carried out for all alloy samples using XRD, EPMA, electron backscatter diffraction (EBSD), and SEM with BSE imaging analysis. Comparisons between experimental and calculated results for all investigated alloys were in good agreement as presented elsewhere [6]. We present here only the results obtained for one typical sample, Mo–18Si–9B–32.5Ti. The equilibrium phases presented in this sample are calculated to be T1 + T2 + bcc + D8$_8$ from 1000°C to 1800°C, as shown in Fig. 8.8. This figure gives information not

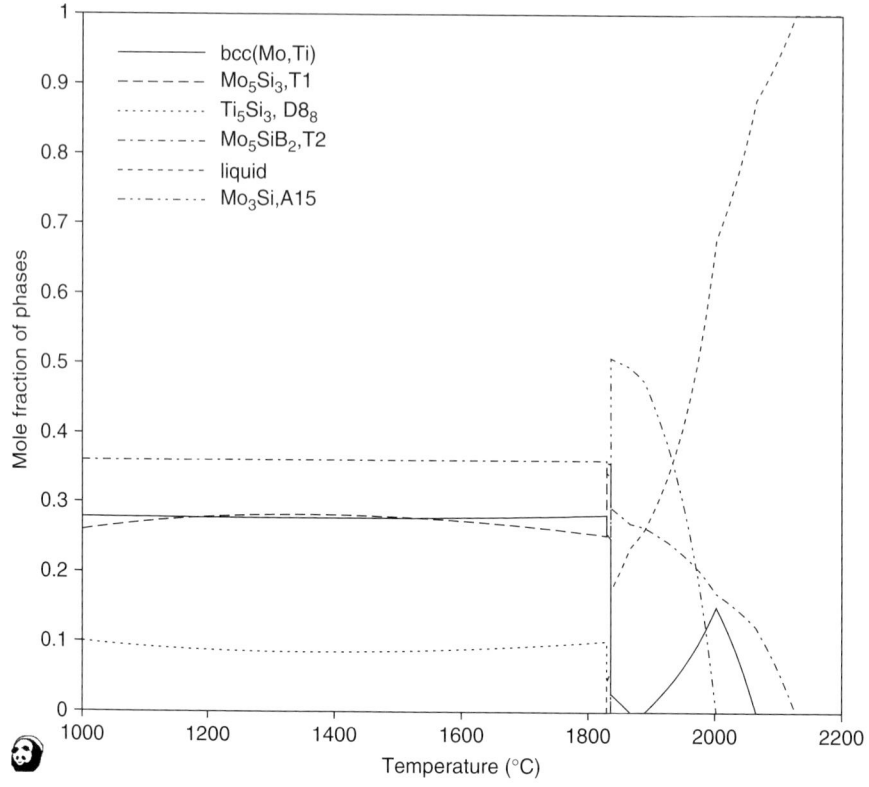

Figure 8.8 Calculated phase fractions as a function of temperature in °C for the sample, Mo–18Si–9B–32.5Ti, in mol%.

only on the phases in equilibrium with each other, but also their amounts. It is evident from this figure that the major phase in the Mo–18Si–9B–32.5Ti is T2 and the minor phase is $D8_8$, with the amounts of T1 and bcc somewhere between these two. BSE, EPMA, and EBSD results show the presence of T1 + T2 + bcc + $D8_8$ in the microstructure of this alloy heated treated at 1200°C and 1600°C. Here we present only the experimental results from samples heat treated at 1600°C. Figure 8.9 shows the BSE image of samples heat treated at 1600°C with the bcc phase exhibiting the brightest contrast, the T2 phase light gray, the T1 phase dark gray, and the $D8_8$ phase black. The BSE images qualitatively confirmed the predictions shown in Fig. 8.9. It should be stated that the crystal structures of the phases were usually identified by XRD. However, it was found that the ternary T1 and $D8_8$ phases were difficult to be discerned by EPMA since their atomic weights are very close to each other in view of the large mutual solubilities of the binary compounds. Furthermore since they were usually present as minor phases (with small volume fractions) in the microstructure, XRD is incapable of differentiating them. However, EBSD was able to identify the crystal structures of these two ternary phases in this alloy as shown in Fig. 8.10(a)–(d).

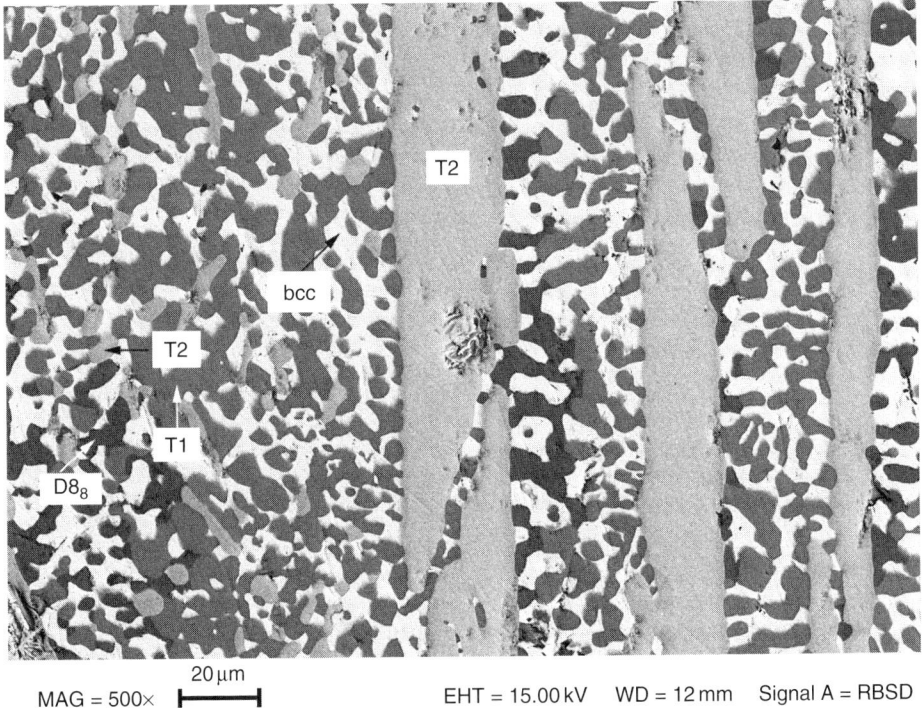

Figure 8.9 BSE image of Mo–18Si–9B–32.5Ti annealed at 1600°C for 150 h.

The EPMA measurements on phase compositions of the Mo–18Si–9B–32.5Ti alloy annealed at 1600°C for 150 h and 1200°C for 50 days are listed in Table 8.2. The concentrations of B and Si were relatively independent of the bulk alloy compositions. Taking the T2 phase as an example, the B and Si concentrations for all samples are 23.5–25.5 mol% and 11–12.4 mol%, respectively, which can be considered as constant values in view of experimental uncertainties. The equilibrium concentrations of Mo and Ti in the T2 phase vary with the overall compositions of the samples. This is also true for the bcc, T1, A15, and $D8_8$ phases. Therefore, only the Ti concentrations of each phase are listed in Table 8.2. The B and Si concentrations were listed for each phase right below the phase name in the same table. The calculated phase compositions, also given in this table, are in accord with the experimentally measured values.

On the basis of the phase equilibrium data alone, the following five multi-phase equilibria: bcc + T2 + A15, bcc + T2 + T1, bcc + T2 + $D8_8$, bcc + T2 + A15 + T1, and bcc + T2 + T1 + $D8_8$, offer the potential to exhibit desirable mechanical properties since they all contain a ductile metallic phase with strong intermetallic compounds. In addition, the two three-phase equilibria consisting of bcc either with T2 and T1 or with T2 and $D8_8$ should also exhibit favorable oxidation resistance due to the higher Si concentrations in the silicides. The phase diagrams of Mo–Si–B–Ti calculated in this study, especially the following multi-phase equilibria, bcc + T2 + A15, bcc + T2 + T1, bcc + T2 + $D8_8$, bcc + T2 + T1 + A15, and

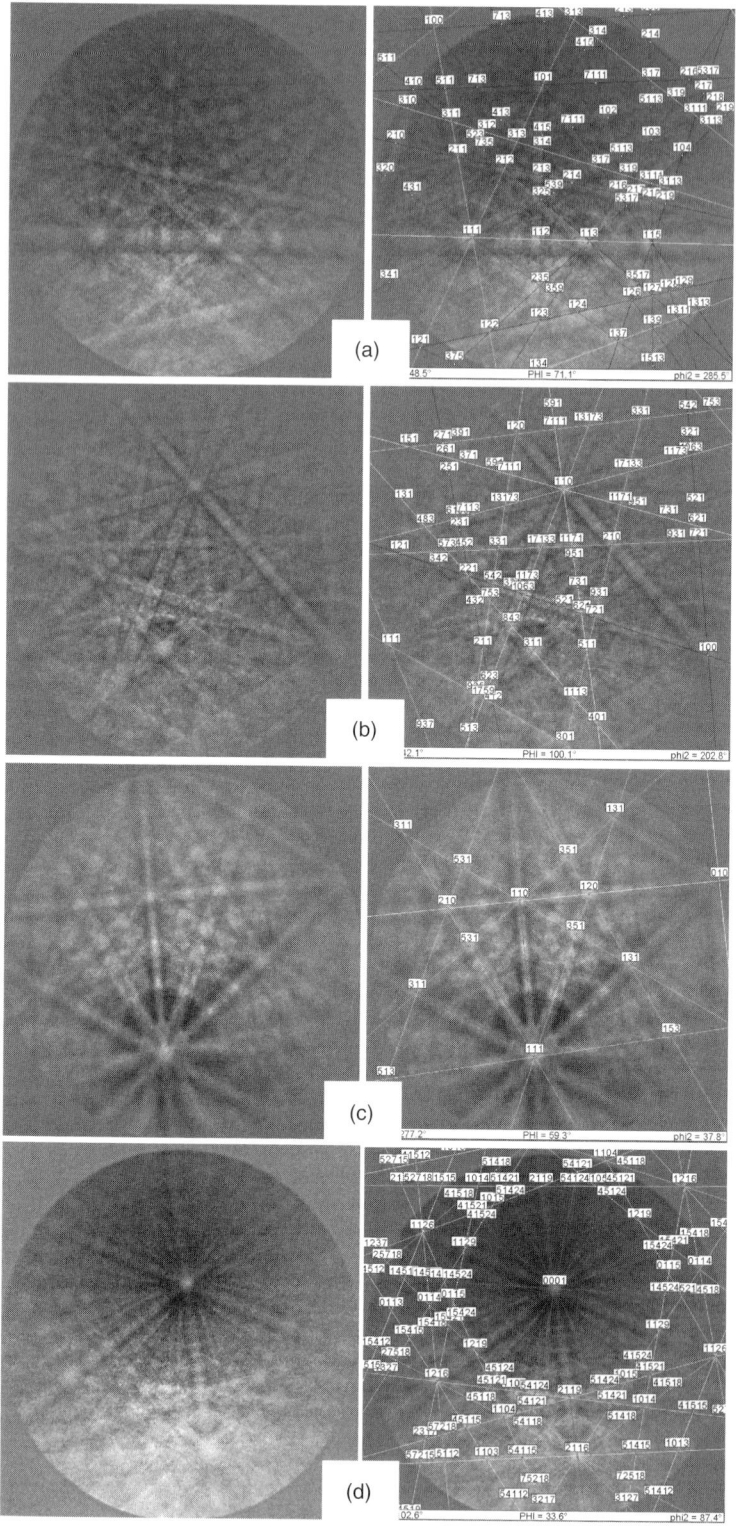

Figure 8.10 EBSD patterns of the phases in Mo–18Si–9B–32.5Ti annealed at 1600°C for 150 h.

Table 8.2 Comparisons between the calculated and the EPMA measured values for the compositions of Ti in mole fractions in each of the four phases for Mo–18Si–9B–32.5Ti annealed at 1600°C for 150 h and 1200°C for 50 days.

Annealing temperature (°C)	T2	bcc	T1	D8$_8$	
	Si = 0.11–0.124 B = 0.235–0.255	Si = 0.016–0.03 B = 0–0.01	Si = 0.35–0.365 B = 0–0.01	Si = 0.35–0.37 B = 0–0.01	
1600°C	0.265 ± 0.02	0.27 ± 0.01	0.423 ± 0.02	0.492 ± 0.01	Experimental
	0.248	0.278	0.416	0.502	Calculated
1200°C	0.26 ± 0.01	0.28 ± 0.02	0.43 ± 0.02	0.50 ± 0.02	Experimental
	0.249	0.261	0.427	0.521	Calculated

bcc + T2 + T1 + D8$_8$, offer wide processing windows to attain optimal microstructures and ultimately the desired mechanical performance. Since the calculated phase diagrams far away from the Mo–Ti rich region were extrapolated from the constituent ternary systems, it is expected that the topological features of the calculated diagrams to be correct but not necessarily the compositions of the component phases in equilibrium with each other. Nevertheless, the calculated phase equilibria could offer an intelligent guide for identifying a few key alloy compositions for further experimental studies. In addition to new phase equilibria found in the Mo–Ti–Si–B system, another important message conveyed here is that thermodynamic modeling provides a powerful tool for studying and visualizing the multiphase equilibria, which otherwise would be a rather challenging task indeed. We choose not to give an example when a new quaternary phase appears since that case is less challenging. Since the phases in equilibrium with a quaternary phase are known, it is possible to obtain its Gibbs energy from those phases in equilibrium with the quaternary one.

4 Concluding Remarks

We presented an efficient and cost-effective way to obtain multicomponent phase diagrams by taking advantage of the success in the use of the phenomenological or CALPHAD approach to calculate phase diagrams, particularly multicomponent ones. It has been known that a thermodynamic description of a multicomponent system can be obtained from those of its lower order systems, normally binaries and ternaries, in most cases. The exceptions are when a new quaternary phase appears or the range of homogeneity of a lower order phase extends significantly to the quaternary compositional space. This means that when thermodynamic descriptions of binaries and ternaries are known, we can readily obtain a description of a multicomponent system. In this chapter we presented two examples to demonstrate this efficient and cost-effective way to obtain a thermodynamic description of Mg–Al–Sr in the Mg–Al rich region and that of Mo–Si–B–Ti in the Mo–Ti rich region. For the ternary system, we identified five alloys for experimentation based on a calculated isotherm using a description extrapolated from those of its binaries. The results obtained from three samples provided the data to develop a final description. The primary phases of solidification obtained from the other two samples are consistent with the calculated liquidus projection. The experimental data reported in a new publication for more than 20 samples are also consistent with the calculated diagrams. For the quaternary system, the composition of a lower order phase extends into the quaternary compositional space. The greater complexity of the phase equilibria involved makes it necessary to go through two cycles, three samples in the first cycle and six additional samples in the second cycle, before a reasonable description was developed. The first three samples were identified based on a preliminary calculated isotherm. The results from these samples enabled developing an improved description. Six samples were next identified, based on a newly calculated isotherm using the improved description, for additional experimentation. Based on the results of these six samples with the other three, a final description was developed. The calculated phase and related diagrams using the final description were validated by a further set of three additional samples. It is worthwhile to point out that the description obtained in the ternary system is restricted to the Mg–Al rich region and Mo–Ti rich region of the quaternary system.

In addition to obtaining complex multicomponent phase diagrams, the thermodynamic descriptions can also provide other information such as thermodynamic factors needed in diffusion calculations and free energy driving forces for phase transformation.

ACKNOWLEDGMENT

We wish to thank Jun Zhu and Hongbo Cao for their critical review of this manuscript and NSF (FRG Grant No. DMR-0309468) and AFRL/ML (Revolutionary High Pressure Turbine Blade Materials Program, Contract no F33615-98-C-2874) through Pratt–Whitney and the Wisconsin Distinguished Professorship for financial support.

REFERENCES

1. K.-C. Chou and Y.A. Chang, *Ber. Bunsen. Phys. Chem.*, 93 (1989) 735.
2. L. Kaufman and H. Bernstein, *Computer Calculation of Phase Diagrams*, Academic Press, New York, NY, 1970.
3. U.R. Kattner, *JOM*, 49(12) (1997) 14.
4. Y.A. Chang, S.-L. Chen, F. Zhang, X.-Y. Yan, F.-Y. Xie, R. Schmid-Fetzer and W.A. Oates, *Prog. Mater. Sci.*, 49 (2004) 313.
5. Y.A. Chang, *Mater. Metall. Trans. A*, 37A (2006) 273.
6. Y. Yang, Y.A. Chang, L. Tan and W. Cao, *Acta Mater.*, 53 (2005) 1711.
7. Y. Yang and Y.A. Chang, *Intermetallics*, 13 (2005) 121.

8. Y. Yang, Y.A. Chang and L. Tan, *Intermetallics*, 13 (2005) 1110.
9. H. Cao, J. Zhu, C. Zhang, K. Wu, N.D. Saddock, J.W. Jones, T.M. Pollock, R. Schmid-Fetzer, Y.A. Chang and Y.A. Chang, *Experimental Investigation and Thermodynamic Modelling of the Mg-rich Corner of the Mg–Al–Sr System*, Z. Metallkd., to be published in 2006.
10. F. Zhang, Y. Yang and F.-Y. Xie, *CompuTherm*, LLC, Madison, WI, 2005, Private communication.
11. P. Liang, H.-L. Su, P. Donnadieu, M.G. Harmelin, A. Quivy, P. Ochin, G. Effenberg, H.J. Seifert, H.L. Lukas and F. Aldinger, *Z. Metallkd.*, 89 (1998) 536.
12. H. Li and Z.-K. Liu, *Poster at International Conference on Phase Diagram Calculations, CALPHAD XXIX*, held on campus of Massachusetts Institute of Technology, Cambridge, USA. June 18–23, 2000. Organized by Larry Kaufman, Sam Allen, and Claude Lupis.
13. C. Wang, Z. Jin and Y. Du, *J. Alloy. Compd.*, 358 (2003) 288.
14. I. Ansara, T.G. Chart, A. Fernández Guillermet, F.H. Hayes, U.R. Kattner, D.G. Pettifor, N. Saunders and K. Zeng, *CALPHAD*, 21 (1997) 171.
15. M.A. Parvez, M. Medraj, E. Essadiqi, A. Muntasar and G. Dénès, *J. Alloy. Compd.*, 402 (2005) 170.
16. Y. Yang, Y.A. Chang, L. Tan and Y. Du, *Mater. Sci. Eng.*, A361 (2003) 281.

CHAPTER NINE

DETERMINATION OF PHASE DIAGRAMS WITH REACTIVE OR VOLATILE ELEMENTS

Cezary Gumiński

Contents

1 Introduction		293
2 Solid–Solid Equilibria		299
2.1	Thermal Analysis	299
2.2	Zone Melting	300
2.3	Microstructural Analysis of Quenched Samples	301
2.4	X-Ray Diffraction	302
2.5	Densitometry	302
2.6	Dilatometry	303
2.7	Interdiffusion	303
2.8	Hardness	304
2.9	Calorimetry	304
2.10	Electrical Resistivity	305
2.11	Superconductivity	305
2.12	Electromotive Force	306
2.13	Coulometric Titration	307
2.14	Galvanic Polarization	309
2.15	Magnetic Susceptibility	310
2.16	Vapor Pressure	311
2.17	Gaseous Thermal Extraction	314
3 Solid–Liquid Equilibria		315
3.1	Weight Loss of a Solid After Equilibration with a Liquid	315
3.2	Chemical Analysis of a Separated Liquid	316
3.3	Chemical Analysis of Quenched Samples	318
3.4	Thermal Analysis	319
3.5	Electromotive Force	320
3.6	Electrical Resistivity	321
3.7	Anodic Oxidation	321
3.8	Magnetic Susceptibility	323
3.9	Densitometry	325
3.10	Enthalpy of Dilution	325
3.11	Kinetics of Alloy Decomposition or Formation	325

3.12	Diffusion Coefficient	327
3.13	Vapor Pressure	327
3.14	X-Ray Absorption Spectrometry	328
3.15	Viscosity	328
3.16	Optical Reflectivity	328
3.17	Corrosion Tests	328
3.18	Motion of Liquid Metal Inclusions in Ionic Crystals	330
4	Liquid–Liquid Equilibria	330
4.1	Chemical Analysis of Separated Liquids	330
4.2	Thermal Analysis	330
4.3	Calorimetry	331
4.4	Densitometry by X-Ray Attenuation	332
4.5	Neutron Transmission	332
4.6	Electromotive Force	332
4.7	Electrical Resistivity	332
4.8	Magnetic Susceptibility	333
4.9	Vapor Pressure	333
5	Solid–Vapor Equilibria	334
5.1	Vapor Pressure	334
5.2	Thermogravimetry	335
6	Liquid–Vapor Equilibria	336
6.1	Vapor Pressure	336
7	Liquid–Fluid Equilibria	337
8	Concluding Remarks	337

1 INTRODUCTION

This chapter is devoted to experimental methods used for determination of phase diagrams with chemically very active elements (alkali and alkaline earth metals, several aggressive non-metals) and easily vaporizing elements (low melting elements). It is not an exhaustive compendium on the fundamentals of the related techniques, but rather a critical assessment of various available methods, their precision, and how to correctly interpret the experimental results.

Independent of the technique applied to an investigation of a system that contains one or more reactive components, their container material must withstand a corrosive attack by an aggressive liquid or gas. The noble metals are generally not suitable because they readily react with the majority of the elements mentioned. Therefore, the selection should be focused on such metals, alloys and ceramics that have a very low solubility in reactive liquid metals and do not form intermetallic compounds with them. In every case, a choice of the container material should take into account: alloy composition, purity of materials used, the temperature range of experiments and possibility of apparatus hermetization.

Degree of purity, both of the sample and the container, is especially important during investigations of alkali and alkaline earth metal alloys. Even small amounts of non-metals (as little as $<0.01\%$) present in either the sample or the container may produce severe corrosive dissolution of the container. For example, an increase of N concentration in Li from 0.0005 to 0.2 at.% increases the solubility of Fe in Li by ~100 times at 1200K with Li_3FeN_2 formed in these conditions being more soluble than pure Fe in Li [1]. The presence of C, H or O in the Fe–Li system has a much smaller effect on the Fe solubility in Li because Li_2C_2, LiH and Li_2O are more stable than analogical compounds formed with Fe. However, in the case of Na, K, Rb and Cs especially, the presence of O may lead to the corrosive dissolution of the metal container. It was experimentally well documented that many complex oxides (e.g., Na_3NbO_4, K_2MoO_4, $RbCrO_2$, Cs_2ZrO_3) are more soluble in the alkali metals than the corresponding pure transition metals [2]. The presence of H in such systems causes embrittlement of containers, and C may be easily leached out from steels. In contrast, N only plays a marginal role in the corrosion of the container metals due to its very low solubility in Na, K, Rb and Cs.

Similar interactions between liquid Mg, Ca, Sr and Ba and solid transition metals perhaps take place in the presence of non-metallic impurities, however, these have not been investigated in detail. For example, the existence of a third allotrope of Sr (hexagonal) between 504K and 896K in earlier studies is now recognized as only a metastable form of Sr induced by a contamination of the Li–Sr system by H or N [3]. Therefore, samples and containers should always be as pure as possible, which is unfortunately neglected in many studies related to the systems mentioned. It should be underlined that all experiments with reactive elements must be performed in an inert gas (Ar, He) atmosphere where concentrations of H, O and H_2O are $<5 \times 10^{-4}$%; and in the case of Li, Mg, Ca, Sr and Ba, N should also be controlled to $<5 \times 10^{-4}$%. A selection of suitable container materials is summarized in Table 9.1 [2,4–16]. Recently, vessels made of synthetic diamond (from chemical vapor deposition) with a volume of a few cm^3 are getting cheaper as their production increases. Chemical resistance of diamond to reactive elements must be confirmed in detailed tests, but it is known so far that diamond interacts with N, O, F and metals – forming carbides only at elevated temperatures [17].

If one starts to investigate a system with one or more readily volatile elements, one should define conditions under which the equilibrium phase diagram will be studied. There are two possibilities: (i) equilibrium phase diagram at 0.101325 MPa (ambient pressure) and (ii) equilibrium phase diagram at an elevated pressure sufficient to keep all components in the condensed state (solid or liquid). The difference between the two types of diagrams is visualized in Fig. 9.1 for the Hg–U system [18]. Although the same intermetallics are formed in both instances, their stability temperature ranges are different. In the first case, Fig. 9.1(a), the experiments were carried out in an open container that was placed in a neutral gas under ambient pressure. An increase of temperature intensifies evaporation of the more volatile component. When the sum of the partial pressures of all components reaches 0.101 MPa then the gas phase over the solid or liquid phase, in absence of neutral gas, contains only atoms or molecules formed within the Hg–U system. In the second case, Fig. 9.1(b), the experiments were performed in a completely closed container with no vapor from the alloy components. When temperature increases above the boiling point of Hg, the pressure inside the container increases significantly, which protects the liquid and solid from decomposition and evaporation. Generally, an increase of pressure up to 10 MPa shifts the temperature of condensed state phase transition less than 1K. Because the

Table 9.1 Selection of container materials suitable to perform equilibrium experiments with selected low melting, easy vaporating and reactive elements in liquid state. Maximum temperatures (in K), below which the container corrosion in static conditions is negligible, are given in the parenthesis. The data were collected from Refs. [2,4–16].

Element	Container material	Remarks
Li	Fe, Cr, ss (750); Ti, Zr, V, Nb, Ta, BN* (1000); Mo, W, Re (1300); W–Re (2000, for short); ZrO_2, ThO_2, BeO, MgO, Y_2O_3 (700)	N enhances corrosion
Na, K, Rb, Cs	Fe, Cr, ss (900); Ti, Zr, V, Nb, Ta (1000); Mo, W (1500); glass (500); ZrO_2, ThO_2, MgO, BeO (800); Mo, W, Re (2500, for short)	O enhances corrosion H and C inadvisable
Mg	Ti, Zr, Hf, Nb, Al_2O_3 (1000); Ta, Mo, W (1250)	N and O inadvisable
Ca, Sr, Ba	Ti, V, Nb, Ta, Cr, Fe, ss (1200); BN (1100); Al_2O_3 (1000); $BaZrO_3$, $CaZrO_3$ (1300)	H, N and O inadvisable
Zn, Cd	Ta (700); Mo, W, Re (1300); quartz, MgO, Al_2O_3, ZrO_2, SiC (1000)	O inadvisable
Hg	ss (500); Nb, Ta (850); glass, quartz, porcelain (650); quartz (1600, for short); Mo, W, Re, corundum (2000, for short)	O inadvisable
Ga	Ta (700); Mo, W, Re (1300); graphite, quartz, Al_2O_3, ZrO_2, MgO (1000)	O inadvisable
In, Tl	Nb, Ta, quartz, graphite, Al_2O_3, ZrO_2, MgO (1000); silica (1200); Mo, W, Re (1300)	O inadvisable
Sn	Mo, quartz (1000); W, silica (1250); Re (1700); graphite (2000)	O inadvisable
Pb	Ni (750); quartz (1000); Nb, Ta, Mo, W (1200); Re (1700)	O inadvisable
P	Glass (500); quartz (800); Al_2O_3, BeO, MgO, ZrO_2 (1200)	O inadvisable
As**	Graphitized quartz (1000); Al_2O_3, BeO, MgO, ZrO_2, silica (1200); graphite (1500)	O inadvisable
Sb	B, graphite, Si (1000); quartz (1100)	O inadvisable
Bi	Nb (800); quartz (1000); Ta, Mo, W, Re, Os (1150); MgO, Al_2O_3 ZrO_2, fused SiO_2 (1250); graphite (1600)	O inadvisable

(*Continued*)

Table 9.1 (Continued)

Element	Container material	Remarks
S	Glass (650); teflon (700); Al, Ta, ss (750); quartz, silica (1100); corundum (2000, for short); quartz (1600, for short)	O inadvisable
Se	ss (500); quartz (1000); silica (1400); corundum (2000, for short); quartz (1600, for short)	O inadvisable
Te	quartz (1000)	O inadvisable
Br, I	Nb, Ta (550); glass (650); teflon, quartz (700); silica (1150)	

ss: Stainless steel.
for short: Used in short investigations of fluidal state in vicinity of the critical point.
*Recommended by Chen et al. [15], but not by Nesper [16].
**The apparatus must withstand pressures higher than 4 MPa to keep as liquid. Alloys with volatile and aggressive components (as As) may be prepared by an equilibration of an alloy containing the volatile element with another metal (or alloy) in which activity of the volatile element is lower. Then the equilibrium is established through the gas phase in a closed container at lower pressure. This way As-Ti alloy was prepared from As-Cu alloy because As is stronger bonded to Ti than to Cu [s 1].

pressure under the experimental conditions of the Hg–U phase diagram determination in the closed container was never higher than 9 MPa, the temperatures were accurate to ±1K in this respect.

It was shown for the Hg–Tl system that a pressure increase up to 4 GPa moved the whole liquidus (both eutectic points and congruent melting of Hg_5Tl_2) almost linearly for about 140K [19]. However, in the case of the As–Te system, the temperature of $\alpha \leftrightarrow \beta$ transformation of As_2Te_3 decreases as the pressure increases. This compound melts congruently up to 0.7 GPa and starts to decompose periteticly at higher pressure [20]. Therefore, every system should be treated individually, and the possibility of new phase formation at a high pressure should be checked.

If a change of free volume over an alloy during investigation is hard to avoid, a change of alloy composition will occur when the temperature changes. With increasing temperature, more volatile component escapes to the gas phase and consequently the condensed part of the alloy is leaner in the more volatile constituent. Some practical suggestions concerning alloy composition control are given in Section 2.1. If one is interested in a fast estimation of the loss of a volatile component upon heating and the maximum pressure that arises upon temperature increases, Table 9.2 collects the equations for saturation vapor pressures of selected liquid elements [10–12,21–23]. Above the boiling points, one may apply the general gas equation to relate various quantities: the amount, volume, pressure and temperature. If stable intermetallics are formed in a system, then the experimentally observed pressure is always lower than that estimated for pure alloy components taken in proper molar fractions.

Table 9.2 also collects the critical parameters for several elements. Over the critical point, one is no longer able to keep an element (or alloy) as liquid and a so-called fluidal phase is formed. In the vicinity of the critical point the metal–insulator phase transition occurs and consequently a metal loses its metallic features [12]. Therefore, the critical parameters set the limit for the liquid metal state.

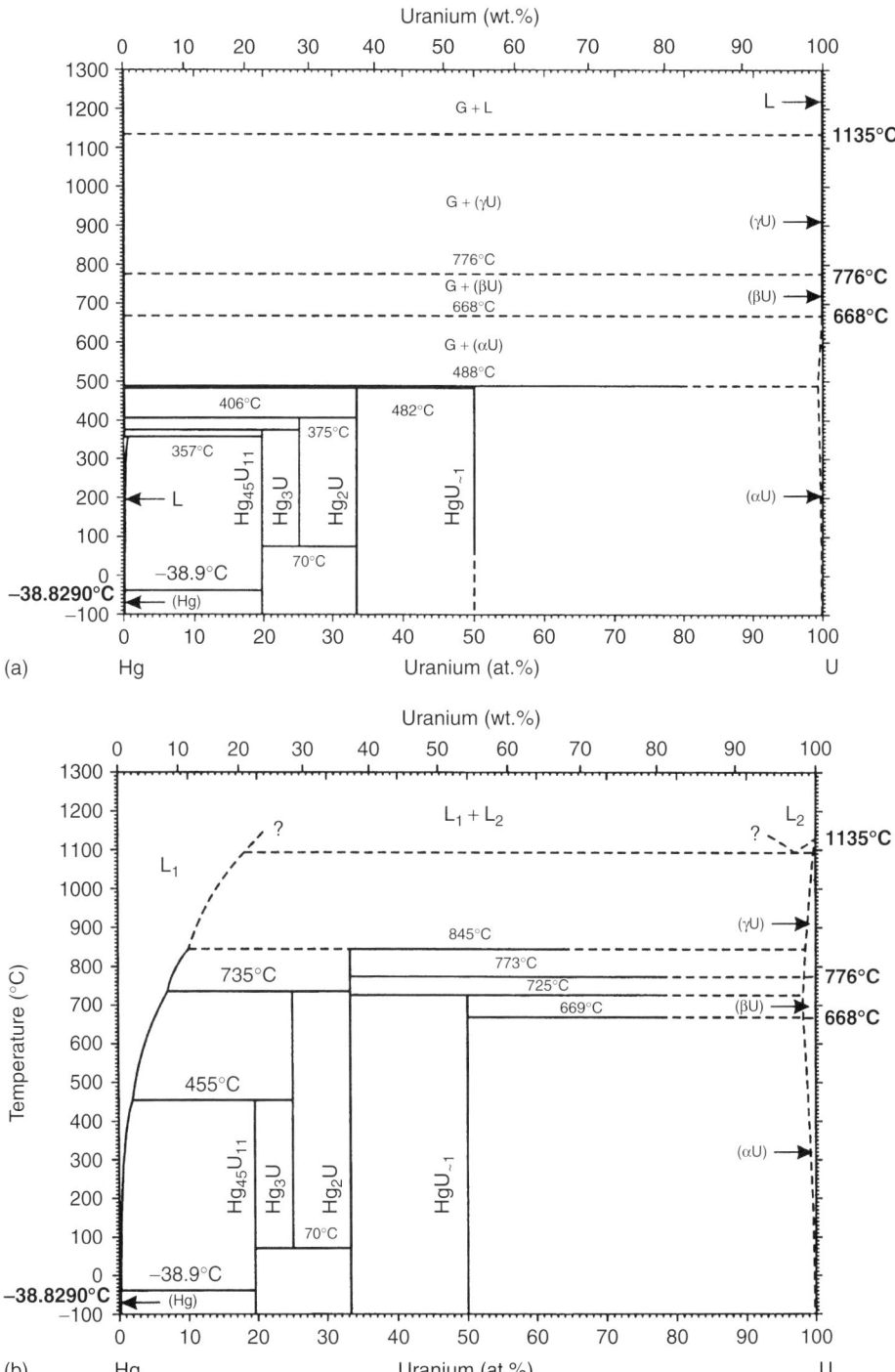

Figure 9.1 Phase diagram of the Hg–U system at 0.101 MPa and under constrained pressure: (a) results obtained with an open container and (b) results with a closed container [18].

Table 9.2 Saturation vapor pressures [21–23] and critical point parameters of selected elements [10–12,23].

Element	Vapor pressure log P (P in Pa)	Temperature range K	Critical pressure MPa	Critical temperature K
Li	$13.415 - 8320/T - 1.0255 \log T$	m.p.–1000	~70	~3200
Na	$13.406 - 5634/T - 1.1748 \log T$	m.p.–700	24.8	2485
K	$13.239 - 4693/T - 1.2403 \log T$	m.p.–600	16.1	2280
Rb	$13.322 - 4275/T - 1.3102 \log T$	m.p.–550	12.5	2017
Cs	$13.238 - 4062/T - 1.3359 \log T$	m.p.–550	9.3	1924
Mg	$14.914 - 7550/T - 1.41 \log T$	m.p.–b.p.	–	–
Ca	$14.575 - 8920/T - 1.39 \log T$	m.p.–b.p.	–	3273
Sr	$14.755 - 9000/T - 1.31 \log T$	m.p.–b.p.	–	3064
Ba	$9.013 - 8163/T$	m.p.–1200	–	3670
Zn	$14.465 - 6620/T - 1.255 \log T$	m.p.–b.p.	–	–
Cd	$14.412 - 5819/T - 1.257 \log T$	m.p.–b.p.	–	–
Hg	$12.480 - 3305/T - 0.795 \log T$	298–b.p.	151	1765
Ga	$13.545 - 14330/T - 0.844 \log T$	m.p.–b.p.	–	–
In	$11.915 - 12580/T - 0.45 \log T$	m.p.–b.p.	–	–
Tl	$12.225 - 9300/T - 0.892 \log T$	m.p.–b.p.	–	–
Sn	$10.355 - 15500/T$	m.p.–b.p.	–	–
Pb	$13.285 - 10130/T - 0.985 \log T$	m.p.–b.p.	–	–
P_4	$9.965 - 2740/T$	m.p.–b.p.	8.3	968
As_4	$9.82 - 6160/T$	600–900	12.5	>1670
Sb	$8.495 - 6500/T$	m.p.–b.p.	–	–
Bi_2	$20.3 - 10.730/T - 3.2 \log T$	m.p.–b.p.	–	–
S	$26.005 - 4830/T - 5.0 \log T$	m.p.–b.p.	20.3	1313
Se	$10.215 - 4990/T$	m.p.–b.p.	38.0	1863
Te_2	$24.415 - 7830/T - 4.27 \log T$	m.p.–b.p.	14.8	2329
Br_2	$19.96 - 2200/T - 4.15 \log T$	m.p.–b.p.	10.3	584
I_2	$23.65 - 3205/T - 5.18 \log T$	m.p.–b.p.	11.6	819

The values for Sb, S and Se are the total pressures of all vapor species; m.p. – melting point; b.p. – boiling point.

2 Solid–Solid Equilibria

2.1 Thermal Analysis

Thermal analysis (TA) is one of the most popular techniques used for construction of phase diagrams. Its fundamentals and related apparatus features may be found in Ref. [24–26] and also in Chapter 5. Nowadays, TA in its classical form is seldom in use because its derivatives, the differential thermal analysis (DTA) and the differential scanning calorimetry (DSC), are much more popular due to their sensitivity, automation, miniaturization and content of valuable information.

In DTA, the temperature difference between a sample and a reference substance (that does not undergo any phase transition during thermal treatment) is recorded by means of thermocouples when both are equally heated or cooled at a constant rate between selected temperatures. Phase transitions are manifested as exothermic or endothermic peaks.

In DSC, the heat flow is precisely recorded and controlled in a calorimeter to keep the sample and the reference at the same temperature, producing a heat flow vs. temperature chart. The temperature at the location of the maximum of a peak recorded under these conditions represents the temperature of a phase transition; and the area under the peak is proportional to the enthalpy of the transition. The heating/cooling rate is especially critical for DSC. If it is too fast, a phase transition is shifted to a higher temperature in heating or a lower temperature in cooling. If the rate is too slow, then the peak is too shallow and the measured enthalpic effect is underrated. More discussion of these effects can be found in Chapters 2 and 5.

Application of DTA and DSC to the investigations of systems with easily volatile elements is not straightforward. To prevent changes of composition during temperature scanning, one must minimize the free space over a sample as much as possible. Sometimes the substances are compressed in containers to pressures higher than the vapor pressures of components, thus requiring the use of pressure vessels. Another possibility is to add an inactive filler to a vessel (e.g., pieces of quartz into a quartz capsule). It is not recommended to perform experiments by evacuating the free volume over the sample; instead, the use of a noble gas filler is a much better solution. Although the compositional changes may be estimated using the data in Table 9.2 and the free space over the sample, one should keep in mind that the fraction of volatile component transferred to the vapor phase changes during the temperature scan. Sufficient time for equilibration (usually longer than the heating/cooling cycle) is required.

If one is not sure about the extent of component evaporation in DTA or DSC, a parallel test with a thermogravimetric device will detect a possible change of alloy composition. If a congruently melting line compound exists with its composition precisely known from crystallographic (or other) considerations and the maximum on the corresponding liquidus line is in agreement with the estimated composition, then component evaporation is rather negligible. It is highly recommended to analyze the condensed part of a sample after a DTA or DSC cycle, which is done best in the course of fast quenching.

An important point, but sometimes neglected by investigators, is the purity of materials used for alloy preparation. If one looks at the experimental phase diagrams of Ca alloys constructed from DTA curves in the seventies and earlier [10], one observes that Ca undergoes either two phase transitions at 598–626K and 883–885K or one transformation at 723–743K. The corresponding thermal arrests are scattered and frequently absent on the heating/cooling curves. Detailed investigation of the Ca-rich part of the Ca–H system [27] finally helped to explain this phenomenon. The formation γ Ca–H phase existing at $0.6 < x_H < 7.3$ at.% and $633K < T < 873K$ is responsible for this perplexity because H is the common impurity present in Ca, and H effectively stabilizes the hexagonal form of Ca in various alloys. The currently accepted αCa ↔ βCa allotropic phase transition occurs at 716K and deviations from it in the absence of sizable solid solubility in Ca reflect the presence of impurities in the Ca–base system investigated. In this context the DTA result of the Ca–Sr system by Ref. [28] with the proposed existence of a hexagonal Ca–Sr solid solution is very likely caused by alloy contamination.

DTA is a very effective method when applied in the composition range 1–99 at.%. In the case of more dilute solid solutions, the results are imprecise. The accuracy of temperature readings depends on the calibration of thermocouples on standard substances. The best precision of the method can be a fraction of a K for temperatures up to ∼400, ∼1K for temperatures up to 1200, and a few K above 1200K. Exact determination of the sample composition depends at first on preparation and then on precision of its chemical analysis. It is suggested to carry out a heating/cooling cycle because the mean temperature of corresponding arrests partially cancels the overheating and undercooling effects. If the arrest temperatures move systematically in subsequent cycles, then partial evaporation of the sample or an interaction of the sample with the container should be suspected. When the components in an alloy differ significantly in their densities, their liquid phases may separate with heavy elements settling at the bottom.

2.2 Zone melting

This technique allows a relatively fast investigation of an entire system with continuously changing compositions; thus the number of experiments may be significantly reduced. The method is recommended for systems with volatile and reactive components since the contact of liquid components with a vessel is quite short. Its principle is illustrated in Fig. 9.2 for the Mg–Pb system [29]. A feed rod is made up of two properly cut-out pieces of metals with a continuously varying triangular shape such that at every section of the rod the sum of component fractions is unity. The rod is placed inside a boat made of an inert material and zone heated by an induction oven or a laser moving along the rod. A movable thermocouple is located very near the molten zone for temperature control. During the movement, the rod melts progressively; the components react in the liquid state and the resulting solid Mg–Pb crystallizes after self-cooling. The rod, after the remelting run, is analyzed in longitudinal and transverse sections to trace properties of new phases formed in the selected composition range. Various methods such as X-ray diffraction (XRD), scanning electron microscopy (SEM), X-ray photoelectron spectroscopy, Auger electron spectrometry, Rutherford back scattering spectrometry (RBS), electron probe microanalysis (EPMA), metallography, electrical resistance, magnetic susceptibility and hardness measurements may be used to identify the phases formed in the cut-off sections. Depending on the remelting and final temperatures, one may obtain information on various regions of

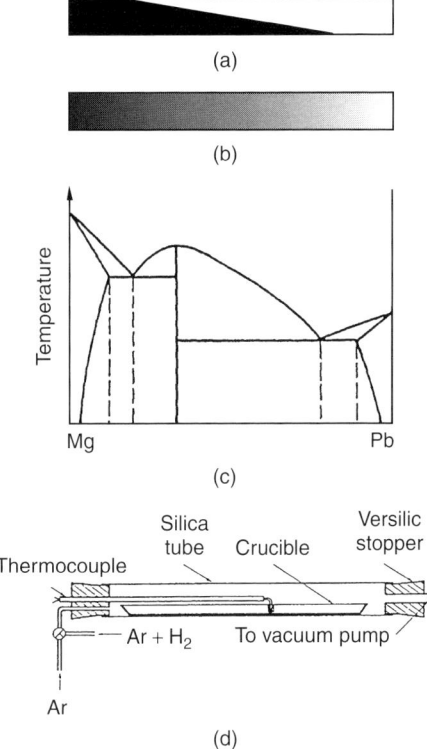

Figure 9.2 The zone melting method for the determination of the Mg–Pb phase diagram: (a) schematic illustration of the feed rods of Mg and Pb before melting, (b) schematic illustration of the feed rod after melting, (c) the Mg–Pb phase diagram and (d) the zone melting device [29].

the phase diagram. It is advised to keep the more dense metal as the upper part of the rod to induce an effective mixing of the metals in the liquid state.

2.3 Microstructural Analysis of Quenched Samples

Sometimes TA is not sufficiently sensitive to define the correct solvus boundaries. In this case, an alloy sample of known composition may be rapidly quenched to room temperature from several equilibration temperatures above and below the estimated solvus. Subsequent examination of the sample using optical or electron microscopy as well as several spectrometric techniques mentioned in the previous paragraph allows one to distinguish grains of a solid solution from mixed phases, as was the case of the solvus line of Hg in Au [30]. For microstructure observation, a sample should be carefully prepared. At first it should be ground, typically with SiC sandpaper with granules of 20–80 μm. Subsequent polishing is then made with the use of diamond, alumina, magnesia, pumice, diatomaceous earth, hematite or limonite pastes with grains of ~1 μm or less which are suspended in oils (in case of active metals) or water (in case of less active elements). Depending on metal activities, the specimens are finally cleaned with pure trichloroethylene, alcohols or water in an inert gas atmosphere.

If microetching of a surface is needed, one may use chemical or electrochemical dissolution methods. Selection of conditions for the electrochemical etching should take into account the formal potentials of the component elements which depend not only on substance but also on electrolyte composition and acidity, solvent, mixing velocity and temperature. Electrochemical etching is not very practical for alkali and alkaline earth metal alloys because they may react with the solutions even without an applied potential and current flow.

Selection of a solution for chemical etching should be based on the chemical properties of the elements. The mildest agent is alcohol (for alkali metals), then in the increasing oxidation strength: water, acetic, phosphoric, hydrochloric, perchloric, sulfuric, nitric, chromic acids and aqua regia. Strong bases or ammonia solutions are applied in special cases. Chemical knowledge of such reactions is very helpful and may be found in books devoted to metallography and chemistry of the elements. The method of ionic polishing of a sample is not recommended for reactive and easily vaporized metals because during an application of a high vacuum to a sample, its surface may decompose.

2.4 X-Ray Diffraction

Solvus or intermetallic boundary lines are frequently determined using XRD, especially when DTA is difficult to perform or insensitive. Two approaches are possible. When a solvus line is steep, several samples of different compositions should be annealed at a selected temperature and rapidly quenched to room temperature for XRD. When a solvus line is of shallow slope, a sample of a selected composition is preferably annealed at different temperatures (above and below the boundary line) and rapidly quenched. Thus, the state of the samples at high temperatures is frozen to room temperature, and may be analyzed in the form of powders by commercially available XRD equipment. Samples may also be analyzed at the equilibration temperatures when a diffractometer is equipped with a hot stage under a protective environment. The resultant crystal structures and lattice parameters are analyzed with respect to composition and temperature. Principles and procedures of XRD may be found elsewhere [31].

A classical example of experiments performed using both approaches is the work of Raynor [32] who exhaustively investigated the In-rich part of the In–Mg system. The boundary lines obtained are widely accepted. The author realized the problem of easy evaporation of Mg from the alloys and compensated for it by small additions of Mg. The samples were sealed in hard glass tubes for annealing. After the XRD measurements, the samples were quantitatively analyzed.

It is a common practice to plot lattice parameters against alloy composition; and a break (a location with a sharp change of slope) on such plots indicates the appearance or disappearance of a phase (see e.g., the change of the "a" lattice parameter ("c" was constant) in the Sn-rich range of the Hg–Sn system in Ref. [33]).

The XRD method may be used over the whole composition range except for very dilute solutions. Precision of the results depends on the quality of XRD investigations, temperature control, sample analysis and persistency of approaching the boundary line with many tests from each side.

2.5 Densitometry

Each intermetallic compound or solid solution, due to its crystal structure and composition, possesses a characteristic gravimetric density. Therefore, at a phase boundary one should observe a change of slope on a plot of alloy density vs. composition. Very often the density is measured with the use of a pycnometer. An alloy piece is weighed in air (m_1), then a neutral liquid filling the pycnometer is weighed before (m_2) and after (m_3) the submersion of the alloy piece in the liquid. Knowing density of the liquid (d_l) one may calculate the alloy density (d_a) from the following equation:

$$d_a = m_1 d_l / (m_2 - m_3) \tag{1}$$

Samples of different compositions are analyzed in the same way and the corresponding density–composition plot is constructed. Although the procedure seems very simple, one should remember that the liquid must not react with the alloy (as is often the case for alkali metal alloys with water or alcohols). The alloy sample must have a uniform microstructure and be free from voids and pores, especially from closed pores not connected to the surface. Some investigators apply a pressure to samples before the measurements. The densitometry measurements are characterized by a relative precision better than 1%.

Density may also be determined using other methods [22] or calculated from the lattice parameters if the crystal structure is precisely known. Aldinger et al. [33] presented a very good example of density measurements of Sn-rich Hg–Sn alloys. The change of slope on the density vs. composition plot defines the γ phase boundary on the Hg–Sn phase diagram.

2.6 Dilatometry

Dilatometry measures the thermal expansion of solids. When a new phase has a different coefficient of thermal expansion (CTE), its formation may be recognized on a dilatometry curve. When two neighboring phases have similar CTEs, the method is not sensitive. Alloy bars with flat ends are heated gradually and their lengths are monitored by means of a micrometer or an optical device. Discontinuity in the CTE vs. alloy composition plot indicates a phase boundary.

Predel and Schwermann [34] investigated the order–disorder transformation within the ω phase of the Cd–Hg system. As expected, the resultant CTE for the ω′ ordered phase ($CdHg_2$) was higher than that of the disordered ω phase. The linear contraction of the polycrystalline $CdHg_2$ rod during ordering was 0.14%. The results obtained by the dilatometric measurements for this system are in agreement with the results gathered by other methods. Precision of this technique is generally no better than several percent.

2.7 Interdiffusion

When a microscopic analysis of samples from quenching or zone melting is not satisfactory, diffusion couples may be used to study the saturation limit of a solid solution. Such experiments were performed by Taylor and Burns [35] for the Sn-rich side of the Hg–Sn system. The investigators contacted a saturated liquid Sn amalgam with solid Sn specimens at constant temperature. After varying time periods of the immersion, the excess of liquid from the specimen surface was blown off. The specimens were further annealed and finally sectioned. Profiles of Hg content vs. Hg penetration depth were obtained from chemical analysis of the individual sections. The saturating Hg concentration was estimated from an extrapolation of the

Hg concentration plateau to the zero penetration depth. The precision of the method depends on the way the chemical analysis was performed. Auger electron, X-ray photoelectronic spectroscopy, microhardness measurements and EPMA can be used to construct the diffusion profiles near the specimen surface.

2.8 Hardness

Cohesive forces in alloys are manifested by their hardness that is measured by commercially available hardness or microhardness testers. The hardness of a series of alloys is typically measured on a selected hardness scale (Brinell, Rockwell, Vickers, Shore) as a function of composition. A change of slope on such a plot corresponds to a phase boundary.

This method is very useful and popular, but not very sensitive. One frequently observes a large scatter of data such that phase diagram information extracted from hardness profiles may not be very reliable. When a sample has a coarse structure with large crystals, a selection of locations for the tests is quite critical. A ball indenter better represents the mean properties of a sample than a cone-shaped indenter.

Calvo and Hierro [36] described microhardness investigations of Hg–Mg alloys. The alloy specimens are sensitive to humidity, therefore they were protected from interaction with environment by an oil film. Distinct changes of the alloy hardness near the compositions $HgMg_2$, Hg_2Mg_5 and $HgMg_3$ enabled the authors to confirm the existence of these compounds in the system.

2.9 Calorimetry

When the enthalpy of alloy formation over an entire composition range is precisely measured in a calorimeter at a selected temperature, some slope change may be observed on the integral enthalpy vs. composition plot. Such slope changes were observed by Kubaschewski and Seith [37] who measured the heats of formation of the Hg–Na alloys in a solution calorimeter at 298K. Although the authors observed three clear breaks at the correct compound compositions, existence of four other compounds formed in this system was not manifested. It is now well known that such a procedure is only effective when a binary system has only a few intermetallics of discriminating stabilities and compositions.

Predel and Schwermann [34] recorded cooling and heating curves in DSC for the Cd–Hg alloys in the vicinity of the order–disorder transformation of the ω phase. The temperatures of the transitions recorded for several samples were scattered within only 2K. These results were in satisfactory agreement with the results obtained by other methods.

Another example of a calorimetric investigation leading to the determination of phase transition temperatures is a measurement of specific heat capacity (c_p) over a selected temperature range. Boiko et al. [38] measured the temperature dependency of c_p for Hg_2Cl_2 and Hg_2Br_2 using an adiabatic calorimeter. The $\alpha \leftrightarrow \beta$ phase transition for each compound was manifested by an abrupt increase of the specific heat, exactly at the phase transition. The experimental accuracy was ±0.1K and no hysteresis was observed. These temperatures were confirmed in the same study by soft mode Raman spectroscopy that may be applied to investigations of salt-like compounds.

2.10 Electrical Resistivity

Electrical resistance can be used to evaluate points on a solvus, limits of an intermediate phase or compound formation in a system. Pure metals and intermetallics (excluding such ionic compounds as AuCs) have relatively low resistance (high conductivity). An addition of other components or impurities to metals increases their resistivity, sometimes to a great extent. There are resistance meters available commercially or one can assemble a Wheatstone bridge where the sample serves as one of its arms. The essential part of a proper apparatus should be thermostatted if one wants to perform measurements at temperatures other than room temperature.

Two experimental approaches are possible: changing composition at a constant temperature or changing temperature for a constant composition. A change of slope of electrical resistivity vs. composition or vs. temperature curves in such experiments reflects either a phase formation at the selected temperature or a phase transition temperature at the fixed composition. The technique is relatively simple and its precision may reach 0.5%. It may be applied to a wide composition range from quite dilute alloys (e.g., 10^{-3} at.%) to pure substances.

Grube and Schaufler [39] complemented their DTA investigation of Li–Tl alloys with electrical resistivity measurements. The experiments were performed in an Ar atmosphere. Alloy bars of exactly known dimensions and compositions were prepared from Li–Tl melts. The solid alloys were conditioned 1 day at a temperature 50K lower than the solidus line and then slowly cooled to room temperature within 3 days. Then electrical resistance of the bars was measured at 10K intervals (2K intervals near phase transition). Figure 9.3 shows the resistance change with temperature for five alloy compositions near the formation of Li_2Tl. Each curve is composed of three segments. The straight lines drawn through the upper break points intersect at the Li_2Tl stoichiometry. Due to the susceptibility of the Li–Tl alloys to undercooling, the thermal arrests corresponding to the peritectoid formation of Li_2Tl at 654K were not accurate (scattered) so the electrical resistance measurements were preferred in this case.

2.11 Superconductivity

By measuring the superconducting transition temperatures of alloys, the solid phase solubility as well as an effect of phase ordering may be traced. Claeson et al. [40] measured

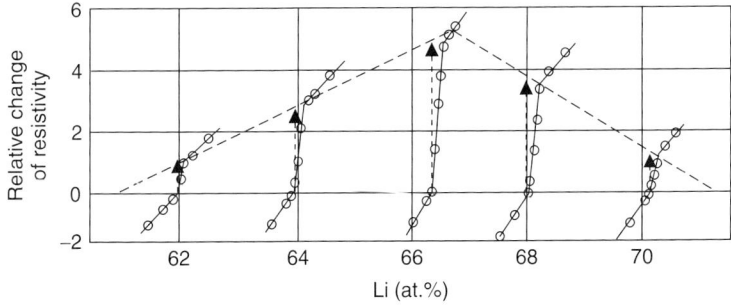

Figure 9.3 Change of the electrical resistance of Li–Tl alloys with increasing temperature in the vicinity of Li_2Tl formation at 654K [39].

the transition temperatures at 0.5–4.2K for many Cd–Hg alloys over the entire composition range. The specimens, encapsulated in quartz tubes were homogenized by annealing at six selected temperatures between 193K and 418K The superconductivity transitions above 1.3K were observed with the aid of an impedance bridge that compared the inductance of a coil containing the sample with that of an empty coil. Both coils were immersed in liquid He. When an alloy became superconducting, the resulting signal of an unbalance was recorded. The bridge was operated at 160 Hz and produced an alternative magnetic field of almost 1 gauss. Temperature was estimated by means of ^{4}He vapor pressure measurement. The measurements below 1.3K were performed by monitoring the self-conductance of a coil containing the alloy sample and submerged in liquid ^{3}He. The magnetic field at the sample was ~0.1 gauss peak to peak at frequency of 95 Hz. The temperature was then determined by measuring the vapor pressure of liquid ^{3}He. The resulting superconductivity transition temperatures were plotted against alloy composition and several break points corresponded to the phase boundaries of (Cd), (Cd) + ω, ω + (Hg) and (Hg) regions (at the temperature of annealing) were determined with a precision of several percentage.

2.12 Electromotive Force

Potentiometry with a galvanic cell allows the measurement of the activity of the most electroactive component of an alloy. The well-known Nernst equation is then applied but we need to appreciate its limitations in practical use. The electrochemical redox system must be reversible which means that reproducible potential readings should be constant within seconds or minutes so that the potential changes reflect the activity changes of the most electroactive element, thus fulfilling the Nernst equation. When concentrations of the redox couple ($<10^{-5}$ at.%) are comparable to the concentrations of impurities, then the potential readings may be fortuitous and un-informative. The measurements ought to be performed in an inert atmosphere to prevent elements such as oxygen from interacting with the base metals. Some easily passived and refractory metals (Al, Ti, Zr, Hf) and their alloys are very difficult to be studied using this method.

Every galvanic cell is comprised of two electrodes which should be kept at exactly the same and constant temperature. A working electrode detects the changes of activity when an alloy composition or the cell temperature is changed. The function of the reference electrode is to keep a constant potential. Both electrodes always have metallic contacts made up of refractory metals; however, solid, liquid or gaseous substances (either of metallic or non-metallic nature) may be involved in the electrode reactions. There are one or two electrolytes between both electrodes, separated by a diaphragm which conducts the ions. Depending on the kind of reaction and the temperature range, the electrolyte may be solid (ThO_2–Y_2O_3, ZrO_2–CaO, glasses, β-alumina, CaF_2, etc.) or liquid (molten KCl–LiCl, Li_2CO_3–Na_2CO_3, $CaCl_2$–CaH_2, $AlBr_3$ in toluene, $MgBr_2$ in pyridine, NaI in ethylamine, $CuSO_4$ in H_2O). Oxides of Th or Zr are electrolytes sensitive for O^{2-}; glasses and β-alumina for light alkali metal ions; CaF_2 for Ca^{2+} or F^-; carbonates for ionic C; hydrides for H^-; sulfides for S^{2-}; salt solutions in aqueous and non-aqueous solvents for specific metal ions. Molten salt electrolytes must be thoroughly dehydrated and used in a completely dry atmosphere.

Fundamentals of electromotive force (emf) measurements and many practical examples are presented elsewhere [41]. This technique is simple, inexpensive and sometimes the most precise (even $\pm\,0.1\%$) when performed in a proper cell. It may be applied over the composition range of 10^{-5} to 100 at.%. Due to the high resistivity of potentiometers, the measured values correspond to true equilibrium conditions.

The emf measurements were applied by Gasior et al. [42] to determine the composition limits of phases formed in the Li–Mg system. Alloys of various compositions were prepared by melting the metals in an inert atmosphere. The following cell was constructed: Li(l)/LiCl-LiF eutectic/Li–Mg(s). The apparatus was thermostatted within ± 0.5 K Constant values of the emf were measured by a potentiometer with a precision of ± 0.1 mV. The results obtained are presented in Fig. 9.4. The limits of the potential plateau define precisely the existence of the (Li) + (Mg) two-phase region.

The formation of intermetallic compounds is manifested in corresponding potential vs. composition curves as steps (the sharper, the more line-like compound it is). If several compounds of very similar compositions exist in a system, one must prepare many alloys over a narrow composition range. In such a case, the technique of coulometric titration outlined in the next paragraph is recommended.

2.13 Coulometric Titration

This method is a combination of electrochemical alloy preparation with subsequent measurements of its potential in the same electrolyte after the electrolysis is stopped. A coulometer with a potentiostat and a high impedance voltmeter are the necessary equipment, and the cell temperature needs to be precisely controlled. The high efficiency of the method comes from the very accurate composition control and the emf

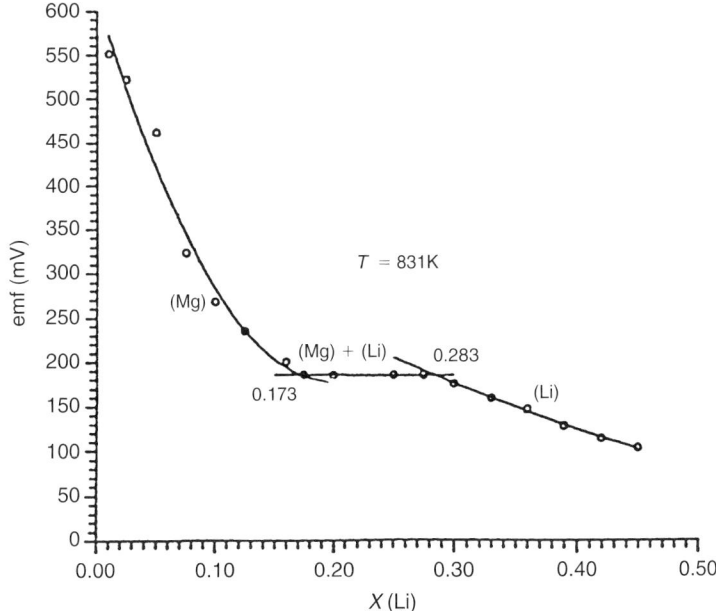

Figure 9.4 Determination of the (Mg) + (Li) phase boundaries from emf measurements [42].

measurements. One may easily increase or decrease the content of the more electroactive component by applying either cathodic or anodic current even in very small steps. Diffusion coefficients of the alloy components should be sufficiently high and the alloy sample should be quite thin so that equilibrium is quickly reached. Equilibrium is practically expressed by the emf stability with time. Either a constant current or constant potential electrolysis mode can be applied. The electric charge is recalculated to the change in alloy composition assuming electrolysis efficiency of 100% (the efficiency should be independently tested):

$$\Delta x_M = (\int_0^t i \, dt)/nF \tag{2}$$

Δx is the mole fraction difference of the alloy tested, i the current, n the number of electrons and F the Faraday constant.

Egan [43] used this method for phase identification and estimation of the homogenity range of Au–Ca compounds at 973K. The following cell was prepared:

$Ca_{0.1}Sn_{0.9}$ (l)/CaF_2 (s)/Au–Ca (s)

The Ca–Sn alloy served as the reference electrode with a well-defined activity of Ca. Because the Au–Ca electrode was composed of a small piece of thin Au foil with electrolytically deposited Ca, the current flow through the cell did not measurably disturb the composition of the much larger reference electrode. If the composition stability of a reference electrode cannot be matched, it will be necessary to use a third auxiliary electrode for the electrolysis step.

The titration curve obtained for the Au-rich side of the Au–Ca system is shown in Fig. 9.5. Due to the high precision of the Ca concentration control in the alloy,

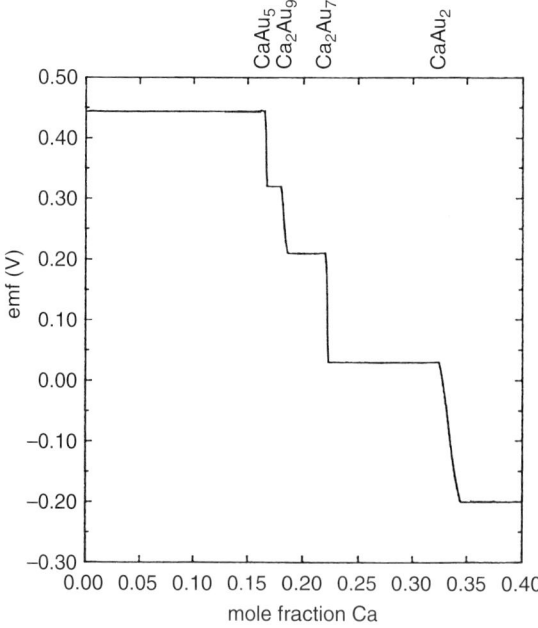

Figure 9.5 Coulometric titration curve of the Au–Ca solid alloys at 973K [43].

it was possible to estimate precisely the homogenity ranges of the compounds formed: Au_5Ca ($0.1655 \leq x_{Ca} \leq 0.1673$), Au_9Ca_2 ($0.1809 \leq x_{Ca} \leq 0.1835$), Au_7Ca_2 ($0.2214 \leq x_{Ca} \leq 0.2226$) and Au_2Ca ($0.3277 \leq x_{Ca} \leq 0.3410$). The homogenity ranges for Na_3Sb, $BiNa_3$ and BiK_3 were evaluated using the same procedure. A perfect agreement of these values with the predictions based on Wagner's model of semi-conducting alloys was observed – Carl Wagner was the inventor of the coulometric titration method.

2.14 Galvanic Polarization

Preparation and subsequent stripping of one component from an alloy by electrolysis can provide information on the stoichiometry of intermetallic compounds. Such electrolysis is reliable if the electrochemical process is reversible (diffusion controlled) and diffusion in a solid is sufficiently fast. To increase the diffusion flux, one can increase the temperature, if possible.

Lai [44] demonstrated oxidation Li-Si alloys at 673K in order to determine the stoichiometry of compounds formed in this system. A Ti fiber mat, impregnated with the alloy, served as the test electrode. Cu and Ni Felt-metal panels impregnated with pure Li served as a counter and reference electrode, respectively. The eutectic melt KCl–LiCl was used as the electrolyte. The test electrode was discharged against the counter electrode using constant current. The test electrode potential changes were recorded with time. The anodic discharge curve was characterized by several plateaus and jumps corresponding to the stoichiometries of the Li–Si intermetallics formed. The stoichiometrics were calculated from the amount of alloy in the test electrode and the quantity of Li oxidized during the discharge stage. The potential jumps observed at the stoichiometries Li_5Si, $Li_{4.1}Si$, $Li_{2.8}Si$ and Li_2Si are in rough agreement with the accepted Li–Si phase diagram. Generally, the Li content in the formulae was somewhat higher than expected, suggesting that a part of the electric charge was consumed by an unidentified side reaction. Investigators quite frequently assume 100% electrolysis efficiency which is not always true and must be tested by independent experiments.

An elegant modification of this method was proposed by Notin and Hertz [45] who prepared and investigated Ag–Ca alloys by electrolysis in a special cell presented in Fig. 9.6. Ca was deposited on sharp tips of both Ag (working) and Fe (reference) electrodes by electroreduction of the solid electrolyte (CaF_2 saturated with CaO). The Ca reacted with Ag to form intermetallics, but it did not react with Fe due to the practical inertness of Fe to Ca. Then potential–time curves for the anodic dissolution of Ca from the Ag–Ca alloy with respect to the reference electrode were recorded. The curves are similar to those shown in Fig. 9.5 for the Au–Ca system (the alloy compositions should be recalculated from the electrolysis times). Depending on the electrodeposition potential, depolarization curves of various extents were observed. The results of Notin and Hertz [45] show a fair agreement with the accepted Ag-rich side of the Ag–Ca phase diagram. The potential differences were successfully used for the determination of the thermodynamic properties of the system. Assuming that the alloy formation occurs in a limited volume and knowing the cathodic and anodic charges, the intermetallic formulae may then be more precisely calculated.

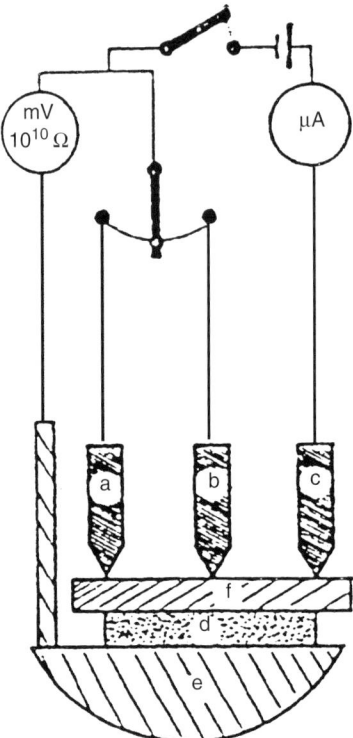

Figure 9.6 Schematic diagram of the galvanic cell: (a) pinpoint Fe electrode, giving the pure Ca/CaF$_2$ potential; (b) pinpoint Ag electrode, alloyed with Ca; (c) pinpoint Fe electrode, serving as auxiliary electrode; (d) large Ni/NiO electrode of constant potential (e) Ni support and (f) CaF$_2$ saturated with CaO, serving as the solid electrolyte [45].

2.15 Magnetic Susceptibility

Phase transitions may introduce abrupt changes of magnetic properties of an alloy when plotted as a function of temperature or composition. The susceptibility is measured by means of a vacuum torsion balance with an electromagnetic compensation. A specimen, placed in a magnetic field H with its gradient dH/dl, interacts with a force F with this field. Taking into account the sample susceptibility χ and its mass m, the following equation is obeyed:

$$F = m \, \chi \, H \, dH/dl \qquad (3)$$

The balancing mechanical force is recorded and the susceptibility calculated.

Figure 9.7 shows the reciprocal χ values plotted against temperature for the Fe-rich side of the Fe–P system investigated by Lorenz and Fabritius [46] using this method. The upper straight line reflects the temperature changes of χ for (γFe) and the bottom straight line for (αFe). The bell-like curves recorded at various P concentrations show the χ changes for (αFe)+(γFe) field of the phase diagram. The points at which these curves intercept the straight lines denote the boundary lines for the (γFe) and (γFe)+(αFe) fields. The method is precise within a few percent.

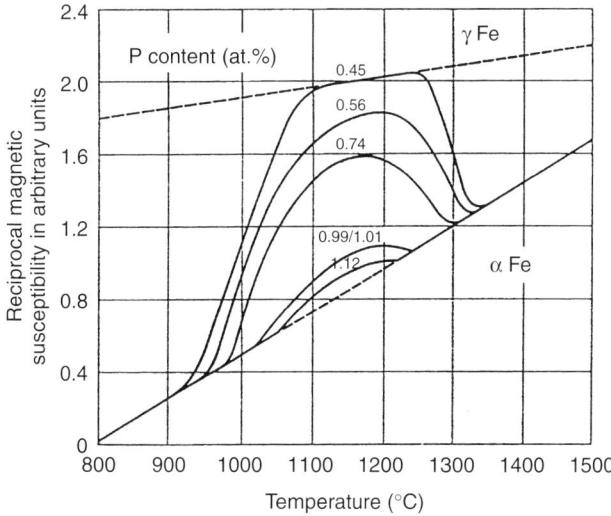

Figure 9.7 Variation of the reciprocal magnetic susceptibilities with temperature for various P contents in solid Fe [46].

2.16 Vapor Pressure

Similar to the emf method, measurements of vapor pressure give precise activity values of a more volatile component of an alloy. When vapor pressures of all components are comparable, the sum of their partial pressures is measured; then the method loses its sensitivity. Advanced techniques allow separation of the total pressure into individual components. The fundamentals of vapor pressure measurements are presented in Ref. [41, s 2] that also sketch the most typical schemes of the apparatus used. There are several experimental approaches within this method: static, dynamic and kinetic. The last technique is not used for easily vaporizing components. Each kind of approach may be further categorized according to the technique used.

In the static method, an alloy under investigation is placed in a closed and isolated container at a constant temperature. If one places the sample in a quartz vessel equipped with a membrane, the membrane position depends on the internal vapor pressure and it may be compensated from outside with a pressure measured with a manometer. This technique is recommended for systems containing aggressive volatile elements (S, Se, Te, P, As, I, Br) in the pressure range of 150–100 000 Pa and the temperature range of 470–800 K.

In similar parameter ranges, the pressure may be measured by an isoteniscope a little bit more precisely. Instead of the manometer membrane, an U-tube (filled with a liquid/melt) is directly bonded to a vessel. Position of the liquid in the U-tube is directly related to the equilibrium pressure inside the container. The liquid (oil, molten salt, molten alloy) is carefully selected according to the alloy properties and the temperature of investigation. It must neither dissolve nor react with the alloy components. Temperature of the sample and the liquid must be the same.

Jangg and Dörtbudak [47] measured the vapor pressure of the Hg–Pt alloys using the isoteniscope. The U-tube was filled with KNO_3–$LiNO_3$–$NaNO_3$ melt

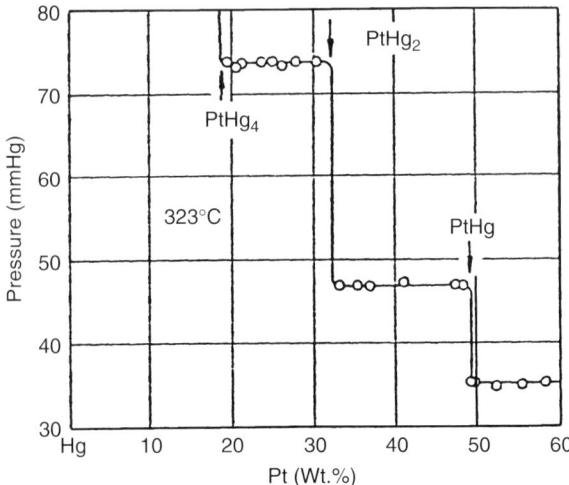

Figure 9.8 Vapor pressures of the Hg–Pt alloys at 596K [47].

(liquid already at ~400K) and the compensating pressure at the other side of the U-tube was measured. Before the experiments, the container was evacuated to a level of 10 Pa. The investigations were performed at fixed compositions of the amalgam. The temperature was gradually changed in the apparatus by an oven. In another set of experiments at selected temperatures, the amalgam composition was changed by controlled Hg distillation in a special device. An isotherm of the vapor pressure vs. Pt content is shown in Fig. 9.8. The pressure jumps correspond to the formation of Hg_4Pt, Hg_2Pt and $HgPt$ intermetallics in the system. Homogenity ranges of these compounds are manifested by the widths of the pressure jumps. The solid solubility of Hg in Pt (maximally 18.7 at.% Hg at 666K) was also determined from the Hg pressure bound. The partial pressure of Pt in this temperature range is incomparably smaller (10^{30} times) than the Hg vapor pressure and may be safely neglected.

The vapor pressure in a closed system may be selectively measured by absorption spectroscopy. One may trace a continuous spectrum or a certain line from an alloy component. The second variant is more sensitive. For instance, Franck [48] measured the Hg partial pressure over solid and liquid In-rich amalgams. The alloys were placed on the bottom of an absorption cell made of quartz. The cell was placed in the optical path of a Hg-vapor lamp and the light absorption, after transmission through the cell, was monitored by a photoelement. The apparatus was primarily calibrated with Hg. Results of the measurements are depicted in Fig. 9.9. The line drawn through the right plateau edges in Fig. 9.9(b) describes the (In) phase above 45°C and the border line of $HgIn_{3.5-15.7}$ phase below 100°C. However, the In-rich boundary of $HgIn_{3.5-15.7}$ and the narrow range of (In) + $HgIn_{3.5-15.7}$ two-phase region are not reflected in these vapor pressure measurements. The range between 88 at.% and pure In below 373K should be more precisely investigated, then another step with a narrow plateau corresponding to the two-phase region should appear. One may also use a radioactive isotope monitored by a suitable detector, for the selective measurement of the vapor pressure of a component in similar experiments.

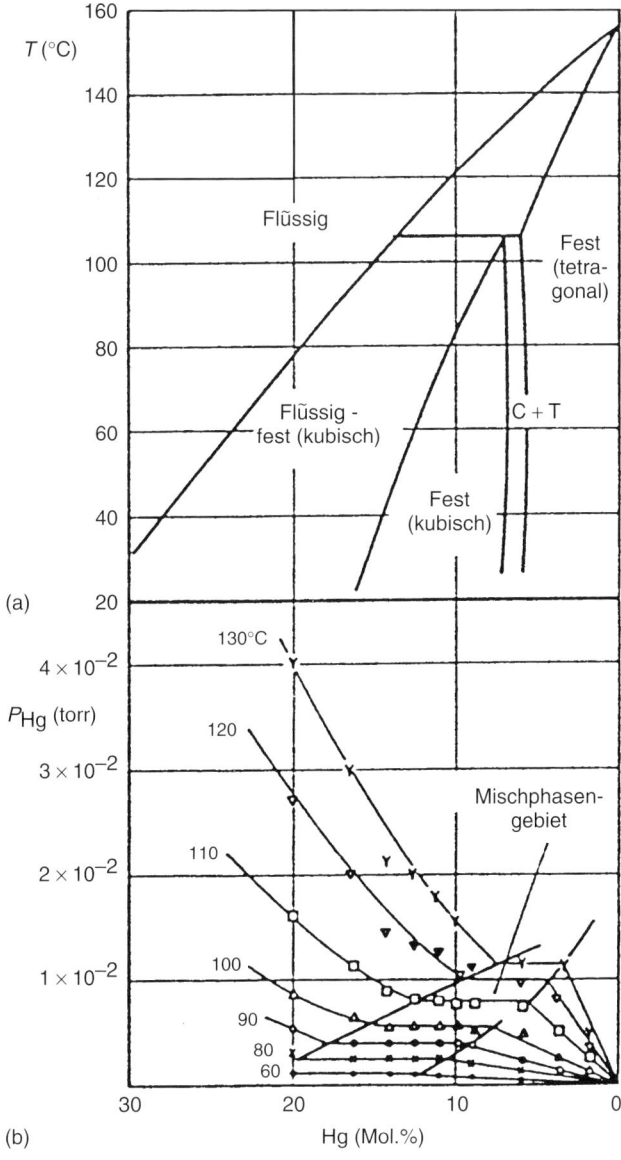

Figure 9.9 The In-rich part of the Hg–In system (a) and the measured vapor pressure isotherms (b) [48].

Vapor pressures over stable compounds are sensitive to polymorphic phase transitions. Using a method based on the substance evaporation rate, Piechotka and Kaldis [49] measured HgI_2 vapor pressures over the temperature range of 323–422 K. A Knudsen cell with an orifice diameter of 100 µm was used. Ion getter and sorption pumps were used to keep the pressure in the chamber at 3×10^{-7} Pa. During HgI_2 vaporization, the pressure in the chamber was about 10^{-6} Pa. A cell containing the sample was placed inside the chamber and aligned with an outside ionization source.

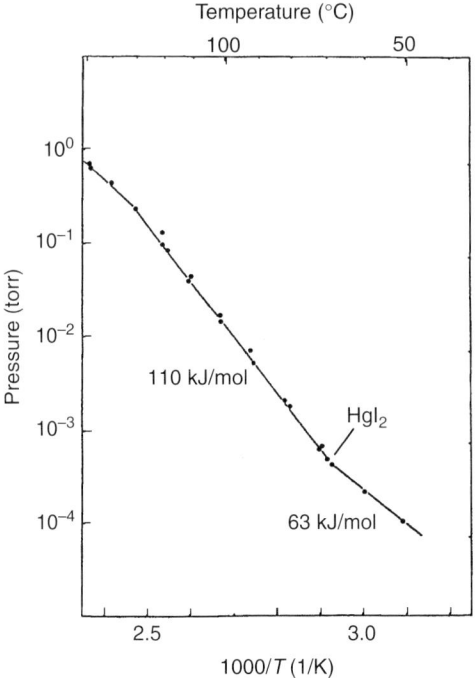

Figure 9.10 Vapor pressure curve of solid HgI_2 vs. reciprocal temperature [49].

The ionized vapor was analyzed using a quadrupole mass spectrometer with a cross-beam source and an off-axis photomultiplier. Decreasing the ionization energy below 18 eV, the HgI_2^+ ion became dominant. The vapor pressures determined for HgI_2 are reproduced in Fig. 9.10 as the logarithm of the pressure vs. reciprocal temperature. Two abrupt changes are observed at 340K and 400K The second one corresponds to the well known $\alpha \leftrightarrow \beta$ polymorphic phase transition of HgI_2. The first one most probably represents a transition of the metastable orange HgI_2 form into αHgI_2. This phenomenon is observable only at the surfaces of HgI_2 crystals.

These various techniques are quite flexible in application due to the wide ranges of temperature (200–2300K and pressure (10^{-6}–10^6 Pa) available as well as their good precision (up to ±0.1%). The phase boundaries found in this way could be precise within ±0.5K and ±0.5%, at the best.

2.17 Gaseous Thermal Extraction

This method is based on remelting of alloys containing a weakly bonded and volatile element. The principle is simple but to do it right, experience and knowledge about the kinetics and equilibria in the system are needed. The remelting may be carried out under vacuum or in an inert gas atmosphere. It may be replaced by a hot extraction technique (especially useful for H) when an alloy is heated for a long time below the melting temperature and all the liberated gas is collected. Amounts of the evolved gas are further analyzed by volumetric or gravimetric means as well as by gas chromatography, thermal conductance (useful for H) or gas spectrometry. Since H diffuses through many metals, the measurement apparatus used must maintain the hydrogen pressures.

This method is frequently used for determination of the solubility of gaseous elements in steels. An example of such an analysis in systems of our interest is the thermal decomposition of NaH [50]. Dissociation of NaH(s) to Na(l) and H(g) in a previously evacuated container is relatively fast when it is carried out at 650–675K. Since the reverse reaction to form solid NaH is very slow, a rapid cooling of the system makes a negligible error. The amount of the liberated H gas was measured by a McLeod gauge. It was confirmed by such a measurement that the H solubility in solid NaH is immeasurably small. The method may be applied in the composition range of $10^{-4}-50$ at.%. Its precision depends on the completeness of the degassing process and the gas analysis.

3 Solid–Liquid Equilibria

Solid–liquid equilibria primarily involve the liquidus and solidus. The liquidus line represents the final dissolution of a solid in a liquid and it is the end of melting. The beginning of melting is named the solidus line. The liquidus may be approached from the supersaturation side after precipitation of a solid phase from a liquid and the solidus can be approached from the undersaturation side after the necessary addition of an element to form a liquid.

3.1 Weight Loss of a Solid After Equilibration with a Liquid

In this method, a solid sample is introduced into a liquid metal pool and allowed to equilibrate in an inert container and atmosphere at a carefully controlled temperature to avoid recrystallization of the solute on the cooler container walls. With the initial mass or volume of the liquid and weight loss of the solid after equilibration precisely known, one can simply determine the solubility value by dividing the weight loss by the amount of the resulting solution. This technique, however, has several limitations. The solid sample must not be disintegrated into fragments during the experiment. The equilibrium between the solid and the liquid should be conclusively reached, that is, the apparent solubility should not increase with time. Stirring the solution or rotating the sample can accelerate the rate of dissolution. The solute must not form any compounds with the solvent and the solvent solubility in the solute metal should be negligible. Therefore, any weight gain of the solute specimen indicates a possible existence of the above-mentioned effect. An indirect form of interaction may also be possible if measurable amounts of the container (or a vessel liner) are dissolved in the solvent metal and subsequently form intermetallics between the solute and the container. It seems that the overrated values of Be solubility in liquid Li (which were obtained in a Fe vessel [51]) are associated with the formation of the Be–Fe intermetallics (most probably Be–Fe is hardly soluble in Li). This reaction makes an apparent additional Be dissolution into the Li phase to compensate for the Be loss caused by the reaction with Fe.

This method was successfully applied to the determination of the solubility of InSb, InAs and InP in liquid In and GaSb, GaAs and GaP in liquid Ga at 470–1670K [52].

The solute crystals were placed in the solvent metals contained in quartz tubes, flushed with H_2 and equilibrated. Then the semiconductor samples were freed from the Ga by dissolution in HCl and from In by dissolution in Hg with subsequent air blast. The clean samples were reweighted with a precision of 0.1 mg. Errors due to the solution of Ga or In in the solids are negligible because their stoichiometries are perfectly maintained. Concerning the possibility of P or As evaporation from the systems, trace deposits of InP and GaP were observed in other part of the apparatus; however, their weight changes were negligible compared to the amounts of the semiconductors dissolved. If, for example, 1 g of solute and 10 g of solvent were used, then the detection limit is 10^{-4} mass % but the precision of the experiment depends mainly on the skills of the investigators and it is typically better than $\pm 0.5\%$. The most convincing results are obtained in the composition range 10^{-2} to 10 at.%.

3.2 Chemical Analysis of a Separated Liquid

Phase separation is the most frequently used procedure in the determination of low solubilities of solids in liquid metals. A solvent is saturated with a solute for a sufficiently long time that the analysis results under the same conditions do not change systematically with time. During the system equilibration the container has to be kept under an inert gas atmosphere and in strictly isothermal conditions (to avoid any particulate formation). Then the saturated solution may be filtered, centrifuged or sampled depending on the integrity/quality of the solute after the equilibration. If the solute is disintegrated, only the filtering and centrifuging are effective ways of separation. The effectiveness of the centrifuging depends on the density difference of the solute and the solvent.

Typical apparatus used for the filtration of supersaturated liquid metals with hardly soluble transition metals is composed of a capsule from an essentially inert metal that is divided by an inert filter. The solute and solvent are placed in one part of the capsule and after sufficient equilibration time at constant temperature the capsule is reverted to allow the saturated liquid to be filtered into the second part of capsule. After cooling, the solidified liquid metal is chemically analyzed for the solute content. The procedure may be modified such that a sampler containing a filter is dipped into the liquid. Instead of filtering, Stanaway and Thompson [53] took liquid Na samples saturated with Mn or Co by piercing the thin wall of a Ni container with a Mo needle to make a 1.5 µm pore.

The method of centrifuging of a suspended solute from a supersaturated liquid has been seldom applied due to the complexity of the apparatus. The centrifuge machine ought to be carefully thermostatted and placed in an inert atmosphere or vacuum. For example, Ginell and Teitel [54] measured the solubilities of Ti, Zr, Nb, Ta, Fe and Ni in liquid potassium (K) using a special L-shaped capsule. The capsule containing the solute and K in its smaller part was welded and heated to 100°C above the equilibration temperature and then reduced to the test temperature while the centrifuge was continuously rotated. The rotation was at first accelerated and then slowly reduced to allow the decanted solution to drain into the larger collector part of the capsule. The solute content was chemically analyzed after cooling the saturated K to room temperature.

Sampling of a saturated liquid with a bucket dipped into the melt is possible if one can assume that the excess solute after equilibration is at the test vessel bottom.

Another possible simplification is the decantation of the upper part of the saturated liquid into a collector by rotating the test container.

Precision of these procedures is between ±0.5% and ±30% and the detectability level is 10^{-5} at.% under the optimal conditions. Many examples of the determination of transition metal solubilities in liquid alkali and alkaline earth metals are collected in critical reviews [2,6].

Atomic absorption spectroscopy or inductively coupled plasma atomic emission spectroscopy are nowadays more frequently used than conventional wet chemical analysis. Very good results are also obtained using radioisotopes [55] or activation analysis in case of light elements (H, B, C, N, O) [56]. Adsorption of the solute metal on another solid is a way of further improvement of the detection limit down to 10^{-7} at.% as was demonstrated by Caputi and Adamson [57] for the solubility determination of PuO_2 in liquid Na with the use of an adsorbing Ni foil. To increase the detectability level, an excess of low melting metal is frequently distilled off. If the distillation is complete, one may simply weigh the residue to find the solute content. For this method to work, the vapor pressure of the solute must be many orders of magnitude lower than that of the solvent.

Another way to analyze the solute in its saturated solution is based on electrochemical oxidation. This requires that the solute be electrochemically more active than the solvent. A difference in standard potentials of the alloy components higher than $200/n$ mV (n = number of exchanged electrons in a reversible process) is required. It is possible to perform a diffusion-controlled electrochemical oxidation of a liquid alloy under the following conditions to find the solute concentration x_M:

1. Chronoamperometry, at constant potential

$$x_M = (i/nFA)(\pi t/D_M)^{0.5} \tag{4}$$

where i is the electrolysis current, F the Faraday constant, A the electrode surface area, t the time of electrolysis and D_M the diffusion coefficient of the solute.

2. Chronopotentiometry, at constant current

$$x_M = (2i/nFA)(\tau/\pi D_M)^{0.5} \tag{5}$$

where τ is the electrolysis time between two potential jumps as shown in Fig. 9.11(a).

3. Chronovoltammetric, at potential linearly changing in time

$$x_M = i_p(RT)^{0.5}/(0.446 F^{1.5} n^{1.5} D_M^{0.5} A v^{0.5}) \tag{6}$$

where i_p is the peak current of the curve shown in Fig. 9.11(c) and v is the electrode polarization rate.

These equations are reasonable for planar electrodes with infinite diffusion fields for both reduced and oxidized forms. In the case of electrodes in the shape of a finite spherical drop or a thin film, more complex equations [58] should be applied. This reference also contains the fundamentals of various derivative techniques. The electroanalytical methods are able to detect concentrations above 10^{-5} at.% and in case of so-called step or pulse techniques (eliminating the capacitance current) even down to 10^{-8} at.%. Precision at optimal concentrations is never better than ±5%.

Ostapczuk and Kublik [59] used chronovoltammetric oxidation of Cu or Ag from their saturated amalgams to determine the solubilities of Cu or Ag in Hg.

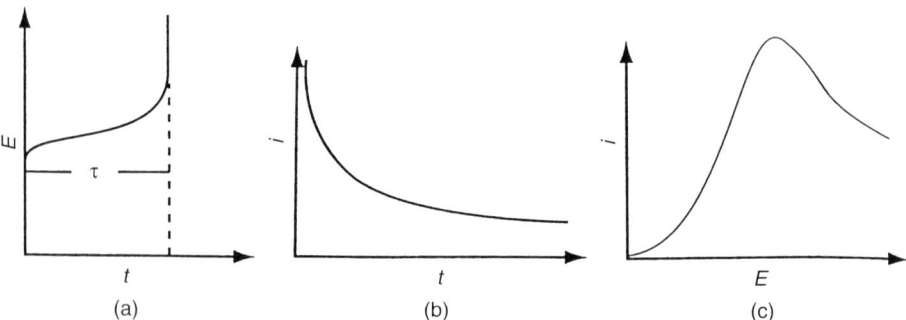

Figure 9.11 Schematic presentations of the chronopotentiometric (a), chronoamperometric (b) and chronovoltammetric (c) curves.

Before the electrolysis, liquid amalgams were separated from Cu_7Hg_6 or $Ag_{11}Hg_{15}$ intermetallics. If one neglects the separation, then during the electrolysis a part of intermetallic crystals may dissolve, thus overestimating the solubility – sometimes even more than an order of magnitude [60].

If a solute is electrochemically less active than the solvent metal, then a special method introduced by Guminski [61] may be used based on the precipitation of hardly soluble compounds with a third metal introduced into the amalgam under a cyclic potentiometric or chronovoltammetric conditions. For example, PdZn is precipitated in Hg when the solubility product of Pd and Zn (4.7×10^{-9} at.%^{2}) is exceeded. If Zn(II) ions are first reduced on saturated Pd amalgam and Hg hanging drop electrodes, and Zn is subsequently oxidized under chronopotentiometric conditions, the difference between Zn oxidation times (Δt) on both electrodes corresponds to the Pd amount being initially dissolved in the electrode drop. If one knows the current of electrolysis (i), the Pd concentration (x_{Pd}) may be easily calculated from:

$$x_{Pd} = i(\Delta t)/2Fv \qquad (7)$$

where v is the volume of both the Hg and Pd amalgam drops.

3.3 Chemical Analysis of Quenched Samples

This paragraph begins description of methods which do not involve phase separation. The equilibrium between a solid and a liquid metal at a high temperature may be fixed by a rapid quench. The system has to be heated gradually and kept at a constant temperature to reach equilibrium. A sudden cooling does not allow the system to segregate and crystallize the excess solute. If one cuts out an internal part of the solidified liquid and performs a chemical analysis for the solute content, one obtains the solubility result. It is not recommended to analyze the whole frozen liquid or its upper or lower parts where the solute crystals may eventually accumulate depending on density.

Bychkov et al. [62] analyzed the solubility of Be, U and several transition metals in liquid Li using this method. Crucibles made of the test metals were equilibrated with Li at 973–1473K and then cooled down to room temperature in less than a minute. The solute metals were determined spectrophotometrically (as colored complex compounds) after a chemical dissolution. The detection limit was 10^{-4} at.% and the precision of the analysis ±10%, of the nominal value at the best.

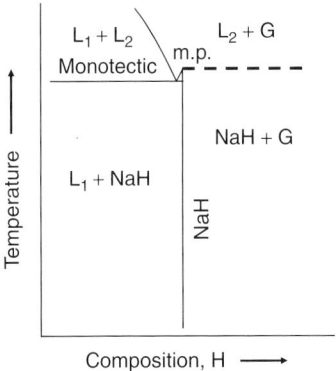

Figure 9.12 Schematic presentation of the central part of the Na–H phase diagram.

3.4 Thermal Analysis

The DTA and DSC are frequently used for the determination of a liquidus line. The relatively high vapor pressure of easy-boiling elements requires the use of tightly closed containers that can withstand elevated pressures and high temperatures. When an alloy contains elements of very different densities (e.g., Au and Rb), their mixing is only possible in a common TA but not in DTA. Another advantage of TA is its ability to accurately determine phase transitions with temperatures less than 25K apart, as demonstrated by Gaweł [63] during investigations of quasi-binary metal telluride systems. Errors in the liquidus temperature estimations are generally larger than those for the eutectic reactions. TA is effective in the concentration range of 1–99 at.%. In case of lower concentrations, a related technique named cryometry may be applied. Then temperatures of melting are measured with high precision by differential thermometers.

As pointed out in the case of the solid–solid equilibria, the free volume over an alloy containing volatile components must be minimized. A pressure developed at an elevated temperature enriches the specimen in the less volatile component and changes the overall composition of the alloy; this can cause misinterpretation of the results. An example is the determination of melting points (m.p.) of alkali metal hydrides. Several measurements were performed [64], but only the determination by Skuratov et al. [65] seems to be faultless. Figure 9.12 shows schematically the central part of the Na–H phase diagram near the m.p. of NaH. It was established that H vapor pressure over NaH reaches 0.101 MPa at 692K, 215K below the m.p. of NaH. Therefore, if one heats even perfectly pure NaH at a normal pressure it is gradually decomposed to gaseous H and liquid Na. H escapes from the condensed system and the thermal arrest observed by DTA reflects in fact the monotectic Na–NaH reaction but not the m.p. of NaH. A reliable result of the melting is obtained when the test capsule is additionally filled with gaseous H at a pressure of 10.7 MPa. This pressure is equal to the decomposition pressure of NaH at its m.p. NaH is virtually a line compound of ionic-like nature with negligibly small solubility of either H or Na. The difference between the m.p. of NaH and the temperature of the Na–NaH monotectic reaction amounts to 2–3K. One may comment that this is within the experimental

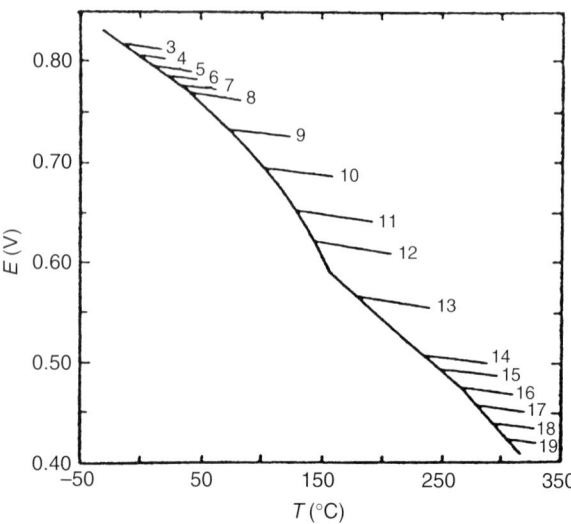

Figure 9.13 The emf of Na/Na$^+$ in a β-alumina/Hg–Na cell vs. temperature for fixed Na compositions $0.0350 \leqslant x_{Na} \leqslant 0.2500$ (denoted by increasing numbers 3–19) [66].

error, but this difference is an important fact and ought not to be neglected. Further increase of the pressure to 20.7 MPa increases the m.p. of NaH another 2K. Similar behavior was also observed for KH, RbH and CsH.

3.5 Electromotive Force

This method is frequently used for liquidus determination because it is relatively simple and precise. Moreover, it may be applied over a very wide concentration range from 10^{-5} at.% to almost pure solute. The related experiments are carried out at either constant composition (changing temperature in steps in a certain range) or constant temperature (changing a sample composition). A possible change of alloy composition due to evaporation of the more volatile component becomes then relatively small.

Figure 9.13 presents results of emf experiments performed by Sun and Cao [66] in the cell: Na/β-alumina/Na–Hg. The solid β-alumina is a specific conductor for Na$^+$ ions. Every experiment was performed at a fixed Na amalgam composition (the increasing Na contents are denoted by numbers) prepared with an accuracy of $\pm 0.2\%$ according to coulometric electrolysis. Temperature was changed in steps and the points of abrupt change on the potential–temperature curves corresponded to the Hg–Na liquidus temperatures for the selected samples with an accuracy of $\pm (0.1$–$0.8)$K.

If one performs emf measurements at a constant temperature while changing the alloy composition, one observes a linear dependence of the potential vs. the logarithm of the concentration. When the concentration surpasses the solubility limit, the potential becomes independent of the alloy concentration as shown in Fig. 9.14 for the measurements of Tl amalgams. However, comparing Fig. 9.14 to the Hg–Tl phase diagram, one may expect some small plateaus related to the (αTl) and (βTl) two-phase equilibria on the Tl-rich side. Such changes may be lower than the detectability with emf measurements (± 0.1 mV).

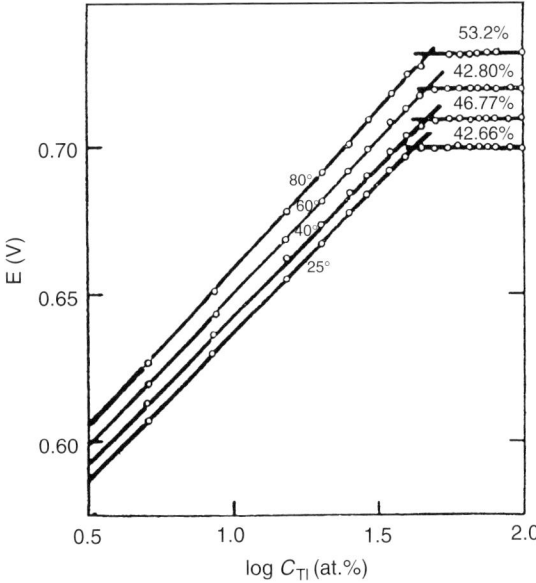

Figure 9.14 The emf of Hg–Tl/0.1 mol dm^{-3} TlClO$_3$, 0.9 mol dm^{-3} NaClO$_4$/1 mol dm^{-3} NaCl, Hg$_2$Cl$_2$, Hg cell vs. Tl content in the amalgam at temperatures 298–353 K [s 3].

3.6 Electrical Resistivity

This method has often been used for the determination of the liquidus of the alkali metal-rich sides of phase diagrams with moderately soluble metallic and non-metallic elements. The typical scheme of apparatus is shown in Fig. 9.15 [67]. The resistivity was measured with a capillary glass tube that is a part of a stainless steel loop filled with a liquid metal. The capillary serves as a part of the Wheatstone bridge. The liquid metal circulation was forced by pumps. The resistivities of the solutions were investigated under either isothermal (resistivity vs. composition) or isocompositional (resistivity vs. temperature) conditions. Results of such investigations for the Al–Li system are presented in Fig. 9.16 [67]. The saturating concentrations of Al are clearly indicated by the break points on the resistivity vs. temperature curves.

When a system is investigated under isothermal conditions, the resistivity of liquid solutions increases linearly with addition of the solute up to its saturation point; and further addition of the solute no longer changes the resistivity. The technique has a precision of ±0.5%. It is only applicable to systems for which the solute and solvent have relatively different resistivity coefficients. Other techniques with resistivity measurements are outlined by Ref. [22].

3.7 Anodic Oxidation

When a solute is electrochemically more active than a solvent, one may perform a chronoamperometric oxidation of the solute at a potential more positive than that of the solute but more negative than that of the solvent. When such a process is carried out in a semi-infinite diffusion field (i.e., a large volume of an alloy serves as

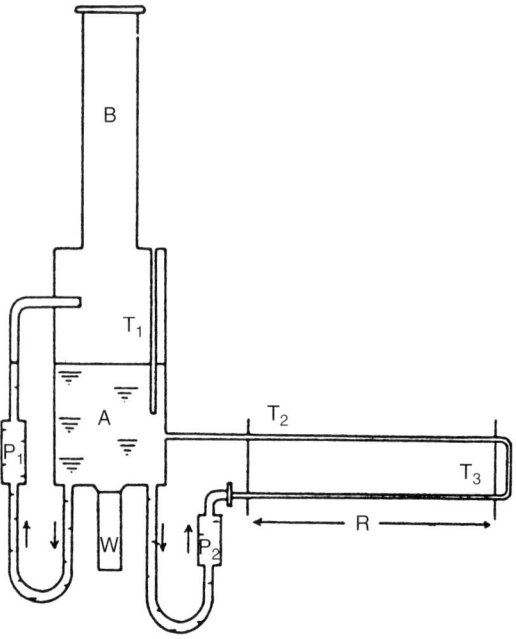

Figure 9.15 Scheme of electrical resistivity measurements apparatus; A: liquid metal reservoir; P_1, P_2: pumps; R: resistivity measurement capillary; T_1, T_2, T_3: thermocouples [67].

the anode), the changes of current with time are dictated by Eq. (4). When the diffusion field in a liquid alloy is limited, or the shape of electrode is not flat, or the alloy is mixed at a constant velocity, then proper equations suitable to such conditions should be selected from the book [58]. The presence of solute crystals in a liquid is manifested by a plateau (independent of current and time). The wider the plateau, the more crystals are present in the heterogeneous alloy. The solubility may then be calculated from the integration of the field under the current–time curve when all crystals are dissolved. It is therefore important to reach a true equilibrium in the heterogeneous alloy before the oxidation step. Presence of any excess solute in the form of microcrystals would result in an overestimation of the solubility result.

An example plot for the determination of Pr solubility in liquid Hg is shown in Fig. 9.17 [68]. The authors applied chronoamperometric oxidation of Pr amalgams of increasing concentration and observed that the Pr oxidation limiting current decreased smoothly with time for the unsaturated amalgam (curve 1) and showed a current plateau for the supersaturated amalgam (curve 2). The solubility estimation from the oxidation of the homogeneous amalgam (after the plateau) is precise $\pm 5\%$ (at the best) and may be applied to diluted alloys ($10^{-3} - 10^{-1}$ at.%).

Shalaevskaya et al. [69] proposed oxidation of a solute covered by a thin layer of a solvent metal. The authors derived an equation relating I, the oxidation current in steady-state conditions, n, the number of electrons exchanged in the electrode process, D_M, the diffusion coefficient of the solute in the liquid solvent, A, the electrode

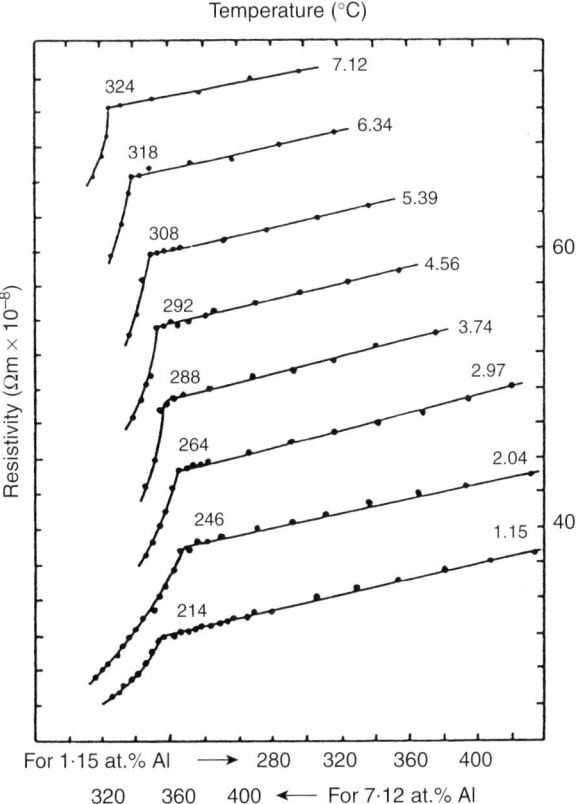

Figure 9.16 Electrical resistivity vs. temperature curves for solutions of Al in liquid Li. Each curve is transposed by 40K from its neighbor [s 4].

area, d, the thickness of the solvent layer, k, the heterogeneous dissolution rate constant and x_M, the saturating solute concentration

$$nFA/I = 1/(kx_M) + d/(D_M x_M) \qquad (8)$$

where F = Faraday's constant.

The steady-state currents were recorded at the various layer thicknesses and the solubilities were calculated from the slope of Eq. (8) using known value of D_M. Applying this method, Shalaevskaya et al. [69] determined the solubilities of Al, Cu and Sn in liquid Hg at several temperatures. The results for Al were slightly too low and those for Cu and Sn too high in comparison with selected solubility data from the literature [60]. The precision of this method was estimated to be ±5% and the range of applicability is $10^{-3}-1$ at.%.

3.8 Magnetic Susceptibility

Zhakupov et al. [70] measured the magnetic susceptibilities of several Ga–Na alloys encapsulated in Ta tubes. The susceptibility was recorded while the alloy sample

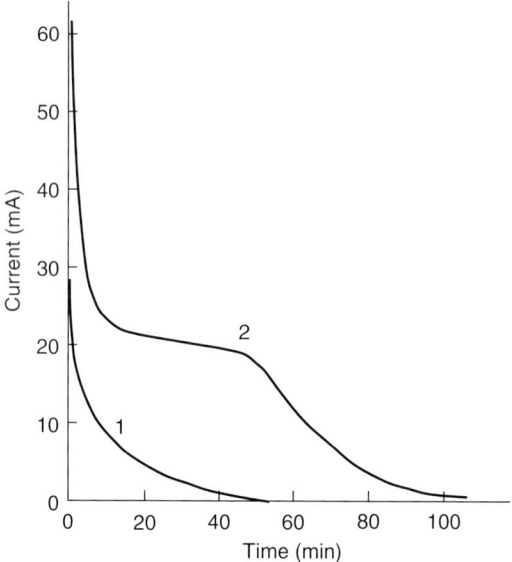

Figure 9.17 Current vs. time curves for the anodic decomposition of Pr amalgam at −0.1 V against saturated calomel electrode at 343K. The Pr amalgam concentration: (1) 0.014 (still homogeneous) and (2) 0.083 at.% (heterogeneous) [68].

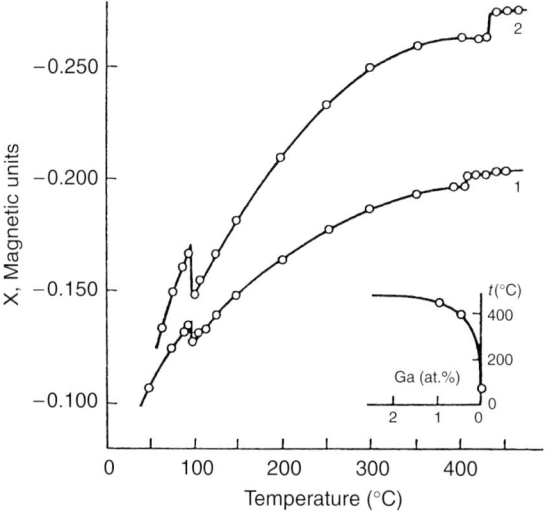

Figure 9.18 Temperature dependence of magnetic susceptibility of Ga–Na alloys with: (1) 0.47 at.% Ga and (2) 0.92 at.% Ga [70].

was heated from room temperature to 1073K in steps of 3–50K. The results for two alloy compositions are shown in Fig. 9.18. The first susceptibility jumps at 371K (98°C) are due to the Na melting and the second steps correspond to the liquidus temperatures. The method is precise within a few K.

It seemed that solubility of Co and Fe in liquid Hg could be easily determined using this method. Unfortunately, due to the easy formation of colloid-like solutions in these amalgams obtained electrochemically by reduction of Co(II) and Fe(II), the solubility values for these metals as measured with magnetic susceptibility were strongly overrated ($\sim 10^6$ times) [71,72].

3.9 Densitometry

Most often crystals of a solute are characterized by a different density than the saturated solution in a liquid metal. If one determines precisely the densities of several solutions below and above the saturation level, one observes a slope change on a density vs. alloy composition curve. The method may be applied in the composition range $10^{-2}-100$ at.% and the precision achieved may be as good as $\pm 0.1\%$, as demonstrated in the case of Zn solubility in liquid Hg by Crenshaw [73]. Densities of supersaturated and diluted Zn amalgams of exactly known concentrations were measured using a glass pycnometer at precisely regulated temperatures (± 0.02K). The Zn saturation level was determined from a slope change on the density–composition plot. The precision achieved was $\pm 0.3\%$, a quite remarkable result. Other techniques based on density measurement are also possible [22].

3.10 Enthalpy of Dilution

When a solid is in equilibrium with a saturated liquid alloy, an addition of a liquid solvent gradually dissolves the solute, resulting in a change of enthalpy of the system. If one performs a titration of the solid–liquid mixture by additions of small amounts of the solvent and measures the accompanying heat effect in a calorimeter, one obtains a calorimetric titration curve where a point of slope change on the enthalpy–composition plot corresponds to the saturation level. Figure 9.19 shows an example of the results obtained for such titrations in the Hg–Tl system [74]. The heterogeneous Tl amalgams were prepared by mixing weighed amounts of the metals. As one may see from Fig. 9.19, the enthalpic effect was, depending on the concentration range, endothermic or exothermic. The extrapolated experimental lines intersect at ~ 43 at.% Tl, in agreement with the assessed Hg–Tl liquidus at 298K [60]. In the case of the Cd–Hg liquidus, the titration effect was endothermic for both undersaturated and supersaturated amalgams, and independent of composition for the Cd homogeneous amalgam.

Precision of such measurements is never better than $\pm 0.5\%$. The technique may be used for metallic systems in which the solubility is higher than 10^{-2} at.% because the enthalpic effect needs to be precisely measured. The dissolution rate of solute ought to be sufficiently high to be reliably traced by a calorimeter.

3.11 Kinetics of Alloy Decomposition or Formation

If a heterogeneous system contains a solute of higher chemical reactivity than the solvent, one may select such oxidizing agent that reacts effectively with only the solute. Kinetics of such a reaction may be recorded by a different instrument. When the reaction is simple and the oxidant is used in excess, the reaction may be treated as pseudo-first order. Its half-decomposition time does not depend on the solute concentration up to the saturation level but then increases with further increase of

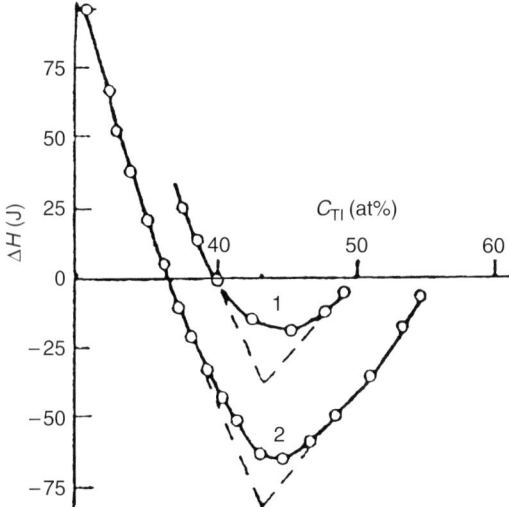

Figure 9.19 Calorimetric titration curves of heterogeneous Tl amalgams at 298K; initial Tl concentrations: (1) 50.0 at.% Tl and (2) 56.0 at.% Tl [74].

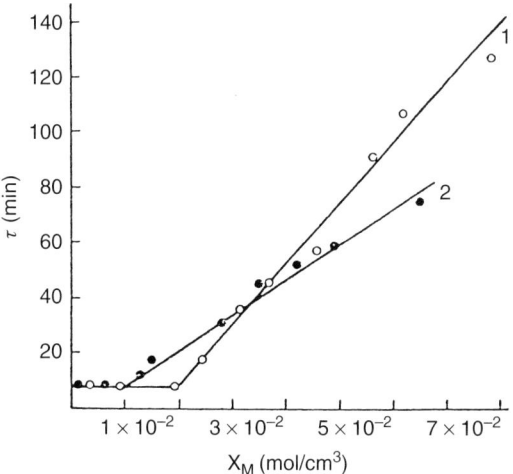

Figure 9.20 Dependencies of half-decomposition times for oxidation of Pr (1) and Nd (2) amalgams on their initial concentrations [75].

the solute concentration in the heterogeneous alloy. The phases in the heterogeneous alloy should dissolve more slowly than the rate of the oxidation reaction.

Using this method, Usenova et al. [75] determined the solubilities of Ce, Pr and Nd in liquid Hg at 298K. The amalgams were prepared by reduction of Ce(III), Pr(III) and Nd(III) (being labeled with radioisotopes) with Na amalgam. Then the lanthanides in the amalgams were oxidized by contacting them with an acetate buffer of high capacity and pH = 3. Periodically collected samples of the buffer solutions were analyzed for the radioisotope contents of the lanthanides. The kinetic curves obtained were analyzed and the half-decomposition time $\tau_{1/2}$ calculated. Figure 9.20 plots $\tau_{1/2}$ values against the initial amalgam concentration for Pr and Nd. The precision of the determination was no better than ±10% and probably cannot be much improved. Rather dilute metallic solutions (10^{-4}–10^{-1} at.%) may be analyzed this way.

Kinetics of reaction between a solute and a solvent may be used to estimate the solubility as shown by Cheburkov and Rozanov [76]. Reactions of O or N with liquid Li were performed in a glass apparatus. Progress of the reactions was monitored in real time under isothermal conditions by manometric measurements of the reacting gas pressure. The kinetic curves were analyzed and the saturation levels of O and N in liquid Li were found from break points when a logarithmic dependence of the kinetic curve was transformed into a linear one. The results so obtained overestimated the solubilities because the procedure had an uncertainty about the degree of oversaturation of the melt with Li_2O or Li_3N before their crystallization started. Likewise, formation of a supersaturated alloy under electrochemical conditions, when solute ions are reduced on a liquid solvent electrode, was used several times to determine the solubility of elements in Hg. However, the results were seldom satisfactory [60].

3.12 Diffusion Coefficient

The diffusion coefficient of a solute by definition is related to diffusion in a homogeneous alloy. In dilute liquid alloys (<0.1 at.%), the coefficients essentially do not depend on the solute concentrations. A supersaturation of such an alloy with the presence of a second phase results in an instant decrease of the diffusion coefficient with increasing solute content. Zakharova et al. [77] performed a chronoamperometric oxidation of Cu from its unsaturated and supersaturated amalgams at a constant temperature. The equilibration time before the Cu oxidation was too short in these experiments. If the precipitation of Cu_7Hg_6 microcrystals did not complete, then the abrupt drop of the diffusion coefficient was observed at a concentration higher than expected. The overestimation of Cu solubility was found to be about 30% higher. Precision of this method is no better than ±5% as is the case for most electroanalytical methods.

3.13 Vapor Pressure

Partial vapor pressure over a homogeneous liquid and heterogeneous solid–liquid alloy reflects the solute activity in both phases. The saturation of the liquid with the solute is manifested by constant vapor pressure upon increase of the solute content. The break point on the vapor pressure–composition plot corresponds to the liquidus point. The experiments are typically performed under isothermal conditions. Their accuracy and sensitivity depend on the system under investigation. The technique is quite popular; its precision is sometimes better than ±1% and it may be applied to rather concentrated alloys (0.1–100 at.%). Figure 9.9 shows the results by Franck [48] on the In-rich amalgams. The left-hand break points on the vapor pressure isotherms correctly reflect the shape of the liquidus line in this composition and temperature range.

3.14 X-Ray Absorption Spectrometry

This method may be applied to light solvent metals that do not absorb X-rays (such as Li) and solute metals that effectively absorb X-rays (transition and heavy metals).

The detectability level (on the order of 10^{-4} at.%) depends on the absorption coefficient and the precision of the method is of a few percent. The technique was elaborated by Beskorovainyi et al. [78] who precisely determined the solubilities of V, Nb, Cr, Mo, Mn, Fe, Co, Ni in liquid Li. These tests were decisive for such difficult systems. Small diameter tubes of the solute metals were filled with Li; the latter was kept inside the tube by means of the surface tension. The tube was placed horizontally in a sealed capsule having a pair of Be blocks placed geometrically on the tube axis. The capsule, filled with He, was heated and equilibrated at selected temperatures. A beam of X-rays passing through the Li containing the dissolved solute was analyzed by means of a Soller spectrometer. The X-ray intensity was measured from both sides of the K-absorption edge of the solute. Saturation concentration of the solute dissolved x_M was calculated from the following equation:

$$x_M = [\mu_{i1} - \mu_{i2}(1-p)]^{-1} \ln[(I_2)^{-1-p} (I_1)^{-1} I_{1o} (I_{2o})^{p-1}] \tag{9}$$

where $p = (\mu_{i1} - \mu_{i2})/\mu_{i2}$ (p values are tabulated in Ref. [78]), μ_{i1} and μ_{i2} are the mass coefficients of X-ray absorption of a solute at two different wavelengths, I_{1o}, I_{2o} and I_1, I_2 are the intensities of the X-rays at two wavelengths which passed through the sample before and after the solute equilibration.

3.15 Viscosity

The appearance of second phase crystals in an oversaturated melt makes a distinct viscosity change and thus may be utilized for an estimation of the liquidus. Savinov et al. [79] measured the attenuated torsional vibrations of a cup and determined the solubility of H in liquid K–Na alloy. They cooled an alloy at a constant rate (0.5K/min) and measured the viscosity. The ratio of the attenuation decrements of the vibration for the alloy containing H to that of the pure alloy was analyzed as a function of temperature. As shown in Fig. 9.21, the abrupt increase of this ratio indicates the saturation temperature of the alloy sample. The results obtained explicitly overestimated the liquidus, probably due either to a side reaction of the melts with the quartz container or to insufficient equilibration time given the alloys before the measurements. The method seems to be applicable to moderate concentrations of solute, and its precision is no better than 10%.

3.16 Optical Reflectivity

Optical reflectivity of surfaces of liquid metallic solutions shows a distinct change when second phase crystals are precipitated from the supersaturated solutions. This allows determination of the liquidus temperature during gradual cooling of a molten alloy. The method was primarily designed to determine precisely the solubility of transition metals in liquid alkali metals [80]. Unfortunately, the related details of this technique have not yet been published.

3.17 Corrosion Tests

Very low solubilities of transition metals in liquid alkali metals were successfully estimated from the extent of transition metal corrosion when dipped in the liquid

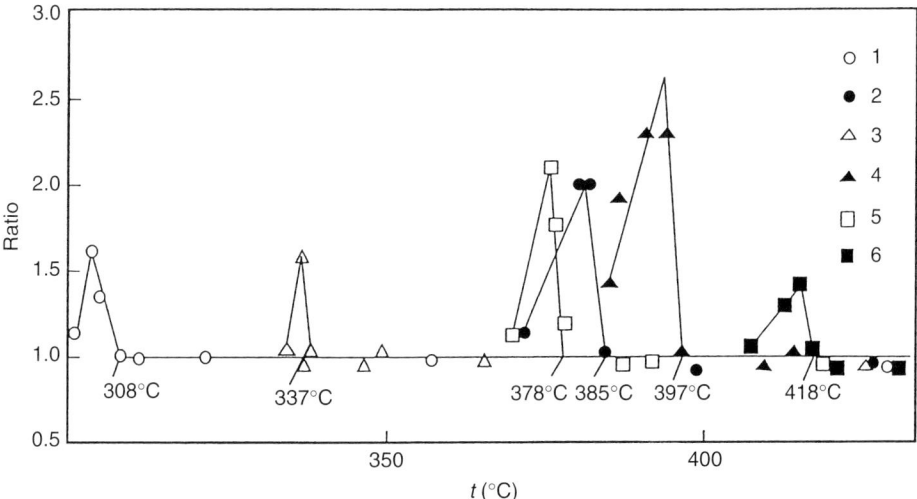

Figure 9.21 Temperature dependence of the ratio of the vibrational attenuation decrements for K–Na alloys containing various H concentrations: (1) 0.0015, (2) 0.0099, (3) 0.0034, (4) 0.0152, (5) 0.0068 and (6) 0.0172 mass % H, respectively [79].

metals. Sample sheets of a transition metal are placed in a metallic loop filled with the turbulently circulating solvent metal. The liquid metal is circulated by electromagnetic pumps, and a temperature gradient in the loop is kept during the experiments. It is very important that the corrosion is simply a dissolution process and that no intermetallics between the solute and the solvent are formed. Under such conditions, the corrosion rate Δm is dependent on many hydrodynamic factors:

$$\Delta m = 0.0481(x_s - x_b) D_M d^{-1}(Re)^{0.75}(Sc)^{0.42} \qquad (10)$$

where x_s and x_b are the saturating and bulk solute concentrations, D_M is the solute diffusion coefficient, d the diameter of the test tube, $Re = vd\upsilon^{-1}$ the Reynolds number, $Sc = D_M^{-1}\upsilon$ the Schmidt number, where v the flow velocity and υ the kinematic viscosity of the solvent. The x_b value is typically much lower than x_s under the experimental conditions, and the former value may be neglected. The dissolution corrosion rate is generally determined as the sample weight loss and it is translated into a decrease of the sample thickness. The diffusion coefficient, if not known, should be calculated from the Sutherland–Einstein equation:

$$D_M = kT(4\pi\mu r_M)^{-1} \qquad (11)$$

where k is the Boltzmann constant, T the absolute temperature, μ the viscosity of the liquid metal and r_M the radius of the solute atom.

Although precision of this procedure is no better than ±10%, the solubility results calculated from the corrosion of V in liquid Li or Na [81] showed quite good agreement with the best available experimental data as well as theoretically predicted values from the cellular model of Miedema et al. [s 5].

3.18 Motion of Liquid Metal Inclusions in Ionic Crystals

This particular method was presented by Geguzin et al. [82] who measured the motion of liquid Li droplets in a LiF monocrystal in a very strong temperature gradient. The Li inclusions in LiF were formed as a result of electric pulse punctures. After the specimen was annealed for 12 h at 973 K, a temperature gradient of 1000 K/cm was applied to the opposite sides of the monocrystal and the speed of motion of the Li droplets was optically monitored. The motion of the Li droplets is due to the LiF dissolution at the hotter side of the cavity and the LiF crystallization at the cooler side. It was assumed that the liquid Li is saturated with LiF at the mean gradient temperature. The corresponding F solubility in Li (x_F) was calculated from the equation:

$$x_F = RT^2 v/(\Delta H_F D_F \Delta T) \tag{12}$$

where R is the gas constant, T temperature, v the speed of the inclusion motion, ΔH_F the partial enthalpy of LiF dissolution in liquid Li, D_F the diffusion coefficient of F in liquid Li and ΔT the temperature gradient along the crystal. The rate of droplet motion did not depend on the dimension of Li inclusion if the latter was between 30 and 70 μm. The ΔH_F and D_F were determined by separate experiments. Although precision of this method is no better than ±10%, the result obtained is of the proper order of magnitude. The method seems to be applicable for concentrations <0.1 at.% if the recrystallization process is sufficiently fast (diffusion controlled).

4 LIQUID–LIQUID EQUILIBRIA

4.1 Chemical Analysis of Separated Liquids

When two liquid alloys of limited miscibility are equilibrated for an adequately long time at a selected temperature, one may determine two points on the immiscibility gap by chemical analysis of each liquid phase. Such an equilibrium may also be frozen by an abrupt quenching.

Smith [83] applied the method to study mutual miscibility within the K–Li system. Both metals were loaded into a Mo crucible and placed inside a sealed container filled with Ar. The system was heated to ~770 K and then cooled down in steps. Samples of the mutually saturated liquid Li (the upper layer) and K (the lower layer) were taken with the use of stainless steel filtering tubes after 4 h of equilibration at a selected temperature. A flame photometric technique was used to analyze the contents of Li and K in the alloy samples. The results obtained were precise within 10%; however, this might be much improved if the sampling and analysis were performed more accurately. For example, Yatsenko and Druzhinina [84] achieved a precision of ±1.5% during analysis of the saturated Ga amalgam by its anodic oxidation with detection of the end point by potentiometry. Ga and Hg show limited miscibility at room temperature.

4.2 Thermal Analysis

Cetin and Ross [85] determined the liquid miscibility gap of the Na–Zn system using TA. The experiments were carried out in a Mo–W container sealed with a Ta lid. The container with the Na and Zn inside was placed in a steel vessel. The sealed system was filled with Ar and subsequently heated to ~1173K. Simple cooling curves were recorded with the use of a calibrated chromel/alumel thermocouple. The thermal arrests corresponding to the points on the miscibility gap boundary were not very strong, but they were in agreement with earlier data on the Zn-rich side; and they were additionally confirmed by simple modeling. There are, however, reservations with respect to this experimental procedure. Serious composition changes may result both from the lack of alloy mixing (the components have very different densities) and from the high temperature to which the alloys were heated (near the boiling points of the components, Na: 1156K and Zn: 1180K while Ar still circulated over the alloys). In fact the results for the Na-rich side of the diagram were found to have solubility values for Zn in Na too high in comparison with the more careful studies of this system.

4.3 Calorimetry

Gaune-Escard and Bros [86] measured the enthalpies of mixing of many Ga–Hg liquid alloys at several temperatures (313–478K) in the range of the limited miscibility of both metals. The authors choose the break-off ampoule technique for the isothermal experiments. Because of the Hg volatility, an Ar atmosphere was maintained in the measurement cell. Both metals were covered with a previously degasified oil layer and did not show any sign of oxidation. The experimental enthalpy of mixing increased for the homogeneous mixtures at both extremes of compositions but it remained constant at compositions inside the miscibility gap, as shown in Fig. 9.22. No plateau of the mixing enthalpy was observed at 478K because the critical temperature of miscibility gap lies 1K below. The method was found to be sufficiently precise ($\pm 2-6\%$), and the results in agreement with other studies.

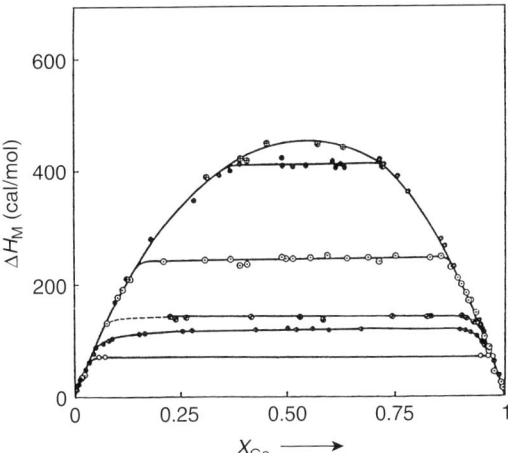

Figure 9.22 Enthalpies of formation of the liquid Ga–Hg mixtures at increasing temperatures: 313K, 353K, 373K, 423K, 467K, 478K (from the bottom) [86].

4.4 Densitometry by X-Ray Attenuation

Abrupt changes of density at the liquid–liquid phase transition may be traced by measuring the attenuation of a primary γ-ray beam that passes through an alloy sample. Chuntonov et al. [87] employed this method to determine several points on the liquid miscibility gap of the Cs–In system. The mixtures of given compositions were heated or cooled in steps during exposure to a γ-ray radiation source of constant intensity, originated from the ^{137}Cs isotope. The radiation intensity that passed through the alloy sample was monitored to construct a intensity vs. temperature plot. Discontinuity points on these plots correspond to temperatures of the liquid–liquid phase transition. The precision of this method depends on the difference in thermal expansion coefficients of heterogeneous and homogeneous alloys.

4.5 Neutron Transmission

This quite expensive technique was used to determine the critical mixing temperature in the Ga–Hg system [88]. The neutron transmission by the Ga–Hg equimolar mixture was measured during a slow and precisely controlled cooling. Due to the different cross-sections of each metal, the thermal neutrons originated from a reactor show different transmissions for the heterogeneous and homogeneous alloys. A very precise determination of the critical temperature (± 0.1K) was possible in this case; however, this temperature was found to be 2K higher than the value accepted from other measurements.

4.6 Electromotive Force

This method was applied to determine the miscibility gap in the liquid phase of the Mg–Na system at 973K [89]. The alloys were prepared by melting weighed amounts of the metals in a high melting borosilicate glass container. The emf of the cell: Mg–Na/borosilicate glass conducting Na$^+$/Na, was measured for different alloy compositions in an Ar atmosphere. Discontinuities on the potential–concentration plot correspond to the saturation of Mg with Na and of Na with Mg. The results obtained were precise within ± 4%.

4.7 Electrical Resistivity

The liquid–liquid phase boundary in the Rb–Se system was investigated at several pressures by Hoshino and Endo [90] using measurements of the electrical resistivity of the melts. The investigations were carried out at pressures up to 110 MPa where the critical mixing and monotectic temperatures matched. The experiments were performed in an internally heated autoclave using a Pyrex glass cell with W electrodes. The homogeneous Rb–Se mixtures (rich in Se) were prepared by rocking the autoclave and keeping the cell at temperature 150K higher than the critical temperature. The resistivity was measured at gradually lowering temperatures. The method of pressure application to the system was not described. The results obtained are presented in Fig. 9.23.

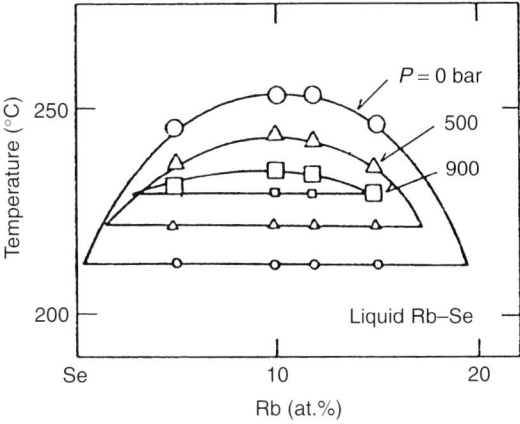

Figure 9.23 Pressure variations of the boundary in the two-phase region of the liquid Rb–Se mixtures [90].

4.8 Magnetic Susceptibility

Because phase transitions in alloys disrupt the monotonic variation of magnetic susceptibility as a function of temperature, slope changes observed on such plots may be used to determine a liquid immiscibility gap. The method was applied to the investigation of the Cs–Ge alloys in the middle composition range [91]. The samples were placed in Mo crucibles that were sealed in vacuum by laser. The susceptibility was measured using the Faraday method on a pendulum balance. The authors obtained reliable results that were in agreement within $\pm 3K$ of the DTA data.

4.9 Vapor Pressure

Using measurements of H vapor pressure over Cs-rich liquid mixtures at several temperatures, Szafranski et al. [92] constructed the essential part of the Cs–H phase diagram. The experiments were carried out in an autoclave made up of a Ni-base superalloy. The autoclave internal space was subdivided into two concentric compartments. The inner compartment contained a thin walled, pure Fe capsule. The capsule was charged with a known amount of liquid Cs under an Ar atmosphere. The capsule was evacuated, and then H gas was introduced into both compartments of the autoclave at a desired pressure. A certain amount of H diffused into the capsule and formed CsH at the proper mole fraction. During this process the H pressure in the outer compartment was gradually reduced, so that the H pressure in both compartments became equal. The amount of H that reacted with the Cs was calculated from the equation of state for H_2. Increasing the H pressure in a stepwise manner, the isotherms of pressure–composition were obtained. The pressures were measured with a strain gauge pressure transducer. The results obtained are presented in Fig. 9.24. The compositions on the limited miscibility gap were estimated from the pressure plateaus. The experimental values have uncertainties of $\pm 2K$, ± 0.05 MPa and $\pm(3-8)\%$, in the composition range studied.

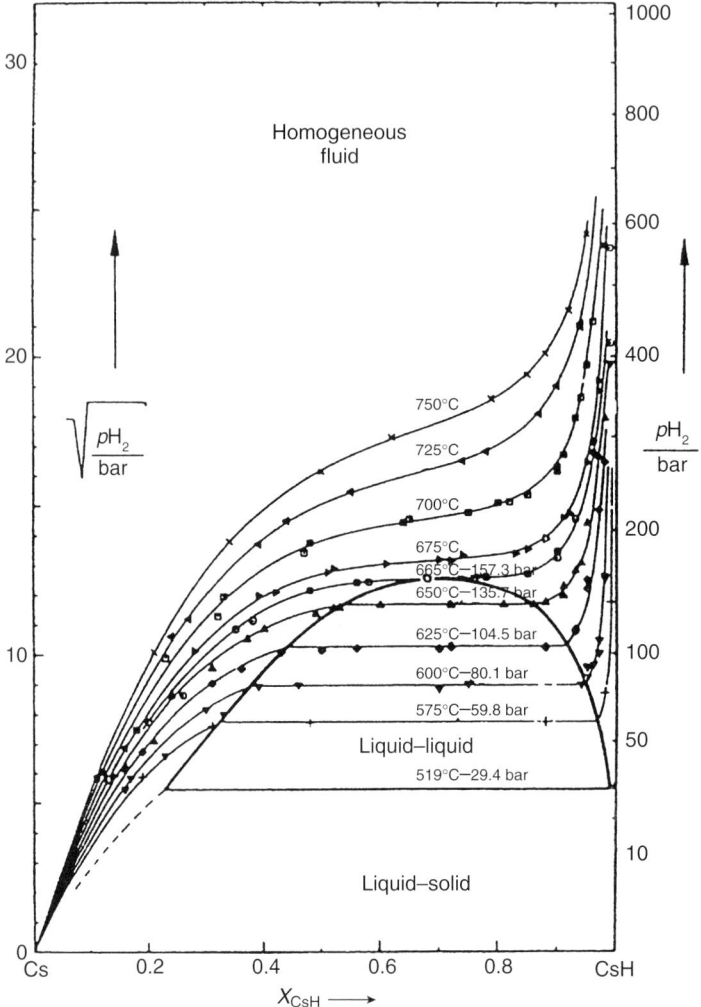

Figure 9.24 Isotherms of pressure of H in equilibrium with the liquid Cs–CsH mixtures vs. their composition [92].

5 SOLID–VAPOR EQUILIBRIA

5.1 Vapor Pressure

This is the key method for the determination of vapor–solid phase equilibria at a constant pressure. Using a Knudsen effusion cell and mass spectrometric analysis of the gas phase, Hilpert [93,94] measured Hg partial vapor pressures of the decomposition reactions of all compounds formed in the Hg–Mg system. Compositions of the alloy samples were carefully controlled by chemical analysis and XRD.

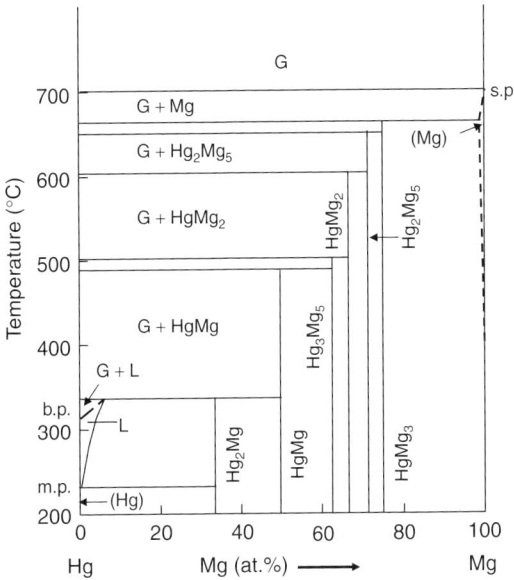

Figure 9.25 The Hg–Mg phase diagram at 1 Pa constructed from the experimental vapor pressure data.

Temperature dependency of the Hg vapor pressures obtained in these measurements for all compounds allowed the construction of the Hg–Mg phase diagram at 1 Pa (this pressure is essential in the production of light bulbs). The vapor pressures for pure Hg and pure Mg were calculated using the equations in Table 9.2. The resulted diagram is shown in Fig. 9.25. Because all Hg–Mg compounds have very narrow homogenity ranges, the temperature steps in the diagram are rectangular. In the case of (Mg) with a probable Hg solubility of ~1 at.%, the temperature should increase smoothly within the limits of (Mg) phase up to the sublimation point of Mg (701°C). Hg boils at 315°C at 1 Pa; and the Hg-rich side of the diagram was based on a widely accepted version.

5.2 Thermogravimetry

Although this technique has a very high potential for phase diagram determination (especially for the gas solubility in melts), it is not frequently used to study solid–vapor equilibria in alloy systems. The equipment is based on a high precision analytical balance with the sample placed in a furnace on one side and a counterbalance on the other side. Changes of the sample weight due to bonding with or expelling of a gaseous component while heating or cooling the sample are recorded by electronic devices connected to the balance.

The experiments performed by Fertonani et al. [95] for the Hg–Pt system did not show the exact temperatures of the decomposition reaction of the intermetallic compounds. Figure 9.26 shows a thermogravimetric curve (solid line) and its derivative (dashed line) for the thermal decomposition of the specimen – a Pt foil covered with electrochemically deposited Hg. In the first step (up to 170°C), the bulk Hg evaporated.

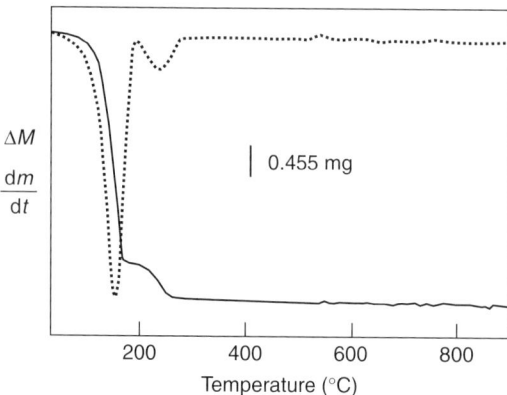

Figure 9.26 Thermogravimetry and derivative thermogravimetry curves for an annealed Pt foil containing electrodeposited Hg; heating rate – 5K/min, N flux – 150 cm^3/min [95].

Between 170°C and 225°C the mass loss of the sample was attributed to the thermal decomposition of Hg$_4$Pt. The next step between 225°C and 275°C was similarly ascribed to the Hg$_2$Pt decomposition. Above 275°C, HgPt should decompose but the authors ascribed it to further decomposition of Hg$_2$Pt formed in the course of side reactions. These experiments were complemented by XRD, energy dispersive X-ray analysis, SEM, optical microscopy and chemical analysis. If we compare these results with the decomposition temperatures of Hg$_4$Pt – 406°C, Hg$_2$Pt – 436°C and HgPt – 471°C at 0.101 MPa that were obtained from precise vapor pressure measurements [47], a serious disagreement emerges. It is well known that equilibria in the Hg–Pt system are reached slowly thus the decomposition reactions perhaps do not represent the temperature changes. Moreover, the use of a constant flow of N$_2$ gas over the samples may decrease the equilibrium partial pressure of Hg (Hg evaporated effectively at ~180°C lower temperature in these experiments than it should be observed at normal pressure).

6 LIQUID–VAPOR EQUILIBRIA

6.1 Vapor Pressure

Vapor pressure measurement is a fundamental method to study the liquid–vapor equilibria that are defined by lines of the beginning and the ending of boiling. Tang et al. [96] used a transpiration method with direct sampling for the investigation of the Hg–K system. Amalgams of different compositions were placed in an evacuated apparatus made of stainless steel. The apparatus was then placed in a furnace, brought up to the isothermal temperature corresponding to a pressure of 0.1013 MPa for each sample. The samples were removed into an evacuated trap and analyzed. The compositions of the condensed vapors and those of the original liquid permitted the establishment of the liquid–vapor phase diagram for the K–Hg system. Figure 9.27 compares the theoretical calculations to experimental data, and shows a satisfactory correlation. The precision of the experimental method is a few percent.

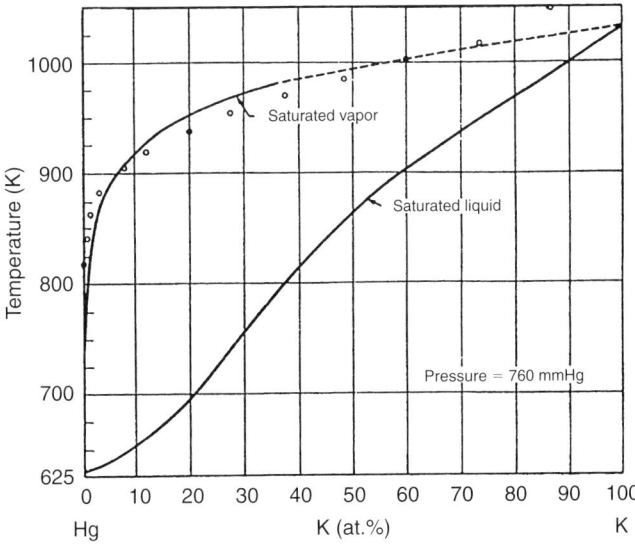

Figure 9.27 Comparison of experimental (points) and predicted liquid–vapor equilibrium data for the Hg–K system [96].

7 LIQUID–FLUID EQUILIBRIA

If one wants to keep a metal as liquid at temperatures higher than its boiling temperature, an increase of pressure over the liquid metal is required. This is well defined on unary phase diagrams for the majority of elements. However, the boundary line of boiling, which separates the liquid and gas, is discontinuous for every element at a critical point (a certain pressure and temperature). The critical point data of several elements are shown in Table 9.2. The phase which exists above the critical point is named a fluidal phase, being neither liquid nor gas. It has been shown experimentally for several metals that near the critical point the density of the metal decreases drastically, which leads to the loss of its metallic properties. The phase transition leading to formation of the fluidal state of pure elements [12, 97, s 6] and amalgams [98] was investigated by means of the following techniques: densitometry, optical absorption, reflectivity, electrical resistance, magnetic properties, neutron diffraction [12], thermopower properties [98], small angle x-ray scattering [s 6] and XRD [97]. The materials used for short time tests in these experiments are shown in Table 9.1. The high pressure cells with specimens are placed in larger autoclaves which are filled with Ar or He under the same pressure as the fluid inside the cell.

8 CONCLUDING REMARKS

This chapter provides a brief survey of most experimental techniques used by various researchers on determination of phase diagrams involving reactive and/or volatile elements. Since many different methods are used, only a brief description of each technique is given. Readers are referred to the original publications for details. Many of the techniques are not widely used and some of them lack sufficient accuracy and precision. Comparison of the results obtained from different techniques allows an assessment of the reliability of these methods. Such an assessment is important both for researchers to select specific techniques for their use and for researchers on phase diagram assessments to decide whether particular data are reliable.

REFERENCES

1. C. Guminski and H.U. Borgstedt, *Arch. Metall. Mater.*, 49 (2004) 529.
2. H.U. Borgstedt and C. Guminski, *IUPAC Solubility Data Series, Metals in Liquid Alkali Metals*, vols. 63 and 64, Oxford University Press, Oxford, 1996, various pages.
3. C. Guminski, *J. Phase Equilib. Diff.*, 25 (2004) 5, and all references therein.
4. E.L. Reed, *J. Am. Ceram. Soc.*, 37 (1954) 146.
5. J.R. Weeks, *ASM Trans. Quart.*, 58 (1965) 302.
6. D.F. Anthrop, US Atom Energy Comm. Rep UCRL-50315, 1967.
7. S.P. Yatsenko, *Galii*, Nauka, Moskva, 1974, various pages.
8. I. Ali-Khan, Corrosion of steels and refractory metals in liquid Pb, in H.U. Borgstedt (ed.), *Material Behavior and Physical Chemistry in Liquid Metal Systems*, Plenum, New York, 1982, p. 243.
9. D.J. DeRenzo, *Corrosion Resistant Materials Handbook*, 4th edn., Noyes Data Corporation, Park Ridge, 1985, various pages.
10. M.E. Drits and L.L. Zusman, *Splavy Shchelochnykh i Shchelochnozemelnykh Metallov*, Metallurgiya, Moskva, 1986, various pages.
11. H.U. Borgstedt and C.K. Mathews, *Applied Chemistry of the Alkali Metals*, Plenum, New York, 1987, various pages.
12. F. Hensel, High temperature–high pressure techniques for the study of fluid electronic conductors, in R. van Eldik, J. Jonas (eds.), *High Pressure Chemistry and Biochemistry*, D. Reidel Publishing Company, Dordrecht, 1987, p. 117.
13. S.P. Yatsenko, *Indii*, Nauka, Moskva, 1987, various pages.
14. W. Zakulski, *Institute of Fundamentals of Metallurgy & Materials – Polish Academy of Science*, Private communication, Cracow, 2005.
15. S.W. Chen, C.H. Jan, J.C. Lin and Y.A. Chang, *Metall. Trans. A*, 20 (1989) 2247.
16. R. Nesper, *Prog. Solid State Chem.*, 20 (1990) 1.
17. S. Evan, Reactivity of diamond surfaces, in G. Davies (ed.), *Properties and Growth of Diamond*, INSPEC, London, 1994, p. 64.
18. C. Guminski, *J. Phase Equilib.*, 24 (2003) 461.
19. J.B. Clark, M.E. Thomas and P.W. Richter, *J. Less-Common Met.*, 132 (1987) 181.
20. V.A. Kirkinskii and V.G. Yakushev, *Izv. Akad. Nauk SSSR, Neorg. Mater.*, 10 (1974) 1431.
21. C.B. Alcock, V.P. Itkin and M.K. Horrigan, *Can. Metall. Quart.*, 23 (1984) 309.
22. T. Iida and R.I.L. Guthrie, *The Physical Properties of Liquid Metals*, Clarendon Press, Oxford, 1993, various pages.
23. *Ullman's Encyclopedia of Industrial Chemistry*, in B. Elvers, S. Hawkins (eds.), VCH, Weinheim, 1989, various pages.
24. W.W. Wendlandt, *Thermal Methods of Analysis*, Wiley, New York, 1986, various pages.
25. B. Wunderlich, *Thermal Analysis*, Academic Press, New York, 1990, various pages.
26. M. Sorai, *Comprehensive Handbook of Calorimetry and Thermal Analysis*, Wiley, Chichester, 2005, various pages.
27. D.T. Peterson and V.G. Fattore, *J. Phys. Chem.*, 65 (1961) 2062.
28. J.C. Schottmiller, A.J. King and F.A. Kanda, *J. Phys. Chem.*, 62 (1958) 1446.
29. R. Cohen-Adad, J.J. Barthelemy and J. Mack, *Compt. Rend.*, 280C (1975) 1477.
30. C. Rolfe and W. Hume-Rothery, *J. Less-Common Met.*, 13 (1967) 1.
31. C.S. Barrett and T.B. Massalski, *Structure of Metals, Crystallographic Methods, Principles and Data*, 3rd edn., Pergamon, Oxford, 1980, various pages.

32. G.V. Raynor, *Trans. Faraday Soc.*, 44 (1948) 15.
33. F. Aldinger, P. Schuler and G. Petzow, *Z. Metallkd.*, 67 (1976) 625.
34. B. Predel and W. Schwermann, *J. Inst. Met.*, 99 (1971) 209.
35. D.F. Taylor and C.L. Burns, *J. Res. Nat. Bur. Stand.*, 67A (1963) 55.
36. F.A. Calvo and M.P. Hierro, *Metallography*, 22 (1989) 97.
37. O. Kubaschewki and W. Seith, *Z. Metallkd.*, 30 (1938) 7.
38. M.E. Boiko, Yu.F. Markov, V.S. Vikhnin, A.S. Yurkov and B.S. Zadokhin, *Ferroelectrics.*, 130 (1992) 263.
39. G. Grube and G. Schaufler, *Z. Elektrochem.*, 40 (1934) 593.
40. T. Claeson, H.L. Luo and M.F. Merriam, *Phys. Rev.*, 141 (1966) 412.
41. O. Kubaschewki, C.B. Alcock and P.J. Spencer, *Materials Thermochemistry*, 6th edn., Pergamon, Oxford, 1993, various pages.
42. W. Gasior, Z. Moser, W. Zakulski and G. Schwitzgebel, *Metall. Mater. Trans. A*, 27 (1996) 2419.
43. J.J. Egan, *High Temp. Sci.*, 19 (1985) 111.
44. S.C. Lai, *J. Electrochem. Soc.*, 123 (1976) 1196.
45. M. Notin and J. Hertz, *CALPHAD*, 6 (1982) 49.
46. K. Lorenz and H. Fabritius, *Arch. Eisenhuttenwes.*, 33 (1962) 269.
47. G. Jangg and T. Dörtbudak, *Z. Metallkd.*, 64 (1973) 715.
48. G. Franck, *Z. Naturforsch. A.*, 26 (1971) 150.
49. M. Piechotka and E. Kaldis, *J. Less-Common Met.*, 115 (1986) 315.
50. C.C. Addison, R.J. Pulham and R.J. Roy, *J. Chem. Soc.*, (1964) 4895.
51. W. Klemm and D. Kunze, Systems of alkali and alkaline earth metals, in *The Alkali Metals*, The Chemical Society, London, 1967, p. 3.
52. R.N. Hall, *J. Electrochem. Soc.*, 110 (1963) 385.
53. W.P. Stanaway and R. Thompson, The solubility of Mn and Co in liquid Na, in H.U. Borgstedt (ed.), *Material Behavior and Physical Chemistry in Liquid Metal Systems*, Plenum, New York, 1982, p. 421.
54. W.S. Ginell and R.J. Teitel, *Trans. Am. Nucl. Soc.*, 8 (1965) 393.
55. G.J. Lamprecht and P. Crowther, *Trans. AIME.*, 242 (1968) 2169.
56. C. Engelmann, in G. Kraft (ed.), *Analysis of Non-metals in Metals*, de Gruyter, Berlin, 1981, p. 135.
57. R.W. Caputi and M.G. Adamson, US Dept Energy Rep CONF-800401-P2 1980; no 18; 62.
58. Z. Galus, *Fundamentals of Electrochemical Analysis*, 2nd edn., Chichester, Horwood, 1994, various pages.
59. P. Ostapczuk and Z. Kublik, *J. Electroanal. Chem.*, 83 (1977) 1.
60. C. Guminski and Z. Galus, in C. Hirayama (ed.), *Solubility Data Series, Metals in Mercury*, Pergamon, Oxford, 1986, various pages.
61. C. Guminski, *J. Less-Common Met.*, 168 (1991) 329.
62. Yu.F. Bychkov, A.N. Rozanov and V.B. Yakovleva, *Atom. Energiya*, 7 (1959) 531.
63. W. Gaweł, Phase studies on the quasi-binary metal telluride systems, in K. Świątkowski (ed.), *Metallurgy on the Turn of the 20th Century*, Polish Academy of Science, Cracow, 2002, p. 413.
64. H.U. Borgstedt and C. Guminski, IUPAC–NIST Solubility Data Series, *Nonmetals in Liquid Alkali Metals*, vol. 75, *J. Phys. Chem., Refer. Data*, 30 (2001) 835.
65. O.A. Skuratov, O.N. Pavlov, V.I. Danikin and I.V. Volkov, *Zh. Neorg. Khim.*, 21 (1976) 2910.
66. C. Sun and Y. Cao, *Acta Metall. Sin. B*, 6 (1993) 256.
67. P.F. Adams, M.G. Down, P. Hubberstey and R.J. Pulham, *J. Less-Common Met.*, 42 (1975) 325.
68. K.Zh. Sagadieva and A.I. Zebreva, Dep Doc VINITI no 1355-77, VINITI, Ivanovo, 1977.
69. V.N. Shalaevskaya, V.A. Igolinskii and G.A. Kataev, Dep Doc VINITI no 588-75, VINITI, Moskva: 1975.
70. Sh.R. Zhakupov, K.A. Chuntonov, G.G. Ugodnikov, S.P. Yatsenko and Kh.O. Shakarov, *Zh. Fiz. Khim.*, 54 (1980) 1023.
71. C. Guminski, in H. Okamoto (ed.), *Phase Diagrams of Binary Fe Alloys*, ASM, Materials Park, 1993, p. 174.
72. C. Guminski, *J. Phase Equilib.*, 14 (1993) 643.
73. J.L. Crenshaw, *J. Phys. Chem.*, 14 (1910) 158.
74. A.I. Zebreva, L.M. Filippova, N.D. Omarova and A.Sh. Gayfullin, *Izv. Vuz. Khim. Khim. Tekhnol.*, 19 (1976) 1043.
75. K.A. Usenova, G.V. Osipova, Sh.D. Krebaeva and R.Sh. Enikeev, *Radiokhimiya*, 16 (1974) 99.
76. V.I. Cheburkov and A.N. Rozanov, *Metall. Metalloved. Chist. Met.*, 7 (1968) 168.

77. E.A. Zakharova, G.A. Kataev, L.A. Ignateva and V.E. Morozova, *Tr. Tomsk Univ.*, 249 (1973) 103.
78. N.M. Beskorovainyi, A.G. Yoltukhovskii, I.E. Lyublinskii and V.K. Vasilev, *Fiz-khim. Mekh. Mater.*, 16(3) (1980) 59.
79. V.B. Savinov, E.E. Konovalov and Sh.I. Peizulaev, *Teplofizicheskie Svoistva Zhidkostei*, Nauka, Moskva, 1970, p. 118.
80. S.P. Garg, Recent developments in determination of high temperature phase diagrams, in C.K. Mathews (ed.), *Thermochemistry and Chemical Processing*, Indian Institute of Metals, Kalpakham, 1992, p. 253.
81. C. Guminski and H.U. Borgstedt, *Z. Metallkd.*, 85 (1994) 771.
82. Ya.E. Geguzin, S.S. Simeonov and V.M. Mostovoi, *Fiz. Tverd. Tela.*, 13 (1971) 100.
83. J.F. Smith, *J. Less-Common Met.*, 35 (1974) 147.
84. S.P. Yatsenko and E.P. Druzhinina, *Zh. Neorg. Khim.*, 6 (1961) 1902.
85. H. Cetin and R.G. Ross, *J. Phase Equilib.*, 12 (1991) 6.
86. M. Gaune-Escard and J.P. Bros, *Thermochim. Acta*, 31 (1979) 323.
87. K.A. Chuntonov, L.Z. Melekhov, A.N. Orlov, G.G. Ugodnikov and S.P. Yatsenko, *J. Less-Common Met.*, 83 (1982) 143.
88. G.D'Abramo, F.P. Ricci and F. Menzinger, *Phys. Rev. Lett.*, 28 (1972) 22.
89. M.F. Lantratov, *Zh. Prikl. Khim.*, 46 (1973) 1982.
90. H. Hoshino and H. Endo, *J. Non-cryst. Solids*, 117 (1990) 525.
91. A.N. Orlov, K.A. Chuntonov, L.Z. Melekhov and A.A. Semyannikov, *Izv. Akad. Nauk SSSR Met.*, (1987), no 6, 172.
92. A.W. Szafranski, E. Keller and E.U. Franck, *Ber. Bunsen. Phys. Chem.*, 96 (1992) 955.
93. K. Hilpert, *Ber. Bunsen. Phys. Chem.*, 84 (1980) 494.
94. K. Hilpert, *Ber. Bunsen. Phys. Chem.*, 88 (1984) 37.
95. F.L. Fertonani, A.V. Benedetti and M. Ionashiro, *Thermochim. Acta*, 265 (1995) 151.
96. Y.S. Tang, C.R. Smith and P.T. Ross, US Atom Energy Comm. Rep ORNL-3605, 1964, p. 110.
97. K. Tamura and S. Hosokawa, *J. Non-cryst. Solids*, 150 (1992) 29.
98. M. Yao, K. Takehana and H. Endo, *J. Non-cryst. Solids*, 156–158 (1993) 807.

[s 1]. J. Romanowska, B. Onderka, G. Wnuk and J. Wypartowicz, *Arch. Metall. Mater.* 51 (2006) 587.
[s 2]. H. Ipser, *Ber. Bunsen. Phys. Chem.* 102 (1998) 1217.
[s 3]. L.F. Kozin, *Tr. Inst. Khim. Nauk Akad. Nauk Kaz. SSR* 9 (1962) 71.
[s 4]. R.J. Pulham, P. Hubberstey and P. Hemptenmacher, *J. Phase Equil.* 15 (1994) 587.
[s 5]. A.K. Niessen, F.R. deBoer, R. Boom, P.F. deChatel, W.C.M. Mattens and A.R. Miedema, *Calphad* 7 (1983) 51.
[s 6]. M. Inui and K. Tamura, *J. Non-cryst. Solids* 312 (2002) 247.

CHAPTER TEN

Phase Diagram Determination of Ceramic Systems

Doreen Edwards

Contents

1	Introduction	341
2	Ex Situ Methods	342
	2.1 Sample Preparation and Equilibration	343
	2.2 Phase and Compositional Analysis	350
	2.3 Identification of New Phases	354
3	In Situ Methods	354
	3.1 Thermal Analysis	355
	3.2 Coulometric Titration	356
	3.3 High-Temperature X-Ray Diffraction	358
	3.4 Thermomicroscopy and Other Optical Techniques	358
	3.5 Oscillation Method of Phase Analysis	358
	3.6 In Situ Electrical, Dielectric, and Magnetic Measurements	359
4	Concluding Remarks	359

1 Introduction

In the broadest sense, ceramic materials are defined as inorganic, non-metallic materials, and thus include borides, carbides, nitrides, chalcogenides, oxides, hydroxides, halides, and other salts. Around twenty thousand experimentally derived phase diagrams of ceramic systems have been compiled in the well-known series *Phase Equilibria Diagrams*, which was known as *Phase Diagrams for Ceramists* prior to 1992 [1]. The series, available in print and electronic formats, is continually updated through the joint efforts of the National Institute of Standards and Technology and the American Ceramic Society. Phase diagrams, which are culled from the archival literature, are evaluated by experts in the field prior to publication in the series.

Searching the entries in *Phase Equilibria Diagrams* for any number of well-studied systems will quickly demonstrate that a system's phase diagram evolves over time. Often, the first experimental diagrams in the open literature are little more than

Department of Materials Science and Engineering, New York State College of Ceramics at Alfred University, Alfred, NY, USA formation diagrams, which indicate the phases that form under known reaction conditions. If the system assumes industrial importance, more sophisticated studies follow. With the continued collection of new and (hopefully) improved data, the constructed diagrams should provide a more accurate representation of the system's equilibrium condition. Nevertheless, it is important to recognize that, as Roth and Vanderah [2] have suggested, "… *it is likely that no diagram is ever completely finished.*"

Like those for any other class of materials, experimental phase diagrams for ceramic systems are based on observations of phase existence or coexistence, reactions, and transitions. Because ceramic systems exhibit a broad range of physical behavior, several experimental techniques, many of them detailed in other chapters in this book, have been used to identify phase relationships. Often, the best understanding of a system develops when multiple techniques are used. Experimental techniques which are suitable for one region of a given phase diagram may not be the best technique for another area. Collecting evidence for phase relationships is just the first step toward constructing a phase diagram. This evidence must be interpreted with an understanding of the Gibbs phase rule, crystal chemistry, and the limitations of the experimental techniques used.

Traditionally, the experimental techniques for probing phase relationships have been broadly characterized as either static or dynamic. Static methods are those in which phase existence (or coexistence) is determined after a sample is equilibrated under constant temperature and pressure. The most commonly used static method involves quenching a specimen to "freeze-in" its high-temperature phase assemblage and allow for room temperature analysis. In dynamic methods, phase changes are monitored as temperature (or pressure) is varied. The applicability of a given method depends largely on the kinetics of the system. For systems with sluggish kinetics, static methods are often preferred. Conversely, dynamic methods are often required for systems with rapid kinetics.

Perhaps a more appropriate view is to classify the techniques ex situ or in situ. The reason for doing this is that many measurements traditionally classified as dynamic are not conducted as temperature (or pressure) is changing. Instead, they are conducted as a series of static measurements, after equilibrium is achieved, at non-ambient temperature and/or pressure conditions.

2 Ex Situ Methods

One of the most commonly used methods for constructing ceramic phase diagrams is the "cook and look" technique. This method involves equilibrating a specimen at high temperature, rapidly cooling to room temperature with the hopes of retaining the high-temperature phase assemblage, and analyzing the sample to determine the phases present and their composition.

Often, initial exploration of a system involves heating samples of known composition at a known temperature for a set period of time, cooling them, and seeing what forms. The process is then repeated with heating to increasingly higher temperatures until the sample melts, volatilizes, or reacts with the container in which it was heated. While the presentation of the resulting phase analysis in the form of a T vs. X diagram is more aptly called a "formation diagram" rather than a "phase diagram", it can provide general information about how the system behaves and can help identify the level of experimental sophistication that will be needed to obtain accurate phase-relationship data.

The primary difference between a formation diagram and a phase diagram is that the latter is constructed using only data collected from samples in a state of thermodynamic equilibrium. Equilibration is often achieved by repeatedly heating and quenching a sample until no further changes are observed. In some systems, the absence of change does not necessarily mean the system is in equilibrium, but rather that the kinetics are very sluggish under the conditions used. One method to assess whether or not equilibrium is achieved is to compare the results obtained using samples with different initial states. For example, consider a system in which a mixture of two components shows no evidence of reaction at 900°C but is known to form a compound at 1000°C. The absence of reaction at 900°C does not necessarily mean that the mixture is the equilibrium phase assemblage, but may simply indicate that the reaction is kinetically limited. If annealing a sample of the compound at 900°C results in the decomposition of the compound into the two separate components, then one can conclude that the mixture is the equilibrium phase assemblage at 900°C. In some systems, it may not be possible to achieve equilibrium in a reasonable length of time.

Several factors affect the accuracy of experimental diagrams derived from "cook and look" data, including: (1) the nature of the system under study; (2) the purity of the starting materials; (3) the techniques used to prepare and equilibrate samples; and (4) the methods used to determine phase assemblage and composition. Because the term "ceramics" encompasses many different types of materials, it is nearly impossible to define a single set of experimental protocols for obtaining good data. Nevertheless, the following discussion describes some common methods used in phase diagram work, with an emphasis on oxide systems.

2.1 Sample Preparation and Equilibration

Specimens for quenching studies can be prepared using a variety of techniques, depending on the nature of the system. In general, specimens used for quenching studies should be near their equilibrium state, or at least have no features that would hinder attainment of equilibrium in a reasonable amount of time. For most systems, this means fine-grained mixtures in which the components are uniformly distributed throughout the specimen.

Solid-state reaction (also called calcining) techniques are commonly used for preparing oxide samples. For oxide systems, starting materials typically include single-cation oxides and/or oxysalts that will provide nominally stoichiometric oxides at the desired temperature. Purity of the starting material should be reported along with any available information regarding the nature and level of minor impurities. If oxides are used as starting materials, they should be dried to remove any adsorbed water and stored in a dessicator prior to use. Oxysalts, such as carbonates and acetates, are often chosen in lieu of oxides when the latter are hygroscopic or prone to reaction with carbon dioxide in air. The oxides of alkali metals and heavier alkaline earth metals are particularly prone to such interactions. For example, barium carbonate is commonly used for determining the phase diagrams of systems in which BaO is one of the components [3,4]. When heated to a sufficiently high temperature, oxysalts decompose to form the corresponding oxide and gaseous products which are released from the sample. If an oxysalt is used as a starting material, thermogravimetric analysis (TGA) should be used to determine its decomposition temperature and the actual oxide yield to allow for batching adjustments. To achieve the desired overall composition, starting powders are weighed with an analytical balance and mixed together. Typically, small (<5 g) batches are made with individual components weighed to the nearest tenth of a milligram. While the use of larger batches would result in a lower level of uncertainty in composition, smaller batches are easier to process using techniques that will retain the overall composition. The starting powders are usually mixed together by grinding in an agate mortar with a small amount of acetone or alcohol. Alternatively, automated milling techniques can be used, but care must be taken to avoid contamination of the sample with grinding media and/or preferential loss of one or more of the starting components. After mixing, the powders are typically pressed into a pellet using hardened steel dies and heated to facilitate reaction and ultimately equilibration. Pressed-powder pellets are preferred over loose powder because they result in more intimate contact of the components, although loose powder can be used to avoid pressure caused by internal grain growth in a sintered body [2].

For slow-reacting oxide systems, solution-based methods may provide more intimate mixing of the component oxides and reduce the times required for equilibration. One of the most common solution-based methods is coprecipitation, which involves the formation of a homogenous liquid solution containing cations in the ratio required for the desired overall composition. Starting materials, usually in the form of oxysalts or metals, are dissolved in water and/or concentrated mineral acids. If oxysalts salts such as carbonates, nitrates, or acetates are used as starting materials, TGA should be conducted to confirm oxide yield. Alternatively, elemental analysis using plasma emission or atomic absorption spectrophotometry can be used to confirm cation content. After forming a homogenous aqueous solution, coprecipitation is accomplished using ammonium hydroxide or by simple evaporation. The products should not be removed by filtration as this can result in preferential loss of one or more components. After drying to remove most of the water, the precipitate is calcined to form a mixed oxide powder. Coprecipitation may not be as effective as more sophisticated solution-based techniques for producing fine grained, homogenous powders, but usually provides a more intimately mixed powder than can be achieved by mechanical mixing. Other solution-based techniques, such as a polymerized-complex method, sol–gel processing, or metal-organic decomposition, can also be used when warranted. In selecting a solution-based method, the primary concerns should be maintaining the desired overall composition and avoiding the introduction of unwanted contaminants.

For compositions with sufficiently low liquidus temperatures, melt processing can be used. For oxide systems, this entails weighing appropriate amounts of the starting materials, typically oxides and carbonates, into a platinum crucible and heating above the liquidus temperature. Intermediate grindings and/or agitation of the melt are used to improve homogeneity. After homogenization of the melt, glass-forming samples are often heated at temperatures slightly below the liquidus to promote the formation of small, uniformly distributed crystals because glass-forming systems can be difficult to equilibrate at much lower temperatures.

When heating the specimen, either to initially form it or to equilibrate it, contamination from the container must be avoided to the extent possible. Noble metal containers (crucibles, setters, etc.) are preferred for many ceramic systems. For samples prepared using solid-state reaction, a less-expensive alternative is to place a thin sacrificial pellet of the same composition as the specimen between the specimen and a refractory-ceramic crucible. To hinder volatilization of the components, the sample can be covered with a powder of the same composition within a closed container. Before and after each heating step, the sample should be weighed to determine if material is being lost, and if it is, to estimate the uncertainty in the overall composition.

Hermetic sealing is sometimes used to prevent volatilization and/or reaction with the atmosphere. Sealed platinum tubes with 0.2 mm wall thickness are reported to withstand internal pressures on the order of 10^6 Pa [2]. To seal a platinum metal tube, the inner surface of the tube is cleaned, and tube is pinched shut using smooth-surfaced pliers to prevent ripping the metal. After pinching, the tube can be sealed with a DC arc welder or a micro torch. Platinum-foil packets can be sealed in a similar fashion. Sealed silica ampules are routinely used because they can be easily evacuated and backfilled with an inert gas prior to sealing. To prepare a sealed ampule, the specimen is placed in a one-end-closed silica tube which is subsequently evacuated and backfilled, if desired. The silica tube is then softened with a torch and pinched shut to provide a hermetic seal. For some systems, it may be necessary to physically separate the sample from the silica tube (ampule) using platinum foil or even a thin layer of carbon deposited on the inner surface of the tube.

Hermetic sealing has been widely used to study ternary metal–metal–oxide systems, such as W–In–O [5], V–W–O [6], and Pr–Fe–O [7]. In such studies, oxygen content is controlled by varying the ratio of one of the metals and its corresponding oxide, such as W and WO_3. One of the drawbacks of the technique is that the oxygen partial pressure in equilibrium with the condensed phases can vary significantly with composition. If a sample equilibrates to produce a high oxygen partial pressure, the overall composition of the condensed phases can be alternated significantly. Thus, isothermal sections constructed from ex situ phase analysis data may not provide an accurate representation of the phase relationships. Nevertheless, the results of such studies can be used in conjunction with other data to construct more accurate representations. For example, Tretyakov et al. [7] combined phase analysis results with data collected for the equilibrium oxygen pressure over $PrFeO_3$ to construct an equilibrium pO_2-vs.-composition diagram for the Pr–Fe–O system at 1300K.

The heating conditions used in phase equilibria studies depend on the nature of the system and the desired objective. If salts are used as starting materials for oxide specimens, the first stage of heating is designed to promote controlled decomposition of the salt, removal of unwanted gaseous byproducts, and solid-state reaction among the remaining oxides. Heating too rapidly or to too high of temperature can result in loss of material through melting and wicking, frothing, and/or volatilization. Wicking occurs when a transient liquid phase flows out of the pressed pellet before reacting to form the equilibrium phases, often fusing the sample to the bottom of the crucible. Frothing occurs when the gaseous byproducts of oxysalts form bubbles in a transient liquid phase. These bubbles can rise, break and eject material away from the pressed pellet.

Once oxysalts are decomposed, the objective of subsequent heating steps is to move the system toward equilibrium, which can involve solid-state reaction, the formation of solid solutions, and/or melting. Intermediate grindings are used to produce fresh surfaces thereby increasing the contact between reacting phases. Clearly, grinding must be done in such a manner so as to retain as much of the original specimen as possible, and thus maintain the overall composition. The objective of the final stage of heating is equilibration under well-defined conditions, followed by quenching to retain the high-temperature phases. In practice, equilibration and quenching are accomplished with varying levels of sophistication which ultimately affect the accuracy of the resulting phase diagram.

A variety of custom and commercial heating systems have been used to prepare ceramic specimens for phase diagram studies. Box and tube furnaces with resistive, nickel–chrome and iron–chrome–aluminum heating elements are the most economical choice for heating at temperatures below 1200°C in air. Furnaces with silicon carbide and $MoSi_2$ resistive elements typically have maximum temperature ratings of 1600°C and 1800°C, respectively, in air. Commercial electrical-resistance furnaces with zirconia elements are now available for heating to 2050°C in air. For firing in vacuum or inert atmosphere, furnaces with graphite heating elements are rated to as high as 3000°C.

Temperature control and measurement, especially during the equilibration stage, is critical to the accuracy of the collected data. Furnaces should be calibrated using melting point standards. Most box furnaces can be calibrated, at best, to ± 10–25°C. Tube furnaces, on the other hand, can be calibrated to ± 2°C using a Pt/Pt–Rh thermocouple near the specimen within the center "hot zone". When heating at very high temperatures, above ~ 1700°C in air or above ~ 2300°C in inert atmospheres, temperatures are typically measured using infrared pyrometry, which can be calibrated to 1% accuracy, at best.

Quenching protocols can also affect the quality of the data. While simply opening a furnace and removing the specimens with tongs is not technically quenching, the technique is often used to generate isothermal sections in the subsolidus region of multicomponent systems. In such cases, the cooling rate can be limited by the thermal mass of the container. One way to increase cooling rate, is to place hot samples on a thermal sink, such as a thick metal plate, or to immerse them in a suitable fluid. For the best data, a vertical quench furnace, such as that described by Roth and Vanderah [2], is recommended.

Figure 10.1 illustrates the key components of a vertical quenching furnace. The heating system is a vertically mounted tube furnace equipped with one or more open-ended refractory cylinders that separate the heating element from the sample. The inner sleeve, or sample chamber, should be large enough so that the sample does not touch the outer wall. The sample is suspended in the predetermined hot zone, near but not touching the measurement thermocouple. Different sample holders can be designed to achieve consistent placement of the sample in relationship to the hot zone and thermocouple and to provide rapid quenching when desired. One such design, illustrated in Fig. 10.1(b), employs a four-hole-bore refractory tube (thermocouple protection tube) that hangs down the center of the sample chamber. Two of the four holes house the measurement thermocouple; the other two house metal-wire hooks. The sample package is placed in a thin wire

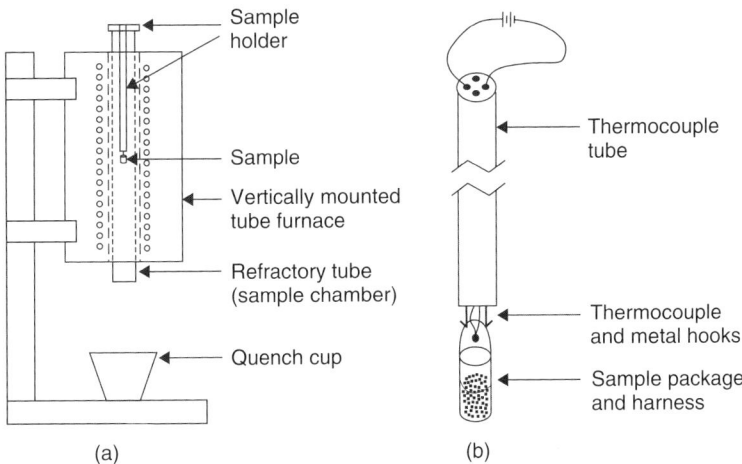

Figure 10.1 Schematic of vertical quench furnace (a) with detail of sample holder (b), after Ref. [2].

harness which hangs upon the hooks. The harness wire is substantially thinner than the hook wire so that a voltage applied across the metal-wire hooks causes the harness wire to melt and the sample package to drop into the quenching fluid.

The function of the quenching medium is to remove heat away from the sample as quickly as possible. Dry quenching in air is frequently used, but may not provide a sufficient quench for some systems. Quenching in cold water is usually adequate, if reactivity with the system is not an issue. If reactivity with water is an issue, forced inert-gas cooling can be used. Direct quenching into liquid nitrogen is not recommended as N_2 gas, formed upon contact with the hot specimen, has poor thermal conductivity [2]. Many older studies report quenching in liquid mercury, which should be avoided because of issues related to toxicity and disposal.

Equilibrating and quenching in controlled atmosphere may be desired for some studies. Figure 10.2 illustrates a modification of the open-air vertical quench furnace to allow for controlled atmosphere equilibration and rapid quenching. In this apparatus, the inner refractory sleeve is closed-off with metal cap equipped with fittings to accommodate incoming and exiting gas. Caps with O-ring compression seals can be used with appropriate cooling, which can be accomplished by wrapping a copper cooling coil near the end of the refractory tube or by using custom-designed caps with embedded cooling channels. If the caps are located sufficiently far from the furnace, forced air cooling is sometimes possible. A sample holder, similar to that described above, is fed through an O-ring compression fitting in the top cap. A dab of silicon caulk, or other similar material, provides an adequate seal between the wires and the four-hole-bore tube. Alternative configurations, which use a draw wire to pull samples from the hot zone of a horizontally oriented furnace, have also been used for phase equilibria studies [8], but these designs do not provide the rapid rate of quenching available with the vertical furnace. Some commercial vacuum/inert-gas systems are equipped with quenching systems, which use helium or other inert gases to cool the work piece more rapidly than possible with ambient cooling, but again not at the rapid rates achieved in a vertical quench furnace.

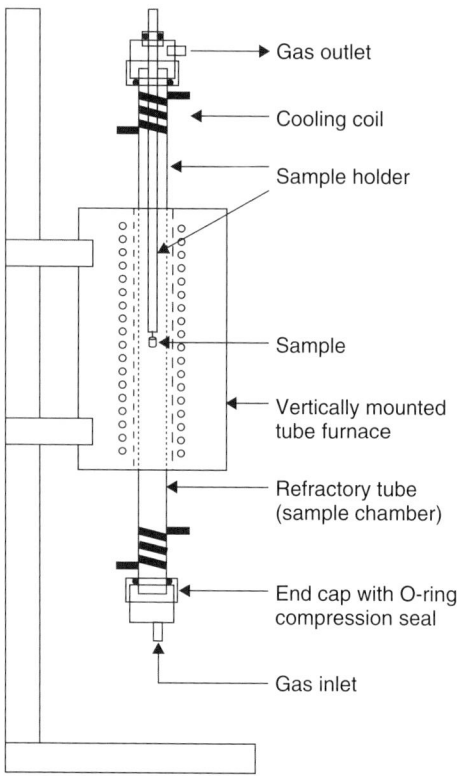

Figure 10.2 Schematic of vertical quench furnace with modifications to allow for controlled atmosphere operation.

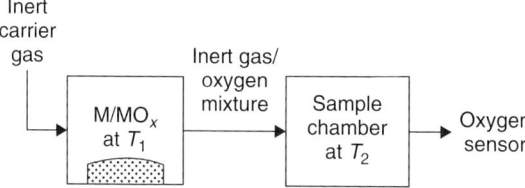

Figure 10.3 Schematic showing the use of metal/metal-oxide buffers to control oxygen partial pressure.

A number of methods have been used to control the partial pressure of oxygen, or pO_2, during equilibration. Oxygen and inert-gases are routinely mixed using mass-flow controllers, for example, to achieve oxygen partial pressures in the range of $1\,\text{atm} > pO_2 > 10^{-6}\,\text{atm}$. Oxygen/inert-gas mixtures with lower pO_2 can be obtained by passing an inert gas over an appropriate metal/metal-oxide mixture. As illustrated in Fig. 10.3, the inert carrier gas is passed across a bed of the metal/metal-oxide mixture held at constant temperature before entering the sample equilibration chamber, often held at another temperature. When heated under isothermal conditions, the metal/metal-oxide mixture will exchange oxygen with the carrier gas until the system's equilibrium pO_2 is established according to the following reaction:

$$M + \frac{x}{2}O_2 \leftrightarrow MO_x \qquad pO_2 = e^{\frac{2\Delta G_{f,MOx}}{xRT}}$$

where $\Delta G_{f,MOx}$ is the molar free energy of formation for the oxide at temperature, R is the ideal gas law constant, and T is absolute temperature. Providing that the kinetics of the exchange reaction are sufficiently rapid and that neither the metal or metal–oxide component becomes depleted, it is possible to supply a continuous flow of an inert-gas/oxygen mixture to the sample of interest. Stainless-steel tubing is required when transporting low pO_2 mixtures to the sample because oxygen can diffuse through the walls of plastic tubing. When using oxygen/inert-gas mixtures, it is best to confirm the oxygen partial pressure of the exiting gas using a zirconia-based oxygen sensor. For low pO_2 levels, it is also necessary to consider whether sufficient oxygen is being delivered to enable equilibration of the sample in a reasonable amount of time. For example, 1 l of gas with an oxygen partial pressure of 10^{-6} atm at 1000K contains less than 1 μg of oxygen. Muan [9] suggested that a pO_2 of approximately 10^{-3} atm is the "practical lower limit for studying the equilibria when oxygen is present only as O_2 molecules in the gas phase."

Buffer-gas mixtures, such as CO/CO_2, H_2/H_2O, CO_2/H_2 are often used instead of oxygen/inert-gas mixtures for maintaining controlled pO_2 levels. In buffer-gas mixtures, the oxygen partial pressure is determined by the ratio of the incoming gases and the temperature near the sample. For example, in a mixture of CO and CO_2, the partial pressure of oxygen will be established by the equilibrium constant, K_{eq}, of the following reaction:

$$2CO + O_2 \leftrightarrow 2CO_2 \qquad K_{eq} \cong \frac{p^2_{CO_2}}{p^2_{CO} p_{O_2}}$$

The equilibrium constant can be calculated from the Gibbs free energy of the reaction, according to

$$\Delta G_{RXN} = -RT \ln K_{eq}$$

where R is the ideal gas law constant and T is absolute temperature. The free energy of reaction can, in turn, be calculated from the free energy of formation of CO and CO_2 as follows:

$$\begin{array}{ll} 2C(graphite) + 2O_2(g) \rightarrow 2CO_2(g) & 2\Delta G_{f,CO_2} \\ \underline{2CO(g) \rightarrow 2C(graphite) + O_2(g)} & \underline{-2\Delta G_{f,CO}} \\ 2CO(g) + O_2(g) \rightarrow 2CO_2(g) & \Delta G_{RXN} = 2\Delta G_{f,CO_2} - 2\Delta G_{f,CO} \end{array}$$

Thus, oxygen partial pressure can be controlled by the ratio of the incoming gases and temperature, according to

$$pO_2 = \left(\frac{pCO_2}{pCO}\right)^2 e^{\frac{\Delta G_{RXN}}{RT}}$$

where ΔG_{RXN} is a function of temperature, that is, the temperature near the sample being equilibrated.

The analysis shown above is only valid for low pO_2 conditions, where the formation of oxygen does not significantly alter the ratio of the incoming gases. For higher pO_2 levels, a slightly more complicated treatment is required. For more detailed discussions about the use of buffer gases and the associated calculations, Muan's paper [9] "Phase equilibria at high temperatures in oxide systems involving changes in oxidations state" and Gaskell's Introductory Thermodynamics Textbook [10] are recommended.

As noted by Muan [9], one of the advantages of using buffer-gas mixtures over oxygen/inert-gas mixtures is that "a mechanism is available by which the oxygen exchange between the gas phase and the condensed phase can take place within a reasonable period of time in spite of a low partial pressure of O_2". One of the potential disadvantages of using buffer gases is that CO_2 and H_2O may react to form carbonates or hydroxides at low temperatures. Moreover, mixtures with high ratios of CO/CO_2 present some health and safety concerns and can result in the deposition of graphite under highly reducing conditions as dictated by the reaction:

$$2CO(g) \leftrightarrow C(graphite) + CO_2(g)$$

Figure 10.4 illustrates the ranges of oxygen partial pressures resulting from the equilibration of CO/CO_2 and H_2/CO_2 gas mixtures. A number of commercial and free-access computer programs are capable of calculating the equilibrated composition of various gas mixtures.

When using buffer gases, accurately measuring the ratio of the incoming gases (including water vapor) and the temperature near the sample is often more important than measuring the oxygen partial pressure of the gas entering or exiting the chamber. In fact, the pO_2 reported by a zirconia-based sensor will not be an accurate measurement of the pO_2 within the sample chamber; instead it will report the pO_2 established at the sensor's operating temperature.

Buffer gases have also been used in the studies of sulfur-containing systems. For example, Grey and Merrit report using mixtures of H_2, CO_2, and H_2S to control the oxygen and sulfur fugacities in their investigation of the Fe–Ti–O–S system [11].

2.2 Phase and Compositional Analysis

Phase and compositional analysis of equilibrated and quenched samples is usually determined through a combination of visual inspection, optical and electron microscopies, and X-ray diffraction.

Simple visual inspection can provide substantial information about the phases present in quenched samples. The color of the sample should be noted. If grains of different phases are sufficiently large and the phases are different colors, the sample may appear spotted. Sometimes even a small amount of a colored phase – at level that lies below the detection limits of X-ray diffraction – can be visually detected in a mixture with one or more white/transparent phases. At temperatures above the solidus but below the liquidus, samples may exhibit a slight rounding of sharp edges or sticking to the sample container. (Note that reaction with the sample container can also result in sticking). Samples that undergo significant cracking or crumbling may signify the existence of an "unquenchable" phase transition.

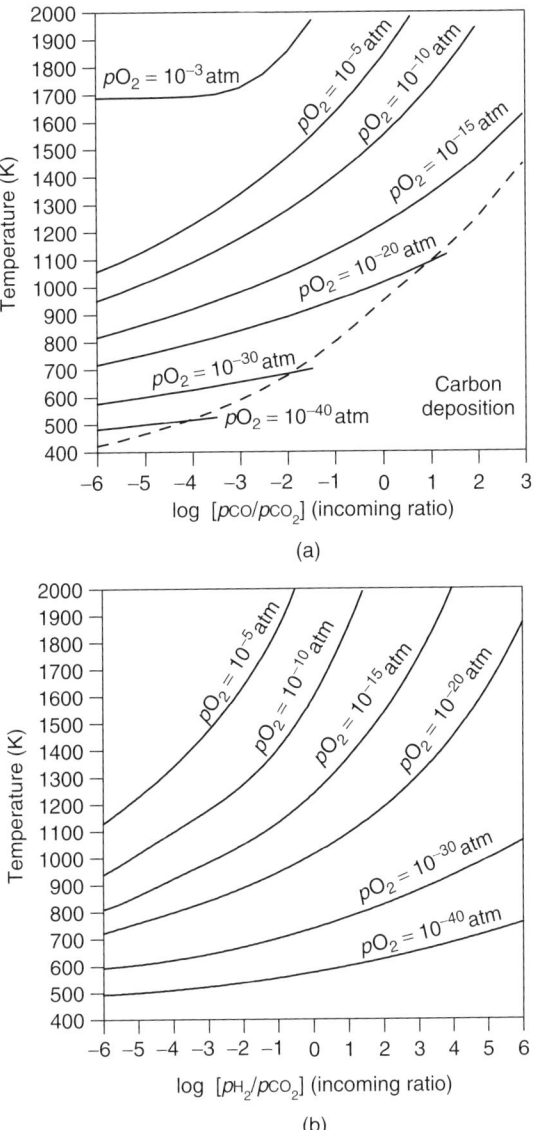

Figure 10.4 Oxygen partial pressures achieved with CO/CO_2 and H_2/CO_2 mixtures at different temperatures and a total pressure of 1 atm. Note that the gas ratios are for incoming gases prior to equilibration.

Optical microscopy using polarized light can also be helpful in identifying small amounts of a second phase. Samples quenched above the liquidus should show no evidence of crystals within a glassy matrix, whereas those quenched from below the liquidus will. Optical microscopy can also help confirm that equilibrium conditions were achieved. For example, equilibrium between a liquid and solid phase is characterized by the presence of small, sharp crystals that are uniformly distributed within a glassy phase. The presence of rounded, corroded, or poorly distributed crystals may indicate that equilibrium was not achieved.

X-ray diffraction is the most commonly used method for determining the phase assemblage of ceramic systems. Most modern diffraction labs have software with search-and-match routines that compare the measured diffraction pattern to those in the powder diffraction file (PDF™) database [12]. While this feature can be extremely helpful for phase identification, it also has limitations because peak position can shift substantially with the formation of solid solutions. Often the best approach to phase identification is to assemble the patterns of all of the known phases in the system under study and to compare them one by one to the sample pattern, keeping in mind that peaks position and relative intensity can change with the formation of solid solutions. Some computer programs are equipped with features which allow one to adjust the lattice parameters of reference patterns to enable better comparison to systems containing solid solutions and to generate theoretical diffraction patterns using known atom positions and probable site occupancy. In the end, each and every peak in the sample pattern must be assigned to a phase and indexed. Peaks that cannot be assigned may represent a previously unidentified phase in the system or a contaminant phase. If the sample was properly equilibrated and no contaminants are present, the number of detected phases in a properly equilibrated sample should not exceed the maximum number allowed by the Gibbs phase rule.

Construction of a phase diagram also requires information regarding the elemental composition of the phases present. If coexisting phases can be separated from each other, which is rare in ceramic systems, methods such as plasma emission or atomic absorption spectrophotometry, can be used to provide compositional information. If coexisting phases cannot be separated, electron probe microanalysis (EPMA) is very useful as long as its limitations are recognized. Quantitative analysis requires the use of standards, which are best if they are chemically similar to the system under study. If possible, the quality of the analyses (and sample preparation technique) should be assessed by measuring the composition of one or more known compounds. Quantitative analysis requires that the specimen has a flat, polished surface and relatively larger grains, that is, larger than the sampling volume of the electron beam. Because it can be difficult to assess whether one is sampling a single grain, it is best to make several measurements on different grains of a given phase within the sample and to report the average and standard deviation of the results along with any criteria used for discarding suspect measurements.

For multiphase samples containing solid solutions, it is sometimes possible to determine the composition of phases in an assemblage using lattice-parameter calibration curves. This entails making a series of phase-pure solid-solution samples and measuring their lattice parameters as a function of composition. The lattice parameters of the phase in a mixed-phase sample can then be compared to the calibration curve to determine its composition. Ideally, the solid solution will obey Vegard's law, changing linearly with changes in composition.

For example, lattice-parameter calibration curves were used to determine the sub-solidus phase relationships in the Ga_2O_3–In_2O_3–SnO_2 system, illustrated in Fig. 10.5 [13]. As shown in Fig. 10.5(b), the substitution of Ga^{3+} by In^{3+} increases the unit-cell volume of beta-gallia (Ga_2O_3) until the solubility limit is reached at ~42%. Similar curves were generated for other solid solutions in the system, including the

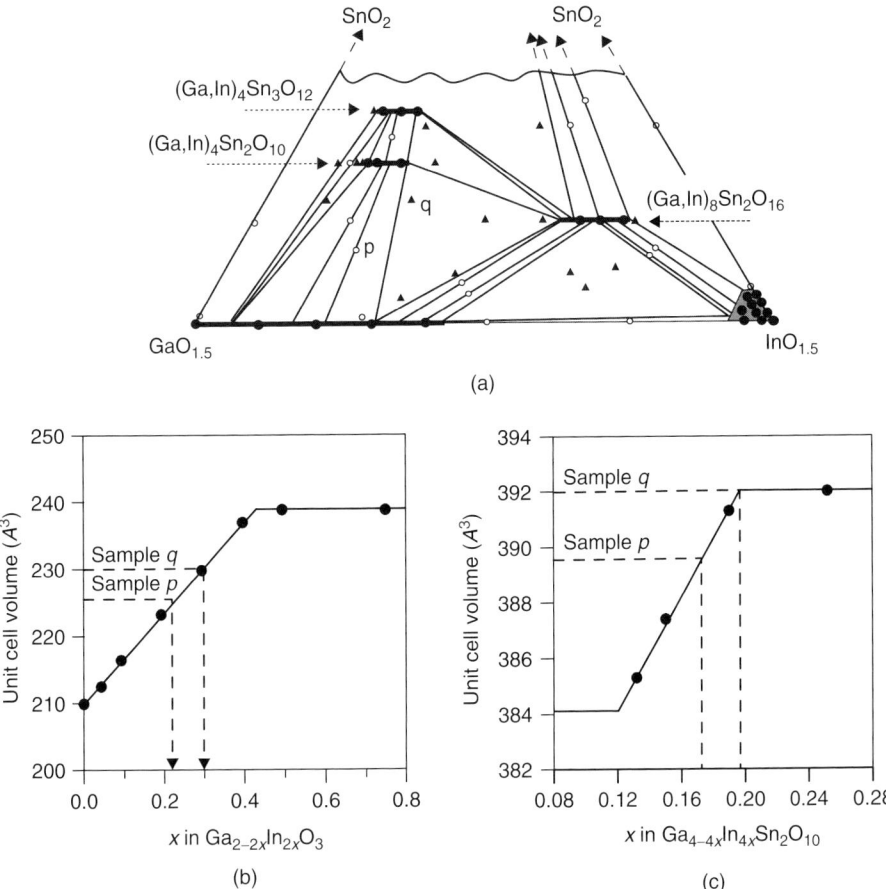

Figure 10.5 A portion of an isothermal section of the Ga_2O_3–In_2O_3–SnO_2 system (a) and graphs showing the compositional dependence of the unit-cell volume of two phases (b and c) after Ref. [13]. By comparing the unit-cell volume of phases in multiphase samples to the lattice-parameter curves it is possible to determine the position of tie lines and compatibility triangles.

$(Ga,In)_4Sn_2O_{10}$ solution shown in Fig. 10.5(c). The tie lines shown between the solid solutions were determined by comparing the unit-cell volumes of the phases in the biphasic samples (shown as open circles in Fig. 10.5(a)) to the calibration curves, as illustrated in Figs. 10.5(b) and (c). This method also allows one to assess the accuracy of the diagram because the law of mass conservation dictates that a tie line connecting two phases should pass through the original composition. Similarly, the overall composition of triphasic specimens, shown as filled triangles in Fig. 10.5(a), must lie within the compatibility triangle suggested by the composition of the coexisting phases.

Because crystal chemistry has such a strong influence on phase stability of oxide materials, it is best to construct ceramic phase diagrams using molar rather than mass compositions. In addition to showing the anion and cation ratios of compounds, molar representations can provide information about the nature of solid solutions. For example, most of the solid solutions in the Ga_2O_3–In_2O_3–SnO_2 system – $(Ga,In)_2O_3$,

$(Ga,In)_4Sn_2O_{10}$, $(Ga,In)_4Sn_3O_{12}$, and $(Ga,In)_8Sn_2O_{16}$ – appear as horizontal lines through single-phase compositions shown as solid circles, thereby illustrating that the solutions form upon the substitution of In^{3+} for Ga^{3+}.

2.3 Identification of New Phases

Accurate interpretation of phase-relationship data often requires an understanding of the crystal chemistry of all observed phases. If a previously unidentified phase is found, an investigation of its structure is warranted. Compositional analysis using the microprobe techniques can provide the information needed to prepare a phase-pure sample, either in single crystal or polycrystalline form. If possible, the elemental analysis should be converted to simple molar ratios of the system's components as this can also provide clues as to the structure of the new phases.

Single-crystal X-ray diffraction studies have been long recognized as the preferred method for structure determination, but Rietveld analysis [14] using X-ray and/or neutron powder diffraction data is increasingly applied to such problems, especially if single crystals are not available. Single-crystal X-ray diffraction studies require crystals on the order of ~0.2–0.5 mm in the largest dimension, which is larger than can be grown using solid-state techniques [2]. Single crystals can be grown, for example, by cooling from the melt or by using a suitable flux if the liquidus temperature is too high. Fluxes are substances, that when mixed with the phase of interest, form a homogenous melt from which the desired phase can grow. Before using flux-grown crystals for structural studies, they should be chemically analyzed to identify possible contaminants which may affect their structure.

Rietveld analysis requires a starting model that includes lattice parameters, correct space group, and approximate atom positions. A number of computer programs are available to assist with pattern indexing. Alternatively, unit-cell information can also be obtained from electron diffraction data obtained from polycrystalline samples using a transmission electron microscopy (TEM). Convergent beam electron diffraction (CBED) can be used to provide symmetry information as can the systematic absence of peaks in diffraction data. To identify approximate atom positions, it is often beneficial to search the International Crystal Structure Database (ICSD) [15] for isostructural phases in related systems. High-resolution TEM imaging can provide information about cation positions. Once a partial structure is known, Fourier difference analysis can lead to starting parameters for anion positions.

3 In Situ Methods

Techniques like differential thermal analysis (DTA), TGA, hot-stage microscopy, and high-temperature X-ray diffraction are sufficiently commonplace that a number of systems are commercially available. Specialized equipment and techniques have been developed to overcome the limitations of the commercially available equipment and to address the peculiarities of some systems. In essence, any physical property (thermal, optical, electric, or magnetic) that changes significantly when a system undergoes a phase transition can be used to monitor that transition.

3.1 Thermal Analysis

Thermal analysis techniques – DTA and TGA – are techniques that are widely used to study phase transitions and reactions in ceramic systems. In DTA, small containers of the sample and a stable reference material are heated (or cooled) while recording the temperature difference between the two as a function of the reference material's temperature. For ceramic systems, alpha–alumina is often used as the reference material because it is stable over a wide range of temperatures. A measured temperature difference between the sample and the reference material signals that the sample is undergoing either a heat-absorbing or heat-liberating processes. In thermogravimetry, sample mass is recorded as function of temperature, which makes this technique useful for studying reactions that involve the release or uptake of gaseous species. Several commercial DTA and TGA systems are available for studies between room temperature at 1500°C in air and controlled atmosphere. Some systems are capable of operation to 2400°C, and other are designed for simultaneous DTA and TGA measurements. Temperature calibration using certified reference materials is required to ensure the highest possible accuracy. The International Confederation for Thermal Analysis has developed a series of reference materials, which are available through the National Institute of Standards and Technology [16].

DTA is a dynamic technique that can provide information about melting, crystallization, chemical reactions, and polymorphic phase transitions. In addition to instrument design and calibration, heating rate and, sample preparation can affect the quality of the data. A heating rate of ~10°C/min is commonly used although a much wider range of heating rates is possible. When slower heating and cooling rates are used, samples are nearer the equilibrium condition, which has obvious advantages for phase equilibria studies but can also result in broad shallow peaks that are more difficult to interpret. Conversely, faster heating rates will result in narrower peaks, but the peak temperature may not represent the equilibrium phase-transition temperature. This is particularly evident in cooling curves of glass-forming samples. To minimize the effects of thermal history, standardized procedures for measuring melting, and crystallization temperature should be followed [17]. Detailed discussion on both DTA and DSC can be found in Chapter 5. Even though the focus Chapter 5 is on metallic systems, the same fundamentals apply to ceramic systems.

DTA sample cups can be made from a variety of different materials. As with quenching studies, contamination of the sample with the container must be minimized, thus platinum containers are commonly used for ceramic systems. Open and sealed containers are available and are often designed for specific commercial instruments. When using sealed containers, one should consider the effects that pressure may have on the temperature of the phase transition. Sample characteristics such as particle size and packing can also influence DTA results. For quantitative analysis, samples are commonly diluted with a reference material. For high-temperature phase equilibria studies, reaction of the reference materials with the sample may be a concern.

Thermogravimetry has been widely used for investigating phase equilibria in metal–oxygen systems [18,19]. Increases or decreases in mass in response to changes in temperature and/or oxygen partial pressure provide evidence for oxidation, reduction, and decomposition reactions involving gaseous products.

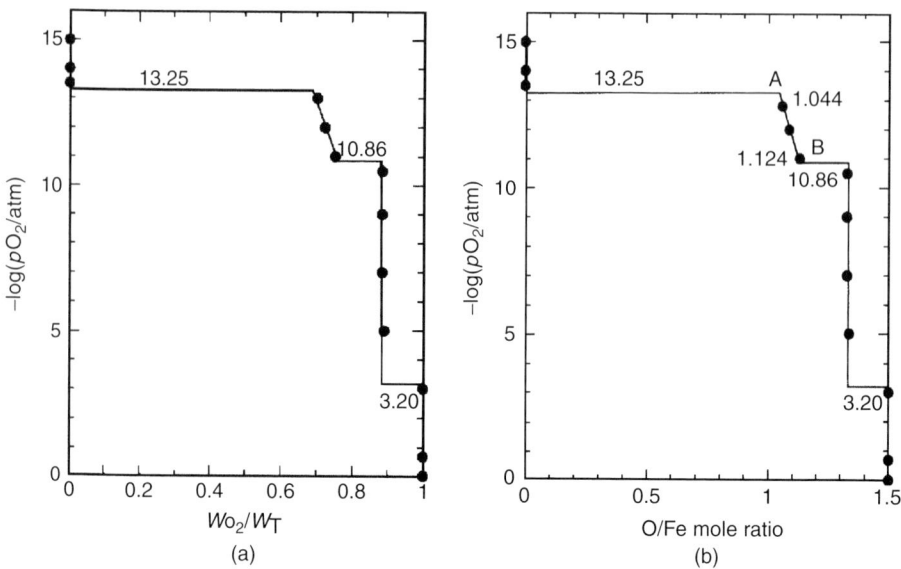

Figure 10.6 Change in the mass of an iron sample in response to changes in partial pressure of oxygen at 1100°C, from Ref. [19].

Figure 10.6 shows an example of how thermogravimetry is used to determine phase relationships in a simple metal–oxide system. In a reinvestigation of the Fe–O system [19], Kitayama et al. heated iron isothermally at 1100°C while increasing the oxygen partial pressure from 10^{-15} to 1 atm using various CO_2/H_2 and CO_2/O_2 gas mixtures [18]. At the $pO_2 = 10^{-15}$ atm, iron is known to exist in its elemental state. Increases in oxygen partial pressure to $10^{-13.25}$ atm resulted in an increase in mass corresponding to the formation of FeO. Further increases in the oxygen partial pressure to $10^{-10.86}$ resulted in a steady increase in mass corresponding to the formation of non-stoichiometric FeO_{1+x}. A rapid increase in mass at $pO_2 = 10^{-10.86}$ signaled the formation of Fe_3O_4, which is stable over $10^{-10.86} < pO_2 < 10^{-3.2}$. The formation of Fe_2O_3 was noted as an increase in mass at $10^{-3.2}$.

For more complicated systems, ex situ X-ray diffraction is often used in conjunction with thermogravimetry to identify the phases present before and after weight changes.

3.2 Coulometric Titration

Coulometric titration is another in situ method used to investigate phase relationships of oxides exhibiting oxygen non-stoichiometry [20–22]. Figure 10.7 illustrates an apparatus for making such measurements. The chamber is a quartz (or alumina) tube, one end of which is sealed with an yttria-stabilized zirconia oxygen sensor/pump and the other end of which leads to vacuum system. After heating under vacuum, the chamber is sealed, and the sample is allowed to equilibrate, as noted by a constant open-circuit potential of the zirconia sensor according to the Nernst equation,

$$E = \frac{RT}{4F} \ln\left(\frac{pO_2}{pO_2^{ref}}\right)$$

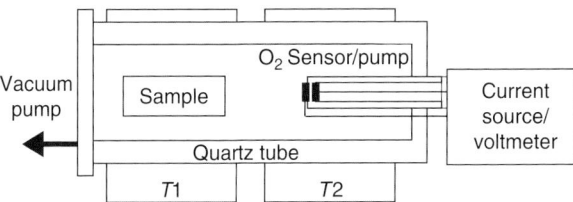

Figure 10.7 Schematic of an apparatus used for making coulometric titration measurements, after Ref. [20].

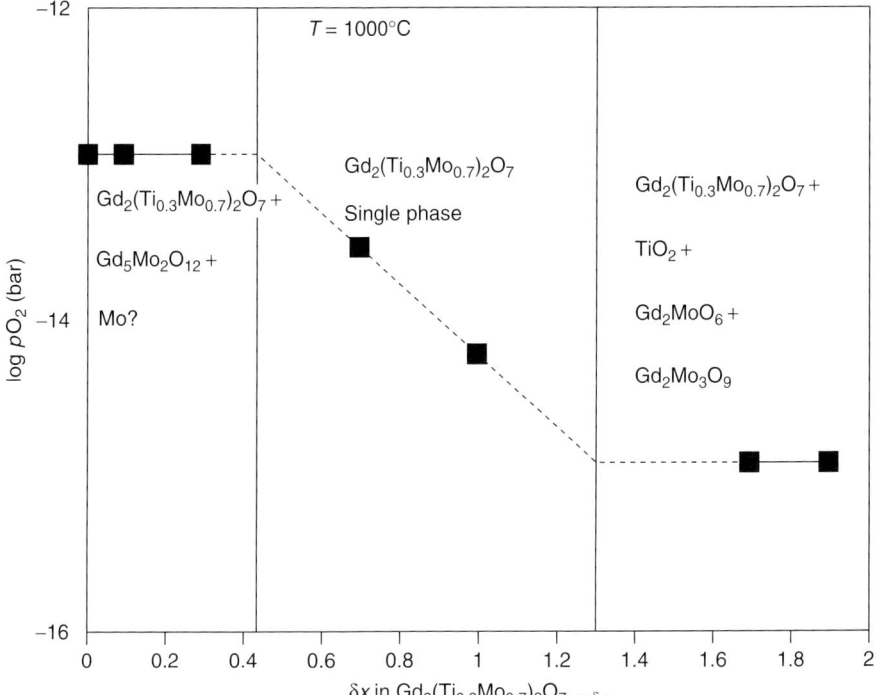

Figure 10.8 A coulometric titration curve for the pyrochlore phase $Gd_2(Ti_{0.3}Mo_{0.7})_2O_{7+x}$, from Ref. [21]. The changes in pressure observed with changing oxygen content, δx, signal the changing oxygen activity in the non-stoichiometric pyrochlore phase. The onset of pO_2 plateaus signals the decomposition of the pyrochlore.

where E is the measured potential, R is the ideal gas law constant, T is absolute temperature, F is Faraday's constant, and pO_2 is the oxygen partial pressure inside the chamber and pO_2^{ref} is the oxygen partial pressure outside the chamber.

After recording the equilibrium oxygen partial pressure of the sample, oxygen is admitted or removed from the system by driving a current through the zirconia, which now acts as an oxygen pump. The amount of oxygen pumped is recorded, and the sample is again allowed to equilibrate. By plotting the equilibrium oxygen pressure as a function of oxygen content, it is possible to extract information regarding the phase relations. As an example, Fig. 10.8 shows a coulometric titration curve for the non-stoichiometric pyrochlore phase $Gd_2(Ti_{0.3}Mo_{0.7})_2O_{7+x}$ [21]. In the single-phase region in the middle of the diagram, the decrease in oxygen partial pressure reflects a decrease in the sample's oxygen activity resulting from a change in its oxygen content, δx. The onset of plateaus at $pO_2 = 10^{-12.9}$ and $pO_2 = 10^{-14.9}$ bar signal the decomposition of the pyrochlore. Ex situ X-ray diffraction methods are often used in conjunction with coulometric titration to identify phases forming after the onset of a pO_2 plateau.

3.3 High-Temperature X-Ray Diffraction

High-temperature X-ray diffraction (HT-XRD) is an extremely useful tool for identifying phases that cannot be retained upon quenching to room temperature and for studying reactions under varying temperature and partial pressure conditions. The technique is often used in conjunction with DTA, which can often provide a more accurate measurement of the transition temperature. Diffraction data can then be collected under isothermal conditions above and below the transition temperature to provided information about the structure of the phases. HT-XRD is also very useful for substances that undergo suspected phase transitions that cannot be detected by thermal analysis. In such cases, diffraction data can be collected dynamically (with changing temperature) using a smaller angular range.

Most high-temperature diffractometers use a thin metal strip to hold and heat the sample, which can lead to large thermal gradients and sample displacement errors that require careful calibration. For phase diagram studies, volume heating is preferred. High-temperature diffractometers with volume heating are now commercially available with operating temperatures as high as 1200°C. A custom-built system designed for controlled atmosphere studies to 1600°C has also been reported in the literature [23].

3.4 Thermomicroscopy and Other Optical Techniques

Hot-stage microscopy, or thermomicroscopy, can be conducted in dynamic or isothermal modes to monitor a variety of physical changes including melting, crystallization, decomposition, oxidation, and even crystal transformations. A number of commercial and custom-made systems, using resistive and laser heating, have been reported [24]. Techniques that use photocells to measure reflected and transmitted light have obvious advantages over direct visual assessment. Thermomicroscopy is often used in conjunction with other techniques, such as DTA [25,26] and high-temperature X-ray diffraction [27].

3.5 Oscillation Method of Phase Analysis

The oscillation method of phase analysis (OPA), described Kaplun and Meshalkin [28–31], monitors crystallization and melting by measuring the harmonic oscillation of a thin plate suspended in a melt. As the temperature is decreased below the liquidus, crystallites form on the thin plate resulting in an abrupt decrease in the oscillation amplitude. After the onset of crystallization, the temperature is increased until the crystals re-dissolve, as noted by an increase in the amplitude of oscillation to its original value. In separate experiments, the metal plate can be removed from the melt so that the forming crystals can be analyzed. One of the reported advantages of the method, compared to DTA, is that measurements can be conducted at very low rates (<1°C/h), or even isothermal conditions, thus enabling the accurate measurement of the liquidus temperatures in systems prone to supercooling. Kaplun and Meshalkin have applied this method to number of ceramic systems, including Li_2O–Cs_2O–B_2O_3, GeO_2–Bi_2O_3 and, BaO–B_2O_3 [28–31].

3.6 In Situ Electrical, Dielectric, and Magnetic Measurements

Because thermally induced polymorphic phase transformations are often accompanied by abrupt changes in the resistivity, permittivity, and/or permeability, in situ measurements of these properties can be used to probe transition temperatures under controlled atmosphere conditions. Conducted in either static or dynamic modes, these measurements can be particularly helpful in identifying phase transitions that cannot be monitored using DTA. As with the other techniques, consideration must be given to temperature measurement, sample contamination, equilibration, etc. However, because electrical properties can be highly dependent on microstructure and defect chemistry, they are seldom used as a stand-alone technique for probing phase equilibria. Instead, they are considered complimentary techniques, which can be used in tandem with DTA [32], TGA [33], and HT-XRD data to provide a more complete understanding of a system.

4 Concluding Remarks

This intent of this chapter was to provide readers with an overview of the experimental techniques most commonly used for constructing ceramic phase diagrams, with a particular emphasis on oxide systems. Because each system has its peculiarities, no one experimental technique, or protocol, can be singled out as the best for ceramic systems. Nevertheless, this chapter has attempted to highlight some common issues that must be considered to obtain high-quality data.

REFERENCES

1. *Phase Equilibria Diagrams CD-ROM database*, The American Ceramic Society, Westerville, OH, USA.
2. R.S. Roth and T.A. Vanderah, in B.V.R. Chowdari, H.-L. Yoo, G.M. Choi, and J.-H. Lee (eds.), *Solid State Ionics: The Science and Technology of Ions in Motion*, World Scientific Publishing, 2004, p. 3.
3. K.W. Kirby and B.A. Wechsler, *J. Am. Ceram. Soc.*, 74 (1991) 1841.
4. W. Wong-Ng and L.P. Cook, *J. Res. Natl. Inst. Stand. Technol.*, 103 (1998) 379.
5. T. Ekstrom, M. Parmentier, K.A. Watts and R.J.D. Tilley, *J. Solid State Chem.*, 54 (1984) 356.
6. T. Ekstrom and R.J.D. Tilley, *J. Solid State Chem.*, 16 (1976) 141.
7. Yu.D. Tretyakov, V.V. Sorokin and A.P. Erastrova, *J. Solid State Chem.*, 18 (1976) 263–269.
8. A.H. Webster and N.F.H. Bright, *J. Am. Ceram. Soc.*, 44 (1961) 110.
9. A. Muan, *Am. J. Sci.*, 256 (1958) 171.
10. D.R. Gaskell, *Introduction to the Thermodynamics of Materials*, 4th edn., Taylor & Francis, New York, 2003.

11. I.E. Grey and R.R. Merritt, *J. Solid State Chem.*, 32 (1980) 41.
12. *Powder Diffraction File TM*, International Centre for Diffraction Data, Newtown Square, PA.
13. D.D. Edwards and T.O. Mason, *J. Am. Ceram. Soc.*, 91 (1998) 3285.
14. R.A. Young, *The Rietveld Method*, Oxford University Press, Oxford, 1995.
15. *International Crystal Structure Database*, Fachinformationszentrum, Karlsruhe, Germany.
16. T. Hatakeyama and Z. Liu, *Handbook of Thermal Analysis*, John Wiley & Sons, 2000, p. 9.
17. ASTM E 794-01, *Standard Test Methods for Melting and Crystallization Temperature by Thermal Analysis*, ASTM International Standards, West Conshohocken, PA.
18. K. Kitayama, *J. Solid State Chem.*, 137 (1998) 255.
19. K. Kitayama, M. Sakaguchi, Y. Takahara, H. Endo and H. Ueki, *J. Solid State Chem.*, 177 (2004) 1933.
20. J.L. MacManus-Driscoll, J.C. Bravman and R.B. Beyers, *Physica C*, 251 (1995) 71.
21. O. Porat, C. Heremans and H.L. Tuller, *Solid State Ionics*, 94 (1997) 75.
22. V.A. Cerepanov, L.Yu. Barkhatova and V.I. Voronin, *J. Solid State Chem.*, 134 (1997) 38.
23. S.T. Misture, *Meas. Sci. Technol.*, 14 (2003) 1091.
24. T. Hatakeyama and Z. Liu, *Handbook of Thermal Analysis*, John Wiley & Sons, 2000, p. 32.
25. H.G. Wiedemann and G. Bayer, *J. Thermal Analy.*, 30 (1984) 1273.
26. B. Forslund, *Chem. Scripta*, 24 (1984) 115.
27. M. Mamiya, T. Suzuki and H. Takei, *J. Cryst. Growth*, 198/199 (1999) 611.
28. A.B. Kaplun and A.B. Meshalkin, *J. Cryst. Growth*, 275 (2005) e1975.
29. A.B. Kaplun and A.B. Meshalkin, *J. Cryst. Growth*, 229 (2001) 248–251.
30. A. B. Kaplun and A.B. Meshalkin, *J. Cryst. Growth*, 167 (1996) 171.
31. A.B. Meshalkin and A.B. Kaplun, *J. Cryst. Growth*, 275 (2005) e301.
32. V.V. Vashook, M.V. Zinkevich and Yu. G. Zonov, *Solid State Ionics*, 116 (1999) 129.
33. T. Tsuji, J. Asakura, T. Yamashita and K. Naito, *J. Solid State Chem.*, 50 (1983) 273.

CHAPTER ELEVEN

DETERMINATION OF PHASE DIAGRAMS INVOLVING ORDER–DISORDER TRANSITIONS

Ryosuke Kainuma,[1] Ikuo Ohnuma[2] and Kiyohito Ishida[2]

Contents

1 Introduction	361
2 ER and Thermal Analysis Methods	362
2.1 ER Method (Resistometric Study)	362
2.2 Thermal Analysis	364
3 Singular Point Method	366
3.1 Origin of Singularity	366
3.2 Examples of SPM	368
4 Concentration Gradient Method	374
4.1 Basics of CGM	376
4.2 Examples of CGM	376
5 Concluding Remarks	380

1 INTRODUCTION

Many kinds of methods, such as X-ray diffraction (XRD), electron diffraction (ED), neutron diffraction (ND), differential scanning calorimetry (DSC) and electrical resistivity (ER), have been utilized to determine the order–disorder (OD) transition temperature T_c and kinetics of long-range ordering (LRO) and short-range ordering (SRO) [1]. It is sometimes difficult to determine the T_c by diffraction techniques, such as XRD and ED, using as-quenched samples, because it is generally hard to suppress the OD transition located at elevated temperatures by quenching. Although in situ examination at elevated temperatures for the diffraction techniques is useful to obtain information on not only crystal structures but also T_c, contamination and oxidation of specimens sometimes hinders the accurate determination of the T_c. On the other hand, both DSC and ER, which are two of the most convenient and powerful techniques, are widely used to determine both T_c and phase equilibria.

---

[1] Institute of Multidisciplinary Research for Advanced Materials, Tohoku University, Sendai, Japan.
[2] Department of Materials Science, Graduate School of Engineering, Tohoku University, Sendai, Japan.

All the above techniques utilize some change of physical properties or atomic configuration with changing temperature in a specimen with a fixed composition in order to determine the T_c. The data obtained by such techniques can be directly plotted on concentration–temperature (C–T) phase diagrams, but not in isothermal sections of phase diagrams of multi-component systems. The diffusion-couple (DC) method is very useful to find and determine unknown phase equilibria at a constant temperature as discussed in detail in Chapter 6. Recently, some new attempts to determine the critical concentration of second-order OD (2O-OD) transition and the two-phase region of phase separation due to first-order OD (1O-OD) transition using DCs have been reported.

In this chapter, the widely used methods such as ER and DSC are introduced first, followed by a detailed illustration of the DC techniques as applied to the determination of the phase stability of ordered phases in isothermal phase diagrams.

2 ER AND THERMAL ANALYSIS METHODS

2.1 ER Method (Resistometric Study)

It is well known that electrical resistance is influenced by OD transition, and the ER method can be used for the determination of the critical temperature of OD transition. Figure 11.1 shows a typical example of ER heating and cooling curves obtained from the Cu–50 at.% Au alloy, as reported by Sprusil and Pfeiler [2]. According to the Au–Cu binary phase diagram [3], this alloy shows an OD transition from a disordered A1 to an ordered L1$_0$ structure at around 410°C. This is the 1O-OD transition which accompanies nucleation and growth of ordered phase domains, and a large thermal hysteresis of about 50°C is observed. In general, the absolute value of

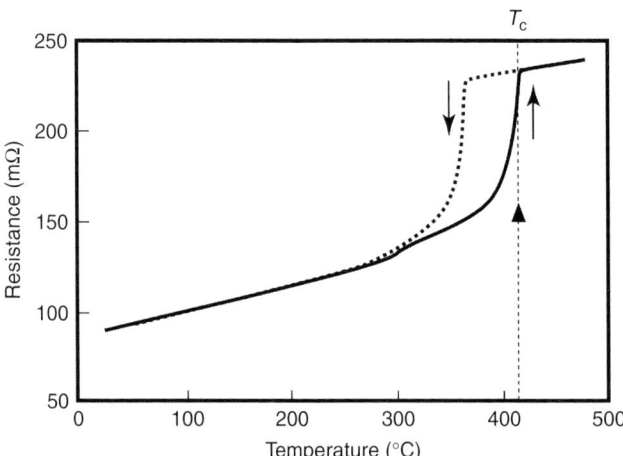

Figure 11.1 Electrical resistance curves showing the 1O-OD transition of a Au–50 at.% Cu alloy reported by Sprusil and Pfeiler [2]. The T_c is defined as the temperature with the maximal gradient in the heating curve.

ER in the ordered phase is relatively lower than that in the disordered phase in the same alloy due to a decrease of electron scattering induced by the more regular periodicity of the ordered atomic configurations, and the critical temperature is usually defined as the temperature with a maximal gradient in the ER heating curve as shown in Fig. 11.1. Here, a cooling curve should not be used for the determination because a supercooling effect due to the nucleation and growth of ordered domains which cannot be suppressed in the usual first-order transitions, such as 1O-OD transition, solidification and precipitation.

In the case of 2O-OD transition, the characteristic features of the ER curve are classified into two types: type I is similar to those in the 1O-OD transition and type II is more complicated, as demonstrated in Figs. 11.2 and 11.3 [4,5], respectively. It can be seen that the ER shows a small and broad peak just below T_c in type II, but has no peak in type I. The origin of this small peak is not clear, but a complicated atomic configuration under a very low degree of order may affect the electron conductivity. This type of ER curve has been reported in $B2/D0_3$ OD transition such as in Fe_3Al and Fe_3Ga [6,7]. In any case, the T_c is usually defined as the temperature with a maximal gradient in the ER curve as well as that in the 1O-OD transition as shown in Figs. 11.2(b) and 11.3(b).

Because the data obtained by the above definition are usually affected by the scanning rate, a low scanning rate should be selected to avoid systematic error.

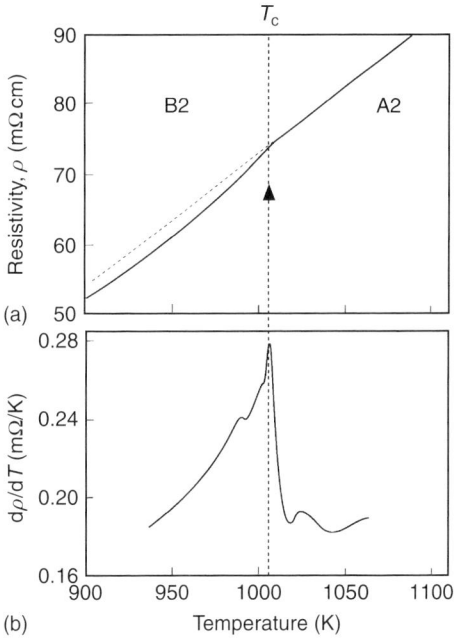

Figure 11.2 Electrical resistance curve (a) and its differential (b) showing a 2O-OD transition of a Fe-50 at.% Co alloy reported by Seehra and Silinsky [4]. The behavior of electrical resistance is similar to that in the 1O-OD transition.

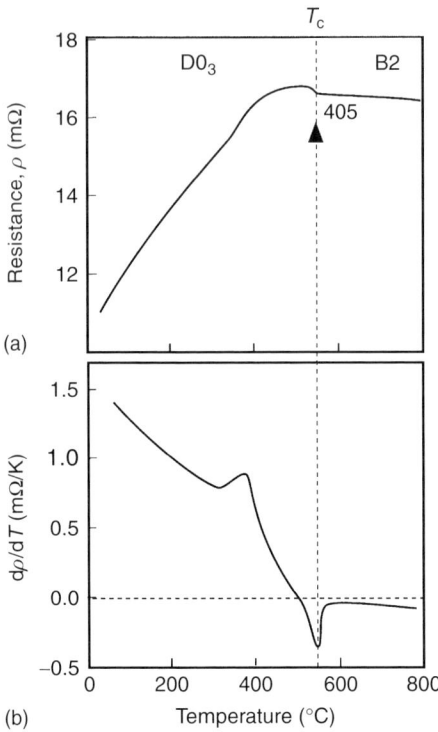

Figure 11.3 Electrical resistance curve (a) and its differential (b) showing a 2O-OD transition of Fe–27 at.% Al alloy reported by Sprusil and Pfeiler [5]. While a small and broad peak appears in the temperature region below T_c, the T_c is defined as the temperature at the minimum point of the negative peak in the differential curve.

2.2 Thermal Analysis

Thermal analysis such as differential thermal analysis (DTA) and DSC is most frequently used for the determination of T_c due to the easy preparation of specimens. The details of this technique are presented in Chapter 5. When thermal analysis is performed for the determination of OD transitions, the small signal from the transition may be one of the most difficult problems. To obtain a large signal, a high scanning rate over 5K/min is generally required. However, it should be considered that the data determined by the thermal analysis at a high scanning rate usually deviate from the real critical temperatures as well as that shown by ER examination. Figure 11.4(a) and (b) shows DSC curves obtained at different scanning rates of 20°C/min and 1°C/min for the same AuCu specimen [8]. Each DSC curve exhibits a large peak corresponding to 1O-OD transition between the $L1_0$ and $A1$ phases, and a large temperature hysteresis due to the first-order transition is detected as shown in Fig. 11.4(a). It is seen that the maximal temperature of the peak in the curve of 20°C/min is higher than that of 1°C/min, which is the effect of the scanning rate. This effect is significant in the 1O-OD transition because this transition proceeds through phase interfaces between the ordered and disordered phases. However, it is shown in Fig. 11.4 that the intersection of a basal line and a tangent at the temperature with a maximal gradient in the peak

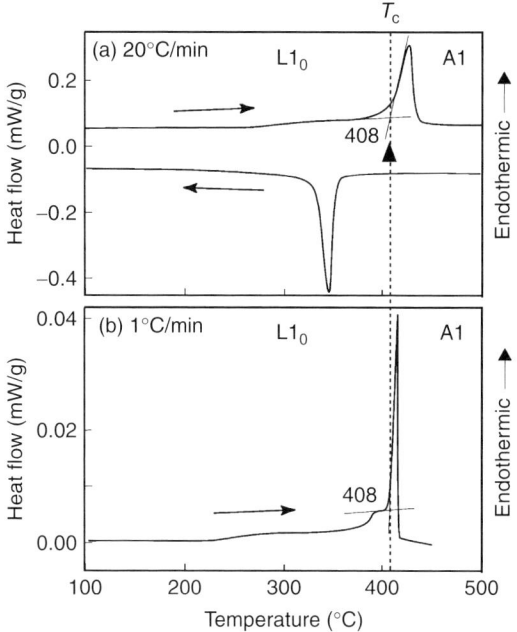

Figure 11.4 DSC curves at scanning rates of 20°C/min (a) and 1°C/min (b) showing a 1O-OD transition of a Au–50 at.% Cu alloy reported by Battezzati et al. [8]. The T_c is defined as the temperature with the intersection of a basal line and a tangent at the temperature with a maximal gradient in the peak, as demonstrated in the figures.

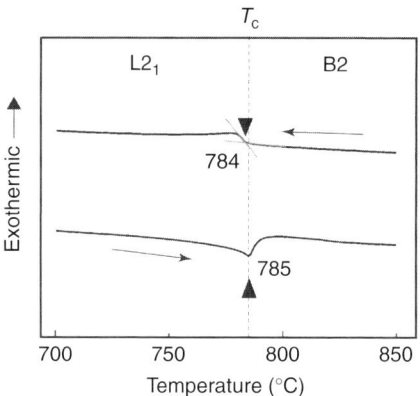

Figure 11.5 DSC curves at scanning rates of 5°C/min showing a 2O-OD transition of a Co–26 at.% Cr–24 at.% Ga alloy [9]. The T_c is defined as the temperature with the minimum point of the λ-shaped peak in the heating curve.

is hardly affected by the scanning rate. This suggests that even at a relatively high scanning rate the critical temperature of the 1O-OD transitions such as A1/L1$_0$ and A1/L1$_2$ is easily estimated by such an intersection.

On the other hand, a continuous ordering reaction, namely, 2O-OD reaction, such as A2/B2 and B2/D0$_3$ (L2$_1$) transitions, usually shows a peak with a λ shape as exhibited in Fig. 11.5 [9]. Because the 2O-OD transition proceeds in a homogeneous fashion without phase boundaries, it ideally has no thermal hysteresis and is not strongly affected by the scanning rate. Critical temperature can usually be defined as the temperature corresponding to a minimum point of the endothermic peak. It is worthwhile noting that the critical temperature determined by the heating curve is usually very close to the temperature at the onset of a peak in the cooling curve as demonstrated by Fig. 11.5. This means that the cooling curve is also viable for the determination of the T_c of 2O-OD transitions.

3 Singular Point Method

The DC method for phase diagram determination is discussed in detail in Chapter 6. After a DC is isothermally annealed at a temperature for an extended period of time, electron-probe microanalysis (EPMA) can be applied to obtain a concentration profile across the phases, which is also called a concentration–distance or concentration–penetration profile. In a concentration–penetration profile involving 2O-OD transitions such as A2/B2 and B2/DO$_3$, singular points sometimes appear at the concentration–penetration curve corresponding to a critical point of the OD transition, as shown in Fig. 11.6 [10]. This phenomenon is the result of the difference in interdiffusivity between the ordered and disordered phases at the critical point and is very useful for evaluating the phase stability of ordered phases. This technique is called the singular point method (SPM) here.

3.1 Origin of Singularity

The flux of interdiffusion is given by Fick's first law. Here, if there is a singular point due to OD transition in a concentration profile, fluxes of both the ordered and disordered phases are expressed as

$$\tilde{J}_O = -\tilde{D}_O \frac{\partial c_{BO}}{\partial x} \tag{1A}$$

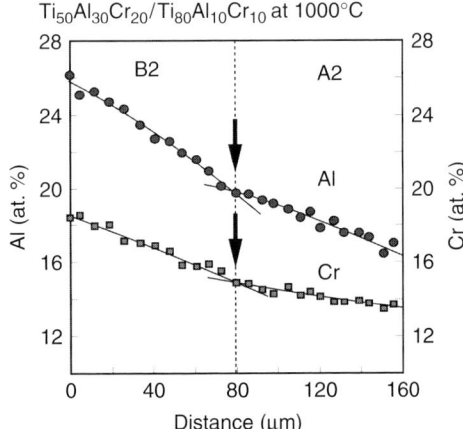

Figure 11.6 Example of a concentration–penetration profile showing a singularity due to 2O-OD transition [10].

$$\tilde{J}_{\mathrm{D}} = -\tilde{D}_{\mathrm{D}} \frac{\partial c_{\mathrm{BD}}}{\partial x} \quad (1\mathrm{B})$$

where $\tilde{D}_{\mathrm{O}}$ and $\tilde{D}_{\mathrm{D}}$ are the interdiffusion coefficients of the ordered and disordered phases, c_{BO} and c_{BD} are the concentrations of solute atom B in the ordered and disordered phases, respectively, and x is the displacement/distance. At the singular point with $c_{\mathrm{BO}} = c_{\mathrm{BD}}$, since $\tilde{J}_{\mathrm{O}} = \tilde{J}_{\mathrm{D}}$ due to mass balance, the following relation is obtained:

$$\tilde{D}_{\mathrm{O}} \frac{\partial c_{\mathrm{BO}}}{\partial x} = \tilde{D}_{\mathrm{D}} \frac{\partial c_{\mathrm{BD}}}{\partial x} \quad (2)$$

Eq. (2) means that the ratio of gradient between the ordered and disordered phases at the singular point in the concentration profile is inversely proportional to that of interdiffusion coefficient, and that the singular point is caused by the difference in interdiffusivity between the ordered and disordered phases in the vicinity of the critical point. The interdiffusion coefficient of a phase in the A–B binary system is given by Darken's equation:

$$\tilde{D} = (n_{\mathrm{A}} D_{\mathrm{B}}^{\star} + n_{\mathrm{B}} D_{\mathrm{A}}^{\star}) \Phi \quad (3)$$

where n_i and $D_i^{\star}$ are the mole fraction and self-diffusion coefficient of element i, respectively, and Φ is the thermodynamic factor [11]. It is known that both the self-diffusivity and thermodynamic factor are affected by the 2O-OD transition. The self-diffusion coefficient continuously changes with temperature or concentration at the critical point, while its Arrhenius plot has a singularity only at the critical point as shown in Fig. 11.7 [12]. On the other hand, Φ is given by

$$\Phi = \frac{\partial \mu_i}{\partial n_i} \cdot \frac{n_i}{RT} \quad (4)$$

where μ_i, R and T are the chemical potential of element i, the gas constant and the temperature, respectively. Eq. (4) means that Φ is proportional to the second-order differential of Gibbs energy, that is, a step should appear at a critical point of the 2O-OD in the Φ vs. the n_i. Figure 11.8 shows the theoretically calculated Φ vs. the Al mole fraction in the Fe–Al binary system on the basis of the cluster variation method (CVM) as reported by Schoen and Inden [13]. It is seen that steps of Φ clearly appear at the critical points of both the A2/B2 and B2/DO$_3$ transitions. Such a step of Φ has also been reported by a theoretical work using the Bragg–Williams–Gorsky (BWG) model [14]. Consequently, the origin of the singularity can be explained by the difference in the interdiffusion coefficient between the ordered and disordered phases as a result of a step of Φ due to the 2O-OD transition. It should be noted that since the interdiffusion coefficient of the higher-ordered phase (e.g., DO$_3$ to B2) is always lower than that of the lower-ordered phase, the higher-ordered phase always shows a concentration gradient larger than the lower-ordered phase at the singular point. In the next section, two examples will be shown where critical

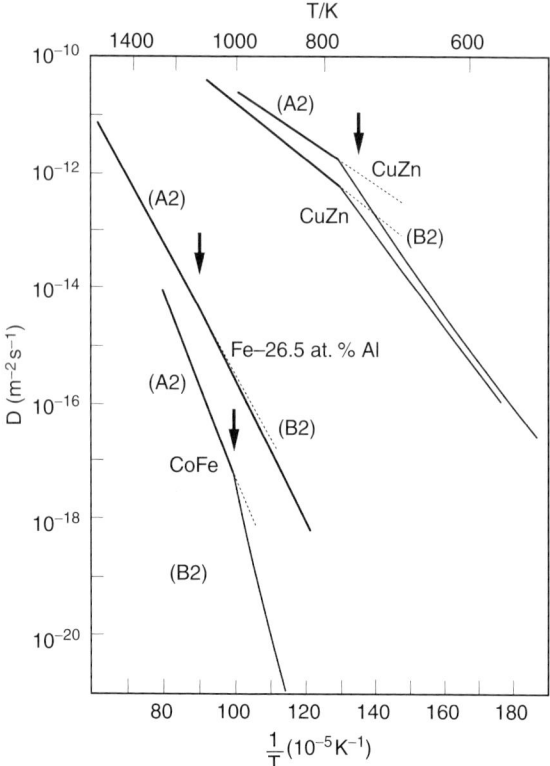

Figure 11.7 Arrhenius plots of the self-diffusion coefficient in the CuZn, FeAl and FeCo alloys involving A2/B2 2O-OD transition [12].

concentrations of B2/L2$_1$ and L1$_0$/D0$_{23}$' 2O-OD transitions were determined using the SPM.

3.2 Examples of SPM

3.2.1 B2/L2$_1$ Transition in Ni–Fe–Ti–Al System [15]

In the Ni–Al–Ti system, which has a phase diagram shown in Fig. 11.9 [15], it has also been reported that the precipitation of the Ni$_2$AlTi (L2$_1$) phase in the NiAl (B2) matrix phase increases the creep strength of NiAl at high temperature. Recently, Kainuma et al. have reported that the B2 + L2$_1$ two-phase regions in the NiAl–Ni$_2$AlTi and Ni$_2$AlTi–NiTi sections narrow with increasing temperature, and that the tricritical points (T_t) are located slightly above the liquidus [15]. On the other hand, only the 2O-OD transition has been reported in the FeAl (B2)–Fe$_2$AlTi (L2$_1$) pseudo-binary section [16]. It is expected, therefore, that the NiAl–Ni$_2$AlTi transition changes to continuous ordering due to the substitution of Fe for Ni.

Figure 11.10 shows a typical concentration profile involving a B2/L2$_1$ continuous ordering [15], where a singular point corresponding to the critical concentration of the transition is clearly detected in the Al and Ti profiles. It can be seen that

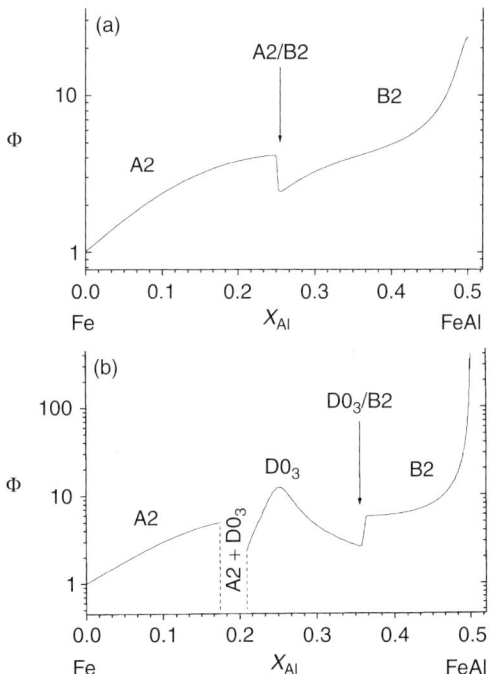

Figure 11.8 Thermodynamic factor for interdiffusion at (a) 1000K and (b) 623K in the bcc phase of Fe–Al system [13].

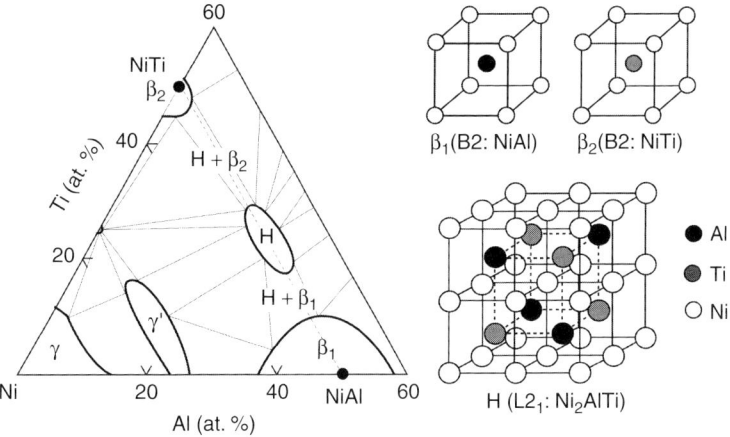

Figure 11.9 The Ni–Al–Ti ternary phase diagram at 900°C [15].

the slope of profile on the left is larger than that on the right, and that the right and the left side phases correspond to the B2 and L2$_1$ ordered phases, respectively. The phase boundary determined by SPM coincides with the data obtained by DSC measurements as shown Fig. 11.11 [15], which confirms the accuracy of the critical point determined using the SPM. All the results in the NiAl–FeAl–FeTi–NiTi section are plotted in Fig. 11.12 [15]. In every isothermal section, the NiAl + Ni$_2$AlTi

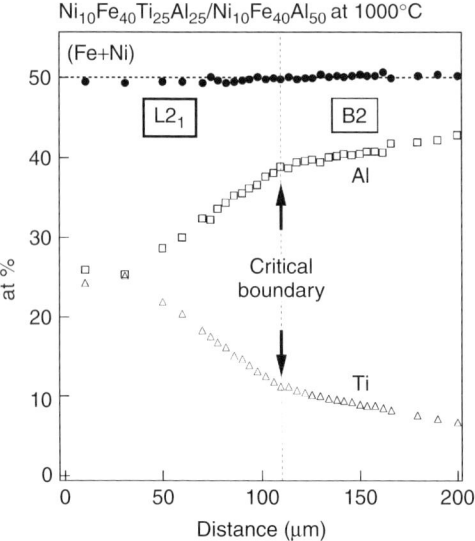

Figure 11.10 Concentration profile obtained from a Ni–40Fe–25Ti–25Al/Ni–40Fe–50Al DC [15].

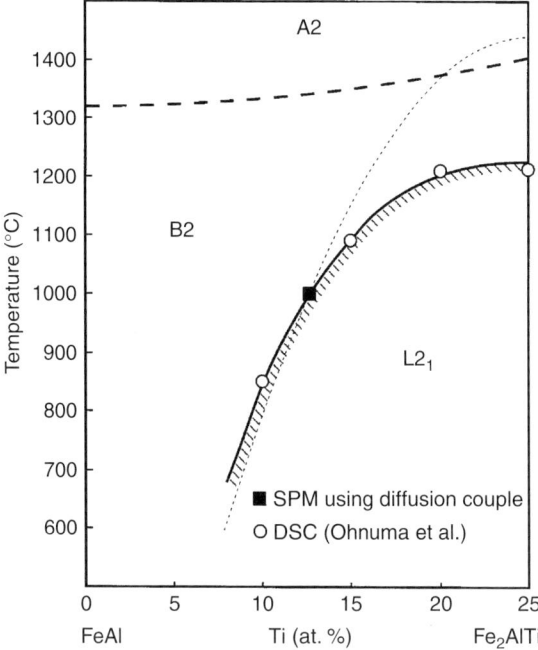

Figure 11.11 Critical temperatures of B2/L2$_1$ 2O–OD transition determined by DSC (open circles) and critical concentration obtained by SPM at 1000°C (closed square) in the FeAl–Fe$_2$AlTi pseudo-binary section [15].

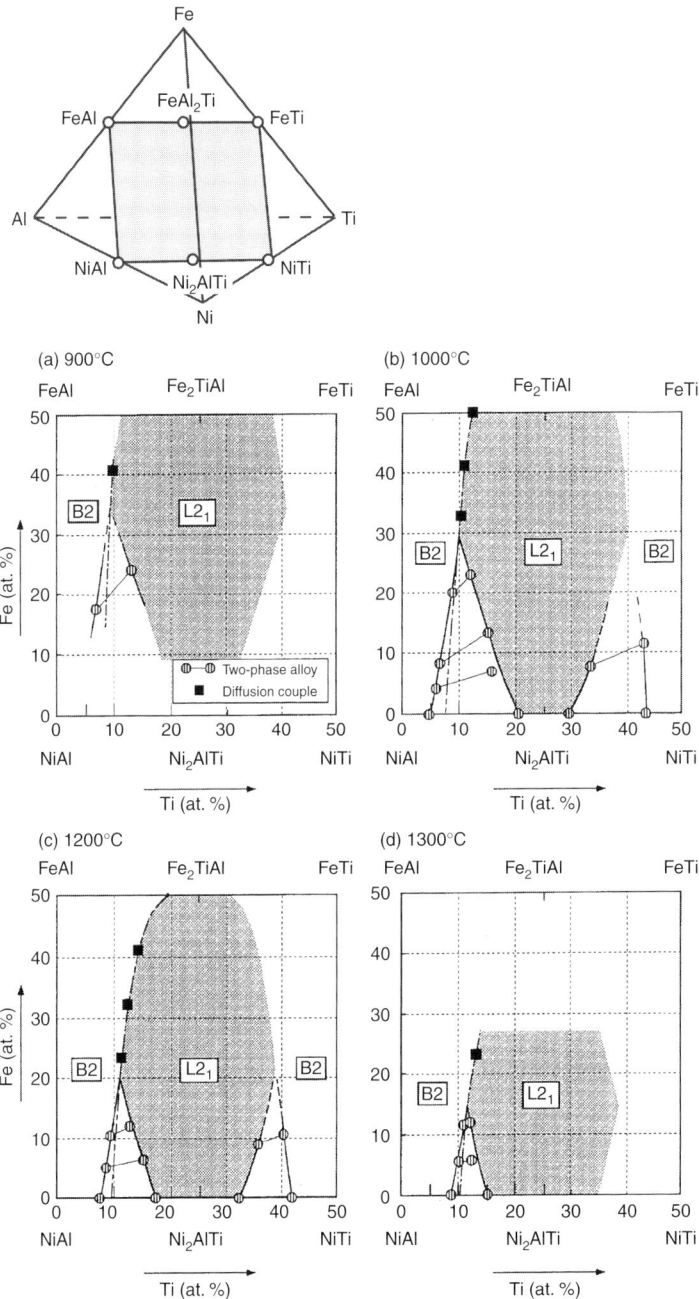

Figure 11.12 Isothermal sections at temperatures ranging from 900°C to 1300°C in the Ni–Fe–Al–Ti quaternary system [15].

two-phase region becomes narrower by the addition of Fe and continues to the critical boundary of the 2O-OD transition, where the two-phase regions were determined by EPMA using two-phase alloys. This excellent agreement of data between the two-phase region and the continuous ordering boundary suggests SPM is a powerful technique for the determination of the OD transition boundary.

3.2.2 L1$_0$/Modified Do$_{23}$ Transition in the Al-Rich Portion of Ti–Al System [17]

Figure 11.13 shows a recent version of the Ti–Al binary phase diagram [18]. In the Ti-rich portion of the Ti–Al system, the authors' group has experimentally examined the phase equilibria of the α(A3), α_2 (Ti$_3$Al: D0$_{19}$), β(A2 or B2) and γ(TiAl: L1$_0$) phases, and has established the existence of an ordered β_2 (B2) phase [19]. The phase diagram in the Al-rich portion has long been under discussion due to the very complicated transitions from the L1$_0$ to highly ordered structures, such as TiAl$_2$ (oC12: Ga$_2$Zr type), Ti$_3$Al$_5$ (oP4: Ti$_3$Al$_5$ type) and Ti$_5$Al$_{11}$ (D0$_{23}$: AlZr$_3$ type). Very recently, Palm et al. [20] have investigated the phase equilibria in the Al-rich, high-temperature region using the DC technique, and suggested from the

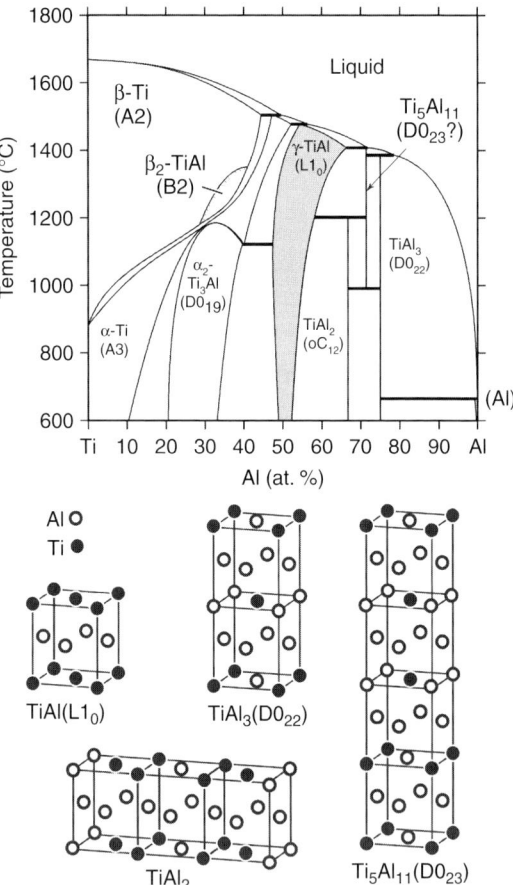

Figure 11.13 The Ti–Al binary phase diagram [18].

results of XRD and transmission electron microscopy (TEM) observation that a second-order transition from $L1_0$ to Ti_5Al_{11} may exist. Neither the composition nor temperature dependence of the critical boundary was determined. In this part, the recent investigation [17] performed by the authors using the SPM is introduced combining with the result of experimental determination of the interdiffusivity.

Figure 11.14 shows the concentration–penetration profile of a Ti–55Al/Ti–75Al DC annealed at 1250°C for 12 h and its corresponding image. Only one boundary was observed in the specimen of 1250°C, and the equilibrium concentrations of each phase appearing in the DC profile can be evaluated by extrapolation from the single-phase region to the phase boundary. The data determined by the DC are in agreement with those by the bulk specimens. A singularity in the concentration–penetration profile of the DC was also detected at about 64.5 at.% Al, as indicated by a double-headed arrow in Fig. 11.14. This suggests that the γ/γ_2 (Ti_5Al_{11}) ordering reaction may be a 2O-OD transition as in the cases of the A2/B2 and B2/L2$_1$

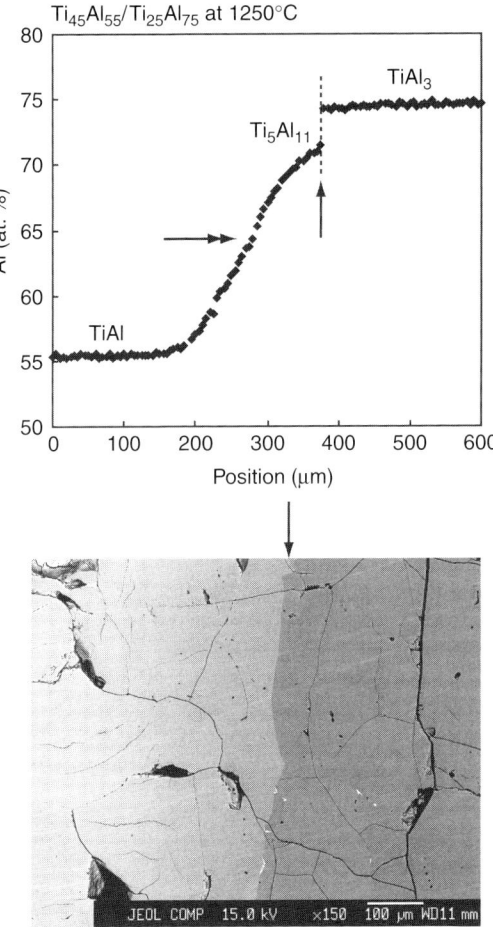

Figure 11.14 Concentration–penetration profile obtained from a Ti–55Al/Ti–75Al DC annealed at 1250°C for 12 h [17].

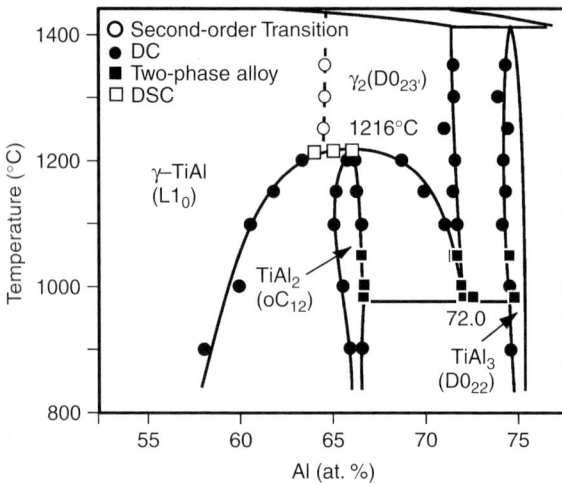

Figure 11.15 Phase diagram in the Al-rich portion of Ti–Al system [17].

transitions. The concentrations obtained at such a singular point are plotted with open circles in Fig. 11.15. It can be seen that all the critical compositions are located at about 64.5 at.%; the critical boundary is drawn as a vertical straight line.

Although the stoichiometric TiAl$_3$ can have a D0$_{23}$ atomic configuration as shown in Fig. 11.13, the configuration of off-stoichiometric alloys with a Ti concentration higher than 25 at.% is slightly complicated. In comparison to the L1$_0$ structure, an atomic configuration with a "modified" D0$_{23}$ (D0$_{23}'$) structure shown in Fig. 11.16 is expected rather than the regular D0$_{23}$ structure shown in Fig. 11.13. The D0$_{23}'$ structure which is an ordered structure of L1$_0$ can be easily obtained by the two-dimensional ordering only in the Ti layer as demonstrated in Fig. 11.16 [17]. The atomic configuration of the D0$_{23}'$ was confirmed by high-temperature XRD (HTXRD) patterns.

As mentioned above, a step in the interdiffusion coefficient vs. concentration (mole fraction) plot is expected at the critical concentration of a 2O-OD transition. Interdiffusion coefficients of the γ and γ_2 phases were extracted from the concentration profiles at 1250–1350°C using the Boltzmann–Matano method. The smoothing of the EPMA data was performed in the same way as that reported by Kainuma and Inden [21]. The compositional dependence of the interdiffusion coefficients of the γ and γ_2 phases is shown in Fig. 11.17 [17]. A step in the diffusivity due to L1$_0$/D0$_{23}'$ ordering exists at about 64.5 at.% Al in each DC at every temperature as indicated by an arrow, which directly corresponds to the singular point in the 1250°C profile shown with the double-headed arrow in Fig. 11.14.

4 CONCENTRATION GRADIENT METHOD

The coherent two-phase microstructure formed by phase separation due to 1O-OD is usually too fine to be directly and quantitatively determined by EPMA. Even after ageing for a long duration, it is difficult to obtain a measurable size of

Determination of Phase Diagrams Involving Order–Disorder Transitions

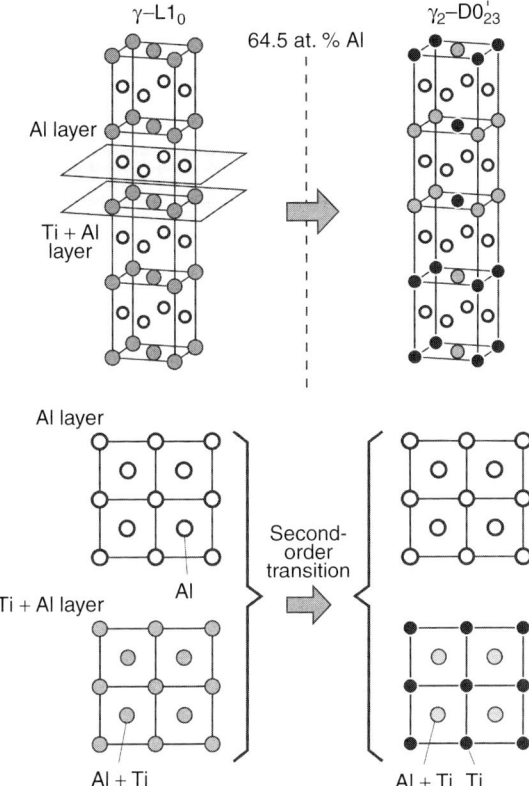

Figure 11.16 OD transition from $L1_0$ to a modified $D0_{23}$ structure [17].

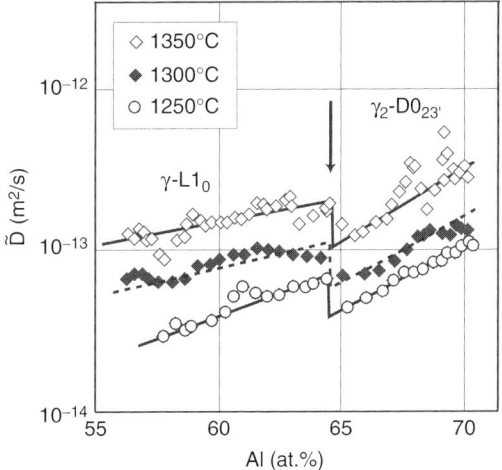

Figure 11.17 Variation of the interdiffusivity with Al concentration in the γ and γ_2 phases [17].

precipitates without loosing the lattice coherence. The concentration gradient method (CGM) was proposed by Miyazaki et al. [22] to determine the coherent two-phase region formed by 1O-OD transition, which is applied to the γ(A1: Ni solid solution) + γ' (L1$_2$: fcc ordered phase) region in the Ni–Al based system in combination with TEM observations and energy dispersive spectroscopy (EDS) analysis of the microstructure appearing in the diffusion–penetration profile of DC specimens. This technique is very useful for the determination of both the two-phase region and the critical point of the 2O-OD transition; and it does not always require the use of TEM and TEM–EDS – only scanning electron microscopy (SEM) and EPMA.

4.1 Basics of CGM

The experimental procedure for the CGM consists of the following four steps:

1. Formation of a continuous concentration profile in a DC specimen by annealing at a high temperature.
2. Formation of precipitation or anti-phase domain (APD) zone in the diffusion layer of the "as-annealed" DC, due to ageing followed by quenching.
3. Microstructural observation in the vicinity of the precipitation or APD zone by TEM or SEM.
4. Construction of a concentration–penetration profile using EDS or EPMA, and determination of the boundary concentrations between the single-phase- (or no APD) and two-phase (or APD) zones.

In the case of the 1O-OD transition, this boundary concentration should correspond to the solubility concentration of the coherent precipitation phase in the matrix phase. It should be noted, however, that the result obtained by CGM is always located in the two-phase region slightly beyond the solubility line at least, because of an incubation time for precipitation. Fortunately, the difference from the solubility line is actually negligible according to our experience.

On the other hand, the critical concentration of 2O-OD transition can be determined with higher accuracy because the APD structure is formed during cooling from the disordered region without nucleation. In the next section, two examples are shown where the A2 + DO$_3$ (B2) two-phase region and the critical concentration of the A2/B2 2O-OD transition were determined using the CGM.

4.2 Examples of CGM

4.2.1 A2 + Do$_3$ Two-Phase Region and A2/B2 Transition in the Fe–Al System [23]

The disordered A2, ordered B2 (FeAl) and D0$_3$ (Fe$_3$Al) phases are formed in the Fe-rich portion of Fe–Al binary system as shown in Fig. 11.18 [24]. In addition, there is a (para/ferro) magnetic transition in the Fe-rich portion and the critical temperature of the magnetic transition ($T_{c\text{-mag}}$) decreases with increasing Al content, which intersects the A2/B2 ordering temperature ($T_{c\text{-A2/B2}}$) in the vicinity of 20 at.% Al. An interesting feature is that $T_{c\text{-mag}}$ drastically decreases in the D0$_3$ phase region. These characteristic features of the stability of the ordered phases in the Fe–Al

Determination of Phase Diagrams Involving Order–Disorder Transitions

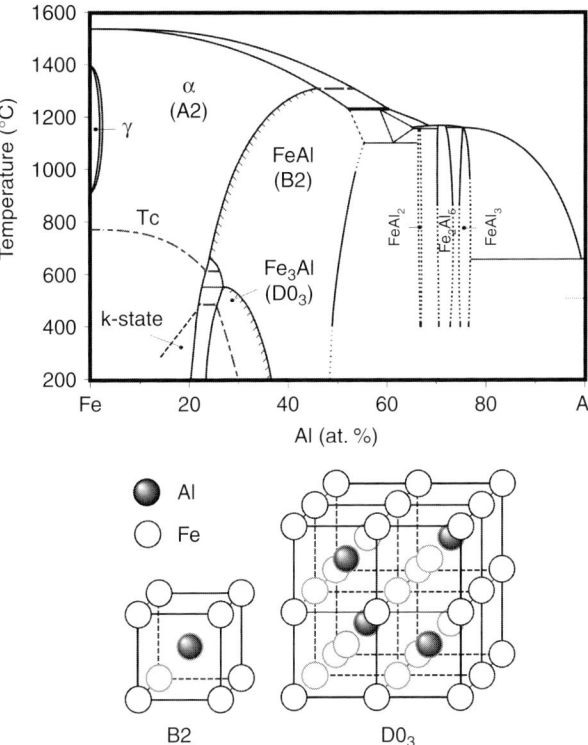

Figure 11.18 Fe–Al binary phase diagram and crystal structures of B2 and D0$_3$ [24].

system suggest that there is a strong interaction between the magnetic and chemical orderings in Fe–Al alloys [25]. Moreover, there is an unknown region called "k-state", in which some anomalies of the electrical resistance, thermal dilatation and magnetic and mechanical properties are observed at very low temperatures below 450 °C in the Al concentration range of 15–20 at.% [26] as shown in Fig. 11.18. Here, accurate determination of the phase equilibrium compositions between the A2 and D0$_3$ phases at low temperatures by the CGM in combination with TEM observation is shown.

DC specimens were prepared by melting pure Al pieces in a Fe-15Al alloy columnar crucible, with dimensions of 10 mm diameter and 10 mm height, having a hole of 3 mm diameter and 5 mm deep in the center. The DC was annealed at 1300 °C for 4 h in a quartz capsule under an argon atmosphere followed by quenching in ice water. Slices of 1 mm in thickness of the as-annealed DC were prepared and then aged at 300–700 °C. TEM specimens with a diameter of 3 mm and a thickness of 0.2 mm were punched out after mechanical thinning of the aged DCs. Thin foils were prepared by the twin-jet method from the disk specimens dimpled in the vicinity of the diffusion zone. Concentration profiles of the DC specimens were determined using SEM with EDS. Figure 11.19(a) and (b) shows the appearance of the DC specimen for TEM observation and the concentration profile along the white points, respectively [23]. It can be seen that a continuous concentration

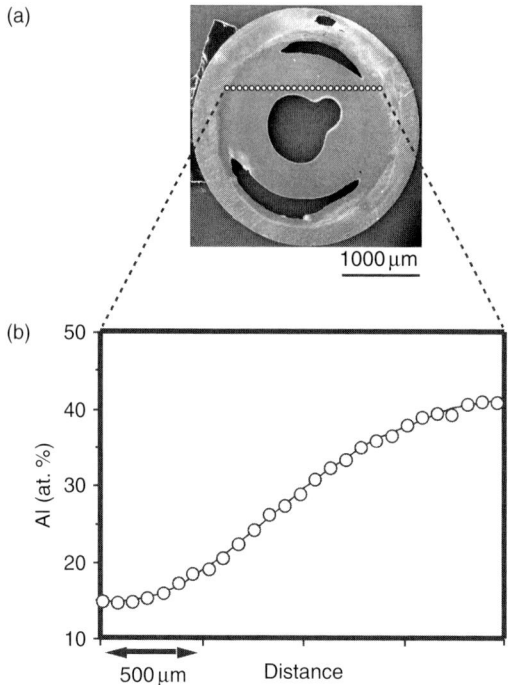

Figure 11.19 (a) SEM micrograph and (b) corresponding concentration profile of the disk specimen for TEM observation cut out of the DC aged at 500°C for 3 days after quenching from 1300°C [23].

profile is obtained by annealing at 1300°C. The shape of this concentration profile taken from the DC quenched from 1300°C did not change by the ageing treatments at the temperatures below 700°C after quenching.

Figure 11.20(a) and (b) shows a SEM micrograph and the corresponding concentration–penetration profiles obtained by SEM–EDS from the thin foil DC specimen annealed at 500°C [23]. The upper and lower regions in the SEM micrograph of Fig. 11.20(a) correspond to the foil and the hole of the TEM specimen, respectively. Although it was confirmed by ED that all regions shown in Fig. 11.20 had the ordered DO_3 structure at room temperature, the dark-field (DF) images of the $\{111\}_{DO_3}$ ordered reflection taken from positions A and B in Fig. 11.20(a) exhibit many small spherical particles as shown in A and B of Fig. 11.20(c), while those from positions C and D show an APD structure as shown in C and D of Fig. 11.20(c). These results suggest that the $A2 + DO_3$ two-phase region exists in the diffusion zone of this DC. It seems easy to determine the $A2/(A2 + DO_3)$ boundary that corresponds to the concentration at the position where the DO_3 particles start to appear. However, it is not easy to distinguish between the DO_3 single-phase and the $A2 + DO_3$ two-phase structures because the microstructure in the $A2 + DO_3$ two-phase region near the DO_3 single-phase region is usually very similar to the APD structure in the ordered DO_3 single phase. It is known that the characteristic features of the APD contrast are useful for distinguishing the DO_3 single phase from the $A2 + DO_3$ two-phase structures. The contrast of the APD boundary of the single

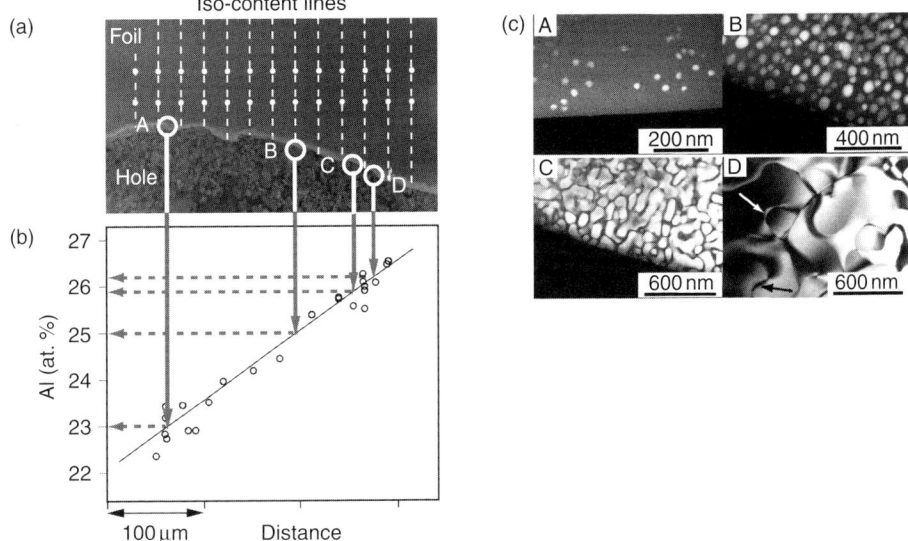

Figure 11.20 (a) SEM micrograph and (b) the corresponding concentration profile in a Fe–Al DC annealed at 500°C for 3 days. (c) TEM DF images with (111) D0$_3$ super-lattice reflection taken from the points (A), (B), (C) and (D) in Fig. 11.3(a), where (A), (B) and (C) show an A2 + D0$_3$ two-phase structure, and (D) exhibits a D0$_3$ single-phase structure [23].

phase in the $\{111\}_{D0_3}$ DF image changes from white to black depending on the reflection condition as indicated by arrows in image D in Fig. 11.20(c), while the APD-like structure in the A2 + D0$_3$ two-phase region does not show such a change – the domain boundary consisting of the disordered phase always has a black contrast as shown in image C in Fig. 11.20(c). It can be concluded from the TEM observation that the A2 + D0$_3$/D0$_3$ boundary in this DC specimen should exist between positions C and D in Fig. 11.20(a). It is also seen in Fig. 11.20(c) that the average size of the D0$_3$ domains in the two-phase region C is considerably smaller than that in the single-phase region D. Figure 11.21 shows the domain size against the Al content obtained by observing the diffusion zone in the DC specimens aged at several temperatures ranging from 350°C to 500°C [23]. In every profile, there exists a step in domain size. This drastic change of domain size is brought about by the difference of growth rate between the APDs in the D0$_3$ single-phase region and two-phase domains in the A2 + D0$_3$ region. In the DC specimen aged at 300°C, only the A2/A2 + D0$_3$ boundary was confirmed by the existence of fine D0$_3$ particles. These results are in good agreement with those determined using the domain boundary contrast.

Figure 11.22(a) and (b) shows a TEM DF image of $\{100\}_{B2}$ and the corresponding concentration–penetration profile obtained by SEM–EDS for the thin foil DC specimen quenched from 700°C to ice water [23]. The left side of the SEM micrograph seems to show many fine APDs, while the right side has no contrast from APDs. This result confirms that the boundary of the A2/B2 OD transition is located at the position indicated by the arrow, where the composition corresponding to boundary is 25.1 at.% Al as shown in Fig. 11.22(b).

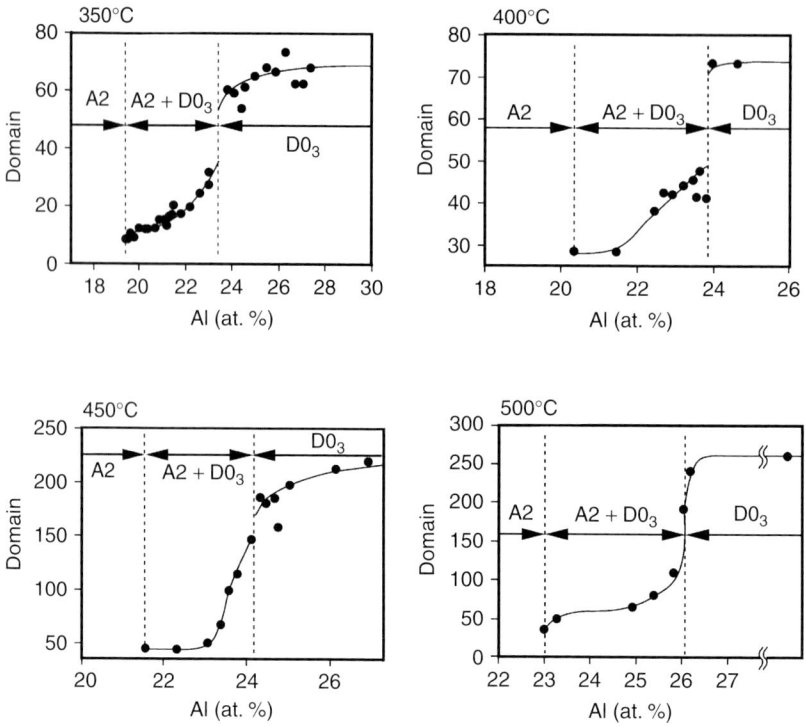

Figure 11.21 The APD size and Al concentration in the DC specimens annealed at 350–500°C [23].

All the data obtained are plotted with solid circles in Fig. 11.23 [23]. As shown in Fig. 11.18, the phase boundaries determined by CGM are in good agreement with those in the assessed phase diagram [24] in the temperature range over 500°C. This indicates that the CGM technique is applicable for the determination of the critical concentration of continuous ordering and the two-phase region formed due to ordering. One of the most important advantages of CGM is that it is able to determine the phase equilibria at low temperatures. As shown in Fig. 11.23, it was firstly confirmed that the Al concentration of the A2/(A2 + D0$_3$) two-phase boundary decreases with decreasing temperature and reaches about 17 at.% Al at 300°C, which suggests that the k-state basically corresponds to the phase separation between A2 and D0$_3$ phases. Finally, it should be emphasized that this method is also applicable for the determination of metastable phase diagram. One example was reported by the present authors for the Fe–Ga binary system [27].

5 CONCLUDING REMARKS

The methods for the determination of phase equilibria involving OD transitions are introduced in this chapter. Less emphasis was placed on the well-known methods such as DSC, ER and XRD. The focus of this chapter was on new techniques, namely, SPM and CGM. The most remarkable feature of SPM and CGM using the

Determination of Phase Diagrams Involving Order–Disorder Transitions

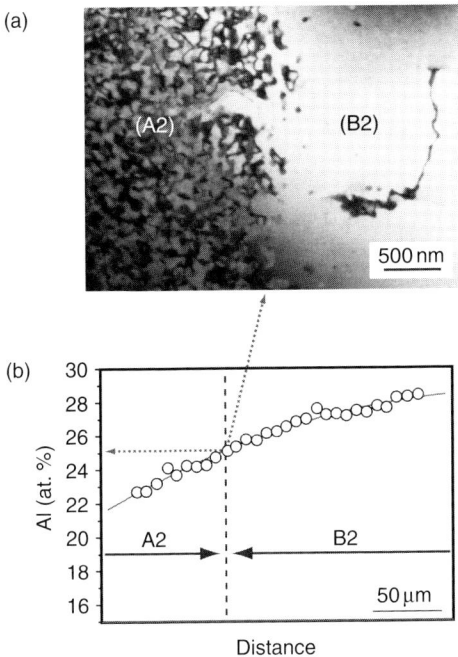

Figure 11.22 (a) The DF image with $(100)_{B2}$ super-lattice reflection taken from the DC aged at 700°C, showing the A2/B2 continuous boundary and (b) concentration profile corresponding to the microstructure in Fig. 5(a) [23].

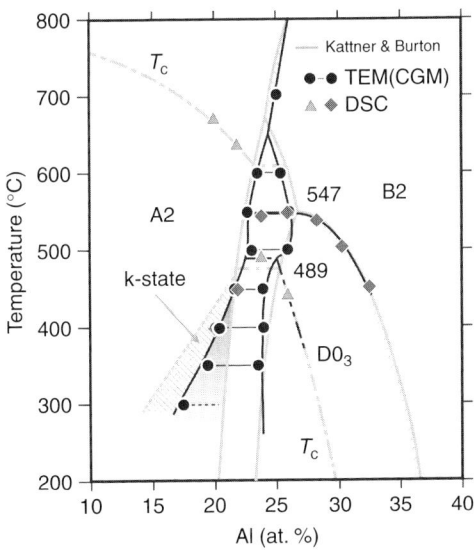

Figure 11.23 Phase diagram of the Fe-rich portion of the Fe–Al binary system [23].

DCs is that they can determine the critical concentration and solubility line of the coherent precipitation at a fixed temperature. This feature is very useful, when isothermal section diagrams are constructed as demonstrated in Fig. 11.12.

Other less widely used methods such as electromotive force (emf) measurements for OD transitions are not discussed in this chapter. Readers are referred to the original publications for details (e.g., Ref. [28]).

REFERENCES

1. O. Dimitrov, in J.H. Westbrook, R.L. Fleischer (eds.), *Intermetallic Compounds Principles and Practice, Vol. 1. Principles*, John Wiley and Sons, New York, 2002, p. 771.
2. B. Sprusil and W. Pfeiler, *Intermetallics*, 5 (1997) 501.
3. H. Okamoto, *Desk Handbook: Phase Diagrams for Binary Alloys*, ASM International, Materials Park, OH, 2000, p. 72.
4. M.S. Seehra and P. Silinsky, *Phys. Rev. B*, 13 (1976) 5183.
5. B. Sprusil and W. Pfeiler, *Intermetallics*, 7 (1999) 635.
6. Y. Nishino, C. Kumada and S. Asano, *Scripta Mater.*, 36 (1997) 461.
7. Y. Nishino, M. Matsuo, S. Asano and N. Kawamiya, *Scripta Metall. Mater.*, 25 (1991) 2291.
8. L. Battezzati, M. Belotti and V. Brunella, *Scripta Mater.*, 44 (2001) 2759.
9. K. Kobayashi, R. Kainuma, K. Fukamichi and K. Ishida, *J. Alloy. Compd.*, 403 (2005) 161.
10. R. Kainuma, I. Ohnuma, K. Ishikawa and K. Ishida, *Intermetallics*, 8 (2000) 869.
11. L.S. Darken, *Trans. Am. Inst. Min. Metall. Eng.*, 175 (1948) 184.
12. R. Kainuma, H. Nakajima and K. Ishida, *High Temp. Mater. Proc.*, 28 (1999) 313.
13. C.G. Schoen and G. Inden, *Acta Mater.*, 46(1998) 4219.
14. T. Helander and J. Agren, *Acta Mater.*, 47 (1999) 1141.
15. R. Kainuma, K. Urushiyama, C.C. Jia, I. Ohnuma and K. Ishida, *Mater. Sci. Eng.*, A239–240 (1997) 235.
16. I. Ohnuma, C.G. Schoen, R. Kainuma, G. Inden and K. Ishida, *Acta. Mater.*, 46 (1998) 2083.
17. R. Kainuma, J. Sato, I. Ohnuma and K. Ishida, *Intermetallics*, 13 (2005) 784.
18. H. Okamoto, *Desk Handbook: Phase Diagrams for Binary Alloys*, ASM International, Materials Park, OH, 2000, p. 46.
19. I. Ohnuma, Y. Fujita, H. Mitsui, K. Ishikawa, R. Kainuma and K. Ishida, *Acta Mater.*, 48 (2000) 3113.
20. M. Palm, L.C. Zhang, F. Stein and G. Sauthoff, *Intermetallics*, 10 (2002) 523.
21. R. Kainuma and G. Inden, *Z. Metallkd.*, 88 (1997) 429.
22. T. Miyazaki, T. Koyama and K. Kobayashi, *Metall. Mater. Trans.*, 27A (1996) 945.
23. O. Ikeda, I. Ohnuma, R. Kainuma and K. Ishida, *Intermetallics*, 9 (2001) 755.
24. U.R. Kattner and B.P. Burton, in H. Okamoto (ed.), *Phase Diagrams of Binary Iron Alloys*, ASM International, Materials Park, OH, 1993, p. 12.
25. I. Ohnuma, O. Ikeda, R. Kainuma and K. Ishida, *Z. Metallkd.*, 89 (1998) 12.
26. R.G. Davies, *J. Phys. Chem. Solids*, 24 (1963) 985.
27. O. Ikeda, I. Ohnuma, R. Kainuma and K. Ishida, *J. Alloy Compd.*, 347 (2002) 198.
28. R.A. Oriani, *J. Electrochem. Soc.*, 103 (1956) 194.

CHAPTER TWELVE

DETERMINATION OF PHASE DIAGRAMS INVOLVING MAGNETIC TRANSITIONS

Ichiro Takeuchi[1] and Samuel E. Lofland[2]

Contents

1 Introduction	383
2 The Basics of Magnetism and the Traditional Methods of Mapping Magnetic Phase Diagrams	384
2.1 Basics of Magnetism	384
2.2 Measurements of Magnetism	385
2.3 Effects of Magnetism on Phase Diagrams	389
3 Combinatorial and High-Throughput Mapping of Magnetic Phase Diagrams	396
3.1 Introduction to the Combinatorial Approach	396
3.2 Combinatorial Mapping of Magnetic Phase Diagrams in Metallic Systems	396
3.3 Combinatorial Mapping of Oxide Systems	404
4 Concluding Remarks	409

1 INTRODUCTION

From the early days of civilization, magnetism, and magnetic materials have always played profound roles in advancing our society. More than a thousand years ago, the Chinese began using the magnetic properties of lodestone to make compasses for navigation. Technologically, magnetic materials represent some of the most important classes of materials. They are used in a wide variety of everyday applications ranging from permanent magnets used in motors and transformers to ferrites used in high-frequency communication devices and multilayer thin film devices used for giant magnetoresistance sensors. The ability to control and modify magnetic properties of these materials is crucial for implementing them in specific applications, and thus, exploring compositional phase diagrams and mapping the compositional dependence of various magnetic properties has always been a central part of development of magnetic materials. Key parameters of magnetic materials include the magnetic transition temperature, saturation magnetization, remnant magnetization, coercive field, and magnetic anisotropy.

[1] Dept. of Materials Science and Engineering and Center for Superconductivity Research, University of Maryland, College Park, MD, USA.
[2] Dept. of Physics and Astronomy, Rowan University, Glassboro, NJ, USA.

In condensed matter physics, magnetic materials have always provided some of the most interesting and intriguing topics of investigation. Long-range ordering in any magnetic system is a collective result of alignment due to various exchange interactions between magnetic ions and/or exchange coupling of magnetic layers at the atomic level. Despite over a century of research, no comprehensive theory of magnetism exists. In many systems, the exact nature of the spin interactions is not understood at the atomic level, particularly in metals. A good example is magnetic semiconductors in which both charge and spin degrees of freedom are to be utilized in the so-called spintronics applications [1]. There is currently an intense worldwide effort to develop magnetic semiconductors with room-temperature magnetism, and there have been reports of many materials in oxide-based systems. But the origin and the mechanism of ferromagnetism in the oxide-based magnetic semiconductors are yet to be understood. Perovskite manganese oxides have been extensively studied by the community and have displayed a rich variety of magnetism related phenomena. Many of these materials are highly correlated systems whose properties are dictated by intricate interplay between several competing forces such as the double-exchange interaction and spin/charge ordering, and subtle change or shift in composition can lead to drastic change in physical properties [2]. Mapping the magnetic phase diagram in such systems has played a key role in uncovering the physics of the materials.

The traditional way of mapping the phase diagram is to make bulk solid samples with compositions in the region of interest and measure the properties one sample at a time. While much information on magnetic properties can be obtained by performing, for instance, magnetization measurements, it is extremely time consuming to make a large number of individual samples to densely cover the compositional range in a multicomponent phase diagram. This is particularly tedious when trying to investigate properties near a critical point.

In comparison, in the emerging field of high-throughput or combinatorial materials science, techniques are being developed where entire (or large fractions of) compositional phase diagrams are being generated and mapped simultaneously in the form of individual wafers/chips [3]. Such composition-spread experiments allow rapid survey of various physical properties as a function of composition. By incorporating appropriate magnetic property mapping techniques, the experiments have the potential to uncover all the salient features of the phase diagrams with just one sample.

This chapter reviews experimental techniques to map magnetic properties of materials across phase diagrams with an emphasis on high-throughput approaches. Traditional methods for investigating magnetic properties across phase diagrams using bulk samples will also be described first with specific examples. This will be followed by a section on high-throughput approaches to magnetic phase diagrams using thin-film techniques. Some of the phase diagrams mapped by the high-throughput techniques will be compared with bulk phase diagrams.

2 THE BASICS OF MAGNETISM AND THE TRADITIONAL METHODS OF MAPPING MAGNETIC PHASE DIAGRAMS

2.1 Basics of Magnetism

Magnetic phase transitions are associated with changes not only in magnetization but in other properties such as heat capacity, resistivity, density, thermal expansion, elastic

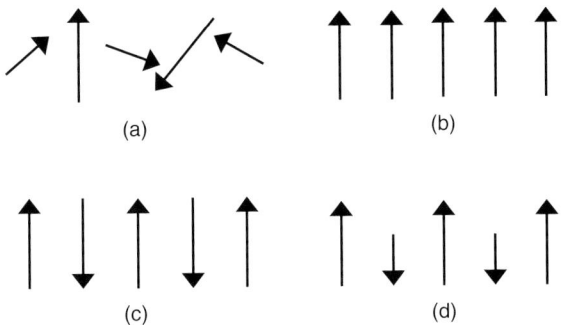

Figure 12.1 Schematic of the spin orientation in various magnetic states: (a) paramagnetic, (b) ferromagnetic, (c) antiferromagnetic, and (d) ferrimagnetic.

moduli, and structure as well. Phase transitions can be discontinuous (first order) or continuous (second order or higher). For most alloys, it is the phases with different crystal structures that dominate the phase diagram; however, in materials that are magnetic, magnetism often plays an important role in the phase diagram, particularly near phase boundaries, just as pressure is important near the melting point of a solid.

From a semi-classical point of view, magnetism arises from the quantum mechanical currents associated with spin angular momentum of electrons, although there can be minor contributions from orbital angular momentum. A single electron has a magnetic moment of one Bohr magneton, $\mu_B = 9.27 \times 10^{-24}$ A m^2.

There are various possible magnetic states, not all of which are necessarily ordered. Among them, paramagnetism, ferromagnetism (ferrimagnetism), antiferromagnetism, and spin glass states are most common (Fig. 12.1). Paramagnetism is a disordered state with no correlation between neighboring spins such as in Cu (Fig. 12.1(a)). Ferromagnetism is an ordered state where all spins are aligned with each other (Fig. 12.1(b)) as in Fe and Co. An antiferromagnet (Fig. 12.1(c)), such as Cr, consists of two or more ordered spin sublattices which are aligned antiparallel such that the net magnetization is zero. In contrast, a ferrimagnet (Fig. 12.1(d)), like magnetite (Fe$_3$O$_4$,) has two or more unequal spin sublattices aligned antiparallel such that there is a net moment. In many ways, a ferrimagnet acts like a ferromagnet. A spin glass is a disordered magnetic state where the spatial average of the magnetization over the sites is zero while the time average of the orientation of any given spin is non-zero, in contrast to a paramagnet where both averages are zero. A spin glass state can arise from a competition between antiferromagnetic and ferromagnetic interactions or from geometric frustration such as triangular lattice with antiferromagnetic coupling between nearest neighbors.

2.2 Measurements of Magnetism

Measurements of magnetic transitions are often done with either a vibrating sample or superconducting quantum interference device (SQUID) magnetometer, which determine the magnetic moment of a sample usually in presence of applied magnetic field H_{app}. Since magnetic field lines always form closed loops, any sample with a finite aspect ratio produces an external dipolar field (Fig. 12.2). This

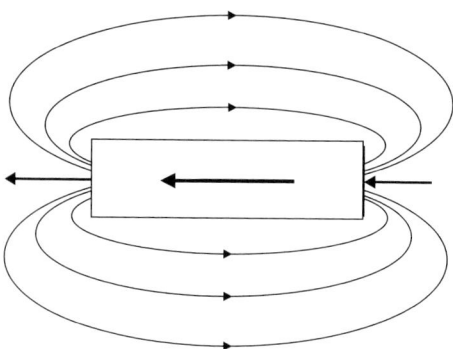

Figure 12.2 Dipolar fields from a magnetic sample oppose the applied field and have to be accounted for.

demagnetizing field H_d opposes H_{app} and is proportional to the volume magnetization of the sample, that is, $H_d = DM$ where D is the demagnetization factor. So the true field sensed by the sample is $H = H_{app} - DM$. Unless the shape of the sample can be represented by an algebraic equation of degree 2 (sphere, ellipsoid, etc.), the demagnetizing field is not uniform. There are numerous analytical and approximate results for various shapes [4].

Each magnetic state has its own characteristic field and temperature dependence. Paramagnets usually show linear field dependence, except near a phase transition. The susceptibility $\chi = M/H$ is the slope of the magnetization curve. Often for magnetic materials, χ can be described by the Curie–Weiss law:

$$\frac{1}{\chi} = \frac{C}{T - \theta} \quad (1)$$

where θ is the Weiss constant and $C = \left| \sum_{i=1}^{n} g_i^2 S_i (S_i + 1) \right|$ is the Curie constant where the sum is over the n magnetic sublattices with spectroscopic splitting factor g_i and spin S_i. The sign of θ usually correlates with the type of magnetic interaction, being positive for ferromagnetic interactions and negative for antiferromagnetic interactions.

A typical field dependence of ferro- and ferrimagnets is shown in Fig. 12.3. Note that the magnetization displays a typical hysteresis which results from the motion of magnetic domains. The domains arise because it is energetically favorable to form them by trading some exchange energy for magnetostatic energy. The magnetic field at which the magnetization changes sign is the coercive field H_c. The remnant magnetization M_r is the magnetization which remains after removal a sufficiently large field to reach the saturation magnetization M_s where all spins are aligned. The field required to reach magnetic saturation is generally a measure of the magnetic anisotropy, except along the preferred direction of spin alignment, the magnetic easy axis.

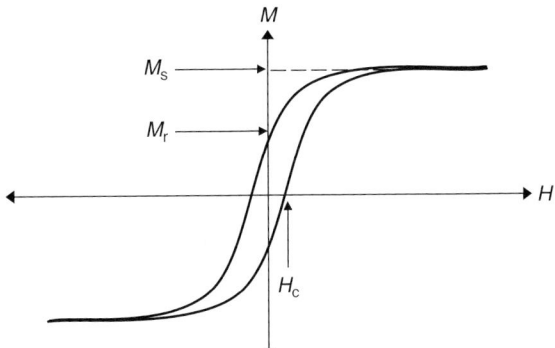

Figure 12.3 Representative field dependent magnetization curve of a ferro (ferrimagnet). H_c is the coercive field, M_r the remnant magnetization, and M_s the saturation magnetization.

At temperatures much lower than the ferromagnetic transition temperature, the Curie temperature T_C, the molar magnetization is given by $M = N_A g S \mu_B$ where N_A is Avogadro's number. It is difficult to precisely locate T_C since magnetization measurements are always performed with some applied magnetic field, and no phase transition takes place because the application of a magnetic field always produces some spin alignment. To resolve this, one uses scaling with the Arrott plots [5] (Fig. 12.4), whereby isotherms of $H/M^{1/\gamma}$ vs. $M^{1/\beta}$ are plotted. With the appropriate values of the critical exponents β and γ, the isotherms near the transition result in a series of parallel lines, and the temperature of the isotherm intersecting the origin corresponds to T_C, providing that one measures along the magnetic easy axis and one carefully corrects for the effects of demagnetization. Arrott plots cannot be employed if the mixtures are inhomogeneous. In such cases, an approximate T_C can be obtained from the steepest slope in the high-field measurements of the temperature dependence of the magnetization, the isochamp.

Antiferromagnets show linear field dependence of the magnetization although there is a characteristic spin-flop transition at suitably high field where the magnetization jumps. At the ordering temperature, the Neel temperature T_N, there is a maximum in the temperature dependence of the magnetization at constant field, the isochamp. However, a peak in an isochamp is insufficient to claim that a material is an antiferromagnet. It is necessary to measure the isochamps for several fields since the temperature dependence of the magnetization of a ferromagnet can also display a peak as a result of magnetic anisotropy [6].

Spin glasses display a distinctive S-shaped field dependence of the magnetization, with no true magnetic saturation reached. In addition, the magnetization has a logarithmic time dependence. The isochamp depends on the history of the sample: cooling in zero field and measuring the isochamp on warming produces a maximum near the spin freezing temperature T_f while cooling in applied field and then measuring the isochamp on warming results in an effectively constant magnetization up to the T_f. Behavior similar to this can be seen in ferromagnets depending upon the applied field and coercive field of the sample, but the time dependence/dynamics is the best way to check for the proper identification of the magnetic state.

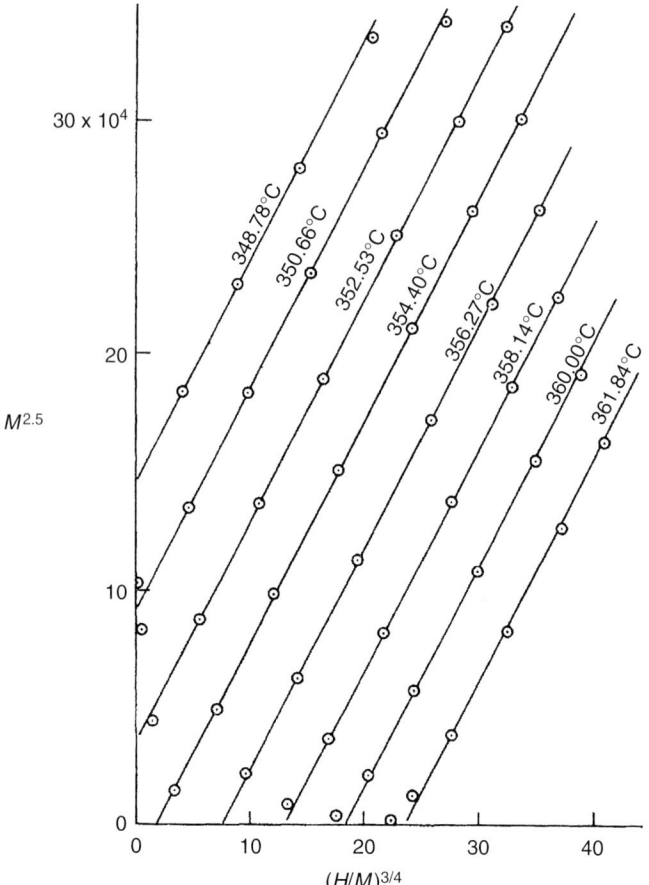

Figure 12.4 Arrott plot of the magnetization of Ni near its Curie temperature. The temperature of the isotherm intersecting the origin corresponds to T_C. From Ref. [5].

Magnetization measurements not only identify the magnetic phase boundary of homogeneous solid solutions but can also be used to identify components and volume fractions of heterogeneous mixtures. From the temperature dependence of the magnetization, one can also analyze phases from multiple ferromagnetic phases. This analysis was first described in detail by Hoselitz [7]. Figure 12.5 shows a schematic of the technique: one assumes that the total saturation magnetization is the sum of the saturation magnetization of the components. Then if the saturation magnetization and T_C of each component are known separately, the relative fractions of the various ferromagnetic components in the sample can be determined [8]. There are several limitations to this method: (1) it is assumed that there are no magnetic interactions between the various components; (2) applying the magnetic field may cause phase changes; and (3) it may be difficult to distinguish between an actual T_C and a structural phase change from a ferromagnetic to a non-ferromagnetic one as in Fe–Co alloys (see below).

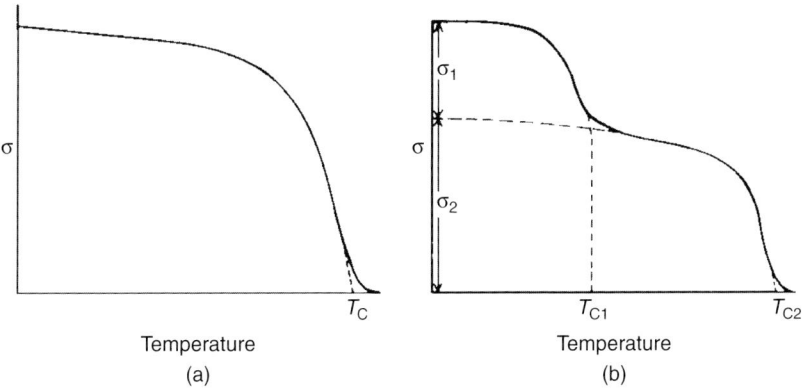

Figure 12.5 Temperature dependence of a homogeneous ferromagnet (a) and a mixture of two ferromagnet phases (b). The T_Cs in (b) identify the phase while the magnetization values determine the relative amounts. From Ref. [8].

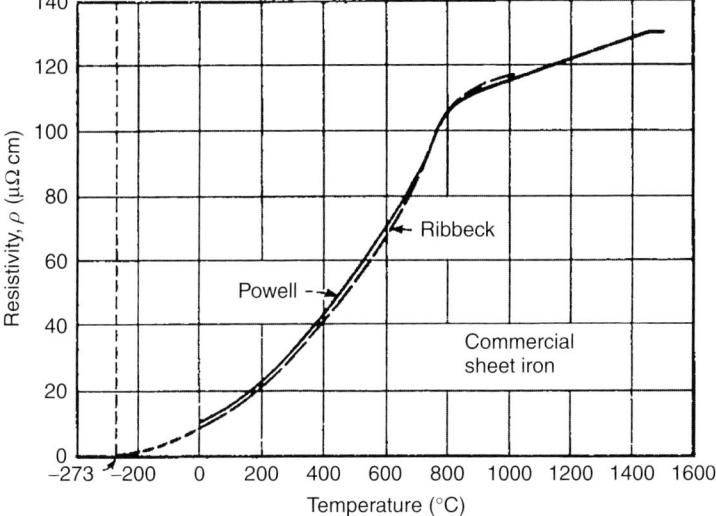

Figure 12.6 Temperature dependence of the resistivity of iron, which shows a kink at T_C. From Ref. [7].

Resistivity can be used as a quick means to identify some magnetic boundaries. Generally, first order transitions (often structural) lead to discontinuities in the resistivity vs. temperature curve as the electronic overlap between neighboring atoms is altered during the transition. Ferromagnetic transitions manifest as kinks in the resistivity (Fig. 12.6), as given by the Fisher–Langer relation [9].

2.3 Effects of Magnetism on Phase Diagrams

Magnetic transitions, with their associated Gibbs energy contribution to the total Gibbs energy of magnetic phases, can have an effect on the topology of phase diagrams. The effects are appreciable when the Gibbs energy of the competing phases are very close. Thus, the magnetic contribution to the total Gibbs energy can make a difference. Under such a condition, the effects of magnetic transitions are manifested in phase diagrams as a change of solubility (phase boundary shape), a formation of a miscibility gap (so-called Nishizawa horn), complex interactions with chemical ordering, and other phenomena. Phase diagrams of a few systems are discussed here to illustrate such effects.

2.3.1 Magnetic Effect on Solubility and Phase Boundary Shape

Many people consider the $3d$ transition elements Fe, Ni, and Co when they think of magnetic materials. Each has a T_C well above room temperature: $T_C = 1040$, 640, and 1380K for Fe, Ni, and Co, respectively. Interest in their binary and ternary phase diagrams was spurned by the discovery of enhanced properties such as nearly zero thermal expansion in Invar ($Fe_{64}Ni_{36}$) and the large magnetization in $Fe_{60}Co_{40}$. For a review, see Ref. [10].

In the Fe–Co system, for less than 70% Co, the alloys display bcc structure at low temperature and fcc at high temperature, Fig. 12.7 [11]. The fcc-Fe is weakly antiferromagnetic, and the magnetic transition temperatures of the γ phase (fcc) alloys have to be relatively low for relatively low atomic fractions of Co. The T_C of the γ phase is below the bcc–fcc boundary; however, they can be mapped out by quenching powders from the fcc phase [12]. On the other hand, the magnetic ordering temperature of the α phase would take place above the fcc–bcc transition (Fig. 12.7) if the structural transition did not occur first. However, the T_C can be reasonably extrapolated from magnetization data [13]. When the T_C of the two phases becomes comparable near 75% Co, the $\alpha + \gamma$ two-phase region drops suddenly [11].

Another appreciable effect of magnetic transitions on phase diagrams is a change in solubility of precipitation when the T_C line intersects the solubility phase boundary

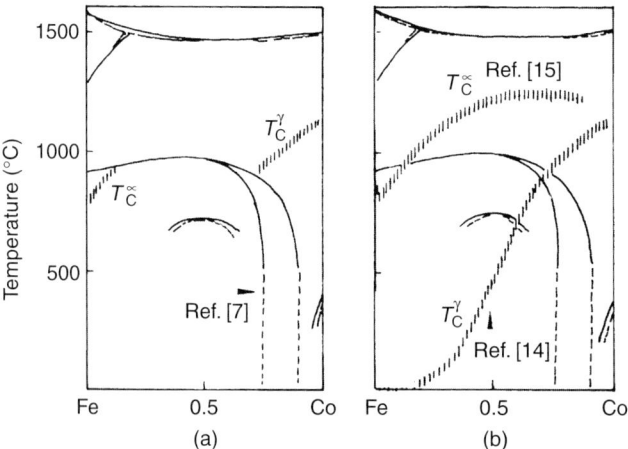

Figure 12.7 Phase diagram of Fe–Co alloys: (a) the equilibrium diagram and (b) metastable continuation of the Curie temperatures. The sudden drop of ($\alpha + \gamma$) two-phase region at around 75 at.% Co is believed to be associated with the comparable magnetic transition temperatures of the α and γ phases. From Ref. [11].

line. At temperatures above that of the intersection point, the solubility is higher in the paramagnetic phase. An abrupt change can be observed in the vicinity of T_C/intersection point; and below that temperature, the solubility in the ferromagnetic phase is much lower. Data for several systems are shown in Fig. 12.8 to illustrate this effect.

2.3.2 Magnetic Transition Induced Miscibility Gap

Magnetic transitions sometimes also promote the formation of a horn-shaped miscibility gap with its critical point (horn tip) sitting on the T_C line. An example is shown in Fig. 12.9 for the Co–Mo binary system [14]. The formation of a miscibility gap (the so-called Nishizawa horn) between the paramagnetic β′ (fcc) phase and the ferromagnetic β (fcc) phase is completely caused from the magnetic transition. Note in this particular case, the miscibility gap is predicted from thermodynamic modeling and has not been explicitly confirmed by experimental observations. The existence of such a miscibility gap was confirmed in the Co–Cr system [15] and is shown in Fig. 2.13(b). It is interesting to note the big difference of Mo solubility in the β′ (fcc) phase at >1143K and in the β (fcc) phase at <1143K. The abrupt change at ~1143K – the approximate temperature where the T_C line intersects the fcc solubility phase boundary – is what is discussed in the above paragraph.

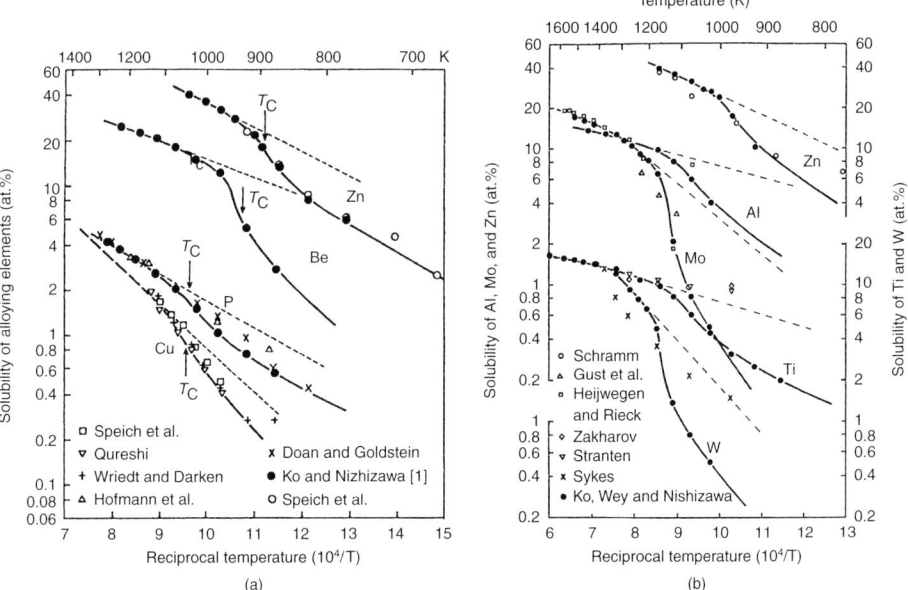

Figure 12.8 Changes in terminal solubility when the Curie temperature (T_C) line intersects the solubility phase boundary [11]: (a) solubility of various elements in bcc-Fe and (b) solubility of various elements in fcc-Co. The extrapolated solubilities based on the high-temperature paramagnetic condition are shown by the dashed lines to illustrate the abrupt change in the vicinity of T_C.

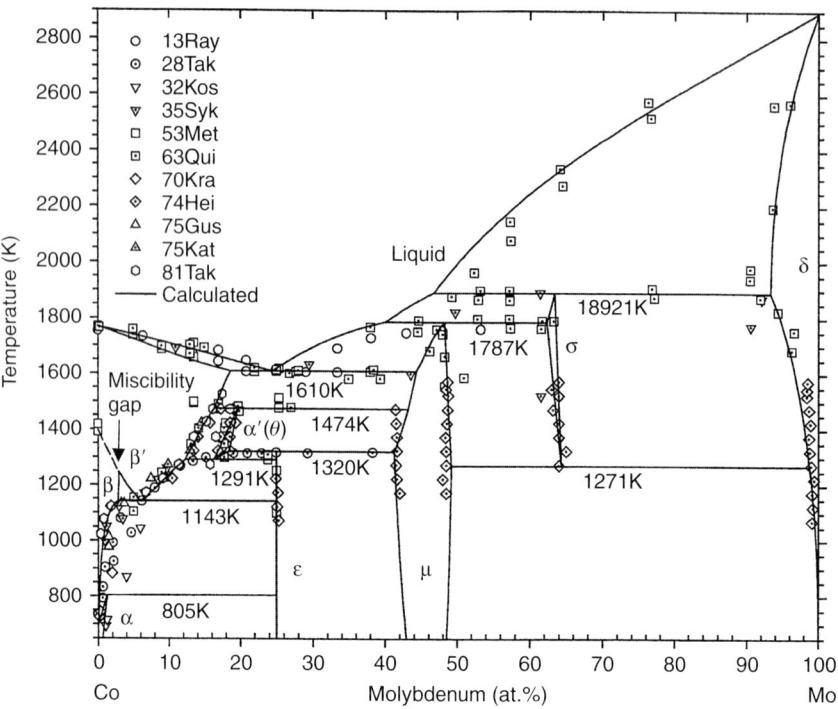

Figure 12.9 The calculated Co–Mo binary phase diagram with experimental data points superimposed [14], showing the formation of a miscibility gap between paramagnetic β′ (fcc) phase and the ferromagnetic β (fcc) phase.

2.3.3 Magnetic Interaction with Chemical Ordering: The Fe–Si System Example

Interest in Fe–Si alloys began in the late nineteenth century when Hadfield [16] found significantly increased hardness in an alloy containing 1.5% Si. By the early 20th century, the magnetic properties of various alloys had been investigated, which was further stimulated by the finding that a few percent Si leads to improved properties over pure Fe: increased permeability, lower hysteresis loss, decreased eddy current losses, and no aging. Thus it became an important new material for use in transformer cores and was one of the first alloy systems where the magnetic phase diagram was mapped out [17].

One of the first complete phase diagrams was obtained by Guggenheimer et al. [18] using the multicomponent magnetic analysis described above. While this helped identify much of the phase diagram into six distinct phases, the actual phase diagram is actually much richer, Fig. 12.10 [19]. For less than 25 at.% Si, the alloys have a bcc structure. This structure undergoes ordering with nearest neighbors or ordering at both first and second nearest neighbor sites (Fig. 12.11(a), inset) [20]. This leads to a very complex phase diagram which was not settled until the 1970s and only when confirmed with calculations [21], which were done with the Bragg–Williams formalism that takes into account nearest and next-nearest neighbor interactions and determines phase diagrams by maximizing the entropy. Such computations

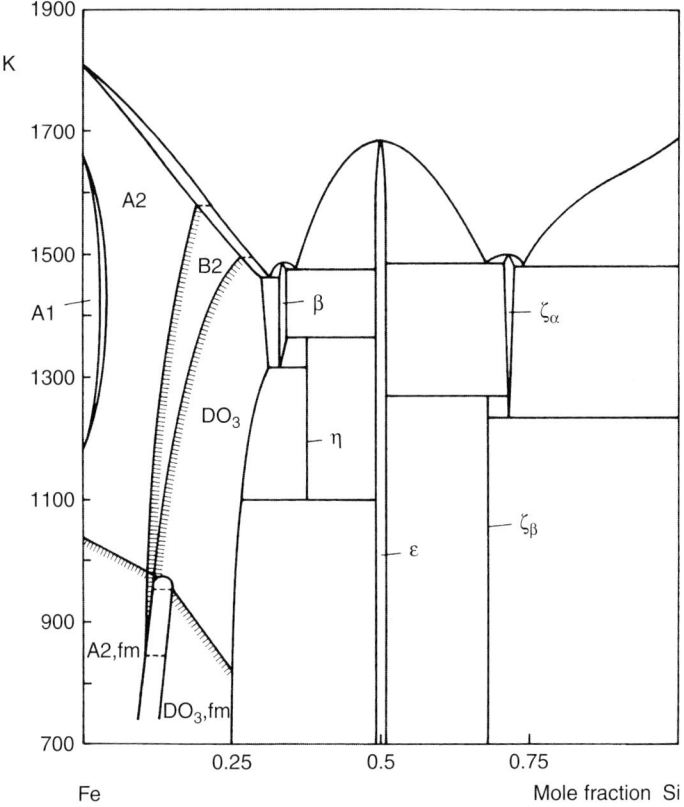

Figure 12.10 The Fe–Si binary phase diagram showing the complex interplay of magnetic transition with chemical ordering in the bcc phase at $<\sim 25$ at.% Si [19]. Details of this interplay are shown in Fig. 12.11.

(Fig. 12.11) were necessary to help distinguish between a heterogeneous mixture over a short-temperature range and continuous magnetic and structural transitions which occur near 12.5% Si [19].

2.3.4 Interplay of Magnetic Transition and Chemical Ordering: The Manganite Example

In the early 1950s, Jonker and van Santen [22] investigated perovskite compounds of the form $La_{1-x}A_xMnO_3$ where A is Ca, Sr, and Ba. $LaMnO_3$ has an antiferromagnetic insulating ground state but alloying with a divalent alkali earth creates mixed valence at the Mn site and leads to a metallic ferromagnet. This was one of the first oxide materials to show ferromagnetism above room temperature. Zener [23] proposed the idea of double exchange where the extra electron on $Mn^{3+}(3d^4)$ can hop to a neighboring $Mn^{4+}(3d^3)$ through the connecting oxygen. Because of strong Hund's rule coupling, which requires the first five electrons to fill the d band with the same spin, the hop can take place provided that moments of the ions are aligned, thus giving rise to enhanced conductivity in the ferromagnetic state.

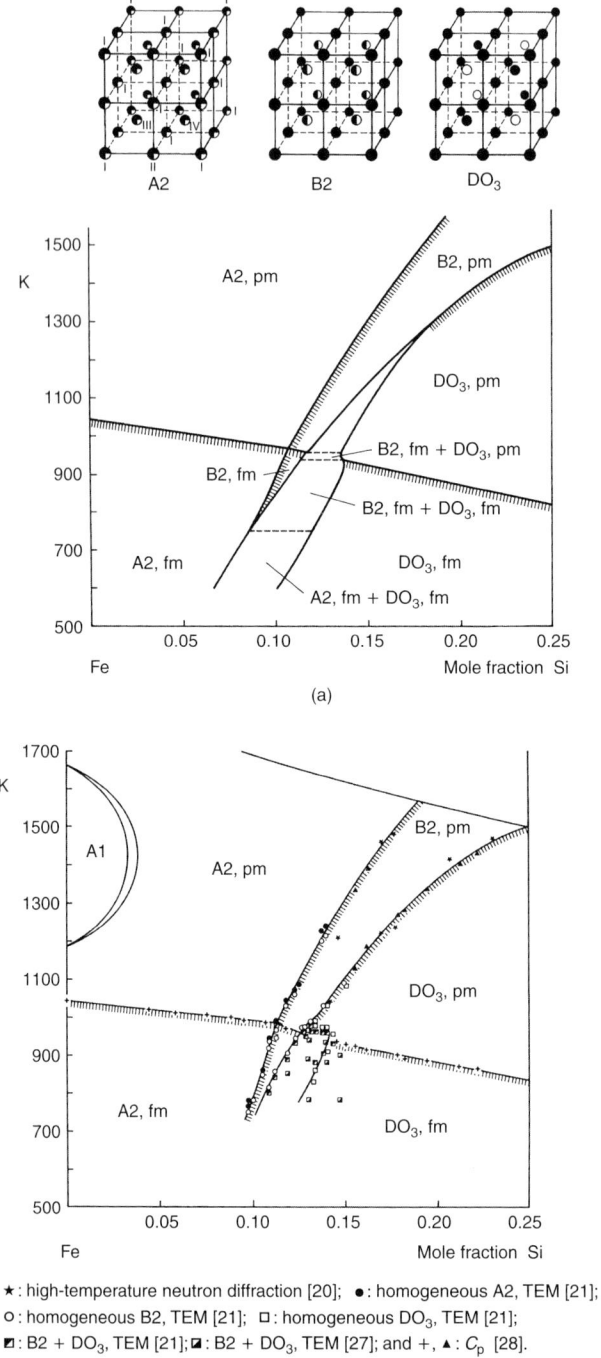

Figure 12.11 Interplay of magnetic transition with chemical ordering in the bcc phase of the Fe–Si binary system [19]: (a) calculated phase diagram using the Bragg–Williams model and (b) measured phase diagram using various techniques including neutron diffraction and transmission electron microscopy. The inset of (a) shows the types of atomic ordering. A2 is disordered, B2 has nearest neighbor ordering and DO_3 has first and second nearest neighbor ordering. From Ref. [19].

This picture was able to describe the basic features of the then-known phase diagram.

There was ongoing research with this set of materials and in 1994 Jin et al. [24] discovered that the resistance could change several orders of magnitude by application of a magnetic field. This effect was dubbed colossal magnetoresistance (CMR) [24], and it led to a redoubled effort to investigate these compounds in detail. Only then it became clear that both the phase diagram and the idea of double exchange were too simplistic. It turns out that the phase diagram is very complex indeed. Consider the Sr-based system (Fig. 12.12), there are several different crystal structures (cubic, rhombohedral, orthorhombic, tetragonal) and many magnetic states: ferromagnetic, canted antiferromagnetic, as well as several different types of antiferromagnetic structures [25,26]. There are even several electronic states: metallic, semiconducting, insulating, and charge ordering [27] where the Mn^{3+} and Mn^{4+} ions form ordered sublattices. The orbital ordering actually has a large impact on the phase diagram.

The new phase diagram (Fig. 12.12) has come about due to new/refined synthesis and detailed measurements using neutron diffraction, synchrotron X-ray diffraction (XRD), and electrical resistivity techniques. For instance, it was long believed that the magnetic transitions were always continuous; however, magnetization and neutron diffraction measurements indicated that the magnetization changed continuously while there are two components with the same crystal structure but different magnetic state: one is ferromagnetic and the other is not [28,29]. The observed magnetic transition arises when the ferromagnetic component changes into the nonmagnetic one and occurs below the actual T_C. The Sr-based compounds can show continuous transitions [30]; however, the Ca system never demonstrates a true second order magnetic transition [31].

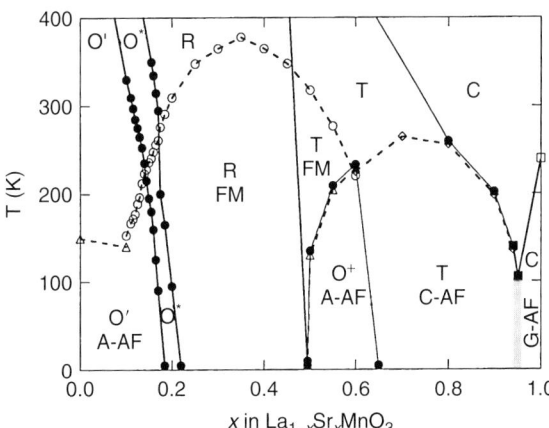

Figure 12.12 Detailed equilibrium $La_{1-x}Sr_xMnO_3$ phase diagram with complex structural, magnetic, and electronic transitions [26]. FM refers to ferromagnet. A-AF, C-AF, and G-AF refer to A-, C-, and G-type antiferromagnets, respectively. The symbols C, T, R, and O refer to cubic, tetragonal, rhombohedral, and orthorhombic symmetries, respectively. Coherent and incoherent Jahn–Teller distortions are present in the O' and O* Pbnm structures, respectively. O^+ refers to a Fmmm orthorhombic symmetry.

3 Combinatorial and High-Throughput Mapping of Magnetic Phase Diagrams

3.1 Introduction to the Combinatorial Approach

The combinatorial approach to materials is a new wave of research methodology which aims to dramatically increase the rate at which new compounds are discovered and improved. In this approach, up to thousands of different compositions can be synthesized and screened in an individual experiment for desired physical properties [3,32–34]. Individual samples containing large number of compositionally varying samples are called combinatorial libraries. This methodology initially started in biochemistry, and within the last 20 years, combinatorial chemistry and high-throughput screening for new drugs and biomolecules have already revolutionized the pharmaceutical and DNA-sequencing industries [35].

The scope of the combinatorial approach is far-reaching, and it can be used to address materials issues at different levels in a wide spectrum of topics ranging from catalytic powders [36] and polymers [37] to electronic and bio-functional materials [33,37,38]. In solid-state applications, the concept and effectiveness of the approach have been demonstrated in successful discoveries of new compounds in a number of key technological areas including optical materials, dielectric materials, and magnetic materials [3,32–34,38,39].

Various thin film synthesis techniques incorporating spatially varying or selective deposition can be used to create combinatorial libraries and composition spreads. There have been significant advances in this area, and at the most sophisticated level, tools for atomic layer controlled lattice engineering such as laser molecular beam epitaxy (LMBE), have been used to fabricate compositionally varying samples [40].

Parallel measurement techniques, scanning probe techniques, and/or sensor arrays are used to rapidly characterize specific physical properties of interest and obtain mapping of the properties across libraries [3,32–34,41]. Making quick and accurate measurements of a specific physical property from small volumes of materials in the libraries often represents a considerable feat of instrumentation. Researchers are taking on this challenge, and in some cases, it has led to invention of new measurement tools.

3.2 Combinatorial Mapping of Magnetic Phase Diagrams in Metallic Systems

There are a number of non-destructive scanning techniques that have been used to obtain mapping of magnetic properties from combinatorial libraries and composition spreads to date [42–47]. As is the case for measuring other physical properties, one of the biggest challenges lies in extracting accurate quantitative values of magnetic properties in such a way so that the value obtained from a spot on a library is the same as what one would obtain from a large scale individual sample of the same composition using a standard "bulk" measurement technique such as a SQUID magnetometer. Thin films can sometimes exhibit properties that significantly deviate from bulk properties. To this extent, phase diagrams mapped using thin-film composition spreads should be, in principle, regarded as "thin-film phase diagrams". But as will be shown below through many examples, the composition–structure–property relationships deduced from thin-film experiments are generally very similar (and sometimes identical) to those obtained from bulk samples. There are also instances where targeted final forms of the applications require thin films. In such cases, thin-film phase diagrams are more relevant than bulk phase diagrams. Alternatively, one can use the diffusion-multiple approach to map bulk phase diagrams in a high-throughput manner [48]. This methodology demands very high spatial resolution, on the order of 1–5 μm, for the measurement techniques.

While there are many important characterization techniques which provide different information regarding the overall magnetic properties of a material, by far the most crucial information for mapping phase diagrams and searching for novel magnetic materials is the magnetization. There are different techniques with which one can obtain magnetic field distribution emanating from thin film combinatorial samples. These include scanning SQUID microscopy [42,43,46], and the scanning Hall probe technique [44].

The magnetic–optical Kerr effect (MOKE) [44,45,47]. Even though field distribution itself does not give the magnetization value, in principle, one can extract magnetization using calibration samples. It is important to remember that the magnetization information obtained from mapping magnetic field distribution from ferromagnetic materials in absence of external applied field is the remnant magnetization.

The MOKE technique is based on the fact that the polarization of light is rotated upon reflection from the surface of a magnetic material and the rotation is directly proportional to the magnetization. MOKE is particularly useful because by sweeping applied magnetic field, one can obtain mapping of the magnetic hysteresis loop from different positions on combinatorial samples, from which one can obtain the saturation magnetization (through calibration) as well as the coercive field values [47].

The scanning Hall probe has also been used to map saturation magnetization values [44]. In order to obtain absolute values of magnetization from measured magnetic field without calibration, an inversion technique was developed which can be applied to any field distribution data [49]. This technique uses a computation algorithm which performs inverse Fourier transform to calculate magnetic pole densities, which in turn is integrated to obtain the value of magnetization. This algorithm has been applied to room-temperature scanning SQUID microscopy data of a variety of magnetic materials libraries [39,42,43]. In particular, the technique has been applied to thin-film composition spreads of ternary metallic alloy systems containing ferromagnetic shape memory alloys. The remnant magnetization values calculated for different compositions on the spread were demonstrated to be in good agreement with values measured by a SQUID magnetometer and a vibrating sample magnetometer for individual large-scale thin films [42].

Prior to the recent surge of activities in the high-throughput experimentation, the idea of utilizing deliberately created composition non-uniformity in co-deposited thin films to obtain large fractions of compositional phase diagrams simultaneously can be traced back to about half a century ago when Boettcher et al. [50] demonstrated the idea for the determination of phase diagrams for the Ag–Pb–Sn and other systems. Some discussion about the history of the combinatorial approach can be found elsewhere [3,35,38,51]. Figure 12.13 illustrates two such systems for fabricating the combinatorial samples [52,53].

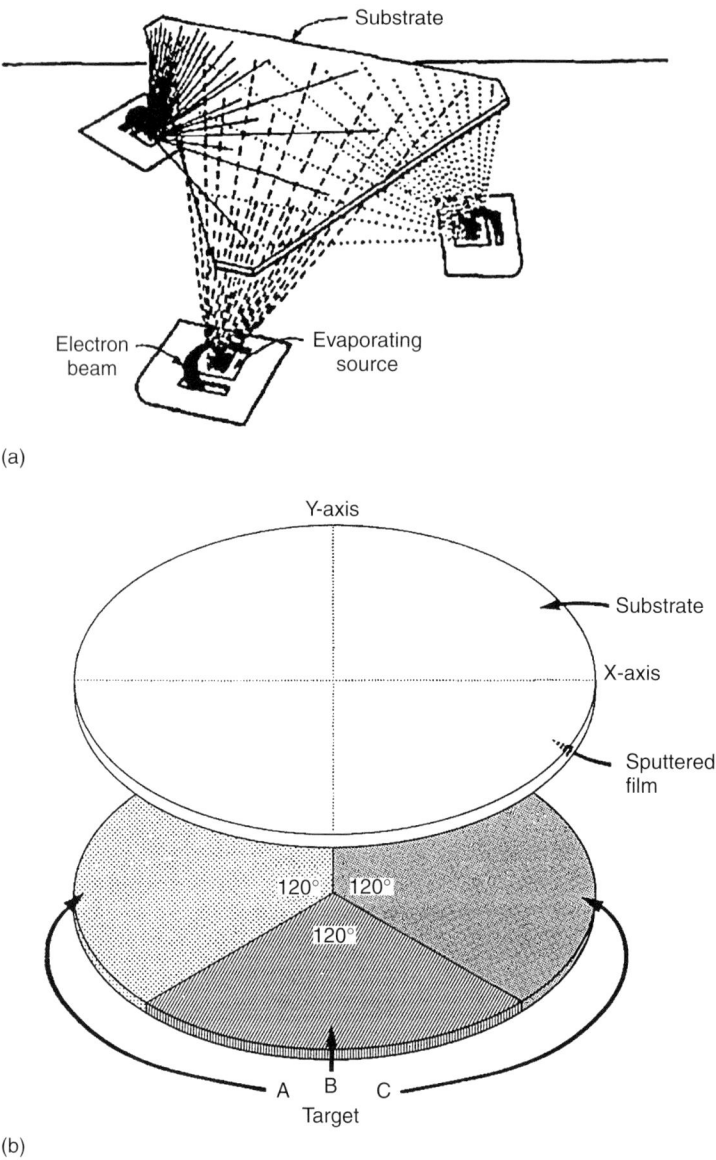

Figure 12.13 Schematic illustration of two methods for fabricating combinatorial thin films: (a) co-evaporation system using electron beam vapor deposition for mapping a ternary phase diagram [52] and (b) co-sputtering system with a three-component sputtering target to map their phase diagrams [53].

In the 1970s, Hanak and co-workers at RCA Laboratory had performed a series of experiments on phase mapping of a variety of materials using co-sputtering techniques [53,54]. Novel materials compositions were actively explored for a number of different functional properties targeting specific technological applications including magnetic recording heads and inductor cores. In one example, Hanak and Gittleman had co-sputtered the ternary system of $(Ni_xFe_y)_{1-x}(SiO_2)_x$ using the segmented target approach [54]. Figure 12.13(b) shows the schematic where three disk segments of Ni, Fe, and SiO_2 were placed on one RF sputtering target electrode. Their analysis of the deposition profile off of the three segments had indicated that a large fraction compositional phase diagram of the pseudo-ternary system could be mapped by this technique. They had performed measurements of resistance, magnetoresistance, permeability, and magnetization. The latter was measured by the magneto-optic Kerr technique. There had also been other instances where magnetic properties have been mapped as a function of composition for co-deposited composition-spread samples [55].

While it is instructive to obtain mapping of magnetic properties alone, in such early experiments, no comprehensive correlation with distribution of structure across phase diagrams was performed. In fact, it is only recently that multi-property mapping techniques have begun to be combined with structural property information, allowing one to obtain composition–structure–property relationships. The key enablers for such an integrated approach are the advent of computer technology combined with sophisticated measurement instrumentation, which provides accurate characterization of various physical properties. In some instances, the measurement techniques can result in precise quantitative values from relatively small volume of samples.

A good demonstration of comprehensive mapping of structural and magnetic properties of a phase diagram using the high-throughput method is the continuous phase diagram experiment by Yoo et al. [56]. The Fe–Ni–Co system was mapped onto a triangular sapphire substrate by ion beam deposition of metal elements in sequence, followed by extensive annealing in vacuum (600°C at 10^{-8} torr) to allow complete interdiffusion of the layers of metallic films and appropriate phase formation.

An in-house X-ray diffractometer was used to obtain structural phase distribution. The 2θ range of 4° (42–46°) sufficed to discern the major features in the phase distribution. For a given structural phase, the X-ray peak intensity allows one to track the compositional range. Regions where there are co-existing phases can be identified by looking for diffraction with peaks arising from multiple phases. It was observed that near the phase boundaries, the full-width half maximum of the peaks tended to increase. Figure 12.14(a) shows the distribution of phases deduced from the chip. They are Ni fcc, Fe bcc, and Ni–Fe alloys with fcc and bcc structures. The result is consistent with the known structural phases in this alloy system.

The magnetic property of the chip was investigated by the MOKE technique (Fig. 12.14(b)). A qualitative saturation magnetic moment of each point of the ternary phase diagram was mapped through the difference in the magneto-optical Kerr rotations at ±50 Oe applied magnetic field. Comparison of the two figures underscores the clear correlation between the phase distribution and the magnetism trend. Namely, the bcc-Fe has the highest magnetization. The observed magnetization variation is also consistent with the known distribution of saturation induction of the Fe–Ni–Co system deduced using the traditional bulk method (Fig. 12.14(c)) [57]. Although the MOKE method does not allow direct measurement of magnetization, by comparing the measured Kerr rotation angle with values obtained from standard methods such as SQUID magnetometry for some calibration samples, in principle, one can convert the mapping of the Kerr angle to mapping of the magnetization value. In this particular experiment, mechanical hardness map was also

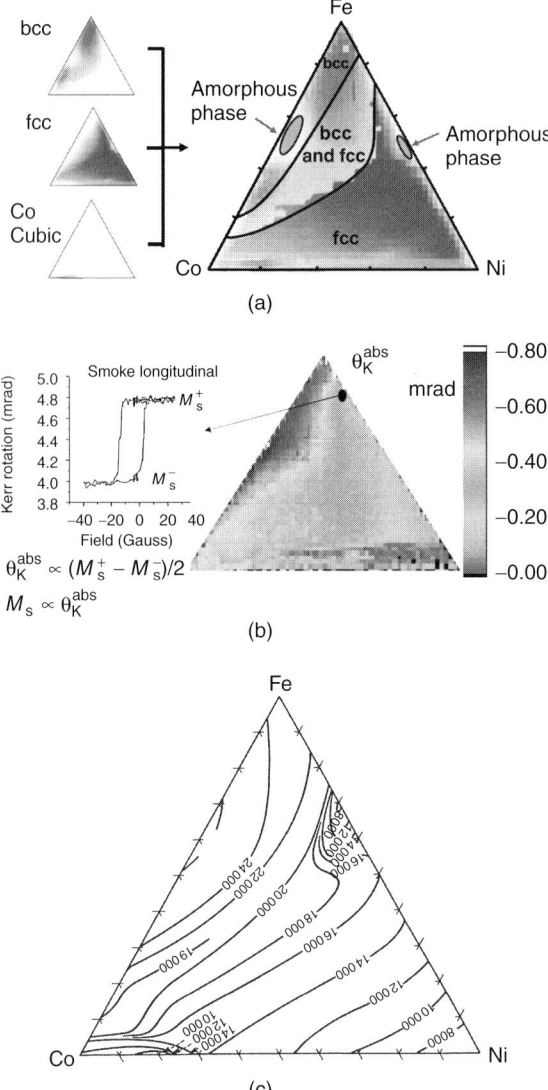

Figure 12.14 Mapping of properties of a Fe–Ni–Co combinatorial thin film library [56]: (a) the Fe–Ni–Co phase diagram obtained from distribution of structural phases deduced from X-ray microdiffraction peaks of fcc, bcc, and "Co cubic" phases; (b) mapping of magnetic property using scanning MOKE measurement. The Kerr rotation angle measured at ±50 Oe applied field is plotted. The left graph shows a typical hysteresis loop. (c) Saturation induction mapped for the Fe–Ni–Co ternary phase diagram using the traditional individual alloys for comparison [57].

obtained using nanoindentation, which revealed that the Fe–Co region containing fcc and bcc or pure bcc crystal structure is hard and that the higher Ni concentration region with fcc symmetry is soft.

Despite the fact that metallurgy has been practiced for many decades, only less than 5% of all possible ternary phase diagrams have been investigated. There are only very limited systems whose magnetic properties are mapped. Recently, systems containing Heusler alloys such as Ni_2MnGa and Co_2MnGe have attracted much attention because of their unique functionalities such as the ferromagnetic shape memory effect and high spin polarization [42,58]. In such alloys systems, it is of great interest to understand how the compositional variation and structural properties affect the magnetic (and other) properties.

The ferromagnetic shape memory effect can occur when an alloy is ferromagnetic and undergoes a reversible martensitic transition. In one investigation, properties of the Ni–Mn–Ga system were mapped in order to map the composition regions which are ferromagnetic shape memory alloys and to understand the origin of the co-existence of ferromagnetism and reversible martensites, which give rise to the shape memory effect [42]. Natural composition spreads were deposited in an ultra high-vacuum magnetron co-sputtering system (with a base pressure in the range of 10^{-9} torr) on 3-inch diameter (100) Si wafers. Three 1.5-inch diameter guns are placed parallel and adjacent to each other in a triangular configuration. The three targets used in the experiment were Ni, Mn, and Ni_2Ga_3. XRD of the fabricated films revealed that they were fiber textured with predominantly (110) orientation normal to the substrate. The wavelength dispersive spectroscopy (WDS) was used to accurately map the composition spread of every wafer. It was confirmed that by adjusting the power applied to each gun and the distance between the guns and the substrate (typically 12 cm), different regions of the ternary phase diagram (as well as the entire ternary) can be mapped out in a controllable manner [59].

A room-temperature scanning SQUID microscope was used for rapid characterization. A numerical algorithm was used to directly convert the field distribution information into quantitative magnetization [49]. The result of this calculation was plotted to obtain a room-temperature mapping of magnetization across the phase diagram (Fig. 12.15(a)). In this figure, the solid triangle-shaped curve marks the compositional region mapped on this particular wafer. It is clear that the most strongly magnetic region stretches from near the middle of the phase diagram toward slightly Ni-rich region. The circle indicates the region surrounding the Heusler composition, Ni_2MnGa, which has been extensively studied [60–62]. This experiment showed that this composition is only one part of a large region which has high magnetization. As one moves away from this general region, the magnetization gets smaller. The peak in magnetization is observed near the center of the phase diagram around the half Heusler composition, NiMnGa. The values of magnetization extracted here is consistent with those of saturation magnetization obtained by a vibrating sample magnetometer on separate individual composition samples at room temperature, and they are 200–300 emu/cc.

In order to map the regions of shape memory alloys and their martensitic transition temperatures, micromachined arrays of cantilevers were fabricated and deposited the composition spreads were directly deposited on the array wafers. Figure 12.15(b) summarizes the magnetic and the martensite phase diagrams deduced from the obtained data. It is evident that there is a large region outside of the near Heusler composition which is both ferromagnetic and reversible martensites. It has been shown that this class of material displays martensitic instability for stoichiometries where the $(s + d)$ electrons/atom ratio is ≈ 7.4 [60,61]. The shaded area in Fig. 12.15(b) covers the region where the $(s + d)$ electrons/atom ratio is 7.3–7.8.

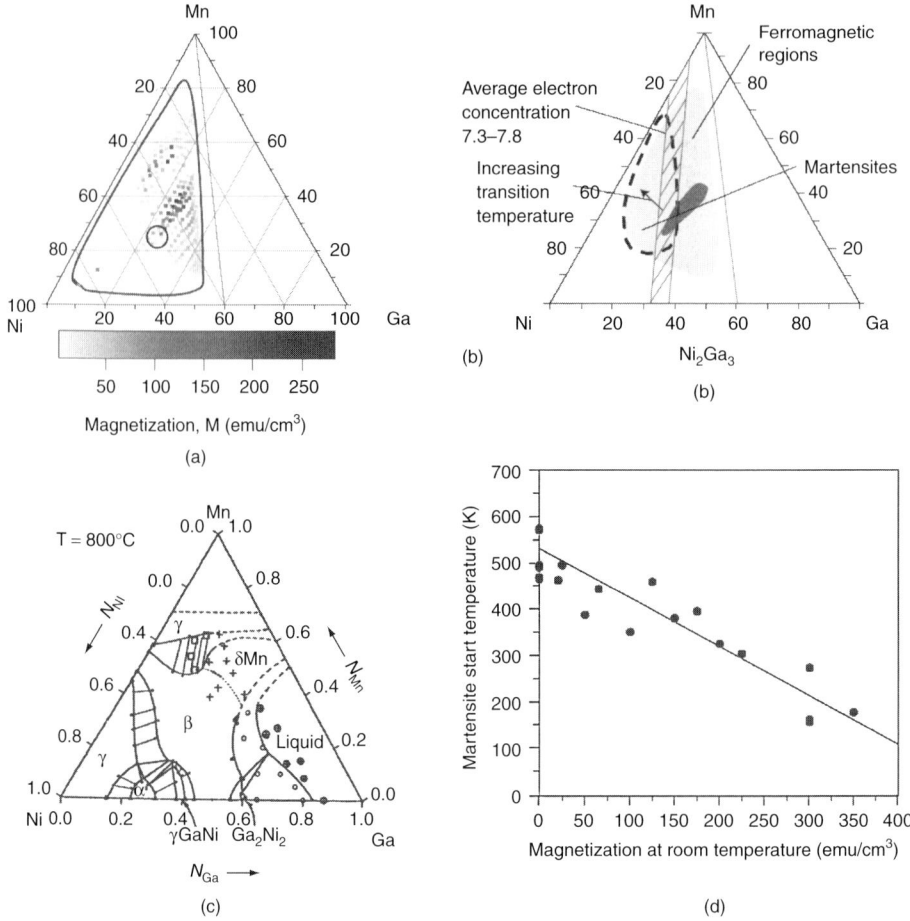

Figure 12.15 Exploration of magnetic shape memory alloys in the Ni–Mn–Ga ternary system [42]: (a) room-temperature mapping of remnant magnetization of the Ni–Mn–Ga system (The region inside the triangular-shaped curve is the compositional region mapped on a composition-spread wafer. The circle marks the compositions near the Ni$_2$MnGa Heusler composition.); (b) superimposed functional phase diagram deduced from magnetic mapping and shape memory alloy mapping (The shaded region has the compositions with average electron/atom ratio 7.3–7.8. The dotted line surrounds the region of shape memory alloys. In the ferromagnetic region, the red area has the highest magnetization.); (c) 800°C isothermal section of the Ni–Mn–Ga ternary system [63]; and, (d) martensitic transformation temperature plotted against room-temperature saturation magnetization (Each data point corresponds to a composition on the spread wafer. The line is a linear fit to the data.).

There is a large overlap between this region and the observed reversible martensite region. The room-temperature magnetization is plotted against the martensitic start temperature in Fig. 12.15(d) for the large compositions range studied here. There is a clear trend between the two parameters: the higher the magnetization, the lower the transformation temperature. We have confirmed from separate measurements that for different samples, the magnetization at room temperature is roughly proportional to the Curie temperature. The plot points to a strong thermodynamic magneto-structural coupling in this system. Such a coupling has previously been

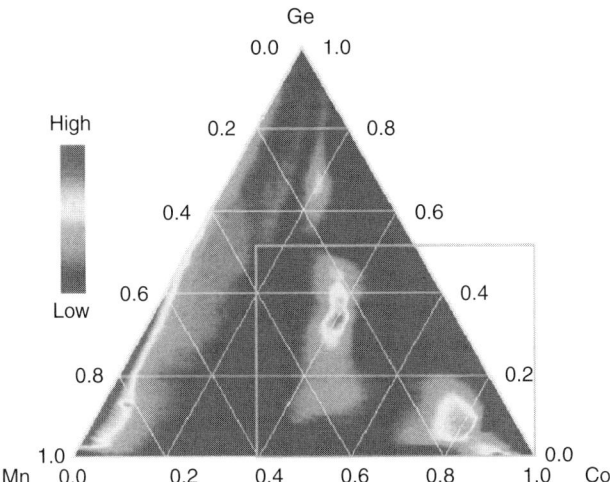

Figure 12.16 Room-temperature magnetic phase diagram of a 25 nm thick Co–Mn–Ge phase diagram sample grown by molecular beam epitaxy (MBE). Image of differential MOKE intensity measured at ±5 KOe, which corresponds to the saturated states of the system. Detailed magnetic behavior of the region inside the box is discussed in Ref. [65].

only observed in a limited range of compositional range near Ni_2MnGa (inside the circle in Fig. 12.15(a)) [62]. This experiment showed that this relationship holds for magnetic alloys with martensitic instability in general. A comparison of Fig. 12.15(b) with an isothermal of the Ni–Mn–Ga system at 800°C (Fig. 12.15(c)) [63], indicates that it is the β phase which largely transforms martensitically at low temperatures. Microdiffraction of the spread indicated that most of the regions have the $L2_1$ structure of the Heusler composition or a distorted $L2_1$ structure.

Co_2MnGe is predicted to be fully spin polarized at the Fermi level. However, to date, the fully spin polarized behavior has not been observed. The possibility of off-stoichiometric effect has motivated the ternary mapping of the Co–Mn–Ge system by Tsui et al. [64,65]. Triangular Ge substrates of several different orientations were used to map the ternary phase diagram using the molecular beam epitaxy technique. Sub-monolayers of Co, Mn, and Ge were deposited sequentially using a combination of moving shadow masks, sample rotation, and source shutters. The differential MOKE imaging technique was used to rapidly characterize the magnetic properties of the entire ternary. Figure 12.16 shows the transverse differential MOKE intensity image of a 15 nm thick continuous ternary phase diagram chip measured at ±5 KOe using a charge-coupled device (CCD) camera. Areas of high intensity correspond to magnetic regions at room temperature with strong magneto-optic coupling.

Three regions in the phase diagram were determined to be strongly ferromagnetic: (1) near pure Co, (2) near the Co-rich region centered around $Co_{0.7}Mn_{0.1}Ge_{0.2}$, and (3) in the middle of the phase diagram around $(Co_{0.5}Mn_{0.5})_{1-x}Ge_x$ for x between 0.2 and 0.6. The area displaying strong signals along the Mn_xGe_{1-x} binary edge for x between 0.6 and 0.9 corresponds to the known antiferromagnetic region. Results are consistent with earlier reports on limited composition ranges inside the ternary system. In order to study the details of magnetism in the $(Co_{1-x}Mn_x)_{0.8}Ge_{0.2}$ range, a separate composition-spread chip focused in this region was synthesized. Reflection of high-energy electron diffraction (RHEED) was performed across the sample together with the MOKE measurement, and it was found that the robust ferromagnetism observed with high saturation magnetization (and high Curie temperature) at $Co_{0.4}Mn_{0.4}Ge_{0.2}$ has strong correlation with the onset of a structural order–disorder transition, which is manifest in an abrupt change in the RHEED intensity [65]. As seen here in several examples, high-throughput mapping of compositional phase diagrams using characterization techniques of different types can go beyond simply mapping phase boundaries and allows one to develop important insight concerning the overall nature of the underlying magnetism.

Accurate Curie temperature–composition dependence can also be mapped from diffusion multiples using magnetic force microscopy (MFM). An example is shown in Fig. 12.17 for the Co–Cr–Mo ternary system [51]. The diffusion multiple is the same one as discussed in Chapter 7. A scanning electron microscopy (SEM) backscattered electron image of the Co–Cr–Mo tri-junction area is shown in Fig. 12.17(b). Superimposed on the image are MFM maps showing magnetic domains in Co-rich compositions of the fcc phase. These MFM maps were taken at ~24°C during the mapping experiments. The topographic atomic force microscopy and MFM images were taken simultaneously. The topographic images (not shown) contained features of the surface topography such as polishing scratches and voids that allowed the exact positions of the MFM maps to be superimposed on the SEM image. The ferromagnetic region of the fcc phase could be easily separated from the paramagnetic region, as shown by the dotted line in Fig. 12.17(b). Electron probe microanalysis (EPMA) along this line gave the compositions at which the magnetic transition took place at 24°C, that is, an iso-Curie temperature line at 24°C, Fig. 12.17(c). By performing MFM mapping at different temperatures (which has not been done), it would be possible to obtain the Curie temperature surface, that is, the Curie temperature–composition relationship for the fcc phase. As a check for reliability, the 24°C Curie temperature composition of the binary Co–Cr system obtained from the MFM and EPMA analysis of the diffusion multiple was evaluated to be ~20 at.% Cr, which agreed with the value (~20 at.%) extrapolated from existing data of the binary system.

3.3 Combinatorial Mapping of Oxide Systems

Since the advent of high-temperature superconductivity in cuprates, there has been an explosion of activities in transition metal oxides in general. A variety of functionalities observed in metal oxides provide a rich platform for exploring novel device applications. Strong correlation in many systems continues to dominate the focus of investigation in condensed matter physics. In magnetism, metal oxides exhibit a myriad of intriguing properties including CMR and charge ordering in manganites to high spin polarization in Fe_3O_4 and long spin relaxation times in ferrites. In addition, there are many metal oxide systems which have traditionally been used as important device materials such as dielectric insulators, ferroelectric materials, and phosphors. Because there are always needs to find better materials with improved physical properties, phase diagram mapping can play a significant role in

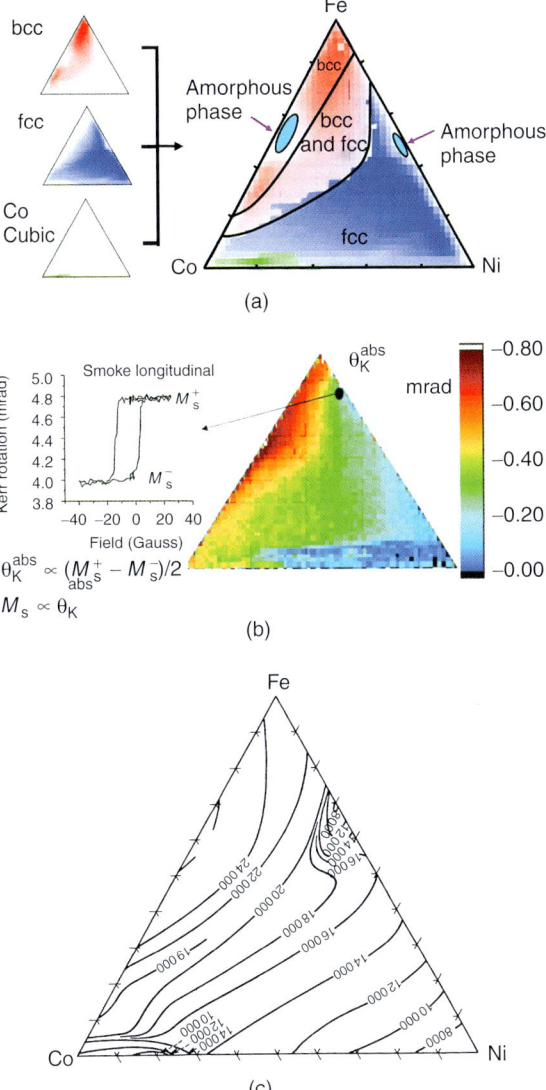

Figure 12.14 Mapping of properties of a Fe–Ni–Co combinatorial thin film library [57]: (a) the Fe–Ni–Co phase diagram obtained from distribution of structural phases deduced from X-ray microdiffraction peaks of fcc, bcc, and "Co cubic" phases; (b) mapping of magnetic property using scanning MOKE measurement. The Kerr rotation angle measured at ±50 Oe applied field is plotted. The left graph shows a typical hysteresis loop. (c) Saturation induction mapped for the Fe–Ni–Co ternary phase diagram using the traditional individual alloys for comparison [58].

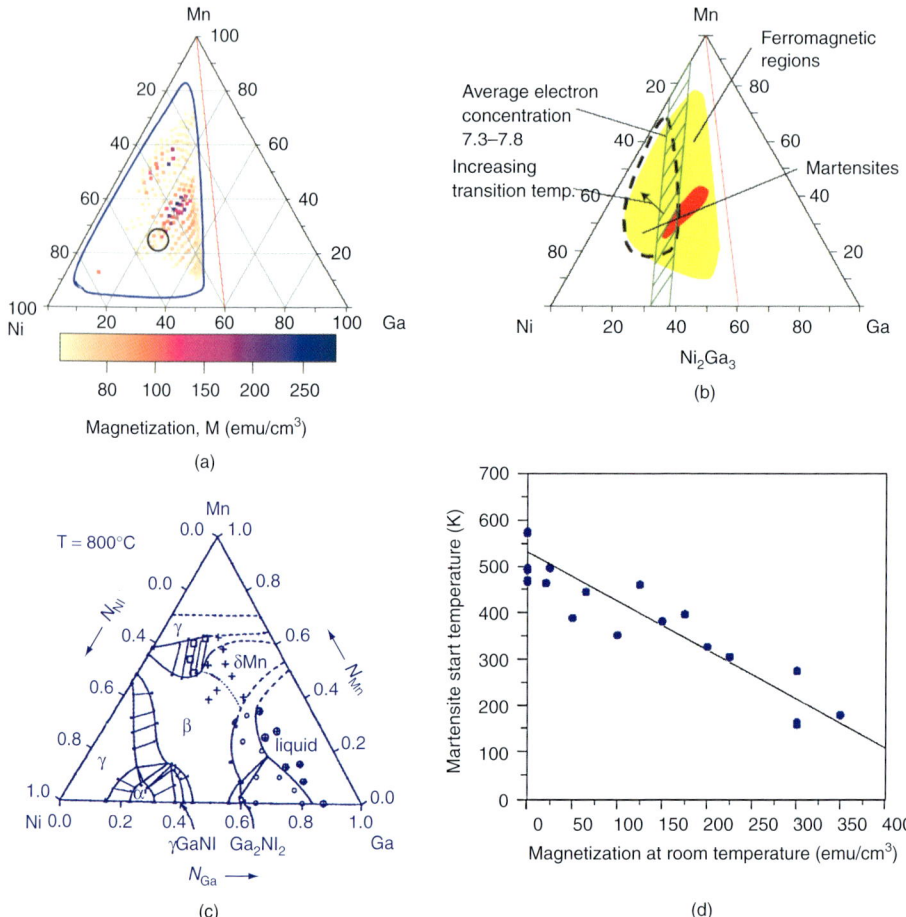

Figure 12.15 Exploration of magnetic shape memory alloys in the Ni–Mn–Ga ternary system [42]: (a) room-temperature mapping of remnant magnetization of the Ni–Mn–Ga system (The region inside the triangular-shaped curve is the compositional region mapped on a composition-spread wafer. The circle marks the compositions near the Ni$_2$MnGa Heusler composition.); (b) superimposed functional phase diagram deduced from magnetic mapping and shape memory alloy mapping (The shaded region has the compositions with average electron/atom ratio 7.3–7.8. The dotted line surrounds the region of shape memory alloys. In the ferromagnetic region, the red area has the highest magnetization.); (c) 800°C isothermal section of the Ni–Mn–Ga ternary system [64]; and, (d) martensitic transformation temperature plotted against room-temperature saturation magnetization (Each data point corresponds to a composition on the spread wafer. The line is a linear fit to the data.).

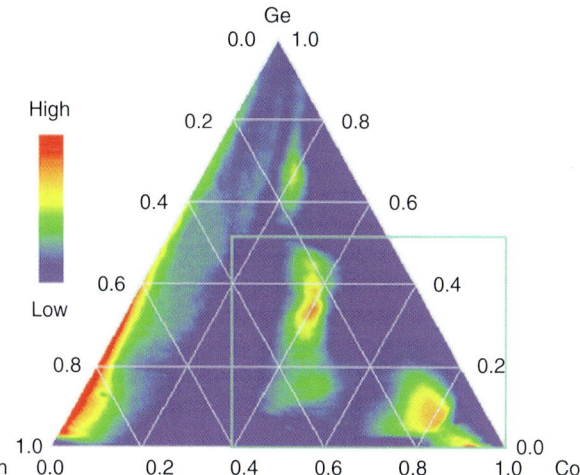

Figure 12.16 Room-temperature magnetic phase diagram of a 25 nm thick Co–Mn–Ge phase diagram sample grown by molecular beam epitaxy (MBE). Image of differential MOKE intensity measured at ±5 Oe, which corresponds to the saturated states of the system. Detailed magnetic behavior of the region inside the box is discussed in Ref. [66].

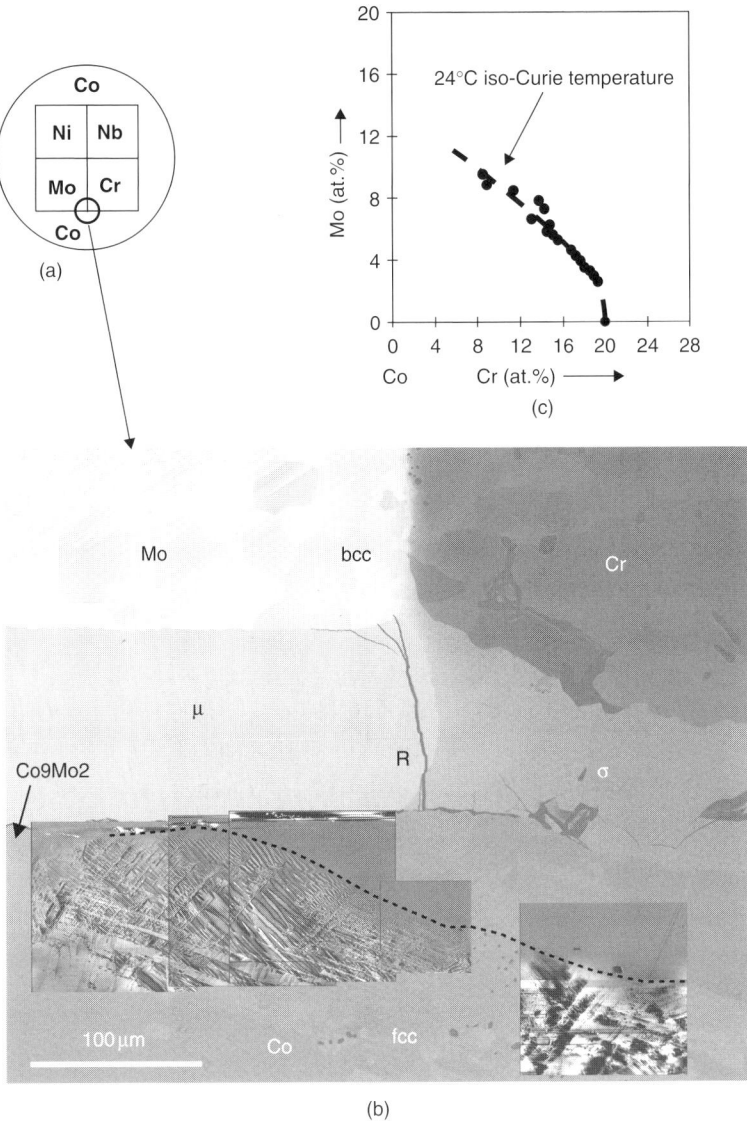

Figure 12.17 Application of MFM to map magnetic domains to obtain Curie temperature–composition relationship for the Co–Cr–Mo ternary system: (a) schematic of a Co–Cr–Mo–Nb–Ni diffusion multiple with the Co–Cr–Mo tri-junction region highlighted; (b) MFM images superimposed on a backscattered electron image of the Co–Cr–Mo tri-junction; and (c) the 24°C iso-Curie temperature compositions of the fcc phase in the Co–Cr–Mo ternary system obtained by tracing the boundary (dotted line in (b)) between the domain/non-domain region with EPMA [51].

development of new metal oxides as well as for investigation of physics. Prior to recent phase diagram mapping experiments on magnetic oxide systems discussed below, there had been reports of ternary mapping of dielectric materials. van Dover et al. have reported on investigation of a large number of pseudo-ternary transition metal oxide systems using co-sputtered composition-spread chips for exploring novel high-κ dielectric materials [66]. Chang et al. have synthesized the first pseudo-ternary composition-spread sample which mapped the $(Ba,Sr,Ca)TiO_3$ phase diagram on a triangular-shaped chip using a thin film precursor technique where amorphous precursor multilayers are deposited at room temperature followed by a series of annealing procedures in order to arrive at crystalline materials via appropriately diffused precursors [67].

Yoo et al. used the precursor thin film synthesis technique for mapping the phase diagram of the $La_{1-x}Ca_xMnO_3$ system on a single chip [68]. The continu-ous phase diagram chip was made by first depositing gradient thicknesses of precursors La_2O_3, Mn_3O_4, and $CaMnO_3$ at room temperature on a (100) $LaAlO_3$ substrate, then annealing the chip at 200°C for several days, 400°C for 30 h, and followed by sintering at 1000°C for 2 h in flowing O_2. A predominantly epitaxial film was obtained on the chip as a result of this process. To map the magnetic phases, a low-temperature scanning SQUID microscopy was performed at 3–7K. To track the qualitative change in electrical properties of the spread, microwave microscopy was utilized at room temperature [69]. The experimental data are summarized in Fig. 12.18.

Correlation between microwave properties and the optical properties (as reflected in the picture of the chip shown in Fig. 12.18(a)) at room temperature (well above the onset of magnetic order) and the magnetic properties at low temperatures are observed as "boundaries" between composition regions where properties change abruptly. The electronic boundaries observed here were found to coincide with those boundaries of orbital orderings predicted by Maezono et al. who describe the spin order in the system as a function of composition using a generalized Hubbard model (shown in Fig. 12.18(b)) [70].

The data were thus analyzed in comparison with the predicted magnetic states for the system. At $x = 0.5$, there is an abrupt change in the optical image where ΔB_z is the minimum. This composition is believed to be displaying the result of ½ charge ordering previously reported in a transmission electron microscopy studies. At $x = 0.93$, where ΔB_z reaches another minimum, the optical image and the microwave microscope data (Fig. 12.18(b)) change in a concomitant manner abruptly. This is believed to be a transition to an electrically insulating state with the antiferromagnetic state. Magnetic domain distributions measured at several spots on the chip by the low-temperature scanning SQUID (Fig. 12.18(d)) are also consistent with the changing magnetic states across the phase diagram.

Tracking magnetic transitions across composition variation is perhaps one of the most important aspects of mapping the magnetic phase diagrams. Simultaneous mapping of the magnetic transition across a composition-spread sample was demonstrated using magneto-optic imaging with an indicator film (MOIF) technique by Turchinskaya et al. [71]. By taking an image of the spread film with an indicator film over it, one can readily detect the ferromagnetic region which would show up as a dark edge due to a stray field in the case of in-plane aligned film. By tracking the changing width and the position of the dark line as a function of temperature, one can map out the Curie temperature of the entire phase diagram. This is schematically illustrated in Fig. 12.19. A $La_{1-x}Ca_xMnO_3$ composition spread was synthesized using the in situ layer-by-layer method described above. Figure 12.20 shows the result of the Curie temperature mapping across this sample (diamonds)

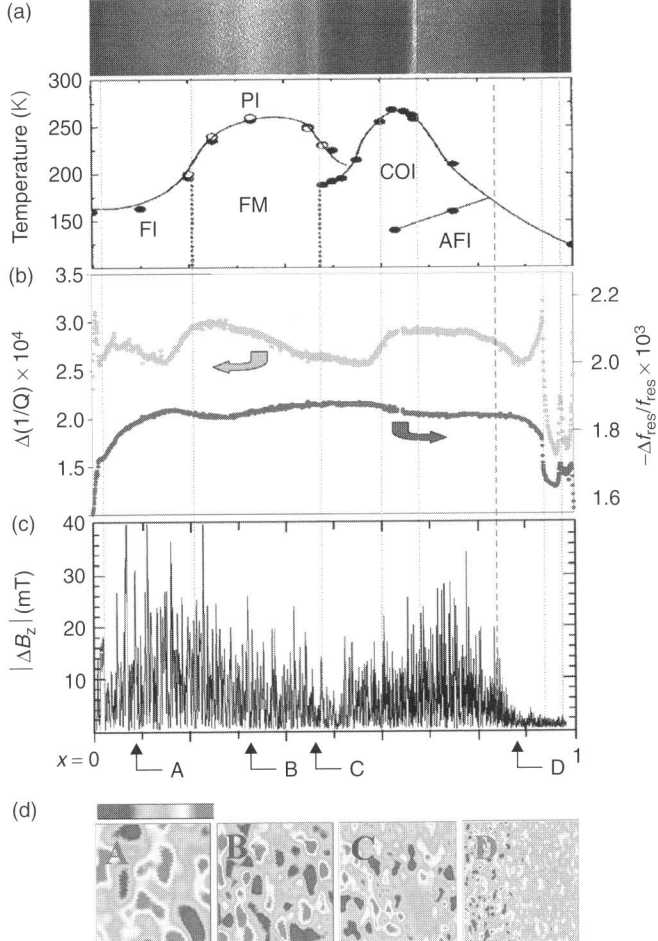

Figure 12.18 Electronic and magnetic transitions in the $La_{1-x}Ca_xMnO_3$ pseudo-binary system studied using scanning evanescent microwave probe (SEMP) and scanning SQUID microscope [69]: (a) a room-temperature CCD photograph of the continuous $La_{1-x}Ca_xMnO_3$ thin film taken under white light and a magnetic phase diagram of the system determined from bulk single crystals (the various states are: paramagnetic insulator (PI), ferromagnetic insulator (FI), ferromagnetic metal (FM), charge-ordered insulator (COI), antiferromagnetic insulator (AFI)); (b) SEMP line-scan profiles of microwave loss $\Delta 1/Q$ (related to conductivity) and relative frequency shift $\Delta f/f$ (measurement performed at room temperature); (c) scanning SQUID line profile obtained at 3K (the oscillation indicates magnetic domains); (d) scanning SQUID images taken at 7K at four different composition regions showing the magnetic domains and the transition from strong to weak magnetization. See Ref. [69] for detailed description of understanding the complex electronic and magnetic transitions from these images and profiles.

superimposed over a previously published bulk phase diagram [2]. Various acronyms indicate different magnetic states. Although significantly shifted compared to the bulk curve, one can clearly see the transition curve from the paramagnetic state to the ferromagnetic state, as well as the sharp composition boundary at $x = 0.5$ between

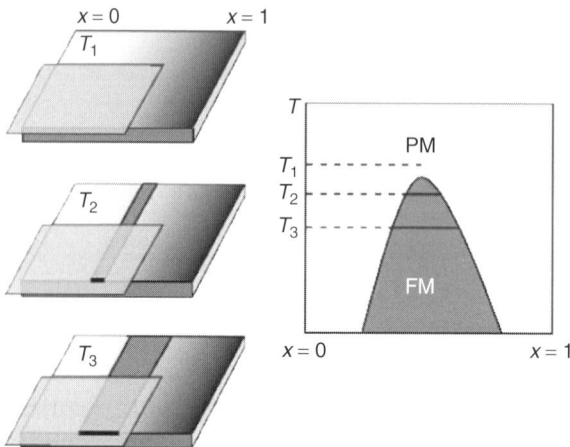

Figure 12.19 Schematic showing how a magnetic phase diagram gets constructed using the MOIF technique [71]. The length and location of dark lines at different temperatures indicate a magnetic phase region on a composition–temperature diagram.

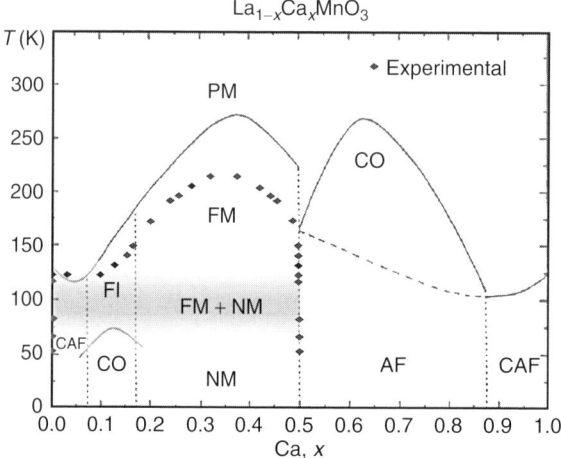

Figure 12.20 Constructed magnetic phase diagram of the $La_{1-x}Ca_xMnO_3$ composition spread on a $SrTiO_3$ substrate [71]. The experimental points (diamonds) are superimposed on the bulk phase diagram from Ref. [2]. PM, FM, NM, AF, CAF, CO, and FI refer to paramagnetic, ferromagnetic, non-magnetic, antiferromagnetic, canted antiferromagnetic, charge ordered, and ferromagnetic insulating states, respectively.

ferromagnetic and the antiferromagnetic state. Thin-film samples are known to exhibit deviated properties compared to bulk samples due to strains and presence of defects, etc. Thus, the shift such as the one seen here is not unreasonable. This experiment clearly shows that one can use this simple technique to map the Curie temperature across composition-spread chips.

4 Concluding Remarks

We have described important aspects of magnetic and structural property measurements aimed at mapping magnetic phase diagrams in different materials systems. In the traditional methods section, we have discussed techniques used for determining magnetic phase boundaries. High-throughput techniques in general are beginning to change the way research is carried out to map a variety of properties across compositional phase diagrams. Magnetic materials are areas where much is being done to demonstrate the utility of such techniques. High-throughput methods frequently use thin film composition spreads where entire phase diagrams can be mapped on individual chips. Although highly dependent on available measure-ment instruments, such spreads are instrumental in rapidly uncovering complex composition–structure–property relationships across phase diagrams. The combinatorial materials science is a rapidly changing field, and we can expect continuing development for years to come.

To perform complete determination of a phase diagram with magnetic transitions requires a number of tasks in materials synthesis as well as characterization. It is important to keep in mind that there are systems where optimum materials processing conditions vary drastically within one phase diagram. In such instances, mapping the entire phase diagram on a single chip is not appropriate. One solution to this is to divide up the diagram into several composition-spread chips to be processed under different conditions. An ideal comprehensive strategy perhaps is to come up with an effective way to combine the high-throughput approach with the traditional bulk method, so that one can determine the phase diagram in a rapid and yet thorough manner.

Acknowledgments

This work was supported by ONR N000140110761, ONR N000140410085, NSF DMR 0094265 (CAREER), NSF DMR 0231291, MRSEC DMR 0520471, and NSF DMR 0503711. I. Takeuchi acknowledges useful discussions with X.-D. Xiang and J.-C. Zhao. S.E. Lofland acknowledges fruitful discussion with S.M. Bhagat.

REFERENCES

1. See for example, *MRS Bulletin*, vol. 28, number 10, (2003), a special issue on new materials for spintronics.
2. S.-W. Cheong and H.Y. Hwang in Y. Tokursa (ed.), *Ferromagnetism vs. Charge/Orbital Ordering in Mixed-Valent Manganites: Colossal Magneto-Resistive Oxides*, Gordon and Breach Science Publisher, Amsterdam, 2000, p. 237.
3. H. Koinuma and I. Takeuchi, *Nat. Mater.*, 3 (2004) 429.
4. W.F. Brown, in E.P. Wohlfarth (ed.), *Magnetostatic Principles in Ferromagnetism*, North Holland, Amsterdam, 1962.
5. A. Arrott and J.E. Noakes, *Phys. Rev. Lett.*, 19 (1967) 786.
6. S.E. Lofland, S.M. Bhagat, K. Ghosh, S. Ramakrishnan and G. Chandra, *J. Magn. Magn. Mater.*, 129 (1994) L120.
7. K. Hoselitz, *Ferromagnetic Properties of Metals and Alloys*, Oxford University Press, Oxford, UK, 1952.

8. A. Berkowitz, in A. Berkowitz, E. Kneller (eds.), *Magnetism and Metallurgy*, vol. 1, Academic Press, New York, 1969.
9. M.E. Fisher and J.S. Langer, *Phys. Rev. Lett.*, 20 (1968) 665.
10. R.M. Bozorth, *Ferromagnetism*, Van Nostrand, New York, 1951.
11. A.P. Miodownik, *Bull. Alloy Phase Diagr.*, 2 (1982) 406.
12. Y. Nakamura, M. Shiga and S. Santa, *J. Phys. Soc. Jpn.*, 29 (1969) 210.
13. G. Inden and W.O. Meyer, *Z. Metallkd.*, 66 (1975) 725.
14. U. Kattner and A. Davydov, *J. Phase Equilib.*, 24 (1003) 209.
15. K. Oikawa, G.-W. Qin, T. Ikeshoji, R. Kainuma and K. Ishida, *Acta Mater.*, 50 (2002) 2223.
16. R.A. Hadfield, *Metallurgy and Its Influence on Modern Progress*, Van Nostrand, New York, 1926.
17. W.F. Barrett, W. Brown and R.A. Hadfield, *Sci. Trans. Roy. Dublin Soc.*, 7 (1900) 67.
18. K.M. Guggenheimer, H. Heitler and K. Hoselitz, *J. Iron Steel Inst.*, 158 (1948) 192.
19. G. Inden, *Bull. Alloy Phase Diagr.*, 2 (1982) 412.
20. G. Inden and W. Pitsch, *Z. Metallkd.*, 63 (1975) 253.
21. G. Inden, *Physica B*, 103B (1981) 82.
22. G.H. Jonker and J.H. van Santen, *Physica*, 16 (1950) 337.
23. C. Zener, *Phys. Rev.*, 82 (1951) 403.
24. S. Jin, T.H. Tiefel, M. McCormack, R.A. Fastnacht, R. Ramesh and L.H. Chen, *Science*, 264 (1994) 413.
25. J. Hemberger, A. Krimmel, T. Kurz, H.-A. Krug von Nidda, V.Yu. Ivanov, A.A. Mukhin, A.M. Balbashov and A. Loidl, *Phys. Rev. B.*, 66 (2002) 094410.
26. O. Chmaissem, B. Dabrowski, S. Kolesnik, J. Mais, J.D. Jorgensen and S. Short, *Phys. Rev. B.*, 67 (2003) 094431.
27. P.G. Radaelli, D.E. Cox, M. Marezio and S.-W. Cheong, *Phys. Rev. B.*, 55 (1997) 3015.
28. E.O. Wollan and W.C. Koeller, *Phys. Rev.*, 100 (1955) 545.
29. C. Kapusta, P.C. Riedi, W. Kocemba, M.R. Ibarra and J.M.D. Coey, *J. Appl. Phys.*, 87 (2000) 7121.
30. S.E. Lofland, V. Ray, P. Kim, S.M. Bhagat, M.A. Manheimer and S.D. Tyagi, *Phys. Rev. B.*, 55 (1997) 2759.
31. C.P. Adams, J.W. Lynn, V.N. Smolyaninova, A. Biswas, R.L. Greene, W. Ratliff, S.W. Cheong, Y.M. Mukovskii and D.A. Shulyatev, *Phys. Rev. B.*, 70 (2004) 134414.
32. I. Takeuchi, R.B. van Dover and H. Koinuma, *MRS Bull.*, 27 (2002) 301.
33. I. Takeuchi, J.M. Newsam, L.C. Wille, H. Koinuma and E.J. Amis (eds.), Combinatorial and Artificial Intelligence Methods in Materials Science, *MRS Symposium Proceedings*, vol. 700, 2002.
34. X.-D. Xiang and I. Takeuchi (eds.), *Combinatorial Materials Synthesis*, Marcel Dekker, New York 2003.
35. See for example, M. Lebl, *J. Comb. Chem.*, 1 (1999) 3.
36. S. Senken, K. Krantz, S. Ozturk, V. Zengin and I. Onal, *Angew. Chem. Int. Edit.*, 38 (1999) 2794.
37. U.S. Schubert and E. Amis (eds.), Macromol. Rapid Comm., Special issue on *Combinatorial Research and High-Throughput Experimentation in Polymer and Materials Research* 24 (1), 2003.
38. X.-D. Xiang, *Annu. Rev. Mater. Sci.*, 29 (1999) 149.
39. Y. Matsumoto, H. Koinuma, T. Hasegawa, I. Takeuchi, F. Tsui and Y.K. Yoo, *MRS Bull.*, 28 (2003) 734.
40. H. Koinuma, *Solid State Ionics*, 108 (1998) 1.
41. R.A. Potyrailo and W.G. Morris, *Rev. Sci. Instrum.*, 75 (2004) 2177.
42. I. Takeuchi, O.O. Famodu, J.C. Read, M.A. Aronova, K.-S. Chang, C. Craciunescu, S.E. Lofland, M. Wuttig, F.C. Wellstood, L. Knouse and A. Orozco, *Nat. Mater.*, 2 (2003) 180.
43. O.O. Famodu, J. Hattrick-Simpers, M. Aronova, K.-S. Chang, M. Murakami, M. Wuttig, T. Okazaki, Y. Furuya and I. Takeuchi, *Mater. Trans. JIM*, 45 (2004) 173.
44. Y.K. Yoo, T. Ohnishi, G. Wang, F.W. Duewer, X.-D. Xiang, Y.-S. Chu, D.D. Mancini, Y.-Q. Li and R.C. O'Handley, *Intermetallics*, 9 (2001) 541.
45. Y.K. Yoo and F. Tsui, *MRS Bull.*, 27 (2002) 316.
46. T. Fukumura, M. Ohtani, M. Kawasaki, Y. Okimoto, T. Kageyama, T. Koida, T. Hasegawa, Y. Tokura and H. Koninuma, *Appl. Phys. Lett.*, 77 (2000) 3426.
47. M.-H. Yu, J. Hattrick-Simpers, I. Takeuchi, J. Li, Z.L. Wang, J.P. Liu, S.E. Lofland, S. Tyagi, J.W. Freeland, D. Giubertoni, M. Bersani and M. Anderle, *J. Appl. Phys.*, 98 (2005) 063908.

48. J.-C. Zhao, M.R. Jackson, L.A. Peluso and L.N. Brewer, *MRS Bull.*, 27 (2002) 324.
49. M. Aronova, Ph.D. Thesis, University of Maryland, 2000.
50. A. Boettcher, G. Haase and R. Thun, *Z. Metallkd.*, 46 (1955) 386.
51. J.-C. Zhao, *Prog. Mater. Sci.*, 51 (2006) 557.
52. K. Kennedy, T. Stefansky, G. Davy, V.F. Zackay and E.R. Parker, *J. Appl. Phys.*, 36 (1965) 3808.
53. J.J. Hanak, *J. Mater. Sci.*, 5 (1970) 964.
54. J.J. Hanak and J.I. Gittleman, *Magnetism and Magnetic Materials – 1972*; 18th Annual Conference, Denver; C.D. Graham, Jr. and J.J. Rhyne, (eds.), AIP Conference Proc., Part 2 (10), New York, 1973, p. 961.
55. R.B. van Dover, M. Hong, E.M. Gyorgy, J.F. Dillon Jr. and S.D. Albiston, *J. Appl. Phys.*, 57 (1985) 3897.
56. Y.K. Yoo, Q. Xue, Y.S. Chu, S. Xu, U. Hangen, H.-C. Lee, W. Stein and X.-D. Xiang, *Intermetallics*, 14 (2006) 241.
57. R.M. Bozorth, *Ferromagnetism 317*, IEEE Press, New Jersey, 1993, p. 165.
58. R.A. de Groot, F.M. Mueller, P.G. Van Engen and K.H.J. Buschow, *Phys. Rev. Lett.*, 50 (1983) 2024.
59. I. Takeuchi, O. Famodu, M. Aronova and J. Hattrick-Simpers, *Proceedings of the Eighth International Symposium on Sputtering & Plasma Processes*, ISSP 2005 Committee, Japan, 2005, p. 201.
60. S.J. Murray, M. Marioni, P.G. Tello, S.M. Allen and R.C. O'Handley, *J. Magn. Magn. Mater.*, 242 (2001) 945.
61. V.A. Chernenko, E. Cesari, V.V. Kokorin and I.N. Vitenko, *Scripta Metall. Mater.*, 33 (1995) 1239.
62. M. Wuttig, L. Liu, K. Tsuchiya and R.D. James, *J. Appl. Phys.*, 87 (2000) 4707.
63. C. Wedel and K. Itagaki, *J. Phase Equilib.*, 22 (2001) 324.
64. Y.K. Yoo and F. Tsui, *MRS Bull.*, 27 (2002) 316.
65. F. Tsui and P.A. Ryan, *Appl. Surf. Sci.*, 189 (2002) 333.
66. R.B. van Dover, L.F. Schneemeyer and R.M. Fleming, *Nature*, 392 (1998) 162.
67. H. Chang, I. Takeuchi and X.-D. Xiang, *Appl. Phys. Lett.*, 74 (1999) 1165.
68. Y.K. Yoo, F. Duewer, T. Fukumura, H. Yang, D. Yi, S. Liu, H. Chang, T. Hasegawa, M. Kawasaki, H. Koinuma and X.-D. Xiang, *Phys. Rev. B*, 63 (2001) 22421.
69. C. Gao, B. Hu, I. Takeuchi, K.-S. Chang, X.-D. Xiang and G. Wang, *Meas. Sci. Technol.*, 16 (2005) 248.
70. R. Maezono, S. Ishihara and N. Nagaosa, *Phys. Rev. B*, 58 (1998) 11583.
71. M.J. Turchinskaya, L.A. Bendersky, A.J. Shapiro, K.S. Chang, I. Takeuchi and A.L. Roytburd, *J. Mater. Res.*, 19 (2004) 2546.

CHAPTER THIRTEEN

DETERMINATION OF PRESSURE-DEPENDENT PHASE DIAGRAMS

Surendra K. Saxena[1] and Yanbin Wang[2]

Contents

1 Introduction	412
2 High-Pressure Devices	413
2.1 Diamond-Anvil Cells	413
2.2 Large-Volume Presses	416
3 Pressure Measurement	418
3.1 Pressure Measurement Using Ruby Fluorescence	418
3.2 Pressure Measurement Using X-Rays	419
4 High Pressure and Temperature	421
4.1 X-Rays at High Pressure	421
4.2 Heating at High Pressure in a DAC	422
5 Examples of Phase Diagrams Determined Using LVPs	431
5.1 Phase Relations in Univariant Systems	431
5.2 Phase Relations in Complex Systems	434
6 Phase Diagrams Using DAC	435
7 Industrial Solids	438

1 INTRODUCTION

High pressure is an important physical variable affecting properties of materials almost as significantly as high temperature. In many instances, a certain process will be possible only because of pressure. Planetary interiors are inaccessible to any direct observation and only high pressure–temperature experiments can provide any information on the nature of planetary interiors. There are myriads of examples showing the effect of pressure leading to formation of important materials. The example of graphite converting to diamond is the one often used in this connection. Pressure alone has resulted in the formation of many novel materials but when pressure is combined with temperature, the results could be astounding. The study of crystal structure at high pressures can reveal the fundamentals of phase transformations and the nature of atomic bonding. High pressure is being used more and more to synthesize new

[1] Center for the Study of Matter at Extreme Conditions, Florida International University, Miami, FL, USA.
[2] Center for Advanced Radiation Sources, The University of Chicago, Chicago, IL, USA.

materials, to study the behavior of existing materials and to tune material physical properties. This chapter will describe the currently used high-pressure techniques in the field of high-pressure science. Although there are relatively few studies involving the simultaneous use of pressure and temperature, the importance of such a combination requires that we also consider the progress made recently in such experimentation. In this chapter, the focus is on techniques that are useful from low to ultra high pressures and which permit the in situ characterization of materials.

Most high-pressure techniques used to date are designed to use in situ determination of properties by using X-ray or neutron-diffraction and Raman spectroscopy. The equipment is created using various anvil devices – multi-anvil and diamond anvil. Pressure is defined as the force applied to a given area. Thus, to increase pressure one can either increase the force being applied or decrease the area. The latter effect is used in these devices; pressure becomes larger as the anvil culet becomes smaller. Thus pressures of many million bars can be created on the tip of a diamond anvil which is 0.02 mm in diameter, restricting the sample size to less than the culet size. Usually pressures up to 100 gigapascals (GPa) can be created with a culet size of 0.3 mm. For larger sample size, one cannot use expensive diamonds and the multi-anvil technique requiring six to eight anvils has to be used. These techniques are described below. In particular, we will describe the diamond-anvil cell (DAC) and the large-volume press (LVP). The sample volume in DAC is small but it is a handy tool for rapid exploration of pressure–temperature space and the identification of new high-pressure phases, through in situ studies of their structure and properties. It can also be used to determine rapidly whether a high-pressure phase remains stable on release of pressure. LVP is not as handy as DAC but it has a very reliable performance and can be easily automated. We already have examples of several novel materials synthesized by using LVP. As we shall describe later, in LVP, the temperatures and synthesis environments can be better controlled than in DAC, but the pressure range rarely attainable to 60–70 Gpa, mostly limited to below 30 GPa. Therefore for industrial studies at low to moderate pressures and high temperatures, LVP is clearly the technique of choice but for planetary studies and for rapid exploration DAC is to be preferred.

2 HIGH-PRESSURE DEVICES

2.1 Diamond-Anvil Cells

Several different types of DAC designs have been described in the literature [1–4]. The cells that use diamonds as anvils are portable devices which may vary in size from a few millimeters to a few centimeters in size. Diamond, in addition to being recognized as the hardest and least compressible material, has the important property of being transparent to most of the spectrum of electromagnetic radiation, including γ-rays, X-rays, portions of ultraviolet, visible, and most of the infrared region. Diamonds are usually modified brilliant design with a table and a culet (Fig. 13.1). High pressure is generated in the DAC by a small force applied to the table and transferring that same force to a small area of the culet (Fig. 13.2). A sample is placed between the polished culets of the diamonds and is contained on the sides

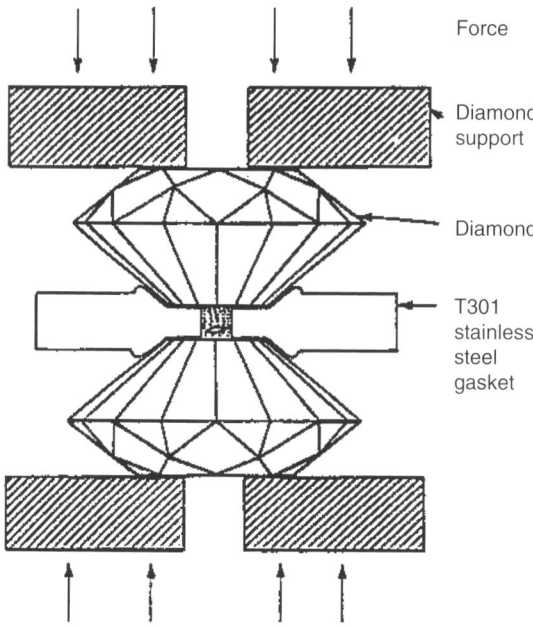

Figure 13.1 Basic principle of a DAC.

Figure 13.2 Two design diamonds showing facets and top culet with a 300 µm diameter.

by a metal gasket. The small sample size is determined by available diamonds for the anvils, which are of the order of 0.2–0.4 mm. Larger natural diamonds are rare and expensive which precludes the use of DAC techniques for creating new materials for industrial use. Scientists at the Geophysical Laboratory, Carnegie Institution of Washington and at the University of Alabama are currently working in enlarging

Figure 13.3 The first opposed DAC at NBS laboratory (photo from the NIST museum display).

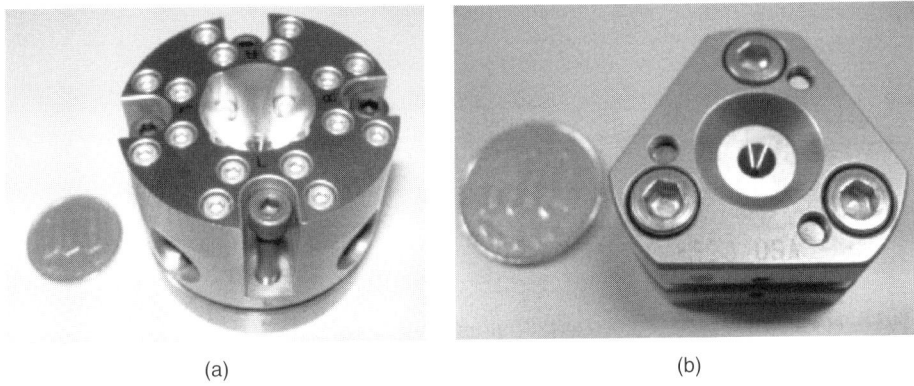

(a)   (b)

Figure 13.4 Two types of DAC from M. Kanzaki, ISEI, Okayama University website. The four-pin cell to the left is based on a piston and cylinder design and can be used for heating the sample with laser.

the diamond size by a microwave plasma vapor deposition technique which could change the future application of the DAC technique.

Several types of diamond cells are available. The first opposed diamond cell device was created at the NBS laboratory during the second half of the twentieth century (Fig. 13.3). This invention of the diamond-anvil high-pressure cell [5] and the development of the optical ruby fluorescence method of pressure measurement [6] led to the subsequently designed DACs and the high-pressure research. The lever-arm design of Weir et al. [5] has undergone some modifications and is used in the Mao–Bell cell (see later). The three-pin Merrill–Bassett cell and three- and four-pin cells (Fig. 13.4) have been designed for various applications requiring a short cell for use in limited space or for double-side laser heating.

Irrespective of which cell is to be used, most opposite anvil cells employ a metal gasket to contain the sample (in some cases boron gaskets have been used – side diffraction for example). The choice of the metal depends on the magnitude of the pressure to be reached. Usually steel gasket is used for moderate pressures up to several GPa. For ultra high pressures, the choice material is rhenium which is quite expensive. Rhenium is also necessary in externally heated diamond cells.

2.2 Large-Volume Presses

LVPs came into use in the 1950s, in an effort to generate high pressures on three-dimensional specimens for synthesis reactions, especially for making diamond (e.g., see Ref. [7], for a review). Significant developmental efforts were carried out in earth science communities in the late 1960s, primarily in Japan, to use LVPs to simulate conditions in earth and planetary interiors [8]. Onodera [9] reviewed various types of LVP apparatus. To our knowledge, the first effort to study materials with the LVP using laboratory X-ray sources started in the 1970s [10,11] leading to the first synchrotron-based LVP facility at the Photon Factory (Japan) in 1983 [12]. Since then, LVPs have become a major player in simultaneous high-pressure, high-temperature research using synchrotron radiation.

LVPs offer sample volumes ranging from about $1\ mm^3$ to about $1\ cm^3$, depending on the pressure range of interest, allowing virtually endless variations in cell design for various in situ physical property measurements combined with X-ray diffraction. There are several technical advantages in using LVP for high-pressure research: (1) resistive heating provides prolonged stable temperatures in excess of 2500K for hours or even days, (2) pressure and temperature gradients are small (typically on the order of 0.1 GPa/mm and 10K/mm, respectively) and well characterized, (3) effects of non-hydrostatic stresses in the sample are small and can be eliminated by heating, (4) with a carefully designed $P-T$ path, a wide pressure and temperature range can be covered in a single experiment, and (5) the large sample volume provides robust counting statistics for powder diffraction for quick data collection in energy-dispersive mode and reliable structural information in monochromatic mode.

Several types of pressure modules are being used at various laboratories for various experimental needs. Here we only discuss some of the most commonly used modules. The first tooling is the well-known cubic-anvil apparatus, or DIA [10], which consists of upper and lower pyramidal guide blocks installed on the heads of the hydraulic rams, four trapezoid thrust blocks, and six anvil holders, as indicated in Fig. 13.5(a). The inner surfaces of the guide blocks form a tetragonal pyramid. Two of the six anvils are along the center line of this pyramid and are fixed opposite to each other on each guide block. The other four anvils are horizontally located on the midpoints of the square edges of the bipyramid, forming a cubic nest bounded by the flat faces of the six anvils. A ram force applied along the vertical axis is decomposed into three pairs acting along three orthogonal directions, forcing the six anvils to advance synchronously toward the center of the cube. X-ray access is through vertical gaps between the side anvils. The diffraction vector is vertical, with possible 2θ angles up to about $30°$.

The second pressure tooling is the so-called Kawai-type apparatus. This double-stage system consists of a first-stage hardened steel cylinder, which is cut into six

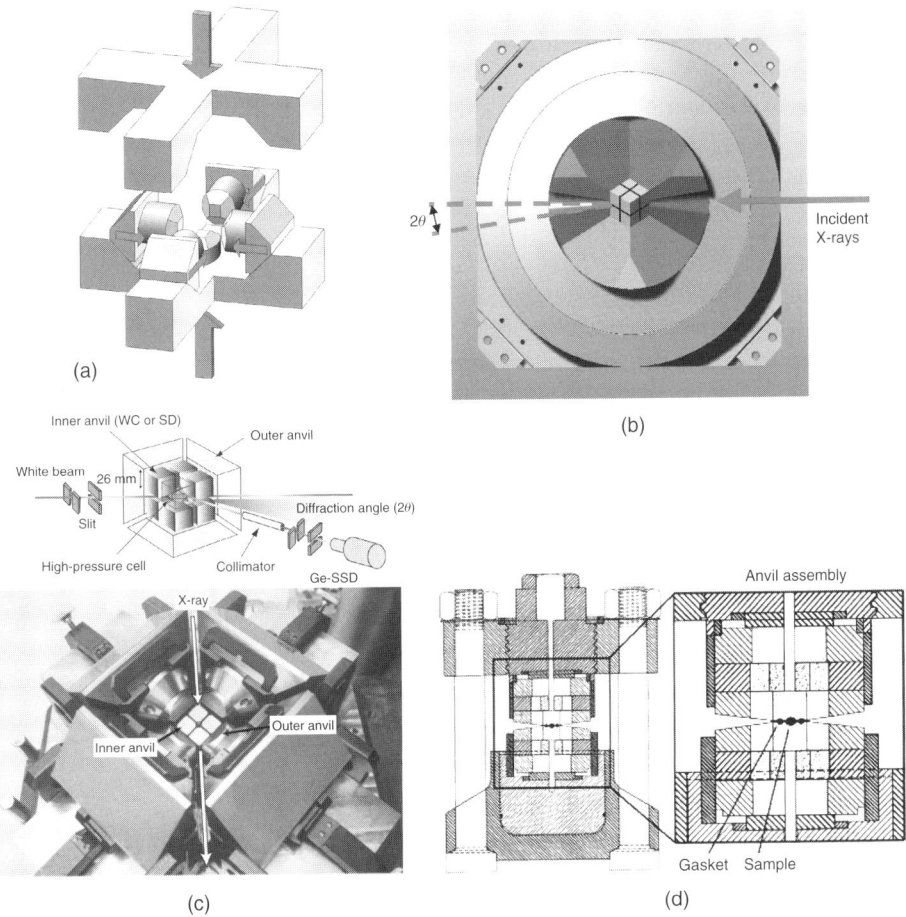

Figure 13.5 Several types of pressure modules for high-pressure experiments: (a) DIA apparatus; (b) T-Cup apparatus; (c) double-stage 6/8 system (SPring-8); and (d) Paris–Edinburgh (PE) cell.

parts enclosing a cubic cavity, with the [111] axis of the cube along the ram load direction. Inside the cubic cavity is the second-stage assembly, consisting of eight WC cubes, which are separated by spacers. Each cube has one corner truncated into a triangular face; the eight truncations form an octahedral cavity in which the pressure medium is compressed. Several variations of such apparatus exist. A miniature version of this apparatus, the T-Cup [13], weighs only a few kilograms with an overall outer diameter of 200 mm and uses 10-mm WC second-stage cubes (Fig. 13.5(b)). The pressure module is compressed by a vertical hydraulic ram, similar to most of the off-line MA8 systems. Diffracted X-rays pass through the gaps between the second-stage cubic anvils in a plane inclined at an angle of 35.3° from the vertical load direction. Conical access notches are made in the first-stage wedges, to allow a range of 2θ angles up to 12°. We will refer to this configuration as Kawai type A, to distinguish it from the configuration described below.

The third configuration is a hybrid system using a set of large DIA anvils instead of the six wedges in the above-mentioned split-cylinder (or sphere) first stage to compress the eight cubes [14]. Figure 13.5(c) shows such a system installed at Beamline BL04B1 at SPring-8 [15]. Hydraulic load in this case is parallel to the [100] direction of the eight-cube assembly, and X-ray access is through horizontal anvil gaps if WC cubes are used. In this case, notches need to be made in the DIA anvils to allow X-ray access for in situ work. With sintered diamond second-stage anvils, pressures in excess of 70 GPa and temperatures over 1700°C are accessible with this technology. This will be referred to as Kawai type B.

Not all LVPs are multi-anvil apparatus. The torroidal cell, for example, operates on the concept of opposed anvils, similar to the DAC. This type of anvil device was first developed in the former Soviet Union [16,17]. Instead of flat anvil tips as in the DAC, both torroidal anvils have a half spherical depression so that the two anvils enclose a semi-spherical cavity. One or several torroidal depressions are made concentric with the central sample cavity for preformed gaskets that are used to minimize extrusion (Fig. 13.5(d)). Besson et al. [18] integrated this device with a compact hydraulic system so that the system is portable, allowing great flexibility. Such integrated torroidal devices are known as the Paris-Edinbough (PE) cell; a PE cell with load of 150 tons weighs only about 25 kg. Several PE cells are currently used at synchrotron sources. The X-ray access is through the gaps between the two anvils and the largest diffraction angle is in the horizontal plane.

The opaque nature of the LVP cell assemblies requires a bright X-ray source for in situ studies. Currently there are a number of LVP facilities at various synchrotron sources worldwide. Most significant multi-anvil activities are in Japan [11,15] and the US [19–23]. At the European Synchrotron Research Facility (ESRF), PE cells are mostly used, with a plan to build a 17.5 MN double-stage system. A similar 17.5 MN system has been constructed at HASYLAB in Germany [24].

Non-in situ techniques such as those based on piston–cylinder design and also the Walker inserts are not reviewed here. For description of these techniques, the reader is referred to other reviews [3].

3 PRESSURE MEASUREMENT

3.1 Pressure Measurement Using Ruby Fluorescence

Pressure may be measured by either the ruby (Cr^{3+} doped Al_2O_3) fluorescence line shift with pressure or by determining the molar volume of a pressure marker by X-ray diffraction. The pressure marker may be any solid with a well-known relationship between pressure and volume. The ruby method requires a laser and a spectrometer and micron sized ruby grains, which can be distributed throughout the sample for checking the uniformity of the pressure. The ruby fluorescence produces R1 and R2 lines (Fig. 13.6) and has been calibrated to 80 GPa, the so-called quasi-hydrostatic ruby scale, by Mao et al. [25]. The method of using markers to calibrate ruby was introduced by Piermarini et al. [26] using the theoretical equation of state (EOS) of NaCl [27] as the pressure standard up to 19.5 GPa at 300K. Mao et al. [28] extended the marker method to the megabar range at 300K. Recent revisions of this scale have been made by Dewaele et al. [29] and Akobuije et al. [30].

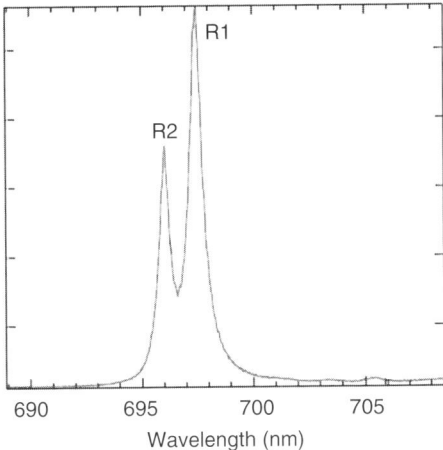

Figure 13.6 The ruby fluorescence lines R1 and R2 for pressure measurement. The lines show a pressure of ~4 GPa.

The latter authors used the P–V–T data of (material) from Dewaele et al. [29], MgO from Zha et al. [31], and their own data on (material) generated recently in quasi-hydrostatic pressure media to 150 GPa. They also considered shock-wave reduced isotherms including corrections for material strength. The new calibration is

$$P = (A/B)\,[(\lambda/\lambda o)B - 1]\ (\text{GPa}) \tag{1}$$

with $A = 1873.4 \pm 6.7$ and $B = 10.82 \pm 0.14$ where λ is the peak wavelength of the ruby R1 line.

In calibrating the ruby with a sample of another material with a known pressure scale, we have to consider the problem of the hydrostatic vs. non-hydrostatic medium. This is a typical problem with all DAC work. If we want to construct a ruby scale for the hydrostatic medium, the ruby and the calibrant should be in a hydrostatic pressure medium to obtain a reliable calibration (this sentence is awkward). At high pressure all materials can develop shear strength, leading to non-hydrostatic conditions. Several fluids can be used to create quasi-hydrostatic conditions. The best high-pressure media are helium, hydrogen, and, possibly, xenon. At low pressures to about 15 GPa, a 4:1 methanol–ethanol mixture has been used in many studies.

3.2 Pressure Measurement Using X-Rays

It is possible to use a solid as an internal standard. NaCl, MgO, and metals principally platinum and gold have been used as X-ray markers. The choice of the solid again depends on the pressure range of the experiments and whether the sample will be heated. The Pt pressure scales proposed by different authors [32,33] are in general agreement. However, gold is problematic because of significant deviatoric stress development. The problem of non-hydrostatic medium is similar to that discussed in the previous section.

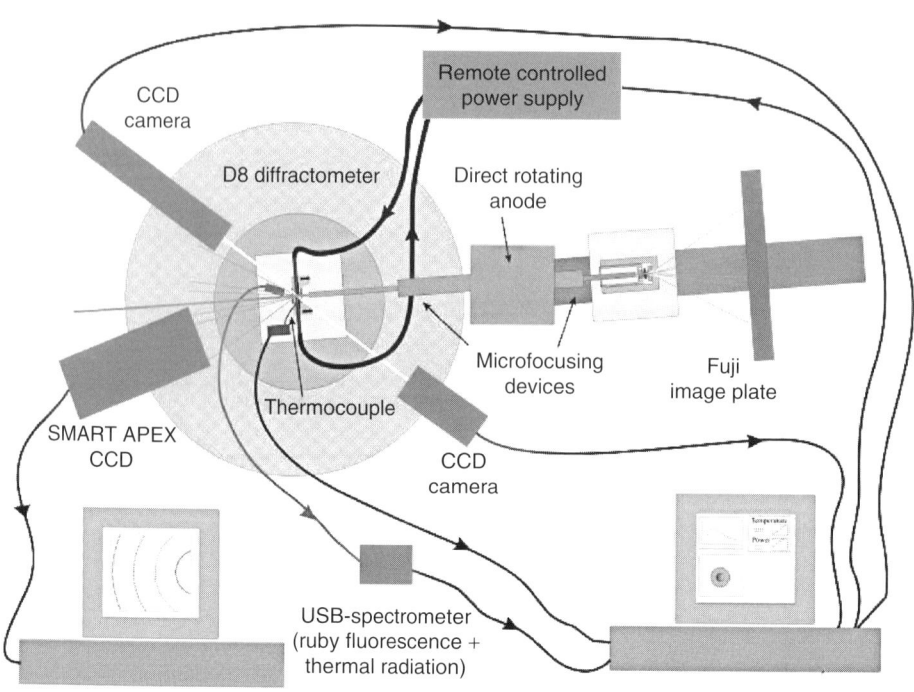

Figure 13.7 X-ray facility for powder diffraction at CeSMEC.

X-ray diffraction measurement of cell constants of solids at high pressure requires a powder diffraction measurement using a high intensity X-ray beam collimated to a size of few microns. Such facilities which use synchrotron radiation are located in a few places around the world. Some user facilities are: Advanced Photon Source at Argonne National Laboratory, European Synchrotron Radiation Facility at Grenoble France, and SPring-8, Japan, [15]. In-house X-ray equipment with high intensity beam is available at many institutions such as the Bayerisches Geoinstitut, Universität Bayreuth, or the one at CeSMEC (Fig. 13.7). For the diamond cells, powdered samples are loaded into a suitable initial diameter hole (size depending on the pressure to be achieved, e.g., a 200 μm hole would get us to a pressure of few tens of GPa, a 30 μm hole could get us to some megabars) of Re gasket confined between bevelled diamonds with suitable sized culets. For the LVPs, powders are first pelletized and then loaded in a capsule inside a heater (made of various electrically conducting materials, such as graphite, $LaCrO_3$, or precious metal tubes). The material choice of the capsule and the heater depends on the chemical reactivity of the sample under the desired P–T conditions. Pressure may then be determined by placing powders with a known EOS (e.g., platinum) as an in situ X-ray standard, either mixed in small proportion with the sample, or as a separate layer in the sample chamber.

Usually for all non-hydrostatic measurements the ruby method can be exploited easily to study the pressure distribution because a few micron size laser beam can be used. With X-ray unless, it is the synchrotron X-ray such pressure determination is not possible (don't quite understand this sentence. Diffraction has been used to measure deviatoric stress). At high temperature sample contamination by reaction of ruby with the sample or with the internal X-ray standard is possible. X-ray method has the advantage that an experiment can be continued without interruption if diffraction data is required over a series of pressure. With increasing pressure, the X-ray signal might weaken. The same can be achieved with ruby for Raman spectroscopy. But the ruby fluorescence peaks also change in quality as pressure of 100 GPa is reached. Chen and Silvera [34] and Akobuije et al. [30] have discussed this problem and elaborated on the method.

4 High Pressure and Temperature

4.1 X-Rays at High Pressure

The availability of synchrotron X-ray has revolutionized the nature and tempo of high-pressure research. We will briefly describe here the facility available at GeoSoilEnviroCARS (GSECARS) at the Advanced Photon Source at Argonne National Laboratory. The high-pressure studies using both DAC and LVP can be performed at two beam lines. Other beam lines at HPCAT facilities (DAC) are becoming available to general users along the same lines. A monochromatic beam is to be usually preferred for the DAC. For LVPs, most experiments prefer energy dispersive diffraction (EDD), although monochromatic diffraction has been gaining popularity. The collection time required for a sample differs from several minutes to few seconds depending on the mode of diffraction, photon flux (~5 × 10^8 bending magnet, ~10^{11} wiggler both at 37 keV), and pressure – the higher the pressure the more exposure time is needed. Beam sizes as small as 5 μm^2 may be achieved at the wiggler beam by focusing. One may opt to use a charge-coupled device (CCD) detector or an image plate. For X-ray diffraction resolution, larger size of image plate is better. For weak signals, image plate could be a better choice because collecting time may be prolonged without dark current accumulation. In case of having poor signal-to-noise ratios, the better dynamic range of IP may help in statistics. For fast data collection, we need to distinguish two cases. One is fast collection only once with a delayed readout. The other is fast data collection continuously. In the case of one-time collection, the benefits of IP still hold. For continuous fast collection, CCD has clear advantages.

Both EDD and angle-dispersive diffraction (ADD) techniques are available for the LVPs, although for determination of phase diagrams EDD technique is sufficient. This is because sample volumes (in the order of 1 mm^3) are several orders of magnitude greater than those in the DAC and counting statistic is rather robust. Typical beam size for LVP experiments is about 0.05–0.1 mm and data collecting time about 2–5 min. An example of the facility setup is shown in Fig. 13.8.

A detailed description may be found on the GSECARS website (www.gsecars.org). Similar experiments may also be performed on beamline ID 30 at the European Synchrotron Radiation Facility (Grenoble, France: DAC) or at beamlines BL04B1 (LVP) and BL22XU (DAC) at SPring-8. These third generation synchrotrons also permit the use of monochromatic radiation to study samples of very small size with area detectors.

Figure 13.8 1000 ton LVP at 13-ID-D of GSECARS. Inset compares with a PE cell at ESRF. Scale is approximately the same.

X-ray powder diffraction data may be processed using a variety of programs. We have found that FIT2D (hammersley@esrf.fr) is a good general purpose data analysis program. The analysis of the integrated X-ray spectra may be done by using the program GSAS and PeakFit 4.0 and by many other programs (e.g., Bruker's TOPAZ). In-house facilities such as that at Bayreuth and CeSMEC can be used at high pressures. The CeSMEC facility consists of a SMART APEX CCD (Bruker) and a Fuji Image Plate (Fig. 13.6). The MacSci rotating anode is used with a 50 Watt generator and 20 amps current. The target is molybdenum. X-rays are focused to a narrow beam of 300, 100 or 50 µm using Australian X-ray Capillary Optics. The high intensity beams permit study of solids to a pressure as high as 50 GPa.

4.2 Heating at High Pressure in a DAC

While the study of high-pressure behavior of fluids and solids has given us results of profound scientific interest, the study of heated compressed materials is opening a window to look at planetary interiors and leading to discovery of new materials. For the DAC, we will consider two methods of in situ heating of a pressed sample; the first is heating the sample with a laser and the second by employing electrical heating devices which may be considered as internal, that is, using a resistively heated wire between the diamonds and external where electrical heaters of various

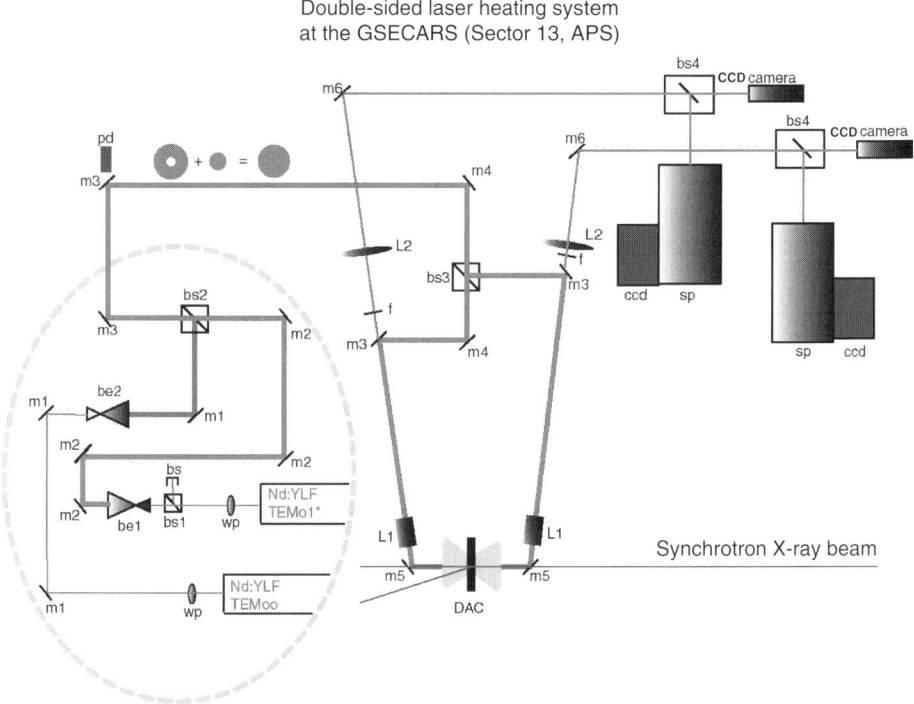

Figure 13.9 A double-side laser heating setup at GSECARS.

kinds may be used. Several such heaters are described by Dubrovinskaia and Dubrovinsky [35].

4.2.1 Laser Heating

Some decades ago Ming and Bassett [36] introduced laser heating in a DAC. The technique is now well established and used at many synchrotron and in-house facilities. Pressurized sample in a DAC may be heated by stabilized Nd:YLF laser (Photonics, 50-Watts CW at 1064 nm, TEM_{00} mode). Temperatures are determined from the thermal radiation of hot sample using spectroradiometry as described below. The use of two-dimensional CCD detector and imaging spectrograph with entrance slit give possibility to measure temperatures along a narrow vertical strip in one measurement with spatial resolution of about 3 μm. In laser heating from one side of the sample, there is a large temperature gradient from the front to back. While this is not a problem for some studies, for example, melting where only surface reaction is involved, it is a definite drawback for determining physical properties of solids where the X-ray samples the whole thickness. Therefore, we use the technique of heating the sample from both sides as developed by Shen et al. [37]. The setup at the GSECARS beam line is shown in Fig. 13.9. At GSECARS, there are two lasers for the double-side heating (Fig. 13.9). The power profiles of the laser is shown in Fig. 13.10. The ideal heating must be homogeneous over as much of

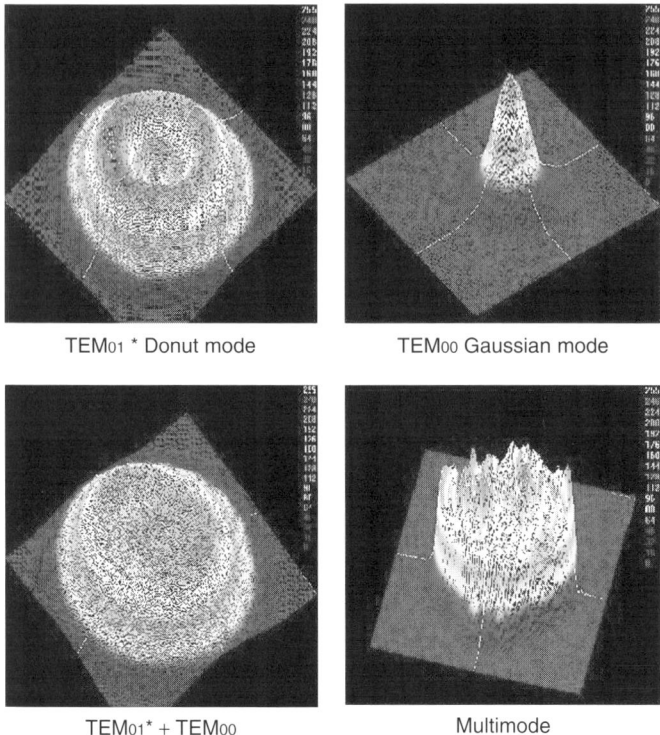

Figure 13.10 The various temperature distribution possibilities when laser modes are combined to achieve a large uniformly heated area.

the sample area as feasible. The CW-mode (continuous wave, as opposed to pulsed mode) laser in a mixed Nd-YLF or Nd-YAG lasers can be used for heating most metals and dark samples. The colorless ceramic powders would require mixing with a small powder of a non-reacting metal powder such as Pt or a CO_2 laser. The ND-YLF or ND-YAG can be used with type I diamonds but the CO_2 laser requires the type II which are double the cost of a type I.

The temperature is measured using the method of spectroradiometry for heated metals and silicates. For correct temperature determination wavelength-dependent emissivity has to be considered when fitting the spectra to the Planck radiation function. After major advances in instrumentation during the last decade, a poor knowledge of high-pressure and temperature emissivities became the main source of systematic errors in temperature measurements in high-pressure studies. These errors may range from few tens to several hundreds of degrees depending on the actual emissivity variation with wavelength and the temperature range of a study. Thermal radiation is collected over a range of wavelengths and the resulting spectrum is fitted with Planck's function. The accuracy of measured temperature requires data on emissivity of a sample as a function of pressure, temperature, and wavelength. In multi-wavelength thermometry the spectral radiance of the target at several wavelengths is curve fitted to Planck's or Wien's equation (Fig. 13.11).

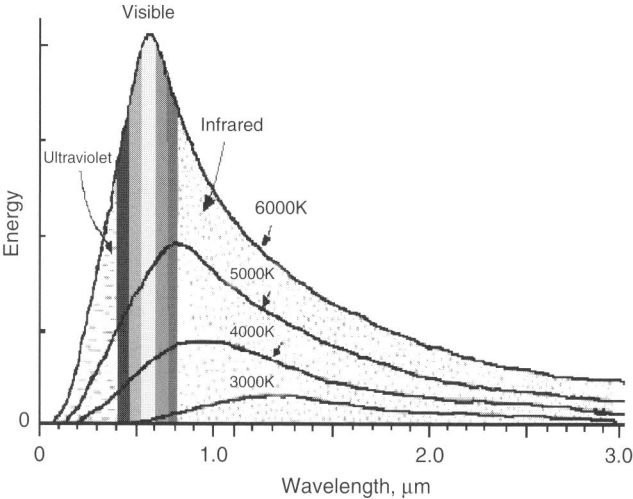

Figure 13.11 Temperature measurements using Planck's black body radiation.

A spectrum of the black body radiation can be described by Planck equation

$$b^0(\lambda, T) = \frac{2\pi hc^2 \lambda^{-5}}{\exp\left(\dfrac{hc}{\lambda kT}\right) - 1} \qquad (2)$$

where $b^0(\lambda, T)$ is the spectral intensity of the black body radiation at wavelength λ and temperature T, h Planck constant, c light velocity, and k Boltzman constant.

For real (gray) body we can write

$$b(\lambda, T) = \varepsilon(\lambda, T) \frac{2\pi hc^2 \lambda^{-5}}{\exp\left(\dfrac{hc}{\lambda kT}\right) - 1} \qquad (3)$$

where $\varepsilon(\lambda, T)$ is the body emissivity at given λ and T.

If $hc/\lambda \gg kT$ (valid up to 5000K at $\lambda < 0.9\,\mu m$), we can use Wien's approximation

$$b^0(\lambda, T) = \frac{2\pi hc^2 \lambda^{-5}}{\exp\left(\dfrac{hc}{\lambda kT}\right)} \qquad (2A)$$

$$b(\lambda, T) = \varepsilon(\lambda, T) \frac{2\pi hc^2 \lambda^{-5}}{\exp\left(\dfrac{hc}{\lambda kT}\right)} \qquad (3A)$$

Table 13.1 Emissivity of some metals and oxides at the melting temperature.

Compounds	Emissivity*		Melting temperature (K)
	ε_{800}	Δ	
W	0.34	−72 100	3685
Pt	0.44	−241 000	2045
Fe	0.28	−782 000	1808
Al_2O_3	0.40	−151 000	2288
ZrO_2	0.29	101 000	2988
Na_2SiO_3	0.21	−102 000	1361

*Emissivity at wavelength λ (in nm) is $\varepsilon_\lambda = \varepsilon_{800} + \Delta(\lambda - 800)/10^9$.

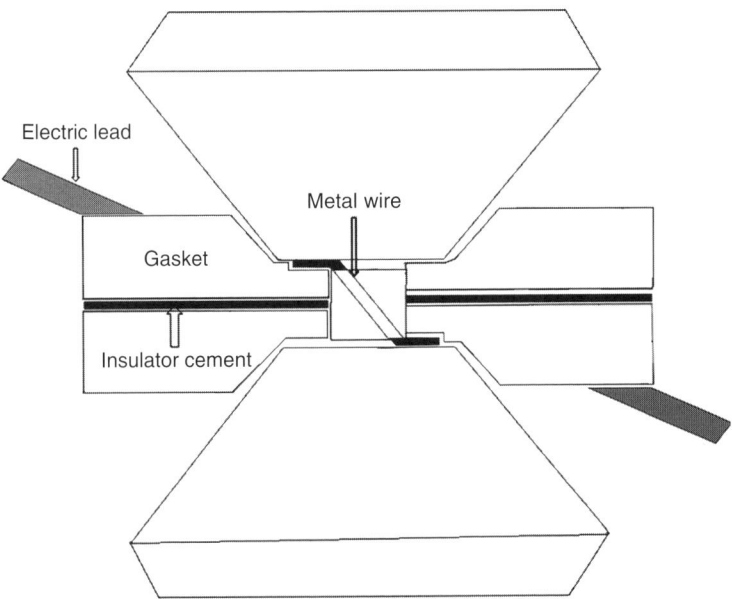

Figure 13.12 Resistive wire heating by Boehler's (1986) design.

Experimentally measured spectrum cannot be described by Eq. (3) or (3A) because it includes effects due to geometry of collecting optics, absorption in lenses, sensitivity of spectrometer, etc., which can affect spectral intensity. To solve this problem, a calibration function from quartz halogen tungsten lamp (QHTL) is usually used. In addition to the available data on emissivity which many times lacks the dependence on wavelength we also determine a temperature calibration using melting temperatures of a series of metals and oxides as shown in Table 13.1. Examples of such measurements may be found in Boehler [38] (Fig. 13.12).

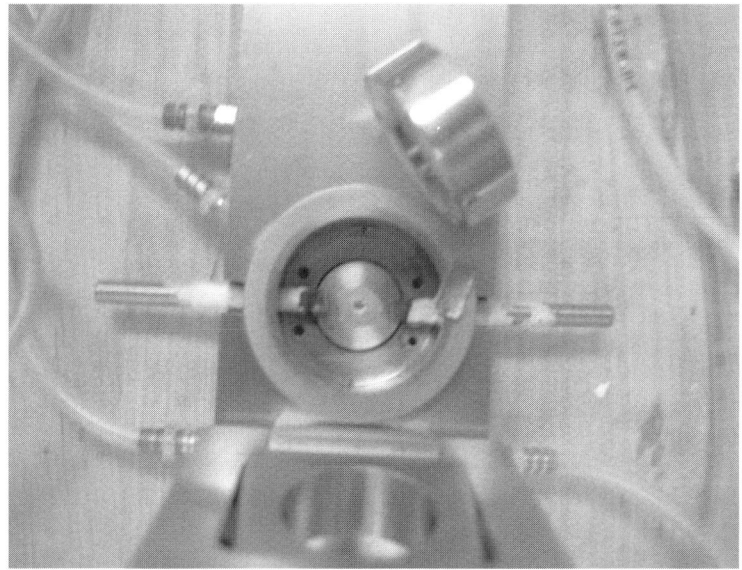

Figure 13.13 A short cell with piston and cylinder for heating a sample with a graphite heater. The two molybdenum rods are used as electric leads. The cell has an inlet for thermocouple and it is water cooled. Pressures to 70 GPa and temperatures to 1500K may be reached.

While laser heating provides an easy method to heat to very high temperatures, there are some serious drawbacks with this method. In most cases, the diamonds must be insulated from the hot spot for long time sustained heating. The insulating material could react with the sample. Examples are the heating of iron with MgO or Al_2O_3. If fluids such as argon are used, at ultra high pressures there may be some diffusion of the fluid in the sample. The inhomogeneity of the sample surface may lead to uneven heating due to different degrees of laser absorption.

4.2.2 Electrical Heating

Electrical heating of the sample may be done in various ways. We may heat the sample by putting a wire through the sample or put a small heater around the diamond seats or wrap the gasket in graphite heater. Electric heating requires time and efforts and is effective in heating a much larger area homogeneously but there are several problems which still need to be solved. Dubrovinskaia and Dubrovinsky [35] recently reviewed the high-temperature in situ techniques quite extensively. In what follows, we have summarized some of the techniques as reviewed by them. For details not covered in this review, we suggest the reader should read their paper.

Resistive Wire Heating The latest resistive wire heating was designed by Zha and Bassett [39] which seems to be the end of a long series of trials by a number of workers with mixed results. Liu and Bassett [40] studied melting of a 5 μm diameter iron wire placed on the diamond culet wrapped in oxide or silicate powder. Since no gasket was used they reached a pressure of about 20 GPa. Boehler [41] used two stainless steel gaskets glued together with a ceramic glue (Fig. 13.13). The gasket assembly is then indented and a sample hole is drilled in the center. Electric current is passed through a metallic wire (several micrometers in diameter) which is in contact with the upper diamond and the upper gasket; it passes through diagonally across the sample chamber reaching the lower gasket. The part of the wire in contact with the diamond remains cold but in the sample hole, where it is surrounded by a ceramic sample, it heats up. Due to the high thermal conductivity of diamonds, the hot spot forms only in the center of the pressure chamber [41,42]. Dubrovinsky and Saxena [43] and Dubrovinsky et al. [44], modified the double-gasket assemblage. A detailed description of this method is to be found in Dubrovinskaia and Dubrovinsky [35]. Zha and Bassett [39] modified the double-gasket assemblage for internal resistive heating by replacing the metal gasket with a non-metallic gasket for holding the heater-sample assemblage. This prevented the failure of the electrical isolation layer under high pressure and pressures up to 10 GPa and temperatures up to 3000K could be reached. Other simpler designs have been used by wire heating with single ceramic covered gaskets [47] for moderate pressures and temperatures with limited success. Dubrovinsky et al. [45] attained significantly higher pressures and temperatures in experiments not requiring the use of in situ characterization by employing a gasket without a sample hole. They conducted an experiment to alloy magnesium and iron at a pressure of 89 GPa and a temperature of 3400°C.

4.2.3 External Electrical Heating

Several types of external heaters have been discussed in the literature. It is essentially a matter of surrounding the gasket and the diamond with a heater. If the gasket is wrapped in a graphite foil, one can reach several hundred degrees but this could take a toll on the diamonds. Ring heaters of various types can be used but, these do not get the samples to temperatures more than few hundreds. Placing the thermocouple securely close to the sample is another problem which is not easy. Of course if one succeeds in doing that, then temperature is easy to measure and the sample can be maintained at a constant value (± 5K at 1500K) for several hours. An important advantage of an externally heated sample is that at $T > 800$K stresses are practically absent and the heating is largely homogeneous [3,46]. For these reasons, the external electrical heating would have been an ideal technique ($T < 1500$K) complementing the laser heating (>1500K) for studying materials at extreme conditions but for several difficulties relating to graphitization of diamond, rapid creep deformation of materials, welding of the gasket to the anvils, and so on [3,47,48]. Several heater designs are described by Dubrovinskaia and Dubrovinsky [35].

A graphite heater may be made by sandwiching the gasket between two flexible graphite layers. Several techniques exist which would permit to bring the electrical leads to the graphite [46,49,50]. The Uppsala–CeSMEC–Bayreuth approach is shown in Fig. 13.13. Two molybdenum rods of 5–6 mm diameter are used to supply the current to the graphite heater. Figure 13.14 shows a Mao–Bell cell with such a heater design with a cooling water jacket at the HPCAT facility as organized by Hans-Peter Liermann. Thermocouple can be suitably inserted. The experiments must be carried out under a flow of argon gas. With such a graphite heater, both the diamonds and the gasket can be homogeneously heated quite easily to 1000K and

Figure 13.14 A Mao–Bell cell designed at CeSMEC being used at HPCAT (Advance Photon Source, ANL, Hans-Peter Liermann). The molybdenum rods are supported by a water-cooling jacket. Thermocouple is shown at the top.

temperatures to 1500K may be obtained but with a certain risk to the diamonds. Details on how to make a heater and the entire setup, are given in the chapter by Dubrovinskaia and Dubrovinsky [35]. They have also reviewed the advantages and the problems. A great advantage is of course the homogenous and sustained heating; the problems are the risk to diamonds, the thinning of the graphite, and the loss of electric contact during the middle of an experiment and the unpredictability of pressure increase or decrease during heating.

Other possibilities of heating a sample under pressure exist. Dubrovinskaia and Dubrovinsky [51] have succeeded in heating a cell (whole body) to 1300K and pressures up to 100 GPa. At CeSMEC, several types of ring heaters have been used for heating to 1000K.

4.2.4 Combination of Electric and Laser Heating

Laser heating may be combined with electric heating with great advantage. Such combined heating could result in less stress for laser-heated samples. This type of heating has been proposed by many but no results have appeared yet.

4.2.5 Heating in LVPs

High temperatures in LVPs are achieved using resistive heaters. Various materials can be used in a variety of forms for heating. Below the diamond stability field, graphite is a convenient heating material. It is easy to machine graphite rods into cylindrical sleeves. Graphite has low X-ray scattering power and hence does not produce strong diffraction lines that may interfere with sample diffraction. For high pressures, $LaCrO_3$ as well as precious metals (Re, Pt, Ta, etc.) are used widely as heating materials. However, all of these latter materials are attenuating to X-rays. Various geometrical tricks have been developed to minimize effects of attenuation on diffraction quality. For example, slots may be cut through a sleeve heater with light materials (such as boron, diamond, and MgO) as "window" inserts, or orient the sleeve heater in such a way that X-rays enter the heater along the cylindrical axis, or use disk heaters and have the X-rays going through the gaps between the two disks. Figure 13.15 shows some examples of cell designs.

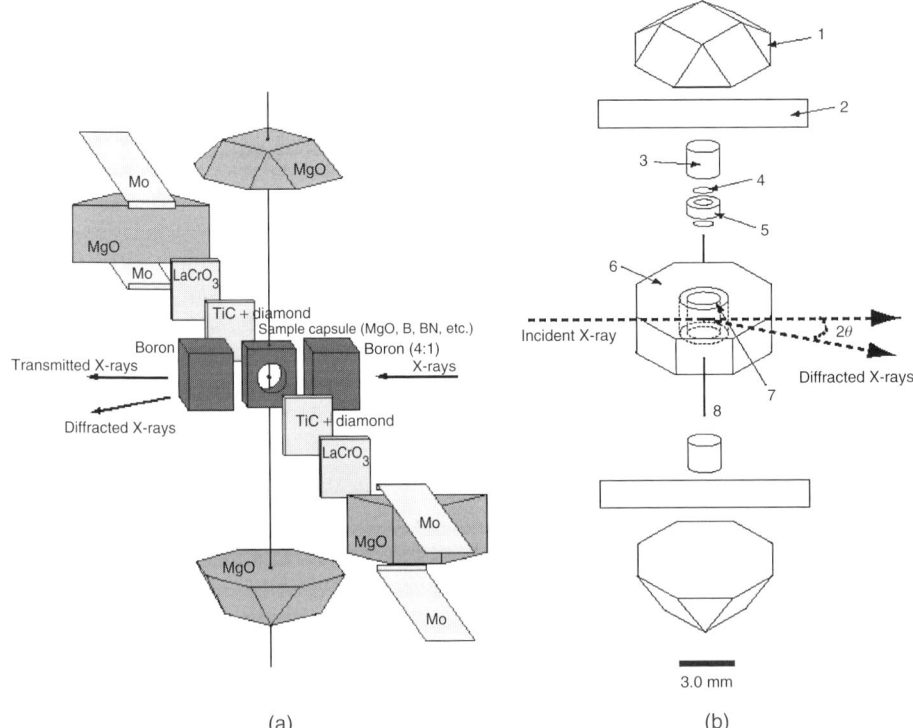

Figure 13.15 Examples of cell assemblies used in LVP for synchrotron work: (a) assembly for T-Cup [54] (b) assembly for double-stage 6/8 system [58].

Temperature is measured by a thermocouple in the LVP. However, pressure effects of thermocouple emf are poorly understood at this stage and are therefore ignored (see, however, Getting and Kennedy [52]). There are some initiatives to examine these effects, either by using neutron absorption resonance (Le Godec et al. [53]) or Johnson noise measurements [54]. These techniques provide potential opportunities to examine pressure effects on thermocouple emf, and quantitative correction schemes.

5 Examples of Phase Diagrams Determined Using LVPs

The LVP has been used for studying phase relations for more than 40 years. Initially, specimens were brought to a desired $P-T$ conditions for reaction and then recovered to ambient conditions, so that the specimens could be examined using X-ray diffraction, electron microprobe, or transmission electron microscopy (TEM). These "ex situ" or quench experiments have provided a wealth of information for minerals and ceramics with quenchable high-pressure phases.

For many systems where high $P-T$ phases are unquenchable, certain in situ techniques must be used. One of the commonly used probes is electrical conductivity. Many solid-state phase transitions are associated with marked change in electrical conductivity, which has been used in LVPs to detect phase diagrams. For example, Bundy [7] used this technique to determine melting curve for Si.

In situ X-ray diffraction experiment in a multi-anvil apparatus is generally considered the most reliable method in determination of high $P-T$ phase relations. In the case of a univariant reaction, stability of phases can be simply judged by detecting growth and diminishing of their diffraction lines. Phase A is more stable than phase B at a given a $P-T$ condition if the diffraction intensities of phase A increase while those of phase B decrease, when the composition does not change during the phase transition. Due to kinetic effects, it usually takes a certain time for the transition to proceed, that is, to be able to observe the appearance and disappearance of the diffraction lines of the phases of interest. For transitions with extremely sluggish kinetics, it is difficult to determine the phase boundary accurately.

In the case where there are more than two freedoms of the reaction (i.e., the phases involved in the transition have different compositions), one must determine compositions of the coexisting phases. This is usually quite difficult because although X-ray diffraction is powerful in determining crystal structure the phases, it is not sensitive to subtle composition variations, especially under high P and T, for the lack of accurate EOS information for the phases involved. Therefore, one must quench the sample by cutting of the electric power, and recover it to ambient pressure. Compositions of the coexisting phases will then be examined by powder X-ray diffraction, electron probe microanalysis, and analytical TEM. Chemical equilibrium of the coexisting phases must be confirmed in a series of experiments at various $P-T$ conditions.

In what follows, we will show a few examples, starting from simple, univariant systems and then review some recent results on more complex systems.

5.1 Phase Relations in Univariant Systems

One of the major advantages of LVPs in studying phase relations and EOS is that P–T paths can be well controlled. Figure 13.16 shows data collected in two experiments on pure iron using the T-10 [55]. Here, different phases can be clearly identified by powder diffraction. The hcp–fcc phase boundary is determined with both forward and reverse reactions, by changing temperature, making this the first reversed phase boundary determination of iron up to 20 GPa. The slope of the phase boundary is 36(3)K/GPa, which is steeper than that reported by Boehler et al. [56] in the DAC (24K/GPa) but is in excellent agreement with the results obtained by cubic-anvil press (Akimoto et al. [57]) (34K/GPa; as estimated from their Fig. 13.2) and the DAC results reported by Mao et al. [58] (35K/GPa). The hcp–fcc boundary and the triple point thus determined place a firm anchor on DAC determined boundaries at higher P and T.

Another example is NaCl. Nishiyama et al. [59] studied the B1–B2 phase boundary to 25 GPa and 2000K, near the melting point of NaCl. They took the advantage of a unique capability of the high-pressure system SPEED-Mk. At

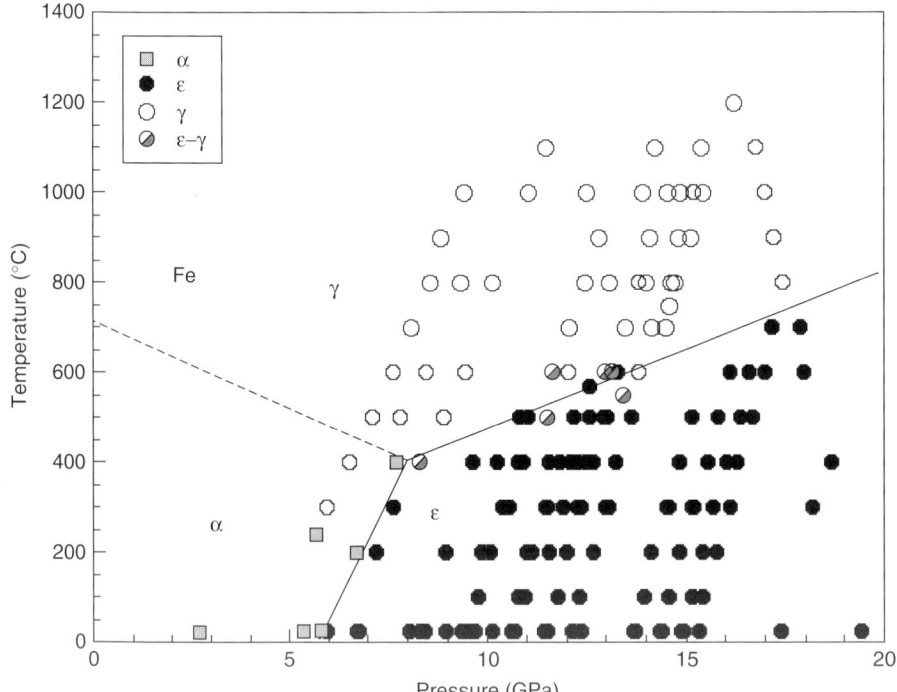

Figure 13.16 Phase diagram of Fe up to 20 GPa and 1200°C, determined in the T-Cup apparatus (Uchida et al. [55]).

SPring-8, that is, the press is able to oscillate over a 10° range during data collection. This ability greatly reduced uncertainty in phase identification near melting, where grain growth generally causes significant preferred orientation. They were able to perform reversals under these conditions and accurately determine the phase boundary (Fig. 13.17).

The phase diagram of Mg_2SiO_4 has been the focus in mineral physics for many years, because the olivine–wadsleyite phase transition is generally considered to be responsible for the 410 km seismic discontinuity in the Earth's mantle. Figure 13.18, taken from Katsura [60], summarizes results of the forsterite– wadsleyite–ringwoodite phase transitions in Mg_2SiO_4. This is a compilation of several groups' work in Japan, spanning more than 10 years. Here, results obtained using synchrotron radiation are presented as heavy solid and dashed lines. Thin solid lines represent previous quench results and dotted lines represent thermodynamic calculations based on calorimetry measurements, both are shown for comparison. Because of different pressure standards used in various studies, Katsura [60] recalculated previously reported pressure values so all the data can be directly compared. The forsterite–wadsleyite boundary determined by Morishima et al. [61], at lower temperatures, shows reasonable agreement with that by Katsura et al. [62] at higher temperatures. Both this boundary and the wadsleyite–ringwoodite boundary determined by Suzuki et al. [63] have steeper Clapeyron slopes (about 3.2 MPa/K for the forsterite–wadsleyite transition and 6.9 MPa/K for the wadsleyite–ringwoodite transition) than previous quench studies indicate (Katsura and Ito [64]), possibly because in quench experiments, pressures were generally calibrated only at a few fixed temperature points. In situ studies have shown complex variation in pressure

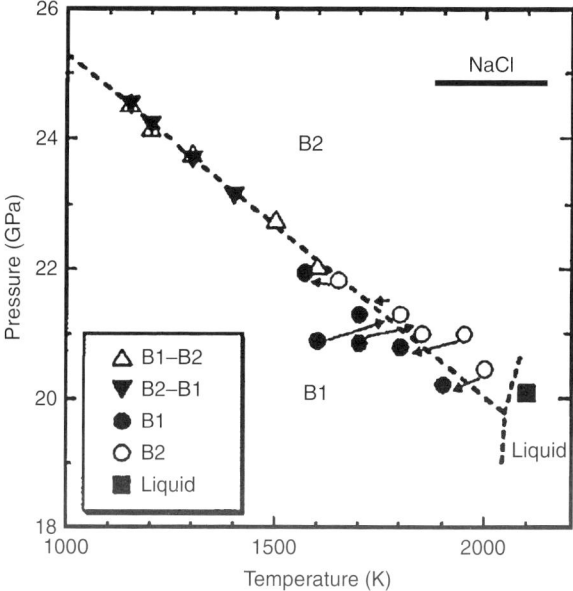

Figure 13.17 Phase boundary between B1 and B2 structures in NaCl (Nishiyama et al. [59]).

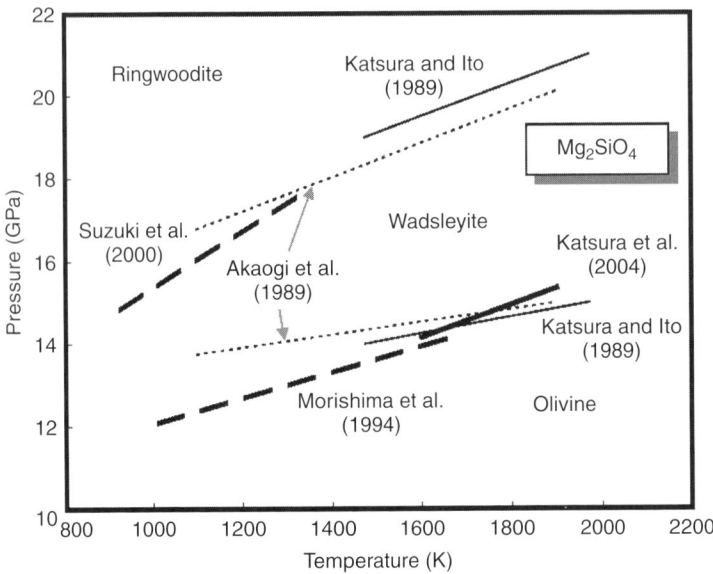

Figure 13.18 Phase boundaries of olivine–wadsleyite–ringwoodite transition in Mg_2SiO_4 determined by in situ X-ray diffraction [61–63], quench method [64], and thermodynamic calculation [65]. The phase boundary by Morishima et al. [61] has been recalculated using the Brown's (1999) scale.

in a multi-anvil experiment, when hydraulic load and temperature are varied. On the other hand, the discrepancy between the in situ phase boundaries and thermodynamic calculation (Akaogi et al. [65]) may suggest systematic errors in transformation enthalpy measurements on small samples at the time.

5.2 Phase Relations in Complex Systems

Katsura et al. [62] studied the binary phase boundary of olivine to wadsleyite in the system of $(Mg,Fe)_2SiO_4$. They used a tube heater, whose cylindrical axis is parallel to the incident X-ray beam. Inside the sample chamber, they placed three samples, each with a different composition (the iron contents varied from 3% to 30% in the starting olivine sample), along with a pressure standard. The pressure scale of Matsui et al. [66] was adopted for MgO. Experiments were carried out at pressures between 12.8 and 14.8 GPa, with a specially designed $P-T$ path to minimize pressure variation while heating the samples to at 1600 and 1900 K. After an extended soaking at a given $P-T$ condition, the sample was quenched and recovered for electron microprobe analysis.

Figure 13.19 shows the binary loops thus determined. The boundaries of the loops are fairly straight, not curve upward as suggested by Helffrich and Wood [67]. In fact there is an indication that the loops actually curve slightly downward, although the precision and compositional range are not sufficient to resolve the curvature. Katsura et al. [62] used a formalism proposed by Stixrude [68] to express

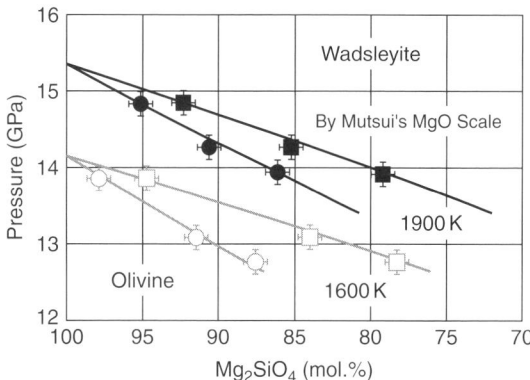

Figure 13.19 Experimental result and phase relations of the olivine–wadsleyite transition in the system $(Mg,Fe)_2SiO_4$ at 1600K and 1900K. The binary loops are drawn using the formalism of Stixrude [68] and Katsura et al. [62].

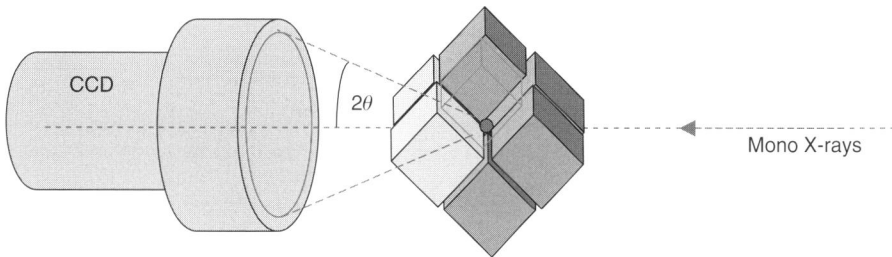

Figure 13.20 Diffraction geometry in angle-dispersive mode using SD anvil as windows. The two light cubes are made of sintered diamond for collecting two-dimensional patterns (Fig. 13.15).

the binary phase loop in P–x space. The parameters involved are transition pressures of end-members Mg_2SiO_4 and a hypothetical Fe_2SiO_4 counterpart and the partition coefficient K^{Ol-Wd}_{Fe-Mg}. They measured the partitioning coefficients and found it to be 0.61(1) at 1900K and 0.51(4) at 1600K. Experimental results on the binary loops at 1600K and 1900K are reproduced using these partition coefficients at transition pressures of 14.15 and 15.35 GPa for Mg_2SiO_4 and 5.9 and 7.2 GPa for Fe_2SiO_4 at 1600K and 1900K, respectively. Thus, the temperature dependence of the transition pressure in Mg_2SiO_4 is fairly large at 4.0 MPa/K. The binary loops generated following Stixrude's (1997) formalism are also shown in Fig. 13.19.

A new technique employing monochromatic diffraction with area detectors in the LVP has been developed for melt curve measurements. Wang et al. [69] introduced two X-ray transparent sintered diamond anvils to the second-stage eight-cube assembly in the T-Cup (Fig. 13.20). This allowed them to collect diffraction Debye rings of the sample over the entire 360° azimuth angle, thereby greatly increasing sampling of the grain population in the specimen. In the study of Si melting curve, they were able to reverse the melting/solidification transformation with a temperature uncertain of ±20K, at pressures up to 14 GPa. They also determined solid phase boundaries in the same experiment (Fig. 13.21).

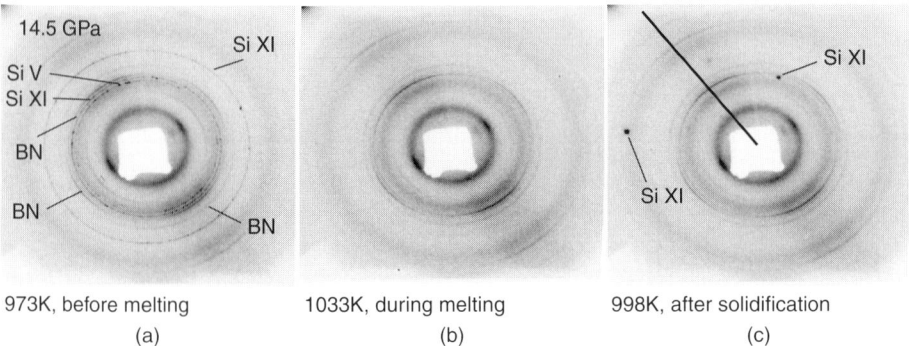

Figure 13.21 Selected two-dimensional diffraction patterns collected at 14.5 GPa and various temperatures before (A), during (B), and after melting (C).

6 Phase Diagrams Using DAC

Geoscientists have used high-pressure techniques to understand the nature of earth materials at high pressures and temperatures. Iron is considered to be the main component of Earth's core and therefore its high-pressure/temperature phase diagram has been a topic of active research. The LVP results were further advanced to ultra high pressure and temperature by laser-heated DAC. Recent reviews of these results have been done by Shen and Heinz [70] and Boehler [71].

The following information from Shen (pers. comm.) gives a modern view of the iron phase diagram. At ambient pressure, the stable phase is α-Fe with a body-centered cubic (bcc) structure. At high pressure it transforms to ε-Fe with hexagonal-close-packed (hcp) structure. At high temperature there is a large stability field for γ-Fe with face-centered cubic (fcc) structure. In several X-ray data particularly those obtained by using external heating a double hcp structure of iron was noted (Fig. 13.22). In laser-heated samples, this has never been found [74,75].

For the melting curve, there is a converging consensus at pressures below 60 GPa, reflected by a narrow uncertainty range in Fig. 13.23. As pressure increases, uncertainties in phase boundaries, including the melting curve, become large as shown by wide bands in Fig. 13.23. The width of the bands represents the scatter in literature data from recent years. Factors causing these uncertainties include those in pressure determinations (neglecting thermal pressure, different EOS, and/or different standard materials), in temperature determinations (large temperature gradient and temporal variation, chromatic aberration in optics), and in sample characterizations (different melting criteria, transition kinetics).

Uncertainties in the melting curve and the γ–ε transition lead to significant variations in the location of γ–ε–liquid triple point. Knowledge of its location is important because it is the starting point used for extrapolation of the melting curve of ε-Fe to core pressures. As shown in Fig. 13.23, the slope of γ–ε transition ranges from 25 to 40K/GPa, placing the triple point between 60 and 100 GPa. Such large uncertainties in the slope of the γ–ε transition mainly arise from the coexisting nature of these two phases in the pressure–temperature range, causing

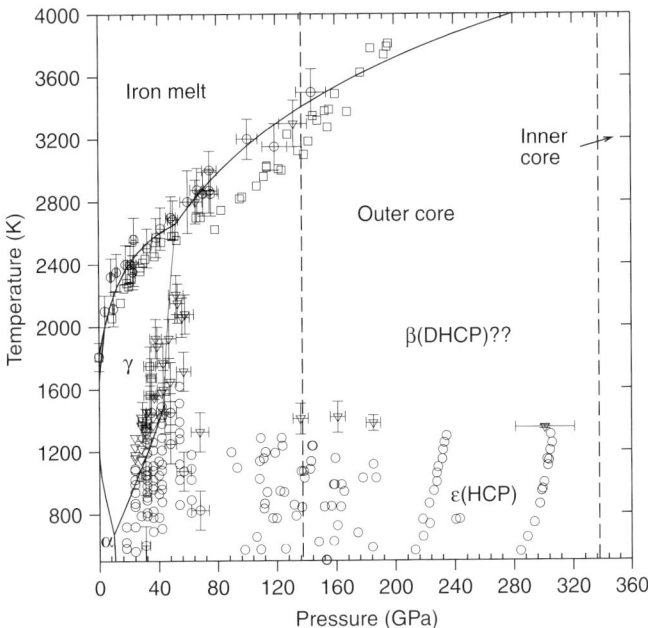

Figure 13.22 X-ray data on phase relations in iron. Only data collected with in situ heating are shown. Open diamonds – fcc; squares – hcp; open circles – fcc (Funamori et al. [78]); circles with plus – fcc/hcp transition from Dubrovinsky et al. [44]; open triangles – β and corresponding hcp phase, respectively. The latter is obtained after cooling to the temperature shown.

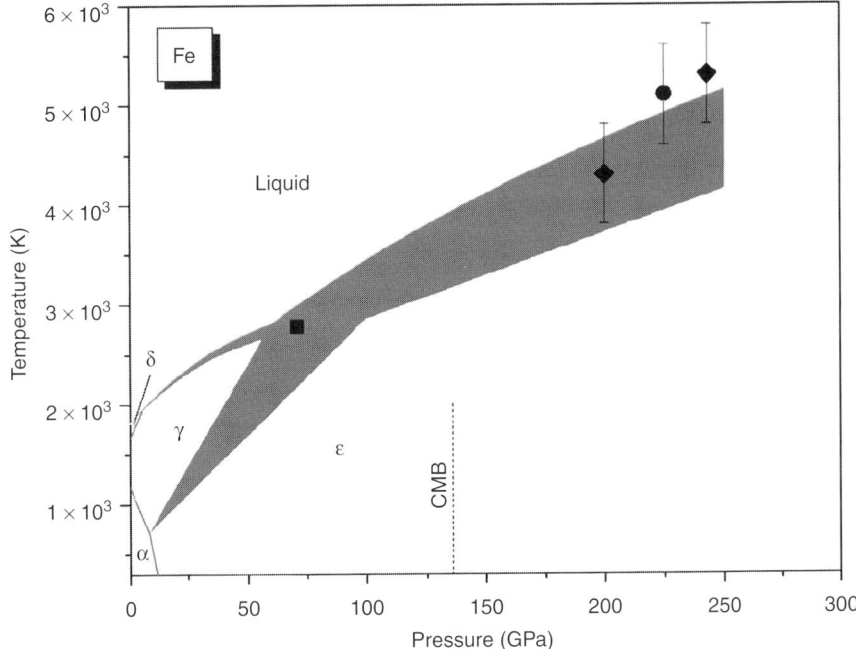

Figure 13.23 Iron phase diagram. Shaded areas represent the range of literature values in recent years with static DAC experiments. Symbols are shock-wave data: square – Ahrens et al. [79], diamonds – Brown and McQueen [76], circle – Nguyen and Holmes [77]. Brown and McQueen's point at 200 GPa was interpreted as a solid–solid transition. All other points are referred as melting.

difficulties in identifying the boundary. The uncertainties are also contributed by the pressure–temperature determination associated with the use of different standard materials and/or different EOS.

The melting data for iron above the triple point are scarce and scattered, reflecting the difficulty level of such experiment. It appears that the early DAC data [72,73] represent a lower bound of the melting curve in this region. The later experimental data falls at higher melting temperatures [74,75]. The shock-wave data [76,77] lie close to the upper bound. Extrapolating the data to the inner core boundary, the melting temperature of iron is between 4800K and 6000K at 330 GPa.

While the use of the LVP in constructing the phase diagrams is well established, the DAC technique has resulted in only partial data. Excellent examples of such work are the studies conducted on the phase $MgSiO_3$-perovskite [80] and magnesiowustite [81]. The geo-scientific interest on these phases is enormous because of their occurrence in Earth's mantle. The difficulty in obtaining such data, which could be useful in creating thermodynamic database, is due to the difficulty of conducting the in situ high-pressure, high-temperature experiments. As discussed elsewhere, long time sustained heating in a DAC reaching temperatures in excess of 1500K can only be achieved with double-side laser heating. Such facilities are rare.

7 INDUSTRIAL SOLIDS

While phase diagrams that provide us with basic data to model natural processes in the planetary interiors, the use of high-pressure techniques to synthesize novel materials with important properties is becoming increasingly common as revealed in an important review by McMillan [82]. While data on several oxides and silicates have been acquired over the years with interesting results, the nitrides are now gaining ground. These nitrides such as the semi-metallic transition metal nitrides (TiN, MoN_x), and nitrides of group-III (B, Al, Ga, In) and group-IV (Si, Ge) elements, when subjected to appropriate pressure–temperature conditions could yield technological materials. Several potentially important new materials [83], which have already been synthesized include phases of Si_3N_4 with a high hardness [84,85] and a new superhard compound c-BC_2N [86]. Several C_3N_4 type phases have to be synthesized [87]. Transition-metal nitrides and carbides are high-hardness technological materials with low compressibility. For example, MoN is found to have a record value of $K_o = 487$ GPa, exceeding that of any solid known. Another property of interest is the low compressibility usually associated with high bulk modulus. Osmium's bulk modulus of 462 GPa exceeds that of diamond's (443 GPa) but it is soft. Dense metal oxides such as the newly synthesized TiO_2 phase has both high bulk modulus and hardness [88].

ACKNOWLEDGMENTS

The work of CeSMEC is supported by grants from National Science Foundation (DMR-0231291) and from Air Force (212600548). Saxena thanks

Guoyin Shen, Leonid Dubrovinsky, and Peter Lazor for supplying the latest information on techniques. Portions of this work were performed at GeoSoil EnviroCARS (Sector 13), Advanced Photon Source (APS), Argonne National Laboratory. GeoSoilEnviroCARS is supported by the National Science Foundation – Earth Sciences (EAR-0217473), Department of Energy – Geosciences (DE-FG02-94ER14466), and the State of Illinois. Use of the APS was supported by the US Department of Energy, Office of Science, Office of Basic Energy Sciences, under Contract No. W-31-109-ENG-38.

REFERENCES

1. A. Jayaraman, *Rev. Mod. Phys.*, 55 (1983) 65.
2. A. Jayaraman, *Rev. Sci. Instrum.*, 57 (1986) 1013.
3. M. Eremets, *High Pressure Experimental Methods*, Oxford University Press, New York, 1996.
4. R.J. Hemley (ed.), *Ultrahigh-Pressure Mineralogy*, Rev. Mineral 37, New York, Mineralogical Society of America 1998.
5. C.E. Weir, E.R. Lippincott, A. Van Valkenburg and E.N. Bunting, *J. Res. Nat. Bur. Stand.*, 63A (1959) 55.
6. R.A. Forman, G.J. Piermarini, J.D. Barnett and S. Block, *Science* 176 (1972) 284.
7. F.P. Bundy, *J. Chem. Phys.*, 41 (1964) 3809.
8. M. Kumazawa, in M.H. Manghnani, S. Akimoto (eds.), *High-Pressure Research: Applications in Geophysics*, Academic Press, New York, 1977, p. 563.
9. A. Onodera, *High Temp. High Press.*, 19 (1997) 579.
10. K. Inoue and T. Asada, *Jpn. J. Appl. Phys.*, 12 (1973) 1786.
11. E. Ohtani, A. Onodera and N. Kawai, *Rev. Sci. Instrum.*, 50 (1979) 308.
12. O. Shimomura, S. Yamaoka, T. Yagi, M. Wakatsuki, K. Tsuji, H. Kawamura, N. Hamaya, O. Fukuoga, K. Aoki and S. Akimoto, in S. Minomura (ed.), *Solid State Physics Under Pressure*, Terra Scientific Publishing Co., Tokyo, 1985, p. 351.
13. M.T. Vaughan, D.J. Weidner, Y.B. Wang, J.H. Chen, C.C. Koleda and I.C. Getting, *Rev. High Press. Sci. Technol.*, 7 (1998) 1520.
14. O. Shimomura, W. Utsumi, T. Taniguchi, T. Kikegawa and T. Nagashima, in Y. Syono, M.H. Manghnani (eds.), *High-Pressure Research: Application to Earth and Planetary Sciences*, TERRAPUB, Tokyo, 1992, p. 3.
15. W. Utsumi, K. Funakoshi, S. Urakawa, M. Yamakata, K. Tsuji, H. Konishi and O. Shimomura, *Rev. High Press. Sci. Technol.*, 7 (1998) 1484.
16. L.G. Khvostantsev, L.F. Vereshchagin and A.P. Novikov, *High Temp. High Press.*, 9 (1977) 637.
17. L.G. Khvostantsev, *High Temp. High Press.*, 16 (1984) 165.
18. J.M. Besson, R.J. Nelmes, G. Hamel, J.S. Loveday, G. Weill and S. Hull, *Physica B*, 180 (1992) 907.
19. D.J. Weidner, M.T. Vaughan, J. Ko, Y. Wang, K. Leinenweber, X. Liu, A. Yeganeh-Haeri, R.E. Pacalo and Y. Zhao, *High Press. Res.*, 8 (1992) 617.
20. D.J. Weidner, M.T. Vaughan, J. Ko, Y. Wang, X. Liu, A. Yeganeh-Haeri, R.E. Pocalo and Y. Zhao, in Y. Syono, M.H. Manghnani (eds.), *High-Pressure Research: Application to Earth and Planetary Sciences*, TERRAPUB, Tokyo, 1992, p. 13.
21. M.L. Rivers, T.S. Duffy, Y. Wang, P.J. Eng, S.R. Sutton and G. Shen, in M.H. Manghnani, T. Yagi (eds.), *Properties of Earth and Planetary Materials at High Pressure and Temperature*, AGU, Washington, DC, 1998, p. 79.
22. Y. Wang, I. Getting, D.J. Weidner and M.T. Vaughan, in M.H. Manghnani, T. Yagi (eds.), *Properties of Earth and Planetary Materials at High Pressure and Temperature*, American Geophysical Union, Washington, DC, 1998, p. 35.
23. Y. Wang, M. Rivers, T. Uchida, P. Murray, G. Shen, S. Sutton, J. Chen, Y. Xu and D. Weidner, in *Proceedings of the International Conference on High Pressure Science and Technology* (AIRAPT-17), *Science and Technology of High Pressure*, vol. 2, 2000, p. 1047.

24. H.J. Mueller, F.R. Schilling and C. Lathe, *19th General Meeting of the International Mineralogical Association (Abstract)*, Kobe, Japan, July 23–28, 2006.
25. H.K. Mao, J. Xu and P.M. Bell, *J. Geophys. Res.*, 91 (1986) 4673.
26. G.J. Piermarini, S. Block, J.D. Barnett and R.A. Forman, *J. Appl. Phys.*, 46 (1975) 2774.
27. D.L. Decker, *J. Appl. Phys.*, 42 (1971) 3239.
28. H.K. Mao, P.M. Bell, J.W. Shaner and D.J. Steinberg, *J. Appl. Phys.*, 49 (1978) 3276.
29. A. Dewaele, P. Loubeyre and M. Mezouar, *Phys. Rev. B*, 70 (2004) 094112.
30. D. Akobuije, D. Chijioke, W.J. Nellis, A. Soldatov and I.F. Silvera, Abstract, SMEC, Miami, 2005.
31. C.-S. Zha, H.K. Mao and R.J. Hemley, *Proc. Nat. Acad. Sci.*, 97 (2000) 13494.
32. J.C. Jamieson, J.N. Fritz and M.H. Manghnani, in S. Akimoto, M.H. Manghnani (eds.), *High-Pressure Research in Geophysics*, Center for Academic Publishing, Tokyo, 1982, p. 27.
33. N.C. Holmes, J.A. Moriarty, G.R. Gatheres and W.J. Nellis, *J. Appl. Phys.*, 66 (1989) 2962.
34. N.H. Chen and I.F. Silvera, *Rev. Sci. Instrum.*, 67 (1996) 4275–4278.
35. N. Dubrovinskaia and L. Dubrovinsky, *Advances in High-Pressure Techniques for Geophysical Applications*, Elsevier, NY, 2005, p. 487.
36. L.C. Ming and W.A. Bassett, *Rev. Sci. Instrum.*, 45 (1974) 1115.
37. G. Shen, H.K. Mao, R.J. Hemley, T.S. Duffy and M.L. Rivers, *Geophys. Res. Lett.*, 25 (1998) 373.
38. R. Boehler, *Mater. Today*, November (2005) 34.
39. C.S. Zha and W.A. Bassett, *Rev. Sci. Instrum.*, 74 (2003) 1255.
40. L. Liu and W.A. Bassett, *J. Geophys. Res. B*, 80 (1975) 3777.
41. R. Boehler, *Geophys. Res. Lett.*, 13 (1986) 1153.
42. R. Boehler, M. Nicol, C.S. Zha and M.L. Jonson, *Physica B*, 139–140 (1986) 916.
43. L.S. Dubrovinsky and S.K. Saxena, *Petrology*, 6 (1998) 535.
44. L.S. Dubrovinsky, S.K. Saxena and P. Lazor, *Phys. Chem. Minerals*, 25 (1998) 434.
45. L.S. Dubrovinsky, S.K. Saxena, F. Tutti and T. Le Bihan, *High Press. High Temp.*, 31 (1999) 553.
46. L.S. Dubrovinsky, S.K. Saxena, F. Tutti and T. Le Bihan, *Phys. Rev. Lett.*, 84 (2000) 1720.
47. D. Schiferl, *Rev. Sci. Instrum.*, 58 (1987) 1316.
48. D.M. Adams and A.G. Christy, *High Press. High Temp.*, 24 (1992) 1.
49. L.S. Dubrovinsky, S.K. Saxena, F. Tutti and T. Le Bihan, *Phys. Rev. Lett.*, 84 (2000) 84–88.
50. S. Rekhi, L.S. Dubrovinsky and S.K. Saxena, *High Press. High Temp.*, 31 (1999) 299.
51. N.A. Dubrovinskaia and L.S. Dubrovinsky, *Rev. Sci. Instrum.*, 74 (2003) 3433.
52. I.C. Getting and G.C. Kennedy, *J. Appl. Phys.*, 41 (1970) 4552.
53. Y. Le Godec, M.T. Dove, D.J. Francis, et al. *Min. Mag.*, 65 (2001) 737–748.
54. M.G. Pepper and J.B. Brown, *J. Phys. E: Sci. Instrum.*, 12 (1979) 31.
55. T. Uchida, Y. Wang, M.L. Rivers and S.R. Sutton, *J. Geophys. Res.*, 106 (2001) 21, 799.
56. R. Boehler, M. Nicol and M.L. Johnson, in M.H. Manghnani, Y. Syono (eds.), *High-Pressure Research in Mineral Physics, Geophys. Monogr. Ser.*, vol. 39, AGU, Washington, DC, 1987, p. 173.
57. S. Akimoto, T. Suzuki, T. Yagi and O. Shimomura, in M.H. Manghnani, Y. Syono (eds.), *High-Pressure Research in Mineral Physics, Geophys. Monogr. Ser.*, vol. 39, AGU, Washington, DC, 1987, p. 149.
58. H.K. Mao, P.M. Bell and C. Hadidiacos, in M.H. Manghnani, Y. Syono (eds.), *High-Pressure Research in Mineral Physics, Geophys. Monogr. Ser.*, vol. 39, AGU, Washington, DC, 1987, p. 135.
59. N. Nishiyama, T. Katsura, K. Funakoshi, A. Kubo, T. Kubo, Y. Tange, Y. Sueda and S. Yokoshi, *Phys. Rev. B*, 68(13) (2003) 134109.
60. T. Katsura, in E. Ohtani (ed.), *Geological Society of America Memoir: Advances in High Pressure Mineralogy*, in press (2006).
61. H. Morishima, T. Kato, M. Suto, E. Ohtani, S. Urakawa, W. Utsumi, O. Shimomura and T. Kikegawa, *Science*, 265 (1994) 1202.
62. T. Katsura, H. Yamada, O. Nishikawa, M.S. Song, A. Kubo, T. Shinmei, S. Yokoshi, Y. Aizawa, T. Yoshino, M.J. Walter, E. Ito and K. Funakoshi, *J. Geophys. Res.*, 109 (2004) B02209.
63. A. Suzuki, E. Ohtani, H. Morishima, T. Kubo, Y. Kanbe, T. Kondo, T. Okada, H. Terasaki, T. Kato and T. Kikegawa, *Geophys. Res. Lett.*, 27 (2000) 803.
64. T. Katsura and E. Ito, *J. Geophys. Res.*, 94 (1989) 15663.
65. M. Akaogi, E. Ito and A. Navrotsky, *J. Geophys. Res.*, 94 (1989) 15671.
66. M. Matsui, S.C. Parker and M. Leslie, *Am. Mineral.*, 85 (2000) 312.
67. G.R. Helffrich and B.J. Wood, *Geophys. J. Int.*, 126 (1996) F7.

68. L. Stixrude, *J. Geophys. Res.*, 102 (1997) 14, 835.
69. C.E. Runge, A. Kubo, B. Kiefer, Y. Wang, T. Uchida, N. Nishiyama and T. Duffy, Abstract, COMPRES Fourth Annual Meeting, June 16–19, (2005), New Paltz, New York.
70. G. Shen and D.L. Heinz, in R.J. Hemley (ed.), *Ultra High Pressure Mineralogy: Physics and Chemistry of the Earth's Deep Interior*, Mineralogical Society of America, Washington, DC, 1998, p. 369.
71. R. Boehler, *Rev. Geophysics*, 38 (2000) 221.
72. R. Boehler, *Nature*, 363 (1993) 534.
73. S.K. Saxena, G. Shen and P. Lazor, *Science*, 264 (1994) 405.
74. G. Shen, H.K. Mao, R.J. Hemley, T.S. Duffy and M.L. Rivers, *Geophys. Res. Lett.*, 25 (1998) 373.
75. Y.Z. Ma, M. Somayazulu, H.K. Mao, J.F. Shu, R.J. Hemley, G. Shen, *Phys. Earth Planet. In.*, 143–144 (2004) 455.
76. J.M. Brown and R.G. McQueen, *J. Geophys. Res.*, 91 (1980) 7485.
77. J.H. Nguyen and N.C. Holmes, *Nature*, 427 (2004) 339.
78. N. Funamori, T. Yagi and T. Uchida, *Geophys. Res. Lett.*, 23 (1997) 953.
79. T.J. Ahrens, K.G. Holland and G.Q. Chen, *Geophys. Res. Lett.*, 29 (2002) 101029.
80. M. Murakami, K. Hirose, K. Kawamura, N. Sata and Y. Ohishi, *Science*, 304 (2004) 855.
81. L. Dubrovinsky, N. Dubrovinskaia, I. Kantor, C. McGammon, W. Crichton and V. Urusov, *J. Alloy Compd.*, 390 (2005) 41.
82. P. McMillan, *Nature Mater.*, 1 (2002) 19.
83. E. Kroke, *Angew. Chem. Int. Edit.*, 41 (2002) 77.
84. A. Zerr, G. Miehe, G. Serghiou, M. Schwarz, E. Kroke, R. Riedel, H. Fuess, P. Kroll and R. Boehler, *Nature*, 400 (1999) 340.
85. A. Zerr, M. Kempf, M. Schwarz, E. Kroke, M. Göken and R. Riedel, *J. Am. Ceram. Soc.*, 85 (2002) 86.
86. V.L. Solozhenko, S.N. Dub and N. Novikov, *Diam. Relat. Mater.*, 10 (2001) 2228.
87. T. Malkow, *Mater. Sci. Eng.*, A302 (2001) 309.
88. L.S. Dubrovinsky, N.A. Dubrovinskaia, V. Swamy, J. Muscat, N.M. Harrison, R. Ahuja, B. Holm and B. Johansson, *Nature*, 410 (2001) 653.

CHAPTER FOURTEEN

THE DETERMINATION OF PHASE DIAGRAMS FOR SLAG SYSTEMS

David R. Gaskell

Contents

1 Introduction	442
2 The CaO–Al$_2$O$_3$–SiO$_2$ System	443
2.1 The CaO–SiO$_2$ System	443
2.2 The Al$_2$O$_3$–SiO$_2$ System	447
2.3 The CaO–Al$_2$O$_3$–SiO$_2$ System	452
3 The CaO–"FeO"–SiO$_2$ System	453
4 The FeO–Fe$_2$O$_3$–SiO$_2$ System	455

1 INTRODUCTION

Metallurgical slags play an important role in the extraction and refining of metals. They are formed when a flux material, added to or included in the charge, reacts with unwanted minerals during extraction or smelting, or with the products of the oxidation of unwanted solute elements during refining. This fluxing produces a low melting complex oxide melt, the slag, which is immiscible with the metal phase and, hence, is easily separated. Efficient fluxing requires knowledge of phase relationships in the slag system, and these are presented graphically best, on phase diagrams.

For example, lime, charged to the iron blast furnace with processed iron ore and metallurgical coke, fluxes the coke ash. North American coke ash contains silica and alumina in the weight ratio 2:1 and a line drawn from this composition on the phase diagram for the ternary system CaO–Al$_2$O$_3$–SiO$_2$ to the CaO apex passes through the composition of a ternary eutectic melt. This melt has the composition 40% CaO–40% SiO$_2$–20% Al$_2$O$_3$ (the so-called 40–40–20 melt) and is liquid at 1265°C. Knowledge of this allows calculation of the quantity of lime to be charged to the blast furnace with any amount of coke to produce the 40–40–20 slag.

School of Materials Engineering, Purdue University, West Lafayette, IN, USA

Oxygen steelmaking involves preferential oxidation of the solutes C, Si and Mn from the hot metal charged to the converter. Liquid iron oxide is formed at beginning of the oxygen blow and this fluxes the silica, formed by oxidation of the silicon. Lime, charged with the hot metal and the scrap, gradually replaces iron oxide as the flux. It produces a slag in the system $CaO-FeO-SiO_2$ which is simultaneously saturated with $3CaO.SiO_2$ and CaO. The composition of this melt is used to calculate the amount of lime that should be charged with any amount of hot metal.

Coppermaking involves separating sulfur and iron from the concentrated copper iron sulfide ore. Sulfur is removed as gaseous SO_2 and iron is removed by oxidation, with silica as a flux, to form an iron silicate slag that is immiscible with the copper-rich sulfide matte. The iron is removed initially in a reverberatory furnace and finally in a Pierce–Smith converter. Oxidation in the converter produces a slag saturated with magnetite and operation of the process is facilitated by knowledge of phase relations in the system $FeO-Fe_2O_3-SiO_2$.

The turn of the twentieth century saw the introduction of new experimental methods for studying phase relationships occurring in materials systems and most of the early work was conducted by petrologists interested in phase equilibria occurring in mixtures of oxides at high temperature. Later work was conducted by metallurgists interested in the phase equilibria occurring in metallurgical slag systems. The new experimental approaches required the development of crucible materials capable of containing mixtures of oxide melts at high temperatures and the means of attaining and measuring these high temperatures. The advantages and disadvantages of these approaches are discussed and the early studies of phase equilibria in the systems $CaO-SiO_2$, $Al_2O_3-SiO_2$, $CaO-Al_2O_3-SiO_2$, $CaO-"FeO"-SiO_2$ and $FeO-Fe_2O_3-SiO_2$ are examined and discussed in this chapter.

2 THE $CaO-Al_2O_3-SiO_2$ SYSTEM

2.1 The $CaO-SiO_2$ System

The turn of the twentieth century saw the introduction of new experimental methods for studying the phase relationships occurring in materials systems. The excitement generated by the ability to construct new phase diagrams is illustrated by the Introduction to the paper by Day et al. on phase equilibria in the system $CaO-Al_2O_3-SiO_2$ [1]:

> "Anyone who has followed the work of the eminent Norwegian Scientist, Prof. J.H.L. Vogt, during the past three or four years, must realize that an extraordinarily effective weapon has come into the service of petrology, the full power of which cannot at once be understood or appreciated. We refer to the methods and established generalizations of physical chemistry. The older science of chemistry has made such strides under these new theories of solutions that we really have little more to do than to apply them ready-made to our own problems, like a smooth and powerful machine tool of guaranteed effectiveness. Mineral solutions, as Bunsen long ago maintained, are after all only chemical solutions over again with slightly different components and a different, a very different, range of temperatures and pressures. There is no need to disparage the difficulties involved in operating at high temperatures and under great pressures; they are very great, probably even greater

than most of us appreciate, but they are certainly not insuperable, and when they are overcome, not only will a new era in the science of petrology have been inaugurated but an important return service will have been rendered to physics and physical chemistry in extending the scope of their generalizations."

Day et al. first studied the components lime and silica. When melted in an electric arc, CaO formed a clear liquid of low viscosity which crystallized readily to form a well-developed cubic structure. At 2000°C pure lime showed no sign of having a high vapor pressure and thus it was assumed that the observed significant volatility of lime, in the carbon arc, was caused by reduction to form the metal vapor, which was immediately reoxidized outside the heated zone. Three measurements of the density of lime, relative to water at 25°C having unit density, gave 3.313, 3.307 and 3.329. Probably this variation was caused by superficial hydration or the absorption of CO_2. The hardness of lime was placed on Moh's scale as being between 3 (calcite) and 4 (apatite).

In 1906, as no previous careful measurement of the melting temperature of silica had been made, conventional wisdom placed the value in the range 1200–2000°C, Day et al. considered that determination of the freezing point was "out of the question" due to the inertness of the viscous melt. A small quantity of finely powdered quartz was placed in a small iridium crucible and heated in an iridium tube furnace. Small fragments of platinum foil were placed on top of the quartz powder and the furnace was slowly heated until the foil was observed to melt. Examination of the charge, after cooling, showed that the crystal grains had inverted to tridymite and that superficial liquefaction had caused them to sinter. Thus, at the melting temperature of platinum, 1720°C, silica showed some evidence of fusion. Further heating for longer periods of time showed that melting occurred at temperatures as low as 1625°C, as measured by an optical pyrometer. Unfortunately an iridium furnace cannot sustain long-term heating, and a platinum furnace is not capable of reaching high enough temperatures. Thus the effort to measure the melting temperature of silica was abandoned and it was assumed that pure silica begins to melt at about 1600°C.

Optical examination identified four compounds in the lime–silica system, three of which exhibited polymorphism; SiO_2, occurring as quartz and tridymite; the metasilicate, $CaO.SiO_2$, occurring as wollastonite at lower temperatures, and pseudowollastonite at higher temperatures; the orthosilicate, $2CaO.SiO_2$, occurring in forms identified as α, β and γ; and one form of CaO.

Day et al. anticipated the existence of eutectics between $CaO.SiO_2$ and SiO_2 and between $CaO.SiO_2$ and $2CaO.SiO_2$. They identified these eutectics by taking small quantities of a number of compositions close to $CaO.SiO_2$, placing them in order on a platinum ribbon and observing the order in which they melted. The eutectics were the first to melt and the compound $CaO.SiO_2$ melted last. A few repetitions of this procedure using intermediate compositions identified eutectic compositions at 46 mol% SiO_2 and at 63 mol% SiO_2. The eutectic between $2CaO.SiO_2$ and CaO, at 32.5 mol% SiO_2, was identified using the same technique with an iridium wire. A maximum melting temperature was observed at 35 mol% SiO_2, which was taken to be that of the anticipated orthosilicate compound, $2CaO.SiO_2$. No other points were obtained up to 2100°C. The phase diagram for the system $CaO–SiO_2$ determined by Day et al. is shown as Fig. 14.1 and the currently accepted phase diagram for this system is shown as Fig. 14.2.

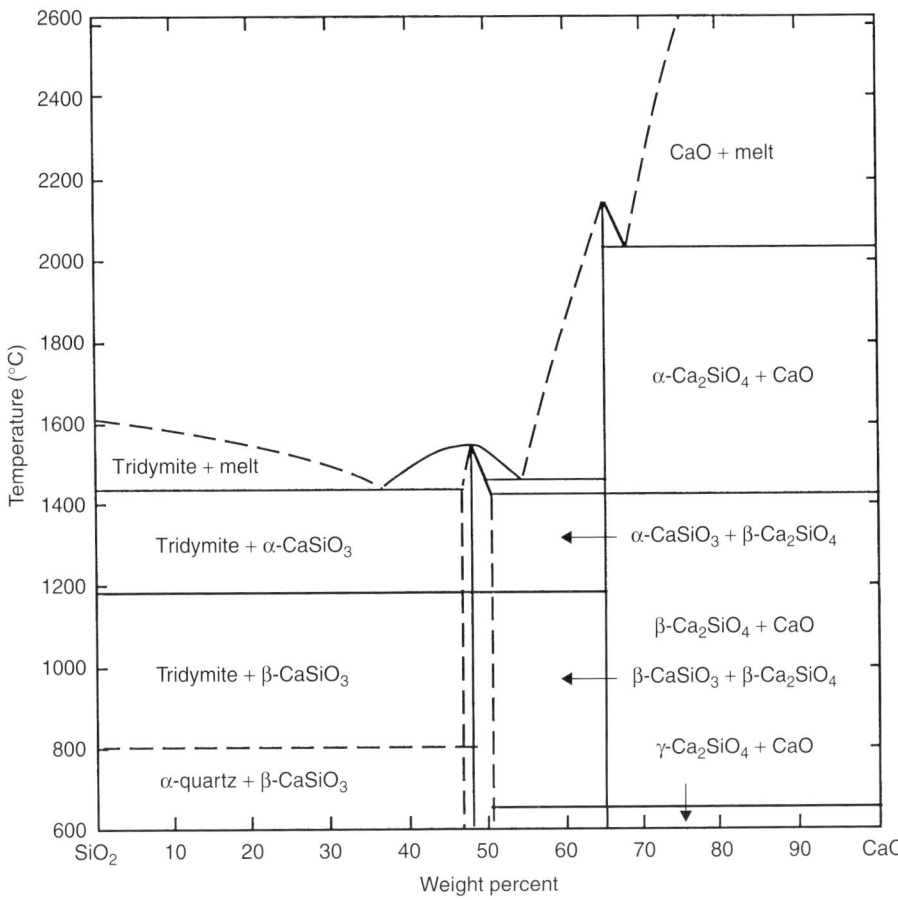

Figure 14.1 The phase diagram for the system CaO–SiO$_2$ determined by Day et al. in 1906.

Day et al. noted that they failed to find the akerman analog, 4CaO.3SiO$_2$, and the tricalcic silicate, 3CaO.SiO$_2$. Failure to find the former led the researchers to speculate that the akermanite mineral is stable only when other substances are present. They further stated that 3CaO.SiO$_2$ owes its "supposed" existence to investigators, who have found it necessary to postulate such a compound, in order to explain the constitution of Portland cement. A survey of the literature showed that no one had ever isolated and described a pure and homogeneous compound of the composition 3CaO.SiO$_2$. Also the authors discussed the problems caused by evaporation of iridium from the Pt–10% Ir furnace coils used for heating. Iridium sublimes slowly at temperatures above 1200°C and contaminated the thermocouples, causing them to give low readings – up to 0.5°C after 1 h heating and up to 50°C after a few weeks of continuous usage. Their remedy for this problem was "simple and sure – use no iridium in the furnace". Other topics of discussion were problems encountered in the measurement of temperature by optical pyrometry.

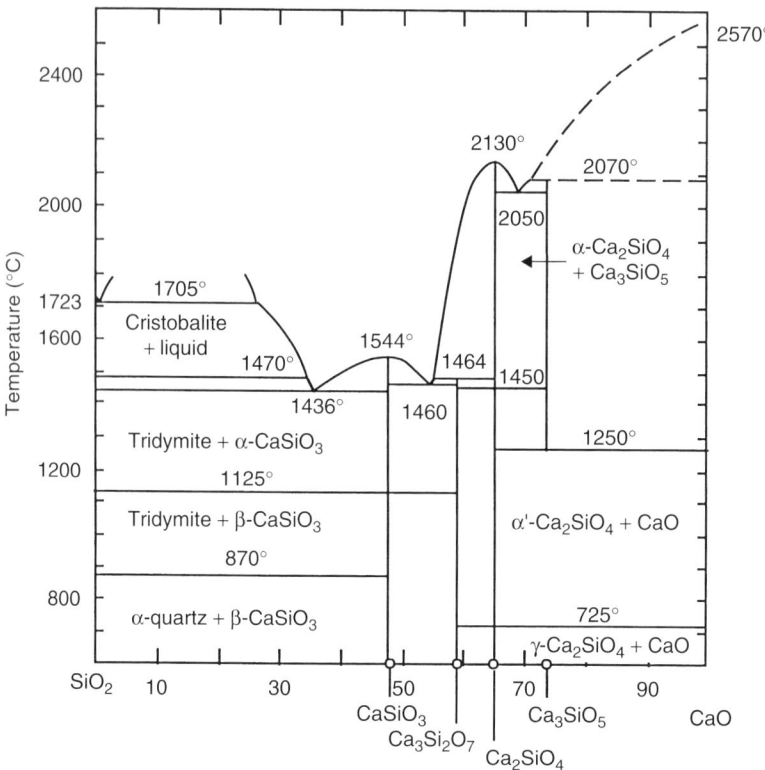

Figure 14.2 The currently accepted phase diagram for the system CaO–SiO$_2$.

In considering study of phase equilibria in the system CaO–SiO$_2$, Welch and Gutt [2] pointed out that dynamic methods, such as thermal analysis, cannot be used for verifying low-temperature decomposition which takes place extremely slowly. Although such decompositions are studied best by quenching, this technique cannot be relied upon to give unequivocal results at very high temperatures. For example, it is not possible to quench liquids of compositions in the range 2CaO.SiO$_2$–3CaO.SiO$_2$ to their corresponding glasses. Thus, a more advanced technique is required, such as one involving the use of a microscope to observe the behavior of a small quantity of melt attached to a thermocouple. By means of a rapidly alternating switch, the thermocouple serves to both heat and to measure the temperature of the melt. The use of a Pt–5% Rh/Pt–20% Rh allowed temperatures of 1800°C to be attained and measured with an accuracy of ±3°C and the use of Pt–20% Rh/Pt–40% Rh allowed temperatures of 1880°C to be attained and read with an accuracy of ±10°C. Attempts were made to use a W–Mo thermocouple in a slightly reducing atmosphere. Although temperatures of 2500°C were attained readily, reaction with the melts precluded the use of W–Mo junctions. The decision was made to use Ir wire as the heating element and to find an alternative thermocouple. The thinnest gauge of Ir, easily obtained, was 0.5 mm diameter. The Pt–Rh alloys previously used as heating elements were only 0.2 mm in diameter and dissipated 6 watts at their maximum temperature. Attainment of the same temperatures with an Ir wire consumed about 25 amperes at 1 volt. Tests showed that this considerable increase in heat dissipation could be tolerated in the gas-tight cell designed for use with the original heating elements, and that there was no risk of damage to the microscope if a 1-inch objective was used for observations. At high temperatures a background source of illumination of greater brilliance was required for viewing crystals in the melt. For temperatures up to 1900°C a group-filament tungsten projector lamp was sufficient but, at the higher temperatures required by Welch and Gutt, a carbon arc lamp was used. A heat-absorbing filter was placed between the arc lamp and the microscope condenser and the observer's eyes were protected by a series of neutral-density filters placed in front of the eyepiece. The use of a disappearing filament pyrometer involved two difficulties. First, over time, a slight clouding of the glass windows of the heater enclosure occurred, causing an appreciable error in the optical pyrometer reading. Second, a more serious error, was caused by a tendency of some melts to form a very thin surface film over large areas of the heating wire. This phenomenon caused a variation in the effective emissivity of the wire. Consequently, even larger and more inconsistent errors in the measurement of temperatures were made and measurement of temperature by optical pyrometry was abandoned.

It was found that an arbitrary temperature scale could be established in terms of the RMS voltage applied to the heater or the primary voltage step-down transformer. This scale was calibrated by observing the voltage required to melt several pure oxides of known melting temperature. One drawback to this procedure was that any deformation of the Ir wire causing cracking, fracture and a change of resistance would vitiate the results. However, it was found that the Ir wire could be rendered ductile by annealing in argon.

After measurements were made by observing the temperature at which the smallest observable remaining crystal hovered unchanged in size at the center of the melt, the authors stated their belief that theirs were the only such measurements made at temperatures of the order of 2000°C. The temperature scale was calibrated at the melting temperatures of $2CaO.SiO_2$, 2130°C, and Al_2O_3, 2040°C. Their results are shown in Fig. 14.3 and earlier measurements, made in the same system, by Rankin and Wright in 1915 [3] are shown in Fig. 14.4.

2.2 The Al_2O_3–SiO_2 System

In their paper on phase equilibria in the system Al_2O_3–SiO_2 Bowen and Greig [4] make reference to earlier work on this system by Shepherd et al. [5]. These latter workers obtained results, shown in Fig. 14.5, which appeared to show a maximum or congruent melting point (1816°C), at the composition of sillimanite, $Al_2O_3.SiO_2$, with a eutectic at 1810°C between sillimanite and corundum, Al_2O_3. Bowen and Greig pointed out that the difference between these temperatures is less than the error of measurement at such temperatures. Thus, the existence of a eutectic cannot be regarded as being proven definitely and the possibility of incongruent melting is not altogether precluded.

Preliminary trials by Bowen and Greig with both natural sillimanite and with a mixture of Al_2O_3 and SiO_2 in the ratio 1:1 showed that, when heated to a high

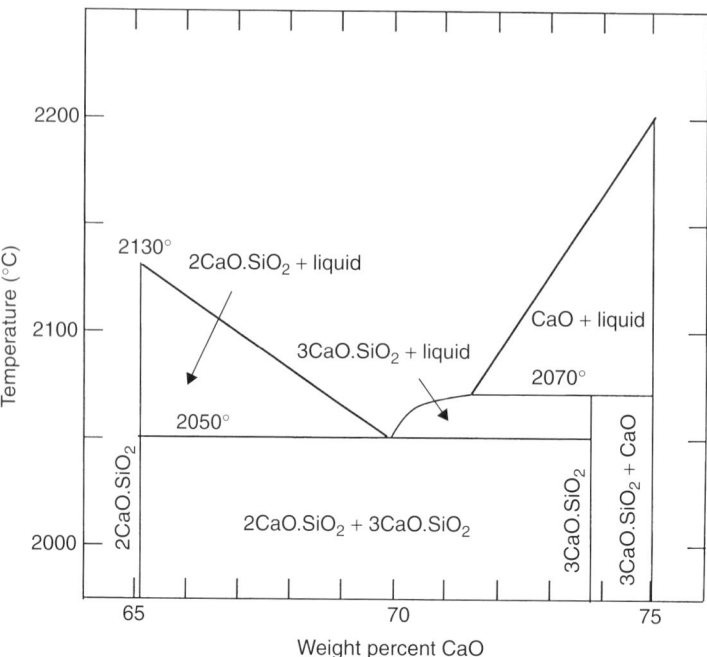

Figure 14.3 Portion of the system CaO–SiO$_2$ determined by Welch and Gutt in 1959.

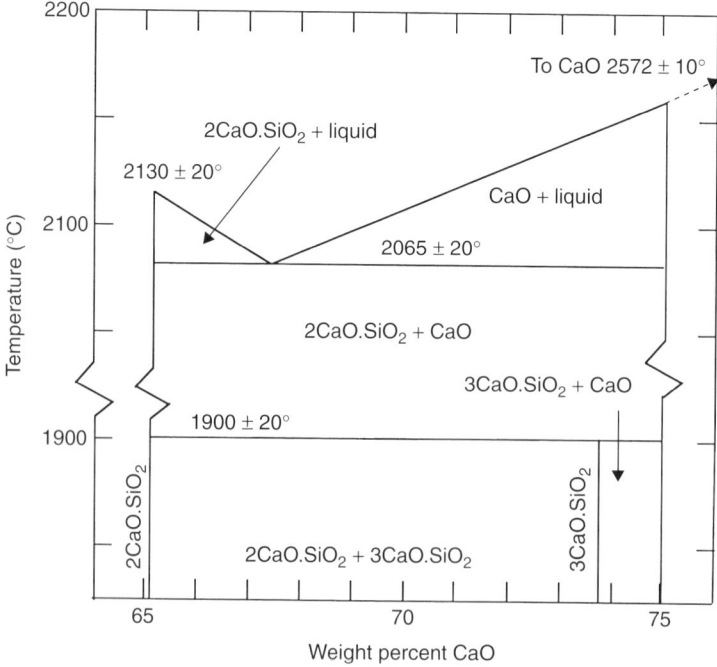

Figure 14.4 Portion of the system CaO–SiO$_2$ determined by Rankin and Wright in 1915.

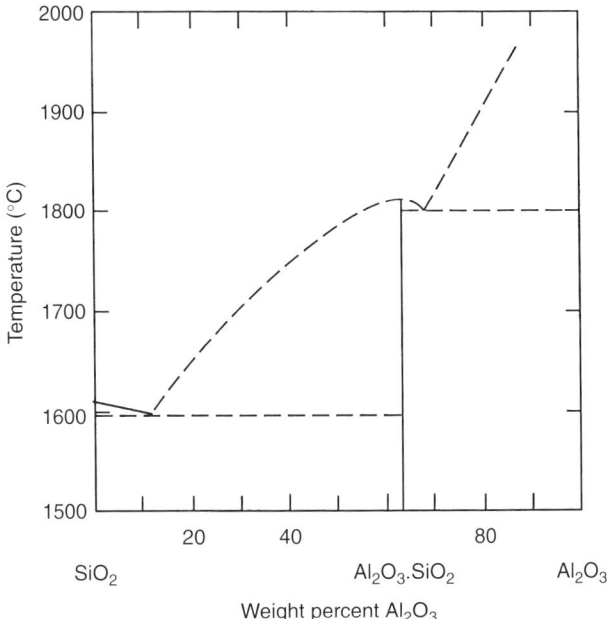

Figure 14.5 The phase diagram for the system Al_2O_3–SiO_2 determined by Shepherd et al. in 1909.

temperature and cooled slowly, good corundum crystals were formed. This demonstrated that a range of temperature exists in which corundum occurs in contact with a liquid in a mixture of the composition of sillimanite. Also, it was found that conditions did not exist in which the 1:1 mixture could be induced to form a homogeneous mass consisting only of crystals corresponding to natural sillimanite. Even when not heated to the very high temperature at which corundum forms, the mass always consisted of two phases, crystals having properties related to those of sillimanite and an interstitial glass. This observation was also made in the earlier work [5]. From this, Bowen and Greig concluded that the glass did not have the same composition as the crystals. If the glass was more siliceous than the overall composition of the mass, then the crystals must be more aluminous. Acting on these indications Bowen and Greig proceeded to determine the composition of the crystals by making up mixtures containing more alumina than the 1:1 ratio. In accordance with their expectations they found that, with the addition of alumina, the amount of crystals increased and that of the glass decreased until, at the composition $3Al_2O_3 \cdot 2SiO_2$, the glass phase disappeared. Then the whole preparation consisted of crystals identical with those in the 1:1 preparation. They concluded that these crystals have the composition $3Al_2O_3 \cdot 2SiO_2$ and are not sillimanite, although they showed a "very remarkable similarity" with sillimanite in optical and crystallographic properties.

In their investigation, Bowen and Greig used the method of quenching, as this allows the determination of the phases present in any mixture at any temperature. They used a furnace wound with Pt–20% Rh wire, which could be operated at 1750°C and in which the temperature was measured with a Pt–Rh thermocouple.

An iridium furnace was used for temperatures higher than 1750°C and the temperature was measured by optical pyrometry. The charges were prepared from specially purified precipitated alumina and quartz ground together in the desired proportion and heated in a gas furnace to near the melting temperature of platinum. The charge was then ground and returned to the furnace. Heating with subsequent grinding repeated five times produced specimens that were either a single homogeneous phase or constituent phases uniformly distributed. The most noticeable feature of the equilibrium diagram determined by Bowen and Greig is that $3Al_2O_3 \cdot 2SiO_2$ is the only compound stable at high temperatures. Specimens prepared with alumina and quartz in the ratio 3:2 were homogeneous aggregates of crystals of one kind at all temperatures up to 1810°C, where melting began. A slight excess of silica caused the formation of some cristobalite below 1545°C and a little glass (liquid) above 1545°C. The 1:1 preparation, $Al_2O_3 \cdot SiO_2$, consisted of the 3:2 compound with cristobalite below the eutectic, 1545°C, and the 3:2 compound with liquid above 1545°C (and below 1810°C). Only one eutectic occurred between $3Al_2O_3 \cdot 2SiO_2$ and cristobalite. The 3:2 compound melts incongruently, breaking up to form corundum and liquid at 1810°C, and the composition is completely molten at temperatures above about 1920°C.

In subsequent studies of phase equilibria in the system Al_2O_3–SiO_2 the 3:2 compound, $3Al_2O_3 \cdot 2SiO_2$, is referred to as mullite, named after the Scottish island of Mull, where it was found to occur in nature.

Aramaki and Roy [6] revisited the system Al_2O_3–SiO_2 and began their paper with the statement that this system "is the most important binary system in high-temperature technology". They presented an extensive review of the literature which showed that, in 1949, a claim was made that single crystals of mullite had been grown using flame fusion. Aramaki and Roy raised the questions; could an incongruently melting compound be formed by flame fusion and does mullite melt congruently? They referred to a claim, made in 1939, that mullite exhibits a range of solid solubility extending from the composition 3:2 to 2:1 and a claim, made in 1951, that mullite does melt congruently. As these claims are at odds with the phase diagram determined by Bowen and Greig, Aramaki and Roy undertook a further study of phase equilibria in this system. They pointed out that SiO_2 can evaporate "quite appreciably" from silicate melts under reducing conditions, causing difficulties when noted that quenching from temperatures higher that 1700°C is difficult because of the lack of the availability of a satisfactory furnace and they considered that the use of sealed systems produces the best results. Consequently, all of their high-temperature data were collected using hermetically sealed Pt–20% Rh or Pt–40% Rh containers of OD 2.5 mm, and length 2.5 cm, which could contain 30–50 mg of sample and be heated to 1830°C. The temperature was measured with an optical pyrometer calibrated against the melting temperature of Pt, 1769°C, and, after equilibration, the containers were quenched by being dropped into mercury or water. Specimens were prepared from α-Al_2O_3 and powdered silica gel and, after quenching, were examined petrographically, by powder X-ray diffraction and by infrared absorption.

They found appreciable solubility of alumina in cristobalite, as indicated by an increase in the lattice constants of cristobalite. The X-ray diffraction data did not give any evidence of the solubility of silica in corundum. Corundum was the only form of alumina found and "real" glasses could be prepared in compositions containing up to 60 mol% Al_2O_3. Invariably, some mullite crystallized in compositions containing more than 50% alumina.

Aramaki and Roy found at least one new synthetic phase of the probable composition $Al_2O_3.SiO_2$, prepared under moderate pressures, 500–5000 bars, at temperatures in the range 500–700°C. Its powder pattern has been mistaken for andalusite, to which it may bear the same structural relation as mullite has to sillimanite. (Note: The composition $Al_2O_3.SiO_2$ exists in three polymorphic forms: sillimanite (discovered naturally in 1824 and named for the American chemist and mineralogist Benjamin Silliman); andalusite (discovered in Andalusia, Spain in peri-aluminous sedimentary rocks); and kyanite.) From the crystallographers point of view [6] "Mullite may be regarded as being a disordered phase of aluminum silicate intermediate between the ordered phases of sillimanite and andalusite".

Aramaki and Roy prepared mullites with compositions ranging from the ratio 3:2 to 2:1 and noted that it appeared that the 2:1 mullite forms metastably at the eutectic temperature due to a structural inheritance from the liquid which favors Al^{3+} in four fold coordination. Although it is metastable from at least 1750°C down, once formed into "relatively" large crystals it resists transformation to the equilibrium form because of the "enormous" activation energy required for the breaking and rearranging of so many Al–Si bonds.

They confirmed earlier observations that heating the sillimanite composition produces mullite plus glass or silica. Indeed, Bragg and Claringbull claim that both sillimanite and andalusite transform to mullite and silica when heated. Kyanite also transforms to mullite and silica when heated in the range 1350°–1380°C. A slightly higher temperature is required for the transformation of andalusite, indicating that andalusite has a smaller Gibbs free energy of formation from alumina and silica, than has kyanite.

Amaraki and Roy first addressed the question: Does mullite melt congruently at equilibrium? Initially they experienced a problem in distinguishing mullite, that formed from the liquid on quenching, from primary mullite present in equilibrium with the liquid at high temperatures. It was found, eventually, that independent observers could separate these two types consistently and the observations clearly showed that mullite melts congruently. However, they experienced a major problem with the equilibrium extent of the solid solution of mullite, finding that the composition of the eutectic point, 67 mol% Al_2O_3 at 1840°C between mullite and corundum, coincided with the composition of the mullite saturated with corundum. This difficulty was removed by considering that the saturated solid solution was metastable and that the composition of the mullite in equilibrium with corundum was 63 mol% Al_2O_3.

Amaraki and Roy thus concluded that mullite melts congruently at 1850°C and forms a eutectic with alumina at 67 mol% Al_2O_3. These phase relationships are shown in Fig. 14.6. They did not find any evidence of equilibrium solubility toward sillimanite but did find that the structure of mullite at any one composition can be ordered or disordered, depending on the heat treatment. Thus X-ray spacing cannot be used to determine the composition of mullite.

Sadanga et al. [8] conducted an X-ray study of mullite and its relationship with the structures of sillimanite and andalusite. They explained the mechanism of transformation of sillimanite and andalusite to mullite and silica in terms of (i) the redistribution of tetrahedrally coordinated silicon and aluminum ions into a

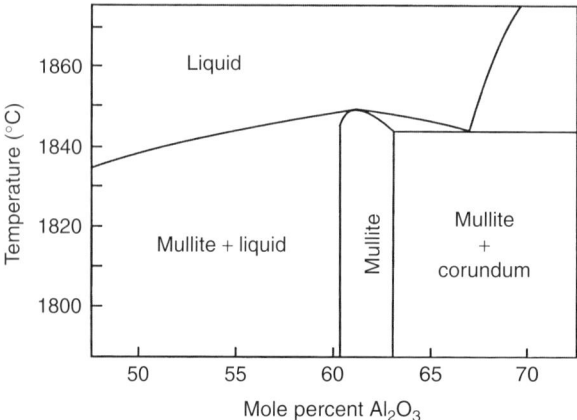

Figure 14.6 The phase diagram for the system Al_2O_3–SiO_2 determined by Aramaki and Roy in 1962.

random way over the same set of positions: (ii) random removal of some of the oxygen atoms shared by the cations, replacement of silicon by aluminum and (iii) shifts of the tetrahedrally coordinated sites into open sites.

2.3 The CaO–Al_2O_3–SiO_2 System

Rankin and Wright published the results of their study of phase equilibria in the system CaO–Al_2O_3–SiO_2 in 1915 [3] and started by accepting the phase diagrams of the constituent binary systems available at that time. They used a phase diagram for the system CaO–SiO_2, that did not show liquid immiscibility at the SiO_2-rich end, Bowen and Greig's phase diagram for the system Al_2O_3–SiO_2 showing the occurrence of congruently melting sillimanite, and used a phase diagram for the system CaO–Al_2O_3 differing from the currently accepted version by identifying $12CaO.7Al_2O_3$ as $5CaO.3Al_2O_3$ and omitting the incongruently melting $CaO.6Al_2O_3$.

Rankin and Wright used three experimental methods: (1) Measurement of the variation of temperature with time during cooling, which identifies ranges of temperature over which primary and binary crystallization occur and identifies the temperatures of eutectic or peritectic decompositions as thermal arrests; (2) holding different compositions at constant temperatures for times long enough to allow reactions to proceed to completion and for crystals to grow to a size large enough to make microscopic examination easier, and (3) location of the positions of the boundaries of the primary phase fields, by determining whether a phase under investigation dissolves in a second solution or crystallizes. This approach involved much trial and error. They used platinum furnaces that could be operated at temperatures up to 1600°C and an iridium furnace for temperatures between 1600°C and 2100°C. The pure constituent oxides were melted in a platinum crucible, solidified and crushed, first in a steel mortar, and then in an agate mortar and remelted. This procedure was repeated three times, to produce a chemically homogeneous product.

All of the experiments were conducted in air, in a platinum or iridium furnace heated by current from a storage battery with a voltage of 110 and a capacity of 3000 ampere-hours. The furnace temperature could be maintained at 1500 ± 2°C

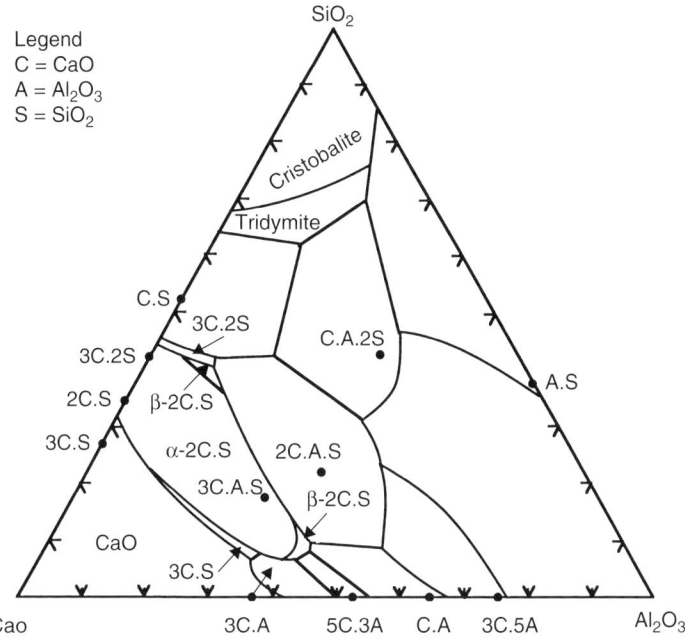

Figure 14.7 Primary phase fields in the system CaO–Al_2O_3–SiO_2 determined by Rankin and Wright in 1915.

for up to 12 h. The work involved 7000 heat treatments and microscopic examinations of 1000 different compositions: The study found three ternary compounds, $CaO.Al_2O_3.SiO_2$ (anorthite), $2CaO.Al_2O_3.SiO_2$ (gehlenite) and $3CaO.Al_2O_3.SiO_2$. The first two of these compounds melted congruently with the third melting incongruently in the primary phase field of α-$2CaO.SiO_2$. The incongruently melting ternary compound does not appear on the currently-accepted phase diagram for the ternary system. Fourteen primary phase fields were found, as were 30 doubly saturated boundary curves at 21 quintuple points (triply saturated melts). These equilibria are shown in Fig. 14.7.

3 THE CaO–"FeO"–SiO_2 SYSTEM

Phase equilibria in the system CaO–"FeO"–SiO_2 have been studied by Bowen et al. [9], who made reference of the importance of melts in this system to metallurgists interested in the behavior of slags. Limitations of temperature confined the study to mixtures less siliceous than the orthosilicate, that is, mixtures with mole fractions of SiO_2 less than 1/3.

Silica, ferrous oxalate (or Fe_2O_3) and $CaCO_3$ mixed in proportion calculated to give the desired product were melted together in a platinum crucible. This procedure caused a small deficiency in iron in the product, due to solution of Fe in the crucible. Samples were then heated in an iron crucible, in an atmosphere of nitrogen. The temperatures of the start of melting, of completion of melting and of any other changes of phase were determined by the method of quenching, combined with microscopic determination of the phases present in the quenched product. Heating in an iron crucible causes some reduction of Fe_2O_3 by the iron crucible as the melt becomes saturated with iron. The authors stated their belief that this may partly balance, or may more than partly balance, the deficiency caused by melting in platinum. Thus it was necessary, in all cases, to make a chemical analysis of each mixture after a run in an iron crucible at a temperature at which a change of phase was found to occur. Repeating this procedure for a series of mixtures allowed construction of the liquidus surfaces in the system.

The ferrous iron contents and the total iron contents were determined by long, involved wet chemical analysis and the difference between the total iron and FeO gave the Fe_2O_3 content. The optical properties of the phases produced were determined using a petrographic microscope. Glass was distinguished readily from crystals and different crystals were distinguished from one another by exact measurements of optical constants. In many cases it was possible to determine the composition of the liquid phase by measuring the refractive index of the glass. Standards were made for calibration, for example, the refractive indices and optic axial angles were measured as functions of composition in wollastonite solid solutions ($CaO.SiO_2$–$FeO.SiO_2$). The authors indicated that the presence of Fe_2O_3 means that, strictly, the system is not a ternary, but for convenience of presentation, the small amounts of ferric oxide were calculated as ferrous oxide and the results are plotted on a ternary diagram.

The results obtained, which are shown in Fig. 14.8, give complete information on the chemical aspects of the attack of a slag upon silica refractories, including those to which some lime has been added.

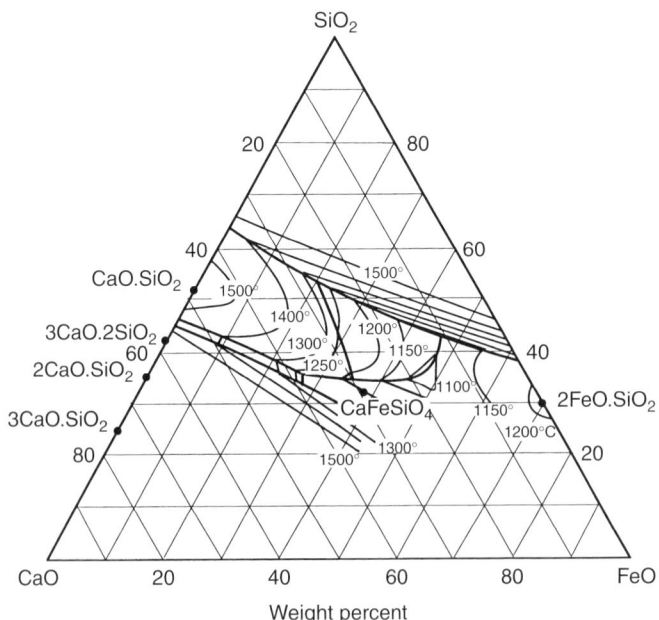

Figure 14.8 The liquidus surfaces in the system CaO–FeO–SiO_2 determined by Bowen et al. in 1933.

4 THE FEO–FE$_2$O$_3$–SIO$_2$ SYSTEM

The thermodynamic properties of iron silicate slags have been determined by Schuhmann and coworkers. In the first of these studies, of the thermodynamics of iron silicate slags saturated with γ-iron, Schuhmann and Ensio [10] melted mixtures of a stock iron silicate with iron oxide or silica in an iron crucible under a CO–CO$_2$ atmosphere at a desired temperature. The iron oxide was produced by oxidizing a bar of armco iron in an electric furnace and collecting the scale, and the orthosilicate, fayalite (2FeO.SiO$_2$), was prepared as the stock material. The mixture of CO and CO$_2$ was bubbled through the melt and analyses were made of the ingoing gas and the mixture after bubbling through it. The composition of the ingoing gas was adjusted, until it became identical with that of the outgoing gas, at which point the final CO$_2$/CO ratio was considered to be the value required for the equilibrium given by the following equation

$$Fe + CO_2 = FeO + CO \tag{1}$$

at the given temperature. Two methods were used to analyze the incoming and outgoing gas mixtures. During the period of adjustment, the composition of the outgoing gas was analyzed continuously by measuring its thermal conductivity. Equilibrium was reached when the ingoing and outgoing gases had the same thermal conductivity. Then the actual composition of this equilibrium gas was determined gravimetrically by passing the gas through a column of ascarite anhydrone, which absorbed the CO$_2$. The remaining CO was passed through a column of hot copper oxide, which oxidized it to CO$_2$, and the CO$_2$ produced was absorbed in a second column of ascarite anhydrone. The CO$_2$ and CO were measured as weights of CO$_2$ and the authors stated that the weight ratio was considered to be the ratio of the partial pressures, assuming ideal gas behavior. This has to be a misstatement, as the ratio of the partial pressures is equal to the ratio of the mole fractions of the gases. The melt was then sampled and analyzed for FeO, Fe$_2$O$_3$ and SiO$_2$. The contents of SiO$_2$ and Fe were determined by standard gravimetric techniques and the Fe^{3+} content was by titration. Thirty-three equilibrium runs were made, each requiring about 7 h. Twenty-three of these runs were made using slags of varying silica content, in the temperature range 1250–1400°C, seven were made using wustite-saturated slags in the range 1250–1350°C and the remaining three with silica-saturated slags in the range 1250–1350°C. The thermodynamic activities of FeO in the melts were calculated from the measured equilibrium ratios of the partial pressures of the gases and the standard change in Gibbs free energy for the reaction given by Eq. (1). It was found that the activity of FeO in a melt of any composition does not vary with temperature. Thus, the liquidus temperatures were identified as those, at which, on cooling, deviations occurred in the measured activities, caused by the compositions of the saturated melts moving down the liquidus lines.

The thermodynamics of ion silicate slags saturated with silica have been determined by Michal and Schuhmann [11], using a technique similar to that described above, employing a silica crucible. Initial attempts to contain the silica-saturated melts in translucent silica crucibles failed because it was found that the melts rapidly penetrated the wall of the crucible. Consequently, double-walled silica crucibles with ground, fused quartz placed in the annular space were constructed, and these successfully contained the melts. The temperatures were measured with a standardized optical pyrometer (which involved corrections in the range 18–24°C). No difficulty was experienced with the deposition of carbon from the equilibrating CO–CO_2 gas mixture.

Attempts to establish the CO_2/CO ratio at values higher than those required for saturation of the melt with magnetite caused the formation of more magnetite with no increase in the ratio of p_{CO_2}–p_{CO}. Consequently, slags with compositions close-saturation with magnetite were melted in platinum crucibles. They were equilibrated with gas mixtures of fixed compositions, quenched, mounted, polished and examined microscopically for primary magnetite. The oxygen potentials of the equilibrating gas mixture were increased gradually, until microscopic examination detected crystals of primary magnetite. The positions of the invariant systems gas:liquid slag:silica: γ-iron and gas:liquid slag:silica:magnetite were determined.

The constitution of the system FeO–Fe_2O_3–SiO_2, as given by the liquidus surfaces of magnetite, wustite and silica (tridymite) in the temperature range 1250–1450°C has been determined by Schuhmann et al. [12]. Combination with other data [10,11] facilitated development of the entire temperature–composition range of stability of iron silicate slags containing both ferrous and ferric ions.

Small charges of solid were heated and melted in platinum crucibles under an atmosphere of nitrogen in the temperature range 1450–1500°C. The melts were then quenched and ground and 76 experimental mixtures were analyzed for FeO, Fe_2O_3 and SiO_2. Five isothermal sections were determined at 1250°C, 1300°C, 1350°C, 1400°C and 1450°C. The melts were equilibrated, quenched and examined microscopically to identify the three-phase triangles and the two-phase lines. Cooling did not prevent crystallization of the melt, but no difficulty was experienced in differentiating the primary crystals from the finer crystals and dendrites formed during cooling. The quenched melts were broken out of the crucibles and crushed lightly. The large grains were mounted in bakelite and polished sections were examined by standard reflected light microscopy.

Primary tridymite and magnetite were readily recognized from their well-developed crystalline habits and from their reflectivities. Primary tridymite was checked by measuring the refractive indices of loose grains. After quenching, wustite grains acquired a light gray color and melts transformed to a dark gray glass or, more commonly, some form of fine structure.

It was found that the weights of the platinum crucibles increased, due to transfer of iron from the melt according to

$$3\text{FeO (mixture)} \rightarrow \text{Fe}_2\text{O}_3 \text{ (mixture)} + \text{Fe (in the platinum)}$$

and, consequently, all of the final equilibrations were conducted in platinum crucibles, that had been used at least once, before with the same mixture of oxides.

The phase equilibria in iron silicates saturated with solid iron are shown in Fig. 14.9 and the liquidus surfaces in the ternary system FeO–Fe_2O_3–SiO_2 are shown in Fig. 14.10.

The Determination of Phase Diagrams for Slag Systems 457

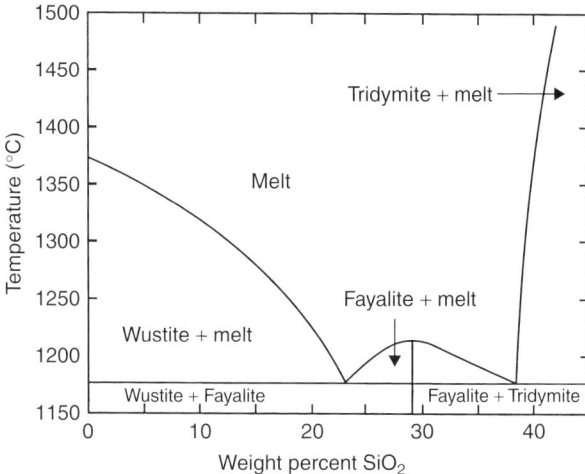

Figure 14.9 Liquidus compositions of iron silicate slags in equilibrium with solid iron determined by Schuhmann and Ensio in 1957.

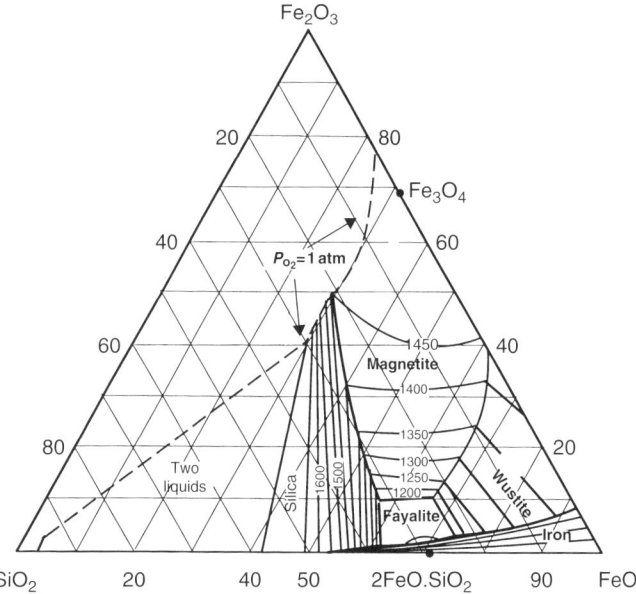

Figure 14.10 The liquidus surfaces in the system FeO–Fe_2O_3–SiO_2 determined by Schuhmann et al. in 1953.

REFERENCES

1. A.L. Day, E.S. Shepherd and F.E. Wright, *Am. J. Sci. (4th Series)*, 22 (1906) 265.
2. J.H. Welch and W. Gutt, *J. Am. Ceram. Soc.*, 42 (1959) 11.
3. G.A. Rankin and F.E. Wright, *Am. J. Sci. (4th Series)*, 39 (1915) 1.
4. N.L. Bowen and J.W. Greig, *J. Am. Ceram. Soc.* 7 (1924) 238.

5. E.S. Shepherd, G.A. Rankin and F.E. Wright, *Am. J. Sci., (4th Series)*, 28 (1909) 293.
6. S. Aramaki and R. Roy, *J. Am. Ceram. Soc.*, 45 (1962) 229.
7. L. Bragg and G.F. Claringbull, *Crystal Structures of Minerals*, Cornell University Press, NewYork, 1965, pp. 190 and 196.
8. R. Sadanga, M. Tokonami and Y. Takeuchi, *Acta Cryst*, 15 (1962) 65.
9. N.L. Bowen, J.F. Schairer and E. Posnjak, *Am. J. Sci. (5th Series)*, 26 (1933) 193.
10. R. Schuhmann Jr. and P.J. Ensio, *J. Metals*, 1951, 401.
11. E.J. Michal and R. Schuhmann Jr., *Trans AIME*, 194 (1952) 723.
12. R. Schuhmann Jr., R.G. Powell and E.J. Michal, *Trans. AIME*, 197 (1953) 1097.

CHAPTER FIFTEEN

DETERMINATION OF PHASE DIAGRAMS FOR HYDROGEN-CONTAINING SYSTEMS

Ted B. Flanagan[1] and Weifang Luo[2]

Contents

1	Introduction	459
2	Phase Diagram Representations of M–H Systems	462
	2.1 Binary M–H Systems	462
	2.2 Multi-component Systems	466
3	Some Useful Rules Relating Phase Diagrams and Reaction Enthalpies for M–H Systems	471
4	Techniques Employed for Phase Diagram Determination of M–H Systems	473
	4.1 Open Systems	474
	4.2 Closed Systems	477
	4.3 Electron Diffraction and TEM	480
	4.4 Magnetic Susceptibility	480
	4.5 Electric Resistivity	480
	4.6 Dilatometry	480

1 INTRODUCTION

The high mobility of interstitially dissolved hydrogen (H) atoms allows equilibrium to be established at relatively low temperatures making these metal–hydrogen (M–H) systems unique in that their phase diagrams can be quite readily determined at low temperatures. The M–H systems also allow the facile measurement of the hydrogen chemical potential from the equilibrium p_{H_2} over the sample. Other metal–gas systems also offer the advantage of being able to determine the gaseous chemical potential from equilibrium pressures but their diffusion is much slower than H and therefore equilibrium can only be established at elevated temperatures. For example, the diffusion constant of H in Pd is $\approx 10^{19}$ times faster than O in Pd at 298K! Even though this chapter concerns the experimental aspects of phase diagram determination for the M–H systems, some theoretical aspects of their phase diagrams are needed because otherwise distinctions between the various degrees of equilibrium vis-à-vis the resulting phase diagrams is unclear.

---

[1] Department of Chemistry, University of Vermont, Burlington, VT, USA
[2] Department of Analytical Material Sciences in the address of Weifang Sandia National Laboratory, Livermore, CA, USA

The Pd–H phase diagram system was employed by Hoitsema and Roozeboom [1] as one of the first demonstrations of the usefulness of Gibbs phase rule but, after this, there was little activity in the determination of M–H phase diagrams until about 50 years ago and even the Pd–H phase diagram was incomplete until 1978 [2]. Oates pointed out [3] that there was little known about the phase diagrams of M–H systems prior to 1950 when McQuillan determined the Ti–H phase diagram [4] from measurements of a series of pressure–composition isotherms (p–r–T) where r = H-to-M atom ratio. After this determination of the Ti–H system, there was some activity in this area due to the interest in metal hydrides as neutron moderators, mainly the group 3B and 4B metals and their alloys and the actinides. The book by Mueller et al. [5] describes the status of these phase diagrams up to 1968.

A series entitled *Hydrogen Metal Systems* edited by Lewis and Aladjem [6] has recently appeared in Solid State Phenomena. The focus was not on phase diagrams, but some M–H phase diagrams were discussed. Recently, Manchester [7] edited an extensive, and very useful, compilation of phase diagrams for pure M–H (binary) systems including some of their structural and thermodynamic properties. The present chapter will emphasize the experimental approaches to M–H phase diagram determination and will not be concerned with the many extant phase diagrams.

The dilute H phase in M–H systems is universally referred to as the α phase and the H-rich phase as the α' phase for miscibility gap systems or as β for structural transformation systems. Both of the latter phases are referred to as hydrides. Any additional higher H content hydride phases are referred as δ, γ, etc. Historically the hydride phase of Pd–H was referred to as the β phase and this is still a common practice. Two broad categories of M–H phase diagrams are miscibility gap and structural transformation systems and these are schematically illustrated in Fig. 15.1 [8]. In the former there is no change in the metal sub-lattice upon hydride formation only a lattice expansion and, consequently, it is possible to pass continuously from the α to the α' phase by passing around the critical point. In structural transformation systems the hydride phase has a different metal sub-lattice than the pure metal and consequently it is not possible to proceed continuously from one, α, to the other, β. Figure 15.1(a) and (b) shows the simplest possible representations for each category and in real systems both may appear together with other ordered hydride phases that may form, especially at low temperatures. It can be seen from Fig. 15.1(c) that the isotherm at T_1 is the same whether the system is a miscibility gap or a structural transformation type. Isotherms over a wide temperature range are required to establish whether or not there is a critical point and consequently if the system is a miscibility gap type. X-ray diffraction (XRD) measurements are the most expedient technique for the determination of whether the system is a miscibility gap or structural transformation type or whether a certain region of the phase diagram is one or the other.

There is, of course, no difficulty in the establishment of the solid phase equilibrium for miscibility gap systems because metal atom diffusion is not needed. For structural transformation systems of binary M–H systems, long-range metal atom diffusion is also unnecessary but only short-range diffusion is needed for local metal rearrangement, and this should also take place at relatively low temperatures. For example, a small local movement of Nb atoms from bcc (α) to fcc orthorhombic (β) can take place below ambient while the H atoms undergo an accompanying

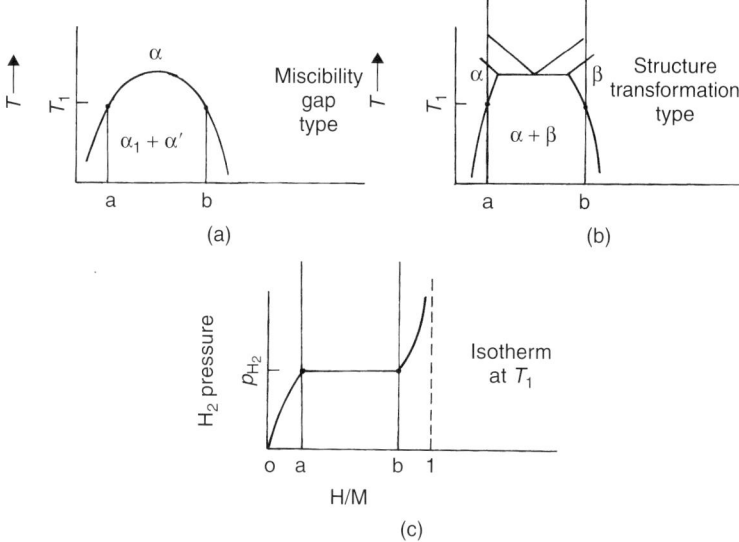

Figure 15.1 Schematic diagrams showing (a) a miscibility gap system, (b) a structural transformation system and (c) an isotherm corresponding either to (a) or (b) at temperature T_1 [8].

ordering. It is quite possible, however, that some of the many H-ordered phases found at *very* low temperature are metastable because the metal atoms may not have sufficient mobility for the short-range diffusion needed at these temperatures.

There are incoherent and coherent phase diagrams for M–H miscibility gap systems. Coherent phase diagrams reflect the unrelaxed stresses arising from the volume expansion of the second, hydride, phase. For incoherent phase diagrams these misfit stresses have been relaxed by dislocations; this review will be limited to incoherent phase diagrams. A theoretical discussion of coherent phases has been given by Wagner and Horner [9] and a review of the Nb–H coherent phase diagram is given by Peisl in Manchester's compendium [7].

Phase diagrams at ultra high p_{H_2} will not be discussed here. Fukai has shown such phase diagrams [10,11] often with a theoretical compositional limit of pure H. Antonov [12] has given plots of T vs. p_{H_2} up to 5 GPa for nine different transition metals based on work at the Institute of Solid State Physics, Russian Academy of Sciences, Russia. High pressure phase diagrams of pure metals and alloys also appear in other publications by the Russian workers (e.g., V. Antonov, E. Ponyatovsky, I. Belash).

Hysteresis is omnipresent in the two solid-phase regions of M–H systems [13] and is usually reflected by a greater plateau pressure for hydride formation than that for decomposition. It also affects the H contents of the co-existing phase boundaries and the dilute/(dilute + hydride) boundary is always greater for hydride formation than for decomposition and the hydride/(hydride+dilute) boundary is also always greater for hydride formation. This means that determinations of the terminal hydrogen solubilities (solvus) will differ between heating and cooling. It seems impossible to avoid the effects of hysteresis in the determination of M–H phase diagrams and, if it is significant, it seems best to take an average value of the boundaries

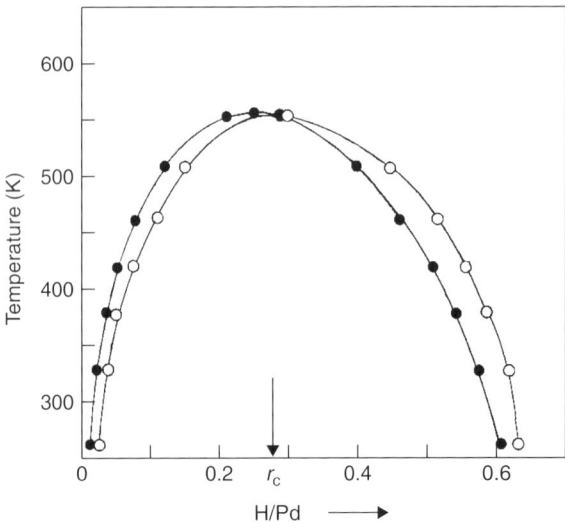

Figure 15.2 The effect of hysteresis on the phase boundaries for the Pd–H system. ○, absorption; ●, desorption [14].

determined during absorption and desorption measurements. An example, of the influence of hysteresis on the phase diagram of an M–H system is shown by Pd–H in Fig. 15.2 [14].

2 PHASE DIAGRAM REPRESENTATIONS OF M–H SYSTEMS

2.1 Binary M–H Systems

For closed, single phase, binary (M + H) systems under only hydrostatic pressure, P, it is necessary to specify three state variables, T, P and $r = $ H/M, atom ratio, to determine G, thus $G(T, P, r)$. It will be assumed here that the hydrostatic pressure P is fixed, and has effectively zero effects on the condensed phases. In open systems generally $P = p_{H_2}$ where p_{H_2} is the H_2 pressure in equilibrium with the condensed phase and P is the overall hydrostatic pressure. It is again assumed that P is effectively zero since p_{H_2} does not increase the hydrostatic pressure sufficiently to influence the condensed phases except at very high p_{H_2}. When P is fixed, a single phase integral property, such as G in a binary M–H system, is a surface in G–T–r space. There will be $G(T, r)$ surfaces for both the dilute and hydride phases and, if there is two-phase equilibrium, these three-dimensional surfaces will intersect. If a tangent plane can be drawn to both of the free energy surfaces in G–T–r space the intercepts will determine the co-existing phase boundaries. When these curves of intersection are projected onto a T plane, the resulting G–r curve will allow two tangent points to be drawn indicative of the two-phase compositions. If, instead, the system were to remain on the G–r curve of the α phase it would have a greater G than if it is on the tangent slope indicative of the co-existence of the two phases. In a binary system (M–H) with two co-existing condensed phases, the three state functions which need to be specified for G are T, r_α, r_β where, r_α, r_β are the phase boundary compositions.

Different phase diagram representations will now be discussed. Often a T–r ($r = $ M/H) representation is employed and then pure H cannot be the limiting concentration but generally the limiting value of r is chosen as an integer corresponding to complete occupation of available interstices. It is convenient to employ r as the concentration variable and it is equivalent to the mole fraction of H in the (vacancy + H) sub-lattice where the metal serves as simply the framework.

In some cases, information about the pure metals assists in the determination of the M–H phase diagrams. For example, Ti and Zr are each allotropic with α (hcp) phases existing from moderate to elevated temperatures and then transitions take place at 1136K and 1156K for Ti and Zr, respectively, to the β (bcc) phase [7]. The existence of the pure metal allotropes must be reflected in the T–r binary M–H phase diagrams because extrapolation to $r \to 0$ must be consistent with the pure allotropic phase transitions. For example, the allotropic transition temperature is lowered from 1136K for pure Ti to 571K at the eutectic composition of dissolved H (Fig. 15.3). The H plays the same role as a solute which, when added to a liquid, lowers its freezing point.

The Zr–H phase diagram is shown in Fig. 15.3 [7] and the Ti–H diagram is similar [7]. It falls into the category of a structural transformation system as shown by comparison with Fig. 15.1(b), although it differs somewhat because of the hcp (α) $\to$ bcc(β) transition of pure Zr and this is reflected in the phase diagram where the ($\alpha + \beta$) two-phase field for the H-containing phases extrapolates at $r \to 0$ to reflect the pure metal $\alpha \to \beta$ transition. This causes the H compositions of both phases to decrease with increase of temperature. H occupies the interstices in a disordered way in both the α and β phases and the β phase of Zr–H is not a hydride phase, as indicated above as the general practice, but a solid solution because the phase retains its bcc structure of pure Zr above the eutectic which is at $r \approx 0.60$ and $T = 819$K. In the ($\beta + \delta$) two-phase field both phase compositions increase in H content with increasing temperature. The δ and ε phases both involve structural changes of the metal sub-lattice, for example, the δ phase is fcc and the ε is fct.

The Zr–H phase diagram (Fig. 15.3) is, of course, a T–r representation and another phase diagram representation employs plots of $\Delta\mu_H = RT \ln p_{H_2}^{1/2}$ or $\ln p_{H_2}^{1/2}$ against T or $1/T$. Such plots have been given for the Pd–H and Pd–alloy–H systems by Kishimoto et al. [15] showing the two-phase region with accompanying single phase iso-H concentration tie lines. The iso-H concentration tie lines in the single α phase intersect the two-phase van't Hoff plot (Fig. 15.4). Similar representations have been given by Toguchi et al. [16] for rare earth–H systems. When a single M–H system has several different two-phase equilibria a more complex $\ln p_{H_2}^{1/2}$–r phase diagram is needed. Representations of several two-phase equilibria lines together with iso-H concentration tie lines in the single-phase regions have also been given, for example, Katz and Gulbransen [17] give such a plot for the Zr–H system showing the lines for the co-existence of the α/β, β/δ and some single α phase iso-H content tie lines. The slopes of these lines of R $\ln p_{H_2}^{1/2}$ against $1/T$ give the corresponding values of ΔH_H and $RT \ln p_{H_2}^{1/2}$ against T give $-\Delta S_H$ or $-\Delta S_{plat}$ when $p_{H_2}^{1/2} = p_{plat}^{1/2}$, where P_{plat} refers to the plateau pressure.

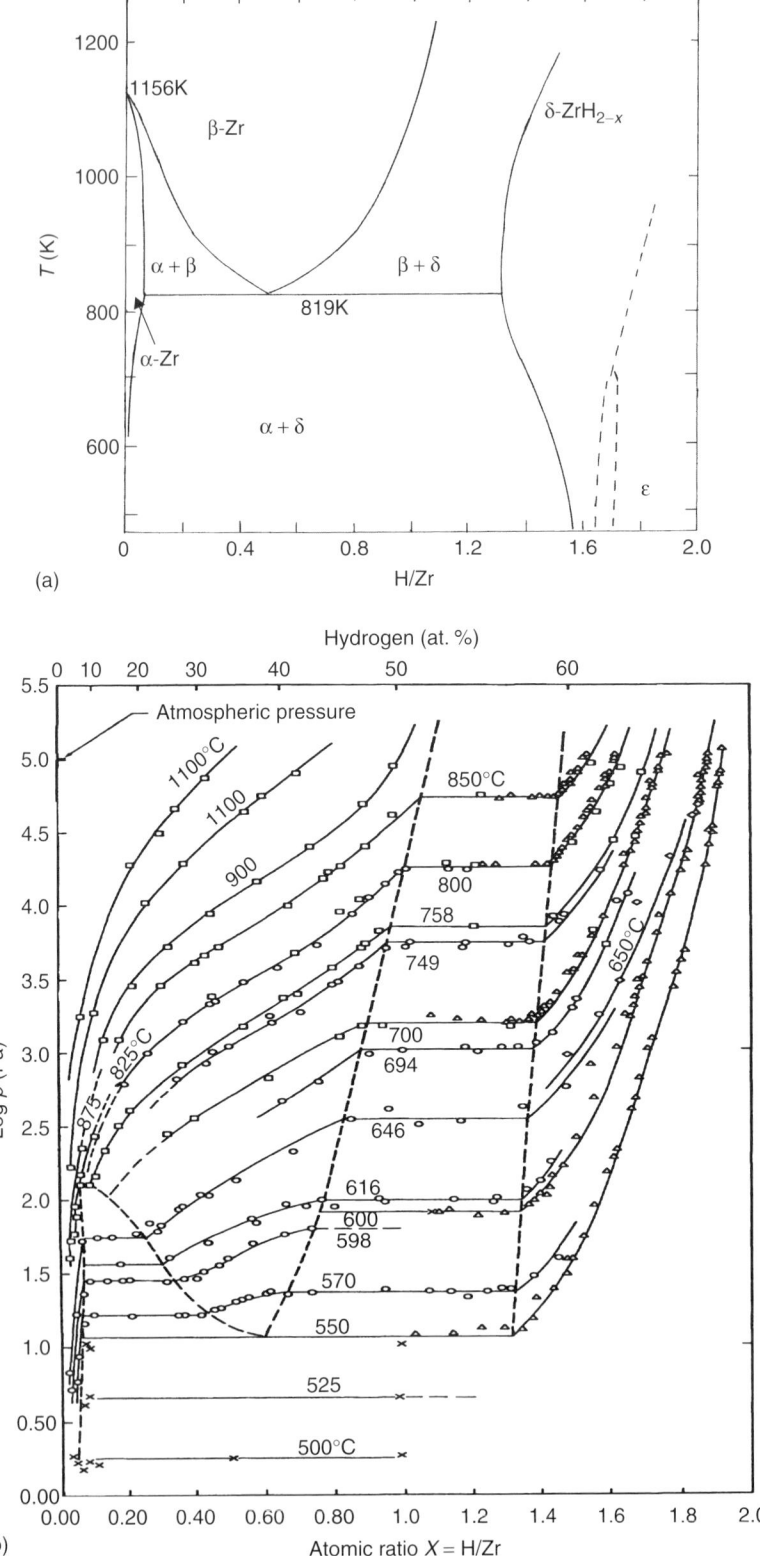

Figure 15.3 Phase diagram of the Zr–H system without (a) and with (b) the pressure isotherms [7].

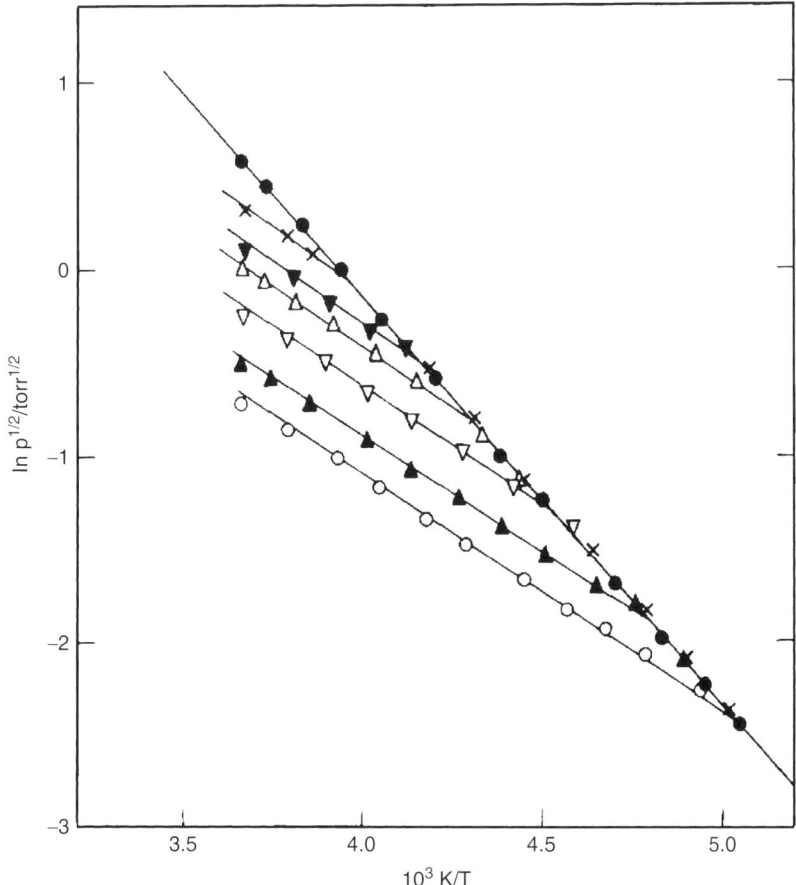

Figure 15.4 Data for the Pd–H system [15]. The $r = H/M$ for the various iso-H concentration lines are: X, $r = 0.0065$; filled upside down triangle, $r = 0.0052$; △, $r = 0.0045$; ▽, $r = 0.0034$; filled right side up triangle, $r = 0.0026$; ○, $r = 0.0021$.

Collections of plots of log $p_{H_2, \text{plat}}$ against $1/T$ for a large number of different M–H systems have been given by several authors (e.g., [18]) and these allow the M–H systems to be readily compared and to allow those with a desired $p_{H_2,\text{plat}}$ to be selected for possible applications.

The ln $p_{H_2,\text{plat}}$ against $1/T$ plots mentioned above are closely related to Ellingham diagrams for O_2 [19] which are plots of $\Delta G° = RT \log p_{O2,\text{eq}}$ against T for a series of stoichiometric oxides where $\Delta G°$ is the standard free energy of formation of the stoichiometric oxide from one mole of O_2 (g) (1 bar) and the pure metal. When there are no phase changes, these plots are essentially linear and a considerable amount of information can be readily gleaned from them. For example, the equilibrium $p_{O_2,\text{eq}}$ value for any temperature for a given oxide equilibrium can be obtained and the relative stabilities of the oxides can be evaluated, for example, whether or not a given one will be reduced by another metal to form its oxide.

Fukai [10] has constructed Ellingham-type plots for a series of metal hydride plateau reactions where, in his construction, the ordinate is $\Delta\mu_H = RT \log p_{H_2,plat}$ rather than $\Delta G°$ because the hydrides are often non-stoichiometric with appreciable H_2 solubilities in the metal phase. These plots give p_{H_2} for a two-phase co-existence plateau at any desired temperature from a nomographic pressure scale along the right-hand ordinate. They also indicate whether a given hydride will decompose in the presence of another metal to form its hydride. The slopes of the plots are all positive and they give $-\Delta S_{plat}$. It is generally thought that ΔS_{plat} is the same for all hydrides but Fukai's plot (Fig. 1.21, Ref. [10]) indicates that this is not true.

Earlier Oates and Flanagan [20] constructed $\Delta\mu_{H_2}$ $(= 2\mu_H - \mu_{H_2}°)$–T phase diagrams for the group VB–H systems where the standard state of H_2 refers to 1 bar and, as in Fukai's plots, μ_H is not in its standard state. In contrast to Fukai's plots, these include single phase tie lines. The $\Delta\mu_{H_2}$–T phase diagrams are constructed from the thermodynamics of the systems together with knowledge of the corresponding T–r phase diagrams and therefore more information is needed than for the T–r diagrams, however, they provide more information and such diagrams can be seen in Ref. [20]. A $\Delta\mu_{H_2}$–T phase diagram is shown here for the Zr–H system in Fig. 15.5 and, since the $\Delta\mu_{H_2}$–T diagrams are less familiar than T–r phase diagrams, this will be explained in some detail.

Iso-H concentration tie lines are shown by dashed lines while the heavy solid lines are two phase co-existence lines (Fig. 15.5). Because the translational entropy of the $H_2(g)$ is lost on absorption, the slopes, $-\Delta S_{H2}$, of the iso-concentration lines are positive and it can be seen that they become increasingly positive (i.e., ΔS_{H2} more negative) as the H concentrations of the iso-concentration lines increase. Except for a section of the $(\alpha + \beta)$ co-existence line, the "plateau" slopes are all positive and therefore the ΔS_{plat} values will also be negative for H_2 absorption.

The nomographic p_{H_2} scale on the inside of the right-hand side axis indicates isobaric paths which are drawn from $T = 0$ and $\Delta\mu_{H_2} = 0$ to the markings on the right-hand axis. Because of the relatively high temperature scale needed for this diagram of Zr–H, it is inconvenient to include 0 K but the nomographic markings have, nonetheless, been drawn at the correct angles to extrapolate to $T = 0$ K and can be used as in an Ellingham plot. The points of intersection of one of these isobaric straight lines with a given phase boundary gives the p_{eq} for those conditions. For example, a straight line originating from the 10^{-4} bar intersects the $(\alpha + \delta)$ boundary at 800 K. The 10^{-4} bar line intersects the iso-H concentration $r = 1.6$ tie line in the single δ phase at 700 K. This type of phase diagram is more useful for M–H systems than the usual T–r one because it gives the equilibrium p_{H_2} in single-phase regions for a given H content and temperature and values of $p_{H_2,plat}$ for two-phase regions.

2.2 Multi-component Systems

The possible nature of the equilibria which can be obtained in multi-component M–H systems will now be discussed because this determines the phase diagrams. Categories of two (solid) phase co-existence regions relevant to multi-component M–H systems have been discussed by Flanagan and Oates [21]. These categories of behavior are: para-equilibrium (frozen metal atom, PE), local equilibrium (LE) and

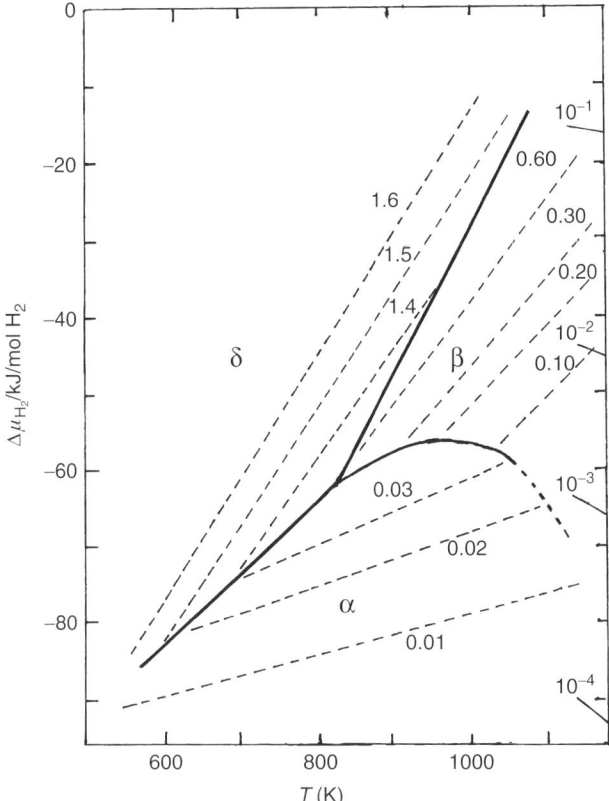

Figure 15.5 $\Delta\mu_{H_2}$–T phase diagram of Zr–H constructed from thermodynamic data given in Refs. [7,36]. The numbers on the dashed tie lines refer to the H/Zr ratios for these iso-concentration lines.

complete equilibrium (CE). For binary systems it does not matter which category obtains with respect to the presence of two-solid phase invariant pressure regions (plateaux) but for multi-component systems it does. Under CE conditions there will *not* be a plateau region for the latter but for PE there will be. For conditions of LE there may or may not be a plateau dependent upon several factors [22].

The progression from PE (pseudo-binary) behavior to CE occurs with increase of temperature and this can be illustrated very nicely by the reaction of H_2 with $CeFe_2$. PE corresponds to the reaction:

$$xH_2 + CeFe_2 = CeFe_2H_{2x} \tag{1}$$

which occurs at moderate temperatures. CE corresponds to

$$yH_2 + CeFe_2 = CeH_{2y} + 2Fe \tag{2}$$

which takes place at elevated temperatures. Figure 15.6 shows the X-ray reflections after reaction of $CeFe_2$ with H_2 at different temperatures [23]. At low temperatures

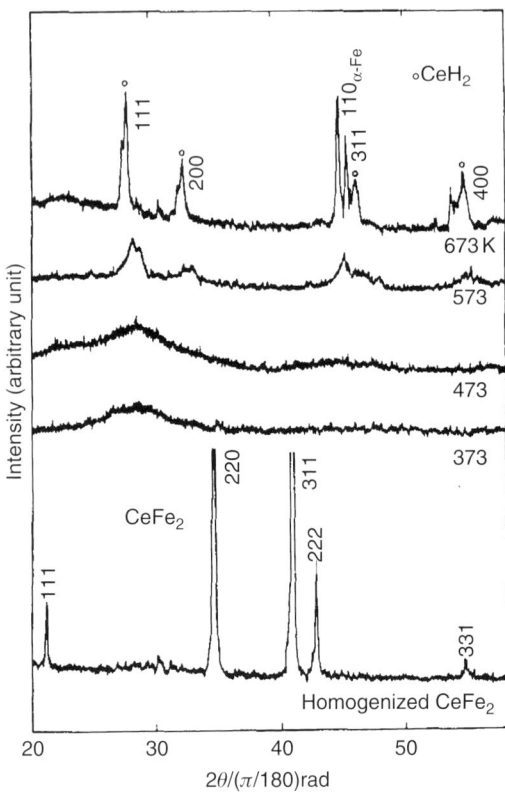

Figure 15.6 XRD patterns of CeFe$_2$ after hydriding at the indicated temperatures [23].

the reflections are from CeFe$_2$H$_{2x}$, at intermediate temperatures reflections are absent because the product is amorphous due to a limited diffusion of the metal atoms (LE) and at high temperatures (CE) the reflections can be identified as CeH$_{2y}$ and Fe according to CE (Eq. (2)).

As illustrated by the CeFe$_2$ example, the H atoms are at equilibrium in all cases because of its great mobility but the metal atoms may or may not be in equilibrium. For ternary systems such as alloys and intermetallic compound–H systems, although the H atoms are in equilibrium, the metal atoms are generally not during two phase co-existence, that is, they exhibit PE. In fact, all intermetallic hydrides, AB$_2$H$_x$, are thermodynamically unstable with respect to decomposition into AH$_x$ and 2B where A is the element which forms the most stable hydride as shown for CeFe$_2$. Therefore phase diagrams for intermetallic–H systems, in contrast to those for pure metals, are generally metastable phase diagrams at moderate temperatures because the intermetallic or alloy is normally assumed not to disproportionate.

Not very many "complete" PE phase diagrams are available for some categories of intermetallic compound–H systems (e.g., AB$_5$–H systems) because disproportionation may become a factor at higher temperatures. For example, the phase diagram of LaNi$_5$–H is still largely unknown partly because the plateau p_{H_2} increases markedly with temperature and partly because of some disproportionation at higher

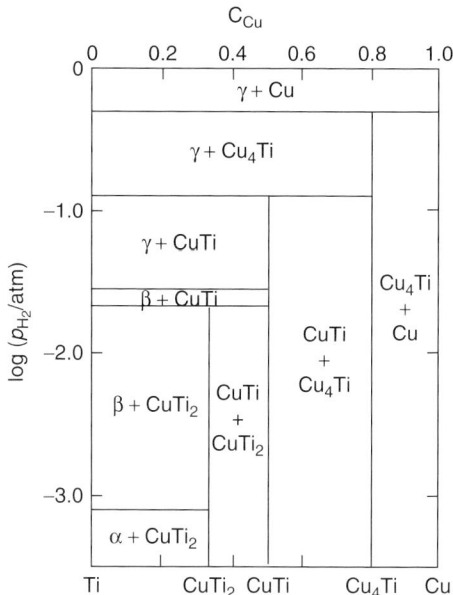

Figure 15.7 Log p_{H_2} vs. X_{Cu} phase diagram at 773K for the Cu+Ti+H system [26,27].

temperatures. From the work of Semenenko et al. [24], it appears as if there is a critical point in this system at about 415K but this has not been confirmed.

It should also be noted that some metal–deuterium (M–D) (metal–tritium, M–T) phase diagrams have also been determined and in a few cases they differ significantly from the M–H diagrams (e.g., V–H(D) [25]). The different isotope phase diagrams will not be discussed here.

For a single-phase ternary system (A+B+H), three state functions needed are T, r, C where $C = n_B/(n_A + n_B)$. For pseudo-binary behavior C is fixed and the system behaves as a binary one but for CE this is not the case and one example of CE, from the work of Kadel and Weiss [26], is the reaction of Ti_2Cu with H_2 in the temperature range from 773K to 1123K where a number of CE two-phase plateau regions were found. The lowest pressure plateau corresponded to

$$Ti_2CuH_{0.2} + 0.175H_2 = \beta\text{-}TiH_{0.4} + CuTiH_{0.15} \qquad (3)$$

where the H contents are approximate. This leads to a plateau since there are four phases and three components because the equilibrium (3) introduces another restriction, $f = \chi - \varpi + 2 = 1$ where f is the degree of freedom, χ the number of components and ϖ the number of phases. A CE phase diagram for the Ti–Cu–H system is shown in Fig. 15.7 at 773K using an isothermal presentation where C_{Cu} is plotted against log (p_{H_2}/bar) and $C_{Cu} = n_{Cu}/(n_{Cu}+n_{Ti})$; the amount of H is not indicated, only its p_{H_2} [27]. The phase diagram is based on the results of Refs. [26,28]. For any point within a two-phase region in the diagram a horizontal tie line can be drawn indicating the co-existing phases and the lever rule gives their proportions. A plateau pressure is found whenever log (p_{H_2}/bar) at a fixed C_{Cu}

crosses a horizontal line. It can be seen that as p_{H2} is increased for $C_{Cu} = 0.33$, a plateau is reached at log $(p_{H_2}/\text{bar}) = -1.67$ which corresponds to reaction (3). When log (p_{H_2}/bar) is increased further to -1.55 another plateau occurs corresponding to the following reaction:

$$\beta\text{-TiH}_z \rightarrow \gamma\text{-TiH}_z$$

where the CuTiH$_{0.15}$ remains unchanged during the reaction. When log (p_{H2}/bar) increases to -0.88, another plateau obtains corresponding to the following reaction:

$$4\text{CuTiH}_{0.15} + \gamma\text{-TiH}_z \rightarrow \text{Cu}_4\text{Ti} + 4\gamma\text{-TiH}_{z'}$$

The final plateau at log $(p_{H2}/\text{bar}) = -0.30$ corresponds to the reaction:

$$\text{Cu}_4\text{Ti} + 4\gamma\text{-TiH}_{z'} \rightarrow 4\text{Cu} + 5\gamma\text{-TiH}_{z''}$$

If the p_{H2} were to be increased over CuTi ($C_{Cu} = 0.5$) instead of CuTi$_2$, only two horizontal lines would be intersected indicating plateaux (773K) according to Figs. 15.3 and 15.4 in Ref. [24] which show these plateaux in the isotherm. A similar CE phase diagram has also been drawn for the important La–Ni system [29].

In contrast to the Cu–Ti–H or Ce–Fe–H systems where only stoichiometric intermetallic compounds are considered, for non-stoichiometric intermetallic compounds such as ZrMn$_{2-x}$ and substitutional alloys such as Pd$_{1-y}$Pt$_y$ the metal compositions of the co-existing phases can differ under CE. Schematic phase diagrams such as Fig. 15.8 can be drawn [30] for explaining the consequences of PE and CE. For CE the chemical potentials of each component are equal to each other in the different phases. In the Fig. 15.8, r is plotted against C_B at a constant temperature. The continuous nearly horizontal lines are the phase boundaries for CE and the solid lines within the two-phase field are tie lines of equal chemical potential for the H and metals. When r increases starting at $C_B = C_o$ within the dilute phase, α, tie lines of constant, and increasing, μ_H, will be crossed until the hydride phase first appears at point 1 where the composition of the hydride phase is given by 2.

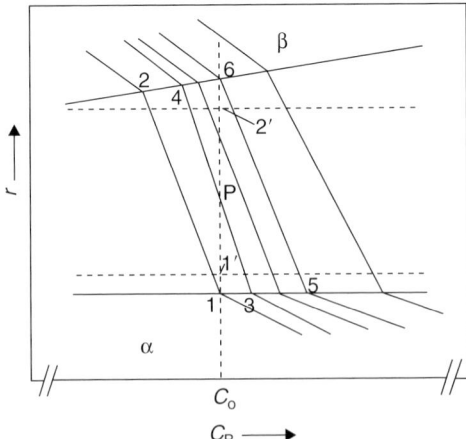

Figure 15.8 Schematic plot of r vs. $c_B = n_B/(n_B+n_A)$ for a ternary system [30].

μ_H must increase with r because of the thermodynamic stability requirement that $(\partial \mu_H / \partial r)_T > 0$. The solid lines within the two-phase field are tie lines along which each chemical potential, μ_H and μ_A and μ_B must be equal in the co-existing phases, for example, $\mu_A(\alpha) = \mu_A(\beta)$. When r increases within the two-phase field, tie lines are cut within the two-phase field with increasing r as in Fig. 15.8 and μ_H must increase with r. Upon reaching 6, the composition of the β phase is 6 and the α phase, 5. Thus CE for non-stoichiometric intermetallics or substitutional alloys does not lead to a two-phase plateau p_{H_2}. Upon further increase of r in the β phase from 6, μ_H increases at C_o as tie lines are crossed at greater r values. An example of CE for a $Pd_{1-y}Pt_yH_x$ alloy it was found that a solid solution $Pd_{0.75}Pt_{0.25}$ fcc alloy heated at 673K and $p_{H_2} \approx 10$ MPa separated into Pd-rich and Pd-poor phases each with different H contents [31] showing that CE conditions obtained.

The PE phase diagram is also shown in Fig. 15.8 by the dashed lines and the vertical one is a tie line of constant μ_H and μ_M where $M = (n_A + n_B)$; the dashed lines crossing it are the phase boundaries for the hydride (β) and dilute (α) phases. Hillert [32] has shown that the PE phase boundaries must lie within the CE ones. When hydrogen is added to the system of composition $C_B = C_o$, μ_H increases in the dilute phase until the hydride phase first appears at $1'$ and when r is increased within the two-phase field, the system must remain on the dashed vertical tie line where the phase compositions are fixed and μ_H is constant until $2'$ is reached when all of the dilute phase has been converted to the hydride phase and then the system remains on the vertical dashed line with μ_H increasing with r in the hydride phase. This is pseudo-binary behavior and the dashed line within the two-phase region is the p_{H2} plateau and this leads to a T–r phase diagram with a two-phase co-existence region.

Another possible representation for ternary systems is to employ triangular phase diagrams. Some workers have employed this approach for alloys such as Pd–M–H [33]. Rosenhall [34] was the first to employ a triangular phase diagram for Pd–Ag–H where the components were taken to be PdH, Pd and Ag; this phase diagram is shown in Fig. 15.9 (298K) where a two-phase region is shown ending at about $X_{Ag} = 0.25$ which is reasonable. The triangular diagrams of Sakamoto et al. [33] employ Pd, M and H as the components. Some of the Pd–M–H [33] phase diagrams shown in Ref. [33] are metastable ones, for example, it is known that Pd–Rh–H separates into Pd-rich and Pd-poor phases at elevated p_{H_2} and $\geqslant$350K [35] and this is not reflected in the diagram.

3 SOME USEFUL RULES RELATING PHASE DIAGRAMS AND REACTION ENTHALPIES FOR M–H SYSTEMS

Compared to systems consisting only of metals, it is relatively easy to measure enthalpies and entropies of solution or evolution of one of the components, H, in M–H systems. The measurement of an equilibrium p_{H2} gives $\Delta\mu_H$ and its temperature variation at a given r value gives ΔH_H or in the plateau regions, the van't Hoff plots of $p_{H2,plat}$ give ΔH_{plat}. Reaction calorimetry can also be employed to determine ΔH_H values as a function of r. This relative ease of determination makes relationships between the enthalpies and the phase diagrams quite useful. It should be recalled that

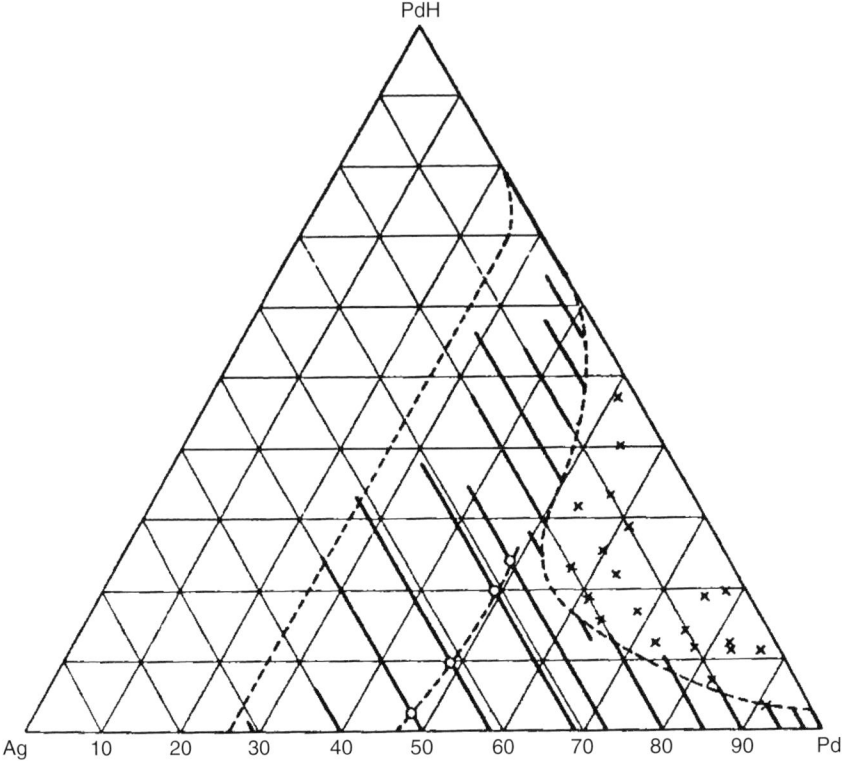

Figure 15.9 Ternary phase diagram (298K) for the Pd+Ag+H system [34]. X, regions showing two lattice parameters (two-phase region); heavy lines indicate regions with only one lattice parameter (single-phase region); ○, region where the Pd+Ag+H alloys have a minimum in electrical resistance.

generally the phase diagrams and thermodynamics are intimately connected because the former can, in principle, be obtained from the latter.

For a three condensed phase co-existence region such as at the eutectic temperature of the Zr–H system (Fig. 15.3), Speiser (in Ref. [5]) has given the following rule,

$$(r_\delta - r_\alpha)[d \ln p^{1/2}(\delta \to \alpha)/d(1/T)] = (r_\beta - r_\alpha)[d \ln p^{1/2}(\beta \to \alpha)/d(1/T)] + (r_\delta - r_\beta)[d \ln p^{1/2}(\delta \to \beta)/d(1/T)]$$

or, upon substitution by ΔH_{plat},

$$(r_\delta - r_\alpha)\Delta H_{plat}(\alpha \to \delta) = (r_\beta - r_\alpha)\Delta H_{plat}(\alpha \to \beta) + (r_\delta - r_\beta)\Delta H_{plat}(\beta \to \delta) \tag{4}$$

Wang and Olander [36] and Zuzek et al. in Ref. [7] employed this equation in order to determine the eutectic composition of Zr–H, however, these two groups of investigators obtained somewhat different values because of different choices of the ΔH_{plat}.

Luo and Flanagan [37] found the following equation to be quite useful for determining the compatibility of the slopes of T–r phase diagrams with the reaction enthalpies as a function of r

$$\Delta H_{\text{plat}} = \Delta H_{\text{H,PB}} + (\partial(\Delta\mu_H/T)/\partial r)_T (dr_{\text{PB}}/dT)(-T^2)$$

or upon substitution and rearrangement,

$$(dr_{\text{PB}}/dT) = (\Delta H_{\text{H,PB}} - \Delta H_{\text{plat}})/[T^2((\partial\Delta\mu_H/T)/\partial r)_T] \quad (5)$$

where the subscript PB refers to a phase boundary and thus (dr_{PB}/dT) is a measure of how a phase boundary composition changes with temperature. For instance, if (dr_{PB}/dT) is zero (i.e., a vertical phase boundary) then $\Delta H_{\text{plat}} = \Delta H_{\text{H,PB}}$ and, if this is not found experimentally, either the enthalpies or the phase boundary shape is incorrect. $(\partial(\Delta\mu_H/T)/\partial r)_T$ must always be positive because it is a thermodynamic stability condition. Eq. (5) applies for any boundary in a T, r phase diagram.

With reference to the $\alpha/(\alpha + \beta)$ and $\alpha/(\alpha + \delta)$ phase boundaries in the Zr–H system just above and below the eutectic temperature at $r = 0.1$ (Fig. 15.3), (dr_{PB}/dT) can be predicted from the known ΔH values. The employment of Eq. (5) for such a situation is relatively reliable because the values of $(\partial(\Delta\mu_H/T)/\partial r)_T$ and $\Delta H_{\text{H,PB}}$ will be the same above and below the eutectic temperatures at each boundary since they both lie in the *same* α phase region. The plateau values are: $\Delta H_{\text{plat}} (\alpha \rightarrow \beta) = -40.9$ and $\Delta H_{\text{plat}} (\alpha \rightarrow \delta) = -87$ kJ/mol H and, as noted, $\Delta H_{\text{H,PB}}$ will be the same for both boundaries (i.e., -49.3 kJ/mol H) [7]. If these values are inserted into Eq. (5), it is found that (dr_{PB}/dT) will be negative for the $\alpha/(\alpha + \beta)$ boundary and positive for the $\alpha/(\alpha + \delta)$ boundary. The phase diagram (Fig. 15.3) agrees with these results and it predicts that the magnitude of (dr_{PB}/dT) is greater for the latter boundary which is also as observed.

Eq. (5) can also be applied in the same way to calculate (dr_{PB}/dT) just above and below the eutectic temperature at the $\delta/(\delta + \beta)$ and $\delta/(\delta + \alpha)$ boundaries (Fig. 15.3), however, $\Delta H_{\text{H,PB}}$ is not as well established for the δ phase, compared to the α phase value of $\Delta H_{\text{H,PB}}$. When the available thermodynamic values for these two boundaries are inserted into Eq. (5), then $(dr_{\text{PB}}/dT) > 0$ for the $\delta/(\beta + \delta)$ boundary and also positive for the $\delta/(\alpha + \delta)$ boundary. The latter is probably incorrect and this is probably due to the magnitude of $\Delta H_{\text{H,PB}}$ in this region which is close to that for $\Delta H_{\text{plat}} (\alpha \rightarrow \delta)$. In any case (dr_{PB}/dT) should be rather vertical for this and for the $\delta/(\beta + \delta)$ boundary as found (Fig. 15.7).

The slopes of the two phase boundaries $\beta/(\beta + \alpha)$ and $\beta/(\beta + \delta)$ near the eutectic composition and temperature can also be predicted from Eq. (5) and the slopes are $+$ and $-$ for the two boundaries, respectively, in agreement with the phase diagram and the slope for the $\beta/(\beta + \delta)$ boundary is greater than that of the $\beta/(\alpha + \beta)$ boundary also in agreement with the phase diagram (Fig. 15.3).

4 TECHNIQUES EMPLOYED FOR PHASE DIAGRAM DETERMINATION OF M–H SYSTEMS

Either open or closed thermodynamic systems can be employed for the determination of M–H phase diagrams. In closed systems phase boundaries can be located by varying the temperature at a given r to obtain T–r phase diagrams. A closed system requires sealing the hydrogen into the M–H system so conditions of iso-H concentration will obtain during the heating or cooling. This is, of course, easiest for a very stable hydride with a low p_{H_2}.

An open system requires that the M–H system be in equilibrium with an H reservoir which can supply H at a given p_{H_2} or its equivalent, a given electrochemical potential. In an open system phase diagrams in a either T–r or $\Delta\mu_H$–T representation can be obtained.

The behavior of metals toward hydrogen can, in some cases, be strongly affected by impurities. For example, the hydrogen solubilities in VB metals are strongly affected by interstitial impurities such as O, N and C. Walter and Chandler [38] found that when they employed less pure Nb (99%), for example, the temperature of the triple phase line ($\alpha - \alpha' - \beta$, see below) was 338.5K and composition was H/Nb = 0.515 whereas for purer Nb (99.5%), these values changed to 341.3K and H/Nb = 0.585. Workers at KFA (Julich) took great precautions to eliminate impurities especially interstitial ones in the group VB metals and they found that this triple phase line was at 361K and H/Nb = 0.52 values which differ significantly from the results for "pure" Nb used by Walter and Chandler. Wenzl and Welter [39] described details of the procedure used by them for the purification of Nb; V and Ta should be purified similarly. The lower temperature phases in these Group VB–H systems can be obtained by cooling a sample with an appropriate r value from a higher temperature because it is impossible to prepare them directly from the metal and H_2 because other phases will form first and the system will remain in such a metastable state. The behavior of Zr and Ti toward H_2 is also strongly affected by interstitial impurities such as O [40].

It should be pointed out that information about a phase diagram such as the extent of two-phase regions and other information can be determined from, for example, cooling/heating curves and/or differential thermal analysis (DTA), but, in the absence of knowledge of the structures of the various phases, a phase diagram is incomplete. Therefore it is important that the structures be identified and this is, perhaps more difficult for M–H systems than "all-metal" systems because of the variety of ordered H phases which can form at low temperatures.

4.1 Open Systems

4.1.1 P_{H_2}–r–T Technique

In an open system an H reservoir is in contact with the sample and $\Delta\mu_H$ can be measured isothermally as a function of r from the p_{H_2} in equilibrium with the M–H condensed phase using

$$\Delta\mu_H (r, T) = \tfrac{1}{2}\Delta\mu_{H_2} = RT \ln p_{H_2}^{1/2} \qquad (6)$$

Such p_{H_2}–r or $\Delta\mu_H$–r isotherms can be determined over a temperature range and they can be measured either by volumetric or gravimetric methods. The volumetric method where the H uptake/evolution is obtained from p_{H_2} changes in known volumes is the most widely employed and such an apparatus is commonly referred to as a Sieverts apparatus. The volumetric approach is quite adequate when the H content changes are relatively large for small pressure changes. Several different computer-controlled Sieverts' systems are now commercially available. If there is a small error in the volumes, however, every H content is affected, that is, its effect is cumulative and this is especially serious in regions where the H changes are small and the pressures are large, as in pure, single-phase hydride regions. Pressures can be recorded either during absorption and or desorption of hydrogen where the H content is determined from the ideal gas law and the p_{H_2} changes in known volumes, however, for p_{H_2} greater than ≈ 20 bar at moderate temperatures, corrections should be made for the H_2 non-ideality especially if a series of absorptions/desorptions are carried out because errors due to uncorrected non-ideality will be additive. The gravimetric method is superior for regions of the isotherms where the H changes are small and the p_{H_2} is large. Gravimetric methods must allow for the effect of buoyancy. Balances are available which operate to ≈ 50 bar for example, the "Intelligent Gravimetric Analyzer" [41]. A different gravimetric technique is the quartz microbalance where the weight is derived from the frequency shift of a vibrating quartz disc. A problem has been reported with this technique [42] which is believed to be due to the effect of changes in the binding of the sample to the quartz disc on its frequency, on the other hand, it appears to have been successfully employed elsewhere for the determinations of isotherms [43].

It was shown by Clewley et al. [44] that it was quite possible to obtain equilibrium between H_2 and Pd or Pd alloys below about 393K by using an all-metal, Hg-free system. Prior to then (~1973), this was not possible because of the use of glass systems containing Hg and grease which poison the sample surface by inhibition of the $H_2 \leftrightarrow 2H$ (surface) reaction. With an all-glass system Wicke and Nernst [45] were able to obtain equilibrium between H_2 and Pd and its alloys at about 323K and higher through the use of so-called H transfer catalysts, for example, Cu power. The introduction of transfer catalysts can be avoided, however, by employing metal systems where equilibrium can be obtained more rapidly and at lower temperatures.

From the p_{H_2}–r isotherms a T–r phase diagram can be constructed because any constant p_{H_2} region, a plateau, indicates a two condensed phase co-existence and the beginning and the end of the plateau corresponds to the phase boundaries for that co-existence region at that temperature and, if it is a miscibility gap system, the critical point can be located from the temperature where the plateaux disappear. This requires isotherm measurements over a wide temperature range. For example, the T–r phase diagram for Pd–H was determined in this way at an early date, 1933, by Brüning and Sieverts [46] from desorption isotherms and the critical point was located. A low temperature portion of the Pd–H phase diagram, where p_{H_2} is too small to measure, was more recently discovered by other means [47]. It was clear that there was an anomaly at about 50K based on the specific heat, electrical resistivity and internal friction behavior in this region. This phase was characterized by single crystal neutron diffraction [47] and was an ordered PdH_x phase with $x = 0.8$. An example of the structure transformation type is shown in Fig. 15.3(b) for the Zr–H binary system.

McQuillan [4] employed p_{H_2}–r–T measurements for the first detailed phase diagram determination of the Ti–H system. For a system such as Ti–H, which has quite low p_{H2} at ambient temperatures, a phase diagram determining in this way is limited to relatively high temperatures ≥ 773K which, of course, aids in attaining gas–solid equilibrium which will be sluggish at lower temperatures with an all-glass system as employed by McQuillan. He located the ($\alpha + \beta$) and the ($\beta + \gamma$) co-existence regions from the pressure invariant regions. A eutectic was predicted where the α, β and δ phases co-exist and which was subsequently verified. Haag and Shipko [48] extended p_{H_2}-r-T measurements on Ti–H down to 573K. A very detailed, up to date, account of this phase diagram is given by Manchester [7].

Another approach using p_{H_2}-r-T methods is to keep r essentially constant and follow the p_{H_2} changes while varying the temperature. In other words, with respect to p_{H_2}, this is an open system although r itself does not change appreciably. If the system is initially in a single-phase region, any discontinuities in these plots with decrease/increase of temperature indicate the intersection with a two-phase co-existence region. Experimentally relatively large samples should be employed and the equilibrium p_{H2} should be relatively low, but measurable, so that r is kept effectively constant. This technique was employed by Kishimoto et al. [15] for Pd–H and $Pd_{0.095}Ag_{0.05}-$, $Pd_{0.09}Ag_{0.10}-$, $Pd_{0.95}Ag_{0.05}-$ and $Pd_{0.09}Ag_{0.10}$–H alloys where, because the $|\Delta H|$ is relatively small, the temperatures had to be low to maintain a low p_{H2} in order to keep r constant. For Pd–H the temperature range was from 200K to 273K where the pressures were 1–500 Pa and for the alloys the ranges were slightly different, dependent upon the specific alloy. At each "fixed" concentration (e.g., H/Pd = 0.0065) the sample is cooled and p_{H_2} measured at temperature intervals until its intersection with the plateau van't Hoff plot and upon further cooling the data follow the van't Hoff plot. The intersection is the solvus or terminal H solubility at the temperature of the intersection (Fig. 15.5). Other intersections at different temperatures map out the $\alpha/\alpha+\alpha'$ phase boundary. Similar characterizations have been made for systems with large $|\Delta H|$ (e.g., rare earth-H systems) where because of the great stability of these hydrides the temperatures must be high [16] and the p_{H2} are still relatively low.

For measurements of p_{H_2} as a function of r, thermal transpiration effects can be appreciable if the sample and pressure gauge are at different temperatures. McQuillan neglected to make the appropriate corrections for the low p_{H_2} range in his initial investigation of the Ti–H system but subsequently realized its importance, and made the necessary corrections at low p_{H_2} [49]. Appropriate corrections must be made for this effect especially if the pressure gauge is at ambient temperature and the connecting tubing narrow, and p_{H_2} small. Bennett and Tompkins [50] give an equation for making such corrections and the Wallbank and McQuillan paper [49] should be referred to where they correct their data for Ti–H. Knudsen [51] gives the equation $P_1/P_2 = (T_1/T_2)^{\gamma}/2$ where subscript 1 refers to the true equilibrium pressure at the sample temperature T_1 and the subscript 2 to the measured pressure at the temperature of the gauge at T_2 and $\gamma = (1+d/\lambda)^{-1}$ where d and λ are the diameter of the connecting tube and the mean free path, respectively.

4.1.2 E–r–T Technique

Instead of measuring equilibrium p_{H_2} as a function of r, the metal's electrode potential may be measured in the appropriate solution where H is introduced into the sample by the electrochemical discharge of H^+. The relation between E_{eq} and p_{H_2} is given by

$$-nFE_{eq} = \tfrac{1}{2}RT(\ln p_{H_2}/p°_{H_2}) = \Delta\mu_H \qquad (7)$$

where E_{eq} is the electrode potential of the M–H system with respect to an electrode of known potential. Kirchheim [52] has employed a mixture of two volumes of glycerin and one volume of concentrated phosphoric acid as the solution because its viscosity inhibits the loss of H_2 from the sample especially by slowing the diffusion of dissolved O_2 up to the sample surface. As a reference electrode, calomel was employed. For Pd and Pd alloys electrochemically charged at relatively low currents ($\approx 5\,\text{mA/cm}^2$), 100% of the discharged H^+ enters the sample at E_{eq} values corresponding to $p_{H_2} < 1$ bar [52]. The electrochemical method must give the same results as the gas phase approach but for temperatures >350K it is inconvenient for the determination of $\Delta\mu_H$–r relationships. When equilibrium p_{H_2} are very small, it has an advantage compared to the p_{H_2}–r–T approach because of the logarithmic relationship, that is, p_{H_2} can be so small as to be difficult to measure while E_{eq} is still quite measurable. The electrochemical method is most suitable for Pd and its alloys or for metals which have been coated with thin layers of Pd which can be reversible hydrogen electrodes. For example, Boes and Züchner [53] measured H_2 isotherms electrochemically for V, Nb and Ta in the temperature range 275–343K where these metals were coated with a thin layer of Pd prior to the electrochemical determinations. The p_{H_2} for these Group VB metals can be very small (e.g., at 275K) for the first plateau of the Nb–H system is 1.3×10^{-6} Pa. This would be nearly impossible to measure using the p_{H_2}–r–T method but is readily measured electrochemically.

4.1.3 Reaction Calorimetry

Increments of H_2, δn_H, are added to/or removed from an M–H sample in this method and the accompanying δq measured and, if small increments of H are used, $(\delta q/\delta n_H) = \Delta H_H$. Usually twin cell, heat leak calorimeters have been employed where the heat of reaction leaks into the adjacent part of the calorimeter mainly via the thermopile and this heat is detected by the temperature imbalance between the twin cells sensed by a thermopile. One of the cells contains the sample and the other a non-H absorbing material similar in size and heat capacity as the sample. When ΔH_H is plotted as a function of r, information about the phase diagram can be obtained as explained above and in the paper by Luo and Flangan [37] because the dependence of ΔH_H on r must be compatible with the phase diagram. In addition, of course, these calorimetric measurements will provide thermodynamic information about the system and, it should be recalled, there is an intimate relationship between the thermodynamics and the phase diagram because the former leads to the latter. A pioneer in employing simultaneous reaction calorimetric and isotherm measurements is Kleppa [54]. More recently, Dantzer and Millet [55] has constructed a computer-controlled calorimetric system and has also considered in detail some of the accompanying gas transfer problems.

4.2 Closed Systems

In a closed system r is fixed during a given set of measurements and p_{H_2} ($\Delta\mu_H$) cannot be measured. This has the advantage that gas/solid equilibration is not needed which may, in any case, be very slow at low temperatures but information can still be obtained for these closed systems from the physical changes during cooling or heating. These methods are most suitable for very stable M–H systems which can be charged with H to a definite r and then heated or cooled whilst maintaining the fixed H content. The r contents can be changed and the process repeated until a region of a phase diagram has been determined. Any technique which can detect abrupt changes in the system such as a first order phase change can be used to determine phase diagrams of M–H systems but some techniques are, of course, more sensitive and provide more information than others. Often such samples can be exposed to the atmosphere while making measurements. The amount of H in the sample can be determined by a higher temperature loading at a known p_{H_2} provided that the appropriate isotherm is available. After loading, the sample can be quickly cooled which effectively seals the H in the metal. H can be introduced electrochemically and its amount determined by weighing the sample although this, perhaps, introduces more error than the gas loading technique. Another method of analysis is by vacuum extraction of the hydrogen. Wenzl and Welter [39] describe this procedure for Nb–H. The H-containing Nb is inserted into an UHV apparatus, which originally has a vacuum of $\approx 10^{-4}$ Pa, and then heated to about 1200K evolving all of the H which is then measured with the appropriate pressure gauge in a known volume.

4.2.1 Thermal Analysis

This method has been extensively employed at lower temperatures for the group VB–H systems which form very stable hydride phases [25]. Sharp thermal signals from DTA occur during cooling or heating when one phase transforms into another or into two phases such as a eutectoid transition during cooling or vice versa during heating. Signals which extend over a range of temperatures result from cooling an H-containing sample in a single phase such as the β phase of the Zr–H system at $r = 0.2$ to intersect the (α + β) two-phase region which continues until the eutectic temperature is reached. The thermal signal is spread over this temperature range. The smaller is r in the β phase, the sharper will be the signal.

Differential scanning calorimetry (DSC) can also be employed and the areas of the signals arising from the cooling or heating when one phase transforms into another or into two phases can be used to determine the ΔH of the transition. Pressure DSC equipment (PDSC) are now available where the sample can be hydrided in situ and the temperature scans can be carried out at fixed p_{H_2}. PDSC equipment can generally operate either above or below 1 bar H_2.

The Nb–H phase diagram is illustrated in Fig. 15.10 in the T against r representation after Schober [56] and Kobler and Welter [57]. This diagram is shown because of its myriad of low temperature phases (Fig. 15.10(b)) whose presence have been detected mainly by DTA and magnetic susceptibility [57]. Some of the phases in Fig. 15.10(b) are tentative and Schober [56] has questioned the existence of the η and θ phases.

4.2.2 Diffraction Methods

XRD can only determine the lattice type and dimensions of the metal sub-lattice in M–H systems. Lattice parameter measurements generally employ powder samples which is particularly convenient for most intermetallic compounds which disintegrate into powder when hydrided. The Bond technique [58] can be used to obtain very accurate lattice parameters. The determination of phase diagrams by XRD involves measurements with a series of samples containing known amounts of H and a large number may be needed. It must be possible to make the XRD measurements at different temperatures in order to generate a phase diagram.

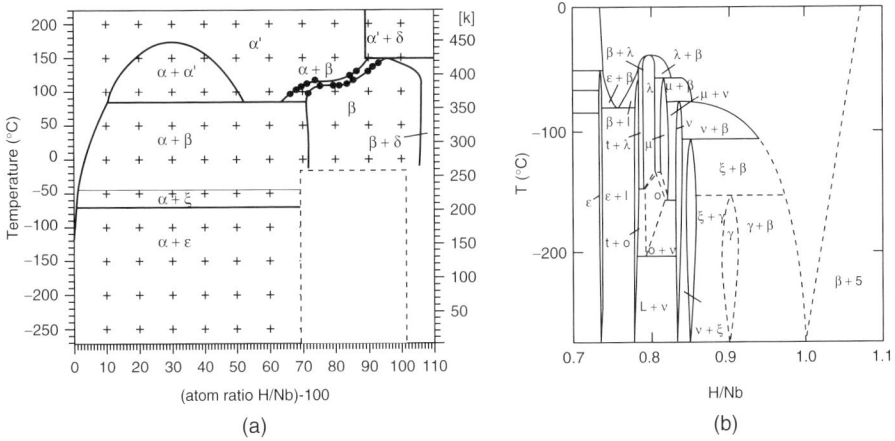

Figure 15.10 The Nb–H phase diagram (a) from Ref. [59] and (b) from Ref. [60].

The first appearance of a second lattice parameter in the dilute phase indicates the onset of hydride formation, that is, the dilute/(dilute + hydride) phase boundary, and when the H content is increased further until one of the parameters disappears which indicates the limit of the two-phase region, the hydride/(hydride + dilute) phase boundary. This method was employed by Walter and Chandler [38] for the Nb–H system. It is relatively easy to detect two-phase regions with XRD but more difficult to accurately delineate the extent of the two-phase co-existence region.

Wise and co-workers [59] employed XRD isothermally for the determination of the critical compositions of a series of fcc-Pd alloy–H systems. They electrochemically loaded the alloys with H. In single-phase regions the lattice parameter, a_o, is a continuous function of r and in two-phase regions there are two lattice parameters which should have constant values across the two-phase plateau. In this technique if the H contents of the alloys are within the two phase miscibility gap, two fcc lattice parameters are apparent. By measuring the lattice parameters of the dilute and hydride phases at a series of X_M (atom fraction metal) values for M = Sc, Ti, Fe, Y, Zr, Ag, In, Ce, Eu and Gd they determined the approximate critical compositions of the alloying elements where hydride formation no longer takes place at 298K. Sakamoto et al. [33] used the same approach for a series of Pd-rich alloys with M = B, Sn, Ti, Zr, Nb, Ta, Pt, Ru and Rh.

It is also possible to employ an open XRD system where r can be varied systematically by changing p_{H_2} while the sample is directly connected to the diffractometer. In most such set-ups the temperature can also be varied. This is an efficient way to investigate a system because one sample suffices. In the apparatus employed by Gavra and Murray [60] the powder sample was in a capillary tube and the accessible p_{H_2} range was 0–5 MPa and temperature range 190–575K. Their sample set-up has the advantage of being mounted temporarily in any multi-use diffractometer. A larger sample can be placed somewhere else in the apparatus to simultaneously measure the p_{H_2}–r isotherm. This XRD method is versatile because it is not limited to samples which are very stable since they are enclosed in the desired H_2 atmosphere.

As is well known, it is difficult with XRD to detect H (D) in metals. Neutron scattering does not depend on the number of electrons (atomic number) but on scattering from the nucleus. Neutron diffraction is well suited for D because its coherent to incoherent scattering ratio is much greater than that for H making it suitable for structural determinations. Neutron diffraction is essential for the characterization of structures of low temperature phases where the D atoms are ordered in the metal sub-lattice. In some cases there is sufficient neutron scattering from H to be detected by neutron diffraction, for example, Somenkov et al. characterized the H sub-lattice in the β phase of Nb–H [61]. Neutron scattering has proven to be an invaluable tool for the study of M–H systems because in addition to the determination of the structures, the H vibrational frequencies and the diffusion constants can be determined.

Many low temperature phases of Nb–H (D) have been characterized by neutron diffraction (Fig. 15.10). For example, the ε phase, Nb_4H_3, has been characterized by both neutron diffraction [61] and transmission electron microscopy (TEM) [62]. It represents an ordering of vacancies starting from the ordered β phase within an orthorhombic unit cell of Nb.

4.3 Electron Diffraction and TEM

In conjunction with TEM, electron diffraction can be carried out and this has the advantage that selected areas can be examined, ≈2 μm; selected area diffraction (SAD) is very useful because this size of the area examined corresponds approximately to that of typical domains in poly-crystalline samples which allows single crystal domains to be examined within a multi-crystalline material. The scattering of electrons from the H in electron diffraction is large enough to observe reflections due to H ordering [62]. Examination of the surfaces is also useful for the identification of phases [25] because the different phases exhibit different morphologies.

4.4 Magnetic Susceptibility

Magnetic susceptibility has been employed by Kobler and Welter [57] to follow phase transitions of hydrogenated Nb and Ta but it is not widely employed.

4.5 Electrical Resistivity

This is frequently employed for the determination of the solvus or terminal H solubilities. The electrical resistivity changes slope upon entering or leaving a two-phase region during cooling or heating [63]. Welter and Schöndube [64] employed this method for the Nb–H system.

4.6 Dilatometry

Very accurate dilatometric techniques are now available in which the length of the sample can be monitored with temperature changes and this will indicate any first order phase changes [52].

ACKNOWLEDGEMENTS

The authors are indebted to Professor W.A. Oates and Professor C.-N. Park for explaining many features of the thermodynamics of M–H systems.

REFERENCES

1. C. Hoitsema and H.W.B. Roozeboom, *Z. Phys. Chem.*, 17 (1895) 1.
2. I. Anderson, C. Carlile and D. Ross, *J. Phys. C. Solid State Phys.*, 11 (1978) L381–L384.
3. W.A. Oates, *J. Less-Common Met.*, 130 (1987) 339–350.
4. A.D. McQuillan, *Proc. Roy. Soc.*, A204 (1950) 309–323.
5. W. Mueller, J. Blackledge and G. Libowitz (eds.), *Metal Hydrides*, Academic Press, New York, 1968.
6. Hydrogen Metal Systems, F. Lewis and A. Aladjem (eds.), *Solid State Phenomena*, Trans Tech Publ., Zurich 49–50, (1996) and 73–75, (2000).
7. F.D. Manchester (ed.), *Phase Diagrams of Binary Hydrogen Systems*, American Society for Metals, Metals Park, Ohio, 2000.
8. P. Rudman, *Int. J. Hydrogen Energ.*, 3 (1978) 431–447.
9. H. Wagner and H. Horner, *Adv. Phys.*, 23 (1974) 587–637.
10. Y. Fukai, *The Metal–Hydrogen System*, Springer-Verlag, Heidelberg, 1993.
11. Y. Fukai, *J. Less-Common Met.*, 172–174 (1991) 8–19.
12. V. Antonov, *J. Alloy. Compd.*, 330–332 (2002) 110–116.
13. T. Flanagan, W. Oates and C.-N. Park, *Prog. Solid State Chem.*, 23 (1995) 291–363.
14. W. Oates and T. Flanagan, *J. Less-Common Met.*, 92 (1983) 131.
15. S. Kishimoto, W. Oates and T. Flanagan, *J. Less-Common Met.*, 88 (1982) 459–468.
16. K. Toguchi, M. Tada and Y. Huang, *J. Less-Common Met.*, 88 (1982) 469–478.
17. O. Katz and E. Gulbransen, *J. Chem. Ed.*, 37 (1960) 533–536.
18. G. Sandrock, *J. Alloy. Compd.*, 293–295 (1999) 877–888.
19. H. Ellingham, *J. Soc. Chem. Ind. (London).*, 63 (1944) 125.
20. W. Oates and T. Flanagan, *Scripta Met.*, 17 (1983) 983.
21. T. Flanagan and W.A. Oates, in L. Schlapbach (ed.), *Hydrogen in Intermetallic Compounds I*, Springer-Verlag, Berlin, 1988, p. 49.
22. C.-N. Park and T. Flanagan, *Ber. Bunsen. Phys. Chem.*, 89 (1985) 1305.
23. K. Aoki, T. Yamamoto, Y. Satoh, K. Fukamichi and T. Masumoto, *Acta Metall.*, 35 (1987) 2465–2470.
24. K. Semenenko, V. Malyshev, V. Petrova, L. Burnasheva and V. Sarynin, *Izv. Adad. Nauk. SSSR Inorg. Mater.*, 13 (1978) 2009–2013.
25. T. Schober and H. Wenzl, in G. Alefeld, J. Volkl (eds.), *Hydrogen in Metals II*, Springer-Verlag, Berlin, 1978.
26. R. Kadel and A. Weiss, *Ber. Bunsen. Phys. Chem.*, 82 (1978) 1290.
27. W.A. Oates and T. Flanagan, *J. Less-Common Met.*, 100 (1984) 299–305.
28. M. Arita, R. Kinaka and M. Someno, *Metall. Trans. A*, 10 (1979) 529.
29. W.A. Oates and T. Flanagan, *Mater. Res. Bull.*, 16 (1981) 3235.
30. W.A. Oates and T. Flanagan, *Scripta Met.*, 17 (1983) 983.
31. T. Flanagan, J. Clewley, H. Noh, J. Barker and Y. Sakamoto, *Acta Metall.*, 46 (1998) 2173–2183.
32. M. Hillert, *JKA-Jernkotoret. Ann.*, 136 (1952) 25.
33. Y. Sakamoto, K. Baba and T. Flanagan, *Z. Phys. Chem. Neue. Fol.*, 158 (1988) 223–235.
34. G. Rosenhall, *Ann. Phys. New York.*, 24 (1935) 297–324.
35. H. Noh, J. Clewley, T. Flanagan and A. Craft, *J. Alloy. Compd.*, 240 (1996) 235–248.
36. W.-E. Wang and D. Olander, *J. Am. Ceram. Soc.*, 78 (1995) 3323–3328.
37. W. Luo and T. Flanagan, *J. Phase Equilib.*, 15 (1994) 20.
38. R. Walter and W. Chandler, *Trans. Met. Soc. AIME*, 233 (1965) 762.
39. H. Wenzl and J.-M. Welter, *Curr. Topics in Mat. Sci.*, 1 (1978) 603–657.

40. S. Yamanaka, T. Tanaka and M. Miyake, *J. Nucl. Mater.*, 167 (1989) 231–237.
41. P. Stonadge, M. Benham, D. Ross, C. Manwaring, I. Harris, *Z. Physik. Chem.*, 181 (1993) 125–131.
42. R. Feenstra, D. de Groot, J. Rector, E. Salomons and R. Griessen, *J. Phys. F. Met. Phys.*, 16 (1986) 1953–1963.
43. K. Papathanassopoulos and H. Wenzl, *J. Phys. F Met. Phys.*, 12 (1982) 1369–1381.
44. J. Clewley, T. Curran, T. Flanagan and W.A. Oates, *J. Chem. Soc. Faraday Trans. I*, 69 (1973) 449–458.
45. E. Wicke and G. Nernst, *Ber. Bunsen. Phys. Chem.*, 68 (1964) 224–235.
46. H. Brüning and A. Sieverts, *Z. Phys. Chem.*, 163A (1933) 409.
47. I. Anderson, D. Ross and C. Carlile, *Phys. Lett. A*, 68 (1978) L381.
48. R.M. Haag and F.J. Shipko, *J. Am. Chem. Soc.*, 78 (1956) 5155–5160.
49. A. Wallbank and A. McQuillan, *Trans. Faraday Soc.*, 71 (1975) 685.
50. M. Bennett and F.C. Tompkins, *Trans. Faraday Soc.*, 53 (1957) 185.
51. M. Knudsen, *Ann. Phys.*, 31 (1910) 633.
52. R. Kirchheim, *Prog. Mater. Sci.*, 32 (1988) 261–325.
53. N. Boes and H. Züchner, *Ber. Bunsen. Phys. Chem.*, 80 (1976) 22–27.
54. O.J. Kleppa, *Ber. Bunsen. Phys. Chem.*, 87 (1983) 741.
55. P. Dantzer and P. Millet, *Rev. Sci. Instrum.*, 71 (2000) 142.
56. T. Schober, *Hydrogen Metal Systems*, vol. 1, F. Lewis and A. Aladjem, (eds.), *Solid State Phenomena*, Trans Tech Publ., Zurich 49 (1996) 331–422.
57. U. Kobler and J.-M. Welter, *J. Less-Common Met.*, 84 (1982) 225–235.
58. W.L. Bond, *Acta Crystallogr.*, 13 (1960) 814.
59. M.L.H. Wise, J.P.G. Farr and I.R. Harris, *J. Less-Common Met.*, 41 (1975) 115–127.
60. Z. Gavra and J. Murray, *Rev. Sci. Instrum.*, 57 (1986) 1590.
61. V. Somenkov, V. Petunin, S. Shil'stein and A. Chertkov, *Sov. Phys. Crystallogr.*, 14 (1970) 522.
62. T. Schober, *Phys. Status. Solidi. A.*, 80 (1976) 107.
63. M. Pershing, G. Bambakidis, J. Thomas and R. Bowman, *J. Less Common Met.*, 75 (1975) 207–222.
64. J.-M. Welter and F. Schöndube, *J. Phys. F Met. Phys.*, 13 (1983) 529.

CHAPTER SIXTEEN

Miscellaneous Topics on Phase Diagrams

J.-C. Zhao[1] and Jack H. Westbrook[2]

Contents

1 Introduction	483
2 Ever-Increasing Interplay of Modeling and Experiment	484
3 Phase Diagrams for Functional Materials	485
4 High-Throughput Approaches to Phase Diagram Determination	486
5 Low-Temperature Phase Diagrams	487
6 Phase Diagram Determination: A Never-Ending Task	488
7 Some Suggestions	489
Further Reading	490
General Phase Diagram Introduction	490
Specimen Purity and Preparation	490
Diffusion as Related to Phase Diagrams and Phase Diagram Determination	490
Interplay of Ordering, Magnetic Transitions, and Phase Equilibria	491
Ceramic Phase Diagram Determination	491
Theoretical/Modeling Work	491
The *CALPHAD* Approach	491
Miscellaneous Recommended Papers on Phase Diagram Determination	491
Some Current Phase Diagram Compilations	492
Elemental and Metallic Systems	493
Ceramic Systems	493
Other Miscellaneous Compilations/Books	493
Some Useful Websites	493

1 Introduction

The 15 prior chapters of this book have covered many different experimental methods for phase diagram determination, some useful information on the correct phase diagram features based on thermodynamics, and some basics on phase transformation kinetics that are important to phase diagram determination and assessment.

[1] GE Global Research, Niskayuna, NY, USA
[2] Brookline Technologies, Ballston Spa, NY, USA

Still, there are a few important topics that have not been covered. This chapter tries to tie up a few loose ends by briefly discussing these topics, including the ever-increasing interplay of modeling and experiment, low-temperature phase diagrams, and the existing compilations of phase diagrams. A recommended list of further reading materials is also provided.

2 EVER-INCREASING INTERPLAY OF MODELING AND EXPERIMENT

A very positive and important trend in phase diagram determination is the ever-increasing interplay of modeling and experiment. In Chapter 8, Chang and Yang illustrated the application of calculations of phase diagram (*CALPHAD*) modeling in efficient determination of phase diagrams. By putting together an approximate thermodynamic description based on the constituent binary systems, an approximate ternary or high order phase diagram can be calculated with assumptions that the solubilities of the third or higher order elements in the binary intermetallics are negligible and there are no ternary or higher order compounds. A few selected alloys can be strategically selected to provide key experimental information to maximize the importance of these data points in improving the thermodynamic parameters of the system. The key experiments are designed to find out whether ternary and higher order compounds exist and to examine the solubility of the third or higher order elements in binary or ternary intermetallics. Such information is used as input for the next iteration of thermodynamic modeling. This approach will become more widely used in the future.

Alloys are becoming increasing complex, for example, Ni–base superalloys routinely contain 8–11 elements for balance of various properties such as yield strength, ductility, creep strength, fatigue resistance, and environmental resistance. For such alloys, the phase equilibria need to be studied in multicomponent systems, far beyond the basic binary and ternary phase diagrams. Thermodynamic modeling such as the *CALPHAD* approach becomes imperative. More importantly, with known thermodynamic parameters, thermodynamic calculation can provide multicomponent phase equilibrium data (phase formation, phase compositions, volume fractions, etc.); can define the driving force for precipitation modeling; the segregation, specific heat, and latent heat data for solidification modeling; and thermodynamic factors for constructing the diffusion coefficient matrices for multicomponent alloys. Thermodynamic databases are also linked to phase field and other models to predict microstructure evolution and certain properties in alloys.

The binary and ternary phase diagrams are essential experimental data for reliable assessment of the thermodynamic parameters for prediction of multicomponent phase equilibria. Since the ultimate goal is to obtain the thermodynamic parameters for the Gibbs energy functions of various phases – not phase diagrams per se, it makes perfect sense to have an iterative process as shown by Chang and Yang in Chapter 8 for the determination of phase diagrams. Such an approach is also practiced by other groups.

CALPHAD modeling has also helped to reveal/predict phase diagram features that otherwise would be hard to be aware of or could be easily missed during experimental investigations. A defining case is the magnetic transition-induced miscibility gap (Nishizawa horn), and an example is shown in Fig. 2.13(b) for the Co–V binary system. Such a miscibility gap was predicted from thermodynamic modeling long before definite experimental confirmation was made [1]. A similar miscibility gap is also shown in Fig. 12.9. Without thermodynamic modeling, it would have been hard to imagine such a miscibility gap and to understand the phase diagram features and materials behavior around such a region.

Thermodynamic modeling is the foundation upon which many incorrect phase diagram features are uncovered and discussed in Chapter 3. The advancement of thermodynamic modeling made such assessments possible.

The coupling of theoretical modeling and experimental investigations has also been very important in understanding complex phase diagram features involving both ordering and magnetic transitions. The example discussed in Fig. 12.11 for the Fe–Si binary system clearly illustrates this point [2]. Without the modeling insight, it would have been difficult to pinpoint the phase diagram features.

Quantum mechanical, especially density functional theory (DFT), calculations are widely used to calculate: the heat of formation (at 0K), phonon dispersion (thus estimation of the specific heat), elastic modulus, electronic density of state, and other parameters. The accuracy of such calculations becomes very impressive, especially for the heat of formation for intermetallic compounds at 0K or the ground state energy of these compounds. A combination of quantum mechanical calculations with data mining can predict the crystal structure of intermetallic compounds in binary systems with very high reliability. With further advancement, it may be able to do so for ternary systems in the future. This would be extremely valuable for ternary and higher order phase diagram determination, since one of the key tasks is to find out whether ternary or higher order intermetallic compounds exist in a certain system. Reliable prediction would make a big difference in reducing the experimental work. Some references related to this subject can be found in the recommended reading list.

3 PHASE DIAGRAMS FOR FUNCTIONAL MATERIALS

Functional materials such as magnetic, magnetocaloric, magnetoelastic, spintronic, thermoelectric, piezoelectric, and dielectric materials are becoming an increasingly important part of our daily life for sensing, actuation, computing, energy conversion, communication, and so on. These materials take advantage of their unique physical, chemical, and structural properties and very often involve complex interplays of structural, magnetic, and electronic transitions with external or induced stress field, magnetic field, electrical field, and thermal field. It is not surprising that their related phase diagrams are often very complex. One can take a look at the phase diagrams of $LaMnO_3$–$SrMnO_3$ and $LaMnO_3$–$CaMnO_3$ (Figs. 12.12 and 12.20) to appreciate the complexity of such systems. Since such systems are becoming increasingly important to our modern society, the phase diagram community needs to position itself to face the challenges of measuring such complex systems.

Many functional materials are used in the thin-film format in real applications. In this case, a non-equilibrium state and/or a constrained state very often exist, and the substrate on which the film is deposited can have a very strong effect on the phase formation in the film. This opens new opportunities for fine-tuning the structure–property relationships of thin-film functional materials; at the same time it complicates the phase diagram studies. One needs to consider the effects of the substrate, especially the coherency strain. One example is the work of Li et al. [3] on the $PbTiO_3$–$PbZrO_3$ system. The phase diagrams are significantly different under none, positive, or negative strain constraint conditions relative to the substrate. A more detailed "phase diagram" under different substrate strain conditions for the $PbZr_{0.53}Ti_{0.47}O_3$ composition was constructed by Choudhury et al. [4]. Phase field modeling greatly helps to define the overall topology of these phase diagrams (these serve as more examples of the importance of the interplay of theory and experiment).

4 HIGH-THROUGHPUT APPROACHES TO PHASE DIAGRAM DETERMINATION

The legendary William Hume–Rothery once said that it took a master's degree student's effort to determine but one isothermal section of a ternary phase diagram. This is why only about 5% of possible metallic ternaries, and fewer than 0.1% of all quaternary phase diagrams, have been even partially investigated [5]. Attempts to accelerate phase diagram determination can be traced back to at least half a century ago. Boettcher et al. [6] in 1955 deposited thin films with composition gradients and used electron diffraction to identify the phases in order to construct the phase diagrams. They tested the methodology using the Ag–Sn and Au–Sn binary systems, and then mapped the phase diagram of the Ag–Pb–Sn ternary system. The phase diagram so determined compares well with the equilibrium phase diagram calculated from the *CALPHAD* approach [7,8]. A decade later, Kennedy et al. [9] used a similar approach to map the Cr–Fe–Ni ternary phase diagram. They evaporated pure Cr, Fe, and Ni from three different ingots using electron-beam physical vapor deposition (EB–PVD) to create a thin film with varying ternary compositions on a substrate. They then cut the thin film on the plate into small pieces to perform compositional analyses and phase identification. In this case, the substrate was heated to 760°C and all the phases formed were equilibrium ones. The Cr–Fe–Ni isothermal section at 760°C determined in this way has an overall topology similar to the equilibrium phase diagram, but there are appreciable differences in the locations of the phase boundaries. Some of these differences may be related to the error in composition determination from X-ray fluorescence (XRF).

Even with state-of-the-art thin-film deposition and phase analysis techniques, the thin-film-based combinatorial approaches for phase diagram mapping are not as straightforward as one might think. One difficulty arises from the fact that crystal structure identification needs to be performed at very localized compositions in order to construct the two- or three-phase regions. Failing to do so often results in poorly defined phase boundaries. In addition, strong textures often form in the deposited films, making crystal structure identification more difficult, especially for complex crystal structures. One also needs to worry about whether equilibrium phases are formed in the thin films, especially when the thin-film substrates are not heated to the temperature of interest during deposition. Sometimes amorphous thin films are produced which are easier to transform to the equilibrium phases than are non-equilibrium crystalline phases. For instance, several annealing trials may be needed to produce the equilibrium phases. The interaction of the substrate with the thin films is also a challenge, especially for alloy systems that contain elements with strong affinity for interstitials (C, N, and O).

The thin-film approach is especially useful for discovering new functional materials and for mapping composition–structure–property relationships for thin films. Since many functional materials and devices are used in the thin-film format in the final product, this approach has tremendous application. In such applications, the equilibrium phase diagrams/conditions are not really as relevant as for structural materials. A brief introduction of the thin-film approach to the determination of phase diagrams involving magnetic transitions can be found in Chapter 12.

Hasebe and Nishizawa [10] as well as Jin [11] introduced the little-known "ternary diffusion couple" technique for efficient ternary phase diagram determination. Jin and his associates applied this technique to determine many ternary phase diagrams (e.g., Ref. [11–14]). The methodology is described in more detail in Chapter 7. The diffusion-multiple technique is an extension of the diffusion-triple technique by combining several diffusion couples and triples into a single sample. A careful examination of the reliability of the diffusion triples/multiples for phase diagram determination was made by comparing the phase diagrams determined from diffusion multiples with those determined from equilibrated alloys [15]. The good agreement indicates the high reliability of the diffusion triples/multiples for equilibrium phase diagram determination. It is now possible to achieve orders of magnitude increases in efficiency without sacrificing the quality of data.

The elegant temperature gradient diffusion-couple technique [16–19] and the wedge-shaped diffusion-couple technique [20–21] are other examples of high-throughput techniques for phase diagram determination. These techniques are very important to clarify the phase relations in several metal–nitrogen phase diagrams. Wider application of these techniques is strongly encouraged.

The floating zone method [22] can be applied to study solid–liquid phase equilibria in a high-throughput fashion, which complements other diffusion-couple/multiple techniques that are only applicable to solid–solid phase equilibria. Liquid phase diffusion couples (both transient and equilibrium) have also been explored to determine phase diagrams with good success [23].

5 LOW-TEMPERATURE PHASE DIAGRAMS

This topic has been briefly discussed in Chapter 2, but it is worthwhile to explore it again. There have been sporadic reports of *room temperature* phase diagrams for high melting point elements. The practice started several decades ago (e.g., Ref. [24] for Co–Fe–W), but unfortunately it still persists nowadays (e.g., Ref. [25] for Co–La–Ti). Individual alloys were usually melted and homogenized at a very high temperature, which is a good practice. They were subsequently step-wise heat treated to, or simply heat treated at, an intermediate temperature, for example, 650–800°C for 45 days for the Co–La–Ti alloys, and then slowly furnace cooled to room temperature. Then X-ray diffraction (XRD) or other techniques were used to identify the phases in the alloys to establish the phase equilibria. Phase diagrams determined in this way should not be called "room temperature phase diagrams"; instead, they should be regarded as phase diagrams at the temperatures of the final heat treatments. In addition, it is always a better practice to quench the samples from the temperatures of final heat treatments to room temperature to avoid the complication of further phase transformations during the slow furnace cooling. All the "room temperature phase diagrams" reported in the literature for systems containing elements with melting points higher than 1000°C should be regarded with suspicion and should actually be considered as phase diagrams at the temperatures of the respective final heat treatments. Slow diffusion and phase transformation kinetics would never allow alloys made up of high melting point elements (e.g., Co–Fe–W) to reach equilibrium at ambient temperature. For Co–Fe–W alloys, even equilibria at 500°C are already extremely difficult to achieve in bulk samples, not even mentioning at ambient temperature.

There may be a way to reach low-temperature phase equilibria for high melting point alloys. The formation of nano-scale intermetallic compounds and solid solutions at ambient temperature using the so-called "metallurgy in a beaker" [26,27] methodology is very encouraging. This method can be tracked back to at least 45 years ago [28] when Kulifay synthesized 28 compounds including intermetallics, antimonides, arsenides, tellurides, and a few ternaries and non-stoichiometric compounds. His method consists of the reduction of aqueous solutions containing the elements by addition to aqueous hydrazine or hypophosphorous acid solutions at 100°C and atmospheric pressure. The recent work of Schaak et al. [26,27] demonstrated similar synthesis of several high melting point binary and ternary intermetallic compounds such as $AuCu_3$, $FePt_3$, and CoPt as well as solid solutions such as Ni–Pt. They synthesized the compounds and alloys at room temperature in a solution and then crystallized them at 300°C. It is remarkable to be able to form such high melting point equilibrium intermetallics and solid solutions at 300°C. A benchmark study using this methodology for low-temperature phase diagram determinations will be very valuable to see whether wider composition spaces can be explored. The methodology may provide critical data for low-temperature phase stability for *CALPHAD* modeling.

For alloys containing both substitutional and interstitial elements, such as steels with carbon or nitrogen, sometimes a para-equilibrium can be reached before the (full) ortho-equilibrium at low temperatures and/or short annealing times [29,30]. Only the interstitial elements reach equilibrium under the para-equilibrium condition, since they are very mobile even at low temperatures and/or short annealing times. Only sufficiently high temperatures and/or long annealing times would allow the substitutional elements to reach the full ortho-equilibrium. Close attention should be paid to the equilibrium conditions when one determine phase diagrams at relatively low temperatures for alloys containing both substitutional and interstitial elements.

6 PHASE DIAGRAM DETERMINATION: A NEVER-ENDING TASK

In a sense, phase diagram determination is somewhat a never-ending task, especially for complex systems. As Roth and Vanderah [31] suggested, "… it is likely that no diagram is ever completely finished". One can appreciate this point by simply looking at the Ti–Al binary system. The binary Ti–Al phase diagram determination can be traced back to 1923 [32]. There are several assessments published already, including those of Murray [33], Kattner et al. [34], and most recently Schuster and Palm [35] which was based on work from 284 references on experimental determination (with an additional 50 references on thermodynamic modeling and 37 on theoretical calculations)! If one can sum up the total research spending on the Ti–Al phase diagram determination, it would be multi-million US dollars for sure. The phase diagram will surely be further refined with time.

The never-ending nature of phase diagram determination comes from several factors: the time and resource constraints dictate the focus of a particular research to a region of the phase diagram of immediate interest; the availability of equipment at the research facility dictates the type of measurements one can perform; some subtle crystal structure changes may be missed by a research group without advanced facilities or relevant knowledge; certain measurement procedures may miss subtle electronic, magnetic, or ordering transitions; and specimen purity and conditions may introduce discrepancies that need further investigation to sort out; and so on. It simply takes a lot of time and effort to get everything and every detail right, even for a binary system.

It is a healthy attitude to realize that most phase diagrams are incomplete and that many uncertainties and errors exist in many published phase diagrams. When one draws a phase diagram from experimental observations, lots of guesswork and intuition/judgment go into how to place the phase boundaries. This is simply because no one can afford to test every possible composition and temperature within a phase diagram. One has to extrapolate and interpolate between the experimental observations using the phase rule, thermodynamics, or simply similarity to analogous systems. It is not hard to imagine the existence of lots of errors in existing phase diagrams, especially ternary phase diagrams.

7 SOME SUGGESTIONS

One of the best suggestions for someone new in phase diagram determination is to look for similar phase diagrams and learn how those were determined, especially the detailed experimental procedures. For instance, if one wants to determine a phase diagram of the Ti–N binary system, one can look at the work that has been performed for systems such as Zr–N, Nb–N, and Hf–N.

The assessments published in the *Bulletin of Alloy Phase Diagrams* and the *Journal of Phase Equilibria (and Diffusion)* are great resources to get information about specific systems. These assessments not only contain phase equilibria data, but also thermodynamic data, crystal structure data, metastable phases reported, lattice parameters, magnetic property data, etc. They are very valuable resources.

The other recommendation is to read the various relevant chapters in this book and items from the recommended list of further reading materials tabulated below. The list was compiled by the authors of this chapter and it is not meant to be comprehensive, but to provide some key references we deem important.

REFERENCES FOR FURTHER READING

General Phase Diagram Introduction
- H. Baker, in *ASM Handbook. Vol. 3: Alloy Phase Diagrams*, Section 1, ASM International, Materials Park, OH, 1992, pp. 1.1–1.30.
- D.R.F. West and N. Saunders, *Ternary Phase Diagrams in Materials Science*, 3rd edn., Maney Publications, London, 2002.
- F.N. Rhines, *Phase Diagrams in Metallurgy*, McGraw-Hill Book Co., New York, 1956.
- G. Massing and B.A. Rodgers, *Ternary Systems*, Dover Publications, New York, 1960.
- A. Prince, *Alloy Phase Equilibria*, Elsevier, Amsterdam, 1966.
- B. Predel, M. Hoch and M. Pool, *Phase Diagrams and Heterogeneous Equilibria: A Practical Introduction*, Springer, Berlin, 2001.
- F.A. Hummel, *Introduction to Phase Equilibria in Ceramic Systems*, Dekker, New York, 1984.
- K.H.J. Buschow, R.W. Cahn, M.C. Flemings, B. Ilschner, E.J. Kramer, S. Mahajan and P. Veyssière (eds.), *Encyclopedia of Materials Science and Technology*, Elsevier, Oxford, UK, 2001.
    T.B. Massalski, *Phase Diagrams*, p. 6842.
    A.P. Miodownik, *Phase Diagrams, Calculation of*, p. 6850.
    G. Effenberg, *Phase Diagrams: Data Compilation*, p. 6857.
    H.J. Seifert, *Phase Equilibria in Fe–C, Fe–X, Fe–X–C, and Fe–C–X_i*, p. 6861.
    J.S. Kirkaldy, *Phase Equilibria, Thermodynamics of*, p. 6867.

Specimen Purity and Preparation
- J.F. Smith, Requisites for reliable determination of phase equilibria, *Mater. Sci. Eng.*, 48 (1981) 1.
- R.D. Shull, Phase diagram sample preparation, *Bull. Alloy Phase Diagr.*, 4 (1983) 5.
- J.P. Abriata, On sample requirements for the correct determination of stable equilibrium phase diagrams, in D. Farkas and F. Dyman (eds.), *Alloy Theory and Phase Equilibria*, ASM International, Metals Park, OH, 1986, p. 69.
- O.N. Carlson and J.F. Smith, Effects of interstitial impurities on phase equilibria, *Bull. Alloy Phase Diagr.*, 8 (1987) 208.
- J.W. Herchenroeder and K.A. Gschneidner, Jr. Stable, metastable and nonexistent allotropes, *Bull. Alloy Phase Diagr.*, 9 (1988) 2.
- J.F. Smith, Factors affecting the validity of both calculated and experimental phase diagrams, *J. Phase Equili.*, 13 (1992) 235.

Thermal Analysis Applied to Phase Diagram Determination (Adapted from Chapter 5)
- P.D. Garn, *Thermoanalytical Methods of Investigation*, Academic Press, New York, 1965.
- M.E. Brown, *Differential thermal analysis and differential scanning calorimetry, Introduction to Thermal Analysis; Techniques and Applications*, Chapman & Hall, New York, Chapter 4, 1st ed., 1988.
- T. Gödecke, Ableitung des kristallisationspfades in ternären gusslegierungen [Delineation of crystallization paths in ternary as-cast alloys], *Z. Metallk.*, 92 (2001) 966 [In German].
- M.J. Starink, Analysis of aluminium based alloys by calorimetry: quantitative analysis of reactions and reaction kinetics, *Inter. Mater. Rev.*, 49 (2004) 191.

Diffusion as Related to Phase Diagrams and Phase Diagram Determination
- J.B. Clark, Conventions for plotting the diffusion paths in multiphase ternary diffusion couples on the isothermal section of a ternary phase diagram, *Trans. Met. Soc. AIME.*, 227 (1963) 1250.
- F.J.J. van Loo, Multiphase diffusion in binary and ternary solid-state systems, *Prog. Solid State Chem.*, 20 (1990) 47.
- A.A. Kodentsov, G.F. Bastin and F.J.J van Loo, The diffusion couple technique in phase diagram determination, *J. Alloy. Compd.*, 320 (2001) 207.
- A.D. Romig, Jr., Thermodynamic considerations in the analysis of phase stability: the role of interfacial equilibrium in the determination of phase diagrams by X-ray microanalysis techniques, *Bull. Alloy Phase Diagr.*, 8 (1987) 308.

- J.E. Morral, C. Jin, A. Engström and J. Ågren, Three types of planar boundaries in multiphase diffusion couples, *Scripta Mater.*, 34 (1996) 1661.
- J.-C. Zhao, Reliability of the diffusion-multiple approach for phase diagram mapping, *J. Mater. Sci.*, 39 (2004) 3913.
- J.S. Kirkaldy and D.J. Young, *Diffusion in the Condensed State*, Institute of Metals, London, 1987.
- H. Wiesenberger, W. Lengauer and P. Ettmayer, Reactive diffusion and phase equilibria in the V–C, Nb–C, Ta–C and Ta–N systems, *Acta Mater.*, 46 (1998) 651.
- W. Lengauer, M. Bohn, B. Wollein and K. Lisak, Phase reactions in the Nb–N system below 1400°C, *Acta Mater.*, 48 (2000) 2633.

Interplay of Ordering, Magnetic Transitions, and Phase Equilibria
- G. Inden, The effect of continuous transformations on phase diagrams, *Bull. Alloy Phase Diagr.*, 2 (1982) 412.
- C. Colinet, G. Inden and R. Kikuchi, CVM calculation of the phase diagram of b.c.c. Fe–Co–Al, *Acta Metall. Mater.*, 41 (1993) 1109.
- A.P. Miodownik, The effect of magnetic transformations on phase diagrams, *Bull. Alloy Phase Diagr.*, 2 (1982) 406.

Ceramic Phase Diagram Determination
- A.M. Alper, *Phase Diagrams: Materials Science and Technology*, Vol. I–VI, Academic Press, New York, 1970.
- A.M. Alper, *Phase Diagrams in Advanced Ceramics*, Academic Press, New York, 1995.
- R.S. Roth and T.A. Vanderah, Experimental determination of phase equilibria diagrams in ceramic systems, in B.V.R. Chowdari, H.-L. Yoo, G.M. Choi and J.-H. Lee, (eds.), *Solid State Ionics: The Science and Technology of Ions in Motion*, World Scientific Publication Co., Singapore, 2004, p. 3.

Theoretical/Modeling Work
- S. Curtarolo, D. Morgan and G. Ceder, Accuracy of ab initio methods in predicting the crystal structures of metals: a review of 80 binary alloys, *CALPHAD*, 29 (2005) 163.
- C.C. Fischer, K.J. Tibbetts, D. Morgan and G. Ceder, Predicting crystal structure by merging data mining with quantum mechanics, *Nature Mater.*, 5 (2006) 641.
- P. Villars, K. Cenzual, J. Daams, Y. Chen and S. Iwata, Data-driven atomic environment prediction for binaries using the Mendeleev number. Part 1. Composition AB, *J. Alloy. Compd.*, 367 (2004) 167.
- J. Hafner, C. Wolverton and G. Ceder, Toward computational materials design: the impact of density functional theory on materials research, *MRS Bull.*, 31 (2006) 659.

The *CALPHAD* Approach
- L. Kaufman and H. Bernstein, *Computer Calculation of Phase Diagrams*, Academic Press, New York, 1970.
- H.A.J. Conk, *Phase Theory: The Thermodynamics of Heterogeneous Equilibria*, Elsevier, Amsterdam, 1983.
- M. Hillert, *Phase Equilibria, Phase Diagrams and Phase Transformations – Their Thermodynamic Basis*, Cambridge University Press, Cambridge, UK, 1998.
- N. Saunders and A.P. Miodownik, *CALPHAD (Calculation of Phase Diagrams): A Comprehensive Guide*, Pergamon Press, Oxford, UK, 1998.
- K. Hack (ed.), *The SGTE Casebook – Thermodynamics at Work*, The Institute of Materials, London, 1996.
- Y.A. Chang, S.-L. Chen, F. Zhang, X.-Y. Yan, F.-Y. Xie, R. Schmid-Fetzer and W.A. Oates, Phase diagram calculation: past, present and future, *Prog. Mater. Sci.*, 49 (2004) 313.

Miscellaneous Recommended Papers on Phase Diagram Determination
- C.J. Altstetter, Metal-oxygen systems, *Bull. Alloy Phase Diagr.*, 5 (1982) 543.

- R. Ferro, G. Cacciamani and G. Borzone, Remarks about data reliability in experimental and computational alloy thermochemistry, *Intermetallics*, 11 (2003) 1081.
- R.G. Faulkner, Segregation to boundaries and interfaces in solids, *Inter. Mater. Rev.*, 41 (1996) 198.
- J.H. Greenberg, Topology of P-T-X phase diagrams of binary systems. Polymorphism, metastable states and high-pressure equilibria, *Mater. Sci. Eng.*, R16 (1996) 223.
- J. Klingbeil and R. Schmid-Fetzer, Ter-Quat: approximate prediction of ternary and quaternary phase diagrams concerning stoichiometric phases, *J. Phase Equili.*, 13 (1992) 522.
- G. Effenberg and R. Schmid-Fetzer, *Critical Evaluation of Ternary Phase Diagram Data*, Notes for Authors, 3rd revised edition, 2000.

SOME CURRENT PHASE DIAGRAM COMPILATIONS

Elemental and Metallic Systems
- D.A. Young, *Phase Diagrams of the Elements*, University of California Press, CA, 1991.
- A.T. Dinsdale, SGTE data for pure elements, *CALPHAD*, 15 (1991) 317.
- E.Yu. Tonkov and E.G. Ponyatovsky, *Phase Transformations of Elements under High Pressure*, CRC Press, Boca Raton, Florida, 2004.
- T.B. Massalski and H. Okamoto (eds.), *Binary Alloy Phase Diagrams*, 2nd edn., ASM International, Materials Park, OH, 1990.
- H. Okamoto, *Desk Handbook: Phase Diagrams for Binary Alloys*, ASM International, Materials Park, OH, 2000.
- H. Okamoto, *Phase Diagrams of Dilute Binary Alloys*, ASM International, Materials Park, OH, 2002.
- B. Predel (ed.), *Landolt-Börnstein Series*, O. Madelung (Editor-in-Chief), vol. IV/5: *Phase Equilibria, Crystallographic and Thermodynamic Data of Binary Alloys*, Subvolumes A to J, Springer, New York, 1991–1998, vol. IV/12, 2006.
- M.E. Kassner and D.E. Peterson (eds.), *Phase Diagrams of Binary Actinide Alloys*, ASM International, Materials Park, OH, 1995.
- L. Tanner and H. Okamoto (eds.), *Phase Diagrams of Binary Beryllium Alloys*, ASM International, Materials Park, OH, 1987.
- P.R. Subramanian, D.J. Chakrabarti and D.E. Laughlin (eds.), *Phase Diagrams of Binary Copper Alloys*, ASM International, Materials Park, OH, 1994.
- H. Okamoto and T.B. Massalski (eds.), *Phase Diagrams of Binary Gold Alloys*, ASM International, Materials Park, OH, 1987.
- F.D. Manchester (ed.), *Phase Diagrams of Binary Hydrogen Alloys*, ASM International, Materials Park, OH, 2000.
- C.E.T. White and H. Okamoto, *Phase Diagrams of Indium Alloys and their Engineering Applications*, ASM International, Materials Park, OH, 1991.
- H. Okamoto (ed.), *Phase Diagrams of Binary Iron Alloys*, ASM International, Materials Park, OH, 1993.
- P. Nash (ed.), *Phase Diagrams of Binary Nickel Alloys*, ASM International, Materials Park, OH, 1991.
- S.P. Garg, M. Venkatraman, N. Krishnamurthy and R. Krishnan (eds.), *Phase Diagrams of Binary Tantalum Alloys*, Indian Institute of Metals, Calcutta, India, 1996.
- J.L. Murray (ed.), *Phase Diagrams of Binary Titanium Alloys*, ASM International, Materials Park, OH, 1987.
- S.V.N. Naidu and P.R. Rao (eds.), *Phase Diagrams of Binary Tungsten Alloys*, Indian Institute of Metals, Calcutta, India, 1991.
- J.F. Smith (ed.), *Phase Diagrams of Binary Vanadium Alloys*, ASM International, Materials Park, OH, 1989.
- J.H. Westbrook (ed.), *Moffatt's Handbook of Binary Phase Diagrams*, 5 loose-leaf volumes, 2006, Genium Publ., Amsterdam, New York, 2006.
- Yu.V. Levinsky and G. Effenberg (eds.), *Pressure Dependent Phase Diagrams of Binary Alloys*, ASM International, Materials Park, OH, 1997.
- G. Petzow and G. Effenberg (eds.), *Ternary Alloys: A Comprehensive Compendium of Evaluated Constitutional Data and Phase Diagrams*, 15 volumes, Wiley-VCH, Weinheim, Germany, 1989–1996.

- P. Villars, A. Prince and H. Okamoto (eds.), *Handbook of Ternary Alloy Phase Diagrams*, 10 volumes, ASM International, Materials Park, OH, 1995.
- P. Rogl and G. Effenberg (eds.), *Phase Diagrams of Ternary Metal–Boron–Carbon Systems*, ASM International, Materials Park, OH, 1998.
- A.A. Prince, G.V. Raynor and D.S. Evans (eds.), *Phase Diagrams of Ternary Gold Alloys*, The Institute of Metals, London, 1990.
- V. Raghavan, *Phase Diagrams of Ternary Iron Alloys*, Parts 1–6, ASM International, Materials Park, OH and The Hindian Institute of Metals, Calcutta, India, 1987–1993.
- K.P. Gupta (ed.), *Phase Diagrams of Ternary Nickel Alloys*, Parts 1 and 2, Indian Institute of Metals, Calcutta, India, 1990.
- V. Raghavan, S.P. Garg, M. Venkatraman, N. Krishnamurthy and R. Krishnan (eds.), *Phase Diagrams of Quaternary Iron Alloys*, Indian Institute of Metals, Calcutta, India, 1996.
- G. Effenberg and S. Ilyenko (eds.), *Landolt-Börnstein Series*, O. Madelung (Editor-in-Chief), Vol. IV/11: *Ternary Alloy Systems. Phase Diagrams, Crystallographic and Thermodynamic Data*, Subvolumes A to C, Springer, New York, 2004–2006.

Ceramic Systems (A Total of 20 Volumes Published by NIST and American Ceramic Society – Not All of Them Listed Here)
- R.S. Roth and T.A. Vanderah, *Phase Equilibrium Diagrams*, American Ceramic Society, Westerville, OH, 2005, 14 volumes (also available on CD-ROMs).
- Phase Diagrams for Ceramists, various (eds.), American Ceramic Society, Westerville, OH, Vol. 1–12 and annuals 1991–1993.
- A.E. McHale, *Phase Equilibria Diagrams for Ceramists: Borides, Carbides, and Nitrides*, American Ceramic Society, Westerville, OH, 1994.
- J.D. Whitler and R.S. Roth, *Phase Diagrams for High T_c Superconductors*, American Ceramic Society, Westerville, OH, 1997.
- T.A. Vanderah, R.S. Roth and H.F. McMurdie, *Phase Diagrams for High T_c Superconductors II*, American Ceramic Society, Westerville, OH, 1997.
- O.B. Fabrichnaya, S.K. Saxena and P. Richet, *Thermodynamic Data, Models and Phase Diagrams in Multicomponent Oxide Systems*, Springer, New York, 2006.

Other Miscellaneous Compilations/Books
- R. Koningsveld, W.H. Stockmayer and E. Nies, *Polymer Phase Diagrams: A Textbook*, Oxford University Press, Oxford, UK, 2001.
- S.A. Morse, *Basalts and Phase Diagrams: An Introduction to the Quantitative Use of Phase Diagrams in Igneous Petrology*, Krieger, Malabar, Florida, 1994.
- E.Yu. Tonkov, *High Pressure Phase Diagrams: A Handbook*, 3 volumes, CRC Press, Boca Raton, Florida, 1996.
- M. Zhao, *Theory and Calculation of P-T-X Multicomponent Phase Diagrams*, Nova Science Publications, New York, 1996.
- A.I. Gusev, A.A. Rempel and A.I. Magel, *Disorder and Order in Strongly Non-Stoichiometric Compounds*, Springer, New York, 2006.
- V.M. Glazov and L.M. Pavlova, *Semiconductor and Metal Binary Systems Phase Equilibria and Chemical Thermodynamics*, Springer, New York, 1989.
- T. Gasparik, *Phase Diagrams for Geoscientists*, Springer, New York, 2003.
- E.G. Ehlers, *The Interpretation of Geological Phase Diagrams*, Dover, Mineola, New York, 1987.
- D. Behrens, H. Engels and R. Eckermann (eds.), *Phase Equilibria and Phase Diagrams of Electrolytes*, Scholium International, Port Washington, New York, 1991.

Some Useful Websites
- ASM Phase Diagrams Center: http://www.asminternational.org/asmenterprise/apd/default.aspx
- MSIT Workplace: http://www.msiwp.com/bookstore/phasediagram.html
- The Pauling File: http://crystdb.nims.go.jp/

- SGTE Binary Phase Diagram Collection: http://web.mse.kth.se/dct/pd/
- NIST (Ceramic) Phase Equilibria Diagrams Database: http://www.nist.gov/srd/nist31.htm
- Landolt-Börnstein: http://www.springer.com/west/home/laboe/group+iv+physical+chem?SGWID=4-10117-12-146645-0&teaserId=48348&CENTER_ID=95859 http://www.knovel.com/knovel2/Toc.jsp?BookID=958

REFERENCES

1. K. Oikawa, G.-W. Qin, T. Ikeshoji, R. Kainuma and K. Ishida, *Acta Mater.*, 50 (2002) 2223.
2. G. Inden, *Bull. Alloy Phase Diagr.*, 2 (1982) 412.
3. Y.L. Li, S. Choudhury, Z.K. Liu and L.-Q. Chen, *Appl. Phys. Lett.*, 83 (2003) 1608.
4. S. Choudhury, Y.L. Li and L.-Q. Chen, *J. Am. Ceram. Soc.*, 88 (2005) 1669.
5. P. Villars, in J.H. Westbrook and R.L. Fleischer (eds.), *Intermetallic Compounds: Principles and Practice*, Wiley, New York, 1995, p. 227.
6. A. Boettcher, G. Haase and R. Thun, *Z. Metallk.*, 46 (1955) 386.
7. J.-C. Zhao, *Prog. Mater. Sci.*, 51 (2006) 557.
8. U.R. Kattner, Private communication to J.-C. Zhao, 2005.
9. K. Kennedy, T. Stefansky, G. Davy, V.F. Zackay and E.R. Parker, *J Appl. Phys.*, 36 (1965) 3808.
10. M. Hasebe and T. Nishizawa, in G.C. Carter (ed.), *Application of Phase Diagrams in Metallurgy and Ceramics*, Vol 2, NBS Special Publications, No. 496, Gaithersburg, MD, 1978, p. 911.
11. Z. Jin, *Scand. J. Metall.*, 10 (1981) 279.
12. J.-C. Zhao and Z. Jin, *Z. Metallk.*, 81 (1990) 247.
13. Z. Jin and C. Qiu, *Metall. Mater. Trans.*, 24A (1993) 2137.
14. J.-C. Zhao, Z. Jin and P. Huang, *Scripta Metall.*, 22 (1988) 1825.
15. J.-C. Zhao, *J. Mater. Sci.*, 39 (2004) 3913.
16. W. Lengauer, *J. Solid St. Chem.*, 91 (1991) 279.
17. W. Lengauer and P. Ettmayer, *Adv. Mater. Proc.*, 144(5) (1993) 116.
18. W. Lengauer, *J. Alloy. Compd.*, 179 (1992) 289.
19. W. Lengauer, *Acta Metall. Mater.*, 39 (1991) 2985.
20. H. Wiesenberger, W. Lengauer and P. Ettmayer, *Acta Mater.*, 46 (1998) 651.
21. W. Lengauer, M. Bohn, B. Wollein and K. Lisak, *Acta Mater.*, 48 (2000) 2633.
22. M.Th. Cohen-Adad, M. Gharbi, C. Goutaudier, R. Cohen-Adad, *J Alloy. Compd.*, 289 (1999) 185.
23. O. Ikeda, I. Ohnuma, R. Kainuma and K. Ishida, *Intermetallics*, 9 (2001) 755.
24. W. Köster, *Arch. Eisen.*, 5 (1932) 431.
25. Q.L. Liu, J.J. Nan, J.K. Liang, F. Huang, Y. Chen, G.H. Rao and X.L. Chen, *J. Alloys Comp.*, 307 (2000) 212.
26. R.E. Schaak, A.K. Sra, B.M. Leonard, R.E. Cable, J.C. Bauer, Y.-F. Han, J. Means, W. Teizer, Y. Vasquez and E.S. Funck, *J. Am. Chem. Soc.*, 127 (2005) 3507.
27. B.M. Leonard, N.S.P. Bhuvanesh and R.E. Schaak, *J. Am. Chem. Soc.*, 127 (2005) 7326.
28. S.M. Kulifay, *J. Am. Chem. Soc.*, 83 (1961) 4916.
29. M. Hillert and J. Ågren, *Scripta Mater.*, 50 (2004) 697.
30. M. Hillert, in L.H. Bennett, T.B. Massalski and B.C. Giessen (eds.), *Alloy Phase Diagrams*, Mater. Res. Soc. Symp. Proc., 19 (1983) 295.
31. R.S. Roth and T.A. Vanderah, in B.V.R. Chowdari, H.-L. Yoo, G.M. Choi and J.-H. Lee (eds.), *Solid State Ionics: The Science and Technology of Ions in Motion*, World Scientific Publishing, 2004, p. 3.
32. E. van Erckelens, *Metall. Erz*, 20 (1923) 206.
33. J.L. Murray, in J.L. Murray (ed.), *Phase Diagrams of Binary Titanium Alloys*, ASM, Metals Park, OH, 1987, p. 12.
34. U.R. Kattner, J.-C. Lin and Y.A. Chang, *Metall. Trans.*, 23A (1992) 2081.
35. J.C. Schuster and M. Palm, *J. Phase Equili. Diff.*, 27 (2006) 255.

Index

A
Ag–Cu phase diagram, 120–121
Ag–Fe–Ti phase diagram, 234
Ag–H–Pd phase diagram, 472
Ag–Pr phase diagram, 80, 81
Al–B phase diagram, 80, 81
Al–Be–Si phase diagram, 131–136
Al–C–Si–Ti system, 137, 139–140, 143
Al–Cr–Nb phase diagram, 45, 46
Al–Cr–Ti phase diagram, 270
Al–Cu–Fe phase diagram, 195
Al–Fe phase diagram, 376–381
Al–Mg–Sr phase diagram, 275–282
Al–Mn phase diagram, 82, 147, 148
Al–Mn–Si phase diagram, 135, 137, 138
Al–Ni–Ti phase diagram, 369
Al–Ni–Ti–Fe phase diagram, 371
Al_2O_3–SiO_2 phase diagram, 449–452
Al_2O_3–CaO–SiO_2 phase diagram, 443–452
Al–Pu phase diagram, 82, 83
Al–Se phase diagram, 83
Al–Ti phase diagram, 372–374
Alloy preparation, 109
   arc melting, 110
   high-temperature melting of alloys, 109–110
   induction furnace melting, 111–112
   powder metallurgy method, 112–113
Analytical electron microscopy (AEM), 226–227, 244
Angle-dispersive diffraction (ADD), 421
Annealing, isothermal, 41, 42
Arc melting, 110–112, 140
ASTM standards for DTA and DSC, 154–155
Au–La phase diagram, 83–84
Au–U phase diagram, 84, 85
Auger electron spectroscopy (AES), 226

B
Back (solid) diffusion, 201
B–Bi phase diagram, 86
Binary alloys
   DTA data analysis for, 213
   enthalpy *vs.* temperature relations, 211–213
B–W phase diagram, 86, 87
B–Si–Mo–Ti phase diagram, 282–289

C
Ca–Eu phase diagram, 87, 88
Ca–H phase diagram, 19
Calcining techniques, 343
Calculation of phase diagrams, *see* CALPHAD
Calorimetry, 152, 201, 304, 331, 477
CALPHAD, 2, 7–10, 38, 48, 109, 140, 171, 201, 273, 289, 484, 486, 488, 491
$CaMnO_3$–$LaMnO_3$ phase diagram, 395
CaO–Al2O3–SiO_2 system, 452–453
CaO–SiO2 system, 443–447
CaO–"FeO"–SiO2 system, 453–454
Ce–Pr phase diagram, 87, 89
Ceramic systems, phase diagram determination of
   ex situ methods, 342
      identification of new phases, 354
      phase and compositional analysis, 350–354
      sample preparation and equilibration, 343–350
   in situ methods, 354
      coulometric titration, 356–358
      high-temperature X-ray diffraction, 358
      oscillation method of phase analysis (OPA), 358–359
      electric, dielectric, and magnetic measurements, 359
      thermal analysis, 355–356
      thermomicroscopy and other optical techniques, 358
C–Ge phase diagram, 87, 88
Charles' Law and Van't Hoff relationship, 59–61
Clausius–Clapeyron equation, 125

Co–Cr phase diagram, 36–38
Co–Cr–Mo phase diagram, 261–269
Co–Fe–Ni phase diagram, 400
Co–Mo phase diagram, 38, 39, 392
Co–Mo–Ni phase diagram, 270
Co–V phase diagram, 13, 15, 16, 39, 40
Combinatorial approach, 384, 396–400, 486
  libraries, 396
Compilation, phase diagrams of, 6, 7, 17, 492–493
Concentration gradient method, 374–375
  basics of, 376
  examples of, 376–380
Continuous cooling transformation (CCT), 26
Convergent beam electron diffraction (CBED), 354
Cooling rate, 23, 26–30, 33, 35, 36, 40, 115, 123, 126, 134, 139, 143, 157, 169, 195, 212, 280, 299, 346, 355
Cooling transformation
  in Co–Cr, 36–38
  in Co–Mo and Co–V, 38–40
  in Fe–Ni, 35–36
  formation of metastable phases during, 27–30
  shifting of transformation-start temperature with, 26–27
Coppermaking, 443
Coulometric titration, 307–309, 356–358
Cr–Nb–Si phase diagram, 43
Cr–Ni phase diagrams, 89, 90
Cr–Ni–Ti phase diagram, 230
Cr–Ni–V phase diagram, 236
Critical point, 11, 12, 73–74, 76, 89, 99, 296, 298, 337, 366–369, 376, 384, 391, 460, 469, 475
Crystal structure identification, 46, 140–145, 257, 486
Cu–Hf phase diagram, 90–92
Cu–Nb phase diagram, 17–19
Cu–Nd phase diagram, 130–131
Cu–Ni–Sn phase diagram, 41, 42
Curie–Weiss law, 386

D
Diamond–Anvil Cell (DAC), 413–415, 418–419, 421–423, 432, 436–438
Darken's equation, 367

Decomposition kinetics, of high temperature phase, 43–44
Degenerated phase equilibrium, 148–149
Degrees of freedom, 201
Densitometry, 302–303, 325, 332, 337
Density functional theory (DFT), 140, 485
Dielectric (measurement), 359
Differential scanning calorimetry, see DSC
  power-compensating, 157, 203
Differential thermal analysis, see DTA
Diffraction, see also X-ray diffraction, 6
  electron, and TEM, 480
  methods, 478–480
Diffusion, back (solid), 201
Diffusion couple, 23, 24, 35–40, 46, 47, 123, 222–244, 246, 255, 259–262, 266, 267, 269, 270, 303, 345, 362, 487, 490
Diffusion couple technique, 22–24, 35–40, 46–47, 222
  error sources in experiments with, 236–244
  experimental procedures
    analytical techniques and specimen preparation, 226–227
    preparation of diffusion couple specimens, 225–228
  general principles of, 223–225
  in Ag–Fe–Ti system, 231–234
  in Ni–Cr–Ti system, 228–231
  phase formation and isothermal phase transformation kinetics in, 46–47
  variations of, 228–236
Diffusion multiple, 2, 43–46, 244, 246–271, 397, 404, 405, 487, 491
Diffusion-multiple approach, 246–247
  considerations for design, 250–255
  EPMA profiling, 257–259
  error sources and reduction, 266–269
  extraction of equilibrium tie lines, 259–266
  fabrication for, 248–250
  imaging examination and phase analysis, 256–257
  phase formation, 46–47
Diffusion path, 224, 225, 228, 231, 233–236, 240, 244
Diffusion zone morphology, 224, 230–231
Dilatometry, 5, 125–126

DSC, 32, 123–125, 148–149, 151–221, 280, 299, 304, 319, 355, 361–362, 364–365, 369–370, 380, 478
DTA, 26, 32, 115, 123–126, 128, 130–136, 147–149, 170, 151–221, 299–300, 302, 305, 319, 333, 354–355, 358–359, 364, 474, 478
  calibration, 164
    temperature calibration, 166
    enthalpy and heat capacity calibration, 169
  data analysis for binary alloys, 169–170, 192–194
  DTA response, 192
  eutectic reactions *vs.* peritectic reactions, 191–192
  general behavior for binary eutectic system
    comparison to experiment, 174, 176
    derivative of enthalpy *vs.* temperature curve and their relation to DTA curves, 171–176
  problems with liquidus determination on heating
    details of computed behavior of alloy on melting, 182, 183
    DTA peak temperature, 183–185
    failure to completely melt, 185–186
    general DTA curve analysis, 180–183
    resolution of difficulties using temperature cycling near the liquidus, 182, 183, 185
    small liquidus solidus separation, 182, 184
  problems with solidus determination on heating
    effect of hold time prior to melting, 177–179
    errors caused by using extrapolated melting onset, 179–180
    incipient melting point *vs.* solidus, 177
  supercooling problem with liquidus determination on cooling onset of freezing, 186–188
  simulation of DTA response for alloys with supercooling, 189–191
  slope of DTA curve on initial freezing, 188–189
DTA data analysis for ternary alloys, 194
  Al–6% Cu–0.5% Fe, 197–198
  Al–20% Cu–0.5% Fe, 195–197

  Al–rich corner of Al–Cu–Fe, 194–195
Dynamic method, 122–123

E
EBSD, 248–249, 256–257, 268–269, 285–286
EDS, 227, 256–258, 268, 278, 376–379
Electrical conductivity measurement, *see* electrical resistivity, 126–128
Electrical resistivity/resistance, 22, 40, 114, 127–128, 300, 305, 307, 321–332, 337, 346, 359, 361–364, 377, 384, 389, 395, 472, 480
  measurement apparatus, 322
  of liquid–liquid equilibria, 330, 332
  of solid–liquid equilibria, 315
  of solid–solid equilibria, 299, 315
Electromotive force (emf), 3, 307–308, 311, 320–321, 332, 382, 431
Electro-discharge machining (EDM), 248–249
Electron backscatter diffraction, *see* EBSD
Electron diffraction, 6, 257, 354, 361, 404, 480, 486
Electron probe microanalysis, *see* EPMA
  diffusion couples measurement, 226–227
  diffusion couple specimens preparation for, 227–228
  profiling in diffusion-multiple approach, 257–259
Electron propagation, in diffusion couple method, 237
Energy dispersive spectrometry/spectroscopy (EDS), *see* EDS
  diffusion couples analysis, 227
Energy dispersive diffraction (EDD), 421
Energy dispersive X-ray microanalysis (EDX), *see* EDS, 116–117, 129, 131, 135, 147
Enthalpy, 153–158, 163–165, 169–174, 178–184, 191–194, 201–208, 211–220, 299, 304, 325, 330–331, 434
  change and phase transition, 123–124
  od dilution, 325
  of mixing, 105
  of pure metal *vs.* temperature curves and DTA response in DTA data analysis, 164–166
  of transition, 299

Enthalpy (*Contd.*)
  vs. temperature curve and their relation to DTA curves, derivative of, 174–176
  vs. temperature curves for binary alloys, 171–174
Entropy, 5, 58–59, 61, 64–65, 68, 70, 74, 76, 80, 90, 97, 124, 392, 466
of fusion, 58
EPMA, 17, 24, 37–39, 116–118, 135, 140–142, 149, 222, 226–229, 232, 236, 237, 239, 244, 246–249, 256–269, 276–278, 285–287, 289, 300, 304, 352, 366, 372, 374, 376, 404, 405
Equilibrated alloy method, for phase diagrams determination, 23, 35, 36, 38, 109–149
  alloy preparation
    arc melting, 110
    high-temperature melting of alloys, 109–110
    induction furnace melting, 111–112
    powder metallurgy method, 112–113
  dynamic method (by heating and cooling experiments) and, 114, 122–129
  examples in Al–Be–Si ternary system, 131, 133, 134–135, 136
    in Al–Mn–Si ternary system, 135, 137, 138
    in Cu–Nd binary system, 130–131, 132–133
    multicomponent phase diagram and Al–C–Si–Ti system, 137, 139–140
  homogenization heat treatment, 113–114
  pitfalls of
    identification of degenerated phase equilibrium in, 148–149
    inconsistency between result from DTA measurement and from XRD and microscopy in, 147–148
    verification of establishment of true equilibrium in, 145–147
  static method (by isothermal experiments) and, 114, 115–122
Equilibrium phase diagrams, 1, 3–5, 7, 13, 23–24, 42, 44, 46, 228, 238, 244, 255, 294, 486–487, 490
Equilibrium states, 24–26, 201

Eutectic (reaction), 32, 113, 114, 116, 117, 127, 131, 134, 135, 148, 149, 154, 160, 161, 166, 170, 171, 173–183, 191–194, 196, 198, 202, 204, 211–217, 250, 251, 296, 307, 309, 319, 442, 463, 472, 473, 475, 478
variant, 204
vs. peritectic reactions, for binary alloys, 191–192
Eutectoid (reaction), 12, 13, 31, 32, 42, 43, 58, 63–65, 90, 261, 267, 478
Evaporation, 111, 112, 115, 124, 145, 163, 226, 251, 254, 294, 299–300, 302, 313, 316, 320, 344, 398, 445

F
Fayalite, 455
Fe (pressure-dependent) phase diagram, 432, 437
Fe–C phase diagram, 4, 5
Fe–Mo–Ni phase diagram, 270
Fe–Ni phase diagram, 36
Fe–Pd phase diagram, 12, 14
Fe–Sc phase diagram, 14
Fe–Si phase diagram, 393–394
FeAl–Fe$_2$AlTi phase diagram, 370
FeO–Fe2O3–SiO2 system, 455–457
First order phase transitions, *see* phase transition, 10–11, 16, 125, 325, 362–381, 480
Fluidal phase, 296, 337
Frothing, 345

G
Ga–I phase diagram, 15
Ga–Mn–Ni phase diagram, 402
Galvanic polarization, 309–310
Ga$_2$O$_3$–In$_2$O$_3$–SnO$_2$ phase diagram, 352–354
Gibbs, J. Willard, 3–5
Gibbs energy, 3, 8, 10–13, 16, 17, 24–26, 201, 202, 274, 276–277, 284, 289, 367, 389–390, 484
Gibbs phase rule, *see* phase rule
Gibbs–Thomson effect, 202
Global equilibrium, 202

H
H–Cu–Ti phase diagram, 469
H–Na phase diagram, 319–320
H–Nb phase diagram, 479

Index   499

H–Pt phase diagram, 462
H–Zr phase diagram, 464
H–Zr phase diagram, 467
Heat capacity, 4, 124–125, 155, 161,
        164–165, 169, 171, 181, 202, 205,
        208, 304, 384, 477
Heat flux–DSC (HF–DSC), 124–125,
        152–156, 199, 201–202, 207
    ASTM standards for, 154–155
    calibration and DTA signal from pure
            metals
        fixed point (pure metal) enthalpy vs.
                temperature curves and DTA
                response, 164–165, 166
        quantitative enthalpy and heat capacity
                calibration, 168–169
        temperature calibration: choice of
                onset temperature, 168
        temperature calibration: effect of
                instrument thermal lags on onset
                determination, 166–167
    factors for proper technique, 169
    ideal reference material, 164
    instruments and operation
        control thermocouple, 157
        DTA/heat-flux DSC vs. power
                compensating (true) DSC, 157
        heat transfer between system
                components, 156–157
        millivolt or Kelvin signal, 158
        plotting signal vs. temperature or time,
                158–159
        variations among instruments, 155
    measurements and DTA, during alloy
            melting and freezing, 153–154, see
            also DTA data analysis for binary
            alloys
    samples
        atmosphere, 161
        crucible selection/reaction, 161, 162,
                163
        evaporation, 163
        inert powder cover, 161
        initial metallurgical state of alloy
                samples, 163–164
        lid, 161
        mass, 159–160
        powder samples, 160
        shape, 160
Heating rate, 30–36, 123, 125, 133, 134,
        148–149, 155–157, 160, 163, 165,
        167, 168, 174–176, 181, 183–185,
        189, 192, 205–210, 336, 355
Heating transformation
    in Co–Cr, 36–38
    in Co–Mo and Co–V, 38–40
    in Fe–Ni, 35–36
    shifting of transformation-start
            temperature with, 30–35
Hermetic sealing, 345
HF–DSC, see Heat-flux DSC measurements
Hg–In phase diagram, 312–314
Hg–K phase diagram, 337
Hg–Mg phase diagram, 335
Hg–U phase diagram, 297
High–temperature melting, 109–110
High–throughput (method/approach), see
        combinatorial, 247, 258, 396–397,
        399, 404, 409, 486–487
Homogenization heat treatment, 113–114
Hot isostatic pressing (HIP), 248–249, 251,
        253–255
Hot-stage microscopy, 358

I

ICSD (International Crystal Structure
        Database), 354
Improbable phase diagrams, 57–59
    examples of, 79–100
Incipient melting, 176, 202
Induction furnace melting, 110–112, 114,
        130
Induction melting, see Induction furnace
        melting
Interface undercooling, 202
Interstitial solutes, 119, 202
Invariant reaction, 5, 6, 12–14, 19, 32, 46,
        55–56, 61, 68, 100, 114, 116–117,
        134–137, 147–149, 154, 194–202,
        204, 215
    types, 13
Inverse miscibility gaps, 59, 76, 97, 99
Ir–Rh phase diagram, 92, 93
Isopleth, 41, 135
Isothermal phase transformation kinetics,
        41–46
Isothermal section, 43–46, 82, 115, 122,
        135–137, 139, 145, 146, 224,
        230–231, 234, 236, 242, 246, 250,
        252, 255, 259–260, 263–270, 277,
        345–346, 353, 362, 369, 371, 382,
        402, 456, 486, 490

K

Kawai-type apparatus, 416–417
Kelvin signal, 158
Kinetics, 22–48, 123, 145, 147–148, 154, 157, 191, 199–201, 206, 235, 239, 250, 274, 314, 325, 327, 342–343, 349, 361, 431, 436, 483, 488

L

Latent heat, 202
Lattice parameter method, 121–122
LaMnO$_3$–SrMnO$_3$ phase diagram, 395
Lever rule, 5, 12, 202
Lever rule solidification, 202
Liquid–fluid equilibria, 337
Liquid–liquid equilibria
  calorimetry, 331
  chemical analysis of separated liquids, 330
  densitometry by X-ray attenuation, 332
  electrical resistivity, 332
  electromotive force, 332
  magnetic susceptibility, 333
  neutron transmission, 332
  thermal analysis, 330–331
  vapor pressure, 333–334
Liquidus, 2, 17–19, 32, 44, 53, 55, 56, 58–72, 79–86, 89–96, 99, 102–104, 154, 171, 174, 176, 180–186, 188, 190–197, 203, 211, 212, 217, 279, 281, 283, 290, 296, 299, 315, 319–321, 324–328, 344, 350–351, 354, 358–359, 368, 454–457
  abrupt change of slope in, 70–72
  asymmetry of, 65–68
  at pure element end, 61–63
  at melting point, 102
  eutectoid temperature and sharpness of, 63–65
  projection of Al–Mn–Si system, 138
  straight line, 102–104
Liquidus determination
  heating problems with, 180–186
  supercooling problem with, 186–191
Liquidus projection, 137, 138, 279, 281, 283, 290
Liquid–vapor equilibria, vapor pressure, 336–337
Local equilibrium, 23–24, 26, 46, 152, 201, 203, 223–224, 228, 235, 246, 259, 263, 266, 466

in diffusion couples, 223–224
Lu–Th phase diagram, 92, 93
Large volume press (LVP), 413, 416, 418, 420–423, 430–432, 435–438

M

Magnetic analysis, 128–129
Magnetic susceptibility, 300, 310–311, 323–325, 333, 478, 480
Magneto–optical Kerr effect (MOKE), 397, 399, 403
Magnetism
  basics of, 384–385
  effect on phase diagrams, 389–395
  measurements of, 385–389
Metallographic analysis, *see* Metallography
Metallography
  and phase diagrams, 5–7
  for phase diagram determination, 115–116, 117
Metastable equilibrium/phase diagram/sate, 5, 24–26, 30, 33, 36, 55–56, 58–59, 61, 65, 70, 77–80, 87, 92, 114, 145, 147, 203, 238, 271, 294, 314, 380, 390, 451, 461, 468, 471, 474, 489, 490, 492
Metastable melting point, 77–80, 92
Metastable phase, 27–30
Metatectic (reaction), 12, 13
Mg–Pb phase diagram, 301
Mg–Sb phase diagram, 92, 94
Mg$_2$SiO$_4$ (pressure dependent) phase diagram, 434–435
Microsegregation, 203
Miscibility gap, 37–38, 40, 46, 59, 62, 73–76, 89, 92, 96–97, 99, 104, 330–333, 390–392, 460–461, 475, 479, 485
Mn–Y phase diagram, 94, 95
Mo–Ru phase diagram, 94, 95
Mo–N–Si phase diagram, 240–243
Monotectic (reaction), 12, 13, 19, 55, 319, 332
Monotectoid (reaction), 12, 13, 55,
Mullite, 450, 451
Multicomponent alloys, melting and freezing of
  aluminum alloy 2219, 213–218
  Udimet 700, 218–220

Index

501

N
N–Nb–Ni phase diagram, 240
NaCl (pressure dependent) phase diagram, 433
Nb–Si phase diagram, 42
Neutron diffraction, 6, 16, 337, 361, 394–395, 413, 475, 480
Nd–Pu phase diagram, 96
Nishizawa horn, 38, 40, 390, 391, 485
Ni–Sr phase diagram, 96, 97
Np–Zr phase diagram, 97, 98
Nucleation during supercooling, 203

O
Off-stoichiometric melting, 104
Olivine–wadsleyite phase transition, 433
Onset temperature, 166–169, 203, 208–211
Optical microscopy (OM), 116
Optical reflectivity, 328
Order-disorder, ordering (transition/transformation), ordered, 16, 41, 113, 127–128, 303–305, 361–382, 384–385, 387, 390, 392–395, 404, 406–408, 451, 460–461, 474, 480, 485, 489, 491, 493
Oscillation method of phase analysis (OPA), 358–359
Oxygen steelmaking, 443
Oxysalts, 343, 344, 345, 346

P
Partial pressure, 110, 240, 242, 243, 294, 311, 312, 327, 334, 345, 348–351, 355, 358, 455
Paris–Edinburgh (PE) cell, 417
Pd–Zr phase diagram, 97, 98–99
Peak intensity method, 119–120
Peak signal, 203
Peak temperature, 158–159, 183–185, 203
Peritectic (reaction), 5, 12, 13, 17–19, 61, 65, 83, 92, 113, 114, 116, 117, 131, 148, 154, 170, 191–194, 198, 203, 204, 213, 215, 296, 452
  vs. eutectic reactions, 191–192
Peritectoid (reaction), 12, 13, 59, 97, 305
Phase, definition, 203
Phase boundary(ies), 4–6, 9, 12, 14, 16–17, 22–26, 32–42, 45, 52–60, 63, 68–70, 72, 86–97, 103–105, 109, 114, 117–119, 120–122, 131, 136–137, 226, 236–237, 244, 256–257, 260–261, 266, 278, 302–307, 314, 332, 366, 369, 373, 380, 385, 388–391, 399, 404, 409, 431–436, 461–463, 466, 470–476, 479, 486–490
  lattice parameter method for, 121–122
  peak intensity method for, 119–120
  two-phase fields, 68–70
Phase diagrams, 484–485, 488–489
  criteria for generating reliable data for, 8–10, 17–20
  experimental considerations, 17–20
  generalized unary, 10, 11
  industrial use of, 2
  low-temperature, 487–488
  metallography and, 5–7
  miscibility gaps and, 73–76
  off-stoichiometric melting and, 104
  Roberts-Austen, William Chandler and, 3–5
  thermodynamic constraints effect on, 10–17
  twentieth century developments, 5–7
  with syntectic reactions, 76–77
  X-ray diffraction and, 6
Phase diagram determination
  binary alloys, 130–131
  differential scanning calorimetry (DSC), 124–125
  differential thermal analysis (DTA), 123–124
  dilatometry, 125–126
  electrical conductivity measurement, 126–128
  electron probe microanalysis (EPMA), 117–118
  energy dispersive X-ray microanalysis (EDX), 116–117
  magnetic analysis, 128–129
  metallography for, 115–116, 117
  optical microscopy (OM), 116
  phase boundaries for, 22–23
  phase transformation kinetics role in, see Phase transformation kinetics
  quaternary alloys, 137, 138, 139–140
  scanning electron microscopy (SEM), 116
  ternary alloys, 131–137
  X-ray diffraction (XRD), 118–119

Phase diagram determination, by diffusion
  couples, 222
  error sources encountered in
    experiments with, 236–244
  experimental procedures
    analytical techniques and specimen
      preparation, 226–227
    preparation of diffusion couple
      specimens for EPMA, 227–228
    techniques for diffusion couple
      preparation, 225–226
  general principles of, 223–225
  variations of, 228–236
Phase diagram determination, by
  equilibrated alloys
  alloy preparation
    arc melting, 110
    high-temperature melting of alloys,
      109–110
    induction furnace melting, 111–112
    powder metallurgy method, 112–113
  crystal structure identification of new
    phases, 140–145
  dynamic method (by heating and cooling
    experiments) and, 114, 122–129
  examples
    in Al–Be–Si ternary system, 131, 133,
      134–135, 136
    in Al–Mn–Si ternary system, 135, 137,
      138
    in Cu–Nd binary system, 130–131,
      132–133
    multicomponent phase diagram and
      Al–C–Si–Ti system, 137, 139–140
  homogenization heat treatment,
    113–114
  pitfalls of
    identification of degenerated phase
      equilibrium in, 148–149
    inconsistency, 147–148
    verification of establishment of true
      equilibrium in, 145–147
  static method (by isothermal
    experiments) and, 114, 115–122
Phase diagram determination, high
  throughput approaches to, 486–487
Phase diagram determination, using
  diffusion-multiple approach
  considerations for design, 250–255
  EPMA profiling, 257–259
  error sources and reduction, 266–269
  extraction of equilibrium tie lines,
    259–266
  fabrication for, 248–250
  imaging examination and phase analysis,
    256–257
Phase diagram determination of ceramic
  systems
  ex situ methods, 342
    identification of new phases, 354
    phase and compositional analysis,
      350–354
    sample preparation and equilibration,
      343–350
  in situ methods, 354
    coulometric titration, 356–358
    electric, dielectric, and magnetic
      measurements, 359
    high-temperature X-ray diffraction, 358
    oscillation method of phase analysis
      (OPA), 358–359
    thermal analysis, 355–356
    thermomicroscopy and other optical
      techniques, 358
Phase diagrams for hydrogen-containing
  systems, determination of, 459–462
  representation of M–H systems
    binary M–H systems, 462–466
    multicomponent systems, 466–471
Phase diagrams, correct and incorrect
  apparent five-phase equilibrium, 100, 101
  apparent four-phase equilibrium, 100,
    101
  off-stoichiometric melting, 104
  pointed liquidus, 102, 103
  straight line liquidus, 102, 103–104
  guidelines for judging validity of phase
    diagrams, 56–79
  improbable
    examples of, 79–100
    situations, 57–59
  phase rule violations, 53–56
Phase diagrams, pressure-dependent,
  determination of
  heating at high pressure in a DAC, 422
  combination of electric and laser
    heating, 430
  external electrical heating, 428–429
  heating in LVPs, 430–431
  laser heating, 423–426
  high-pressure devices
    diamond-anvil cells (DAC), 413–416

large-volume presses (LVP), 416–418
industrial solids, 438
phase diagrams determination using LVPs, 431
phase diagrams determination using DAC, 436–438
phase relations in complex systems, 434–435
phase relations in univariant systems, 432–434
pressure measurement
  using ruby fluorescence, 418–419
  using X-rays, 419–421
X-rays at high pressure, 421–422
Phase diagrams and reaction enthalpies for M–H systems, rules relating to, 471–473
Phase diagram determination, for crystal structure phase identification, 140–145
Phase diagram determination of M–H systems, techniques employed for, 473
  closed systems, 477–478
    diffraction methods, 478–480
    thermal analysis, 478
    dilatometry, 480
    electric resistivity, 480
    electron diffraction and TEM, 480
    magnetic susceptibility, 480
  open systems
    E–r–T technique, 476–477
    P_{H_2}–r–T technique, 474–476
    reaction calorimetry, 477
Phase diagrams for functional materials, 485–486
Phase diagrams involving magnetic transitions, determination of
  combinatorial and high-throughput mapping of magnetic phase diagrams
    in metallic systems, 396–404
    of oxide systems, 404–408
  magnetic effect on phase diagrams, 389–395
  measurements of, 385–389
Phase diagrams involving order-disorder transitions, determination of
  concentration gradient method, 374–376
    basics of, 376
    examples of, 376–380

ER (electrical resistivity) method, 362–363
singular point method
  examples of, 368–374
  origin of singularity, 366–368
  thermal analysis, 364–366
Phase precipitation, from quenched alloys, 41–46
Phase rule, 4, 10–12, 51, 53, 55, 56, 70, 109, 145, 147, 201, 223, 342, 352, 460, 489
  to binary, ternary, and higher order systems, 11–12
  violations, 53–56
Phase transformation kinetics, for phase diagram determination
  during cooling and heating experiments, 24–26
    in Fe–Ni, 35–36
    fcc–hcp phase equilibria in Co–Cr, 36–38
    fcc–hcp phase equilibria in Co–Mo and Co–V, 38–40
    formation of metastable phases during cooling, 27–30
    shifting of transformation-start temperature with cooling rate, 26–27
    shifting of transformation-start temperature with heating rate, 30–35
  during isothermal experiments
    kinetics and phase formation in diffusion couples/multiples, 46–47
    precipitation of phases from quenched alloys, 41–46
Phase transitions
  enthalpy change and, 123–124
  first-order, 10–11, 16, 125, 325, 362–381, 480
  olivine-wadsleyite, 433
  second-order, 13, 16, 55, 124–125, 362–381
Powder metallurgy method, 112–113
Power-compensating (true) DSC, 157, 203
Precipitation of phases from quenched alloys, 41–46
Pt–Zr phase diagram, 99–100

Q
Quartz halogen tungsten lamp (QHTL), 426
Quasi-equilibrated diffusion zone, 238

Quaternary alloys, phase diagram
determination, 137, 138, 139–140
Quaternary Mo–Si–B–Ti system, 282–289
Quenching method and phase equilibria, 115

R
Reaction calorimetry, 477.
Recalescence, 187, 190, 191, 203
Resistivity, *see* also electrical
resistance/resistivity, 41, 305, 307,
321–323, 332, 359, 361–364, 377,
384, 389, 395, 475, 480
Rietveld analysis, 354
Roberts-Austen, William Chandler, 3–5
Rutherford backscattering spectroscopy
(RBS), 226, 300

S
Scanning electron microscopy, see SEM
Scanning transmission electron microscopy
(STEM), 17
Scheil–Gulliver solidification, 171, 172,
174, 176–179, 190–200, 203,
212–220
Scheil solidification, *see* Scheil-Gulliver
solidification
Second-order transitions, 13, 16, 55,
124–125, 362–381
  magnetic transition, 16
Secondary ion mass spectroscopy (SIMS), 226
SEM, 47, 116–118, 131, 134–135, 140, 147,
249, 253, 256–257, 261–262, 276,
285, 300, 336, 376–379, 404
Si–Ti phase diagram, 46, 47
Sieverts apparatus, 474
Singular point method
  examples of, 368–374
  origin of singularity, 366–368
Solidification path, 194–195, 203
Solid–liquid equilibria
  anodic oxidation, 321–323
  chemical analysis of a separated liquid, 316–318
  chemical analysis of quenched samples, 318
  corrosion tests, 328–329
  densitometry, 325
  diffusion coefficient, 327
  electrical resistivity, 321

electromotive force, 320
enthalpy of dilution, 325
kinetics of alloy decomposition or
formation, 325–327
magnetic susceptibility, 323–325
motion of liquid metal inclusions in
ionic crystals, 330
optical reflectivity, 328
thermal analysis, 319–320
vapor pressure, 327
viscosity, 328
weight loss of a solid after equilibration
with a liquid, 315–316
X-ray absorption spectrometry, 327–328
Solid–solid equilibria
  calorimetry, 304
  coulometric titration, 307–309
  densitometry, 302–303
  dilatometry, 303
  electrical resistivity, 305
  electromotive force, 306–307
  galvanic polarization, 309
  gaseous thermal extraction, 314–315
  hardness, 304
  interdiffusion, 303–304
  magnetic susceptibility, 310
  microstructural analysis of quenched
samples, 301–302
  superconductivity, 305–306
  thermal analysis, 299–300
  vapor pressure, 311–314
  X-ray diffraction, 302
  zone melting, 300–301
Solid–state reaction techniques, 343
Solidus, 2, 31, 33, 35, 55–56, 58–61, 68, 82,
89, 92, 99, 104, 109–110, 113–114,
123–124, 126–127, 154, 164, 171,
176, 180–186, 192, 203, 211, 251,
305, 315, 350
  determination on heating, problems
with, 177–180
  line, 315
Solid–vapor equilibria
  thermogravimetry, 335–336
  vapor pressure, 334–335
Static method, 114, 115, 117, 122
Supercooling, *see* also Undercooling, 134,
154, 160, 168, 183, 185–191, 194,
196, 203–204, 359
Superheating, 30, 32, 33, 194
Syntectic (reaction), 12, 13

Syntectic reactions, phase diagram with, 76–77

T
TEM (transmission electron microscopy), 117, 256–257, 354, 373, 376–379, 381, 394, 431, 480
Temperature coefficient of electrical conductivity (TCEC), 127–128
Ternary alloys
  DTA data analysis for, *see* DTA data analysis for ternary alloys
  phase diagram determination, 131–137
Ternary Mg–Al–Sr system, 275
  experimental method, 276
  experimental results and discussion, 277–282
  thermodynamic models, 276–277
Ternary systems, diffusion zones and, 224
TGA (Thermal Gravimetric Analysis), 183, 335–336, 344, 354–355, 359
Thermal analysis, 152–153, 299–300, 319–320, 330–331, 355–356, 364–366, 478, *see also* Differential @3:thermal analysis
Thermal conductivity, 157, 161, 204
Thermocouple, 4, 153, 156–159, 163–164, 166–169, 181, 183, 187–191, 197–199, 205–210, 299–301, 322, 331, 346–347, 420, 428–431, 445–446, 449
Thermodynamic constraints, on phase diagrams, 10–17
Tie lines, 9, 24, 46, 55, 118, 122, 204, 224, 228, 233, 244, 259–260, 262–267, 272, 353, 463, 466–467, 469–471
  extraction of equilibrium, in diffusion multiple approach, 259–266
Thermogravimetry, *see* TGA, 335–336
Time-temperature–transformation (TTT) diagrams, 41
Titration, coulometric, 307–309, 356–358
Transformation-start temperature, shifting of
  with cooling, 26–27
  with heating, 30–35
Transition reaction, 204
Transuses, *see* Phase boundaries

U
Undercooling, *see also* Supercooling, 23, 26, 32, 33, 135, 202, 300, 305, 363
Udimet 700 (alloy), 218–220

V
Van't Hoff relationship and Charles' Law, 59–61
Vapor pressure, *see also* Partial pressure, 3–4, 17–19, 110, 161, 163, 251, 296, 298–299, 306, 311–314, 317, 319, 327, 333–336, 444
Variant eutectic reaction, 204
Vegard's law, 352
Vertical section, *see* Isopleth, 41, 122, 135, 136
Viscosity, 328

W
Wavelength dispersive spectrometry (WDS), *see also* EPMA, 117, 227, 258, 401
Wicking, 345

X
X-ray absorption effect, 237
X-ray diffraction, see XRD, *see also* Diffraction
  high-temperature, 358
  inconsistency between results from DTA, SEM/EDX and, 147–148
  and phase diagrams, 6
X-ray absorption spectrometry, 327–328
XRD, 6, 24, 38–40, 48, 109, 114, 115, 117–120, 123, 130–132, 134, 135, 139–142, 144, 147–149, 257, 276, 280–281, 285–286, 300, 302, 334, 336–337, 358–359, 361, 373–374, 380, 395, 401, 460, 468, 478–480, 487

Z
Zero degree C, phase diagrams below, 105
Zone melting, 300–301